Industrial Hygiene Control of Airborne Chemical Hazards

Related Titles

L1339 *Physical Hazards of the Workplace* by Larry Collins

L1506 *Handbook of Chemical Risk Assessment: Health Hazards to Humans, Plants, and Animals*, Three Volume Set by Ronald Eisler

L1620 *Occupational Health and Safety Management: A Practical Approach* by Charles Reese

L1536 *Handbook of Environmental Health, Fourth Edition, Volume I: Biological, Chemical, and Physical Agents of Environmentally Related Disease* by Herman Koren

L1547 *Handbook of Environmental Health, Fourth Edition, Volume II: Pollutant Interactions in Air, Water, and Soil* by Herman Koren

L1606 *Biological Risk Engineering Handbook: Infection Control and Decontamination* by Martha Boss

Industrial Hygiene Control of Airborne Chemical Hazards

William Popendorf

CRC Taylor & Francis
Taylor & Francis Group
Boca Raton London New York

CRC is an imprint of the Taylor & Francis Group,
an informa business

Published in 2006 by
CRC Press
Taylor & Francis Group
6000 Broken Sound Parkway NW, Suite 300
Boca Raton, FL 33487-2742

No claim to original U.S. Government works
Printed in the United States of America on acid-free paper
10 9 8 7 6 5 4 3 2 1

International Standard Book Number-10: 0-8493-9528-3 (Hardcover)
International Standard Book Number-13: 978-0-8493-9528-4 (Hardcover)

informa
Taylor & Francis Group
is the Academic Division of Informa plc.

Visit the Taylor & Francis Web site at
http://www.taylorandfrancis.com

and the CRC Press Web site at
http://www.crcpress.com

Conversion Factors

To convert from the initial unit on the left to the new unit on the right, *multiply* the value in the units on the left by its "equivalence." For example, 10 $cm^2 \times 0.155 = 1.55$ $inches^2$.

To convert from the initial unit on the right to the unit on the left, *divide* the value in the units on the right by its "equivalence." For example, 1 $inch^2/0.155 = 6.45$ cm^2.

AREA

1 $centimeter^2$	=	0.155	$inch^2$
1 $inch^2$	=	6.4516	$centimeter^2$
1 $meter^2$	=	10.7636	$feet^2$
	=	1.19591	$yard^2$
1 hectare	=	2.47	acres
1 acre	=	43,560	$feet^2$

DENSITY

1 $gram/cm^3$	=	62.4	lb/ft^3
1 $grain/ft^3$	=	2288	mg/m^3

ENERGY

1 Btu	=	0.2520	kcal
1 Joule	=	0.23901	calorie
	=	9.480×10^{-4}	British thermal unit
1 kcal = 1 Cal	=	3.9683	Btu
1 Therm	=	10^5	Btu
1 kWh	=	11,000	Btu
1 bbl crude oil	≈	5.8×10^6	Btu
1 lb bitum. coal	≈	12,500	Btu
1 ft^3 natural gas	≈	1050	Btu
1 lb hardwood	≈	8600	Btu

FLOW

1 ft^3/min	=	28.317	L/min
1 m^3/min	=	35.3147	ft^3/min
1 m^3/sec	=	2118.9	ft^3/min

LENGTH

1 centimeter	=	0.3937	inch
1 inch	=	2.5400	centimeters
1 foot	=	0.3048	meter
1 meter	=	3.2884	feet
	=	1.09361	yard
1 kilometer	=	0.6214	mile (statute)
1 statute mile	=	5280	feet
1 nautical mile	=	6080	feet

MASS

1 grain	=	1/7000	pound
1 gram	=	0.035274	ounce
1 ounce	=	28.350	grams
1 pound	=	453.6	grams
1 kilogram	=	2.2046	pound (lb)
1 ton	=	2000	pounds
1 metric ton	=	2204.6	pounds
	=	1000	kilograms

POWER

1 ampere	=	1	coulomb/sec
1 kWatt	=	1.341	horsepower
1 horsepower	=	6356	″wg × ft^3/min
	=	33,000	ft × lb/min

PRESSURE

1 atmosphere	=	1.01325	bar
	=	14.696	psi (pounds per $inch^2$)
	=	407.60	inch H_2O at 21°C
	=	101513	Pascal
1 Pascal	=	0.1450×10^{-3}	psi
	=	0.00400	inch H_2O

TEMPERATURE

°C	=	5 × (°F − 32)/9
°F	=	(9 × °C/5) + 32
°K	=	°C + 273.15 ≈ °C + 273
°R	=	°F + 459.67 ≈ °F + 460

VELOCITY

1 meter/min	=	3.281	ft/min
1 mile/hour	=	88	ft/min
1 meter/sec	=	196.85	ft/min

VOLUME

1 cm^3	=	0.061	$inch^3$
1 $foot^3$	=	28.317	liters
1 liter	=	0.0353147	$foot^3$
	=	0.2642	gallon
1 gallon	=	0.1337	$foot^3$
1 m^3	=	35.3147	$feet^3$
1 barrel	=	42	gallons

List of Abbreviations

List of abbreviations, acronyms, notations, and symbols used in this book (and equation in text where specific)

General abbreviations

A	Area (e.g., surface area, face area, duct flow area) except in Antoine equation. (Equation 2.10)
ACH	Air changes/hour (Equation 20.15)
AHDD	Annual heating degree day (ca. Equation 17.9)
a.k.a.	Also known as
ASL	Above sea level (ca. Equation 2.4)
C	Concentration, usually in mg/m^3
℄	Centerline
CMD	Count median diameter (ca. Equation 3.3 to Equation 3.6)
D	A large diameter (as in a duct)
d	A small diameter (as in a particle)
d*f*	*Manual of Industrial Ventilation*'s "density factor" (see Equation 16.8)
$\mathcal{D}$	Diffusivity (Equation 5.2 and Equation 5.5)
EDR	Environmental dilution ratio (Equation 5.16 and Equation 5.17)
EL	Generic exposure limit (TLV, PEL, etc.)
F	Force (as in Equation 2.3 and Equation 3.7)
FanSP	Fan static pressure (Equation 16.4 and Equation 16.5).
FanTP	Fan total pressure (Equation 16.1 and Equation 16.2).
G	Contaminant generation rate in g/min or moles with subscript (Equation 5.3 and Equation 5.13 and Chapter 20)
H_i	Henry's law coefficient (Equation 6.13)
$h_{\text{something}}$	Head loss due to "something" (Section 14.I.4)
K_{hood}	DallaValle factors (Equation 13.6)
K_{mix}	Design mixing factor, $K_{\text{room mixing}}$ (Equation 19.2)
K	*Manual of Industrial Ventilation*'s abrupt contraction loss factor
L	*Manual of Industrial Ventilation*'s tapered contraction loss factor
LEV	Local exhaust ventilation
LFL	Lower flammable limit (same as LEL)
m	mass (e.g., g or kg)
MMAD	Mass median aerodynamic diameter (Equation 3.17).
MMD	Mass median diameter (Equation 3.6)
MW	Molecular weight of a chemical
n	Any number or a specific number of moles
N_{Reynolds}	Reynolds number (Equation 12.3)
NIOSH	National Institute for Occupational Safety and Health, a U.S. governmental research agency
OSHA	Occupational Safety and Health Administration, a U.S. government enforcement agency
P	Pressure (molecular, static, ambient, or local but not necessarily the "normal" pressure)
$P_{\text{partial of } i}$	The partial pressure of generic chemical "i"
P_{vapor}	Vapor pressure. $P_{\text{vapor, } i}$ is P_{vapor} when chemical "i" is in a liquid mixture (see Equation 6.7)
ppm	Parts of a chemical per million parts of air (ppm_ℓ refers to ppm of a liquid)
Q	Volumetric flow rate (ft^3/min or m^3/sec)
R	Universal gas constant (Equation 2.16)
R	*Manual of Industrial Ventilation*'s regain factor for duct expansions

Rel.Y_i	Relative Y_i if air is disregarded (Equation 6.19)
Rel.Haz	Fraction of the total VHR (Equation 6.21)
SP	$P_{\text{inside a duct}} - P_{\text{outside the duct}}$ (Equation 11.14 through Equation 11.16)
T	Temperature
t	Time
TP	$(P + \text{VP})_{\text{inside duct}} - P_{\text{outside duct}}$ (Equation 11.28)
TUC	Time of useful consciousness (Table 4.3)
V	Velocity, such as air velocity
VHI	log (VHR) (Equation 8.11)
VHR	Vapor hazard ratio = P_{vapor}/TLV (Equation 8.4)
Vol	Volume (as in air or room volume)
VP	Velocity pressure (Equation 11.21)
X	Distance (e.g., in front of a hood in Equation 13.2)
X_i	Molar fraction of "i" in a liquid (Equation 2.5 and Equation 6.3)
Y_i	Molar fraction of "i" in air (Equation 2.6 and Equation 6.15)
Z	Vertical distance (Figure 13.30)

Abbreviated subscripts and superscripts

aero	Aerodynamic (Equation 3.13)
BZ	Breathing zone (Equation 5.17 to Equation 5.21)
i	The ith component in a group or mixture
ℓ	In a liquid state (cf. airborne)
moles	Moles or based on moles (cf. based on mass)
n	The last (or nth) of a series (see "i" above)
net	After adjusting for buoyancy (Equation 3.7)
o	The initial or starting point
w	Weight basis as in pounds (Equation 11.3)
∞	After a long time; steady state (when $_{\text{subscripted}}$)
∞	Infinitely large dilution (when ^{superscripted})

Greek symbols

Δ	Change or difference between two conditions
γ	Activity coefficient (Equation 6.8)
κ	Generic proportionality constant (e.g., Equation 5.2)
η	Dynamic or absolute viscosity (gm/cm/sec)
ν	Kinematic viscosity (cm^2/sec)
ρ	Density, mass / volume (see also ℓ subscript).
τ	Characteristic time or the time of one room air change (Equation 19.13)

Abbreviated Latin phrases

ca.	*Circa*, about, or approximately
cf.	*Confer* or compare with or compared to
e.g.	*Exempli gratia* or for example
i.e.	*Id est* or that is or in other words
viz.	*Videlicet* or namely (before a list)

Respirator acronyms (Chapter 22)

Most of these terms can be found in paragraph (b) of OSHA's Respiratory Protection Standard (29 CFR 1910.134).

ADA	Americans with Disabilities Act. See www.usdoj.gov/crt/ada/adahom1.htm.
ANSI	American National Standards Institute. See www.ansi.org.
APF	Assigned protection factor. The workplace level of respiratory protection that a respirator or class of respirators is expected to provide to employees ("who have been properly fitted and trained" is often added).
APR	Air purifying respirator. A respirator with an air-purifying filter, cartridge, or canister that removes specific air contaminants by passing ambient air through the air-purifying element.

ASR	Atmosphere supplying respirator. A respirator that supplies the wearer with breathing air from a source independent of the ambient atmosphere, and includes SARs and SCBA units.
DFM	Dust/fume/mist filter. An old designation for filters with TC-21C-XXX or TC-23C-XXX approval numbers certified under Part 11 "for respiratory protection against dusts, fumes and mists having a time weighted average not less than 0.05 milligrams per cubic meter or 2 million particles per cubic foot."
DOP	Dioctylphthalate. An oil-based aerosolized agent used for quantitative fit testing.
EPF	Effective protection factor. The measured protection provided by a properly selected, fit-tested, and functioning respirator when used intermittently for only some fraction of the total workplace exposure time, i.e., sampling is conducted during periods when a respirator is worn and when it is not worn.
ESLI	End of service life indicator. A system that warns the respirator user of the approach of the end of adequate respiratory protection, e.g., that the sorbent is approaching saturation or is no longer effective.
FF	Fit factor. A quantitative estimate of the fit (or actual protection) of a particular respirator when worn by a specific individual. OSHA uses FF whether associated with a QNFT or QLFT.
HEPA	High efficiency particulate aerosol (filter). A filter that is at least 99.97% efficient in removing monodisperse particles that are 0.3 m in diameter. The equivalent to N100, R100, and P100 particulate filters certified by NIOSH under 42 CFR 84.
IDLH	Immediately dangerous to life or health. An atmosphere that poses an immediate threat to life would cause irreversible adverse health effects, or would impair an individual's ability to escape from a dangerous atmosphere.
MUC	Maximum use concentration. The maximum atmospheric concentration of a hazardous substance from which an employee can be expected to be protected when wearing a respirator. While an MUC can be calculated by multiplying the assigned protection factor specified for a respirator or class of respirators by an appropriate exposure limit for the hazardous substance, employers (and users) must comply with the respirator manufacturer's MUC for a hazardous substance when the manufacturer's MUC is lower than the calculated MUC.
PAPR	Powered air-purifying respirator. An air-purifying respirator that uses a blower to force the ambient air through air-purifying elements to the inlet covering.
PLHCP	Physician or other licensed health care professional. An OSHA term for an individual whose legally permitted scope of practice (i.e., license, registration, or certification) allows him or her independently to provide some or all of the health care services required by paragraph (e) of 29 CFR 1910.134.
PPF	Program protection factor. The measured protection provided by a respirator within a specific respirator program. PPFs are affected by all factors of the program, including respirator selection and maintenance, user training and motivation, work activities, and program administration.
QLFT	Qualitative fit test. A pass-or-fail fit test to assess the adequacy of a particular respirator's fit that relies on the individual's response to the test agent.
QNFF	Quantitative fit factor. Used in this book for the particular fit factor resulting from a QNFT; see also FF.
QNFT	Quantitative fit test. An assessment of the adequacy of a respirator whose protection was determined by numerically measuring the amount of leakage into the respirator.
SAR	Supplied air respirator, a.k.a. an airline respirator. An atmosphere-supplying respirator for which the source of breathing air is not designed to be carried by the user.
SCBA	Self-contained breathing apparatus. An atmosphere-supplying respirator for which the breathing air source is designed to be carried by the user.

SWPF	Simulated workplace protection factor. The protection measured in a controlled setting and in which contaminant sampling is performed while the subject performs a series of set activities. The simulated setting is used to control (or reduce the impact of) many of the variables found in workplace studies, while the activities may be work or exercises that simulate the work activities of respirator users.
WPF	Workplace protection factor. The measured protection provided by a properly selected, fit-tested, and functioning respirator when the respirator is correctly worn and used as part of a comprehensive respirator program. Contaminant measurements are obtained only while the respirator is worn during performance of normal work tasks, i.e., samples are not collected when the respirator is not worn.

Preface

Are you a practicing occupational hygienist who has wondered how to choose a substitute organic solvent that is actually safer to use than the hazardous one your company is currently using? Are you a new hygienist looking for an effective, economical, but perhaps temporary method to reduce exposures before you can convince your employer to spend money on a more permanent solution? Are you an industrial hygiene student expecting to learn about local exhaust ventilation as the so-called "universal" chemical exposure control? Or are you just starting to learn about personal protective equipment such as respirators and gloves? All of these topics and more are covered by this book on *Industrial Hygiene Control of Airborne Chemical Hazards*.

Industrial hygiene is typically taught and viewed in a discrete sequence of anticipation and recognition, evaluation, and control. This book focuses squarely on the control of chemical hazards. Its focus is predominantly on airborne chemical hazards, although many of the principles presented would also apply to airborne biological hazards. While this book is science-based and very quantitative, very little day-to-day IH work is quantitative. Therefore, the book also attempts to introduce the student to the kinds of judgment needed to control exposures in the diverse settings in which people work. The ability of industrial hygienists to extend their knowledge and apply judgment, rather than only plugging data into a formula, is one of several personal rewards of practicing the profession that this book tries to illuminate.

The approach in this book to learning how to control chemical hazards tries to bridge the gap between our existing knowledge of physical principles and mechanisms that underlie the generation and dispersion of airborne chemicals and the wealth of recommendations, techniques, and tools accumulated by prior generations of IH practitioners to control chemical hazards. This book assumes that the reader is academically trained in science and math, has seen at least a small number of manufacturing or other work settings with chemical hazards, but is inexperienced in the selection, design, implementation, or management of chemical exposure control systems. Nonventilation controls rely on knowledge from physical chemistry and basic Newtonian physics that the student should be able to recall with appropriate prompting. While much of ventilation has its origins in engineering, no engineering background is expected from the reader. On the other hand, this book tries to introduce science students to some engineering tools. Where the book is quantitative, of course there are lots of formulae, but the author tries to avoid the vague notation (symbols without meaning) and long derivations that characterize most engineering texts. The chapters in this book have been organized into the following sequence and groups:

The first chapter starts with a brief introduction to IH, the approach taken by the author in teaching control, and an overview of features within the book.

The first two technical chapters introduce some terms and discuss important principles and physical mechanisms that affect gases, vapors, aerosols, and their plumes.

Chapter 4 refers to some of the above principles as it reviews some occupational health and safety regulations and exposure limits from a physical perspective rather than the perhaps more familiar toxicological or epidemiological perspectives.

Two chapters discuss the mechanisms affecting liquid evaporation, the behavior of gaseous hazards in four generic exposure scenarios, and nonideal effects within mixtures of liquids that can complicate predictions of their behavior (but not negate them).

Only one chapter is devoted to the important but less science-based technologic, psychologic, and economic dimensions to the workplace in which hygienists make changes.

Two chapters describe applications of these above principles to the control chemical hazards without adding ventilation. The first of these chapters covers chemical or process substitution as a hazard control, and the second covers work practice modifications and other pathway controls.

Chapter 10 introduces terms and concepts used in six later chapters on local exhaust ventilation. The next chapter discusses the physical relationships between air velocity and flow rate and between air's energy and measurable air pressures that underlie the function, design, and operation of ventilation systems. Chapter 12 introduces the practical tools used to measure ventilation velocities and pressures.

Six chapters are devoted to local exhaust ventilation, a staple in the industrial hygienist's arsenal of controls. This sequence of chapters begins with the local exhaust hood, then progresses (as does the air) to the exhaust ducts, to air (emission) cleaners, and to the fan that makes the system work. Two more chapters cover the costs of operating a ventilation system and some aspects on how to keep a system running and protecting workers as well as it should.

Two chapters discuss the ability and limitations of general room and building ventilation to control the accumulation of airborne chemical hazards.

The final two chapters discuss the ability and limitations of protecting employees via administrative controls, personal protective equipment in general, and respirators in particular.

Each technical chapter herein is preceded by specific learning goals. Each learning goal incorporates a key phrase loosely adapted from Bloom's cognitive domains. The idea behind these key phrases is that one does not need to know everything at the same level. Each key phrase implies a level of learning that should be sought at this time in a novice IH's education. Thus, this book is selective in its coverage of some ventilation topics. At the same time, this book goes well beyond any other single text book on the topics of nonventilation controls and receiver controls. Taken as a whole, this book provides a unique, comprehensive tool to learn the challenging yet rewarding role that industrial hygiene can play in controlling chemical hazards at work.

About the Author

William Popendorf is a professor of industrial hygiene at Utah State University. He has been on the board of the American Industrial Hygiene Association and a director of the American Board of Industrial Hygiene. Dr. Popendorf has taught and conducted research for over 30 years and published more than 65 papers and book chapters beginning with pesticide hazards to farm workers 1972 to 1992, inorganic dusts 1978 to 1982, organic dusts from grains and livestock 1982 to 1995, automotive industry foundries and metal working fluids 1987 to 1994, and since 1991 broader reviews. His general interest has been to develop or apply predictive models (many developed in other fields) that describe how physical mechanisms cause (and can be used to control) exposures of workers to organic vapors, hazardous particulate aerosols, and dermally toxic chemicals, with the expectation that such tools will improve the overall practice and knowledge-base of industrial hygiene.

Acknowledgments

We don't get a second chance to learn from our own fatal mistakes. And disabling mistakes are expensive lessons. But we can all learn by reading from and listening to those who have preceded us. I thank many predecessors from many fields of knowledge for passing on their wisdom, and my wife Joyce for encouraging and sustaining me in this contribution to passing on a little more knowledge that might improve the practice of industrial hygiene.

William Popendorf

Table of Contents

1 An Introduction to Industrial Hygiene Chemical Hazard Control

I. AN OVERVIEW OF INDUSTRIAL HYGIENE

The formal definition of industrial hygiene (IH) provides a guide to how the profession is usually taught, learned, and practiced. While this definition may appear at first glance to be just words, to the experienced hygienist each of these phrases has a significant meaning that will be discussed below.

> "Industrial hygiene is both the *science and art*
> devoted to the *Anticipation, Recognition, Evaluation, and Control*
> of those *environmental* factors or stresses
> arising in or from the workplace
> which cause *sickness, impaired health, significant discomfort or inefficiency*
> *among workers or among citizens* of the community."

Certainly, "science" differs from the "art" of any human venture. The need to apply both is crucial in an applied science such as IH. The student who can excel in the science classroom or during an examination is not necessarily creative in applying these concepts, effective at conveying new ideas or at convincing others to change, or comfortable working in an advisory capacity as any industrial hygienist must. On the other hand, someone who is good at selling ideas but does not have the scientific knowledge to make valid (let alone cost-effective) decisions will not succeed as an industrial hygienist either. The best hygienists are good at both the science and the art. (Someone who is or who wants to be good at only one thing, may be trying to enter the wrong field.) Good industrial hygienists have a wide range of both technical and interpersonal skills. It is partly this diversity that makes IH an interesting profession.

The next phrase of the definition that says the profession is devoted to four functions implies that the practice of IH is best viewed as progressing from one of these functions to the next:

Anticipation: the prospective recognition of hazardous conditions based on chemistry, physics, engineering, and toxicology

Recognition: both the detection and identification of hazards or their adverse effects through chemistry, physics, and epidemiology

Evaluation: the quantitative measurement of exposure to environmental hazards and the qualitative interpretation of those hazards

Control: conception, education, design, and implementation of beneficial interventions carried out that reduce, minimize, or eliminate hazardous conditions

IH is typically taught in this same sequence of functions. A course centered on this book would fall squarely into the control function. This list of functions makes IH sound nicely organized if not compartmentalized, but, in fact, a practicing hygienist's day may easily be fragmented into some of each of these functions; and while working at one function, the IH practitioner needs to use or be cognizant of all the other functions and always keep that framework of multiple functions clearly in the back of their mind. To leave any of these functions completely alone is to court professional failure.

The range of stressors with which IH deals is wide, and this also contributes to job satisfaction. Hygienists often group stressors into the categories (with examples) listed below. Although many would argue that the last one is not really an environmental factor, most all would agree that psychologic factors at work can be stressors. This book focuses on the control of chemical hazards and predominantly on airborne chemical hazards, although many of the principles presented also apply to airborne biological hazards.

Chemical:	gases, vapors, dusts, fumes, mists, solvents
Physical:	barometric pressure, temperature, noise, vibration, nonionizing and ionizing radiation
Biological:	bacteria, fungi, parasites, and their toxins (although toxins are also chemicals)
Ergonomic:	the interaction of machine design and operational practices with our human anatomy
Mechanical:	primarily safety (injury, fatality, property damage)
Psychological:	peer pressure, job security and satisfaction, education, and motivation

IH's focus on *factors and stressors arising in* versus *from the workplace* has varied over the years. In its origin, industrial hygiene was a branch of public health that was rooted in the workplace. However, what starts within the boundaries of the workplace often crosses the fence line to enter the surrounding community. Thus, the practice of industrial hygiene has historically oscillated between a narrow focus on employees within the plant to a wider interest in air and water emissions, solid waste, and physical hazards that leave the workplace. The reference in the definition to two settings and later to two groups of people reflects IH's broad responsibilities rather than a narrow focus.

People in both settings can respond to such health hazards in a variety of ways, from fatalities (not explicitly in the definition) to *sickness, impaired health, significant discomfort or inefficiency*, or even outrage. It is interesting to speculate why fatalities on the high end of the response spectrum were left out of the definition of IH. Outrage, expressed as either a personal or social response to a hazard, is a dimension of which the IH profession has only recently become aware. The level of subjective outrage is often irrespective of either the quantitative health hazard or the above implied spectrum, but as Peter Sandman advocates, it is just as measurable and often just as important to the mission and goal of a company or agency as the objective health hazard. This wide spectrum of adverse health effects is also important to keep in mind when trying to set priorities among multiple health hazards, and is a key factor underlying the adage that "not all exposure limits are created equal."

Not all IH work is quantitative. While the root of IH is based on science, very little day-to-day IH work is quantitative. This apparent dichotomy is reflected in a quotation attributed to Albert Einstein: "Not everything that counts can be counted. And not everything that can be counted counts." Not only is much of IH an art, the science of IH has severe limits to its knowledge. Take exposure limits for example: each is our best quantitative indicator of a chemical's toxicity at the time it was adopted. However, most exposure limits have changed at least once owing to new knowledge or interpretations. In fact, we are forced to use judgment in each of the four functions of IH: to anticipate the effects of new chemicals; to recognize a widening spectrum of adverse health responses; to evaluate the links between historic exposures and reported health effects and the statistical uncertainty between current exposure measurements and future health risks; and to control exposures in the diverse settings in which people work.

Industrial hygienists do not go around personally controlling workplace hazards. The practice of IH is a management function that involves multiple tasks and activities from quantitative exposure assessment to human and public relations, the supervision of technical staffs, financial planning and accountability, and documentation (*that means, lots and lots of writing*). Management skills are talked about in some classes but are difficult to learn in the classroom. They build on character but rely on experience. The ability of industrial hygienists to extend their knowledge and apply judgment and management skills, rather than just plugging data into a formula, is another personal reward of practicing the profession.

Yet another personal reward is the satisfaction of doing something in one's own life that can make a positive difference to someone else's life. For industrial hygienists, the someone else is ostensibly the employees, but good preventive health decisions also benefit the employers who might not otherwise make good choices for themselves and the community who might not otherwise have any advocate within the workplace. While the functions of anticipation, recognition, and evaluation are essential to industrial hygiene and are necessary precursors to control, it is only by controlling a hazard that anyone can actually make a positive change. It is only after a control has been implemented that exposures and the risks of adverse health effects are reduced, and we can say that we have made a positive difference.

Some IH jobs such as consulting or OSHA enforcement focus on recognition or evaluation, but rarely provide the opportunity to implement feasible and cost-effective solutions in a specific workplace that affects individual employees. IH researchers usually only see their findings implemented years later via new policies, methods, or technologies. It is primarily the purview of the plant hygienist working as part of a management team who can routinely see the beneficial impact of their work and knowledge. Yet no matter what role we play within the profession, if we all keep our eye on the eventual objective of control, we can all claim a proportion of the satisfaction of the accomplishments of the profession.

II. THE SPATIAL SPECTRUM OF EXPOSURE SCENARIOS

Much of the early part of this book focuses on the physical mechanisms that determine how people are exposed to airborne chemical hazards at work and how these same mechanisms can be used to reduce employee exposure. Exposure starts with a source of a chemical contaminant. The effect of factors controlling the rate that a given source can emit gases and vapors are predictable. Chapter 2 introduces the reader to (or hopefully reminds them of) the basic laws of physics that govern the behavior of gases and vapors. Chapter 3 discusses the basic laws of physics that govern the behavior of aerosols. Chapter 4 weaves the differences between gaseous hazards and aerosol hazards into a discussion of the standards and guidelines upon which decisions to control hazards are based. Chapter 5 and Chapter 6 explain the importance of vapor pressure in determining the rate at which vapors are generated (this intrinsic chemical property contrasts sharply with aerosols whose rate of generation depends almost entirely upon some external process independent of the kind of aerosol being generated).

Once an airborne hazard is generated by whatever mechanism, the pathway of the contaminant can be described by a perhaps surprisingly narrow range of exposure scenarios. Chapter 5 elaborates upon the thesis that virtually every workplace airborne chemical exposure can be categorized into one of the following four scenarios. For those readers who are familiar with the terms near-field, free-field, and reverberant field as they are used in noise, these same concepts pertain to the first three of these four chemical exposure scenarios. The last chemical exposure scenario of vapor equilibrium has no equivalence in the science of noise because noise dissipates literally at the speed of sound and cannot accumulate.

1. *People working within about an arm's length from the source, where its plume begins to form but does not have much time to dissipate.* A plume is nothing more than a highly concentrated current of air such as smoke from a fire (although gaseous plumes are usually not hot). While a person standing upwind of the source may not be in the often invisible core of the plume, turbulence can cause part of the contaminant to migrate

upwind the short arm's distance into their breathing zone. Control of exposures in this near-field is difficult and will focus on substitution with another chemical whose vapors need less dilution and on local exhaust ventilation to deflect the plume into a hood and out of nearby breathing zones.

2. *People working in the plume after it leaves the source by convection and begins to dissipate by turbulent diffusion.* In this scenario, the plume is not locally exhausted, nor is it yet constrained by the walls of the workroom. Control of exposure from a freely dissipating plume (where concentration within the plume will decrease slowly with distance from the source) is limited to personally avoiding the plume (a seemingly simple step if it were visible) or substituting the chemical in use with a less hazardous chemical, as above.
3. *People working where the plume is not completely exhausted and the chemicals recirculate within the airspace.* The concentration of a chemical in this more uniform reverberant field will rise to a level where the rate of contaminant generation will exactly balance the rate at which it is removed by general room air ventilation. General or dilution ventilation can control some airborne hazards in this scenario, but it is usually not the most cost-efficient way to control a chemical hazard.
4. *People working in a room or air space with a source of airborne chemical but little or no ventilation.* Without ventilation, airborne hazards will accumulate. An aerosol will accumulate with no upper limit, but a vapor concentration will eventually be limited either by the amount of liquid available to evaporate or by the liquid's vapor pressure. Work in such confined spaces is the *worst case* scenario.

Once the contaminant becomes an airborne plume, convection (the flow of air) is the primary mechanism causing the plume to move, and turbulent diffusion is the primary mechanism causing the plume to dissipate. Together, convection and turbulence largely control employee exposures in scenarios 1 and 2 above. Thermal buoyancy and sedimentation (particle falling velocities) are two secondary mechanisms that can be important in these scenarios but are more often important in scenario 3. Chapter 2 discusses buoyancy. Chapter 3 discusses aerosols. Chapter 5 discusses the primary source generation mechanisms, elaborates on the above exposure scenarios, and presents methods to predict exposure concentrations in scenarios 1, 3, and 4. Chapter 8 and Chapter 9 focus on how the principles introduced in the previous chapters can be used to reduce the rate at which contaminants are generated at their source. Most of the subsequent chapters focus on either removing the plume from the room (via local exhaust ventilation in scenarios 1 and 2, explained in Chapter 10 to Chapter 18) or diluting the plume within the room (via general ventilation in scenario 3, explained in Chapter 19 and Chapter 20). The final two chapters (Chapter 21 and Chapter 22) discuss how to control personal exposures in any of these scenarios where ambient concentrations are still too high, namely, how to protect employees via limiting their time of exposure or providing respirators and other forms of personal protective equipment. In contrast to learning about other occupational health hazards such as heat, noise, pressure, and radiation that are physically and pathologically unrelated, the topics in this book are closely linked and build upon themselves. Thus, the reader will reencounter the principles of Chapter 2 and Chapter 3 many more times later in this book.

III. THE IH PARADIGM OF CONTROL PRIORITIES

This book is also structured around a long-standing and well-justified paradigm for how industrial hygienists would prefer to control occupational hazards. A paradigm (par̍ədīm̍) is a good example often referred to for guidance as a pattern or model. Rarely do standards or guidelines specify the exact form by which a hazard in the workplace must be controlled. Thus, the professional is generally free to recommend the best way to control each hazard at each workplace. Over time,

TABLE 1.1
The Role of the IH Control Paradigm in the Structure of this Book

Traditional Preferences	Modern Preferences	Specific Control	Chapter No.
Engineering controls	Source controls	1. Substitution	8
		2. Source modifications	9
		3. Work practice modifications	9
	Pathway controls	4. Automation	9
		5. Separation	9
		6. Isolation	9
Administrative controls		7. Ventilation	10–20
	Receiver controls	8. Administrative scheduling	21
Personal protection controls		9. Personal protective equipment respirators, as a particular PPE	22

IH practitioners have developed a philosophic hierarchy of preferred controls referred to as "the industrial hygiene control paradigm." The IH control paradigm has been described in different ways at different times, but the basic concepts embodied in Table 1.1 are always the same. In its essence, the paradigm states that the best way to control a hazard is as physically close to its source as feasible.

The traditional statement: "Engineering controls are preferred over administrative controls are preferred over personal protection."
The modern statement: "Source controls are preferred over pathway controls are preferred over receiver controls."

Controlling the source means to change the origin or intervene in the process that creates the hazard. Proactive intervention means to recognize potential hazards during the design phase of a process or factory (before it is built). Making changes on paper (or in a computer plan), especially to the source of the hazard, is always less expensive than changing an existing hazard. The following "...tions" are all source controls:

1. Substitution — use a less hazardous chemical(s) or process.
2. Modification — change the layout, operating conditions, or work practices.
3. Automation — use robotics or computer-aided manufacturing (CAM).
4. Separation — place the source or employee in different locations.
5. Isolation — enclose major identifiable sources or the employee.

Controlling the pathway for respiratory hazards generally means to intervene in how the contaminant moves through the air from the source to the breathing zone, although the pathway can also lead to the skin (for dermal hazards) and rarely to the mouth (for ingestion hazards).

6. Ventilation is the major pathway control for the airborne route of exposure and can reduce the accumulation of chemicals on working surfaces, which will in turn reduce skin contact and secondary hand-to-mouth doses. General ventilation (such as in a classroom) is not a very cost-effective means to control most airborne hazards. A better means is local exhaust ventilation (LEV) that collects or contains and extracts the contaminant near its source via a hood, ducting, a fan, etc.

Controlling the receiver means to reduce the dose reaching employees through administrative control and personal protective equipment (PPE).

7. Administrative control of a hazard can be achieved through personnel rotation or some other form of schedule adjustment to reduce the time of exposure.
8. PPE for chemical hazards includes clothing, gloves, and respirators that intervene at the last possible moment in the chemical's trip from the source to the absorptive barriers of the body.

Both administration and personal protection are grouped as a receiver control in the modern statement of control preferences because they both require active participation by the exposed employee. While employee participation has some advantages and should be considered a part of any control, reliance upon their consistent cooperation is the least predictable of the three levels of control.

As can be seen in Table 1.1, if they can be implemented, the first two nonventilation controls are the most highly preferred options. Those source and nonventilation pathway controls outlined in Chapter 8 and Chapter 9 rely on the basic scientific principles introduced in the first five technical chapters (Chapter 2 to Chapter 6). Further application of nonventilation controls will evolve as a practicing hygienist develops a broad understanding of manufacturing processes generally, knowledge of the proprietary production conditions specific to their workplace, or/and gains access to commercially available devices, components, and technologies. Since access to such applied knowledge of individual workplaces is much more easily gleaned on the job than from a textbook or classroom, it comprises only a small proportion of this book.

Some of the following paragraphs elaborate on the controls listed in Table 1.1, and in some cases demonstrate how a given control may not always be categorized into just one category.

1. Substitution means changing to a chemical or manufacturing process from one with intrinsically high health and safety risks to one with lower health and safety risks. As will be pointed out numerous times throughout the book, the behavior of volatile chemicals (nominally solvents) is different from that of nonvolatile chemicals (such as solid metals, powders, dusts, and pastes). Both forms of chemicals can be inherently toxic; however, the vapor pressure of volatile chemicals is sufficient for them to become a hazard just sitting passively in the workplace whereas nonvolatile materials have no similar intrinsic property driving them to become airborne.
2. Modification means to change the physical operating conditions without changing the chemical or the manufacturing process. Modifications can include wetting a dust or powder, reducing a solvent's temperature to reduce its vapor pressure (its volatility), and training employees to anticipate and avoid the plume. Surface contamination is a secondary source that can contribute to all three routes of exposure (airborne, dermal, and ingestion); thus, good housekeeping to avoid and to remove surface contamination can be an important control, especially for low-volatility or nonvolatile contaminants. (Some references call training and good housekeeping an administrative control.)
3. Automation (in this context) is the use of some form of robotics to replace a function previously provided manually by an exposed employee. Thus, automation can be viewed as a form of increasing the distance between the source and employee by removing the employee. Similarly, some level of automation is necessary if a previously manually controlled source is to be physically separated or isolated from previously exposed operators. Automation has both advantages and disadvantages that are defined by the technology of automation itself, an often large financial commitment in both the hardware, software, and operator training, and the potential for psychological resistance that can hinder its wholehearted acceptance.
4. Separation means to increase the distance or to change the orientation between the hazard and the employee. To be effective, the change in separation generally needs to be large (a couple of exceptions to this generality will be discussed in Chapter 8). Separation is not a

reliable means of control for an invisible gaseous plume or if the plume's direction is not consistent. (Plumes will be discussed in Chapter 5.) If the plume direction is known or knowable, more control can be achieved by changing the directional orientation between the source and employee than by increasing the distance. The best way for separation to be a reliably effective control (with no caveats) is by complete relocation, which may also require some degree of automation.

5. Isolation means to separate the source and employee by a physical barrier such as an enclosure or at least a partial barrier such as a wall or partition. Isolation (similar to separation) may require some degree of automation. Enclosures usually require some amount of exhaust ventilation. Partial barriers are not as effective as an enclosure and often still require other controls such as ventilation or personal protective equipment.

The above controls all change the work process. However, some managers and employees feel threatened by change. Ventilation leaves the process unchanged, perhaps leading some industrial hygienists to conclude that exhausting the contaminant near its source is the only way to solve a chemical hazard, and terming local exhaust ventilation the "universal chemical exposure control." However, a more objective view clearly places substitution and modification controls at the head of the list. Automation, separation, and isolation leave the nature of the source unchanged but change the ability of the plume to reach the employees' breathing zones. Local exhaust ventilation can remove the plume from the workplace with little physical disruption, but it too has its real costs (money).

6. Ventilation means moving air. Simply mixing a plume into the room's air can dilute a contaminant for a short time but will expose everyone in the room to the diluted chemical. Diluting the plume into general room ventilation (fresh air) also usually causes more people to be exposed to the diluted chemical before it is eventually removed with the excess air leaving the room. Thus, general room or dilution ventilation can only be an effective control in certain limited circumstances. Local exhaust ventilation (removing the plume from the workspace near its source) is versatile and almost always a much more cost-effective control than general room ventilation. However, it should always be considered a secondary choice to source controls; and despite general room ventilation's common presence in nonresidential buildings, its limited ability to control airborne hazards dictates that it should only be considered a supplemental control, placing it even further down the list of preferences.

Receiver controls include administrative control and personal protective equipment. Both are considered the least preferred options by which to control an occupational hazard because they rely on individual behavior rather than strictly physical principles.

7. Administrative control means to reduce the dose of a hazardous agent, believing that the risk of disease is proportionate to dose. Dose can be reduced either by reducing the duration of exposure for each individual (for instance, by employee rotation) and by reducing the number of individuals exposed (for instance, by conducting an intermittent hazardous process only when no or fewer individuals are present). Some people include employee education, supervision, and good housekeeping as administrative controls. The approach taken herein is that employee education and supervision are such basic management elements that they should be used in conjunction with all of the eight controls listed here (although perhaps substitution (where it eliminates the hazard) can be an exception to this generality). Housekeeping (good or bad) is just another way to modify the work practice. Thus, administrative control is restricted herein to mean the control of the duration, frequency, and number of people exposed.
8. Personal protective equipment (PPE) for chemical exposures includes chemical-resistant clothing, gloves, and respirators. Respirators are a common control for airborne chemicals that cannot be controlled by other means; however, because respirators have many

limitations, respirator programs are expensive to manage well. A detailed description of the nature of respirators, their selection, and their limitations will be provided in Chapter 22.

Given the strikingly different exposure scenarios introduced above, it should not be surprising that there is no single way to control all chemical exposures or even any given chemical exposure. One control may be more costly to install or to operate in one setting but not in another setting. Ethnic, educational, and social differences among settings can have as much influence on the success of a given control as the technical issues. Cost-effectiveness also depends upon the corporate culture and market issues affecting the temporal horizon to which the use of the control can be foreseen. A good industrial hygienist will develop a wide arsenal of tools from which to choose the best control in any situation. Even within a chosen control, the hygienist must still choose the parameters that will affect its eventual effectiveness and work within the management policies that will affect its acceptance. The point is that IH is based on science, but the details of control are an acquired art.

IV. THE APPROACH, ORGANIZATION, AND PHILOSOPHY OF THIS BOOK

It is the perennial goal of educators in general (and this author in particular) to try to put "old heads on young bodies." Since learning builds on prior experience, the learning process followed by this book tries to build on prior academic training rather than expecting the student to have had prior field experience.

In attempting to teach chemical hazard control, this book tries to bridge the gap between the basic physical science principles and mechanisms that underlie the generation and dispersion of airborne chemicals and the wealth of chemical hazard control recommendations, techniques, and tools accumulated by previous generations of IH practitioners. This book assumes that the reader is academically trained in science and math, has seen at least a small number of manufacturing or other work settings with chemical hazards, but is inexperienced in the selection, design, implementation, or management of chemical exposure control systems. Thus, most of the technical chapters start with the principles of physical chemistry and Newtonian physics that the student should be able to recall with appropriate prompting.

The emphasis will be on learning *why* more than *what*: the logic being that if you understand why something happens or why two variables are related, you will be better able to understand the formula and how to use the relationship rather than just seeing a formula as a sequence of letters or symbols. Examples of this logic will be pointed out several times throughout this book.

While much of ventilation has its origins in engineering, no engineering background is expected from the reader. As a result, this book may be a science student's introduction to two classic engineering tools: empirical equations[a] and force, mass, and energy balances.[b] These tools

[a] Empirical equations are usually based on a theory or principle but include one or more empirical coefficients or powers found by regression to fit the equation to the quantitative experimental data. The following two examples are from future chapters: Equation 3.11a uses a C_{slip} correction to Stoke's law to predict the falling velocity of a small particle and Equation 5.2 predicts a solvent's vapor generate rate:

$$V_{fall} = C_{slip}(\rho_{net}\, d^2)g/18\eta \tag{3.11a}$$

$$G_{moles} = \kappa_{mass\ tx}[\text{Width}][\text{Length}]^{0.5}\frac{V^{0.6}\mathscr{D}^{0.67}}{\nu^{0.17}}(P_{vapor} - \text{ambient}\ P_{partial}) \tag{5.2}$$

[b] Mass or energy balances yield equations based on the principle that what goes into a system plus what is generated within that system equals what comes out of that system plus what is destroyed or accumulates within the system, and is depicted diagrammatically by Figure 19.1. This concept of balances can be applied to everything from biology, to air or water, to pollutants, and to personal finances.

will not be taught as methods in and of themselves, but they will be used repeatedly throughout the text. The book has many formulae, but the author has tried diligently to avoid using vague notation that is so endemic to engineering texts. The frequent use of long subscripts makes some formulae look longer than they might otherwise appear, but the longer, more explicit notation will hopefully be more meaningful. A one-page list of standard and therefore often less-than-explicit symbols used herein is included near the front of the book.[c] As a textbook, short derivations of equations are included to help lead at least some readers' transition from what they hopefully already know (or at least knew at one perhaps brief time in another class) to an end point that is useful in IH. Useful end points are typically boxed for later reference (see a later explanation of box notation). Industrial hygienists do not typically derive such equations in practice. If you find yourself getting too bogged down in a derivation, try going straight to the end point and then maybe only look back at the derivation as needed to explain its terms.

The succeeding chapters in this book have been organized into the following sequence and groups.

Chapter

2–3: The first two technical chapters cover the physics applicable to airborne contaminants. The first chapter focuses on the physical principles affecting gases and vapors, in particular on vapor pressure and vapor density that underlie the mechanisms of liquid evaporation and buoyancy, respectively. The next chapter covers the principles affecting aerosols. Differences between aerosols and vapors affect both hazard generation and their airborne behavior both within the environment and within the body. The examples used within these two chapters were selected more as learning tools about the physical principles and mechanisms than about specific workplace processes or settings.

4: This chapter uses several of the principles from the previous two chapters as it reviews some occupational health and safety regulations and exposure limits from a physical perspective rather than the perhaps more familiar toxicological or epidemiological perspectives.

5–6: The next two chapters discuss more physical principles: first, how physical principles can be used to predict gaseous hazards in the four universal exposure scenarios; then, how nonideal effects within mixtures of liquids can complicate otherwise simple predictions.

7: This chapter tries to give the reader a flavor for the multiplicity of human dimensions that go into the art of industrial hygiene. Topics covered include the sources of information about technologic processes and alternatives, some of the personal and group dynamics that affect decisions to change, and a brief introduction to the economics constraining change.

8–9: These two chapters cover source controls and the important nonventilation pathway controls. The first chapter is devoted to substitution as a chemical hazard control. The next covers modification and the other pathway controls. Each of these chapters tries to show how the above principles can be applied to control chemical hazards without adding ventilation.

10–12: The first of these transition chapters introduces the topic of local exhaust ventilation. The following chapter discusses the physical relationships between air velocity and flow rate and between energy and measurable air pressures that underlie the function,

[c] A few exceptions to clear notation are made where the author uses a notation already accepted within the profession to avoid later confusion that might result from a reader trying to adapt from a clearer notation to that professional norm.

design, and operation of ventilation systems. The last of these transition chapters introduces the reader to the tools used to measure ventilation velocities and pressures.

13–18: Six chapters are devoted to local exhaust ventilation, a staple of the industrial hygienist's arsenal of controls. This sequence of chapters begins with the local exhaust hood, then progresses (as does the air) to the exhaust ducts, to air (emission) cleaners, and to the fans that make the system work. Two more chapters cover the costs to operate a ventilation system and some management programs to help assure that an installed system keeps on protecting employees.

19–20: The next two chapters discuss the ability and limitations of general room and building ventilation to control the accumulation of airborne chemical hazards created in the third and fourth of the above universal exposure scenarios.

21–22: The final two chapters discuss the ability and limitations of protecting the receiver via administrative controls or personal protective equipment in general, and respirators in particular. Despite their limitations, the IH control paradigm leads one to choose PPE where none of the other controls are feasible; thus, an IH must know how to choose the appropriate PPE and how to manage a PPE program.

This book contains more information than a student needs to learn at one time. The excess will hopefully provide a professional resource after graduation. To guide the student, each of the above chapters is preceded by specific learning goals. Each learning goal is preceded by one of the following notations. The two box notations should correspond to the notation bracketing the material within the text and to the key phrase used within the learning goal. The third notation highlights important concepts that do not necessarily have a specific focal point within the text. Your instructor may modify these learning goals. Trying to read only the portions of text that pertain to these learning goals runs the risk of failing to see their context.

☐ Something that appears in this book within a solid box (not dotted) means that it is important to know thoroughly. The reader should be able to use what is in the box without having to look it up. The author considers information enclosed in a solid box to be either central to or underlying the science of IH. Where the item is an equation, learn it as a concept not just as a formula.

⬚ Dotted boxes are used in this book to highlight important tools or information that the reader may want to find again later (or need to find again quickly) but it is not expected or intended to be memorized. The student should still know how or be able to use the enclosed information with the book open at that page. During an examination is not the time to try to use an equation for the first time.

✍ The *take note* symbol is used where a learning goal summarizes concepts that may not have a single focus in this book that could or should be boxed in either of the above ways. To paraphrase Einstein very loosely, not everything that counts can be boxed.

✓ Rather than highlighting an explicit learning goal, this notation is used within the text to highlight a rule-of-thumb, perhaps a summary, or a trend that is intended to give the student a practical (field-usable) perspective on a topic. A rule-of-thumb is one way "to put an old head on a young body."

Each learning goal also incorporates a key phrase loosely adapted from Bloom et al.'s "cognitive domains."[1] The idea behind these key phrases is that one does not need to know everything at the same level. The implied meaning or level of learning that (at least this author thinks) should be sought at this time in the reader's education is listed below for each key phrase.

Phrase	Implied Meaning
"Know … "	Knowing means more than just memorizing the words or symbols in a box. Knowing includes also understanding the concept and how it might be used in IH practice. This is especially true for boxed equations; they are more than just letters and numbers, and should be viewed as a shorthand method of expressing a concept
"Be able to … "	Ability is a small step short of *to know* as defined above. One goal of *be able* is to be able to translate an IH problem as presented into terms from which the right equation or principle can be applied. Being able to apply concepts to real-world problems lends itself to open-book questions
"Know how … "	Know how to use the information for an IH purpose; usually said of a figure, graph, or equation that definitely does not have to be memorized
"Understand … "	Be able to discuss, evaluate, or interpret the concept. Often related to problems where a nonquantitative or qualitative answer is expected; e.g., what are the factors that determine an outcome? Or, in what direction would the outcome change for a given change in one of its underlying factors?
"Be familiar with… " or "Be aware of … "	Recognize the listed term or concept if it is presented; be able to fit it into or separate it from a list; or at the very most, be able to discuss the term or topic in a short essay question

Chapter 10 to Chapter 18 of this book heavily reference the ACGIH *Ventilation Manual* (often referred to here as simply the *Manual of Industrial Ventilation*).[2,d] While the *Manual of Industrial Ventilation* is a valuable working tool, the industrial hygiene student does not need to know everything in the *Manual of Industrial Ventilation*. Thus, this book is selective in its coverage of the *Manual of Industrial Ventilation*'s ventilation topics. This book also goes well beyond the *Manual of Industrial Ventilation* on the topics of nonventilation and receiver controls. Taken as a whole, this book provides a unique, comprehensive tool to learn the challenging yet rewarding role that IH can play in controlling chemical health hazards at work. The engineering student or specialist who wishes to go beyond the level of ventilation design expected of most competent industrial hygienists may be better served by other texts written by and for engineers such as Heinsohn, Burton, or Goodfellow and Tähti.[3–5]

Admittedly, not every industrial hygienist works very much with ventilation. Few probably design a full ventilation system on their own. However, almost all hygienists will encounter ventilation and will work with people who are in charge of ventilation at their facilities. For better or for worse, students rarely know whether or not ventilation will be a major responsibility in their individual future careers. Numerous past graduates, who as a student thought ventilation was a waste of time, in fact, got their first job because of that particular knowledge. Others unexpectedly found themselves in charge of monitoring and improving ventilation systems at their facility. This book is intended for you, the IH student, and your future, in the expectation that with knowledge the best time of your life is yet to come.

REFERENCES

1. Bloom, B. S., Englehart, M., Furst, E., Hill, W., and Krathwohl, D., *Taxonomy of Educational Objectives: The Classification of Educational Goals Handbook I: Cognitive Domain*, Longmans, Green, New York, 1956.

[d] This book cites the 24th edition (2001) but recent editions such as the 23rd (1998) or 22nd (1995) only differ incrementally.

2. Committee on Industrial Ventilation, *Industrial Ventilation: A Manual of Recommended Practice*, 24th ed., American Conference of Governmental Industrial Hygienists, Cincinnati, OH, 2001.
3. Heinsohn, R. J., *Industrial Ventilation: Engineering Principles*, Wiley, New York, 1991.
4. Burton, D. J., Ed., *Hemeon's Plant and Process Ventilation*, 3rd ed., Lewis Press, Boca Raton, FL, 1999.
5. Goodfellow, H. and Tähti, E., Eds., *Industrial Ventilation Design Guidebook*, Academic Press, London, 2001.

2 Basic Gas and Vapor Behavior

The learning goals of this chapter:

- ✍ Understand why partial pressure is an expression of concentration.
- ☐ Know how to calculate the equilibrium ppm at NTP from vapor pressure $= [P_{vapor}\ \text{mmHg}] \times 10^6/760$.
- ✍ Know the effect of atmospheric pressure on a saturated vapor's ppm and mg/m^3 concentrations.
- ⬚ Be able to calculate a vapor pressure if given an Antoine equation.
- ☐ Know how to approximate the change in vapor pressure with a change in its liquid temperature.
- ☐ Understand Dalton's law, particularly that $P_{partial} = Y_i \times P$. (Of course, know what Y_i is.)
- ☐ Be able to convert between molar fraction (Y_i) and ppm.
- ☐ Know how to convert between ppm and mg/m^3 (usually covered in a previous IH course).
- ✍ Understand the circumstances that determine whether ppm or mg/m^3 will change or remain constant.
- ☐ Know how to calculate the density of any gaseous material at NTP if its MW is known.
- ⬚ Know how to calculate the density of a known vapor relative to the density of air at NTP.
- ✍ Understand how vapor density can affect the vertical movement of a plume leaving a source.

I. GAS AND VAPOR DEFINITIONS

1. Normal temperature for environmental and occupational health purposes is clearly different from the standard temperature of 0°C used for chemistry and physics laboratory purposes. Standard temperature is too cold for human comfort or for safe work without the appropriate clothing. Normal temperature is in the range of human comfort and long-term human habitation without special clothing. Unfortunately, not everyone uses quite the same definition of "normal" temperature.
   a. 25°C or 77°F was assumed to be normal by the people who popularized the common molar volume of 24.45 L. This is a larger volume than the 22.4 L that one mole of an ideal gas occupies at STP like you (hopefully) learned in chemistry because gases expand as they are heated. While 24.45 is one of industrial hygiene's "magic numbers" that just must be learned, the temperature to which it corresponds is probably a bit warmer than most indoor settings or even many outdoor settings (of course, the latter varies by geographic region, season, and time of day).
   b. 21°C or 70°F was assumed to be normal by the ACGIH Ventilation Committee and is used in their *Industrial Ventilation Manual*.[1] This value is a typical thermostat setting in controlled indoor environments, in homes, or institutional buildings and is a nice round Fahrenheit value. (70°F actually corresponds to 21.1°C, but the difference of 0.1°C is inconsequential in field practice.)

c. 20°C or 68°F was assumed to be normal by NIOSH in publications such as their *Pocket Guide to Chemical Hazards.*[2] It is also the minimum temperature specified by the International Building Code that a space-heating system should be able to maintain (IBC Section 1204).[3] This value may be a bit too cool for normal temperatures during the summer season, but it is a nice round metric value (and another (along with 25°C) exact integer conversion between °C and °F). A molar volume at this temperature would be 24.04 L, but this value is not common in the IH literature.

The lack of a universally agreed upon "normal temperature" in occupational health is awkward, but the professional literature has several such nuances. The above range of normal temperatures generally has a small impact on most practical (real world) problems.[a] Any real world problem should state or be defined at a specific temperature. By default (or should a temperature be left unstated), this book will assume that "normal" is 25°C or 77°F.

Temperatures expressed in degrees Celsius or Fahrenheit could be said to be (but are rarely called) "relative" temperatures. Their zero values correspond to the freezing temperatures of water and of a 50:50 salt-to-water mixture, respectively. Each of these relative temperature scales also has a corresponding absolute temperature scale in which each degree is the same size as in their respective relative scales, but the zero value is based on the theoretical absence of all molecular energy. The chemistry student will, no doubt, be familiar with the Kelvin scale (given by Equation 2.1), while the engineering student may have also been introduced to the Rankine scale (given by Equation 2.2).

$$\text{Kelvin}: \ {}^\circ\text{K} = {}^\circ\text{C} + 273.15 = (5/9)\,{}^\circ\text{R} \approx {}^\circ\text{C} + 273 \tag{2.1}$$

$$\text{Rankine}: \ {}^\circ\text{R} = {}^\circ\text{F} + 459.67 = (9/5)\,{}^\circ\text{K} \approx {}^\circ\text{F} + 460 \tag{2.2}$$

Some IH students may remember the use of absolute temperature in degrees Rankine from studying radiant heat stress. All four of these temperature scales will be used herein from time to time.

2. Pressure (P) is the force (F) per unit of surface area (A), as stated in Equation 2.3a. Of course, the force exerted on an object or flat surface is the product by the pressure perpendicular to that surface times its area as stated in Equation 2.3b.

$$\text{Pressure} = P = \frac{F}{A} = \frac{\text{Force}}{\text{Area}} \tag{2.3a}$$

$$\text{Force} = F = P \times A = \text{Pressure} \times \text{Area} \tag{2.3b}$$

Pressure can exist within solids as tension (pulling), compression (pushing), shear (as in friction or sliding parallel to a surface), or torsion (twisting). Fluids create pressure by the inertia of its molecules striking a surface in which it is in contact. One normal atmosphere of air pressure is created by more than 10^{23} molecules impacting on each square centimeter of any surface every second. This static pressure is exerted perpendicular to any surface either immersed in or containing the gas.[b] Air pressure is a part of meteorology, aerodynamics, pneumatics, and ventilation. We will see later in this chapter that pressure is also related to the airborne concentration of a gas or vapor (as in the

[a] For instance, you will soon learn that vapor pressures approximately double every 12°C. Therefore, a difference of 5°C corresponds to approximately five twelfths or a 40% difference in evaporation rates and vapor concentrations. In this author's experience, a predictive difference of this magnitude will not be decisive without generating real exposure data.

[b] In contrast, the friction of air or any other fluid moving over any surface will produce a shear force parallel to that surface.

oxygen that we need to survive or the airborne chemical hazards that IHs are attempting to control). Common units of pressure are listed in Table 2.1 along with the normal values for one atmosphere in these units.[4]

a. If you work in a mountainous region or in the aviation industry, the fact that pressure and air density decrease with the altitude will be important. Equation 2.4 can be used to anticipate the ambient pressure at any habitable altitude above sea level.[5] Equation 2.4d contains a particularly simple set of values to remember.

$$P \text{ at altitude} = P \text{ at sea level} \times e^{(-\text{altitude in feet}/25{,}970)} \quad (2.4a)$$

$$= 29.92 \times 2^{(-\text{feet}/18{,}000)} \text{ inches Hg} \quad (2.4b)$$

$$= 760 \times 2^{(-\text{feet}/18{,}000)} \text{ mmHg} \quad (2.4c)$$

$$P \text{ at altitude} = 2^{(-\text{feet}/18{,}000)} \text{ atmosphere} \quad (2.4d)$$

Equation 2.4 can also be used for altitudes in meters by substituting 7915 m for 25,970 ft or 5500 m for 18,000 ft. These equations will be used with Dalton's law (the effect of ambient pressure on partial pressure will be discussed in Section III) and with the Ideal Gas law (the effect of ambient pressure on air and vapor density will be discussed in Section IV). Changes in air density due to altitude will also affect rotameters, anemometers, human physiology, the calibration of direct reading gas or vapor meters, and some noise meter calibrators.

b. As in temperature, pressures can also be either absolute or relative. An absolute pressure (such as those in Table 2.1) is measured above the zero that exists in a complete vacuum. Because absolute pressure gauges must contain a vacuum as an internal reference, accurate absolute pressure gauges are more expensive and larger than relative pressure gauges. A mercury-filled barometer is a primary absolute pressure gauge. Mechanical absolute pressure gauges are manufactured commercially and will invariably be so noted on their face. For instance, a pressure gauge marked "psia" measures in pounds per square inch absolute (i.e., pressure above the zero of a vacuum). Because thermodynamic properties and chemical reaction rates are dependent upon the absolute pressure and temperature of the reactants, most

TABLE 2.1
One "Normal" Atmosphere in Various Units of Pressure

	English Units	Metric Units
Force per unit area	14.696≈14.7 psi (where psi is pounds per square inch)	101,325 Pa[a] (1 Pascal = 1 N/m^2)
Barometric height[b]	29.921 in. Hg (mercury)	760.0 mmHg
Ventilation[c]	407.60 in. of water	1035.3 cm of water

[a] A "bar" is a pressure term used in meteorology meaning 10^6 dynes/cm^2, 105 N/m^2, 10^5 Pa, or 0.986923 atm. Thus, one normal atmosphere is also 1.01325 bar.

[b] A barometer measures pressure by the height that a liquid (normally mercury) with nothing above it (ideally a vacuum except for the liquid's vapor pressure) can be pushed up into a glass column.

[c] Small differences in pressure (as in ventilation) are often measured by the height of a column of water. Although no one measures atmospheric pressure with a water-filled barometer, the basis for a value of 407.60 in. of water at 21°C (70°F) is documented in an appendix to Chapter 11.

pressure gauges on chemical reactors are in units like psia. The body also responds to, or performs in relation to, absolute pressure, and physical relationships such as Dalton's law and the Universal Gas law use absolute pressure.
The barometric pressure reported with the evening weather (e.g., 29.92 in. Hg) would be an absolute pressure, except that the pressures in weather and aviation reports are adjusted to what the pressure would be at sea level at the reporting location. They are not real except at sea level.

c. Relative pressures are measured in comparison to the ambient atmospheric pressure. A positive pressure relative to the ambient atmosphere is commonly called a "gage" or "gauge pressure," and a relative pressure meter will usually (but not always) indicate its units of measurement with an appended "g" such as "psig," meaning its units are a gauge or relative pressure.[c] Equation 2.5 may be used where one needs to convert between absolute and relative pressures.

$$P_{\text{gauge}} = P_{\text{absolute}} - [\text{the ambient atmosphere in the units of } P] \quad (2.5a)$$

$$P_{\text{absolute}} = P_{\text{gauge}} + [\text{the ambient atmosphere in the units of } P] \quad (2.5b)$$

$$\text{For example, psia} = \text{psig} + 14.7 \text{ psi at NTP} \quad (2.5c)$$

Gauge pressures are particularly important in safety and ventilation. The outward stress on a pipe or pressure vessel is determined by gauge pressure because one ambient atmosphere is always on the outside of the vessel pushing in. Air will flow in response to differences in static pressure. The small pressure differences applicable to ventilation are conveniently expressed by the height difference of two columns of water in a manometer (to be covered in Section II of Chapter 12). The abbreviation "wg is used in the American ventilation literature to mean inches of water gauge. Neither relative pressure nor pressure differences have a normal value (cf., the normal absolute pressures in Table 2.1).

3. When dealing with differences (e.g., $\Delta T = T_2 - T_1$ or $\Delta P = P_2 - P_1$), then one can use either relative or absolute values (you will obtain the same answer). However, when dealing with ratios (e.g., T_2/T_1 or P_2/P_1), then one must use absolute values (or obtain the wrong answer). Choose wisely.

4. Partial pressure is the pressure exerted by the molecules of one gas or vapor that exists in combination with other gases or vapors that comprise the total atmosphere. Thus, the term "partial" means a given pressure is only a portion of the total pressure. Partial pressure is often abbreviated P_i where the subscript i refers to the partial pressure of the generic ith component within a mixture. This book will use the slightly more expressive P_{partial} or occasionally the subscript will be a named chemical. P_{partial} is commonly expressed in units of either mm Hg or atmospheres; however, the reader will soon be shown, via Equation 2.14, that a partial pressure is equivalent to an airborne chemical concentration that can also be expressed in the traditional IH units of mg/m^3 or ppm.
5. Molar fractions can be encountered in either liquid or gaseous mixtures, i.e., when more than one chemical is present. Molar fractions as expressed by Equation 2.6 or Equation 2.7 can be either a proper or decimal fraction, and values should always be entered into an

[c] A negative pressure relative to the ambient atmosphere is commonly called a vacuum pressure. Vacuum pressure can be expressed either as a relative pressure (viz., the ambient atmosphere − the low absolute pressure, wherever it is) usually in terms of head of mercury, e.g., inches or millimeters of Hg or as an absolute pressure in terms such as "torr" (meaning mmHg).

equation as such. One can convert an environmental concentration into a fraction by dividing a percentage (%) by 100 or parts per million (ppm) by 10^6, respectively. Recall that a mole is Avogadro's number of molecules (1.9×10^{23} molecules); thus, both of the following equations are also the ratio between moles within a mixture.

a. The symbol X_i is (by convention) always used to denote the molar fraction in a liquid.

$$X_i = \frac{\text{(Number of } i \text{ molecules within a liquid mixture)}}{\text{(Number of all molecules within a liquid mixture)}} \tag{2.6}$$

Unfortunately liquid mixtures are more commonly expressed as either weight or volume fractions than as molar fractions. The conversion from these other definitions of a mixture's composition into molar fractions will be explained in Chapter 6.

b. Similarly, the symbol Y_i is (again by convention) always used to denote the molar fraction in a vapor or gas.

$$Y_i = \frac{\text{(Number of } i \text{ molecules within an airborne mixture)}}{\text{(Number of all molecules within an airborne mixture)}} \tag{2.7}$$

Because a mole of any two gases or vapors at the same T and P would comprise the same volume, the Y_i molar fraction is also a volume-to-volume ratio. An air concentration expressed as a percent or ppm always implies a molar ratio. Dalton's law (to be discussed in Section III) states that a molar fraction is also a pressure-to-pressure ratio.

The molar fractions of gases and vapors are physically synonymous with ppm, and thus comprise a central topic in industrial hygiene. We will see in Section III how partial pressure, molar fraction, ppm, and mg/m^3 are related via Dalton's law. We will also see in Chapter 6 how physical and chemical conditions in the liquid can affect airborne concentration. In fact, it may help you to remember the X_i and Y_i notation to glance at the X-Y diagrams in Figure 6.4 and think of the airborne Y_i as the dependant variable that varies with the independent liquid X_i variable.

6. A "closed room" is simply an unventilated room (one without either natural or mechanical ventilation). Note that a closed room is not a hermetically sealed room (to construct an airtight room, chamber, or vessel requires special materials and construction methods). Gaps in doors, windows, and even pores in common wall materials allow the air pressure inside any closed room to always stay at the same pressure as the ambient conditions outside the room. Thus, if gases or vapors are released from a source inside a closed room (such as by evaporation), the added airborne molecules will cause some leakage of molecules from inside the room to seep outside so that the total pressure inside the room stays at whatever the ambient atmospheric pressure is outside the room. Similarly, if the air temperature inside the room increases (or decreases) causing a change in air density, some air will escape from the room (or enter into it) to keep the pressure inside the room equal to the ambient atmospheric pressure outside the room. Ventilated rooms maintain a very small pressure difference between inside and outside the room; less than 0.01% of an atmosphere is enough to cause ventilation air to move (to be discussed in Chapter 11 and Chapter 20).
7. A gas is the airborne state of a chemical whose liquid is so volatile that its vapors cannot reach an equilibrium with its liquid (in comparison to a vapor that forms an equilibrium with its liquid state).[d] If such a chemical were present in an open container in a closed

[d] A more technical definition of a gas is a chemical whose vapor pressure at normal room temperature exceeds one atmosphere; however, "vapor pressure" will not be defined for a couple more paragraphs.

room, all of the liquid would evaporate. In fact, if the liquid from of a gas were poured into an open container at room temperature, it would evaporate so fast that it would appear to boil because the ambient pressure is less than the gas's equilibrium partial pressure. If a gas is to be stored at normal temperature, it must be stored under a pressure of more than one atmosphere (e.g., in a gas storage cylinder). Alternatively, if the gas is to be stored under normal pressure as a liquid, it must be stored in a well-insulated container at a temperature below its boiling point. (Such very low temperatures are called cryogenic.) The use of the term "gas" herein will also apply to a chemical that is so volatile as to form a gas.

8. A vapor is the airborne state of a chemical which, if a sufficiently large amount of liquid were released into a closed room at normal temperature, would not completely evaporate but rather would reach an equilibrium with its liquid. By this definition, the molar fraction of a vapor that is in equilibrium with its liquid source must be less than one (otherwise it would be a gas).[e] A chemical engineering model to be introduced in Chapter 5 pictures the vapor molecules immediately above an evaporating liquid source as if they are in equilibrium with its evaporating liquid, although the vapors will start to be diluted as soon as they leave that source. While gases and vapors differ in the conditions at which they will reach equilibrium, each airborne molecule of a vapor and gas acts independently from each other molecule. Thus, airborne gases and vapors will behave in identical ways.
9. Vapor pressure (P_{vapor}) is the partial pressure exerted by the airborne molecules of a compound that is in equilibrium with its liquid. To be in equilibrium, the rate that molecules are leaving the liquid (evaporating) must be equal to the rate that airborne molecules are being reabsorbed back into the liquid. The resorption rate depends upon the rate at which that chemical's airborne molecules strike each unit surface area. That rate depends upon the concentration of airborne vapor molecules (moles/m^3) at a given T. The rate that molecules are hitting a surface at a given temperature determines the pressure exerted by those molecules, viz., its vapor pressure (see definition 2 above).

   The equivalence between vapor pressure (or any partial pressure) and concentration in the units common to IH (mg/m^3 and ppm) is simply applied chemistry. The chemistry student learns that a mole of any gas or vapor will occupy 22.4 L at a standard temperature of 0°C and one atmosphere of pressure (760 mmHg). At a normal temperature of 25°C, this molar volume is 24.45 L. Dalton's law (to be described in Section III and Equation 2.13) states that the fraction of the molecules in a mixture and the fraction of a mixture's total volume are both based on the ratio of its P_{vapor} to the total atmospheric pressure (normally 760 mmHg). The mass of a mole of any chemical will equal that chemical's molecular weight. (Still chemistry, right?) Thus, the concentration in mg/m^3 of any vapor in equilibrium with its liquid can be predicted as that fraction ($P_{vapor}/760$) of its molecular weight ($\times 10^3$ to be in mg) in a molar volume ($\div 10^{-3}$ to be in m^3). Equation 2.8 is industrial hygiene.

$$C_{equilibrium}(\mathrm{mg/m^3}) = \frac{P_{vapor}\mathrm{mmHg} \times \mathrm{MW} \times 10^6}{760\ \mathrm{mmHg} \times 24.45} = 53.82 \times P_{vapor} \times \mathrm{MW} \quad (2.8)$$

✍ By virtue of its definition, a chemical's vapor pressure is always equivalent to its absolute airborne concentration in terms of mg/m^3, given by Equation 2.8, no matter what the ambient pressure actually is. In contrast, the ratio of vapor molecules to air molecules

[e] In contrast to a gas, the more technical definition of a vapor is gaseous state of a chemical whose vapor pressure at normal temperature is less than one atmosphere ("vapor pressure" is defined in the next paragraph).

(expressed, for example, in terms of ppm) will change when the density of the air molecules changes as with air temperature, altitude, or the weather independent of the volatile chemical's vapor pressure or absolute concentration. While Equation 2.8 is based on a definition with constant physical parameters, and thus will give the same answer at any ambient pressure, the result of Equation 2.9 will depend upon the pressure in a given setting. As you will see, the conceptual simplicity of Equation 2.9 will encourage its use throughout this book even though the ambient pressure is not always the "normal" 760 mmHg.

$$\text{ppm}_{\text{equilibrium}} = \frac{P_{\text{vapor}} \times 10^6}{P_{\text{ambient}}} = \frac{P_{\text{vapor}}\ \text{mmHg} \times 10^6}{760\ \text{mmHg}} \quad \text{at NTP} \tag{2.9}$$

a. Vapor pressure (P_{vapor}) is a direct indicator of a chemical's volatility. A chemical with a high vapor pressure is highly volatile, meaning it will evaporate rapidly. A chemical with a low vapor pressure is less volatile. However, all materials have a vapor pressure. Even solids or high molecular weight liquids have a very low vapor pressure ($\ll$ 1 mmHg) that can still be important if the material is sufficiently toxic (e.g., liquid mercury) and any given chemical's vapor pressure will increase with its liquid temperature, as will be discussed in Section II of this chapter.
b. P_{vapor} is the maximum concentration at which a chemical can exist as a vapor at a specified temperature. When the concentration is at a chemical's vapor pressure, the vapor is said to be saturated. Thus, the term "saturated vapor" is equivalent to a vapor and liquid being in equilibrium. The only way that an airborne chemical can be more concentrated than its vapor pressure is if it were present as an aerosol. When a saturated vapor concentration is cooled, the vapor will start to condense into either a visible fog (what some might call steam, or what an IH should call a mist) or a fume (solid particles).
c. For a whole closed room to come into equilibrium with its liquid, enough liquid must initially be present to generate enough vapor molecules to fill the entire enclosed airspace at its vapor pressure concentration (with some residual liquid left over). The criterion for "enough liquid" will be discussed in Section VII of Chapter 5. Examples include a shipping, storage or dispensing container that has just been opened, or a spill inside a poorly ventilated (closed) room.
d. P_{vapor} is also an important predictor of the airborne concentration in ventilated workplaces or in a plume whether indoors or outdoors. A vapor is always considered to be in equilibrium right at its air-liquid boundary. Therefore, P_{vapor} is the concentration that always exists right at any liquid's surface. Fortunately, the vapors in which most people work are much more dilute than their vapor pressure (with the notable exception of water vapor at high humidity). However, because the dominant mechanism that dilutes hazardous plumes in most workplaces is independent of the kind of vapors in the plume, a vapor's concentration anywhere in a room is an at least theoretically predictable fraction of its vapor pressure concentration at its source.

Vapor pressure has been abbreviated as P_0, P^0, P_V, P_{VP}, P_{sat}, or even VP in some other chemistry, industrial hygiene, and engineering texts, such as those listed in Table 2.2. Unfortunately for IH students, VP is used in both the industrial hygiene and ventilation literature to denote velocity pressure, a term that is related to the kinetic energy of moving air (the term is first used in Chapter 10 but will be formally defined in Chapter 11) and used extensively both when designing ventilation systems (e.g., sizing ducts, selecting a fan, etc. in Chapters 14 and 16) and later when assessing their performance (Chapter 18).

The more literal abbreviation P_{vapor} will be used herein to help in clarity for the reader and to differentiate it further from the abbreviation VP for velocity pressure.[f]

II. VAPOR PRESSURE CHANGES WITH LIQUID TEMPERATURE

The temperature of a solvent can vary independently from the air temperature into which its vapors disperse. Such differences in temperature can occur, for example, when the solvent resides on a surface (e.g., a machine or a working or walking surface) whose temperature is different from the surrounding air. (Such surfaces are usually hotter than room temperature; cooler temperatures are possible but rare in the workplace.) The effect of temperature on vapor pressure will be explained below.

1. Temperature is a result (and an indication) of the kinetic energy of individual molecules. The more kinetic energy that a molecule has, the more likely it is to leave the surface of a liquid and become airborne. The greater the tendency for molecules to become airborne, the greater will be their equilibrium vapor concentration. Thus, heating a liquid will increase its vapor pressure, and cooling a liquid will decrease its vapor pressure. Conversely, a single vapor pressure (gleaned from say a chemical's MSDS or a handbook) is only applicable at a single temperature. For a single vapor pressure to be useful, the reference temperature for that vapor pressure must be known, and because not everyone uses the same normal temperature, a P_{vapor} value without a known reference temperature can be misleading.
2. The variation of vapor pressure with its liquid temperature is nonlinear. That is, P_{vapor} changes more rapidly than the change in liquid temperature. While P_{vapor} at normal temperature varies widely among chemicals (values of P_{vapor} in the appendix to this book span eight orders of magnitude), the relative rate at which vapor pressures change with temperature is strikingly similar among chemicals.

   ✓ As a rule-of-thumb for organic solvents, P_{vapor} will approximately double from that at NTP for an increase in liquid temperature of approximately 12°C or 21°F (with a 95% variability in P_{vapor} of $\pm 70\%$).

   The $\pm 70\%$ means that extrapolations beyond one or two doublings in pressure are not very precise.
3. The generic Antoine equation (developed *ca* 1888) is a common, reasonably accurate (errors are usually within $\pm 1\%$), but not unique, mathematical description of the relationship between P_{vapor} and the liquid temperature for an individual chemical. A common form of the Antoine equation is shown in Equation 2.10, but be aware that Antoine equations come in several mathematical variations.

$$\log[P_{vapor}] = A - \frac{B}{(C+T)} \quad \text{the generic Antoine Equation} \quad (2.10)$$

Such variations include the natural logarithm ($\ln[P_{vapor}]$) instead of the common logarithm ($\log[P_{vapor}]$), T as °C instead of °K (which would shift the range of most C coefficients from positive to negative, respectively), or the C coefficient may be omitted altogether (the latter case will be discussed *ca* Equation 2.11). The "Instructions for Using the Vapor Pressure Table" in Appendix B of Perkins (1997) provides some examples of these variations.[6] However, all of these variations are mathematically equivalent and each

[f] The notation P_{vapor} implies the vapor pressure of a pure chemical, whereas the notation $P_{vapor,i}$ will be used extensively in Chapter 6 to denote the vapor pressure of the *i*th component within a liquid mixture of volatile chemicals.

TABLE 2.2
Some Published Sources of Vapor Pressure Data Listed in Chronological Order

Single Temperature	Antoine Coefficient	Publication Citation
√	√	Dreisback, R.R.: *Physical Properties of Chemical Compounds*[7–9]
	√	Boublik, Fried, and Hala: *The Vapor Pressures of Pure Substances*[10]
	√	Ohe, S.: *Computer Aided Data Book of Vapor Pressures*[11]
	√	Reid, Prausnitz, and Sherwood: *The Properties of Gases and Liquids*[12]
	√	Yaws: *Handbook of Vapor Pressure*[13]
√	√	Perkins: *Modern Industrial Hygiene*[6]
√		Howard and Meylan: *Handbook of Physical Properties of Organic Materials*[14]
√		*CRC Handbook of Chemistry and Physics*[4]

is easy to use as long as the user notes and uses the right form pertinent to the individual source of data. Each source of Antoine coefficients listed in Table 2.2 clearly shows the form of the equation they use.

Each set of Antoine coefficients listed in Table 2.2 is also presented with a temperature range over which it is purported to be accurate. Because much of the pressure–temperature data was generated for use in industrial distillation, refining, and reaction processes, the temperature range may be higher than the temperatures normally of interest to IH. Extrapolation beyond those limits will result in degraded accuracy that may still be acceptable for IH purposes but should be so noted in any of your written documentation. A search across multiple sources may be necessary to find a set of coefficients that covers, or better approximates, your temperature range of interest. Of these sources, Perkins' IH text is good for Antoine coefficients in the normal temperature range. Howard and Meylan probably have the most extensive chemical coverage of single point P_{vapor} data at 25°C, plus it has molecular structure diagrams and Henry's constants (the latter is to be discussed in Chapter 6).

Example 2.1

Find the vapor pressure at 12°C intervals above and below 25°C (viz., at 13 and 37°C) using the following Antoine coefficients from Ref. [10] for the common log of vapor pressure of toluene in mm Hg and temperatures between 0 and 50°C.

$$A = 6.93325 \qquad B = 1335.533 \qquad C = 218.8$$

The P_{vapor} values in the right columns of data shown below are the antilog of the $\log[P_{\text{vapor}}]$ values shown in the left columns of data.

$$\log[P_{\text{vapor}}] = A - \frac{B}{(C+T)} \qquad P_{\text{vapor}} = 10^{\log P_{\text{vapor}}}$$

$$\log[P_{\text{vapor}}] \text{ at } 13°\text{C} = 6.93325 - \frac{1335.533}{(218.8+13)} = 1.1717 \qquad P_{\text{vapor}} \text{ at } 13°\text{C} = 10^{1.1717} = 14.8 \text{ mmHg}$$

$$\log[P_{\text{vapor}}] \text{ at } 25°\text{C} = 6.93325 - \frac{1335.533}{(218.8+25)} = 1.4542 \qquad P_{\text{vapor}} \text{ at } 25°\text{C} = 10^{1.4542} = 28.5 \text{ mmHg}$$

$$\log[P_{\text{vapor}}] \text{ at } 37°\text{C} = 6.93325 - \frac{1335.533}{(218.8+37)} = 1.7122 \qquad P_{\text{vapor}} \text{ at } 37°\text{C} = 10^{1.7122} = 51.6 \text{ mmHg}$$

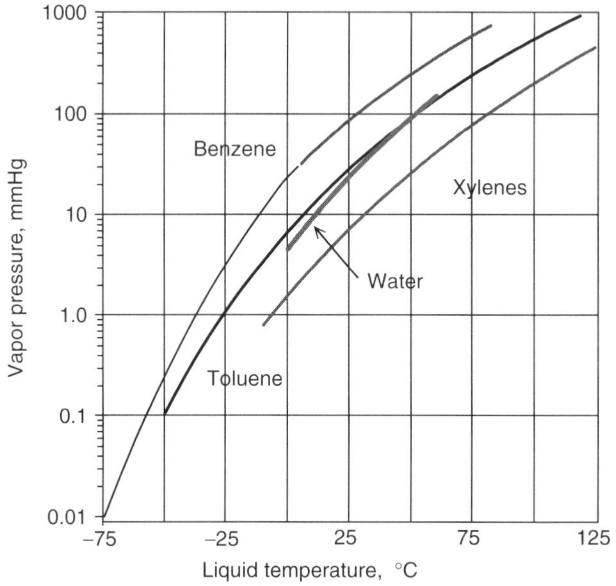

FIGURE 2.1 Antoine results for a few selected solvents.

The above temperatures and vapor pressures for toluene can be found in Figure 2.1 along with the curves for its two adjacent homologous partners of aromatic solvents and water. Each curve in Figure 2.1 constitutes what is sometimes called a "saturation line." Notice how similar the vapor pressure of water is to the vapor pressure of toluene (at least within this temperature range). In the examples above, the P_{vapor} of toluene increases by ratios of 1.93 and 1.81 each time the liquid temperature increases by 12°C. These ratios are close to the rule-of-thumb factor of 2 × per 12°C pointed out above. The pattern of the ratio of toluene's vapor pressure decreasing from 1.93 × over the 12°C below 25°C to 1.81 × for the 12°C above 25°C reflects the subtle pattern predicted by the mathematical form of the Antoine equation that is visible in Figure 2.1.[g] This gradual flattening of the log[P_{vapor}] curve indicates that extrapolating a 2 × change for more than one 12°C increment will soon result in an error that overestimates the actual vapor pressure and its associated exposure hazard.

In contrast (but probably of less practical importance), the antilog of Equation 2.10 (shown below) indicates that the absolute rate at which vapor pressure increases with

[g] The mathematically adventuresome could find the relative change in P_{vapor} vs. T by actually using your half-forgotten calculus to differentiate the Antoine Equation with respect to temperature. The temperature change required for the vapor pressure to double [ΔT] can then be found by first noting that

$$\frac{P_2}{P_1} = 2 \approx \frac{P_1 + \frac{\delta P}{\delta T}\Delta T}{P_1}$$

Finding ΔT via

$$\Delta T = \frac{P_{vapor}}{(\delta P_{vapor}/\delta T)}$$

is much easier than going through the long process used in Example 2.1 either for any particular set of Antoine coefficients (such as the calculated doubling temperature for toluene of 13.4°C) or for many organic solvents as was done to produce the above mean doubling temperature rule-of-thumb.

temperature (mm Hg per degree) actually increases with temperature. This latter pattern is not visible on the semi-log plot shown in Figure 2.1 but can be seen by the shape of the saturation line on any psychrometric chart used for heat stress analysis and comfort ventilation, which is in fact a linear plot of the vapor pressure of water vs. temperature ("100% relative humidity" in Figure 20.6 is an example of such a saturation line).

$$P_{\text{vapor}} = 10^{A-(B/(C+T))} \quad \text{or} \quad e^{A-(B/(C+T))} \quad \text{(the antilog of Equation 2.10)}$$

The case study reported by Franklin et al. (2000) presents a classic example of not anticipating the effect of temperature on vapor pressure, exposure levels, and employee hazard.[15] They almost serendipitously uncovered seven painters who suffered rather severe respiratory (dyspnea and asthma) and dermal (rash) symptoms from exposure to vapors from a polyurethane paint with a hexamethylene diisocyanate (HDI) hardener. Normally the vapor pressure of HDI is less than 0.05 mm Hg (depending upon the concentration of its biuret form); however, these particular employees were instructed to start painting a boiler before it had been given adequate time to cool. Its vapor pressure at over 300°F was close to boiling, and exposures would have been approximately 10,000 × greater than normal. Even without being "too hot to handle," you should now be able to anticipate and avoid such acute hazards.

4. The vapor pressure of metals also varies with temperature. The *CRC Handbook of Chemistry and Physics* is one good source for such data.[4] Molecules of molten metals evaporate just like those of a solvent, but when hot metal vapors cool they condense into fumes rather than liquid mists. The hotter the metal, the more fumes are formed. The vapor pressure of metals seems to follow the somewhat simpler Clapeyron equation shown below (the temperature in the denominator must be absolute).

$$\log[P_{\text{vapor}}] = A - \frac{B}{(T_{\text{absolute}})} \quad \text{the Clapeyron Equation} \qquad (2.11)$$

The Clapeyron equation is rooted in thermodynamics and historically preceded the Antoine equation, to which a third coefficient was added to extend the temperature range over which it will accurately predict the vapor pressure of organic solvents. The mathematical difference between Equation 2.10 (the Antoine equation) and Equation 2.11 (the Clapeyron equation) is not important at this time; practically speaking, only thermodynamicists and chemical engineers can tell them apart.

5. A chemical's boiling point [bp] or boiling temperature is a common but often imprecise surrogate for its vapor pressure at normal temperature. A chemical will appear to boil whenever its vapor pressure equals or exceeds the ambient atmospheric pressure. The boiling point listed on a chemical's MSDS or a similar tabulation is the temperature at which its vapor pressure equals 760 mm Hg. Thus, a listed boiling point is a chemical's boiling temperature at sea level. A chemical's vapor molar fraction will equal one at its boiling point. Figure 2.2 plots the vapor pressure at 25°C vs. its boiling temperature for over 400 chemicals. The agreement looks quite good for those chemicals whose boiling point is close to the 20 to 25°C normal temperature range. However, the further that a chemical's boiling point is above this normal temperature range, the greater will be the effect of deviations from the above $2 \times P_{\text{vapor}}$ per 12°C rule-of-thumb. The result is a spreading of the data toward the right side of Figure 2.2. The magnitude of the potential errors that can result from trying to predict a chemical's P_{vapor} from its boiling temperature must be interpreted in comparison to the logarithmic scale on the *Y*-axis. Thus, boiling temperatures much above 100°C or 200°F (375 K) can only predict a

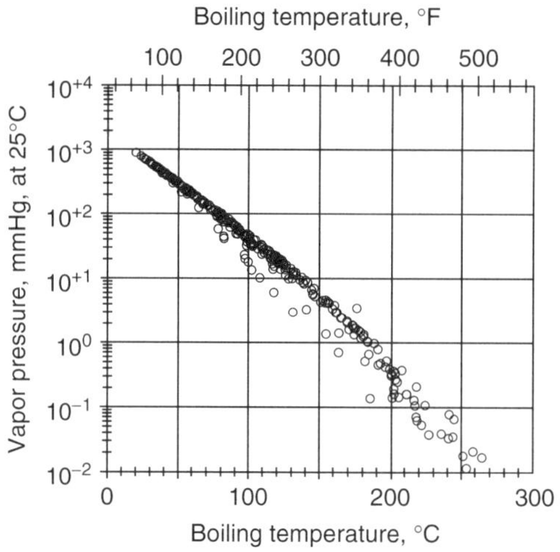

FIGURE 2.2 Plot of vapor pressure at 25°C as a function of boiling temperature at 760 mmHg.

chemical's vapor pressure at normal temperature within about an order of magnitude (a range of 10 ×).

It can also be important to remember if you work in a mountainous region or the aviation industry, that a chemical will physically appear to boil whenever the ambient atmospheric pressure is reduced to equal a chemical's vapor pressure at normal temperatures. Thus, a chemical's apparent boiling temperature will be reduced at higher altitudes (see Example 2.4 and Example 2.5). However, simply reducing the ambient pressure does not increase a chemical's vapor pressure, nor cause it to evaporate more rapidly than at sea level.[h] While knowing a chemical's boiling temperature at a given altitude is rarely a useful IH topic, Equation 2.4 could be used with the Antoine equation to make that prediction.

III. IH USES OF DALTON'S LAW

1. In 1801, John Dalton reported his findings that the total pressure of a mixture of gases equals the sum of its partial pressures. His "law" is stated mathematically in Equation 2.12. Recall that partial pressures are abbreviated as P_{partial} and the total pressure is abbreviated as just P without subscripts. When Dalton's law is applied to IH, the sum equals the ambient P which may or may not be the normal 760 mm Hg due either to the altitude (as per Equation 2.4) or to changes in the weather.[i]

$$P[\text{total pressure}] = P_{\text{ambient}} = \text{sum of } P_{\text{partial}} = \Sigma P_{\text{partial}} = \sum_{i=1}^{i=n} (Y_i \times P_{\text{ambient}}) \quad (2.12)$$

Hygienists commonly use the following corollary to Dalton's law: "The partial pressure (P_{partial}) of each component of gas or vapor in a mixture is proportional to its molar

[h] Although one can imagine that the agitation caused by boiling (should that occur) will increase a liquid's evaporation rate, that phenomenon is not documented.

[i] The reader is reminded of the list of symbols provided in the front matter of this book as a potentially useful resource.

fraction in that gaseous mixture."

$$\boxed{P_{\text{partial}} = Y_i \times P_{\text{ambient}}} \tag{2.13a}$$

or

$$\boxed{Y_i = \frac{P_{\text{partial}}}{P_{\text{ambient}}}} \tag{2.13b}$$

These relationships are valid because each molecule in the air anywhere near normal pressure is largely independent of all the other molecules surrounding it. Statistical thermodynamics has since shown that the pressure that gaseous molecules exert on a surface depends upon the number of molecules and their kinetic energy; kinetic energy is determined by the temperature but not by the kind of molecules striking that surface. Thus, each kind of gaseous molecule creates its own partial pressure in proportion to the fraction of molecules that each component comprises in the total mixture of airborne molecules.

2. One can combine the above definitions of partial pressure and molar fraction with Dalton's law to show that the various expressions of concentrations shown in the following list are in fact equivalent. Physicists would probably be most comfortable with the first or second terms (recall that the second term is just a special case of the first); chemists might relate easily to the second and third terms; and industrial hygienists tend to use the fourth and fifth terms.

Term or Expression	Symbol	Units
Partial pressure	P_{partial}	mmHg
Vapor pressure	P_{vapor}	mmHg
Gaseous molar fraction	Y_i	Unitless (P_{partial}/P) or atmospheres
Gaseous parts per million	ppm_i	Parts per million ($Y_i \times 10^6$)
Mass per volume concentration	C_i	mg/m^3

Dalton's law (Equation 2.13) relates the first two terms to the third. The third and fourth terms are related by their definition stated mathematically in Equation 2.14a. IH students are likely to be already acquainted with the conversion from ppm to C in mg/m^3 and back again, as stated in Equation 2.15a and Equation 2.15b, respectively. In fact, all of these terms are equivalent as stated in Equation 2.14b. These terms should be thoroughly understood as physical concepts, not just as symbols in an abstract equation. The same principle that underlies Dalton's law (the independent nature of airborne molecules) underlies each of these equations. This principle, combined with the concept of vapor pressure, was also the basis for equating vapor pressure and concentration embodied in Equation 2.8 and Equation 2.9.

$$\frac{P_{\text{partial}}}{P} \times 10^6 = Y_i \times 10^6 = \text{ppm}_i \tag{2.14a}$$

$$\boxed{\frac{P_{\text{partial}}}{P} \times 10^6 = Y_i \times 10^6 = \text{ppm}_i = C_i \frac{24.45}{\text{MW}_i} \text{ at NTP}} \tag{2.14b}$$

3. Equation 2.15a (a part of Equation 2.14b) offers a classic example of how to understand an equation conceptually. One way to view this equation is that ppm is a ratio of volume-to-volume, mole-to-mole, and pressure-to-pressure but has no relationship to mass. Whereas, units of mg/m^3 in C clearly includes the mass of the chemical being described. In order to convert from mg/m^3 to ppm, one must conceptually remove the mass component in C_i by dividing it by the chemical's molecular weight. The 24.45 (actually a gas's molar volume in liters) can be viewed as just the memorable proportionality constant that is always on the other side of the ratio from MW_i.

$$\boxed{\text{ppm}_i = C_i\,(\text{mg/m}^3)\,\frac{24.45}{\text{MW}_i}\ \text{at NTP}} \tag{2.15a}$$

To make the inverse conversion as shown by Equation 2.15b, one must conceptually insert mass into the dimensionless ppm by multiplying ppm_i by the MW_i, with 24.45 still on the other side of the ratio.

$$\text{at } 25°\text{C and 760 mm/Hg}\quad \boxed{C_i\,(\text{mg/m}^3) = \text{ppm}_i\,\frac{\text{MW}_i}{24.45}} \tag{2.15b}$$

Another way to view these equations conceptually is by noticing that MW/24.45 is the concentration or density of grams per liter of the chemical as a pure gas or vapor. The actual concentration is the ppm fraction of that pure concentration (you do not have to remember how the 10^6 gets cancelled out in the g to mg and L to m^3 conversions in Equation 2.15). Understanding a concept embodied within an equation allows one to be able to reconstruct and use the equation without having to memorize it as just a series of letters and numbers.

4. Practicing hygienists can use the concept of partial pressure and especially vapor pressure often. The following examples use these concepts to demonstrate how to convert between vapor pressure and concentration. The first example is of much more practical value to an IH than the second.

Example 2.2

What will be the equilibrium vapor concentration of *n*-hexane in a closed room at normal temperature and sea level pressure?

In this example, i = hexane = C_6H_{14}. Its vapor pressure is 151 mm Hg at 25°C (as found for instance in the Vapor Hazard Ratio tables attached as an appendix to this book). Specifying that the room is closed (meaning "poorly ventilated but not sealed") is necessary to assure that, as the evaporating vapor molecules add to the air molecules already existing within the room, as many excess molecules can leak out of the room as necessary to keep the total P inside the room at the same normal one atmosphere as the 760 mm Hg air outside the room. The $C_{\text{saturation}}$ for hexane could also have been found using Equation 2.8, but is shown here in three steps as a demonstration.

$$Y_i = \frac{P_{\text{partial}}}{P} = \frac{P_{\text{vapor}}}{P} = \frac{151}{760} = 0.1987\ (\text{a unitless fraction}) \qquad \text{(using Equation 2.13b)}$$

$$\text{ppm}_i = Y_i \times 10^6 = 198{,}700\ \text{ppm (a fraction with units)} \qquad \text{(using Equation 2.14a)}$$

$$C_i = \text{ppm}_i = \frac{\text{MW}_\text{i}}{24.45} = 198{,}700\frac{86.2}{24.45} = 700{,}474 \approx 700{,}000\ \text{mg/m}^3 \qquad \text{(using Equation 2.15b)}$$

This equilibrium concentration is approximately 3975 times higher than the TLV® for hexane.

EXAMPLE 2.3

IHs do not have much need to go from ppm all the way to mm Hg, but physiologists like to report partial pressures. Thus, if you wanted to talk to a physiologist about the effect of carbon dioxide at its TLV, you might want to find its partial pressure in mm Hg at 5000 ppm at sea level.

Here, the subscript $i = CO_2$. The same equations can be used in approximately the reverse order.

$$Y_i = \frac{\text{ppm}_i}{10^6} = \frac{5000}{10^6} = 0.005 \text{ (again } Y_i \text{ is just a unitless fraction)} \qquad \text{(using Equation 2.14a)}$$

$$P_{\text{partial}} = Y_i \times P = 0.005 \times 760 = 3.8 \text{ mmHg} \qquad \text{(using Equation 2.13a)}$$

The last two examples involve two conditions at which vapor pressure equals the ambient atmospheric pressure. Both conditions are extreme and of more academic than practical interest, but they will test your understanding of the concept of vapor pressure and partial pressures that underlies Equation 2.8 and Equation 2.13.

EXAMPLE 2.4

What will eventually happen when a solvent continues to be heated until its vapor pressure equals the ambient pressure? Eventually, when P_{vapor} rises to equal the ambient pressure, Y_i will equal one and the liquid will appear to boil. A liquid's temperature when this happens at normal P is called its "boiling temperature" or its "boiling point." Fortunately for employee health, organic solvents are rarely boiled.

EXAMPLE 2.5

What will happen to a solvent when the total ambient pressure is reduced to equal P_{vapor}? The most common instance in which ambient pressure is reduced is as one goes up in altitude. Recall that P_{vapor} is unaffected by the ambient P, but any time the ambient P is decreased enough to equal the liquid's P_{vapor}, Y_i will again equal one and the liquid will appear to boil even at a normal room temperature. Water at normal body temperature would appear to boil at an altitude of approximately 62,800 ft. Fortunately (again), no one works unprotected at such altitudes.

IV. IH USES OF THE IDEAL GAS LAW

1. The Ideal Gas law is commonly expressed as Equation 2.16. While the Ideal Gas law may not be the most accurate predictor of how gas pressure, volume, and temperature are related (compared for instance to the van der Waals equation of state), it is adequately accurate for dilute gases such as air and for virtually all IH purposes.

$$\text{The Ideal Gas law: } P \times \text{Vol} = n \times R \times T \qquad (2.16)$$

where

P = absolute pressure of the gas (usually total pressure although the law can also be used to predict the behavior of individual components using their partial pressures).

Vol = volume of air (e.g., L, m^3, or ft^3). The symbol "Vol" is used herein to avoid confusion with the common use of V in the fields of IH and ventilation to mean velocity.

n = number of moles of a gas (n = mass in grams/MW).

T = absolute temperature in either °K = °C + 273 or degrees Rankine °R = °F + 460.

R = the universal gas constant in units that match those used for P, Vol, and T such as 0.082 atm × L/°K × mole or 0.730 atm × ft^3/°R × mole. Luckily, a hygienist does not need to memorize any value for R.

2. Despite its underlying importance, Equation 2.16 is not boxed herein because IHs rarely use this form of the Ideal Gas law and almost never have to use one of the above values of R. Notice that because $P \times \mathrm{Vol}/T$ for a given parcel of air or other gas always equals the constants $n \times R$, the value of $P_1\mathrm{Vol}_1/T_1$ in one condition (subscripted as condition "1") must equal the same ratio in another condition (subscripted as condition "2"). Thus, IHs find the form of the Ideal Gas law expressed in Equation 2.17 to be much more useful. The word absolute is included within the subscripts of Equation 2.17 as a reminder to avoid using °C, °F, or a gauge pressure in such ratios (see Section I.3).

$$\frac{P_{1,\text{absolute}}\mathrm{Vol}_1}{T_{1,\text{absolute}}} = \frac{P_{2,\text{absolute}}\mathrm{Vol}_2}{T_{2,\text{absolute}}} \tag{2.17}$$

The next two sections of this chapter will focus on using this more useful form of the Ideal Gas law to solve two different kinds of IH problems. The first kind of problem is to find or predict the concentration of a contaminant in a given parcel of air whose volume or density will change due to changes in P and/or T, especially to or from NTP conditions. The second kind of problem is to predict the ratio of the density of two different parcels of air, especially a contaminated plume in normal air.

While solving Equation 2.17 for P at some elevated T has safety implications (e.g., to anticipate a tank or reactor rupture), these variables are almost never unknown values within the workplace. The pressure within the workplace is usually the same as the ambient pressure, it is measurable or at least predictable, and the ambient pressure at work rarely comprises a health hazard (although high ambient pressure under water or low pressure in aviation or high mountain work can comprise a health hazard).[5] While workplace temperatures can change due to a nearby heat source, solar heating, or a seasonal pattern, the change in ambient temperature within a workplace is not the result of changes in air P or Volume (although again heat and cold can comprise a health hazard).[16]

Nonetheless, a hygienist can use a known change in P and/or T within one of the modified forms of Equation 2.17, shown as Equation 2.18 to find or predict the air volume while air sampling (usually covered within another course) or the change in the air volume while it moves through an exhaust ventilation system (the latter problem is addressed in Chapter 11, specifically in Equation 11.6).

$$\mathrm{Vol}_2 = \mathrm{Vol}_1\left(\frac{P_{1,\text{absolute}}}{P_{2,\text{absolute}}}\right) \times \left(\frac{T_{2,\text{absolute}}}{T_{1,\text{absolute}}}\right) \tag{2.18a}$$

and

$$\frac{\mathrm{Vol}_2}{\mathrm{Vol}_1} = \left(\frac{P_{1,\text{absolute}}}{P_{2,\text{absolute}}}\right) \times \left(\frac{T_{2,\text{absolute}}}{T_{1,\text{absolute}}}\right) \tag{2.18b}$$

For example, Equation 2.18a can be used to calculate the gaseous molar volume at any nonnormal T and P (condition 2) from the molar volume of any gas at NTP (condition 1), which the IH student should by now know is 24.45 L, as shown below.[j]

$$\text{molar Vol[l]} = 24.45 \times \frac{760 \times (°\mathrm{C} + 273)}{P(\mathrm{mmHg}) \times 298} = 24.45 \times \frac{29.92 \times (°\mathrm{C} + 460)}{P(\text{inch Hg}) \times 537} \tag{2.18c}$$

[j] The molar volume is actually the value for Vol/n when $n = 1$, and technically, as Perkins (1997) points out,[6] a chemical's molar volume also decreases as its molecular weight (MW) increases (up to 5% for solvents with a MW of 75 to 100 and 10% for solvents with an MW over 150), but IHs learn and almost always use this one "magic" number.

Another use of Equation 2.17 is to find or predict a new density (mass per unit of volume) or change in the density of a given parcel of air as expressed within Equation 2.19. Please notice how the ratios in Equation 2.18 and Equation 2.19 are inverted because volume is in the denominator of density.

$$\rho_2 = \rho_1 \left(\frac{P_{2,\text{absolute}}}{P_{1,\text{absolute}}} \right) \times \left(\frac{T_{1,\text{absolute}}}{T_{2,\text{absolute}}} \right) \tag{2.19a}$$

and

$$\frac{\rho_2}{\rho_1} = \left(\frac{P_{2,\text{absolute}}}{P_{1,\text{absolute}}} \right) \times \left(\frac{T_{1,\text{absolute}}}{T_{2,\text{absolute}}} \right) \tag{2.19b}$$

A related but different form of Equation 2.19 will be used in Section VI to anticipate or explain how the temperature and molecular weight of different parcels of air can affect density. A difference between a plume's density and the ambient air's buoyancy can cause that plume to travel vertically.

V. THE EFFECTS OF *T* AND *P* ON AIRBORNE CONCENTRATION

The airborne concentration of a contaminant can be expressed as either a molar concentration (ppm_i), a mass concentration (C_i as mg/m^3), or partial pressure (P_{partial} as mmHg). Dalton's law, the Ideal Gas law, and the effect of temperature on vapor pressure (e.g., the Antoine Equation) all work together to determine how ppm_i, C_i, and P_{partial} each do or do not change when either T and/or P differ from normal. Which of these nominally equivalent expressions of concentration changes and which remains constant depends intimately upon what T and/or P changes and when it occurs. The three scenarios described below could apply to a small spill that will completely evaporate, a gas leaking at a constant rate, evaporation at a fixed rate, or a process producing an aerosol at a fixed rate. The same concepts will have other applications in Chapter 5 and are used to predict the effect of external changes or purposeful modifications to a source in Chapter 9. The examples that follow these descriptions demonstrate why it is important to take a moment or two to think about the setting before leaping to a conclusion.

1. What concentration would result from heating or expanding a parcel of air already contaminated by a prior event or process? Whenever a given (existing) parcel of both air and contaminant is subjected to an increase in T or a decrease in P, the ratio of different molecules within that parcel to each other will stay constant. Therefore, the ppm of a previously contaminated air parcel of air is unaffected by these changes in T and P.

$$\text{ppm}_2 = \text{ppm}_1 \tag{2.20a}$$

However, an increase in T or a decrease in P will expand the molar volume that all the molecules in that parcel of air occupy. As a result, the number of molecules within each liter or cubic meter of air and the corresponding mass of a contaminant per cubic meter (C in mg/m^3) will decrease. If one already knows the mass concentration in condition 1, then one can use Equation 2.20b to find the new concentration by directly applying the Ideal Gas law.

$$C_2 = C_1 \frac{P_{2,\text{absolute}} \times T_{1,\text{absolute}}}{P_{1,\text{absolute}} \times T_{2,\text{absolute}}} \tag{2.20b}$$

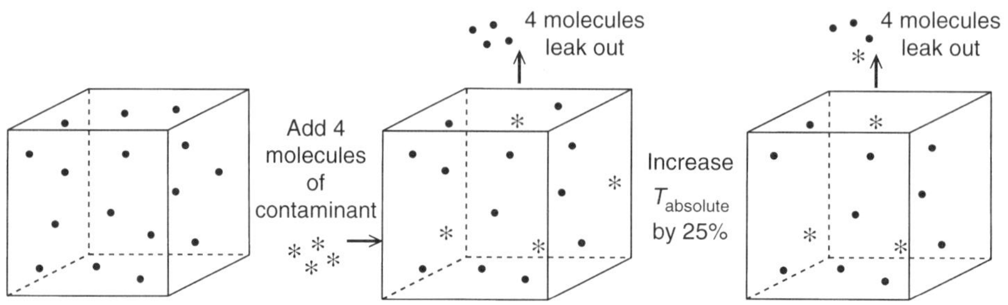

FIGURE 2.3 The effect of adding four molecules of a contaminant, heating an already contaminated parcel of air, and finding the same ppm as predicted by Equation 2.20a.

If one only knows the initial ppm ratio in condition 1 (and because ppm stays constant), one can use Equation 2.20c to find the C in mg/m^3 at any T and P. (Equation 2.20c is actually a modified form of Equation 2.15b derived by using Equation 2.18c to calculate the molar volume at condition 2.)

$$C_2 = \text{ppm}\frac{\text{MW}_i \times P_2(\text{atm}) \times 298}{24.45 \times 1\ \text{atm} \times T_2[°\text{C}+273]} = \text{ppm}\frac{\text{MW}_i \times P_2(\text{atm}) \times 537}{24.45 \times 1\ \text{atm} \times T_2[°\text{F}+460]} \quad (2.20\text{c})$$

This scenario is depicted in Figure 2.3. The cube on the left depicts a small parcel of air containing 16 molecules at NTP (this parcel could be scaled up to a small room that contains many moles at NTP). As four molecules of a contaminant (depicted by stars) are released into that parcel, four molecules of air (depicted by dots) must leak out to keep the pressure constant. The condition before heating would look like the center cube. In this state, four out of the 16 molecules in that parcel would be contaminant, $Y_i = 25\%$ or 250,000 ppm, and the mass concentration would be the mass of four contaminant molecules per cubic volume. Heating that already contaminated air by 25% would drive off 25% of all the molecules (as predicted by Equation 2.19b) of which 25% (or one out of four) are the contaminant. The result is a parcel in which three out of the 12 molecules remaining in the parcel are the contaminant, Y_i is still 250,000 ppm contaminant (as predicted by Equation 2.20a), but the mass concentration is now only three contaminant molecules per cubic volume (as predicted by Equation 2.20b).

2. Would the same result occur from cooling a parcel of air already contaminated by a prior event or process? No, because when an enclosed parcel of air is cooled, pure (uncontaminated) air would enter the enclosure (e.g., room). The amount of contaminant would be unchanged resulting in the same C (mg/m^3) concentration, but the ppm concentration would decrease as will be predicted by Equation 2.21b due to more air molecules being present within the cooled enclosure. The interesting (but not practical) net result of repeatedly heating and cooling a room of contaminated air is to eventually drive all of the contaminant out of the room.
3. What concentration would result from adding a fixed amount of a contaminant into air already at a nonnormal T and/or P? As you might expect, the answer is exactly the opposite from that above. Whenever a given mass of chemical contaminant is diluted into a volume of air at any T and P specified by the setting, the C calculated as mass per unit of volume such as mg/m^3 will stay constant. Equation 2.8 states that a liquid's equilibrium concentration (C due to its vapor pressure) will also be constant independent of changes in the ambient air density due to air temperature, pressure, or altitude.

$$C_2 = C_1 = \frac{\text{mass}}{\text{volume}} \quad (2.21\text{a})$$

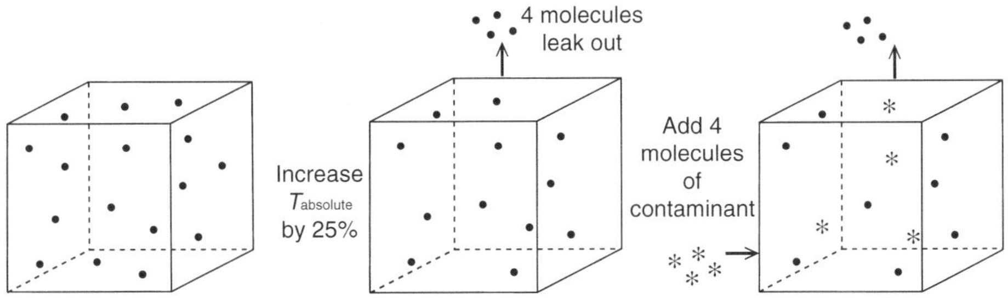

FIGURE 2.4 The effect of heating a parcel of air to drive off some air molecules, then adding contaminant to that already heated parcel to yield the same C as predicted by Equation 2.21a.

However, because T and/or P change the molar volume (as per Equation 2.18) and consequently the density of the air into which the contaminant is diluted, the same number of moles of contaminant will now be diluted into a different number of moles of air. Thus, both Y_i and ppm_i will change.

Analogous to Equation 2.20b, if one already knows the ppm in condition 1, then one can use Equation 2.21b to find the new ppm by again directly applying the Ideal Gas law.

$$ppm_2 = ppm_1 \frac{P_{1,absolute} \times T_{2,absolute}}{P_{2,absolute} \times T_{1,absolute}} \tag{2.21b}$$

If one only knows the mass concentration in condition 1, then one can use Equation 2.21c to find the ppm in condition 2 (again analogous to Equation 2.20c).

$$\begin{aligned} ppm_2 &= C_1 \frac{24.45 \times 1 \text{ atm} \times (T_2(^\circ C) + 273)}{MW_i \times P_2(\text{atm}) \times (298)} \\ &= C_1 \frac{24.45 \times 1 \text{ atm} \times (T_2(^\circ F) + 460)}{MW_i \times P_2(\text{atm}) \times 537} \end{aligned} \tag{2.21c}$$

This scenario is depicted in Figure 2.4. The cube on the left depicts the same small parcel of air containing 16 molecules (or moles) at NTP. The center cube depicts the effect of heating that parcel by the same 25% which would again drive off 25% of the molecules but they are all air (dots). When four molecules (or moles) of a contaminant (depicted by stars) is added to the cube, another four molecules of air are driven out. The result in the final cube is again four molecules of contaminant mass concentration as without heating (as predicted by Equation 2.21a), but now four out of the 12 molecules in that parcel would be contaminant, $Y_i = 33\%$ or 333,333 ppm (as predicted by Equation 2.21b).

4. Can the concentration of the contaminant change independent of the air's T and/or P? The answer is certainly yes. The concentration will change any time the amount of contaminant being added or generated changes. For instance, Section II and Equation 2.10 explained how a change in the liquid temperature (e.g., due to heating the liquid container or the surface onto which a liquid is applied) will cause its corresponding vapor pressure to change. Any change in P_{vapor} will affect the equilibrium concentration ($C_{equilibrium}$ or $C_{saturation}$) and the mass concentration right at the source, as stated in Equation 2.8.

$$C_{sat}(\text{mg/m}^3) = \frac{P_{vapor} \times MW \times 10^6}{760 \times 24.45} = 53.82 \times P_{vapor} \times MW \text{ (repeat of Equation 2.8)}$$

Similarly, because the amount of dilution caused by a given environment is independent of the vapor being diluted and of the air's T and P, any lower concentration resulting from

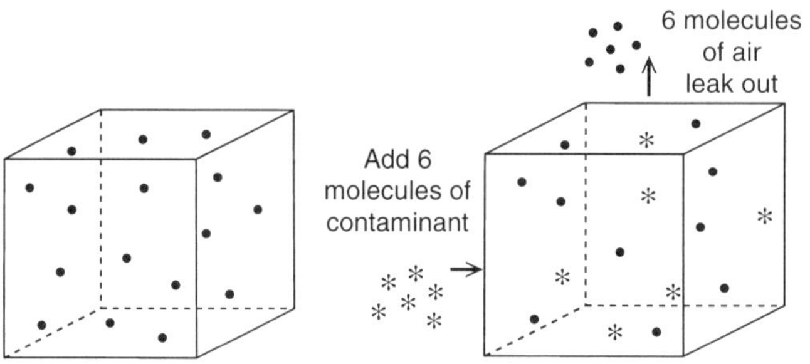

FIGURE 2.5 The effect of increasing a contaminant's vapor pressure for the same air T and P.

the dilution of this equilibrium concentration at the source will only change in direct proportion to either a change in the contaminant's vapor pressure or two chemicals' molecular weights, as stated in Equation 2.22a.

$$C_2 = C_1 \times \frac{\mathrm{MW}_2 \times P_{\mathrm{vapor},2}}{\mathrm{MW}_2 \times P_{\mathrm{vapor},1}} \tag{2.22a}$$

And because the process of dilution is independent of the chemical's molecular weight, a change in molar concentration for two similarly diluted vapors is only proportionate to a change in their vapor pressures, as stated in Equation 2.22b.

$$\mathrm{ppm}_2 = \mathrm{ppm}_1 \times \frac{P_{\mathrm{vapor},2}}{P_{\mathrm{vapor},1}} \tag{2.22b}$$

Because vapor pressure is practically the only chemical-specific parameter to affect evaporation rate, and because the mechanisms that determine a plume's trajectory and dispersion as it leaves its source affects all molecules in a similar way, Equation 2.22a and Equation 2.22b apply both to the change in the vapor pressure of one chemical or to any other diluted vapor emanating from the same source.[k] The effect of increasing a chemical's vapor pressure by 50% is depicted in Figure 2.5 by an increase in the number of contaminated molecules being released into the parcel from four to six. The result at the same air T and P (still 16 total molecules within the parcel) is both a 50% higher mass concentration (as predicted by Equation 2.22a) and a 50% higher ratio of stars to dots (Y_i or ppm as predicted by Equation 2.22b).

Based on the above four scenarios and differing predictive equations, it should be clear that when confronted with a changing or a nonnormal T and P problem, the first step is to recognize what stays constant in your setting. Only with that knowledge can one determine into which of the above four scenarios to place a given problem and which equations to apply.

Example 2.6

Suppose that a direct fired heater is known to emit 100 ppm of carbon monoxide (CO) at 500°F. What is the mass of CO released in these emissions that needs to be diluted by room air?

[k] Chapter 5 will discuss evaporation rate (e.g., Equation 5.3). While physical or environmental factors can affect the rate at which a solvent will evaporate, Equation 2.22 effectively predicts how a contaminant's emission rate will change quantitatively with changes in vapor pressure. The emitted vapors will be diluted by both air turbulence and the amount of mixing between the vapors and the fresh air passing through the room from general ventilation. Thus, vapor pressure is also one predictor of the molar concentration of the contaminant anywhere within a room.

Because mass is the product of concentration times volume, in order to answer this question we will need to measure the volume of hot air emitted by the heater and to know the mg/m^3 concentration of CO in these emissions. Thus, this problem involves finding the mass concentration of CO in already hot air, scenario 1 above, i.e., using Equation 2.20b with the MW of carbon monoxide set at 28.01 g/mole.

$$C_i = \text{ppm}_i \times \frac{\text{MW}_i \times 537}{24.45(°\text{F} + 460)} = 100 \times \frac{28.01 \times 537}{24.45(500 + 460)} = 64\,\text{mg/m}^3 \qquad \text{(using Equation 2.20b)}$$

EXAMPLE 2.7

Suppose that metal parts that are slightly wet with hexanes are sent through a drying tunnel, the consumption rate of hexanes ($\rho_\ell = 0.656$ g/mL) is known from purchase records to be 1.5 L/h. Room air enters the drying tunnel, is heated, and then blown past the parts to be dried. Measurements show an exhaust gas flow rate of 580 ft^3/min. Dividing a mass of 16.4 g/min into a volume of 16.4 m^3/min predicts a vapor concentration of 1000 mg/m^3 in the exhaust gas. But what is the ppm at the nearly 200°F exhaust gas temperature? Is it likely to be flammable? (Hexane's LFL is 1.1% at NTP but may be as low at 0.9% at 200°F; see a discussion on the effect of temperature on LFL in Section III.1.c in Chapter 4.)

This problem involves a constant amount of contaminant evaporating into hotter than normal air. This is setting 2 above, and Equation 2.21b applies. The MW of hexane is 86.2 g/mole (see Example 2.2).

$$\text{ppm}_{\text{hexane}} = C_{\text{hexane}} \times \frac{24.45(°\text{F} + 460)}{\text{MW}_{\text{hexane}}(537)} = 1000 \times \frac{24.45 \times 660}{86.2 \times 537} = 349\text{ ppm which is } \ll 1\%$$

EXAMPLE 2.8

Suppose instead of heating the metal parts in the above problem, that the conveyer were slowed down and the parts were allowed to dry at normal room temperature. What would you anticipate would be the hexane concentration in either mg/m^3 or ppm if the conveyer speed were slowed in proportion to the change in hexane's vapor pressure from 200°F to 77°F but the air flow rate were unchanged?

The vapor pressure of *n*-hexane at 25°C (as in Example 2.2) is 151 mm Hg. While the vapor pressure of *n*-hexane at 200°F could be calculated from an Antoine equation, we will assume that all of it evaporated at a vapor pressure of 760 mm Hg as the metal reached its boiling temperature of 156°F.

$$C_2 = C_1 \times \frac{P_{\text{vapor},2}}{P_{\text{vapor},1}} \text{ in the same setting} = 1000\,\text{mg/m}^3 \times \frac{151}{760} \approx 200\text{ ppm} \qquad \text{(using Equation 2.22a)}$$

Because the evaporation rate is proportional to vapor pressure, one could predict that the conveyor system would need to be slowed in proportion to the 151/760 = 20% ratio of vapor pressures, i.e., to one fifth the speed in order to achieve the same drying effect at a much lower temperature. Since this speed would slow the rate that pieces would pass through the drying tunnel by this amount, this information could also help determine whether this would be acceptable to production rates.

EXAMPLE 2.9

What is the maximum concentration of hexane vapor that can occur at normal temperature on campus in Logan, Utah, at an altitude of 4600 ft above sea level? How does this concentration differ (if any) from the maximum concentration at sea level found in Example 2.2?

Because Equation 2.8 shows that a vapor's mass concentration is unaffected by altitude, this problem can be solved as a constant addition of contaminant into air already at a nonnormal altitude, setting 3 above. However, this problem can also be solved as the special case of saturated vapor in setting 4 using Equation 2.22b. Again, the vapor pressure of hexane is 151 mmHg. The ambient P at Logan's altitude found using Equation 2.4d is used in the denominator of Equation 2.22b.

$$P = 2^{(-\text{altitude}/18{,}000)} \times 760 = 2^{(-4600/18{,}000)} \times 760 = 0.8377 \times 760 = 637\,\text{mmHg} \quad \text{(using Equation 2.4d)}$$

$$\text{ppm}_i = \frac{P_{\text{vapor}} \times 10^6 \times (^\circ\text{F} + 460)}{P_{\text{ambient}} \times (537)} = \frac{151 \times 10^6}{637} = 237{,}000\,\text{ppm} \qquad \text{(using Equation 2.22b)}$$

The 237,000 ppm found above is 19% larger than the 198,700 ppm found at sea level in Example 2.2 because the air is less dense at a higher altitude. While the mass concentration of hexane molecules that are generated by a vapor pressure of 151 mmHg is the same at all altitudes, any given P_{vapor} represents a larger fraction of the lower atmospheric pressure and air density at a higher altitude. Thus, as altitude increases, the Y_i and ppm_i of the same vapor in an otherwise identical setting will increase.

VI. THE EFFECTS OF *T* AND MW ON PLUME DENSITY

Airborne contaminants will generally move with the air in which they are diluted or suspended. However, a difference between the density of a plume and the density of the ambient air surrounding that plume creates a force called buoyancy that can cause the plume to take a somewhat different path.[1] The effect of buoyancy on the vertical trajectory of a plume will depend on the ratio of the two densities and on the air's horizontal velocity and turbulence. This section will show that buoyancy can be caused by a difference between the two parcels' temperature and/or molecular weight but never by a difference in air pressure.

1. Density is simply mass divided by volume. Notice in Equation 2.23a that the typical units of gas or vapor density (g/L) differ from those for most solids and liquids (g/mL) by approximately three orders of magnitude. This book will attempt to differentiate these two densities by always using just plain ρ when referring to a gas or vapor density in g/L vs. ρ_ℓ when referring to the density of solids or liquids in g/mL. Thus, a ρ notation without the ℓ should be assumed to be the gaseous density in g/L.

$$\text{Gas density} = \rho\,(\text{g/L}) = \frac{\text{mass (g)}}{\text{volume (L)}} \qquad (2.23a)$$

   A simple and memorable, albeit somewhat limited, way to quantify gas density is to recall that the mass of one mole of a gas (i.e., a chemical's molecular weight in grams) occupies one molar volume (i.e., 24.45 at NTP). Thus, the density of any gaseous material of known MW at NTP can be found by using Equation 2.23b. This boxed equation should be learned, but again, it is best learned conceptually as a tool rather than just as a couple of mathematical symbols.

$$\frac{\text{the mass of one mole}}{\text{one molar volume}} = \boxed{\rho_{\text{at NTP}} = \frac{\text{MW}_{\text{known}}}{24.45}} \qquad (2.23b)$$

   Equation 2.23c combines the above definition at NTP with Equation 2.18 to yield the gas density in the more general case where conditions are not normal. Units of *P* in mm Hg or inches of mercury could be substituted for atmospheres as long as they are used in both the numerator and the denominator.

$$\rho = \frac{\text{MW}_{\text{known}} \times P(\text{atm}) \times 298}{24.45 \times 1\ \text{atm} \times (^\circ\text{C} + 273)} = \frac{\text{MW}_{\text{known}} \times P(\text{atm}) \times 537}{24.45 \times 1\ \text{atm} \times (^\circ\text{F} + 460)} \qquad (2.23c)$$

2. The limitation to Equation 2.23 will be encountered when one attempts to determine the MW of a mixture of air and a toxic contaminant. In principle, the molecular weight of any

[1] The effect of buoyancy on individual molecules would be slow (much like that of molecular diffusion). In contrast, the effect of buoyancy acting on a parcel of air (such as on a plume) is much faster (much like convection).

mixture ($MW_{mixture}$) can be found by adding the mass contributed by each component that makes up the mixture, again using Dalton's law, as shown in Equation 2.24. This is an important principle but not very useful by itself.

$$MW_{mixture} = \frac{\Sigma(Y_i \times MW_i)}{\Sigma(Y_i)} \approx \frac{\Sigma(Y_i \times MW_i)}{1(\text{or } 100 \text{ if } Y_i \text{ is in } \%)} \quad (2.24)$$

EXAMPLE 2.10

Use Equation 2.24 to find the MW of a mixture whose molar percentages ($Y_i \times 100$) are 78% nitrogen (MW = 28) and 21% oxygen (MW = 32).

$$MW_{mixture} = \frac{(0.78 \times 28) + (0.21 \times 32)}{0.78 + 0.21} = \frac{28.56}{0.99} = 28.85 \text{ g/mole} \quad \text{(using Equation 2.24)}$$

In IH practice, the molecular weight of a hazardous mixture is rarely known (and as you will read, not commonly calculated). However, pure air is one obviously common mixture of gases that is well studied and whose molecular weight is known. The major components of dry standard air shown in Table 2.3 are applicable to all altitudes in which humans can live or work without special respirators (and even well into the range in which we can survive only with a respirator and pressure suit). This table also shows how Equation 2.24 can be used to calculate the molecular weight of standard dry air (MW_{air}). Notice that 28.964 is slightly larger than the approximation in Example 2.10. The normal range of humidity ($MW_{water} = 18$) can reduce the molecular weight of air by 0.1 to 0.2 g.

On the other hand, all of us should be interested to know that data recorded monthly since 1958 on a high peak in Hawaii (shown in Figure 2.6) indicate that carbon dioxide is accumulating in the Earth's atmosphere at a rate of more than 1 ppm per year. The Mauna Loa data also show a consistent seasonal fluctuation in CO_2 with peaks in May and lows in September or October. The CO_2 since 2002 has been consistently above 370 ppm! Although the cumulative effect of such an annual increase in carbon dioxide on the molecular weight of air can be seen from Equation 2.24, Table 2.3, and Example 2.10 to be more negligible than the effect of water, its effect on global warning is not so clear but could be substantial.

TABLE 2.3
Chemical Composition of Standard Dry Air[17]

Chemical Component	Molecular Weight MW_i	Y_i (%)	$Y_i MW_i$
Nitrogen (N_2)	28.0134	78.084	21.8740
Oxygen (O_2)	31.9988	20.9476	6.7030
Argon (A)	39.9480	0.934	0.37312
Carbon dioxide (CO_2)	44.0098	0.0314	0.0138
Neon (Ne)	20.179	0.00182	0.00037
Helium (He)	4.0026	0.000524	0.00002
Methane (CH_4)	16.0425	0.0002	0.00003
Krypton (Kr)	83.80	0.000114	0.00010
Hydrogen (H_2)	2.0158	0.00005	0.00000
	Sum of molar fractions = $\Sigma(Y_i)$ =	99.9997	
	Molecular weight via Equation 2.24 =		28.964

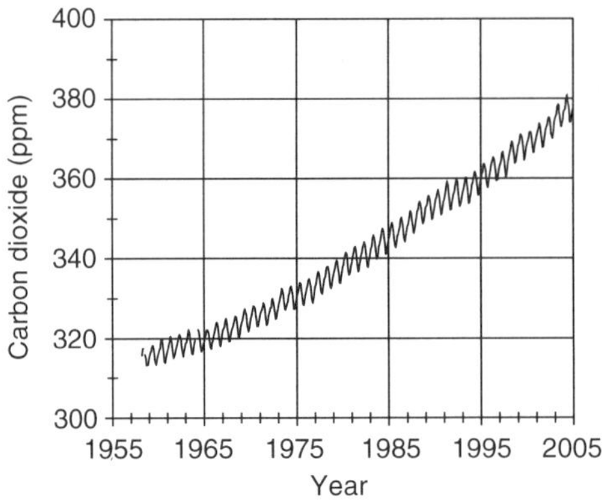

FIGURE 2.6 Monthly carbon dioxide data from Keeling and Whorf, 2005.[m,18]

EXAMPLE 2.11

What is the normal density of air? Recall that Equation 2.23b can be used to find the density of any pure gas of known MW at NTP. Although the MW of a mixture is rarely known, the results of Table 2.3 for pure air creates one of those rare times.

$$\rho_{\text{air at NTP}} = \frac{\text{MW}_{\text{air}}}{24.45} = \frac{28.964}{24.45} = 1.1846\ \text{g/L or about } 1.2\ \text{g/L} \qquad \text{(using Equation 2.23b)}$$

3. Knowing the density of pure air at NTP is a good academic question, but an IH is more likely to ask how does the density of a contaminated plume at another molecular weight, and perhaps at another temperature, compare the density of pure air at NTP? From Equation 2.23c, one can see that pressure, temperature, and molecular weight can all affect gas density. However, for the purposes of calculating a plume's density, it is safe to assume that the entire room or any vicinity near an outdoor source is at the same ambient air pressure. For instance, we will see in the ventilation chapters how a difference in air pressure as small as 0.01 in. of water pressure (out of a total of approximately 407 in. of water in 1 atm) can cause an air velocity of up to 400 fpm (2 m/sec or 4.5 m/h). Since the velocity of room air is usually much slower than that, we can conclude that the ratio of pressures within a room, workplace, or plume will be less than $0.01/407 = 3 \times 10^{-5}$. Such a small difference within the ambient atmospheric pressure will not affect the density of contaminants or contaminated air in a plume any differently than it affects the ambient air in which the plume is being dispersed.

   Thus, the two practical reasons that the density of a plume could be different from the density of the ambient air are because either the plume's molecular weight or its temperature differs significantly from the ambient air's. Rather than just any two molecular weight and temperature conditions (the generic condition 1 vs. condition 2 shown in Equation 2.25a), IHs are most commonly interested in the ratio between a contaminated and/or a hot plume and pure air at normal temperature (25°C), as shown in Equation 2.25b. Notice that when looking at the ratio of densities in Equation 2.25 that the

[m] Atmospheric CO_2 concentrations (ppmv) derived from *in situ* air samples collected at Mauna Loa Observatory, Hawaii (elev. 13,680 ft) by C.D. Keeling, T.P. Whorf, and the Carbon Dioxide Research Group of the Scripps Institution of Oceanography, University of California, La Jolla, as of May, 2005. See cdiac.esd.ornl.gov/trends/co2/sio-mlo.htm.

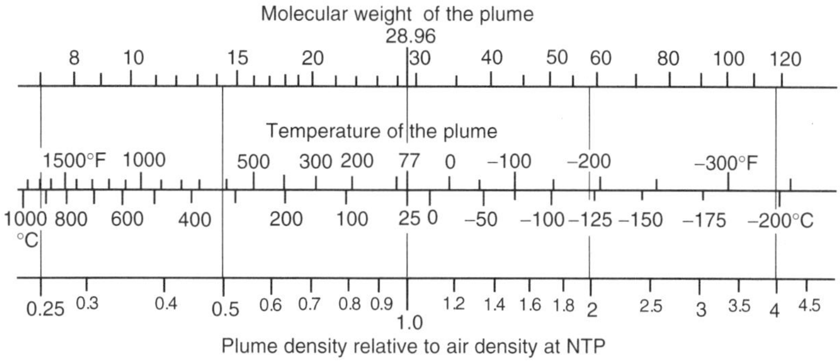

FIGURE 2.7 The effect of the plume's molecular weight (on the top scale) or temperature (on the middle scale) on plume density (on the bottom scale) are aligned vertically in this diagram. For instance, a plume with a molecular weight of 60 or at a temperature of −200°F would both be approximately twice as heavy as air.

values of P (whatever it is locally) and 24.45 L/mole in Equation 2.23c drop out.

$$\frac{\rho_2}{\rho_1} = \frac{\text{MW}_2 \times (°\text{C}_1 + 273)}{\text{MW}_1 \times (°\text{C}_2 + 273)} = \frac{\text{MW}_2 \times (°\text{F}_1 + 460)}{\text{MW}_1 \times (°\text{F}_2 + 460)} \tag{2.25a}$$

$$\frac{\rho_{\text{plume MW at } T}}{\rho_{\text{air at 25°C}}} = \frac{\text{MW}_{\text{plume}} \times 298}{28.964 \times (°\text{C}_{\text{plume}} + 273)} = \frac{\text{MW}_{\text{plume}} \times 537}{28.964 \times (°\text{F}_{\text{plume}} + 460)} \tag{2.25b}$$

Figure 2.7 shows the effect of temperature and molecular weight on a plume's density relative to the density of normal room air at 25°C determined using Equation 2.25b. The effect of temperature will equal the effect due to molecular weight at values in Figure 2.7 that align vertically. For instance, a plume with an MW of 60 will have the same density as a plume at −200°F. The MW of a plume that would behave equivalent to a plume of a given temperature (or *vice versa*) can be found using Equation 2.25c. Of the two variables, the temperature of a plume is more likely to differ from ambient air by a greater ratio than will its molecular weight, temperature is considerably easier to measure, and its effect on a plume is more intuitive for most people to envision.

$$\text{equivalent} \frac{\text{MW}_{\text{plume}}}{\text{MW}_{\text{air}}} = \frac{298}{(°\text{C}_{\text{plume}} + 273)} = \frac{537}{(°\text{F}_{\text{plume}} + 460)} \tag{2.25c}$$

4. Hot processes are very common in industry. Almost everyone has seen smoke rise from a fire. Smoke will rise because it is hotter and less dense than the surrounding air. An equation will be presented in Chapter 13 to predict the velocity of such a plume as a function of its temperature. If a heated plume is not initially captured and removed via local exhaust ventilation, it will continue to rise and cool. Indoors, such a plume will disperse slowly while it rises until it encounters either a ceiling or a thermal inversion layer, where it will disperse horizontally and perhaps recirculate back into the workspace. Eventually, either the contaminant will slowly escape from the room via general ventilation (exposure scenario 3 from Chapters 1 and 5) or its vapor will reach an equilibrium with its source (exposure scenario 4). The latter unventilated scenario is often hazardous to health. Any plume (whether visible or not) that is less dense than the surrounding air will tend to rise in the same manner as smoke.

Cold plumes are rare in the workplace. Vapor plumes will experience a small amount of cooling because the process of evaporation is endothermic. The heat of evaporation must be absorbed from the ambient environment and thus can cool both the liquid source and its vapor plume. Examples of this cooling include the effect on one's body due to the evaporation of perspiration and the cooling of air passing through a swamp cooler. The higher the solvent's volatility, the more rapidly cooling will occur. A pool of a highly volatile solvent such as ethyl ether ($P_{vapor} = 537$ mmHg), methylene chloride ($P_{vapor} = 435$ mmHg), carbon disulfide ($P_{vapor} = 358$ mmHg), or 1,2-dichloroethylene ($P_{vapor} = 331$ mmHg) can be 40°F (20°C) below room temperature, but evaporation will cool most solvents by less than 20°F (10°C), and a plume's temperature is likely to be cooled much less than 10°F (5°C) below room temperature. Just as the low density in an initially hot plume will cause an upward plume velocity, a high density would cause a downward velocity. Substantially the same effect would occur whether the plume were colder that the surrounding air or were contaminated by a vapor whose molecular weight was greater than the MW of air.

The following examples of air densities at three nonnormal temperatures are intended to help the reader visualize the effect of a plume's density on its behavior implied by Equation 2.25b. The resulting small density ratios will be converted into a percent difference using Equation 2.26.

$$\boxed{\%\ \text{difference} = 100 \times [\text{ratio} - 1]} = 100 \times \left[\frac{\text{numerator} - \text{denominator}}{\text{denominator}}\right] \quad (2.26)$$

Example 2.12

Compare the density of steam at the temperature of boiling water to that of air.

$$\frac{\rho_{\text{plume at 100°C}}}{\rho_{\text{air at 25°C}}} = \frac{\text{MW}_{\text{plume}} \times 298}{\text{MW}_{\text{air}} \times 100 + 273} = \frac{298}{373} = 0.799 \qquad \text{(using Equation 2.25b)}$$

Using Equation 2.26 shows that $100 \times (0.799 - 1) = -20\%$ or a ratio of 0.799 is 20% less dense than normal air. Recall how steam from a boiling pan on your stove tends to rise but can easily be blown sideways or even downward if you happen to have a stove-top exhaust hood (popularized by Jenn-Air).

Example 2.13

Compare the density of a fog that can sometimes be seen in a highly humid setting drifting down the outside of a glass of ice-cold water to the density of air. Assume the fog is at 0°C.

$$\frac{\rho_{\text{plume at 0°C}}}{\rho_{\text{air at 25°C}}} = \frac{\text{MW}_{\text{plume}} \times 298}{\text{MW}_{\text{air}} \times 0 + 273} = \frac{298}{273} = 1.091 \qquad \text{(using Equation 2.25b)}$$

Using Equation 2.26 shows that $100 \times (1.091 - 1) = 9\%$ more dense than normal air. This fog is even less different from air than was steam (albeit more dense rather than less dense).

Example 2.14

Compare the density of an easily visible fog that will form when dry ice is placed in water to the density of air. Assume this fog is at the −103°C or −153°F sublimation temperature of CO_2.

$$\frac{\rho_{\text{plume at }-103°\text{C}}}{\rho_{\text{air at 25°C}}} = \frac{\text{MW}_{\text{plume}} \times 298}{\text{MW}_{\text{air}} \times -103 + 273} = \frac{298}{195} = 1.528 \qquad \text{(using Equation 2.25b)}$$

And again, using Equation 2.26, shows that $100 \times (1.528 - 1) = 53\%$ more dense than normal air. Recall how dry-ice fog will flow down onto a table or, if formed as a stage effect, will flow down-stage into the audience or orchestra pit. This density difference is significant.

VII. DENSE VAPORS

1. While any version of Equation 2.25 works fine and well for either a pure chemical or a mixture at a known molecular weight, vapors cannot (by definition) be present at 100% purity and the MW of a vapor–air mixture is not easily known.[n] One could use Equation 2.24 and the process shown in Table 2.3 to find the mixture's molecular weight ($MW_{mixture}$), but that is way too long to be practical. A much simpler method (but not yet the simplest) is to treat standard air as one component and a toxic vapor as the other component of a mixture. This principle (along with Dalton's and the Ideal Gas laws) was used to derive Equation 2.27a.

$$\rho_{\text{mixture}} = \sum_{i=1}^{i=n} (Y_i\ \rho_{\text{pure}\ i}) + (Y_{\text{air}}\ \rho_{\text{air}}) \tag{2.27a}$$

where

Y_i = the molar fraction of each contaminant.

$\rho_{\text{pure}\ i}$ = the density of each gaseous contaminant i calculated as if it were a pure gas using Equation 2.23b or 2.23c (even though a pure vapor is, by definition, not possible).

Y_{air} = the remaining fraction of the atmosphere that is not the contaminant (i.e., $1 - \Sigma Y_i$ as shown in Equation 2.27b).

ρ_{air} = either 1.1846 g/L at NTP or calculated using Equation 2.23c for any other T and P.

Equation 2.27b uses the further simplification that whatever is not air is the contaminant that an IH typically wants to measure or predict.

$$\rho_{\text{mixture}} = \sum_{i=1}^{i=n} (Y_i\ \rho_{\text{pure}\ i}) + \left(\left(1 - \sum_{i=1}^{i=n} Y_i \right) \rho_{\text{air}} \right) \tag{2.27b}$$

Similar to cancellations that occurred to yield Equation 2.25 the molar volume (whether at NTP or any other T and P) in both the $\rho_{\text{pure}\ i}$ and ρ_{air} values will cancel out of Equation 2.27b, yielding Equation 2.28.

$$\frac{\rho_{\text{mixture}}}{\rho_{\text{air}}} = \frac{\sum_{i=1}^{i=n} Y_i \text{MW}_i}{\text{MW}_{\text{air}}} + \left(1 - \sum_{i=1}^{i=n} Y_i \right) \tag{2.28}$$

If the mixture only has only one toxic vapor component (defining i = that one vapor), then Equation 2.29 is almost as simple as this prediction can get to answer the common question, "How much more dense than air is a vapor of some specified concentration of 'dimethyl-chicken wire'?"

$$\frac{\rho_{\text{one vapor}}}{\rho_{\text{air}}} = 1 + \left(Y_{\text{vapor}} \times \left[\frac{\text{MW}_{\text{vapor}}}{\text{MW}_{\text{air}}} - 1 \right] \right) \quad \text{for any one vapor} \tag{2.29}$$

[n] By definition, a chemical's vapor pressure (P_{vapor}) is its maximum vapor concentration at a given temperature, and the P_{vapor} of any vapor under normal conditions must be less than one atmosphere (otherwise it would be a gas), and because P_{vapor} for a vapor must be less than one atmosphere, its Y_i must always be less than unity. The rest of that total atmosphere is air. Therefore, a vapor plume will always be a mixture of the vapor and air.

Finally, recall that a vapor's maximum concentration is limited to its saturation vapor pressure. It follows from Equation 2.9 that the maximum Y_{vapor} is limited to P_{vapor}/P. (If that statement is not obvious by now, please go back and review Equation 2.9 and Equation 2.14) Thus, a chemical's maximum ρ_{vapor}/ρ_{air} depends not only upon the chemical's MW_{vapor} but also upon its vapor pressure (which determines the maximum Y_{vapor}). Notice that ρ_{vapor}/ρ_{air} will reduce to the ratio of MW_{vapor}/MW_{air} for gases where Y_{vapor} within the plume can equal one or where $P_{vapor} = P_{ambient}$. Equation 2.30 (again for i = one vapor) can answer the more specific common question, "How much more dense than air can a vapor of 'dimethyl-chicken wire' possibly be?"

$$\frac{\rho_{sat.\ vapor}}{\rho_{air}} = 1 + \left(\frac{P_{vapor}}{\text{ambient } P} \times \left[\frac{MW_{vapor}}{MW_{air}} - 1 \right] \right) \quad \text{for one vapor at saturation} \tag{2.30}$$

Example 2.15

Compare the mixture density for saturated benzene vapor at NTP to that for air.

One can tell by comparing the $MW_{benzene} = 78$ with the $MW_{air} = 28.964$ that the density of a hypothetically purely gaseous state of benzene would be 78/28.96 = 2.7 × more dense than air. However, benzene cannot form a pure vapor at NTP because its vapor pressure is only 95 mm Hg and its maximum $Y_{benzene}$ at NTP = 9.5/760 = 0.0125. Equation 2.30 is the simplest way to find the saturated vapor density.

$$\text{maximum} \frac{\rho_{sat.\ vapor}}{\rho_{air}} = 1 + \left(\frac{95}{760} \times \left[\frac{78}{28.964} - 1 \right] \right) = 1.211 \quad \text{(using Equation 2.30)}$$

Again, 1.211 equals +21% using Equation 2.26. Thus, the maximum density of benzene vapor is as much more than air as steam from boiling water is less dense than air (Example 2.12). A density difference of approximately 20% compared to air is capable of causing a vertical plume velocity of perhaps 20 fpm, but any horizontal surface below the source (e.g., the liquid or wetted surface, the surface supporting the liquid's container, or the floor or ground) will prevent, or at least impede, the downward velocity of such a plume.

Example 2.16

Find the mixture density for 1.2% benzene vapor in room air, the lowest concentration at which a benzene vapor will burn (its LFL). Again, a lot of optional solutions are possible, but Equation 2.29 is the easiest and most appropriate solution when Y is not a saturated vapor.

$$\frac{\rho_{one\ vapor}}{\rho_{air}} = 1 + 0.012 \times \left(\frac{78}{28.964} - 1 \right) = 1.020 \quad \text{(using Equation 2.29)}$$

A ratio of 1.02 means that the vapor is only 2% more dense than air. It turns out that the lowest vapor concentration at which most combustible vapors can be ignited and burn is around 1 to 2%. For benzene, this 1.2% is almost a tenfold decrease in the concentration below its 12.5% equilibrium vapor concentration. In other words, the vapors leaving the liquid benzene must be diluted just over tenfold to no longer be a flammable concentration. At this point, its plume density is only 2% higher than the density of the surrounding air. The density of such a marginally LFL concentration is capable of causing a downward plume velocity of perhaps 5 to 10 fpm which is less than most random room air currents. Thus, one can conclude that by the time the saturated benzene vapor has dissipated approximately tenfold to its lowest combustible limit, the plume is no longer significantly more dense than the air around it. Or stated in a more generalized rule-of-thumb, ✓ *most flammable vapors are only approximately 2 to 4% heavier than air and will generally move with the air around it.*

EXAMPLE 2.17

Find the mixture density for 1000 ppm benzene vapor in room air.

$$\text{The specified } Y_{\text{benzene}} = 1000 \text{ ppm}/1{,}000{,}000 \text{ ppm} = 0.001 \qquad \text{(using Equation 2.14)}$$

$$\frac{\rho_{\text{one vapor}}}{\rho_{\text{air}}} = 1 + 0.001 \times \left(\frac{78}{28.964} - 1 \right) = 1.00169 \qquad \text{(again using Equation 2.29)}$$

$$\text{Percent difference} = 100 \times (1.0017 - 1) = 0.17\% \text{ heavier than air} \qquad \text{(again using Equation 2.26)}$$

Almost all gas and vapor exposure limits are less than 1000 ppm and most are in the 0.1 to 10 ppm range. Thus, as another rule-of-thumb, occupationally hazardous vapors are less than 0.1% heavier than air. The nearly neutral buoyancy of vapors at less than 1000 ppm means that virtually no vertical forces act on a plume diluted to a concentration in the range of occupational health exposure limits, and such contaminants will follow the air in which they are diluted.

2. Only a limited range of chemicals in a limited range of conditions can create a vapor whose plume is significantly greater than air. The data in Figure 2.8 and Figure 2.9 were generated using Equation 2.30 for 257 organic solvents and commercial products that have a TLV and whose P_{vapor} at 25°C is less than 760 mmHg (thus, the chemicals are at least somewhat common in industry and each forms a vapor rather than a gas). Each circle is the ratio of the density of a chemical's saturated vapor concentration relative to the density of uncontaminated air at 25°C. The straight line is a hypothetical density ratio attributed solely to its molecular weight as if it could exist at a pure (100%) concentration (which is only true of gases but not of vapors). The only difference between these two figures is in their vertical scales. Figure 2.8 shows density ratios up to 15 where it is easier to see the divergence between the line of hypothetical densities based only on molecular weight and the real saturated vapor densities that are limited by their vapor pressures. The vertical scale in Figure 2.9 only goes up to a ratio of four in order to enhance the view of the density ratios of real vapors that tend to cluster near one. None of the ratios for these chemicals exceed four.

   A few lessons can be drawn from these figures. First, the density of most of these vapors is much less than the density that NFPA and others attribute solely to the ratio of the molecular weights of the pure vapor or gas to air (viz., MW/29 in NFPA 325[19]).

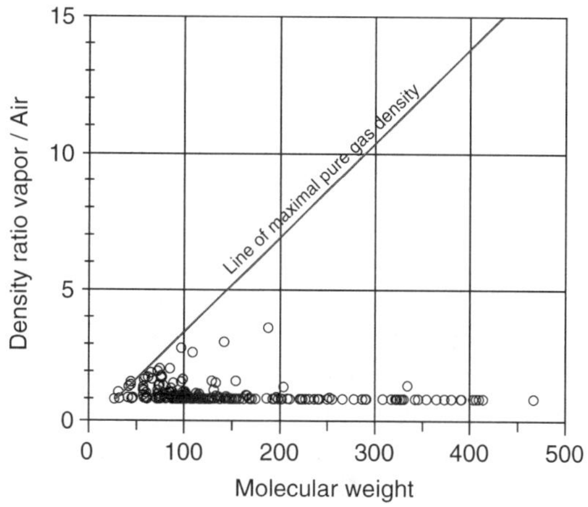

FIGURE 2.8 The density of real saturated vapors relative to that of clean air, highlighting their divergence from hypothetical "pure" vapors.

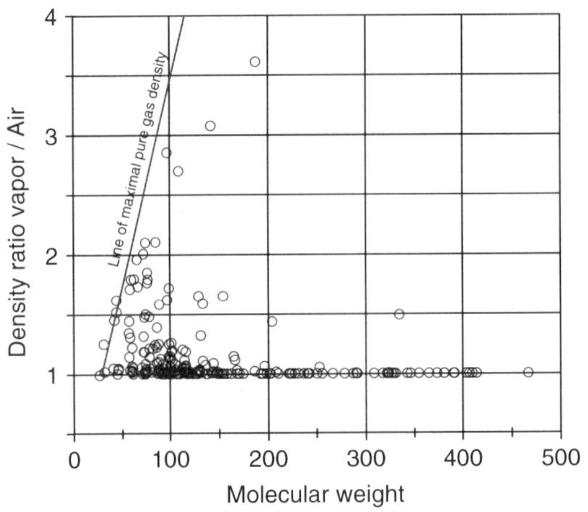

FIGURE 2.9 The density of the same real saturated vapors relative to that of clean air presented on a reduced vertical scale.

The tendency for a chemical's vapor pressure to decrease as its molecular weight increases keeps the maximum density of even a saturated vapor from continuing to increase with its MW. Figure 2.9 shows that the highest vapor densities occur for chemicals with a molecular weight in the region of 70 to 80 g/mole. Lighter chemicals do not add much molecular weight to the mixture, and heavier chemicals cannot exist at high vapor molar fractions.

Second, the range of the density of most vapors is not very broad. Only approximately 2% of the vapor densities of all the chemicals in Figure 2.8 are more than twice as dense as air.

Third, the density of most vapors is really close to that of normal air. The maximum density of over 80% of the vapors in Figure 2.8 is within 20% of the density of air, and just over 50% of these are within 1% of air's density. Recall that the density due to a vapor's molecular weight can be equated to that due to a lower temperature. For instance, a plume that is 20% more dense than normal air due to heavy vapors (found on the bottom line in Figure 2.7) will behave just like a plume that is approximately −20°C or −5°F (found by going straight up to the middle line in Figure 2.7). Looking at the above statistics another way, only approximately 18% of the vapors plotted in Figure 2.8 can possibly be that dense, and buoyancy can have little effect at a density ratio of only 1%.

A fourth lesson is not obvious from these figures, but the initial density of a saturated vapor near just above a liquid source (whatever it is) will decrease rapidly as the plume is diluted. The densities in Figure 2.10 are those of the same chemicals that are shown in Figure 2.9 but diluted by a factor of 10. Virtually none of these minimally diluted vapor densities exceeds that of clean air by more than 20%. Even the density of a heavy gas (that could initially be present at $Y = 100\%$) will rapidly approach that of air when it is diluted tenfold (as represented by the dashed line in Figure 2.10).

A few other points of comparison are offered. The vapors of solvents in the typical 1 to 5% flammable range are roughly 2 to 10% more dense than clean air. The vapors of a chemical whose flash point is above room temperature cannot even reach an ignitable concentration at room temperature; therefore, such vapors below their LFL are even closer to air's density. And vapors at concentrations below 1000 ppm (which is 10 to 100 × greater than most occupational exposure limits) are never dense enough to flow to and collect near the floor under any practical circumstance.

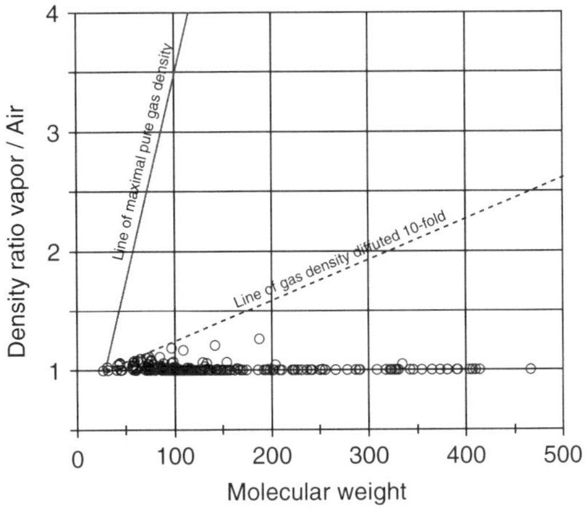

FIGURE 2.10 The density of vapors diluted tenfold compared to that of clean air.

In general, the density of room temperature vapors is not important in rooms with occupants, machinery, or ventilation that each typically cause general air currents in excess of approximately 40 or 50 fpm. Looking ahead at data to be presented in Chapter 5, air currents will rapidly dilute any saturated vapor pressure concentration by at least 1000-fold within about one arm's length from its liquid origin or source. Turbulence caused by someone standing upwind of a source will easily dilute the saturated vapor pressure concentration at the liquid surface by at least 10^4 by the time the vapors reach their breathing zone.

3. So when are dense vapors important? A dense vapor from a liquid already on the floor or ground (e.g., from a liquid leak or spill) needs an external mechanism such as a draft or wind to generate some turbulence and begin the dilution process. Examples of common dilution mechanisms include ventilation, human activity, wind, or even thermal currents. Without an external mechanism (such as in a closed, unventilated, and/or unoccupied room or outdoors on a cool, still night), a dense vapor can accumulate on the floor or ground, fill a walled-space vertically or spread horizontally, and even run downhill by gravity like a liquid. Perhaps the plume will flow down a vent, a stairway, or a slope to an ignition source, or accumulate enough to create an acute inhalation hazard or even an oxygen deficiency.

However, if the liquid evaporates somewhere above the ground, the requisite falling motion of an initially dense plume would start the dilution process and change the density from that depicted in Figure 2.9 to that depicted in Figure 2.10. Once diluted, a molecule's weight will not cause it to separate from the air or to accumulate near the ground. In other words, if a dense vapor does not start out on the bottom, it will not get there on its own.

The *Manual of Industrial Ventilation*'s Figure 3.2 and the ventilation control in VS-75-30 illustrate this point. The former shows that it is inappropriate to place an exhaust hood near the floor in a production facility, but the latter shows an exhaust hood near the floor in a paint storage room. Several sections within the *International Mechanical Code* (part of most local building code requirements in the U.S.A.) specify that the inlet to a local exhaust ventilation system for flammable vapors and gases shall be located not more than 18 in. (457 mm) above the floor.[20] Although this building code is a categorical requirement, the recommendation is probably not beneficial when mechanical ventilation or human activity is mixing the room air. The flow and dispersion of dense plumes have not been well studied except in the case of liquefied natural gas (LNG)

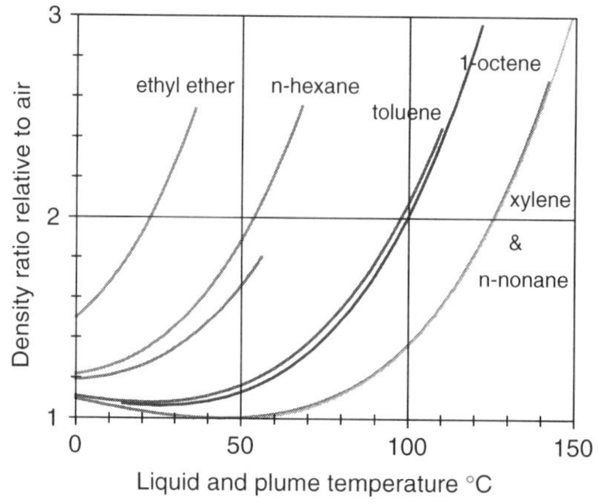

FIGURE 2.11 The effect on saturated plume density of heating some organic solvents.

storage facilities; while the molecular weight of natural gas (mostly methane) is less than air, the cold temperature of LNG makes its vapors very dense.[21] For the more technically inclined reader, Webber (1991) wrote a nice overview of cryogen hazard modeling.[22]

4. An academically interesting, but relatively rare situation in practice (except for vapor degreasers to be discussed in Section VII. 5 of Chapter 4) would occur if a moderately high MW solvent were heated. Heat would tend to drive the vapor pressure up as predicted by the Antoine equation (Equation 2.10) making the plume more dense, but a hotter plume would tend to make the plume less dense as predicted by the T_{room}/T_{plume} temperature ratio (in absolute degrees) in the Equation 2.19b version of the Universal Gas law. Which change would predominate?

   The data shown in Figure 2.11 were derived using Equation 2.25b assuming that the plume temperature is equal to the vapor temperature. Each chemical in Figure 2.11 seems to follow the same pattern. As long as the solvent temperature is well below its boiling point, its vapor density drifts downward linearly as the plume gets hotter. As the temperature approaches each solvent's boiling point, the exponential influence of temperature associated with the Antoine equation starts to predominate over the linear influence of absolute temperature. As a result, the plume's density will eventually increase but still appears to stay well below that predicted by molecular weight alone.

   The actual behavior of an unconstrained heated solvent plume is difficult to predict because the mechanisms that control molecular and turbulent dispersion turn out to be related to, but are somewhat different from, those mechanisms that control thermal cooling. A large, hot plume may cool rapidly enough to create a supersaturated vapor and form visible steam (or what would appear to be steam if the solvent were water). Eventually, any plume will be diluted below its saturation concentration, and the steam (mist) will re-evaporate and disappear. Thus, the behavior of the now diluted and cooled vapor plume might be determined at least as much by its temperature as by its molecular weight. Should you ever encounter such a scenario, try using a smoke tube (described in Section I of Chapter 12) to follow its plume. The pattern may surprise you.

PRACTICE PROBLEMS

1. Envision someone having to enter a series of railroad tank cars in order to remove residual (unevaporated) chemicals on a normal (25°C) day near sea level. This scenario

satisfies the criteria to create a vapor–liquid equilibrium inside the tank car. Estimate the vapor concentration above each of the following chemicals by converting mm Hg into ppm. The last column asks for the unitless ratio of the concentration inside of each tank car to its TLV (viz., ppm/ppm). The $[P_{vapor}]$ and TLV data for all of these chemicals is in the Vapor Hazard Ratio tables in the back of your book. One chemical is completed as an example.

Chemical	Vapor Pressure (mmHg)	$[P_{vapor}]$ (ppm)	TLV (ppm)	Equilibrium ($[P_{vapor}]$/TLV)
(a) Iso-pentane (not *n*-pentane)				
(b) Cyclohexane				
(c) Octane				
(d) Cyclohexanol	0.8	1053	55	21
(e) Naphthalene				
(f) *p*-Nitrochlorobenzene				
(g) Phthalic anhydride				

2. Suppose that the tank cars with residual octane were not cleaned until late in the afternoon when the temperature had climbed to 37°C (12°C above normal) or 98.6°F. What would you expect to be the octane concentration at that point based on a rule-of-thumb of how vapor pressures for organic chemicals change with temperature?

Chemical	P_{vapor} (mmHg)	Concentration (ppm)	TLV (ppm)	Equilibrium ($[P_{vapor}]$ / TLV)
Octane				

3. Use the following parameters for the Antoine equation to find a more accurate prediction of the vapor pressure of octane at 37°C.

$\ln [P_{vapor}] = A - (B/(°K + C))$	A =	B =	C =	Valid Range °C
n-Octane	15.9426	3120.29	−63.63	24 to 135

Chemical	P_{vapor} (mmHg)	Concentration (ppm)	TLV (ppm)	Equilibrium ($[P_{vapor}]$ / TLV)
Octane				

As a check of your calculations, this Antoine equation should yield a P_{vapor} that is between 5% and 6% lower than the approximate results in Q.2 using Equation 2.26.

4. Use Dalton's law to find the partial pressure of oxygen in mm Hg at sea level on a 25°C normal day, knowing that it comprises 20.948% of clean air. (The TLV uses mm Hg oxygen to define an oxygen deficiency.)

ans. = 159 mm Hg

5. Assume that the pressure in your car tires is 35 psig in the morning when the tire is at 50°F or 10°C. Remember that the Ideal Gas law is based on absolute pressure, while tires are measured as a gauge pressure. See Table 2.1 for normal atmospheric pressure.
   a. What is the absolute pressure inside the tire in the morning?

   ans. = 49.7 psia

b. What is the tire pressure while driving at highway speeds on a summer afternoon when the tire temperature might be 150°F or 65.5°C? (A tire is usually not more than approximately 50°F above ambient.)

ans. = 44.7 psig

6. Assume that your refrigerator is set at 37°F or 2.8°C, and that when you open the door to look for a snack, enough cold air is replaced by room air to raise the air temperature inside to 57°F or 14°C. The moment you close the refrigerator, the air inside starts to cool from 57°F back down toward 37°F. The cooling causes the pressure inside the refrigerator to start to drop below normal ambient pressure.
   a. What is the minimum pressure that could occur inside the closed refrigerator when the air again reaches 37°F or 2.8°C if there were no leaks?

   ans. = 14.127 psia

   b. What is the pressure difference at that point between the ambient pressure and the pressure inside the closed refrigerator? (Can you hear the air leaking into your refrigerator or freezer at home when you close the door? Try to open your freezer just a few seconds after you close it.)

   ans. = 0.569 psig

   c. If the refrigerator door were 30 in. wide by 48 in. tall, what force would the above pressure difference create on the door tending to hold it closed? (It is a good thing that the door seals leak!)

   ans. = 819 lb

7. How would the density of a hypothetically pure ($Y = 100\%$) xylene vapor compare to normal air density? The CAS# for xylenes is 1330-20-7, MW = 106.16, and P_{vapor} at NTP = 8 mm Hg (cf., iso-, meta-, or para-xylene).

   ans. 3.666 times the density of air or 267% more dense than air

8. How would the density of xylenes' vapor at its saturated vapor pressure at 25°C compare to normal air density?

   ans. 1.028 times the density of air or 2.8% more dense than air

9. Use Figure 2.7 (or some other means) to estimate the plume temperature of air that would have the same density as each of the above plumes due to their molecular weight. Think about how plumes at temperatures with which you are familiar behave. On which if any of these plumes is buoyancy likely to have a significant effect in a workplace? And, would that vertical trajectory (if any) be upward or downward?

   ans. equivalent temperatures of −192°C and 17°C

REFERENCES

1. *ACGIH Industrial Ventilation: A Manual of Recommended Practice*, 24th ed., American Conference of Governmental Industrial Hygienists, Cincinnati, OH, 2001.
2. *Pocket Guide to Chemical Hazards*, NIOSH, DHHS (NIOSH) 97–140, 3rd printing, 2004. (Also on-line at www.cdc.gov/niosh/npg/npg.html).
3. 2003 International Building Code®, International Code Council, Falls Church, VA, 2004, Section 1204.
4. Lide, D. R., Ed., *CRC Hand book of Chemistry and physics*, 83rd ed., CRC Press, BocaRaton, FL, 2002.
5. Popendorf, W., Barometric hazards, In *The Occupational Environment: Its Evaluation and Control*, 2nd ed., DiNardi, S. R., Ed., American Industrial Hygiene Association, Farifad, VA, pp. 647–667, chap. 25, 2002.
6. Perkins, J., *Modern Industrial Hygiene*, Van Nostrand Reinhold, New York, 1997.
7. Dreisback, R. R., *Physical Properties of Chemical Compounds*, Advances in Chemistry Series, *Am. Chem. Soc.*, I(15), 1955.

8. Dreisback, R. R., *Physical Properties of Chemical Compounds*, Advances in Chemistry Series, *Am. Chem. Soc.*, II(22), 1958.
9. Dreisback, R. R., *Physical Properties of Chemical Compounds*, Advance in Chemistry Series, *Am. Chem. Soc*, III(29), 1961.
10. Boublik, T., Fried, V., and Hala, E., *The Vapor Pressures of Pure Substances*, Elsevier Scientific Publishers, New York, 1973.
11. Ohe, S., *Computer Aided Data Book of Vapor Pressures*, Data Book Publishers, Tokyo, 1976.
12. Reid, R. C., Prausnitz, J. M., and Sherwood, T. K., *The Properties of Gases and Liquids*, 4th ed., McGraw-Hill, New York, 1987.
13. Yaws, C., *Handbook of Vapor Pressure*, Vols. I–III, Gulf Publ. Co., Houston, TX, 1994.
14. Howard, P. H. and Meylan, W. M., Eds., *Handbook of Physical Properties of Organic Materials*, CRC Press, Boca Raton, FL, 1997.
15. Franklin, P. J., Goldenberg, W. S., Ducatman, A. M., and Franklin, E., Too hot to handle: an unusual exposure of HDI in specialty paints, *Am. J. Ind. Med.*, 37(4), 431–437, 2000.
16. Ramsey, J. D. and Bishop, P. A., Hot and cold environments, In *The Occupational Environment: Its Evaluation and Control*, 2nd ed., DiNardi, S. R., Ed., American Industrial Hygiene Association, Farifax, VA, pp. 613–644, chap. 24, 2002.
17. *U.S. Standard Atmosphere*, NOAA, National Oceanic and Atmospheric Administration, National Aeronautics and Space Administration, United States Air Force, NOAA-S/T 76-1562, Washington DC, Oct. 1976.
18. Keeling, C.D., Whorf, T.P., and the Carbon Dioxide Research Group of the Scripps Institution of Oceanography, University of California, La Jolla, as of May, 2005.
19. NFPA 325, *Guide to Fire Hazard Properties of Flammable Liquids, Gases, and Volatile Solids*, National Fire Protection Association, Quincy, MA, 1994.
20. International Mechanical Code, International Code Council, Falls Church, VA, 1999.
21. Hartwig, S., Ed., *Heavy Gas and Risk Assessment*, D. Reidel Publ., Dordrecht, Holland, 1980.
22. Webber, D. M., Source terms, *J. Loss Prev. Process Ind*, 4, 5–14, 1991.

3 Basic Aerosol Behavior

The learning goals for this chapter:

✍ Be able to classify a given aerosol as a fume, mist, dust, or fiber.
✍ Understand the mechanisms of impaction and sedimentation and how they affect aerosol behavior in the environment and in the respiratory tract.
✍ Be familiar with the mechanism of centrifugation as used in aerosol samplers and air cleaners.
✍ Understand the difference between a normal and log-normal distribution and which applies to aerosols.
☐ Know how to use $d_{\text{aero}} = d_{\text{Stokes}}\sqrt{\rho}$ to convert between a Stokes diameter and an aerodynamic diameter.
✍ Know where and how to access particle size data and how to find or calculate a particle's falling velocity.
✍ Be able to use a figure or table to determine the percent inhalable, thoracic, or respirable fraction for a given particle aerodynamic diameter.

I. HOW AEROSOLS DIFFER FROM GASES AND VAPORS

Aerosols are unlike vapors in at least three aspects: they lack an intrinsic property (such as vapor pressure) to become airborne, they are much larger than the air molecules in which they are suspended, and aerosols can be generated in many sizes (diameters).

1. Because powders, dusts, and solid materials have no intrinsic ability (such as vapor pressure) that would cause them to become airborne spontaneously, the generation of most aerosols relies on some external mechanism.[a] Grinding, crushing, or dropping can both break up a solid material and propel small particles into the air. Scraping, shaking, or blowing can aerosolize an already existing powder. Spraying or splashing can aerosolize both a relatively nonvolatile liquid and any solid suspended within the liquid. Because solid materials lack an intrinsic source-generating property, source control of aerosols is limited to either modifying the external mechanism that generates the particles or reducing the intrinsic susceptibility of a given dust, powder, or fiber to become aerosolized (as by wetting or increasing the particle's size). Wetting or increasing a particle's size is not feasible in many circumstances. Thus, most aerosol controls focus on the external mechanism.
2. Once airborne, aerosol particles have a much larger mass than the air molecules in which they are suspended. Appreciating the magnitude of this difference between their masses might help the reader to understand, or even anticipate, some aerosol behaviors, such as why aerosol particles cannot be accelerated or turned as rapidly as air can. As shown near the top of Table 3.1, the ratio of the mass of a 1 μm dust

[a] The exception are molten metals and other high temperature liquids that form fumes. Their vapor pressure at high temperatures causes these materials to evaporate. Rapid cooling causes such vapors to condensed forming fumes or smoke.

TABLE 3.1
Analogies to the Ratio of the Mass of a Dust Particle ("Large *d* Mass") to the Mass of an Air Molecule ("small *d* mass")

	Diameter (mm)	ρ (g/cm³)	Mass (g)	Ratio $\frac{\text{large } d \text{ mass}}{\text{small } d \text{ mass}}$
1 μm dust particle	0.001	2	1.05×10^{-12}	2.2×10^{10}
Nitrogen molecule	0.4×10^{-6}		4.67×10^{-23}	
Bowling ball (16 lb)	218	1.33	7264	5.3×10^{7}
Piece of pencil lead	0.5	2.1	0.00014	
Bowling ball (16 lb)	218	1.33	7264	2.9×10^{3}
A ping pong ball	38.1	0.085	2.46	

particle to the mass of an individual nitrogen molecule in air is approximately 2×10^{10} to 1.

Envisioning a larger scale analog to such a large ratio is a bit of a challenge. For instance, this ratio is over a million times greater than the ratio of the mass of a bowling ball to a ping-pong ball. It is still a thousand times greater than the ratio of the mass of a bowling ball to a 0.5 mm long by 0.5 mm diameter piece of mechanical pencil lead (graphite). If these small amounts of graphite were traveling at sonic speed (like air molecules), over 10^{16} of them would be hitting each side of that bowling ball every second. Since they are hitting all sides of the ball, those hits would not keep the bowling ball suspended, but they are just enough to slow the speed at which the ball would fall.[b]

Similarly, air molecules and turbulence can keep aerosol particles suspended for a long time, but eventually even small (meaning inhalable) aerosol particles will fall out of the air. The relative difference in mass will again affect aerosol particles if the air is accelerated or changes its direction of motion rapidly. The same force that can accelerate molecules rapidly cannot change the path of the much larger particles as rapidly, and a suspended aerosol will lag behind the path of the air. The difference in their paths can cause particles to not enter into a sampling or respiratory orifice; to separate within and be collected by an impaction air sampler, an air cleaning filter, or a cyclone pre-selector; to not be inhaled deeply into ones lung; and/or to settle and be deposited there.

3. Mensuration formula (such as Equation 3.1 and Equation 3.2 embedded within Table 3.2) allow one to use a particle's diameter to calculate its surface area and internal volume. (Mensuration is "the algebra of measuring.") Most natural aerosol particles (such as those in Figure 3.1) are somewhere between a sphere and a cube, so the ratio of $\pi/6$ (approximately one half) between their respective surface areas or volumes turns out not to cause much practical difference. For instance, a particle's behavior will be shown later to vary with the square of its diameter; so the maximum error in estimating an aerosol's diameter as if it were a cube when it might be a sphere only varies with the square root of $\pi/6$, or approximately $\pm 15\%$. Such a difference can be important in detailed research work, but it has only a small effect when applied to industrial hygiene controls.
4. Now we will use the cube's width as an approximation of a sphere's diameter to generate some data that should begin to explain why and how an aerosol particle's diameter is related to its physical behavior, to its fate in the environment, to whether and where it will

[b] To be more technically correct, one should assume that the molecules and pencil leads have equal thermodynamic temperatures; in which case the pencil leads would be traveling slower by the square-root of their mass ratios and it would take approximately 10^{20} hits per side for the little pieces of lead to create the equivalent of one atmosphere of pressure.

TABLE 3.2
Mensuration Formulas for Aerosol Particles Approximated by Spheres or Cubes

	For a Sphere	For a Cube	The Ratio of a Sphere to a Cube
Surface Area	$4\pi r^2 = \pi d^2$ (3.1a)	$6d^2$ (3.1b)	$\pi/6$
Volume	$4\pi r^3/3 = \pi d^3/6$ (3.2a)	d^3 (3.2b)	$\pi/6$

be deposited within a lung, and even to the toxicity of that deposit. Imagine for a moment cutting up a 1 cm cube into smaller and smaller particles. Start by cutting each edge of the cube into ten smaller pieces; that will make 1000 particles. Then repeat that process on each smaller cube, producing a series of progressively smaller and smaller uniform-sized cubes. If the cuts are perfect, the total volume of material and the total mass of all the particles will stay the same as the one whole cube with which you started. Table 3.3 shows how the total number of particles (N) increases in proportion to the diameter of the particles cubed, and their total surface area increases in direct proportion to their diameter. Meanwhile, each particle's individual mass decreases in accordance with Equation 3.2b and its individual falling velocity (V_{falling}) decreases in accordance with Stokes law and Equation 3.11 (to be described in Section IV.2).

The falling velocity of particles increases with the square of their diameter. Falling velocity determines the time that an aerosol particle can stay airborne and the time needed for it to be deposited by sedimentation within the lung's alveoli. The falling velocity is also related to the efficiency with which an aerosol particle can enter into an orifice, such as a sample collector or our nostrils. Thus, particle diameter (squared) determines the fraction of an aerosol that can first be inhaled and that can reach progressively deeper portions of the lung.

a. Particles above 50 μm in diameter will fall rapidly from the air before they can travel far from their source, and not all that pass by the orifices of the respiratory tract (nose and/or mouth) can enter. However, large particles can easily land on one's skin and

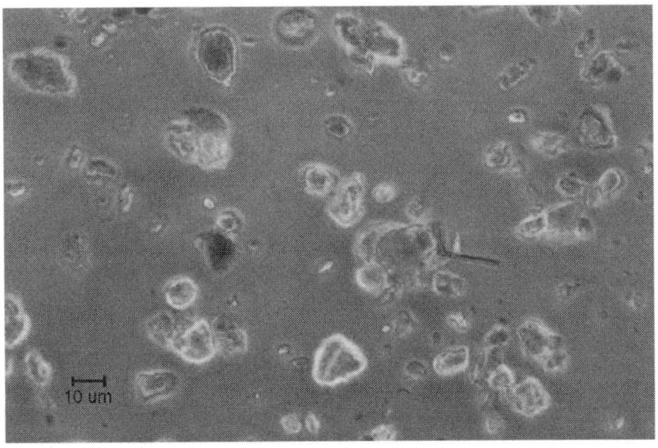

FIGURE 3.1 A photomicrograph of a soil-derived aerosol.

TABLE 3.3
Hypothetical Examples of the Diminution of One Large Cubic Particle into Smaller Uniform Cubic Particles

Individual Width (d)	Total N	Total Surface Area (m^2)	Individual Mass (g)	Individual $V_{falling}$ (cm/sec)	Time to Fall 200 μm (sec)
1 cm	10^0	0.0006	1	2200	0.000000009
1 mm	10^3	0.006	1×10^{-3}	600	0.00003
100 μm	10^6	0.06	1×10^{-6}	50	0.0004
10 μm	10^9	0.6	1×10^{-9}	0.6	0.033
1 μm	10^{12}	6	1×10^{-12}	0.007	2.8
0.1 μm	10^{15}	60	1×10^{-15}	0.0002	100
0.01 μm	10^{18}	600	1×10^{-18}	0.000014	1400

The same concept would also apply to spherical particles.

can carry chemicals that can be absorbed through intact skin (e.g., organophosphate insecticides) or cause dermatitis.

b. Among those particles that are inhaled, less than half of those between 50 and 10 μm in diameter will pass beyond the throat. Those larger than 10 μm fall too fast (or stop too slowly) to reach the deep lung easily, and among those small enough to reach the deep lung, those smaller than approximately 1 μm do not fall fast enough to be deposited within the short time that one's inhaled breath is within the alveoli (ca. 0.75 sec). Diffusion does not begin to be important until particles are smaller than approximately 0.1 μm in diameter. Thus, it turns out that particles near 0.3 μm in diameter are the most difficult to remove from the air, either by a filter or by our lungs.

c. A given mass of small aerosol particles will comprise a geometrically larger number and linearly larger surface area than an equal mass of larger particles. A greater surface area probably increases the toxicity of that given mass of small particles. In addition, each solid particle deposits considerably more mass onto one location or cell than would be deposited by distributing that mass as individual molecules uniformly across the lung's tissues.

II. AEROSOL DEFINITIONS

1. The following terms characterize the physical nature of an aerosol and, indirectly, its source:

   Dust: Solid particles that are formed by the fragmentation or diminution of larger particles into diameters that are small enough to stay airborne. Dusts smaller that approximately 0.1 μm are difficult to generate, and those larger than approximately 100 μm in diameter will not stay airborne long.

   Fume: Solid particles that are formed by condensation from a high temperature vapor, such as from a molten metal or smoke. Fumes form at an initially very small diameter (ca. 0.01 μm); and although they will aggregate into larger particles, they still rarely get larger than approximately 0.5 μm.

   Mist: Any airborne liquid particles. A water mist in the form of steam, fog, or a fine spray is a common example, but mists of an organic solvent or even mercury can be formed.

Mists smaller than approximately 1 μm are hard to generate, and mists larger than approximately 100 μm will not stay airborne long.

Fiber: Elongated particles whose length-to-diameter ratio (called an "aspect ratio") is at least 3:1. A fiber's aerodynamic behavior is determined mostly by its diameter (vs. its length).

Bioaerosol: Particles of biologic origin (plant, food, fungi, bacteria, etc.). Bioaerosols may be either organic dusts or mists of solutions containing microorganisms.

2. Two terms are used to characterize the statistical dispersion or variability among the diameters of all of the particles within an aerosol (or within a powder that could become an aerosol).

   A population of particles of equal (or nearly equal) diameters is said to be monodisperse. Monodisperse aerosols almost never occur naturally. In fact, researchers have to work hard to make a monodisperse aerosol and to keep an aerosol's diameter from growing by agglomeration.

   A population of particles of unequal or varied diameters is said to be polydisperse. Most natural aerosols are polydisperse. The quantitative indicator of how widely dispersed the diameters within a given aerosol is the geometric standard deviation (GSD). The next main section of this chapter (Section III) contains further discussions on polydisperse aerosol statistics.

3. Two terms are used to characterize an aerosol particle's diameter (either individually or collectively). These two diameters are generally not equal, but either one can be calculated from the other if the particle's density is known (by methods to be discussed in Section V).

   A Stokes diameter (d_{Stokes}) is the real diameter of a spherical or nearly spherical particle. A particle's Stokes diameter closely approximates its real diameter because most real particles approximate a sphere (unless they are fibers or some other convoluted shape).

   An aerodynamic diameter (d_{aero}) is the hypothetical diameter of a spherical particle of unit density that behaves just like a given particle. A particle's aerodynamic diameter increases (or decreases) in proportion to the square root of the ratio of its real density to unit density ($d_{\text{aero}} = d_{\text{Stokes}} \times \sqrt{2}$).

4. The following mechanisms affect aerosol particles in the environment, in common air cleaners, in aerosol sampler collectors, and/or within the respiratory tract.

   Interception is the physical entrapment of particles while flowing through a narrow passage (e.g., the collection of large particles as air flows through or between filter fibers). Interception usually only affects very large particles, and as a result is rarely a primary collection mechanism on air sampling filters, in air purifying respirators, or within the respiratory tract.

   "Inertial separation" is a broad term encompassing impaction, centrifugation, and other separation processes that rely on the difference in the inertia between particles and air molecules. If air is accelerated laterally (turned), the much larger inertia of particles will cause them to lag behind and/or cross the path followed by the turning air molecules (air paths are called "air streams").

   a. An impactor rapidly turns air from its original direction through at least 90°. Impaction is caused by the inability of large particles to turn as rapidly as the air as it passes around an obstruction, resulting in contact with and deposition onto the obstruction. Impaction is the principal mechanism in impactor air samplers, in venturi air scrubbers, and in most filters,[c] and is a significant mechanism within the upper respiratory tract.

---

[c] Impaction caused by tortuous paths within filters is more important than interception, although the performance of filters used on air purifying respirators is also enhanced by electrostatic forces.

b. A cyclone causes the air to travel several revolutions inside a tapered cylinder. Centrifugation is the mechanism by which the centrifugal force from the radial acceleration of the air into a circular pattern causes the particles to move outward (radially) across or perpendicular to the direction of air flow. Centrifugation is the principal mechanism operating within cyclone preselectors and cyclone particulate air cleaners, causing at least some particles to make contact with and collect on the wall of the centrifuge. From the perspective of particle dynamics, centrifugation is closely related to impaction, just less abrupt.

c. Inertial separation also occurs as particles are drawn into a human respiratory orifice, an air sampling orifice, or a local exhaust hood causing fewer particles to enter into the orifice than are in the ambient air outside that orifice. If the orifice through which the air enters has a small diameter or a high entry air speed, another disregard fraction of particles that enter the orifice will impact on its entry walls due to entry turbulence. These "entry loss" mechanisms underlie the definition of an inhalable aerosol (to be discussed in Section VI of this chapter).

Gravitational separation is a broad term encompassing elutriation, sedimentation, and any other effect that results from the falling or vertical movement of aerosol particles relative to the air in which they are suspended.

d. Elutriation is the selective separation of one material from another. Thus, an elutriator is an air sampler that utilizes the differential falling speed of aerosol particles to remove the larger sized portion of an ambient aerosol that would otherwise not reach deeply into the lung before it collects the remaining smaller particles. A vertical elutriator is used to collect cotton dust, and the MRE horizontal elutriator is used in British industry to collect respirable sized particles.

e. Sedimentation is the deposition of particles onto a surface due to their vertical falling velocity. Sedimentation is the most important mechanism affecting aerosol deposition within the lower or pulmonary respiratory tract where the air is moving slowly or not at all. It can also cause large particles to fall rapidly out of the air onto skin or other horizontal surfaces close to a source. Undesirable sedimentation can happen when trying to move dusts and even larger debris through long, horizontal local exhaust ducts; the buildup of large particles can clog ducts that do not have enough velocity to keep such material suspended.

5. The following mechanisms affect aerosol particles more slowly than inertial and gravitational separation and are therefore usually of secondary importance to them.

Evaporation is an important mechanism affecting volatile mists. Evaporation will cause the diameter of individual mist droplets to decrease (mist particles may still be called droplets even when they are too small to actually drop), allowing an initially large droplet to penetrate more easily into the respiratory tract and perhaps decreasing its size to the point of leaving only any suspended or dissolved solids initially within the liquid to become a suspended dust. Ferron and Soderholm estimated the times for the evaporation of 10 to 100 μm water droplets to range from 1 to 30 sec, respectively, in 30 to 80% relative humidity conditions, respectively.[1]

Agglomeration, aggregation, and coagulation are all largely interchangeable terms that relate to the growth in an aerosol's diameter by totally inelastic collisions between individual particles (particles that stick together when they bump into each other). Such growth in particle diameter usually precedes their removal by sedimentation or impaction (especially for fumes that start out too small to be affected by most aerosol removal mechanisms other than filtration or electrostatic or thermal precipitation).

Small particles can be transported by diffusion in a concentration gradient similar to molecular diffusion. Diffusion is not important for particles more than approximately 0.1 μm in diameter, and even then it is not important over large distances. About the only time diffusion of aerosols is important is for the aggregation of fumes into larger particles, for penetration of particles less than 0.1 μm in diameter through a filter, or for deposition of such small particles within the lung.

6. The following mechanisms only affect aerosol particle deposition in certain air cleaners and specialized aerosol sample collectors.

   Electrostatic precipitation (ESP) is the transference of electrical charges onto aerosol particles and their subsequent collection onto an oppositely charged collecting surface. ESP is a common air emission control at power plants and has been used as a specialized air sampling technique. In both cases, ESP is effective at collecting small particles. Filters can either be impregnated with a resin or made from polarized fibers to generate minute electrostatic fields that will greatly improve the collection of small particles (e.g., in respirator filters).

   Thermal precipitation can occur as an aerosol passes between two surfaces at different temperatures. The temperature difference creates a difference in the kinetic energy of air molecules hitting the two sides of each particle that propels particles toward, and can eventually cause deposition onto, the cooler collection surface. Thermal precipitation is a rarely used air sampling technique that works well for very small particles. Thermal precipitation also causes dust and soot to collect on the cool, reflective light shade or luminaire around a hot light bulb that will decrease its illumination.

III. PARTICLE DIAMETER DISTRIBUTIONS

The diameters of particles within most aerosols (and many other environmental measurements for that matter) appear to be log-normally distributed. When a log-normal distribution is plotted on a linear scale (as shown at the top of Figure 3.2) the data appear asymmetric or skewed with the higher values (those to the right of their central or most common value) spread out over a wider

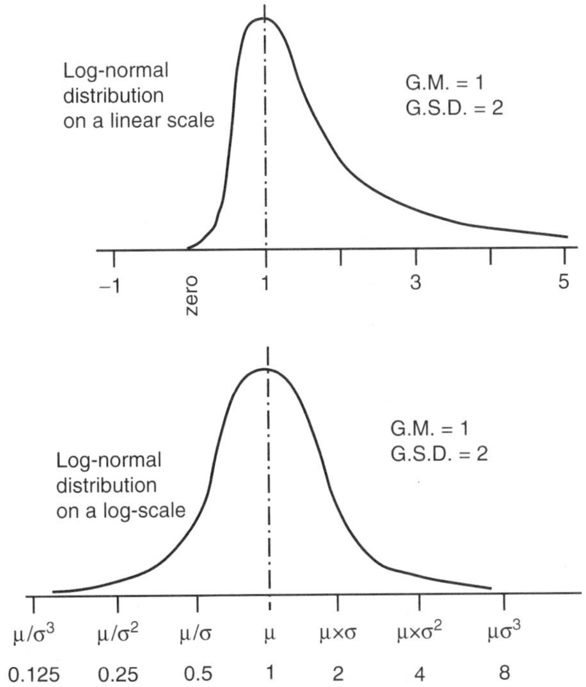

FIGURE 3.2 Depiction of a log-normal distribution plotted on a linear and logarithmic X axis.

range than the lower values (those to the left of their center or most common value). No values are less than zero, another characteristic of a log-normal distribution. A third and probably the most useful attribute of log-normal distributions is that the logarithms of log-normally distributed values are "normally" distributed (as shown at the bottom of Figure 3.2). Thus, the full range of characteristics and statistical tests applicable to normally distributed data can be performed on the logarithms of log-normally distributed data. For instance, just as normally distributed data appear as a straight line on a linear-probability plot, log-normally distributed data appear as a straight line on a log-probability plot such as depicted in Figure 3.3.

1. Log-normal statistical parameters can be derived from a log-probability plot or calculated as shown below from the logarithms of the measurements. The symbol X in these equations represents any generic value being measured. For instance, each X_i can be the diameter of an aerosol particle.

$$\text{Geometric Mean (GM)} = \text{antilog [the average of all of the log}[X_i]\text{ values]} \quad (3.3)$$

Either common or natural logarithms may be used, as long as one uses the same function to take the antilog. "Antilog" means to raise the value in question to the power of the base of the logarithm, either 10 (for common logarithms) or e = 2.71828... or using the "exp" button on many hand calculators (for natural logarithms). Notice that by taking the antilog of the log of X_i values, the units of the geometric mean are the same as the units of whatever is being measured. For instance, if one measures the particle diameters in units of micrometers (μm), then the GM would be in micrometers or μm. The GM for aerosols may also be abbreviated as either d_g, CMD, or MMD for a geometric mean diameter, the count median diameter (CMD), or mass median diameter (MMD), respectively (the median is the center value, the midpoint, or the point in the distribution at which 50% of all measurements are above and below it).

$$\text{Geometric Standard Deviation (GSD)} = \text{antilog [standard deviation of all log}[X_i]\text{ values]} \quad (3.4)$$

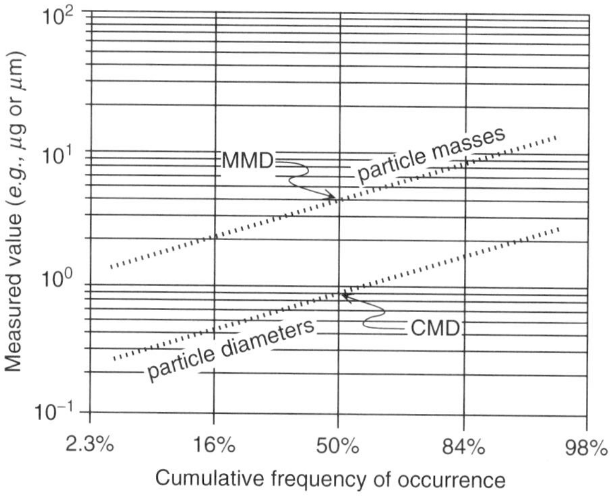

FIGURE 3.3 A depiction of a generic cumulative probability plot of aerosol count and mass.

Recall from basic statistics theory that a predictable portion of a normal population will lie within (±) any given number of standard deviations from the mean. Recall also that adding two logarithms is like multiplying by their antilogs; similarly subtracting is like dividing.[d] Thus, predictable portions of a log-normal population will lie within the same number of multiples of the GSD times the GM. The ratios in Equation 3.5 are each one multiple of the GSD, and (similar to normal statistics) each multiple increases the percentile position within the log-normal distribution by one probit.[e]

$$\mathrm{GSD} = d_{84\%}/d_{50\%} = d_{50\%}/d_{16\%} = d_{16\%}/d_{2.2\%} \tag{3.5}$$

The subscripts in Equation 3.5 refer to the percentile position. For instance, "$d_{84\%}$" means the diameter that is larger than 84% of all particles. Thus, the GSD is always equal to or greater than one. The GSD for a monodisperse aerosol will (by definition) be one (or very close to one). The GSD for all natural samples (or those not purposely created to be one size) will always be greater than one, making them polydisperse. The GSD for natural aerosols is often approximately two or a little bit more.

2. An aerosol's geometric mean diameter and GSD can be determined by counting the number of particles within predefined size ranges. Such a geometric mean is called a CMD. The CMD and GSD parameters were traditionally determined (meaning way back when the author was young) by collecting a lightly loaded aerosol sample onto a filter, visually counting approximately 200 to 300 particles within selected size ranges using a light microscope, plotting the cumulative frequency of particles within each size category on log-probability paper, and fitting a line through the data. The CMD is the diameter of 50% of the particles (where the fitted line passes through fifty cumulative percent of all particles), as shown in Figure 3.3. The GSD is the slope of that line determined as the ratio of two diameters one probit apart using Equation 3.5.[2,3]
3. The mass median diameter or MMD can be determined by measuring the distribution of particle masses separated and collected into a series of size categories by their inertia such as within a cascade impactor.[4] If density is uniform across particle sizes, an aerosol's distribution by mass also represents the distribution of particle volume, and the GSD of aerosol diameters by mass will be identical to the GSD of its diameters by count, as depicted in Figure 3.3. However, because the volume of each spherical particle is proportional to d^3, the MMD is larger than the CMD by a ratio that is a function of the GSD as given by the Hatch–Choate Equation, shown below in both its common and natural logarithmic forms[5]:

$$\mathrm{MMD} = \mathrm{antilog}\{\log[\mathrm{CMD}] + 6.91 \times (\log[\mathrm{GSD}])^2\} \tag{3.6a}$$

$$\mathrm{MMD} = \mathrm{antiln}\{\ln[\mathrm{CMD}] + 3 \times (\ln[\mathrm{GSD}])^2\} \tag{3.6b}$$

Equation 3.6b can be further simplified by taking its antilog as described for Equation 3.3 and Equation 3.4 (and perhaps after reviewing the logarithmic relationships described in footnote [d]), yielding Equation 3.6c.

$$\mathrm{MMD} = \mathrm{CMD} \times \exp(3(\ln[\mathrm{GSD}])^2) \tag{3.6c}$$

The examples of distributions given in Table 3.4 show how much larger the MMD can be than its corresponding CMD as the value of the GSD increases. The GSD for natural aerosols

[d] For review: $\mathrm{Log}(A \times B) = \log(A) + \log(B)$, $\mathrm{Log}(A/B) = \log(A) - \log(B)$, and $\mathrm{Log}(A^B) = B \log(A)$. The same applies for natural logarithms. The reader is also reminded (for later use) of the following relationships, $A^B \times A^C = A^{B+C}$ and $(A^B)^C = A^{B \times C}$.

[e] A "probit" or probability unit is the portion of a population spanned by one standard or geometric standard deviation.

TABLE 3.4
Examples of Using the Hatch–Choate Equation

MMD	$= 1.10 \times$ CMD	if GSD $= 1.2$
	$= 1.64 \times$ CMD	if GSD $= 1.5$
	$= 4.23 \times$ CMD	if GSD $= 2.0$ ⇦ A common GSD
	$= 12.4 \times$ CMD	if GSD $= 2.5$
	$= 37.4 \times$ CMD	if GSD $= 3.0$
	$= 319 \times$ CMD	if GSD $= 4.0$

tends to be slightly above two, but a GSD can exceed three or even four near a source such as grinding or excavating and can be less than two if the particles were manufactured with an ability to control their diameter. Test your understanding of the Hatch–Choate equation by checking to see if you can obtain the same ratio of MMD/CMD for any value of GSD in Table 3.4 that you chose.

Thus, how one interprets a reported median particle size depends qualitatively upon whether the median was measured by counting the number within each size range (yielding the CMD) or by analyzing the masses within each size range (yielding the MMD or MMAD). The magnitude of the difference depends greatly upon how polydisperse the aerosol is (i.e., the value of its GSD). Make sure you know which median diameter is being referred to in written reports.

IV. MODELING PARTICLE BEHAVIOR

The first subsection below explains how a basic model of aerosol behavior can be developed. This particular model is developed based on sedimentation, but its general result (discussed in the second subsection below) also applies to the inertial effects previously described in Section II.4.

1. The first step in building this model is called a "force analysis" where all of the forces on a particle are identified and characterized mathematically. Newton's law of motion states that the acceleration of a body is proportional to the net force acting on that body. Similarly, where a body (a particle in this case) has either no motion or a constant motion, the forces must balance each other and equal zero. This analysis will be made on a body falling at a constant velocity where the forces balance (similar in engineering philosophy to mass and energy balances that the reader will encounter in later chapters).

   An aerosol particle falling in air is affected by two forces: gravity pulling it down and drag slowing it up, as depicted in Figure 3.4. Intuitively, you might imagine that drag will get larger as the particle falls faster. If a particle were to be released into the air with no initial downward velocity, the force of gravity would cause the particle to start to fall (downward). The faster it falls, the greater is the force of drag pushing the particle up. Eventually a falling body (any body) will reach a steady state velocity where the upward force of drag will equal the downward force of gravity and the speed will be constant.

   a. Gravity produces a constant downward force on a particle equal to its mass times the gravitational acceleration, its "weight." Mass in turn is equal to the body's density times its volume (see Equation 3.2b for a mensuration formula for the volume of a sphere).

$$F_{\text{down}} = \text{mass} \times g = \text{density} \times \text{volume} \times g = \rho_{\text{net}}(\pi d^3/6)g \tag{3.7}$$

where

$g =$ the acceleration of gravity $= 980$ cm/sec^2;
$d =$ particle diameter in centimeters;
$\rho_{\text{net}} = \rho_{\text{particle}} - \rho_{\text{fluid}}$ which for aerosols is approximately ρ_{particle} as explained below.

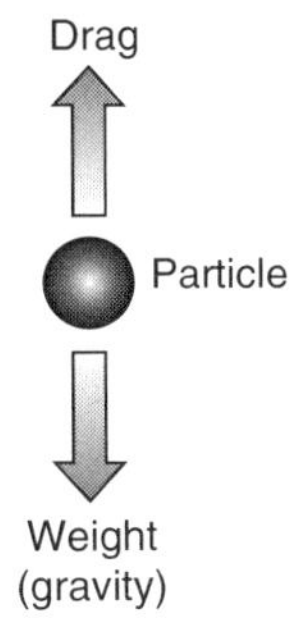

FIGURE 3.4 Forces acting on a falling particle.

The force of buoyancy can, on rare occasions, play a role in the downward force. Buoyancy would also oppose the force of gravity by an amount equal to the weight of the volume of fluid displaced by the body. However, buoyancy for a particle can only be important if its density is similar to (or less than) the density of its suspending media. Because the density of most dusts and mists (generally between 1 and 5 g/cm^3) is much greater than the density of air (found in Example 2.11 to be approximately 0.0012 g/cm^3), the force of buoyancy on aerosols is usually disregarded and $\rho_{net} \approx \rho_{particle}$.[f]

b. Friction from the air flowing around a falling particle produces the upward force of drag on the particle. The following formula for drag on a sphere at slow speeds, referred to as Stokes law (Ca., 1851), is accurate for particles between approximately 0.5 and 100 μm in diameter.[g]

$$F_{up} = 3\pi\, \eta\, d\, V \tag{3.8}$$

where η is the fluid's dynamic or absolute viscosity or 1.8×10^{-4} gm/cm $\times$ sec for air near normal temperature. V is the velocity at which the particle is moving through the air or other fluid, cm/sec. (IHs never use the value of viscosity in practice, but it will be encountered again in Section V.2 of Chapter 12 to describe air flow in ventilation.)

Particles smaller than 1 μm start to act as if they can slip between air molecules and therefore have less air drag, fall faster, and can cross air flow streamlines more rapidly than expected. Cunningham (ca. 1910) conducted a theoretical analysis of the effect of the mean free path of air molecules on particle behavior.[6] This concept evolved over approximately 50 years into various forms of an empirical slip correction factor (C_{slip}) that can be inserted as a denominator into Stokes law to predict the reduced drag for small particles. While most forms of what came to be called the "Cunningham slip correction factor" involve an exponential term (such as Equation 3.9a), the linear forms of C_{slip} in Equation 3.9b and Equation 3.9c are accurate to within $\pm 7\%$ down to approximately 0.01 μm

[f] Two exceptions are pointed out where buoyancy is important. One is in laboratory research in which bubbles are generated with a mixture of air and helium in exactly the right proportion for the buoyancy of the gas to completely neutralize the weight of the surrounding bubble. The other is a solid with the same density as water or other liquid in which it is suspended.

[g] Technically, Stokes law is accurate for Reynolds numbers less than approximately 1. A Reynolds number is the ratio between inertial forces and viscous forces within a fluid flowing past a body that will be revisited in Chapter 12. The Reynolds number of a 100 μm particle falling through air is right about 1 (see Figure 3.5).

(covering the range of most fume and smoke particles).[7,8]

$$C_{slip} = 1 + \frac{\lambda}{d}\left[2.514 + 0.800 \exp\left(-0.55\frac{d}{\lambda}\right)\right] \quad (3.9a)$$

$$C_{slip} = 1 + [(6.65 \times 10^{-4})T/d] \qquad \text{for } 1 \leq d \leq 0.01\ \mu\text{m} \quad (3.9b)$$

$$C_{slip} = 1 + (0.195/d) \qquad \text{for NTP and } 1 \leq d \leq 0.01\ \mu\text{m} \quad (3.9c)$$

where C_{slip} is unitless, λ is the mean free path of air (0.0065 μm), T is air temperature in °K, and d is the particle diameter in μm. For air at 20°C, C_{slip} can be approximated by Equation 3.9c.

2. When the force of drag equals the force of gravity, the falling velocity becomes constant at a speed called a body's "terminal falling velocity" (abbreviated $V_{falling}$ herein). The sum of these forces is expressed quantitatively by Equation 3.10. When a body falls at a constant velocity, by definition its acceleration must equal zero. Applying Newton's famous "$F = m \times a$" formula, if acceleration equals zero, the sum of the vertical forces up and down must also equal zero.

$$\Sigma F_{vertical} = F_{down} - F_{up} = 0 \quad (3.10a)$$

Thus, when an aerosol particle falls at its terminal falling velocity ($V_{falling}$), the vertical force of gravity pulling it down (defined by Equation 3.7) balances or equals the force of drag pushing it up (defined for Stokes particles by Equation 3.8 and modified for slippage by Equation 3.9).

$$F_{down} = F_{up} \quad (3.10b)$$

$$\text{Gravity} = \text{Drag} \quad (3.10c)$$

$$\rho_{net}(\pi d^3/6)g = 3\pi \times d \times \eta \times V_{falling}/C_{slip} \quad (3.10d)$$

$$\rho_{net}(d^2)g = 18 \times \eta \times V_{falling}/C_{slip} \quad (3.10e)$$

And Equation 3.10e can be solved for $V_{falling}$, a Stokes particle's terminal falling velocity.

$$V_{falling} = C_{slip}(\rho_{net}d^2)g/18\eta \quad (3.11a)$$

On the rare occasion that one actually needs to calculate such a result, it is convenient to notice that g (gravity), 18, and η in Equation 3.11a are all either actually or at least practically constants.[h] Equation 3.11b results from inserting values at NTP for each of these constants into Equation 3.11a.

$$V_{falling}\ (\text{cm/sec}) = 0.003025\ \rho_{particle}\ \text{d}^2 \times \left(1 + \frac{0.195}{\text{d}}\right) \text{ where } d \text{ is in } \mu\text{m} \quad (3.11b)$$

The slip correction factor (C_{slip} from Equation 3.9c is the term in parentheses in Equation 3.11b) can be assumed to be unity for all particles over approximately 1 μm,

[h] The value of η depends upon the fluid and its temperature. The viscosity of air increases approximately 10% per 100°F (the reverse of how the viscosity in liquids behave), but it too can be assumed to be constant in workplaces of IH interest.

where the correction drops to below 1.2. The particle density (the $\rho_{particle}$ term in Equation 3.11b) depends upon the nature of the aerosol. The density of a mist of water is 1 g/cm^3. Its density is still near one even if several percent of a suspended or dissolved powder are in the water. The density of solids found in handbooks range roughly from 0.3–0.8 for wood to 1.5–3 g/cc for minerals and 2 to 8 g/cm^3 for most metal dusts. However, the density of agglomerated metal fume particles is much less than the density of the metals from which they form because of internal voids. Calculating a particle's falling velocity is even easier if one already knows its aerodynamic diameter (d_{aero} or just d_a in many texts) because its density is one.

$$V_{falling}(\text{cm/sec}) = 0.003025 \times d_{aero}^2 \times \left(1 + \frac{0.195}{d_{aero}}\right) \text{ where } d_{aero} \text{ is in } \mu\text{m} \quad (3.11c)$$

Values for $V_{falling}$ are also shown graphically near the bottom of Figure 3.5. For those unfamiliar with aerosols, it could also be very useful to peruse the array of typical particle diameters in the upper half of this diagram. The falling velocities for particles less than 100 μm in diameter in this figure were calculated using Equation 3.11a with Equation 3.9a used for C_{slip} and a density of 1 g/cm^3. Other equations (to be mentioned in Section IV.4) were used for larger diameter particles. Equation 3.11d will allow one to calculate the settling velocity for a different density.

$$V_{falling} = (V_{falling} \text{ from Figure 3.5}) \times \rho_{net} \quad (3.11d)$$

EXAMPLE 3.1

Determine $V_{falling}$ for a 5 μm dust particle (where $C_{slip} \approx 1$) whose density is 4 g/cm^3.

The reader has their choice from among three ways to determine a particle's falling velocity. One could either use Equation 3.11a, use Equation 3.11b, or use Figure 3.5 and Equation 3.11d. Using Equation 3.11a is probably the longest way:

$$V_{falling} = \frac{4\text{ g}}{\text{cm}^3} \times \frac{(0.0005^2)\text{ cm}^2}{18} \times \frac{\text{cm sec}}{1.8 \times 10^{-4}\text{ g}} \times \frac{980\text{ cm}}{\text{sec}^2} = 0.30\text{ cm/sec (using Equation 3.11a)}$$

Equation 3.11b is in a dotted-box because it is simpler to use and adequately accurate for IH purposes.

$$V_{falling} = 0.003025 \times 4 \times 5^2 = 0.30\text{ cm/sec} \quad \text{(using Equation 3.11b)}$$

Using Figure 3.5 is only an easy way to determine $V_{falling}$ if the particle's density happens to be near 1 g/cm^3. When the particle's density differs from 1 g/cm^3 (such as >3 g/cm^3), one might be better off with an equation. Notice that the settling velocity in the diagram for a 5 μm particle lines up with the "8" between 10^{-2} and 10^{-1} which is really 8×10^{-2}. Using this value in Equation 3.11d with a particle density of 4 g/cm^3 yields just about the same falling velocity as above.

$$V_{falling} = (V_{falling} \text{ from Figure 3.5}) \times \rho_{net} = 0.08 \times 4 = 0.32\text{ cm/sec} \quad \text{(using Equation 3.11d)}$$

3. The criteria for what constitutes a significant falling velocity depends upon the local air currents and the distance to the ground or other surface. Horizontal air current velocities in typical mechanically ventilated rooms are generally at least 20 to 40 fpm, often 100 to 200 fpm (1 to 2 m/sec), but rarely as high as 1000 fpm. Room air currents may also be vertical in a given location (especially near a warm

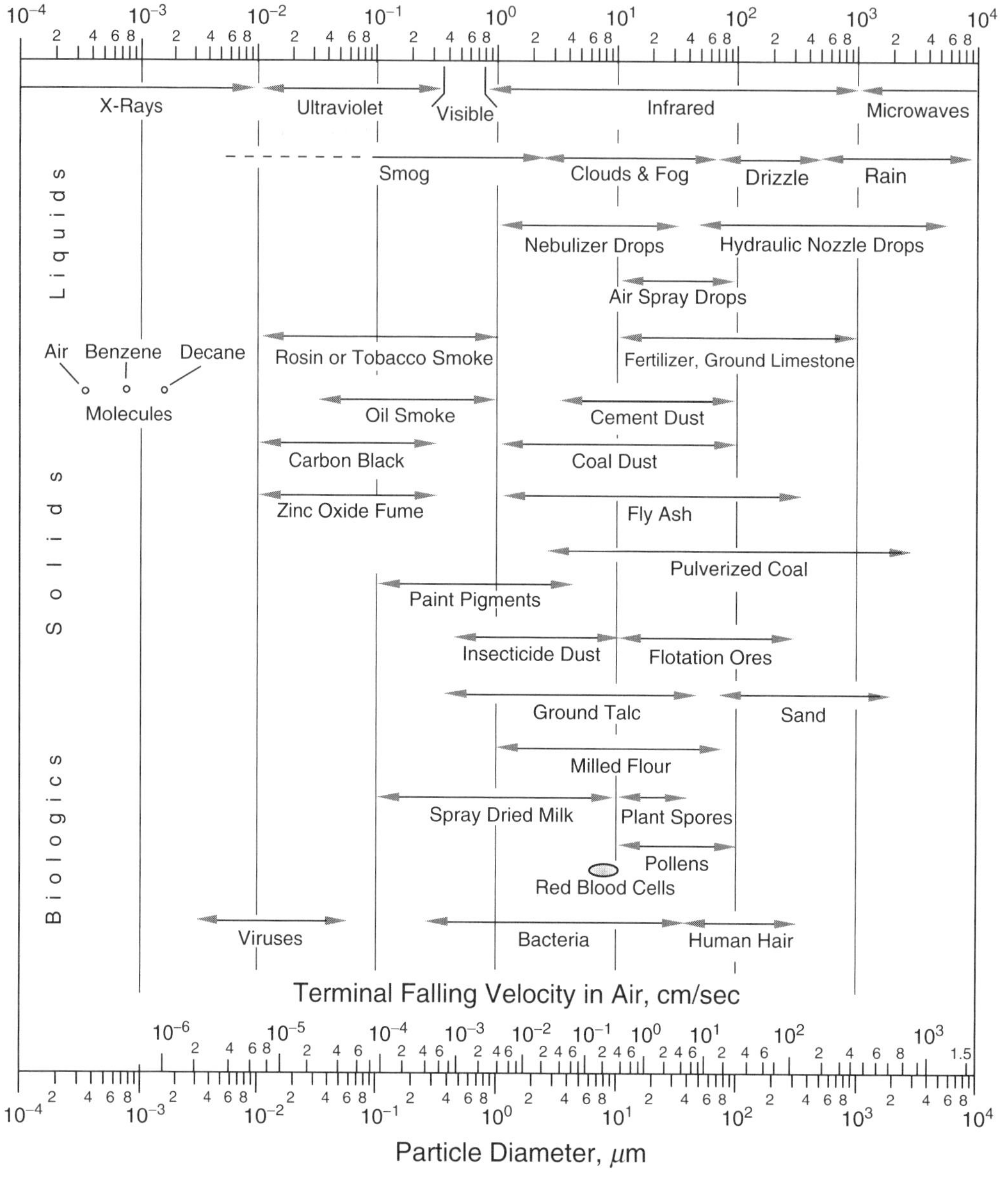

FIGURE 3.5 Characteristics of aerosol particles. V_{fall} values are for $\rho = 1$. (*Source*: Adapted from Lapple, C. E., *Stanford Res. Inst. J.*, 5, 322–325, 1961; Engineering Staff of George D. Claytone and Associates, *The Industrial Environment — Its Evaluation and Control*, NIOSH Publications, Cincinnati, Ohio, chap. 43, 1973; Goodfellow, H. and Tähti, E., Eds., *Industrial Ventilation Design Guidebook*, Academic Press, London, 2001.)

or hot source) and will usually have momentary vertical components due turbulence. Particles will fall any time their falling velocity exceeds the local upward air velocity. Figure 3.6 was generated using the above particle falling velocities, an assumed falling distance of 1 m, and various horizontal air velocities. One can use Figure 3.6 (after adjusting for particle density) to see how far an airborne particle (or plume) will travel horizontally in various wind conditions while falling a vertical distance of one meter. Particles larger than approximately 100 μm will usually fall

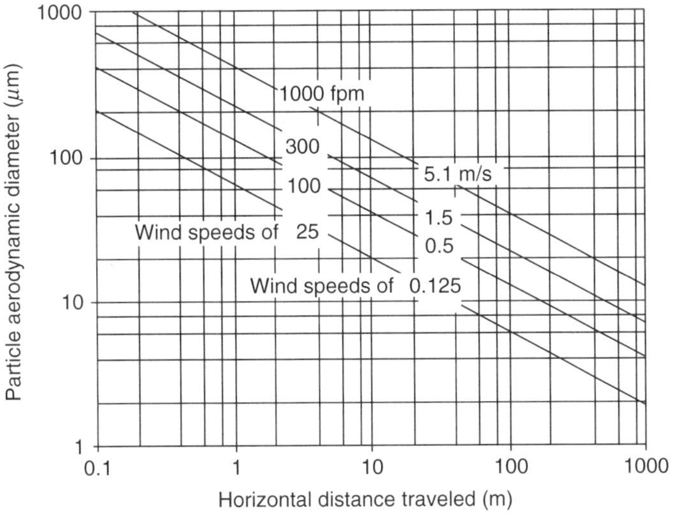

FIGURE 3.6 The horizontal distance a particle can travel while falling one meter.

out of the air in close proximity to their source (within about 1 to 4 m in typical room air currents). On the other hand, particles smaller than approximately 50 μm have a good chance of staying airborne for more than 10 m. One can often see a gradient of particle deposition on the floor near a source as a function of particle size. Further down the scale, sedimentation has almost no effect on particles 10 μm or less in diameter. Once these small particles leave the vicinity of their source, their dispersion will be similar to that of gases and vapors.[12] However, Section VI will show that even small aerosol particles will behave differently from gases and vapors in the small dimensions of the lung.

4. Although this chapter focuses on Stokes law, it may be useful some time in the future to know that not all particles obey Stokes law. Stokes law only predicts aerosol behavior accurately for particle diameters between approximately 0.5 μm and 100 μm (a range that, fortunately, covers most aerosols that are a respiratory hazard except for fumes). Predicting the faster falling velocity of smaller particles can be accomplished fairly easily by adding a slip correction factor to Stokes law (such as shown in Equation 3.11). However, predicting the slower falling velocity of larger particles (such as rain, rocks, or sky divers) requires one or more different models for drag.[i]

   The falling velocity of particles is plotted in Figure 3.7 over a wide range of diameters. The straight dashed line indicates particle falling velocities based only on Stokes law. The solid lines use the slip factor for small particles and other formulas for larger particles. The bottom solid line duplicates the V_{fall} tabulated in Figure 3.5 for $\rho = 1$ g/cm^3. The solid lines above it correspond to the higher

[i] Stokes flow applies to spheres with a Reynolds number < 1. Newtonian flow applies to a Reynolds number $> 10^3$, and Intermediate flow applies to Reynolds numbers between the above two conditions. Thus, one of the following equations for the upward force on any particle due to drag can be equated to the downward force from gravity in Equation 3.7[13]:

For Stokes flow: $F_{drag} = 3\pi\eta dV/C_{slip}$ applies to falling particles with aerodynamic diameters < 100 μm.

For Intermediate flow: $F_{drag} = 2.31\ \rho_{fluid}^{0.4}\ \eta^{0.6}\ (dV)^{1.4}$

For Newtonian flow: $F_{drag} = 0.55\ \pi\ \rho_{fluid}\ d^2V^2$ applies to falling particles with aerodynamic diameters > 1 mm.

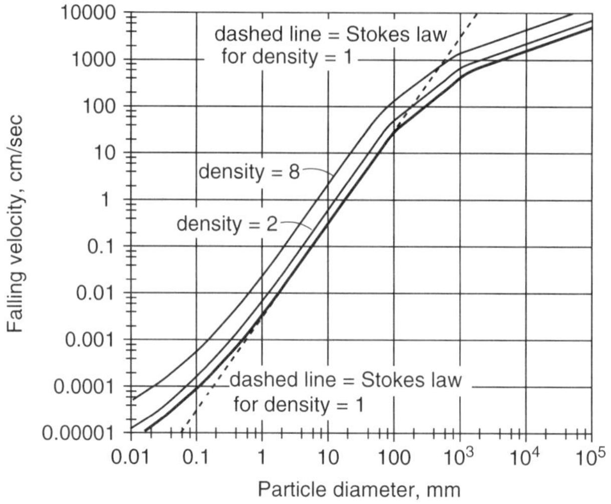

FIGURE 3.7 Particle falling velocities calculated using multiple models.

falling velocities for particle densities of 2 and 8 g/cm^3, respectively. The effect of C_{slip} can be seen as the deviation from Stokes law in the lower left corner of this figure. Deviations from Stokes law on the upper right corner of Figure 3.7 occur as the air passing the spherical particle transitions from laminar to turbulent flow. Stokes Law will overestimate the falling velocity of particles greater than 100 μm because they have more drag in turbulent flow than in laminar flow conditions (which is fortunate for sky divers but not for airplanes, their fuel consumption, and our airline ticket prices).

V. STOKES AND AERODYNAMIC DIAMETERS

Stokes diameter and aerodynamic diameter are two common ways to measure or define an aerosol. In short, a particle's Stokes diameter is physically measured (e.g., by a visual microscope), while a particle's aerodynamic diameter is based on its behavior. In principle, one can choose whether to measure an aerosol's Stokes diameter and/or its aerodynamic diameter, but hygienists are often given one but want the other. Thus, we sometimes need to be able to convert from one to the other.

1. A Stokes diameter (d_{Stokes}) is the physical diameter of a spherical particle that obeys Stokes law.

When viewed and sized with a microscope, most nonfibrous particles turn out to be roughly spherical. As long as a particle's diameter is within the 0.01 to 100 μm range applicable to Stokes law with a slip correction (Equation 3.11), a nonfibrous particle's Stokes diameter approximates its real diameter (abbreviated as just "*d*"). However, because Stokes law incorporates a particle's density, one can neither estimate a particle's behavior without knowing its density nor estimate a particle's Stokes diameter by measuring its behavior without knowing its density. For instance, to estimate a particle's falling behavior using Equation 3.12, one needs a value for particle density ρ:

$$\text{Stokes diameter} = d_{\text{Stokes}} = \sqrt{(V_{\text{falling}} 18\eta / C_{\text{slip}} \rho g)} \tag{3.12}$$

where ρ is the density of the material comprising the aerosol.

2. The aerodynamic diameter (d_{aero}) is the size of a spherical particle of unit density (1 g/cc).

Spherical aerosol particles of unit density can be generated in a laboratory for research and testing purposes. Any water mist is a particle of unit density (however, because water will rapidly evaporate, a water mist is not temporally stable unless the air is at 100% relative humidity). Because all unit density spheres in the 0.01 to 100 μm diameter size range will obey Stokes law, any aerosol's aerodynamic diameter can be determined by measuring one of its behaviors that can be predicted based on Stokes law just by assuming it has unit density. For instance, an aerodynamic diameter can be determined by measuring a particle's V_{falling} using Equation 3.13 which assumes $\rho = 1$ g/cm^3.

$$\text{Aerodynamic diameter} = d_{\text{aero}} = \sqrt{(V_{\text{falling}} 18\eta / C_{\text{slip}} g)} \tag{3.13}$$

An aerosol's aerodynamic diameter is more important to an IH than is its Stokes diameter because we are really interested in an aerosol's behavior. The behavior of an aerosol can be measured without knowing its density if one knows its aerodynamic diameter. Similarly, a particle's behavior can be predicted without knowing its density if its aerodynamic diameter is known.

3. One can sometimes choose to measure either a particle's Stokes diameter or its aerodynamic diameter. At other times, you are given one but want the other. To convert or relate a particle's Stokes diameter to its aerodynamic diameter, one has to know the particle's density (ρ) and its aerodynamic behavior. While various aerodynamic behaviors can be used, Equation 3.14 equates a particle's falling velocity as defined by its Stokes diameter in Equation 3.12 to its falling velocity as defined by its aerodynamic diameter in Equation 3.13 (recall that d_{Stokes} approximates a nonfibrous particle's "d").

$$V_{\text{falling}} = 1\ \text{g/cm}^3 (d_{\text{aero}}^2)\ g/18\eta = \rho\,(d_{\text{Stokes}}^2)\ g/18\eta \approx \rho\,(d^2)\ g/18\eta \tag{3.14}$$

Since the constants "g/18η" affect both definitions equally, they can be canceled out of Equation 3.14, leaving the simple and memorable form of these aerosol equations.

$$d_{\text{aero}}^2 = \rho\, d_{\text{Stokes}}^2 \quad \text{or} \quad \boxed{d_{\text{aerodynamic}} = d\sqrt{\rho}} \quad \text{for } \rho \text{ in g/cc} \tag{3.15}$$

(Note that the apparent units resulting from $d_{\text{aero}} = d\sqrt{\rho}$ in Equation 3.15 cannot be taken literally because the density of $\rho = 1$ g/cm^3 pertaining to an aerodynamic diameter is implied but not shown for simplicity. Just keep in mind that the values of d will be correct if the units of ρ are either g/cm^3, g/mL, or a dimensionless specific gravity, but will be incorrect if ρ is in other units such as g/L where water is 1000 g/L or pounds/cubic foot where water is 62.4 lb/ft^3.)

Thus, the difference between an aerodynamic diameter and a real or Stokes diameter always depends upon the particle's density (more specifically upon the square root of its density). Particles made of materials denser than water ($\rho > 1$ g/cm^3) will behave as if they are bigger than they look (bigger than their real diameters). And particles that are less dense than water ($\rho < 1$ g/cm^3) will behave as if they are smaller than they look (as if they are smaller than their real diameters). The collection efficiency of commercial aerosol samplers is almost always reported in terms of aerodynamic diameter, as is the ability of particles to penetrate through filters, air cleaners, or into the respiratory tract (to be discussed in Section VI). One can measure an aerosol's real diameter or its aerodynamic diameter, but predicting one from the other will always require knowing the particle's density. In fact, Equation 3.15 might correctly suggest that measuring both a particle's real diameter $d \approx d_{\mathrm{Stokes}}$ and its aerodynamic diameter d_{aero} is one way to find its density as $\rho = (d_{\mathrm{aero}}/\mathrm{d})^2$.

4. The relationships predicted by Stokes law, Equation 3.11, and Equation 3.15 have broad applications. Any process that creates an aerodynamic force on particles whose diameters are between approximately 0.01 and 100 μm, will follow Stokes law. The common processes that create an aerodynamic force include elutriation and sedimentation (due to gravitational falling), impaction (due to the inertial forces created in a cascade or virtual impactor or a scrubber), and centrifugation (due to the centrifugal force created in a cyclone). Elutriation and sedimentation due to a particle's falling velocity (as previously discussed) are applicable to size selective aerosol sampling, particle deposition within the deep lung, and air flow in long, horizontal ducts. Impaction due to a particle's stopping distance and centrifugation are applicable to size selective sampling, to particle deposition within the upper respiratory tract, and to air flow in elbows, fans and some air cleaning devices. Respiratory hazards will be discussed in the last section of this chapter, duct flow in Chapter 14, and air cleaning devices in Chapter 15.

   a. A particle's stopping distance can predict an aerosol particle's behavior as it leaves a source at a high velocity or as it approaches an obstacle such as an impactor plate (used in air samplers and air cleaners), a branch in the respiratory tract, or a sharp edge of a local exhaust ventilation hood. Stopping distance is applicable any time the drag force on a particle exceeds either its weight in the vertical axis or zero in the horizontal axis. Such force imbalances occur when a particle is thrown into the environment (say from a grinding wheel) or when the air transporting an aerosol turns quickly (such as in an impactor or other abrupt air deflection).

      The stopping distance given by Equation 3.16 was found by integrating the drag force due to air friction over the time it takes for a particle to loose its initial speed (V_{initial}). V_{initial} might be the horizontal velocity at which a particle leaves a source. Or V_{initial} might be a particle's velocity in an impinger just before the air makes an abrupt turn.

$$\text{Stopping distance} = \rho d^2/18\eta \times V_{\mathrm{initial}} \qquad (\text{for } 1 \leq d_{\mathrm{aero}} \leq 100\ \mu\mathrm{m}) \qquad (3.16)$$

      Figure 3.8 shows the stopping distance in air for particle sizes up to 1 mm and for initial velocities ranging from 50 to 1000 fpm (0.25 to 5 m/sec). The stopping distance of a particle 10 μm or less at any of the initial velocities shown will be less than approximately 0.01 cm (0.1 mm). Such short stopping distances allow small particles to go pretty much wherever the bulk flow of the air takes them. However, even these small stopping distances can become important within the confines of an air cleaner, an air sampler, or the respiratory tract. On the other hand, stopping distances approach or exceed 1 cm for particles greater than 100 μm in diameter and/or initially traveling faster than 1000 fpm. Such large stopping distances cause particles to impact upon and accumulate around the edges of any aspirating orifice such as the inlet of an air sampler (not covered in this text), slots cut into the face of a slotted local exhaust hood (to be covered in Section IV of Chapter 13), or even someone's nose or mouth (to be covered in Section VI). Particles in even higher size and speed ranges present more of an eye safety hazard than they do a respiratory health hazard.

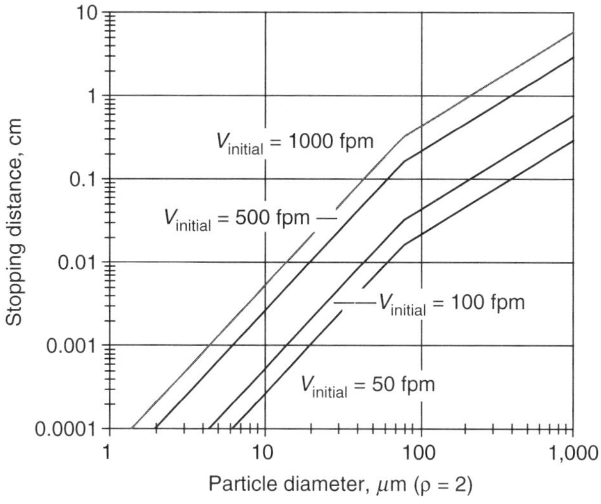

FIGURE 3.8 Stopping distances for particles at various initial velocities (V_{initial} in Equation 3.16).

b. The cascade impactor is a common air sampler used to separate an aerosol into several narrow ranges of aerodynamic diameters as they impact on one of a series of plates or "stages." At each stage, either the air velocity and the particle's stopping distance will increase and/or the distance between plates will decrease, causing each stage to collect successively smaller aerodynamic diameter particles than the previous stage. The collection efficiency of each stage is often characterized by the aerodynamic diameter at which 50% of the particles will be collected (referred to as the 50% cut-point). From the distribution of masses collected on such a series of stages and their previously calibrated cut-point diameters, one can determine an aerosol's mass median aerodynamic diameter (MMAD). Just as a particle's aerodynamic diameter is related to its Stokes or real diameter by the square root of its density in Equation 3.15, an aerosol's MMAD is related to its MMD described in Section III.3) by the square root of the particles' density, as shown in Equation 3.17.

$$\text{MMAD} = \text{MMD}\sqrt{\rho} \tag{3.17}$$

Obviously, converting between an aerosol's MMD and MMAD requires that the particle's density is known. An aerosol's MMD is also related by the Hatch–Choate equation (Equation 3.6) to its CMD or d_g that could be determined by counting and sizing an array of individual particles.

c. Centrifugal acceleration equals the tangential velocity squared over its radius of curvature. If a cyclone is built with a radius of b, then the term V_{inlet}^2/b can be substituted for the gravitational acceleration in Equation 3.11a to again use Stokes law, this time to predict a particle's terminal centrifugal velocity as shown in Equation 3.18.

$$V_{\text{centrifugal}} = C_{\text{slip}}(\rho_{\text{net}}d^2)V_{\text{inlet}}^2/18\eta b \tag{3.18}$$

One can solve Equation 3.18 for the diameter of a particle that will fall fast enough to reach an outer wall (and thus be removed from the air stream) in the time that the air makes some number of revolutions around the cyclone. The result (to be given as Equation 15.2) can be used to predict the performance of a cyclone for either air sampling or air emission control.

5. On the other hand, the $\sqrt{\rho}$ equivalence between aerodynamic behavior ($d_{\text{aerodynamic}}$ or MMAD) and its Stokes diameter or MMD does not apply to every situation. For instance, sample collectors or air cleaners that rely on diffusion, ESP, or thermal precipitation follow different laws. It is in fact this exception that makes such samplers particularly well suited to collect very small particles. Mechanical entrapment is another exception that does not rely on particle density. Mechanical entrapment generally only affects large particles or the dust that cakes up on air cleaning filters. Most air sampling filters collect particles by impaction, and most respirator air purifying filters collect particles by a combination of impaction and ESP rather than by entrapment.

VI. AEROSOL DIAMETERS AND HUMAN HEALTH

An aerosol's hazard is strongly dependent upon where in the respiratory system it will be deposited (along with its inherent toxicity at the point of deposition). Particle deposition within the respiratory system depends on its aerodynamic behaviors of stopping distance and falling velocity. Particle deposition has been described as if it occurs in three or four conceptual anatomical compartments; four are shown in Table 3.5. Of course in reality, deposition occurs as a continuum, and such compartmentalization is only useful to simplify the description of a complex, dynamic process.

1. The most important mechanism affecting particle penetration into and deposition within the upper respiratory system varies with depth into the lung due to changing airway diameters and air velocities.

   The high inertia of particles with an $d_{\text{aerodynamic}}$ greater than 30 μm prevents them from keeping up with the air at it responds to comparatively rapid curvilinear accelerations while flowing into a orifice. For instance, at least 50% of particles larger than 50 μm are separated from the air that enters the nostrils (less so for air entering an open mouth). Thus, inhaled air has fewer particles (a lower aerosol concentration) than the ambient air before it enters into those orifices. This inertial separation mechanism also applies to aspirated air samplers (samplers with a sampling pump) and to hoods with small openings (especially slotted hoods).

   A combination of the moderate inertia of particles with $d_{\text{aerodynamic}}$ between 10 and 30 μm and the high speed of air flowing through the curving tortuous path within the nose causes virtually all of these particles to be separated and deposited in the head's airways or nasopharyngeal (NP) region due to a combination of impaction and centrifugation. Comparatively less deposition will occur in the pharynx if one inhales through their nose than will occur if one inhales through their mouth; however, by the time the air passes through the next region the net effect is largely the same.

   The tracheo-bronchial region does not have the high speeds and accelerating turns of the NP region, the branches are not abrupt enough to create high rates of impaction, nor are the passageways small enough to allow much elutriation. However, the mechanisms of impaction and elutriation are sufficient to cause some deposition within the mid-size aerosol particles ca. $5 < d_{\text{aerodynamic}} < 15$ μm. Particle deposition within the passage-ways of this thoracic region tends to cause small airway or asthmatic types of symptoms.
2. The modest dimensions and air speed within the brochiolar region are not sufficient to cause a great deal of particle deposition within this region. However, some deposition of particles around 10 μm will occur, comprising what are called "thoracic aerosols."
3. Thus, only particles with diameters below about 10 μm can penetrate into the pulmonary (P) region of the lung. The enormous number of respiratory bronchiols and alveolar ducts within the pulmonary region creates a very low velocity precluding further inertial effects.

TABLE 3.5
Physical Parameters of the Respiratory Tract for the Respiratory Parameters Described in the Legend

	Number of Passage Ways	Range of Diameters (mm)	Total Average X-area (cm^2)	Average Air Velocity (mm/sec)	Other Anatomical Features
Head airways region (a.k.a. naso-pharyngeal [NP] region, extra thoracic region, or anterior nasal passages)					
Nostrils (nares)	2	8	1	12,000	Hairs (coarse filter for large particles)
Nasal cavities	2	4–6	0.4	30,000	Turbinates; cilia and goblet cells (olfaction)
Pharynx	1	~30	7	2000	Cilia and goblet cells in upper pharynx
Larynx or epiglottis	1	—	—	—	Flow area through vocal cords varies
Tracheo-bronchial region [the TB region (not the abbreviation of tuberculosis)]					
Trachea	1	15–18	2.2	6000	Cartilage; cilia and goblet cells
Upper bronchi and lower bronchi	2–300	12–2	2.5–8	5000–1500	Eight divisions; cilia; some smooth muscle; decreasing cartilage; fewer goblet cells
Bronchiolar region (sometimes combined with the tracheo-bronchial region)					
Upper bronchioles and lower bronchioles	500 to 1.5×10^4	1.7–0.6	20–50	1000–300	Another seven divisions; mostly smooth muscle; ciliated cells only; sedimentation starts
Terminal bronchioles	3.2×10^4	0.5	60	200	One division; ciliated and secretory Clara cells
Gas exchange region (a.k.a. the alveolar-interstitial or pulmonary [P] region)					
Respiratory bronchioles	8×10^5	0.5	1550	8	2–3 divisions, some containing alveoli
Alveolar ducts	4×10^6	0.4	5000	2	Squamous cells with multiple branches
Alveoli	4×10^8	0.4	5×10^5	<0.1	Highly vascular squamous cells

Air velocities were based on a moderate work rate with a 24 l/min RMV and a 1.2 l tidal volume at 20 breaths/min, yielding a 1.2 l/sec peak flow rate and 12 m^3 inhaled over 8 h from Tables 6 and 8 in Refs. [14] and [15], which is slightly higher than the 22 l/min from the earlier classic Report of the Task Group on Lung Dynamics (Ref. [16]).

TABLE 3.6
The Deposition Rates of Aerosols within the Human Pulmonary Region

$d_{aerodynamic}$ (μm)	$V_{falling}$ (mm/sec)	Falling Distance in 0.75 sec (μm)	% Reaching P Region × % Deposited	% Deposition within P Region
10.0	3.1	2310	1.3 × 100	1
7.0	1.5	1140	9 × 100	9
5.0	0.78	588	30 × 100	30
3.5	0.39	292	50 × 83	42
2.0	0.13	99	90 × 28	25
1.0	0.036	27	100 × 8	8
0.5	0.010	8	100 × 2	2

The combination of low turbulence and small vertical dimensions of alveoli allow sedimentation to become the dominant deposition mechanism. A particle's falling distance is the product of falling velocity times time. Therefore, sedimentation efficiency is a function of $V_{falling}$ (set by the particles' aerodynamic diameter), the average residence time of an inhaled breath (set by the breathing or work rate), and the diameter of the alveoli (set by human anatomy). The falling velocities predicted by Stokes law are listed in Tables 3.3 and 3.6. A residence time of 0.75 sec is assumed for breathing at a moderate work rate (see the legend to Table 3.5). Terminal bronchioles are ca. 700 μm in diameter, while alveoli are 300 to 400 μm in diameter.

As the particle size decreases, penetration through the upper regions of the respiratory tract increases but deposition within the alveoli due to sedimentation decreases. The decreased inertial force on smaller particles allows them to reach the alveoli more easily, but the decreased gravitational force on the smaller particles slows the deposition of those that do penetrate. The combined effects yield the nominal pulmonary deposition efficiencies shown in Table 3.6. For instance, only 30% of the 5 μm particles are expected to reach the P region (70% will be removed from the air in the upper respiratory regions). But since their falling distance in three quarters of a second is greater than the 350 μm diameter of the average alveoli, 100% of the 30% of the 5 μm particles that do reach the alveoli will be deposited there by sedimentation. In this analysis, the combined effect of increasing penetration and decreasing pulmonary sedimentation reaches a maximum of approximately 42% deposition at an $d_{aerodynamic} = 3.5$ μm.

Particles between 1 and 0.2 μm in diameter penetrate into the alveoli very well but are too small to be deposited efficiently by sedimentation and too big to be affected by diffusion. Diffusion does not start to affect particles until $d_{aerodynamic} < 0.2$ μm (not shown in Table 3.6), which will increase their likelihood of coming into contact with and sticking to wet lung tissue within the upper airways. Therefore, deposition within the entire respiratory system is at a minimum for particles near 0.2 μm.

4. The following working definitions of aerosols is written from an industrial hygiene perspective, is based on Appendix D of the ACGIH TLVs®, and is compatible with the above aerodynamic behavior of aerosols and the structure of the human respiratory tract.[17]
   a. *Total aerosol* is a generic rather than a technical term meaning all of the particles that are present in the ambient air whether or not they can enter the respiratory orifices of the body or of a sampling device. Any aerosol sample collected without a preselector is commonly referred to in the literature as a "total aerosol sample." However, the inertia of large particles is so much greater than that of air that they cannot follow the same path as the air is being pulled into an orifice of either a sampler, the body, or a

local exhaust hood; they pass by the orifice or impact on its edges and are therefore not collected. These entry losses make actually collecting all particles (the total aerosol) difficult if not impossible.[18–20]

b. *Inhalable aerosol* is the technical definition of that portion of the total aerosol that can enter the human respiratory orifices. The ACGIH definition of inhalable ("SI[d]" in Appendix D of their TLV booklet) states that a constant 50% of all $d_{aerodynamic} > 100\ \mu m$ can be inhaled and the fraction increases exponentially for particles $<100\ \mu m$ according to Equation 3.19. Equation 3.19a is the TLVs book version; Equation 3.19b uses a ratio of diameters. For example, 77% of particles 10 μm and 90% of particles 8.5 μm in aerodynamic diameter are inhalable.

$$\mathrm{SI}(d) = 50(1 + \exp[-0.06\, d_{aero}]) \pm 10 \tag{3.19a}$$

$$\mathrm{SI}(d) = 50(1 + \exp[-d_{aero}/16.67]) \pm 10 \tag{3.19b}$$

Inhalable aerosol sampling is broadly applicable to aerosols that are hazardous when deposited anywhere in the respiratory tract. It is particularly pertinent for agents that affect the naso-pharyngeal region, especially the nose. Examples of such agents include the following:

1. Chemical irritants (e.g., acid mists and several metals) that affect the highest region of the respiratory system.
2. Nuisance dusts of all sorts (now called "Particulates Not Otherwise Specified" in the TLV booklet or abbreviated as "PNOS") which can simply overwhelm the respiratory system.
3. Dermally absorbable chemicals (e.g., organophosphate pesticides can be desorbed from dust by perspiration then absorbed through the skin; the same process can happen with mucus membranes within the NP passages).

c. *Thoracic aerosol* is that portion of particles small enough to be inhaled and pass through the naso-pharynx, but just large enough to be significantly deposited by elutriation in the small conductive passages of the T–B region. The current ACGIH definition of percent thoracic ("ST[*d*]" in Appendix D of their booklet) is formulated in terms of a fraction of the inhalable fraction.[j]

$$\mathrm{ST}(d) = \mathrm{SI}(d) \times [1 - \mathrm{F}(x)] \tag{3.20}$$

where F(x) is the standardized normal cumulative probability function of

$$x = \frac{\ln(d_{aero}/11.64\ \mu m)}{\ln(1.5)}$$

The implications of their stated value of $\Gamma = 11.64\ \mu m$ in Equation 3.20 must be interpreted in conjunction with the inhalable fraction (SI[*d*] from Equation 3.19). The result of these two functions is that 50% of 10 μm particles are considered thoracic (the diameter at which the collection efficiency is 50%).

Thoracic aerosol sampling is particularly applicable for agents that are hazardous when deposited anywhere within the lung airways and the gas-exchange region. The best example of such an agent is cotton dust but it should include any other asthma inducing agent like endotoxin.

[j] Values of the cumulative probability function can be obtained within Windows® spreadsheets by such functions as @NORMSDIST in QuatroPro®.

d. *Respirable aerosol* is the portion with even smaller diameters that can reach the deepest pulmonary region of the lung. The ACGIH definition of respirable ("SR[*d*]") is also described by a cumulative log-normal function with the same geometric deviation of 1.5 as in the thoracic aerosol above but with a cut point of 4.0 μm. (This part of the definition was changed in 1992 from one with a 50% cut point of 3.5 μm that is still used by OSHA 1910.1000 Table Z-3.)

$$SR(d) = SI(d) \times [1 - F(x)] \tag{3.21}$$

where F(x) is the standardized normal cumulative probability function of

$$x = \frac{\ln(d_{aero}/4.25\ \mu m)}{\ln(1.5)}$$

Note again that a 50% cut point of 4.0 μm results when their $\Gamma = 4.25\ \mu$m is multiplied by the inhalable fraction ("SI[*d*]"). The performance of a 10-mm Door-Oliver cyclone preselector operated at 1.7 lpm still closely matches this slightly changed criterion.

Respirable aerosol sampling is applicable for agents that are hazardous when deposited in the gas-exchange region of the lung (the alveoli) including the following:

1. Fibrogenic dusts such as quartz or beryllium.
2. Radioactive aerosols where the pulmonary tissues are the target organ.
3. Particles with adsorbed irritant gases that would normally irritate the upper respiratory tract (e.g., an aerosol that contains or has absorbed SO_2, formaldehyde, or ammonia).

Table 3.7 and Figure 3.9 summarize the above size selective sampling performance criteria. Each criterion describes the fraction of the total aerosol (expressed as an aerodynamic diameter [d_{aero}]) that would be included in its definition. The 50% cut points for thoracic and respirable aerosols are shown in bold within Table 3.7. The inhalable

TABLE 3.7
The Three Fractions (%) of a Total Aerosol that Are Defined by the ACGIH TLV Committee

Aerodynamic Diameter (μm)	Inhalable SI (*d*)	Thoracic ST (*d*)	Respirable SR (*d*)
100	50.0	3.0×10^{-6}	2.0×10^{-13}
50	52.5	0.009	3.0×10^{-8}
30	58.0	0.6	0.00004
20	65.0	6.0	0.004
15	70.0	19.0	0.07
10	77.0	50.0	1.3
7	83.0	74.0	9.0
6	85.0	80.0	16.0
5	87.0	85.0	30.0
4	89.0	89.0	50.0
3	92.0	92.0	74.0
2	94.0	94.0	91.0
1	97.0	97.0	97.0
0.5	98.5	97.7	97.7
0.1	99.7	99.7	99.7

These data are also plotted in Figure 3.9.

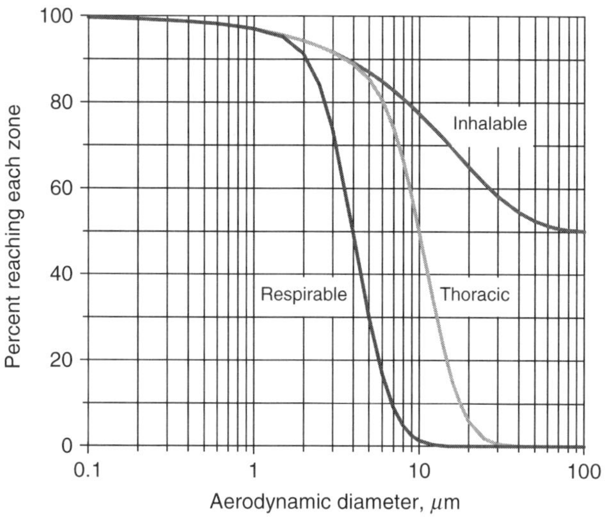

FIGURE 3.9 Three functional fractions of a total aerosol as defined by the ACGIH TLV Committee plotted in terms of the fraction that reaches each portion of the respiratory tract. These curves also correspond to the data in Table 3.7.

aerosol does not really have a cut point because 50% of all particles larger than 100 μm (not shown in either this table or figure) are inhalable aerosols.

PRACTICE PROBLEMS

1. In each of the following cases (except for the first two), choose a mid-range particle diameter from Figure 3.5. The first and last cases are solved as examples.
   a. What is the typical size of particles you might find in the following settings?
   b. What is the terminal gravitational settling velocity of the above particle if it had the density shown?
   c. How long would the particle take to fall out of the air if it were generated or released 1.5 m (5 ft) above the ground (near a worker's breathing zone)?

Setting	a. Particle Diameter (μm)	Particle Density (ρ)	b. Settling Velocity (cm/sec)	c. Falling Time (sec)
a. The smallest ZnO welding fumes (fresh)	0.01	4	1.5×10^{-5}	1.0×10^{7}
b. The largest ZnO welding fumes		4		
c. Wettable powder insecticide (dust)		2		
d. Ground talc		2		
e. Milled flour dust		1		
f. Pollen		1		
g. Pieces of human hair		1		
h. Drizzle		1		
i. Rain	2500	1	10^{3}	0.15

FYI: 10^3 cm/sec. 1800 fpm

2. What is the aerodynamic diameter, settling velocity, and the percent respirable fraction of a 2 μm insecticide spray aerosol that is approximately 99% water? Either of the four versions of Equation 3.11 could solve this type of problem. Try using a couple of different versions (see also Q. 3) to see which you like the best.
   a. These particles' density = _____ g/cm^3 or g/mL
   b. Aerodynamic diameter = _____ μm
   c. Settling velocity = _____ cm/sec ans. = 0.013 cm/sec
   d. Percent respirable = _____ % from Table 3.7 or Figure 3.9.

3. What is the aerodynamic diameter, settling velocity, and percent respirable fraction of a 2 μm brass (70% Cu, 30% Zn) dust particle of density 8.5 g/cm^3? The above solution(s) also apply to this problem.
   a. Aerodynamic diameter = _____ μm
   b. Settling velocity = _____ cm/sec ans. = 0.11 cm/sec
   c. Percent respirable = _____ % from Table 3.7 or Figure 3.9.

REFERENCES

1. Ferron, G. A., and Soderholm, S. C., Estimation of the times for evaporation of pure water droplets and for stabilization of salt solution particles, *J. Aerosol Sci.*, 21(3), 415–429, 1990.
2. Fraser, D. A., *Sizing Methodology, The Industrial Environment — Its Evaluation and Control*, NIOSH Publications, Cincinnati, Ohio, chap. 14, 1973.
3. Peterson, C. M., Aerosol sampling and particle size analysis, Air Sampling Instruments, 5th ed., American Conference of Governmental Industrial Hygienists, Cincinnati, OH, Sec., 1978.
4. Sioutas, C., Measurement and presentation of aerosol size distributions, Air Sampling Instruments, 9th ed., ACGIH Worldwide, Cincinnati, OH, chap. 6, 2001.
5. Hatch, T. and Choate, S., Statistical description of the size properties of non-uniform particulate substances, *J. Franklin I.*, 207, 369–387, 1929.
6. Cunningham, E., On the velocity of steady fall of spherical particles through a fluid medium, *Proc. Roy. Soc.*, A-83, 357–365, 1910.
7. Mercer, T. T., *Aerosol Technology in Hazard Evaluation*, Academic Press, New York, 1973.
8. Hines, W. C., *Aerosol Technology: Properties, Behavior and Measurement of Airborne Particles*, Wiley, New York, 1999.
9. Lapple, C. E., The size of common aerosols, *Stanford Res. Inst. J.*, 5, 322–325, 1961.
10. Engineering Staff of George D. Claytone and Associates, Control of industrial stack emissions, *The Industrial Environment — Its Evaluation and Control*, NIOSH Publications, Cincinnati, Ohio, chap. 43, 1973.
11. Goodfellow, H. and Tähti, E., Eds., *Industrial Ventilation Design Guidebook*, Academic Press, London, 2001.
12. Bémer, D., Callé, S., Godinot, S., Régnier, R., and Dessagne, J. M., Measurement of the emission rate of an aerosol source — comparison of aerosol and gas transport coefficients, *App. Occup. Environ. Hyg.*, 15(12), 904–910, 2000.
13. Bird, R. B., Stewart, W. E., and Lightfoot, E. N., *Transport Phenomena*, Wiley, New York, 1960.
14. ICRP Publication 66, Human respiratory tract model for radiological protection, *Health Phys.*, 24(1–3), 1–201, 1994.
15. Lippman, M., Size selective health hazard sampling, Air Sampling Instruments, 9th ed., ACGIH, Cincinnati, OH, chap. 5, 1999.
16. Task Group on Lung Dynamics, Deposition and retention models for internal dosimetry of the human respiratory tract, *Health Phys.*, 12, 173–207, 1966.
17. Threshold Limit Values for Chemical Substances and Physical Agents, *ACGIH Worldwide*, Cincinnati, OH, 2005.

18. Agarwal, J. K. and Liu, B. Y. H., Criterion for accurate aerosol sampling in calm air, *Am. Ind. Hyg. Assoc. J.*, 41, 191–197, 1980.
19. Beaulieu, H. J., Fidino, A. V., Arlington, K. B., and Buchan, R. M., A comparison of aerosol sampling techniques: "open-" versus "closed- face" cassettes, *Am. Ind. Hyg. Assoc. J.*, 41, 758–765, 1980.
20. Buchan, R. M., Soderholm, S. C., and Tillery, M. I., Aerosol sampling efficiency of 37 mm filter cassettes, *Am. Ind. Hyg. Assoc. J.*, 47, 825–831, 1986.

4 Chemical Exposure Control Criteria

The learning goals of this chapter:

- ✍ Know how the terms LFL, flash point, and vapor pressure are related to each other.
- ✍ Understand the basis for occupational exposure limits and the Occupational Safety and Healthy Act (OSHA) definition of oxygen deficiency.
- ✍ Be aware of OSHA ventilation regulations.
- ✍ Know how to use the ANSI Z9.1 dichotomous hazard potential tables to classify an open-surface tank hazard. (These results will be used in Chapter 13 to determine its minimum ventilation rate.)

This chapter discusses health and safety standards that affect industrial hygiene (IH) control practices. The criteria upon which these standards are based are categorized here into either health, safety, or good work practices. For the purposes of this chapter, health standards are based on protecting the health of people, i.e., based on human toxicity. Safety standards are based on preventing combustion, fire, and the rare potential for gases and vapors actually to explode. A few standards appear to be based on neither health nor safety criteria; the criteria for oxygen deficiency will be presented as a particularly good example of a standard set simply on good work practices. The criteria for the major OSHA ventilation standards are also categorized herein as health or safety.

I. PERFORMANCE AND SPECIFICATION STANDARDS

Health and safety standards and guidelines can be usefully categorized into two broad forms. Health standards tend to be performance standards. Safety standards tend to be specification standards.

1. Performance standards set goals or levels that are believed to be protective in some way. The following IH exposure limits are classic examples of performance standards and guidelines:
   a. TLVs® (Threshold Limit Values are recommended exposure limits that generally require only voluntary compliance).
   b. PELs (OSHA enforceable exposure limits; “standards”).
   c. OSHA action levels (limits that when exceeded require certain defined protective actions be taken).
   d. Other chemical exposure limits (e.g., corporate recommendations, government agency limits).

   The OSHA general duty clause PL 94-596, Section 5(a) is also performance based with its own explicit criteria:

   > Each employer shall furnish to each of his employees employment and a place of employment which are free from recognized hazards that are causing or are likely to cause death or serious physical harm to his employees.[1]

The purpose of the 1969 OSHA Act (stated in Section 2) is "to assure so far as possible every working man and woman in the Nation safe and healthful working conditions and to preserve our human resources." Section 6(b)(5) of the Act goes on to say that standards dealing with toxic materials or harmful physical agents shall be those

> which most adequately assures, to the extent feasible ... that no employee will suffer material impairment of health or functional capacity even if such employee has regular exposure to the hazard dealt with by such standard for the period of his working life.

The section ends with the statement, "Whenever practicable, the standard promulgated shall be expressed in terms of *objective criteria* and of the *performance* desired." (emphasis added)

Implicit in this simple statement is the sometimes questionable assumption that a threshold of exposure does, in fact, exist below which no adverse effects will occur. Whether or not a threshold exists, it is certainly true that current toxicological methods cannot reliably predict an exposure threshold for a working lifetime. Likewise, our epidemiological methods suffer from a lack of comprehensive data base linking exposure levels to enough occupationally exposed people for a cause-and-effect relationship effect to be detectable. A philosophically more achievable although less politically acceptable goal would be to avoid unacceptable risks of health effects.

TLVs have a split approach. Their general policy statement on the uses of TLVs and BEIs merely states "The Threshold Limit Values (TLVs) and Biological Exposure Indices (BEIs®) are developed as guidelines to assist in the control of health hazards." The ACGIH statement in the introduction to the chemical substances is considerably more ambitious: "Threshold Limit Values (TLVs) refer to airborne concentrations of substances and represent conditions under which it is believed that nearly all workers may be repeatedly exposed day after day without adverse health effects. Because of ... individual susceptibility, however, a small percentage of workers may experience discomfort at or below the threshold limit; a smaller percentage may be affected more seriously ... by development of an occupational illness." One must read the criteria documentation quite closely to even hope to discern what incidence of disease might be expected at the assigned exposure limit.[2]

There is some evidence that the incidence of adverse health effects at the TLV is not consistent. Roach and Rappaport presented a provocative view of the variability in the incidence of adverse effects and the margin of safety (or lack thereof).[3] Figure 4.1 plots the fraction of the studied population with an adverse health response at the exposed concentration on the *X* axis vs. the ratio of their exposure level to the eventually adopted TLV on the *Y* axis. The data are from the 1986 *Documentation of the TLVs* and are given in appendices to the original article. The authors argue that the scatter in the percentages of the studied population that experienced adverse responses at exposure levels near the TLV indicates an historic inconsistency in the incidence of adverse health effects acceptable to the TLV committee. The number of times and the sometimes high percentages of the studied population that experienced adverse responses at exposure levels equal to or below the concentration at which the TLV committee set the TLV suggests that at least some TLVs are not as protective as is often assumed.

2. Specification standards define the way certain things will be done, constructed, or built in measurable terms. Specification standards are more abundant in safety than in health. Two examples of specification standards used by OSHA include standard guard rail heights around holes and elevated platforms shall be 42 in. (1910.23[e]) and working clearances around access to live parts operating at up to 600 V may not be less than 3 ft (§1910.303[g]).

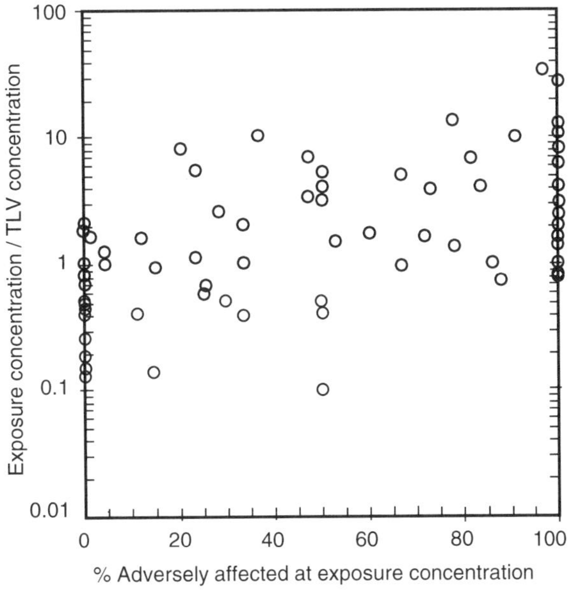

FIGURE 4.1 Exposure expressed as a multiple of the adopted TLV vs. the percent of individuals adversely affected. (*Source:* From Roach, S.A. and Rappaport, S.M., *Am. J. Ind. Med.*, 17, 727–753, 1990, Reprinted with permission of Wiley-Liss Inc., a subsidiary of Wiley.)

3. While health standards tend to be performance standards and safety standards tend to be specification standards, exceptions exist in both categories.

   One exception of a safety criterion that is a performance standard is the lower flammable limit (LFL) or lower explosive limit (LEL). These terms (LFL and LEL) are essentially interchangeable (as will be discussed in Section III of this chapter). OSHA 1910.106(a)(31) requires employers to maintain measured ambient air concentrations below 25% of the LFL or below the PEL, whichever is less (some states limit such concentrations to 20%). LFLs are always higher than PELs or TLVs, usually much higher (e.g., by 10 to 1000 times, as will be discussed in Section V).

   On the other side, most ventilation standards are based on health but written as specification standards. A good example is 1910.252 that specifies local exhaust ventilation at a rate of at least 100 fpm shall be provided when welding in a space of less than 10,000 ft^3 per welder, in a room having a ceiling height of less than 16 ft, or in a confined space or where cross-ventilation is significantly obstructed.

II. HEALTH CRITERIA FOR CHEMICAL CONTROL

1. The following sources of airborne exposure limits are listed in the order of generally increasing concentrations, meaning that where multiple exposure limits exist for the same chemical, allowed levels of exposure will increase as one goes down the list. Only the first (lowest) seven limits are based on protecting the health of people. The last two limits are based on safety and good work practices, respectively, but are included in this list to demonstrate their place in the general progression from lower to higher exposure limits.
   a. A *corporate recommendation or other stated policy limit* must always be at least as stringent as permitted by OSHA where they have set limits; however, chemical manufacturers or users can (and some do) recommend limits for their own products, proprietary chemicals, or intermediates where no other limits have been set. As manufacturers of commercial chemicals bear a significant product liability

responsibility, corporate exposure limits set by manufacturers tend to be conservative. Where a company or other agency chooses to set their own limit for a contaminant present or inadvertently formed but not sold (e.g., dioxin), there is less motivation to be conservative.

b. NIOSH *recommended exposure limits* (RELs) were initially developed (as suggested within PL 94-596, the OSHA Act) as independent recommended exposure limits to be forwarded to OSHA who actually initiates and manages the standard setting process.[5] However, OSHA is under no obligation to act on those recommendations and, in fact, has rarely acted to propose RELs as a standard. RELs tend to be less than TLVs.

c. Two nongovernmental organizations recommend *8-TWA* (8-h time-weighted average) exposure limits. TLVs are recommended by the ACGIH TLV Committee.[2] TLVs are not legally enforceable except under the Walsh–Healy Act for government contractors or in support of a general duty clause violation (see g below). Workplace environmental exposure levels (WEELs) are recommended by the AIHA WEEL committee.[6] The WEELs process was developed to set airborne exposure limits for chemicals for which there are no existing guidelines. Neither the WEELs nor the TLVs use a consensus standard setting process.

d. *OSHA action levels* are specified for some chemicals regulated in 29 CFR 1910 with subparts above 1000. The original basis for using one half of the 8-h PEL for an Action Level came from analyses of the statistical variability in the airborne concentration levels that occur in many workplaces.[7] Exposures above the action level are not an OSHA violation, but they usually require the employer to conduct some form of additional monitoring.

e. *8-TWA* permissible exposure limits (*PELs*) are also 8-h time-weighted average concentration limits that are legally enforceable by OSHA (cf., TLVs and WEELs which are not). All of the PELs in OSHA Table Z-1 are in fact the 1968 TLVs that were existing federal standards at the time the OSHA was passed. Since the TLV list has expanded and many individual TLVs have been reduced since 1968, there are now fewer PELs than there are TLVs, and about half of the PELs are less protective than their corresponding TLVs. (Copper fume in Table Z-1 is probably the only exception where TLV is higher than its PEL.)

f. Fluctuations in exposure during an 8-h day are expected. Short-term exposure limits such as *STELs, ceiling limits, acceptable ceiling concentrations,* and *acceptable maximum peaks* all allow exposures for a short portion of a workday to higher limits than the 8-TWA, TLVs, or PELs. STELs are found only in TLVs. Short-term TWAs are found only in the AIHA's WEELs. Acceptable ceiling concentrations and acceptable maximum peaks are old ANSI standards that were adopted as OSHA PELs in Table Z-2 of 29 CFR 1910.1000 because they were existing consensus standards in 1969 when OSHA was created. Those TLVs and PELs marked as C in their respective lists are ceiling limits that should never be exceeded. For chemicals that do not have a STEL, the TLV committee recommends a generic TLV excursion limit based on statistical control rather than toxicity.[a]

g. The *general duty clause* Section 5(a) of the OSHA Act allows OSHA to enforce over-exposures to chemicals that do not have a PEL, and perhaps may not even

[a] For chemicals without a STEL, the introduction to the TLVs recommends "Excursions in worker exposure levels may exceed three times the TLV–TWA for no more than a total of 30 min during a workday, and under no circumstances should they exceed five times the TLV–TWA, provided that the TLV–TWA is not exceeded."[2]

have a TLV, but present a clearly recognizable hazard to employees. The vigor with which the general duty clause has been enforced has varied with the political climate overseeing OSHA, but in general, enforcement requires a clear and present danger.

PL 94-596, Section 5(a): "Each employer shall furnish to each of his employees employment and a place of employment which are free from recognized hazards that are causing or are likely to cause death or serious physical harm to his employees."

A parallel employee duty clause, Section 5(b) has never been enforced: "Each employee shall comply with occupational safety and health standards and all rules, regulations, and orders insured pursuant to this Act which are applicable to his own actions and conduct."

h. Fire and explosions are safety hazards (see the next section under Safety Criteria) rather than health hazards but would otherwise fall into the sequence about here because the concentration at which a chemical will burn is almost always greater than where it will be a health hazard.

i. The concentration of an airborne contaminant necessary to displace ambient oxygen down to its minimum allowable level (called an oxygen deficiency) is the highest of any of the above limits. (see Table 4.4)

2. In order to compare the exposure limits of gaseous and vapors with those of aerosols, a list of 52 chemicals representative of each group was assembled (see Table 4A.1 in the appendix of this chapter). The chemicals in each group were equally apportioned among those whose exposure limit was based primarily upon their carcinogenic effects, their ability to irritate the respiratory tract, or their ability to cause systemic effects elsewhere (no anticholinergic chemicals, e.g., insecticides were included). As the TLV exposure limits for aerosols are only given in terms of mg/m^3 and *are* not convertible to ppm, all gas and vapor TLVs were converted into mg/m^3 via Equation 2.12c before the summary data in Table 4.1 were calculated.

   As can be seen in Table 4.1 and summarized graphically by Figure 4.2, exposure limits for aerosols in all categories are considerably lower than those for gases and vapors, meaning that aerosols appear on average to be over 100 times more toxic per mass than gases or vapors.[b] Part of the reason for this ratio is the difference in the physical diameters between aerosol particles and gas or vapor molecules and the corresponding difference in the localized dose to respiratory tissue. The physical behavior of aerosols that also partly underlies this difference was covered in the previous chapter.

3. A quite different point that can be introduced here is that each chemical's exposure limit as well as its vapor pressure is an intrinsic characteristic of the individual chemical. The amount by which the chemical's plume is diluted by ventilation and distance is a characteristic of the environment. Mentally separating the chemical characteristics from the environmental characteristics is a useful strategy not only to help extrapolate the recognition and evaluation of hazardous conditions from one workplace to the next but also to predict the effectiveness of certain controls. This point will be explained in more detail and stressed as a learning lesson in Chapter 5 and used in Chapter 8.

[b] The probability that the two overall categories are statistically similar using a log-normal t-test with equal variance is $\ll 0.00001$; the probability using a normal t-test with unequal variances is <0.0001. These distributions are different.

TABLE 4.1
The Geometric Mean and Geometric Standard Deviation (in parentheses) of Selected 1999 TLV Values Expressed in Units of mg/m³

	Geometric Mean [mg/m³] (Geometric Standard Deviation)				Ratio of Gaseous to Aerosol TLVs
	n	Gas and vapor TLVs	*n*	Aerosol TLVs	
Carcinogens	5	0.82 (10.5)	6	0.022 (5.1)	37:1
Irritants	23	114 (12.9)	23	2.2 (4.7)	52:1
Systemic toxins	24	125 (4.7)	23	0.51 (8.2)	245:1
Overall	52	78 (12.3)	52	0.68 (9.9)	115:1

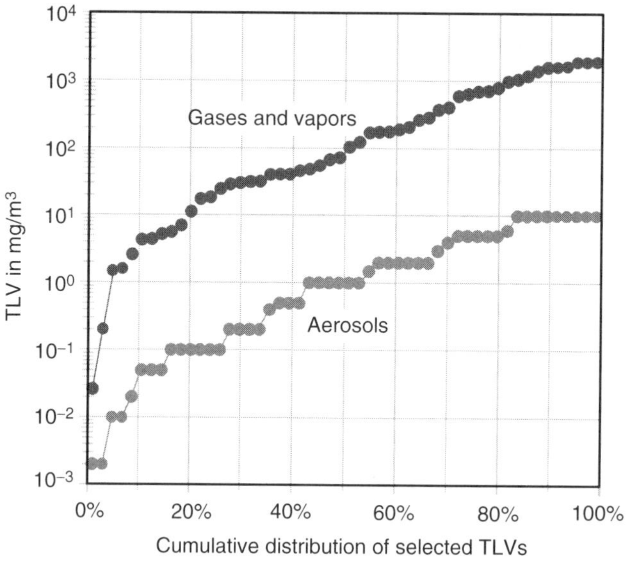

FIGURE 4.2 Composite distribution of 52 aerosol and 52 gaseous TLVs.

III. SAFETY CRITERIA FOR CHEMICAL CONTROL

The principal historic concern of gas and vapor hazards at work was to prevent their combustion. Combustion and fire are safety criteria because of their acute nature and their impact on property (as well as people). Much like the above list of various health exposure limits, each of the following safety limits is an intrinsic characteristic of the individual chemical. The solvent's temperature and the amount by which its vapors are diluted by distance or ventilation are characteristics of the environment.

1. Lower Flammable Limit (LFL) or Lower Explosive Limit (LEL) is the lowest concentration (usually expressed as a percent) needed for an airborne contaminant to support or sustain combustion *if it were ignited.* A spark or open flame will not ignite

vapors at a concentration below its LFL because there are too few burnable molecules present (or they are too far apart) to sustain combustion. Although the widely used term LEL means lower explosive limit, it does not really mean that such vapors will explode (in fact, few vapors will actually explode if ignited). LEL's popularity is perhaps just because it is easier to say than LFL. For practical purposes, the two terms are synonymous.

a. Each chemical has its own LFL. Most LFLs are in the range of just over 1 to about 6%. Only about 31 LFLs are listed in 29 CFR 1910.94, although more are listed in OSHA's technical link at www.osha-slc.gov/SLTC/smallbusiness/sec8.html. Other sources of LFL data include the NIOSH/OSHA *Pocket Guide to Chemical Hazards*,[5] Appendix B of the ACGIH *Ventilation Manual*,[8] and Perkins' *Modern Industrial Hygiene*.[9] The LFL for almost 900 chemicals are listed in the *CRC Handbook of Chemistry and Physics*.[10]
b. OSHA, e.g., 29 CFR 1910.106(a)(31) and 1910.124, requires employers to maintain the airborne concentration of any contaminant below 25% of its LFL or below its PEL, whichever is less. Section 1910.146 prohibits entry into a confined space if flammable or combustible gas or vapor concentrations exceed 10% of their LFL. Continuous or direct reading combustible gas monitors may be set to alarm at 10% of the LFL. The $4\times$ or $10\times$ reduction (to 25 and 10%, respectively) is appropriate because the concentration will always decrease with distance from the actual source of a gas or vapor; if the concentration is 10 or 25% of the LFL at the combustible gas detector, it will be more concentrated and potentially ignitable closer to the source.
c. Any given chemical's LFL will decrease slightly as its vapor's temperature increases. Data for several chemicals in the NFPA *Fire Protection Handbook* show each chemical's LFL decreases relative to its LFL at normal temperature by roughly 10% per 100°F.[11] Thus, acetone's 2.5% LFL at 25°C is reduced almost 50% at 500°F to an LFL of 1.3% (close to $2.5 - 5\times 0.25$). The *Manual of Industrial Ventilation* (Section 2.5) recommends decreasing any LFL to 0.7 of its normal value at any temperature above 250°F (which is equivalent to a one-time 30% reduction, rather than an incremental proportion). The actual decrease in LFL means that an initially safe vapor could become ignitable at the same concentration if it were heated sufficiently. This phenomenon is used to advantage in a thermal oxidizer or after-burning air cleaner and could happen by accident.
d. Not every organic solvent will burn. The multiply-substituted chloro- and fluorocarbon solvents are particularly common examples. Their intrinsic fire safety was the rationale for their popularity and widespread use before their toxicity was adequately recognized. Examples of nonflammable solvents include carbon tetrachloride (CCl_4), chloroform ($CHCl_3$), perchloroethylene ($Cl_2C{:}CCl_2$), and dichlorodifluoromethane (CCl_2F_2). On the other hand, some chlorinated solvents, such as the dichloroethanes, dichloroethylenes, and trichloroethanes are flammable; and some, such as trichloroethylene are so difficult to ignite that they are sometimes considered nonflammable.

2. Upper flammable limit (UFL) or upper explosive limit (UEL) is the maximum gas or vapor concentration that will sustain combustion. The same comments regarding explosiveness vs. flammability that were made for the LEL vs. LFL also apply to UEL vs. UFL. People or sources who use LEL tend also to use UEL, and again, the two terms are practically synonymous. At a concentration above the UFL, there is not enough oxygen in the air (or the oxygen molecules are too far apart) to keep a flame burning. UEL values can usually be found in the same sources that list LEL or LFL values. However, the UFL is rarely an important parameter to industrial hygienists because vapors are naturally diluted as they move further from the source; and if a plume starts out above its UFL, the gas or vapor

concentration will *always* be in the flammable range between the UFL and LFL somewhere further from the source. Should an ignition occur out in the flammable region, the flame front would rapidly propagate back to the source. Not a pleasant experience.

3. Flash point (abbreviated as either F.P., f.p., or fl.pt.) is the minimum temperature of a liquid that can produce enough vapor to be ignited. If an otherwise flammable or combustible liquid is below its flash point, the concentration of its vapors is below its LFL and they cannot be ignited. As the liquid is heated, its vapor pressure will increase. When the liquid's vapor pressure reaches its LFL, it will burn if ignited. As the concentration at the liquid surface always equals its P_{vapor}, the flash point is also the temperature at which the chemical's vapor pressure equals its LFL. Equation 4.1 shows the relationships between flash point, LFL, and vapor pressure mathematically.

$$P_{vapor} \text{ at flash point} = \frac{\text{LFL as a \%}}{100} \times 760 \text{ mmHg} = Y_{at\ LFL} \times 760 \tag{4.1}$$

These relationships can be seen graphically in Figure 4.3, in which the vapor pressure for toluene increases as a function of liquid temperature in accordance with the Antoine equation. Below toluene's 4.5°C or 40°F flash point, its vapor pressure is less than its LFL.

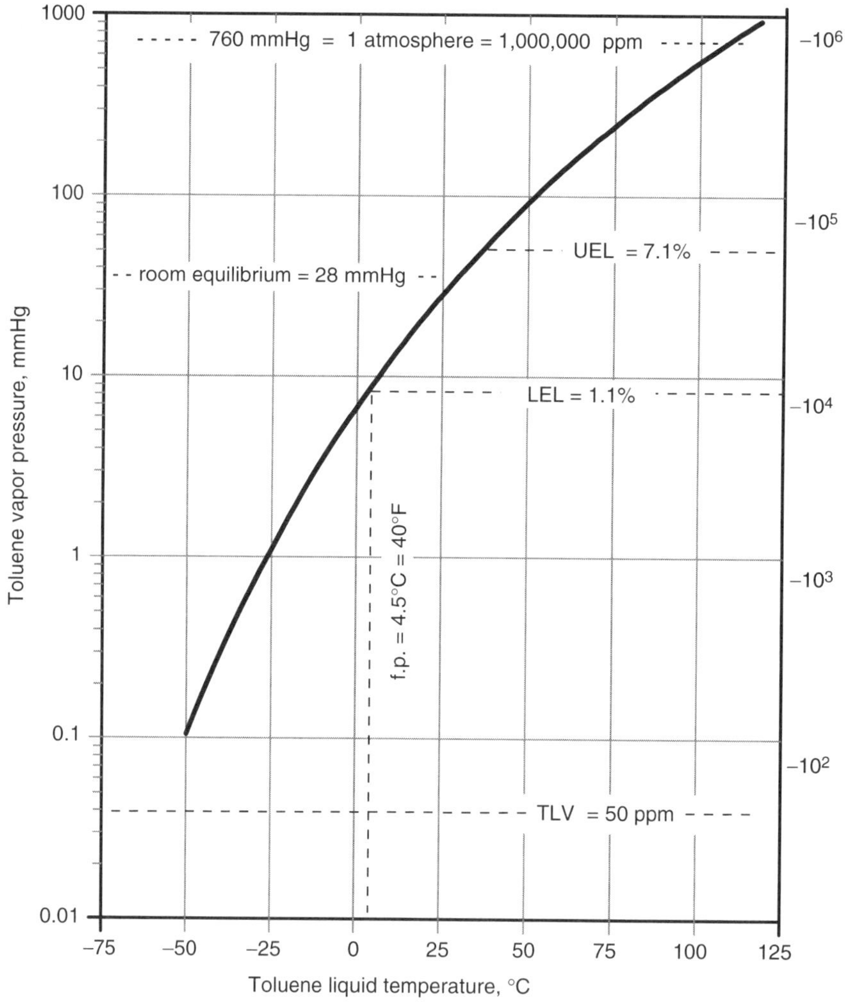

FIGURE 4.3 Schematic relationship between P_{vapor}, LFL, and flash point.

When the liquid toluene's temperature reaches its flash point, its P_{vapor} is 8 mmHg. That 8 mmHg divided by 760 mmHg is 1.1% which, *not* coincidently, is toluene's LEL or LFL. At 25°C (77°F), toluene's vapor pressure is 28 mmHg, corresponding to 36,840 ppm or 737 times higher than its TLV (in other words, at NTP the vapors at the liquid surface must be diluted by a ratio of 737 to reach an acceptable occupational exposure level). At 111°C, toluene's vapor pressure is 760 mmHg or the boiling point of toluene at sea level.

The same relationships apply to all flammable solvents just at different percent concentrations and temperatures. While it is helpful to understand these relationships, it is essential to know that a chemical's flash point is the temperature (in either °F or °C) at which the vapors above the liquid surface will sustain combustion if ignited. Flash points are listed in most of the same sources that list LFLs except OSHA 1910.94.

a. The flash point is much more useful in the field than LFL because the room or liquid temperature is much easier to measure, to predict, or even to sense than is the vapor concentration or its percent of the LFL at any given location (even the most concentrated vapor in the region near the source). The following table provides a broad guide to the implications of knowing the chemical's flash point, the setting's temperature, and the type of ventilation that is present.

<table>
<tr><th>Liquid Temperature</th><th>Unventilated Room</th><th>Ventilated Room</th></tr>
<tr><td>>Flash point</td><td>Ignitable throughout the room unless > UFL</td><td>Ignitable somewhere in the plume</td></tr>
<tr><td><Flash point</td><td colspan="2">Not flammable anywhere</td></tr>
</table>

If a liquid temperature is above its flash point, the vapor concentration right at the liquid source will always exceed its LFL, and therefore the vapor will always be ignitable, either at the source or somewhere in the diluted plume. If a liquid temperature is below its flash point, the vapor concentration is below its LFL everywhere; it will not burn even if an ignition source is present. Therefore, flammability hazards can be controlled either by keeping the liquid temperature below the chemical's flash point or by choosing a chemical with a flash point greater than the liquid temperature. (Normally the liquid will be at or near the ambient temperature, but that is not always the case if, for instance, the liquid were heated by the process.)

b. A liquid's flash point can be measured by two laboratory methods: one is called the closed cup method, and the other is the open cup method. The open cup would seem to be more applicable to conditions typical of ignitable vapors in the workplace; however, closed cup data are more common and, in fact, specified by OSHA 1910.106. The difference between the two methods can be a technical annoyance, but the two data points for any given chemical are generally close enough together that either one could be used for safety hazard assessment and planning.

c. Flash points are also related to the terms flammable and combustible, used above generically. Technically they are different. OSHA 1910.106(a)(18) incorporates an NFPA classification scheme for flammable and combustible liquids, summarized in Table 4.2. The NFPA defines three classes of liquids that differentiate a liquid as either flammable from combustible based on its f.p. Some classes of liquids have subclasses differentiated by boiling point (b.p.).

Flammable liquids (all Class I liquids) are more dangerous than combustible liquids because they have a lower flash point than Class II or III liquids. At normal working temperatures of 68 to 77°F, most Class IA and IB chemicals are ignitable. However, keep in mind that any combustible liquid can be ignited if it is heated sufficiently above normal working temperatures.

TABLE 4.2
NFPA Classifications and Terminology of Flammable and Combustible Liquids[11]

Class	Flash Point (f.p.) °F	Subclass Based on f.p. and b.p., °F A	B	C	Nomenclature or Terminology
I	<100	f.p. <73 & b.p. <100	f.p. <73 & b.p. ≥ 100	73 ≥ f.p. <100	Flammable
II	100 ≤ f.p. 140				Combustible
III	≥140	140 ≥ f.p. <200	f.p. ≥ 200		Combustible

4. Autoignition temperature (or sometimes just referred to as the ignition temperature) is the minimum temperature that will cause the vapor–air mixture to ignite spontaneously. Typical ignition temperatures of 500 to 1000°F virtually never exist in the occupied workplace. Thus for IHs, the autoignition temperature is the least important of any parameter related to combustion. However, it is important to know that a chemical's autoignition temperature is much, much greater than its flash point. It can also be important in some special work settings to know that the published ignition temperature determined in the laboratory can differ from values found in industrial settings because of catalytic effects.

IV. OXYGEN DEFICIENCY CRITERIA

1. Oxygen deficiency (a low concentration of airborne oxygen) is a special environmental hazard that has developed its own terminology related to its toxic effects.
   a. Hypoxia is a generic term for the physiologic shortage of oxygen at the cellular level whether due to an ambient oxygen deficiency or to a body's inability to inhale, absorb, transport, or metabolize oxygen. Hypoxia is a measurable deficiency within the body that may or may not cause clinical symptoms. The earliest symptoms (those at the lowest level of hypoxia) are decreased physiologic performance (e.g., visual acuity or maximum sustainable exertion rate). As ambient and physiologic concentrations of oxygen decrease, symptoms will progress to euphoria, incoordination, loss of consciousness (asphyxia), and death.
   b. Chemical asphyxiants block the body's ability to utilize oxygen in some way. The following three examples of chemical asphyxiants each block the utilization of oxygen in a different way.
      1. Carbon monoxide (CO) blocks the blood's ability to transport oxygen.
      2. Hydrogen cyanide (HCN) blocks the transport of oxygen into the cells.
      3. Hydrogen sulfide (H_2S) and acrylonitrile ($CH_2{=}CHCN$) block oxygen metabolism by mitochondrial cytochrome oxidase within the cells.
   c. Simple asphyxiants are toxicologically inert gases that can only affect health after their vapors displace enough air to reduce the oxygen partial pressure to create hypoxia with no other toxic effect. The presence of any gaseous air contaminant will displace the oxygen and nitrogen that comprise normal air (except of course if the contaminant *is* oxygen or nitrogen in which case, it will still displace the other major component of air). An inert gaseous contaminant that accumulates, for example, in a poorly ventilated area or confined space will displace the air and can create an ambient oxygen deficiency. N_2, H_2, He, CH_4, C_2H_4, C_2H_6, C_3H_8 (but not CO_2) are simple asphyxiants.

A few other chemicals, such as some Freons, could also be considered simple asphyxiants because they are toxic only above concentrations at which they would be an asphyxiant.

Exposure limits to simple asphyxiants are not specified explicitly within either the PELs or TLVs. However, both limits do specify minimum concentration of oxygen to which someone should be exposed in an oxygen-deficient atmosphere. Therefore, a method to calculate the maximum allowable concentration of contaminants from the minimum allowable percent of oxygen will be developed below that will eventually allow a comparison to be made among the concentrations at which a chemical will exceed its health-based exposure limit, will exceed its lower limit of flammability to become a fire hazard, and will displace enough oxygen to become an asphyxiant hazard.

2. Recall that Dalton's law states that our atmospheric pressure is simply the sum of its molar parts as shown in Table 2.2. The major constituents (99.03%) of normal air are nitrogen ($Y_{N_2} = 78.08\%$) and oxygen ($Y_{O_2} = 20.95\%$). For use in the following equations, the subscript N_2 etc will be used to include all of the other *natural* components of pure air in addition to nitrogen such as argon, carbon dioxide, etc. listed in Table 2.2. Taken together, $Y_{N_2\,\text{etc}} = 79.04\%$ at any humanly survivable altitude (until well above 100,000 ft). Thus, the ratio of nitrogen etc. to oxygen is a fixed constant of 3.773 as long as the contaminant is not N_2 and O_2 is neither being emitted, absorbed, nor consumed.

$$Y_{N_2\,\text{etc}}/Y_{O_2} = 0.7904/0.2095 = 3.773 \quad (4.2a)$$

or

$$Y_{N_2\,\text{etc}} = 3.773\ Y_{O_2} \quad (4.2b)$$

Keep in mind that the assumption of a fixed ratio of N_2 to O_2 is only valid for oxygen deficiency caused when a chemical contaminant displaces oxygen. Processes such as microbial metabolism, combustion, and even corrosion can consume oxygen from the air of a confined space over time. When oxygen is consumed, the remaining air will not have the above ratio of N_2 to O_2 and *the quantitative parts of the rest of this discussion will not apply*. Similarly, adding nitrogen to air to keep oxygen out (a process called inerting) will also invalidate the assumed ratio and the following discussion.

a. If a contaminant is present, the Y_{contam} is one part of the sum that totals one atmosphere.

$$Y_{O_2} + Y_{N_2\,\text{etc}} + Y_{\text{contam}} = 1 \quad (4.3a)$$

or

$$Y_{\text{contam}} = 1 - (Y_{O_2} + Y_{N_2\,\text{etc}}) \quad (4.3b)$$

b. By mathematically substituting the amount of nitrogen etc. from Equation 4.2b into Equation 4.3, one can solve either for the Y_{O_2} at a measured concentration of the contaminant (as shown by Equation 4.4a) or for the Y_{contam} at a set value of Y_{O_2} based on health or other criteria (as shown by Equation 4.4b).

$$Y_{O_2} = (1 - Y_{\text{contam}})/4.773 = 0.2095 \times (1 - Y_{\text{contam}}) \quad (4.4a)$$

or

$$Y_{\text{contam}} = 1 - (4.773\ Y_{O_2}) \quad (4.4b)$$

Equation 4.4 allows one to calculate the molar fraction of oxygen is present at any contaminant concentration. While the molar factions in Equation 4.2 will be constant with altitude, the partial pressures will decrease with altitude (as predicted by Equation 2.4d) as shown in Equation 4.5c and Equation 4.6c. Equation 4.5 yields P_{O_2} for any known amount of contaminant in terms of Y_i (typically converted from % or ppm of i).

$$P_{O_2} = 760 \times (1 - Y_{contam})/4.773 \quad \text{at sea level} \quad (4.5a)$$

$$= 159 \times (1 - Y_{contam}) \quad \text{at sea level} \quad (4.5b)$$

$$P_{O_2} = 159 \times 2^{(-ft/18{,}000)} \times (1 - Y_{contam}) \quad (4.5c)$$

Equation 4.6 yields the corresponding Y_{contam} and P_{contam} for a known partial pressure of oxygen.

$$Y_{contam} = 1 - (P_{O_2}/159) \quad \text{at sea level} \quad (4.6a)$$

$$Y_{contam} = 1 - (P_{O_2}/(159 \times 2^{-ft/18{,}000})) \quad (4.6b)$$

$$P_{contam} = 760 \times 2^{(-ft/18{,}000)} \times (1 - P_{O_2}/159)) \quad (4.6c)$$

3. The same hypoxic effects can occur whether oxygen is displaced due to the abnormal presence of a contaminant gas or vapor or whether the total pressure of air is reduced due to increasing altitude (or a combination of the two causes). The physiologic effects of hypoxia created by high altitudes (hypobaric pressures) have been well studied in aviation medicine. For instance, the first four columns of Table 4.3 summarize human physiologic responses to altitude found in the classic studies reported by Henderson and Haggard[13]. The two rightmost columns are the

TABLE 4.3
Equating the Health Effects Due to Altitude with Those Due to Airborne Contaminants at Sea Level[12–15]

Altitude (ft)	Ambient PO_2 (mmHg)	Alveoli PO_2 (mmHg)	Health Effects	Eqv. Sea Level YO_2 (%)	Y_{contam} (%)
6000	127	82	Reduced capacity for sustained maximum exertion (beyond 2 min)	17	21
12,000	101	65	Decreased night vision and AMS symptoms[a]	13	37
18,000	79	44	Euphoria, loss of coordination	10	51
Exposure >18,000 is limited by one's time of useful consciousness (TUC)					
20,000	73	40	TUC = 10–20 min	9.6	55
25,000	59	25	TUC = 3–5 min	8	63
30,000	47	21	TUC = 1–2 min[b]	6	71
35,000	37	12	TUC = 30–60 sec[b]	5	77
40,000	29	12	TUC = 15–20 sec[b]	4	82

[a] Benign acute mountain sickness is a constellation of symptoms ranging from discomfort to incapacitation, highlighted by frontal headaches (see Ref. 16).

[b] Complete loss of consciousness can be expected shortly above 30,000 ft.

contaminant conditions at sea level that would create the same oxygen deficiency conditions as clean air at each altitude. The equivalent Y_{O_2} was derived by dividing Equation 4.5c when Y_{contam} equals zero by 760 mmHg at sea level, as shown in Equation 4.7a. The equivalent Y_{contam} was derived by setting the P_{O_2} in Equation 4.5b due to the presence of a contaminant at sea level equal to 760 × the equivalent Y_{O_2} just derived from Equation 4.7a due to altitude and solving for Y_{contam}, as shown in Equation 4.7b.

$$\text{Equivalent } Y_{O_2 \text{ at sea level}} = P_{O_2 \text{ at altitude}}/760 = 0.2095 \times 2^{(-ft/18,000)} \quad (4.7a)$$

$$\text{Equivalent } Y_{\text{contam. at sea level}} = 1 - 2^{(-ft/18,000)} \quad (4.7b)$$

4. Four oxygen deficiency limits are potentially applicable to the workplace. Each of the four limits is described below, and all are plotted in Figure 4.4. By comparing the three oxygen deficiency limits with the rightmost columns in Table 4.3, which describe conditions at sea level that would produce an equivalent effect, one can see that most of these occupational limits for oxygen deficiency are all well short of asphyxiation. Functionally debilitating symptoms will not occur at sea level until $<10\%$ O_2, although measurable mental effects will occur at $<13\%$ O_2. The first effect noted in Table 4.3 (an inability to sustain a maximum level of human exertion) at $<17\%$ O_2 is a response that is not applicable to many jobs.[15,17] As will be shown in the calculations below, the amount of a contaminant that must be present to displace oxygen from a normal of 20.9% down to even 19.5% would have already exceeded all toxicity and most flammability limits. Therefore, rather than viewing most of the recommended oxygen deficiency limits as either a health or safety limit, they are best viewed as simply *a good work practice*. Since ambient air is both

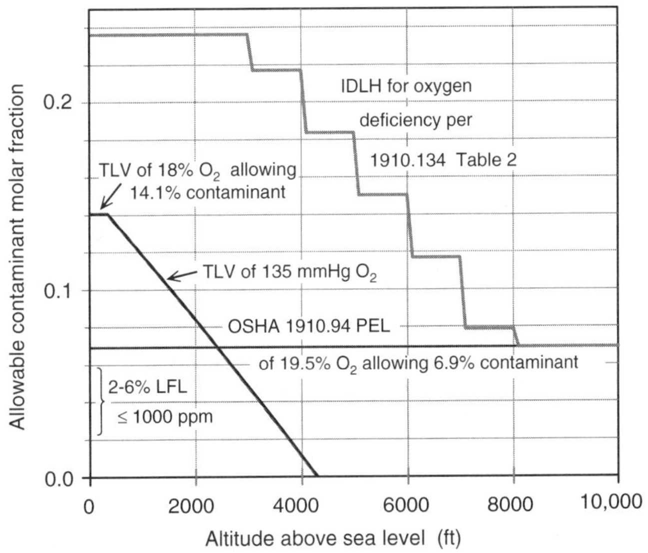

FIGURE 4.4 A depiction of the contaminant concentration necessary to deplete oxygen down to three limits for oxygen deficiency. OSHA 1910.94 for ventilation appears herein as a horizontal line. OSHA 1910.134 definition of an IDLH oxygen deficiency is a stair-step line across the upper right (see Section VI.4 in Chapter 22). The TLV starts as a horizontal line but rapidly transitions to a diagonal line across the lower left as altitude increases.

plentiful and cheap, there are few good reasons why an operator, manager, or anyone should fail to bring enough outside air into a workplace to prevent an oxygen deficiency. The most common reasons to exceed any of these oxygen limits are either that oxygen is purposefully being excluded or the person in charge is ignorant of the hazard or just does not care. Any person who enters a known oxygen-deficient space (or a confined space whose oxygen content is unknown) must be provided with breathing air via an atmosphere-supplied respirator (see Chapter 22).

a. OSHA 1910.94(d)(9), 1910.134, and 1926.57 consistently specify that air is oxygen deficient when the oxygen molar fraction drops to or below 19.5% O_2. This is the most restrictive of the four oxygen deficiency limits. This OSHA requirement is equivalent to a P_{O_2} at sea level of 148 mmHg (found by inserting $Y_{O_2} = 0.195$ into Equation 2.11a) and to the oxygen in clean air at an altitude of only 1860 ft (567 m).[c] The following calculation using Equation 4.4b shows that it would take 6.9% contaminant to displace oxygen down to 19.5%.

$$Y_{\text{contam}} = 1 - (4.773 \times Y_{O_2}) \quad \text{(using Equation 4.4b)}$$

$$Y_{\text{contam}} = 1 - (4.773 \times 0.195) = 1 - 0.931 = 0.069 = 6.9\% \quad \text{(using Equation 4.4b)}$$

Thus, the 19.5% PEL for oxygen effectively restricts any airborne contaminant concentration at sea level to 6.9% before the P_{O_2} will become deficient. This limit is shown as a horizontal line across Figure 4.4. This 6.9% turns out to be above the LFL of nearly all flammable or combustible contaminants.

b. Up until 2004, the ACGIH recommended limit for oxygen deficiency appeared in the introduction to the chemical substances section of the TLV booklet under the label *Simple Asphyxiants — "Inert" Gases and Vapors* as "18% by volume under normal atmospheric pressure (equivalent to a partial pressure P_{O_2} of 135 torr)." Equation 2.11a can again be used to show that a Y_{O_2} of 18% is actually equivalent to 136.8 mmHg and 135 mmHg is actually equivalent to 17.8% oxygen at sea level. This small difference was probably a rounding error ($Y_{O_2} = 18\%$ is equivalent to 135 mmHg at an altitude of about 290 ft (88 m), as shown in Figure 4.4). Equation 4.4b can be used again to show that a 14% contaminant concentration would be required to displace oxygen down to the 18% O_2 TLV (vs. a 6.9% contaminant concentration limit for the OSHA 19.5% oxygen limit).

$$Y_{\text{contam}} = 1 - (4.773 \times 0.18) = 1 - 0.859 = 0.141 = 14\% \quad \text{(using Equation 4.4b)}$$

As the 18% applies explicitly to sea level, the TLV implies a physiological justification to apply that 135 mmHg O_2 oxygen limit to other altitudes. In order to keep a constant 135 mmHg ambient partial pressure of oxygen, Equation 4.6b implies that as the altitude increases less Y_{contam} would be allowed. In fact, if this equation is solved for the altitude at which no exposure is allowed (namely, when $Y_{\text{contam}} = 0$), one would find that this interpretation of the TLV would allow no contaminant exposure at all above 4287 ft. Indeed, any work at an altitude above 4287 ft would violate the implied 135 mmHg TLV, which seems to be an unworkable recommendation for many otherwise industrially productive geographic regions. (Do you suppose that a lack of oxygen in Utah made me want to write this textbook?).

c. Table II of OSHA's respirator standard 1910.134(d)(2) states a series of limits at different altitudes at which OSHA considers an oxygen deficiency to be IDLH

[c] The altitude in feet with the same P_{O_2} as a contaminant at sea level $= 25{,}970 \times \ln(0.2095/Y_{O_2}) = 1863$ ft.

(see Table 22.6 and the stair-step limit in Figure 4.4). Using this definition, an oxygen deficiency would be considered IDLH if it drops below 16% O_2 at or below 3000 ft and when it drops below 19.5% O_2 above an altitude of 8000 ft. Equation 4.4b is used for the third time below to solve for the amount of contaminant allowed at sea level for $Y_{O_2} = 16\%$.

$$Y_{contam} = 1 - (4.773\,Y_{O_2}) = 1 - (4.773 \times 0.16) = 1 - 0.764 = 0.236$$

(using Equation 4.4b)

Thus, if the oxygen deficiency were caused by displacement, a whopping 23.6% contaminant would be allowed at sea level before it would create an IDLH oxygen-deficient atmosphere. You will learn in Chapter 22 that any IDLH atmosphere would require a full facepiece pressure demand SCBA or similar supplied air respirator (SAR) with back-up escape respirator. However, a contaminant above its LFL is already IDLH, and the LFLs of virtually *all* flammable and combustible liquids are all less than 23%. Thus the lesson here is that a flammable contaminant above its LFL will always be IDLH because of its flammability well before it can create an IDLH oxygen-deficient atmosphere by displacing oxygen.

d. ANSI Z88.2-1992 has the only recommendations that approach a health criterion. Its Section 7.3.4.2 states that oxygen is not deficient for purposes of respirator selection until 122 mmHg (equivalent to 16% O_2 at sea level or zero contaminant at an elevation of 7000 ft), and its Section 7.3.4.1 states that IDLH does not begin until 95 mmHg (equivalent to 12.5% O_2 at sea level or zero contaminant at about 14,000 ft). While being the least protective of the four oxygen deficiency limits, the health effects described in Table 4-3 at 7000 ft are still only the reduced capacity to sustain maximum exertion and the higher level symptoms do not reach those upon which an IDLH is traditionally defined (see Section F.4 of Chapter 22).

5. The reader is reminded again that the presence of a gaseous contaminant is not the only way to create an oxygen deficiency. Biologic metabolism, slow combustion, and even slower but sustained corrosion in a confined space can each create an oxygen deficiency by depleting oxygen without adding any contaminant. Oxygen consumption (or an artificially raised nitrogen level) will increase the N_2/O_2 ratio above the 3.773 in normal air, meaning that most of the above formulae do not apply. The amount of oxygen consumed depends completely on the unobserved, uncontrolled, and usually unpredictable chemical transformations going on within the confined space. Therefore, before entering a previously unoccupied confined space where such biological or chemical consumption of oxygen is possible, its air must always be tested for the percent O_2 in addition to testing for the presence of known contaminants and anticipating the presence of unknown chemicals.

V. COMPARISONS AMONG CRITERIA

Examples of exposure limits based on safety (the LFL), health (the TLV or PEL), and oxygen deficiency (%O_2) can be compared in Table 4.4, in which the lowest control criterion for each chemical is highlighted in bold. Note how the LFL is the lowest criterion only for the toxicologically inert gases (simple asphyxiants). The concentration at which any of these chemicals will displace enough oxygen to create a legal oxygen deficiency is above the safety (fire) criteria and well above the health (toxicity) criteria.

As a general statement, health-based exposure limits are usually at least a couple orders of magnitude smaller than safety-based concentration limits for fire protection. Safety-based criteria are somewhat smaller than the criteria set for oxygen deficiency. No adverse health effects should result from work in well less than 19.5% oxygen. Thus, oxygen deficiency regulatory limits are best

TABLE 4.4
Occupational Exposure Limits Based on Safety, Health, and Oxygen Deficiency for Selected Chemicals

Chemical	P_{vapor} (mmHg)	$P_{vapor}/760$ (%)	Safety-Based Criteria			Health-Based Criteria (TLV)		
			LEL (%)	UEL (%)	Fl. Pt. (°F)	mg/m^3	ppm	%
Methane	>760		**5.3**	14.0	−300	19.5% O_2 ⇔ 6.9% contaminant		
Hydrogen	>760		**4.0**	75.0	−436	or 18% O_2 ⇔ 14.1% contaminant		
Acetone	231	30.4	2.5	13.0	0	1780	750	**0.075**
Ethyl acetate	93.7	12.3	2.2	11.4	30	1440	400	**0.04**
Heptane	46	6.1	1.1	6.7	25	1210	400	**0.04**
Ethyl ether	537	70.7	1.9	48	−49	1200	400	**0.04**
Xylene	8	1.1	1.0	7.0	81–90	435	100	**0.01**
Toluene	28.4	3.7	1.2	7.1	39	188	50	**0.005**
Tetrachloroethylene[a]	18.6	2.4	Nonflammable	Nonflammable	Nonflammable	170	25	**0.0025**
1,2-Dichlororethane[a]	78.9	10.4	6.2	15.9	65	40	10	**0.001**
Nitrobenzene	0.245	0.03	1.8	unk.	190	5	1	**0.0001**
Benzene	95	12.5	1.3	7.1	12	1.6	0.5	**0.00005**
Aluminum dust[b]			35 g/m^3	n.a.	n.a.	**15**	n.a.	n.a.

The dotted box within this table was added mostly to highlight the oxygen deficiency data (and not necessarily to make it easier to find on an examination).

[a] Tetrachloroethylene is a.k.a. perchloroethylene. 1,2-Dichloroethane is also known as ethylene dichloride.

[b] Oxidizable dusts can burn in air but at concentrations on the order of g/m^3 vs. health exposure limits in mg/m^3.[18]

viewed as good practice standards rather than health or even safety standards per se. There is always a lot of fresh air somewhere nearby. OSHA requires and common sense dictates that we access it!

VI. OSHA VENTILATION STANDARDS

Once an airborne contaminant is generated, its subsequent direction of movement, dispersion, and arrival in someone's breathing zone are determined largely by the velocity (speed and direction) of the air in which it is suspended. You will learn about control velocity in Chapter 10, how to measure it in Chapter 11, and how to select or specify it in Chapter 13 to design a local exhaust ventilation system to collect or contain the contaminant (and thereby protect employees). Technically, the rate that air flows into a local exhaust hood is an even greater determinant of employee protection but is less easily measured than velocity. The air velocity within any duct intended to transport a contaminant away from the workplace should be more than sufficient to maintain aerosols in suspension but not so high as to cause excessive friction or turbulence (energy losses and noise). Thus, ventilation standards often specify the control velocity, the air flow rate, and the transport velocity within ducts. The examples given in this section and the next should provide the reader with a sense of the specificity of such standards.

1. The OSHA regulations listed below relate to either health or safety concerns as defined above. Most of these regulations are specification standards. As a general guide for chemical hazards, OSHA relies primarily upon performance standards (the PELs contained in Sections 1910.1000 to 1910.1090) for their enforcement actions and uses the specification standards contained within Section 1910.94 to 1910.252 only as a secondary violation. That is, where a PEL is exceeded, violations of one of the following specification standards can add to OSHA's citation authority and potentially to their fine, but a citation for insufficient ventilation is unlikely where chemical exposure or concentration limits are not exceeded

Regulations	Operations	Criteria
a. OSHA 29 CFR 1910 general industry regulations related to ventilation. (An * means that selected details of that regulation are presented later within this chapter.)		
1910.94a	Abrasive blasting	Health
* 1910.94b	Grinding, polishing, buffing (metals)	Health
* 1910.94c	Spray-finishing (basically spray painting)	Safety
1910.103	Hydrogen storage	Safety
1910.104	Oxygen storage	Safety
* 1910.106	Flammable and combustible liquids	Safety
1910.107	Spray finishing and spray booths	Safety
1910.124	Dipping and coating operations in general	Health
1910.125	Dipping and coating operations with organic solvents	Health
1910.252c	Welding, cutting, brazing	Health
1910.307	Electrical requirements in locations defined by 1910.399	Safety
* 1910.399	Definitions of hazardous locations	Safety
b. OSHA 29 CFR 1926 Construction standards		
1926.57	General ventilation	Health
1926.154	Temporary heating devices	Health
1926.353	Welding, cutting, heating	Health
c. OSHA 29 CFR 1915 and 1918 Maritime standards		
1915.32	Toxic cleaning solvents	Health
1915.51	Welding, cutting, heating	Health
1918.93	Atmospheric conditions	Health

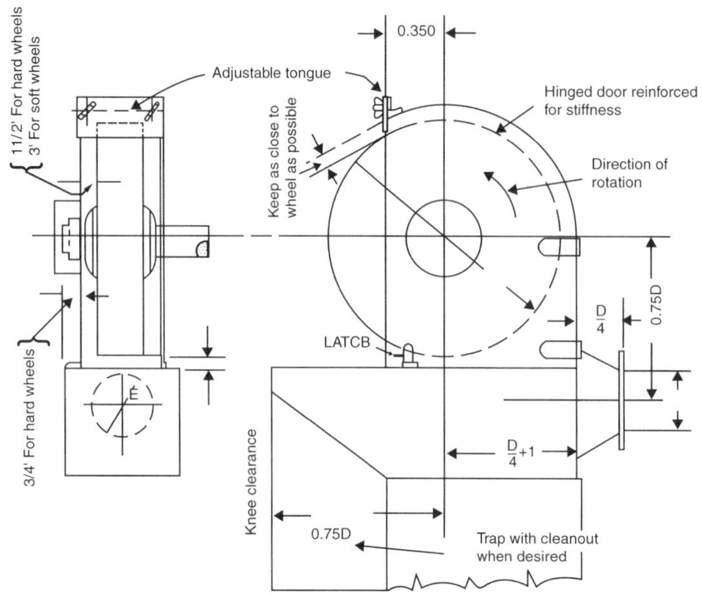

OSHA Table G-5:
Buffing and polishing wheels

Wheel diameter (inches)	Wheel diameter (inches)	Minimum Exhaust Volume (ft^3/min)
Up to 9	2	300
Over 9 to 16	3	500
Over 16 to 19	4	610
Over 19 to 24	5	740
Over 24 to 30	6	1040
Over 30 to 36	6	1200

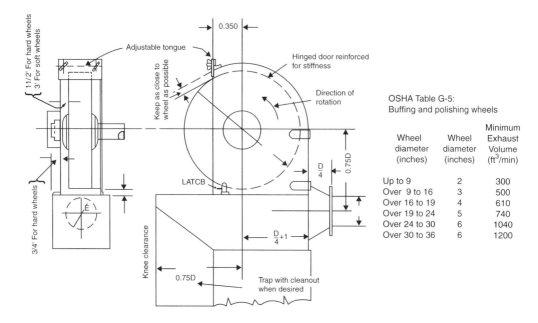

FIGURE 4.5 OSHA Figure G-4 and associated table for standard buffing and polishing hood from 1910.94(b)(5)(v).

2. OSHA 1910.94(b) specifies the exhaust ventilation parameters for grinding, polishing, and buffing operations shown in Figure 4.5 and Figure 4.6 and their accompanying tables.
   a. The minimum exhaust volume (Q or ft^3/min) is specified in the tables in Figure 4.5 and Figure 4.6 as a function of grinding wheel diameter.
   b. The minimum duct velocity (an air velocity within the duct sufficient to keep these contaminants suspended) is specified as 3500 fpm in the main duct and 4500 fpm in branch ducts.
3. OSHA 1910.94(c) specifies local exhaust ventilation parameters for spray-finishing operations (e.g., spray-painting booths such as depicted in Figure 4.7) and include exhaust air velocities, the total air flow rate to be exhausted, and the temperature of make-up air that replaces the exhausted air.
   a. The speed at which air must flow into the open face of the exhaust hood (called the face velocity) is specified to be between 50 and 250 fpm as a function of the hood's cross-draft velocity (the velocity of unplanned air movement laterally past or across the face of the hood).
   b. The total air flow volume (Q in terms similar to ft^3/min of exhaust air flow) must be sufficient to reduce the concentration within the exhaust air stream to 25% of the LFL (usually referred to as LEL in OSHA regulations).
   c. The temperature of the make-up air is specified as equal to or greater than 65°F when the outside air temperature is 55°F or less (unless radiant heat is used). Special provisions are specified for direct (vs. indirect) fired heaters (see Chapter 17).
4. OSHA 1910.399(a) also defines hazardous locations where flammable and combustible materials might be airborne (adopted from Article 505 of the National Electrical Code).[20] The following classes and divisions of hazardous locations are applicable to electrical safety and to the need for and qualifications of intrinsically safe equipment that may be used therein. (Do not confuse these hazardous locations with the somewhat similarly classified flammable and combustible liquids defined above; they are unrelated!)

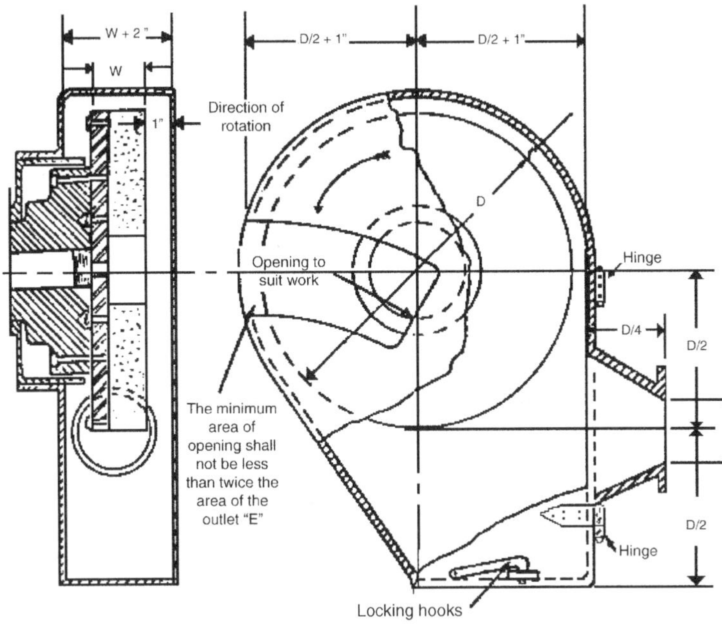

OSHA Table G-6: Horizontal Single-spindle Disc Grinder.

Disc diameter (inches)	Exhaust Volume (ft^3/min)
Up to 12 ...	220
Over 12 to 19	390
Over 19 to 30	610
Over 30 to 60	880

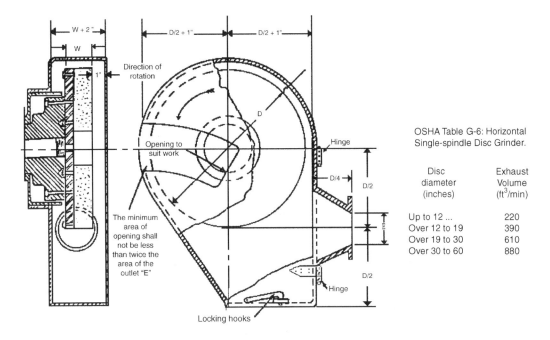

FIGURE 4.6 OSHA Figure G-6 horizontal single-spindle disc grinder exhaust hood and branch pipe connections from 1910.94(b)(5)(vii) and OSHA Table G-6 from 1910.94(b)(3)(iv).

a. Class I hazardous locations are defined by their flammable vapor or gas concentrations:

Division 1: where flammable vapor or gas concentrations may exist either normally, frequently during maintenance or leakage, or accidentally where breakdown or fault might also cause failure of electrical equipment (implying either providing an ignition source or disabling ventilation controls leading to probable vapor accumulation over LFL).

Division 2: where flammable vapor/gas concentrations are normally prevented by an enclosure or mechanical ventilation that could fail.

b. Class II hazardous locations are defined by their combustible aerosol concentrations:

Division 1: where combustible aerosol concentrations may exist either normally, where a mechanical failure might cause such concentrations and an ignition source, or where the combustible dust is electrically conductive, e.g., metals or carbon fiber (used in composites).

Division 2: where combustible aerosol concentrations are not normally present but their deposition or accumulation may cause equipment to overheat or be ignited from an available source.

c. Class III hazardous locations are defined by ignitable fibers:

Division 1: where easily ignitable fibers are handled, manufactured, or used.

Division 2: where easily ignitable fibers are stored or handled except for manufacture.

These are also the classifications to which equipment is tested and certified as intrinsically safe per ANSI/UL 913.[21] Such equipment not only includes that installed within the location but also that which an industrial hygienist might carry, such as air

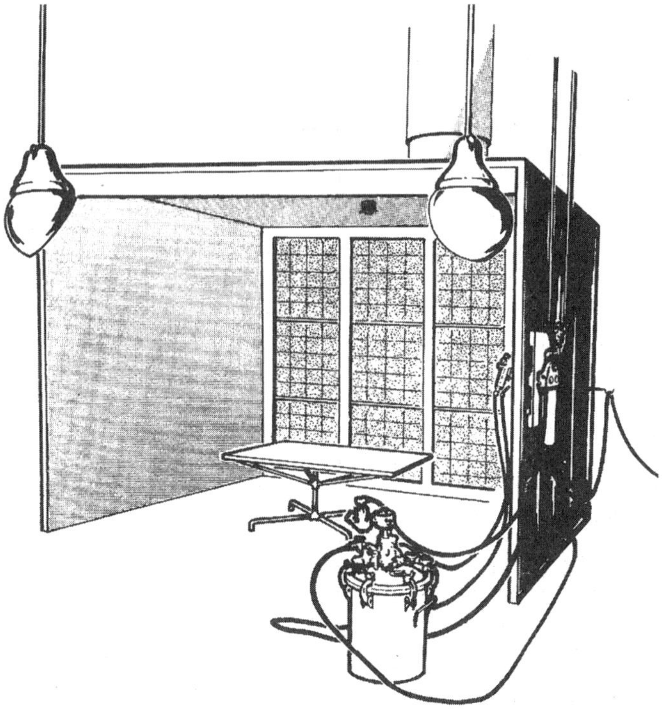

FIGURE 4.7 An example of a walk-in sized spray paint booth with dry-paint removing filters. *Source*: From Patty, F.A., Clayton, G.D. and Clayton, F.E., *Patty's Industrial Hygiene and Toxicology, Volume I, Part A*, Wiley, New York, reprinted with permission of Wiley.

sampling pumps, sound level meters, or gas and vapor detectors. Any piece of intrinsically safe equipment (including instruments and tools used by industrial hygienists) should list the class and category of hazards in which it is certified for use. Check out the label on equipment that you use or have access to.

VII. OPEN-SURFACE TANKS

The following paragraphs first present the background to the ANSI open-surface tank regulations that have been adopted by OSHA. The first two subsections explain the two components of the standard's dichotomous hazard classification scheme. The fourth subsection below explains how these classifications are used to determine ventilation velocities and flow rates (Q); however, such uses in hood design will not be presented in detail until Chapter 13. The final four subsections deal with special open-surface tank conditions.

Open-surface tanks are used for the immersion of materials into liquids to clean, to alter their surface, to add or impart a finish, or to change the immersed material's character. Such operations include washing, electroplating, anodizing, pickling, quenching, dyeing, dipping, tanning, dressing, bleaching, degreasing, alkaline cleaning, stripping, rinsing, and digesting. Prior to 1999, OSHA explicitly included the actual ANSI Z9.1-1971 consensus standard for "Practices for ventilation and operation of open-surface tanks" within 29 CFR 1910.94(d). However, the updated ANSI Z9.1-1991 consensus standard is incorporated only by reference within Section 1910.6 and Section 1910.124.[22] Examples and a short discussion of open-surface tanks also appear in the ACGIH *Manual*, Section 10.70.

ANSI, OSHA, and the *Manual of Industrial Ventilation* all use a dichotomous risk classification scheme to categorize open-surface tank operations. Dichotomous in general means in two parts and in this case risks that are classified by two factors. The two factors used to classify the risk of each tank are its *hazard potential* (designated by a letter from A to D, inclusive) and its *rate of hazard evolution* or *rate of hazard generation* (designated by a number from 1 to 4, inclusive). Thus, any open-surface tank operation can be classified into one of 16 risk classes, numbered A-1 to D-4, inclusive. For example, a class B-3 tank means that its *hazard potential* is B and its *rate of hazard generation* is 3.

1. The chemical's *hazard potential* is categorized as shown in Table 4.5 on a scale of A to D based on either its PEL or TLV (when the chemical has a defined health-based exposure limit) or on its flash point (if the chemical is flammable or combustible). *Hazard potential* is a characteristic of the chemical contained within the tank.
   The following procedures should be used to evaluate the *hazard potential*:
   a. The toxic hazard is determined from the concentration, expressed either in ppm or in mg/m^3, below which ill-effects are unlikely to occur to the exposed employee. The concentrations shall be legal limits where they exist (e.g., PELs set by OSHA) or other recommendations (e.g., TLVs set by ACGIH).
   b. The relative fire or explosion hazard is determined by the closed-cup flash point of the substance in the tank measured in degrees Fahrenheit (or degrees Celsius). Detailed information on the prevention of fire hazards in dip tanks may be found in the Standard for Dipping and Coating Processes Using Flammable or Combustible Liquids.[23] See www.nfpa.org/Codes/
   c. When the exposure limit may be either in ppm or in mg/m^3 or the chemical has both an exposure limit and a flash point, the index classification indicating the greater hazard shall be used (in other words, A takes precedence over B or C; B over C; C over D).
   d. Where the tank contains a mixture of liquids, the ANSI standard says to use the most toxic component (for example, the one having the lowest ppm or mg/m^3 exposure limit) except where that substance constitutes an insignificantly small fraction of the mixture. This approach has the effect of making a health-conservative assumption that the most toxic component comprises the entire mixture. An accurate estimate for the equivalent exposure limit to a mist emanating from a nonvolatile mixture ($EL_{mixture}$) would be based on the additive effects of the liquid component's molar fractions (X_i) and their respective exposure limits (EL_i) as shown in Equation 4.8.

For mists from a nonvolatile mixture:
$$\frac{1}{EL_{mixture}} = \frac{X_1}{EL_1} + \frac{X_2}{EL_2} + \cdots + \frac{X_n}{EL_n} \qquad (4.8)$$

Since vapors do not typically emanate from a volatile mixture in proportion to their liquid molar fractions (Y_i), Equation 4.8 would rarely apply to a volatile mixture (only if one is willing to assume that each component is equally volatile). A more accurate equivalent

TABLE 4.5
Determination of the Hazard Potential from an Open-Surface Tank

	Toxicity Group		Flash Point	
Hazard Potential	**Gas or Vapor (ppm)**	**Mist mg/m^3**	**(°F)**	**(°C)**
A	0–10	0–0.1	—	—
B	11–100	0.11–1.0	Under 100	Under 38
C	101–500	1.1–10	100–200	38–93
D	Over 500	Over 10	Over 200	Over 93

exposure limit to a vapor emanating from a volatile mixture would result from using Equation 4.9; however, Chapter 6 will discuss the more involved calculations needed to find each component's Relative Y_i that would actually be used in Equation 4.9. The following example demonstrates these different approaches.

$$\text{For vapors from a mixture}: \frac{1}{\text{EL}_{\text{mixture}}} = \frac{Y_1}{\text{EL}_1} + \frac{Y_2}{\text{EL}_2} + \cdots + \frac{Y_n}{\text{EL}_n} \qquad (4.9)$$

Example 4.1

What is the equivalent exposure limit for a mixture containing equal molar portions of ethyl chloride and ethylene dichloride? Since the equal portions are in molar terms, we know that each $X_i = 0.5$. The following data will eventually be needed:

$$\text{For } i = 1: \text{ ethyl chloride } P_{\text{vapor}} = 227 \text{ mmHg and TLV} = 100 \text{ ppm}$$

$$\text{For } i = 2: \text{ ethylene dichloride } P_{\text{vapor}} = 78.9 \text{ mmHg and TLV} = 10 \text{ ppm}$$

If one assumes that the entire mixture is the most toxic component, then the assumed exposure limit would be the 10 ppm of ethylene dichloride. If one uses Equation 4.8 as shown below, then the equivalent exposure limit would be 20 ppm.

$$\frac{1}{\text{EL}_{\text{mixture}}} = \frac{0.5}{10} + \frac{0.5}{100} = 0.0505 \Rightarrow \text{EL}_{\text{mix}} = 20 \text{ ppm} \qquad \text{(using Equation 4.8)}$$

If one were concerned enough about the greater volatility of ethyl chloride owing to its higher vapor pressure to read ahead in Chapter 6, one could use Equation 6.19 to find that 74% of the vapors would be ethyl chloride and 26% of the vapors would be ethylene dichloride; then, one could use Equation 4.9 to find that the most accurate equivalent exposure limit for the vapors from this mixture would be 30 ppm:

$$\text{Relative } Y_1 = \frac{X_1 \times P_{\text{vapor of component 1}}}{\Sigma X_i \times P_{\text{vapor of } i}} = \frac{0.5 \times 227}{(0.5 \times 227) + (0.5 \times 78.9)} = 0.74 \qquad \text{(using Equation 6.19)}$$

$$\frac{1}{\text{EL}_{\text{mixture}}} = \frac{0.26}{10} + \frac{0.74}{100} = 0.0334 \Rightarrow \text{EL}_{\text{mix}} = 30 \text{ ppm} \qquad \text{(using Equation 4.9)}$$

Since toxicity and vapor pressure are not correlated with each other, the trend of the three limits in this example to increase incrementally would not necessarily apply to another mixture.

2. The contaminant's *rate of evolution* is categorized as shown in Table 4.6 on a scale of 1 to 4 that characterizes the conditions in which the chemical is being used. As with the *hazard potential*, when different conditions yield different rates, the lowest numerical value shall be used.
   a. The temperature of the liquid in the tank in degrees Fahrenheit or Celsius. The liquid's temperature is a surrogate for the plume's buoyancy and how rapidly it will rise from the source (see earlier discussion of plume density in Section VI in Chapter 2).
   b. The number of degrees that the liquid temperature is below its boiling point. A liquid's degrees below boiling point is a surrogate for vapor pressure (see the earlier discussion of volatility in Section II in Chapter 2 and in footnote a to Table 4.6).
   c. The relative evaporation of the liquid is the time required for a small volume of the liquid in quiescent air at room temperature to evaporate completely. However, the testing procedure is not standardized nor is the reported data consistent (see footnote b to Table 4.6). Although drying time could be advantageous for mixtures if a standard

TABLE 4.6
Determination of Rate of Gas, Vapor, or Mist Evolution

Rate of Evolution	Liquid Temperature °F	Liquid Temperature °C	Degrees below b.p.[a] °F	Degrees below b.p.[a] °C	Relative Evaporation[b]	Gassing[c]
1	Over 200	Over 93	0–20	0 – 11	Fast	High
2	150–200	66–93	21–50	12–28	Medium	Medium
3	94–149	34–65	51–100	29–56	Slow	Low
4	Under 94	Under 34	Over 100	Over 56	Nil	Nil

[a] The relationship between the degrees that a solvent's boiling point is below 25°C and vapor pressure depicted in Figure 4.8 is similar to that between boiling point and vapor pressure depicted in Figure 2.2. The use of degrees below boiling in Table 4.6 allows for some variation in the operating temperature of a solvent; as a solvent's temperature is increased, the spread between its boiling temperature and operating temperature would decrease linearly and its vapor pressure would increase exponentially (according to an Antoine-type of equation, Equation 2.10), pushing its position in Figure 4.8 up and to the left. The rationale for the categories of degrees below boiling in Table 4.6 are unknown, but one can see in Figure 4.8 that they span only a small range of the over 400 solvents plotted in Figure 2.2. The scale in Figure 4.8 is expanded in Figure 4.9 which also shows the corresponding vapor pressure boundaries.

[b] Relative evaporation rate is the time for 100% of a test solution (for example, 2 g) to evaporate according to methods reportedly described by A.K. Doolittle.[24] However (to quote that source), "the methods for determining evaporate rates are, unfortunately, not subject to so nice a control as are the methods of determining boiling ranges. The results are rarely reproducible on the same solvent on different days, but the importance of knowledge of the relative evaporation rates of different solvents is sufficient to offset the disadvantage of lesser accuracy." Indeed, Doolittle's 1935 data differ from his 1954 data,[25] and the "solvent drying times" in Appendix B of the *Manual of Industrial Ventilation* differ from either of Doolittle's.[8] And to top that off, the categories of evolution rate differ between Z9.1-1991 and the *Manual of Industrial Ventilation* as shown below. While evaporation rate is a potentially simple index (and one especially advantageous to mixtures), its application to risk assessment is fraught with uncertainties.

	Fast	Medium	Slow	Nil
Categories per ANSI 1991.	0-3 hours	3-12 hours	12-50 hours	> 50 hours
Categories per the *Manual of Industrial Ventilation*	< 5 hours	5-15 hours	15-75 hours	> 75 hours

[c] Gassing means the formation by either chemical or electrochemical action of bubbles of gas under the surface of the liquid in the tank (not to be confused with boiling). Gassing is particularly common in aqueous solutions where electrochemical reactions result in the formation of hydrogen or oxygen bubbles. The hazard of concern is not the gas *per se* but rather the mist generated when the gas bubbles break the surface. While the rate of gassing can be evaluated for an individual installation, both the ANSI standard and the *Manual of Industrial Ventilation* contain several tables that identify airborne contaminants and dichotomous classifications for many specific open-surface tank operations, from which the following examples were extracted:

High: Bright dip of brass or bronze with high temperature nitric acid. Pickling of steel with sulfuric acid. Chrome plating. Anodizing (a metal oxide) or satin finishing of aluminum.

Medium: Bright dip of brass or bronze with low temperature nitric acid. Pickling of steel with hydrochloric acid. Alkaline cleaning of aluminum (cold). Striking (using a high salt, low metal bath).

Low: Alkaline cyanide plating of zinc. Phosphoric acid dipping of steel. Tin plating from a stannate solution. Stripping galvanizing in sodium hydroxide.

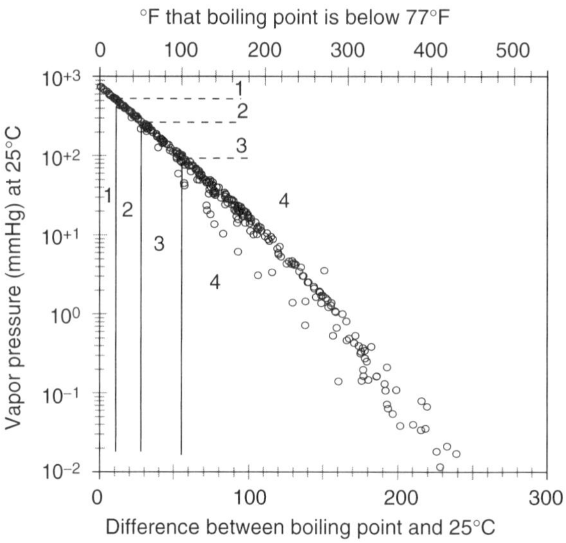

FIGURE 4.8 The relationship between the degrees below boiling and vapor pressure for the chemicals shown in Figure 2.2.

method were (or in fact could be) described, it is currently a poor substitute for vapor pressure and is not likely to be improved.

d. The extent of "gassing" is based on the quantity of bubbles that are observed (or anticipated) to agitate the liquid surface enough to produce a mist (see footnote c to Table 4.6).

3. ANSI Z9.1 goes on to describe ventilation requirements for open-surface tanks when control relies only on ventilation (cf. control by other means in the next subsection). The next version of this ANSI standard will offer the user two options by which to specify the exhaust ventilation conditions for an open-surface tank: the control velocity method or

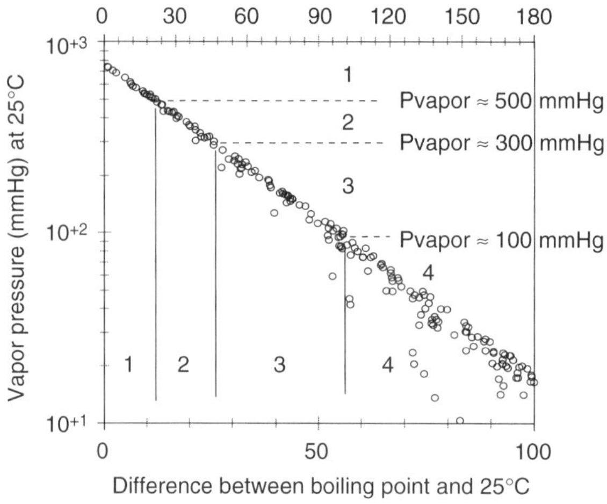

FIGURE 4.9 Narrowed view of Figure 4.8 showing the vapor pressures corresponding to the degrees below boiling categories in Table 4.6.

the Hemeon method. These methods are only mentioned here; a full discussion of them will be deferred until Chapter 13 (after the reader has been introduced to the concept of control velocity and its role and use in the design of hoods). The former is the traditional ANSI method of specifying a control velocity based solely on the dichotomous risk classification from Table 4.5 and Table 4.6 and hood designs to be covered in Section I in Chapter 10. The latter method employs the basic relationship $Q = V \times A$ (to be explained in Chapter 11) where Q is the exhaust volumetric flow rate (ft^3/min or m^3/min), V (fpm or m/sec)] is what Hemeon calls the "critical capture velocity," and A (ft^2 or m^2) is the contoured surface area comprising the outer edge of a hypothetical control zone through which that air must flow en route to the local exhaust hood. These requirements will make more sense by the time you get to Chapter 13.

4. Physical control by means other than ventilation (or as an adjunct to ventilation) can also effectively reduce the concentrations of hazardous materials in the vicinity of the employee. For instance, tank covers can be used to confine the gases, mists, or vapors to the area under the cover even if that space is not exhaust ventilated (some ventilation would help even more). Foams, beads, chips, or other materials that can be made to float on the tank surface will both reduce the liquid surface area available for evaporation and block the formation of a mist from air or gas bubbles breaking on the liquid surface (discussions pertaining to Equation 5.3 will explain the significance of wetted surface area to evaporation rate). Surfactants can also be added to some tank liquids to minimize mist formation by reducing the liquid's surface tension (Section III in Chapter 8 will elaborate on the structure and function of surfactants).
5. A vapor phase degreaser (or often just called a vapor degreaser as depicted in Figure 4.10) is a special class of open-surface tank containing a volatile organic solvent (usually a chlorinated solvent) that normally does not need local exhaust ventilation. This device heats the solvent at the bottom of the tank to create a high concentration of warm vapors. These high density vapors (as discussed in Section VII of Chapter 2) will displace much

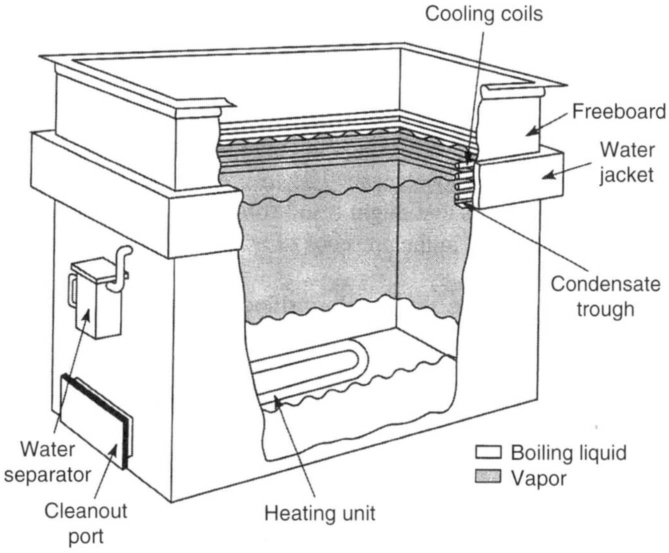

FIGURE 4.10 Schematic view of a vapor degreaser and its key components. Source: From Patty, F.A., Clayton, G.D. and Clayton, F.E., *Patty's Industrial Hygiene and Toxicology, Volume I, Part A*, Wiley, New York, reprinted with permission of Wiley.

of the air as they rise within the tall walls of the degreaser. When these vapors rise to the level of the cooling coils near the open top of a vapor degreaser, they condense into a cold, dense mist that cannot easily escape over the tank's raised freeboard sides. Parts to be cleaned are lowered into this condensation zone where the vapor will also condense as clean solvent onto the parts' comparatively cool surfaces. The accumulated condensate (now contaminated or dirty with oils or grease) will drip back into the bottom of the tank to be recycled as a clean vapor. Variations on this theme have been developed to deal with insoluble or more adherent contaminants, such as adding a hand spray wand to the system, using multiple condensation compartments, or preceding vapor degreasing with an ultrasonic cleaning module to loosen oils and other debris.

Although a top (lid) can be added to a vapor degreaser (sliding is preferred to avoid the updrafts created by a hinged lid), most vapor degreasers are operated as an open-surface tank to facilitate lowering and extracting the parts being cleaned. ANSI Z9.1-1991 addressed several operational issues regarding vapor degreasing tanks that may not be included in the next revision to this standard.

a. For instance, this consensus standard specifies that when chlorinated or fluorinated hydrocarbon solvents (for example, trichloroethylene or Freon) are used, that special precautions are taken to prevent solvent fumes from entering the combustion air of this or any other heater. The combustion of such solvents would form hydrochloric and hydrofluoric acid, respectively.
b. Another specification is that solvent cleaning or vapor degreasing tanks with more than 4 ft^2 of vapor area shall be equipped with well-sealed cleanout or sludge doors located near the bottom of each tank. Periodic cleaning of the heater is necessary for the proper operation of its thermostatic control. The vapors from residual solvent within a deep tank can be acutely hazardous.
c. Procedures to assign the *evolution* of vapors from a vapor degreaser were addressed in a footnote to the equivalent of Table 4.6 to the 1991 ANSI standard. The rate of vapor evolution from the tank into the workroom is not dependent upon the factors listed in Table 4.6, but rather upon a tank's operating procedures such as carryout of vapors from excessively fast action, dragout of liquid by entrainment in parts, contamination of solvent by water and other materials, or improper heat balance. When the operating procedure is excellent, the effective rate of evolution may be taken as 4. When the operating procedure is average, the effective rate of evolution may be taken as 3. When the operation is poor, a rate of 2 or 1 is indicated, depending upon observed conditions. When such equipment is operated properly, emissions are slow to nil, and a vapor degreaser can be a very efficient and safe parts cleaner.

Other discussions on vapor degreasers may be found in OSHA 1919.126 d,[4] the *Manual of Industrial Ventilation* VS-70-20 and VS-70-21,[8] and *Patty's Volume I Part A*.[19]

PRACTICE PROBLEMS

1. What is the LFL and flash point of the following chemicals? You can indicate Fl. Pt. in either °C or °F but not necessarily both. Since various sources of flammability data either differ slightly or do not have all of these chemicals, use the NIOSH Pocket Guide to answer this question, e.g., at www.cdc.gov/niosh/npg/npg.html. Indicate whether each of these solvents is either flammable (F), combustible (C), or not combustible (NC) in the space provided.

	LFL	Flash Point	Flam.?	Formula or/and Other Chemical Name
a. Nitrobenzene	_____%	_____°C or _____°F	_____	$C_6H_5NO_2$
b. Turpentine	_____%	_____°C or _____°F	_____	$C_{10}H_{16}$ (approximately)
c. MIBK	_____%	_____°C or _____°F	_____	$C_6H_{12}O$, methyl isobutyl ketone, hexone, or 4-methyl-2-pentanone
d. Octane	_____%	_____°C or _____°F	_____	C_8H_{10}
e. Toluene	_____%	_____°C or _____°F	_____	C_7H_8, methyl benzene
f. MEK	_____%	_____°C or _____°F	_____	C_4H_8O, methyl ethyl ketone, 2-butanone
g. Acetone	_____%	_____°C or _____°F	_____	C_3H_6O 2-propanone
h. Hexane	_____%	_____°C or _____°F	_____	C_6H_{14} n-hexane
i. "Perc"	_____%	_____°C or _____°F	_____	Perchloroethylene or tetrachloroethylene

2. a. What is the vapor pressure of octane at 25°C from Appendix A in this book?
_____ mmHg
b. Notice that its flash point is 12°C below 25°C or 21°F below 77°F. This is *really close* to the rule-of-thumb for a doubling of a solvent's vapor pressure. Estimate the P_{vapor} at octane's flash point based on its P_{vapor} at 25°C, the above ΔT, and the rule-of-thumb:
_____ mmHg
c. Now express this estimated P_{vapor} at octane's flash point as a molar fraction then as a percent molar fraction:

$$Y_{octane} \frac{P_{vapor} \text{ at the flash point}}{\text{normal atmosphere}} = \_\_\%$$

The P_{vapor} of octane at 13°C using the Antoine equation from Practice problem set 1 is 6.8 mmHg. If you estimated correctly using the right rule-of-thumb, your predicted percent of octane in the air at its flash point should be very close to its 1.0% LFL. This should always be true because the flash point is the temperature at which the vapor is ignitable.

3. a. Acetone or any other vapor would displace a predictable amount of oxygen and nitrogen. If acetone were present at its LFL of 2.5%, what would be the molar fraction of oxygen (Y_{O_2}) in air at NTP?
ans. = 20.4%
b. What is the partial pressure of oxygen (P_{O_2}) in this situation?
ans. = 155 mmHg
c. Would this represent an oxygen-deficient condition as defined by either OSHA or the TLV for oxygen? Recall that the TLV uses the partial pressure of oxygen as its limit.
4. The following data are provided for methylene chloride (CH_2Cl_2 CAS 75-09-2) used in a dip cleaning process that does not release gases or form bubbles within the liquid

NIOSH REL:	Possible carcinogen	MW:		84.9	
TLV:	50 ppm (A3)	P_{vapor}:		435 mmHg at 25°C	
OSHA 1910.1052		BP:	104°F	FRZ:	−139°F
TWA PEL:	25 ppm	LEL:	Nonflam.	UEL:	Nonflam.
STEL:	STEL: 125 ppm for 5 min in any 2 h	Fl.Pt.:		Not applicable	
		Sp. gr:		1.33	

a. If this chemical were to be used in an open-surface tank at normal room temperature, what would be its hazard potential in accordance with OSHA and ANSI Z9.1-1991 (as covered in Table 4.5 of the book)? Fill in each potentially assigned letter; then circle the greatest hazard potential.

_____based on toxicity (health, *use the lowest exposure limit*)

_____ based on flash point (safety)

b. As above, what would be its rate of evolution category (using Table 4.6)? Fill in each potentially assigned number; then circle the highest rate of evolution (lowest number).

_____based on liquid temperature

_____based on degrees below boiling point

_____based on gassing (Reread footnote c to Table 4.6, then reread the question.)

These (or similar) dichotomous ratings will be used much later in Chapter 13 as one way to specify a ventilation control velocity.

APPENDIX TO CHAPTER 4

TABLE 4A.1

The First Part of This Table Lists TLV for Selected Gases and Vapors Expressed as ppm and Converted into mg/m^3 and the Clinical Effects upon which They Were Primarily Based. Vapor Pressure Is Given at 25°C unless Indicated Otherwise. The Second Part of This Table Lists the TLV for Selected Aerosols Expressed Only as mg/m^3 and the Clinical Effects upon which They Were Primarily Based.

		TLV					
Gases and Vapors	**CAS**	**ppm**	**Note**	**Toxicity**	**mg/m^3**	**P_{vapor} mmHg**	**T °C**
Acetic acid	64-19-7	10		Irritation	**24.5**	15.7	
Acetonitrile	75-05-8	40		A4, lung, anox	**67.2**	88.8	
Acetone	67-641	500		A4, irrat.	**1187**	230	
Acrolein	107-02-8	0.1		Irritation	**0.2**	274	
Acrylonitrile	107-13-1	2	Skin	A2, cancer	**4.3**	109	
Ammonia	7664-417	25		Irritation	**17.4**	760	
Benzene	71-43-2	0.5	Skin	A1, cancer	**1.60**	95	
1,3-Butadiene	106-99-0	2		A2, cancer	**4.42**		
Butane	106-97-8	800		Narcosis	**1902**	760	
Carbon tetrachloride	56-235	5	Skin	A2, liver	**31.5**	113	
Carbon disulfide	75-150	10	Skin	CVS,CNS	**31.1**	361	
Carbon monoxide	630-080	25		Anoxia	**28.6**	760	
Chlorine	7782-50-5	0.5		A4, irrit.	**1.5**	760	
Chlorobenzene	108-90-7	10		A3, liver	**46.0**	12	
Chloroform	67-66-3	10		A3, CVS, +	**48.8**	197	
Chlorostyrene	2039-87-4	50		Liver, kidney	**283**	0.96	
Cyclohexane	110-82-71	100		Irritation	**344**	96.9	
Cyclohexylamine	108-91-8	10		A4, irrit.	**40.6**	10.1	
1,4-Dichloro-2-butene	764-41-0	.005	Skin	A2, cancer	**0.026**		
Dichloroethane	75-34-3	100		Liver, kidney	**405**	227	
1,2-Dichloroethane (ethylene dichloride)	107-06-2	10		A4, liver	**40.1**	78.9	
1,2-Dichloroethylene	156-60-5	200		Liver	**793**	331	
Diethyl ketone	96-22-0	200		Irritation	**705**	37.7	
Dimethylformamide	68-12-2	10	Skin	A4, liver	**29.9**		

Continued

TABLE 4A.1 (Continued)

Gases and Vapors	CAS	TLV ppm	Note	Toxicity	mg/m^3	P_{vapor} mmHg	T °C
1,4-Dioxane	123-91-1	20	Skin	A3, irrit.	**72.1**	38.1	
Ethanol	64-17-5	1000		A4, irrit.	**1884**	59	
2-Ethoxyethanol (cellosolve)	110-80-5	5	Skin	Reproduct.	**18.4**	5.3	
Ethylene dichloride	107-21-1	10		A4,liver,nar.	**40.5**		
n-heptane	142-82-5	400		Irritation	**1639**	46	
n-hexane	110-54-3	50		Neuropathy	**176**	151	
1-hexane	592-41-6	30		CNS, irrit.	**103**	186	
Hydrogen sulfide	7783-064	5		Irrit., death	**7.0**	760	
Isopropyl alcohol (IPA)	67-670	400		Irritation	**983**	33	20
Methanol	67-56-1	200	Skin	Neuro., vis.	**262**	127	
Methyl acetate	79-20-9	200	ps	A4, irrit.	**638**	760	
Methyl ethyl ketone (MEK)	78-93-3	200		Irritation	**590**	19.9	
Methyl isobutyl ketone (MIBK)	108-10-1	50		Irritation	**205**	19.9	
Methylcyclohexane	108-87-2	400		Narcosis	**1606**	46	
Methylene chloride (dichloromethane)	75-09-2	50		A3, CNS	**174**	435	
Nitrogen dioxide	10102-440	3		A4, irrit.	**5.6**	760	
p-Nitrotoluene	99-99-0	2	Skin	Anoxia	**11.2**	0.164	
Nonane	111-84-2	200		CNS, narco.	**1049**	4.45	
Octane	111-65-9	300		Irritation	**1401**	14.1	
Perchloroethylene (tetrachloroethylene)	127-18-4	25		A3, irrit.	**170**	18	
Propylene glycol monomethyl ether	107-98-2	100		Irritation	**369**	12.5	
Rubber solvent or naphtha	8030-30-6	400		Irritation	**1587**	4	20
Sulfur dioxide	7446-095	2		A4, irrit.	**5.2**	760	
Toluene	108-88-3	50	Skin	A4, CNS	**188**	28.4	
1,1,2-Trichloroethane	79-00-5	10	Skin	CNS	**54.6**	23	
1,1,1-Trichloroethane (methyl chloroform)	71-55-6	350		CNS	**1910**	124	
Trimethyl benzene (mixed)	25551-13-7	25		Irritation	**123**	2.1	
Vinyl chloride	75-01-4	1	A1	CNS, cancer	**2.6**	>760	
				Geometric mean	77.0		
$n = 52$				Geom. Std. dev.	12.1		

Aerosols	CAS	Note	TLV Toxicity	mg/m^3
Aluminum dust	7429-90-5		Irritation	10
Aluminum fume	7429-90-5		B2, irritation	5
Barium	7440-39-3		Irritation	0.5
Beryllium	7440-41-7		A1, lung cancer	0.002
Boron oxide	1303-86-2		Irritation	10
Cadmium	7440-43-9		A2, cancer	0.01
Calcium carbonate	1317-65-3		Irritation	10
Calcium oxide	1305-78-8		Irritation	2
Cellulose	9004-34-6		Irritation	10
Chromite ore			A1, lung cancer	0.05
Chromium metal and Cr III	7440-47-3		Irritation	0.5

Continued

TABLE 4A.1 (Continued)

			TLV	
Aerosols	**CAS**	**Note**	**Toxicity**	**mg/m^3**
Water soluble Cr VI	7440-47-3		Cancer	0.05
Water insoluble Cr VI	7440-47-3		Cancer	0.01
Coal dust, anthracite			Pulmonary fibrosis	0.4
Cobalt elemental and inorg.	7440-48-4		Asthma	0.02
Copper fume	7440-50-8		Irritation	0.2
Copper dust	7440-50-8		Irritation	1
Cotton dust			Byssinosis	0.2
Grain dust			Irritation	4
Graphite	7782-42-5		Pneumoconiosis	2
Hafnium	7440-58-6		Liver, irritation	0.5
Indium	7440-74-6		Pulmon. edema	0.1
Iron oxide dust	1309-37-1		Pneumoconiosis	5
Iron salts, soluble			Irritation	1
Kaolin	1332-58-7		Pneumoconiosis	2
Lead elemental and inorg.	7439-92-1		CNS	0.05
Magnesium oxide fume	1309-48-4		Irritation	10
Manganese elemental and inorg	7439-96-5		CNS	0.2
Mica	12001-26-2		Pneumoconiosis	3
Molybdenum metal and insol	7439-98-7		Irritation	10
Nickel elemental	7440-02-0		A5, derm., pneumo.	1.5
Nickel soluble	7440-02-0		A4, CNS, irritation	0.1
Nickel insoluble	7440-02-0		A1, lung cancer	0.2
Paraffin wax fume	8002-74-2		Irritation	2
Persulfates (NH_4, K, Na)			Irritation	0.1
Platinum metal	7440-06-4		Irritation	1.0
Platinum soluble	7440-06-4		Asthma	0.002
Portland cement	65997-15-1		Irritation	10
Rhodium metal and insol.	7440-16-6		Irritation	1
Silica amorphous inhalable	61790-53-2		Irritation	10
Silica crystalline quartz	14808-60-7		Pulmonary fibrosis	0.1
Silver metal	7440-22-4		Argyria	0.1
Soapstone total			Pneumoconiosis	6
Sodium bisulfite	7631-90-5		Irritation	5
Talc with no fibers	14807-96-6		Lung	2
Thallium elemental and insol.	7440-28-0		Irritation	0.1
Tin metal and inorg.	7440-31-5		Stannosis	2
Tungsten metal and insol.	7440-33-7		Irritation	5
Tungsten soluble	7440-33-7		CNS, irritation	1
Yttrium metal and comp.	7440-65-5		Fibrosis	1
Zinc oxide fume	1314-13-2		Lung, metal f. f.	5
Zinc oxide dust	1314-13-2		Lung	10
			Geometric mean	0.68
$n = 52$			Geom. std. dev.	9.9

A1, Confirmed human carcinogen based on epidemiologic observation or study; A2, Suspect human carcinogen based on relevant animal data but insufficient evidence to confirm the chemical as a human carcinogen; A3, Animal carcinogen based on positive animal data but is either not confirmed by epidemiological evidence or where the animal data is not relevant to normal levels of human exposure; A4, Not classifiable as a human carcinogen, inadequate data exists on either humans and/or animals to classify it one way or the other; A5, Not suspected as a human carcinogen, based on valid epidemiologic studies which indicate insignificant human risk of cancer.

REFERENCES

1. Public Law 91-596 (the Occupational Safety and Health Act of 1970), 84 STAT. 1590, 91st Congress, S.2193, December 29, 1970, as amended through January 1, 2004.
2. *Threshold Limit Values for Chemical Substances and Physical Agents*, ACGIH Worldwide, Cincinnati, OH, 2005.
3. Roach, S.A and Rappaport, S.M., A critical analysis of threshold limit values, *Am. J. Ind. Med,* 17, 727–753. 1990.
4. The U.S. Code of Federal Regulations, Part Number 1910, Occupational Safety and Health Standards, Chapter. 29.
5. NIOSH, *Pocket Guide to Chemical Hazards*, 3rd printing, *DHHS (NIOSH) 97–140*, Also on-line at www.cdc.gov/niosh/npg/npg.html, 2004.
6. AIHA, *The AIHA 2005 Emergency Response Planning Guidelines and Workplace Environmental Exposure Level Handbook*, American Industrial Hygiene Association, Fairfax, Virginia, 2005.
7. Leidel, N. A., Busch, K. A., and Lynch, J. R., Occupational Exposure Sampling Strategy Manual, *DHHS* (NIOSH) *Publication*, 77–173, 1977.
8. ACGIH, *Industrial Ventilation: A Manual of Recommended Practice*, 24th ed., American Conference of Governmental Industrial Hygienists, Cincinnati, OH, 2001.
9. Perkins, J., *Modern Industrial Hygiene*, Van Nostrand Reinhold, New York, 1997.
10. Lide, D. R. Ed., In *CRC Handbook of Chemistry and Physics*, 83rd ed., CRC Press, Boca Raton, FL, 2002.
11. *NFPA: Fire Protection Handbook*, 2003 ed., National Fire Protection Association, Quincy, MA, 2003.
12. McFarland, R. A. and Evans, J. N., Alterations in dark adaptations under reduced oxygen tensions, *Am. J. Physiol,* 127, 37–50, 1939.
13. Henderson, Y. and Haggard, H. W., *Noxious Gases*, Van Nostrand Reinhold, New York, 1943.
14. Sheffield, P. J. and Heimbach, R. D., Respiratory physiology, In *Fundamentals of Aerospace Medicine*, DeHart, R. L. Ed., Lea and Febiger, Philadelphia, PA, pp. 72–109, 1985.
15. Knight, D. R., Schlichting, C. L., Fulco, C. S., and Cymerman, A., Mental performance during submaximal exercise in 13 and 17% oxygen, *Undersea Biomed. Res,* 17(3), 223–230, 1990.
16. Popendorf, W., Barometric hazards, In *The Occupational Environment: Its Evaluation and Control*, DiNardi, S. R. Ed., 2nd ed., American Industrial Hygiene Association, Fairfax, VA, pp. 647–667, chap. 25, 2002.
17. McArdle, W. D., Katch, F. I., and Katch, V. L., *Exercise Physiology: Energy, Nutrition, and Human Performance*, Lea and Febiger, Philadelphia, 1991.
18. Palmer, K. N., *Dust Explosions and Fires*, Chapman and Hall, London, 1973.
19. Patty, F. A., In *Patty's Industrial Hygiene and Toxicology Volume I, Part A*, Clayton, G. D., Clayton, F. E. Eds., 4th ed, Wiley, New York, 1991.
20. National Electrical Code Softbound Edition (NFPA 70), National Fire Protection Association, Quincy, MA, 2005.
21. ANSI/UL 913, *Intrinsically Safe Apparatus and Associated Apparatus for Use in Class I, II, and III, Division 1, Hazardous (Classified) Locations*, Underwriters Laboratories, Northbrook, IL, 2002.
22. AIHA/ANSI Z9.1, 1991 *Practices for Ventilation and Operation of Open-Surface Tanks*, American Industrial Hygiene Association, Fairfax, VA.
23. NFPA 34, *Standard for Dipping and Coating Processes Using Flammable or Combustible Liquids*, National Fire Protection Association, Quincy, MA, 2003.
24. Doolittle, A. K., Lacquer and Solvents in Commercial Use, *Industrial and Engineering Chemistry* 27(10), 1169–1179, 1935.
25. Doolittle, A. K., *The Technology of Solvents and Plasticizers*, Wiley, New York, 1954.

5 Vapor Generation and Behavior

The learning goals for this chapter:

- ☐ Understand the basic evaporation equation ($G = \text{Geom.Coef.} \times A \times V^{0.5} \times P_{vapor}$).
- ✍ Understand why molecular diffusion is not an important mechanism for dispersing a plume or distributing contaminants in most work places.
- ✍ Be able to categorize a given setting into one of the four universal airborne exposure scenarios.
- ✍ Be able to differentiate between a complete evaporation and an incomplete evaporation problem or setting.
- ☐ Know why the incomplete evaporation concentration is P_{vapor}.
- ☐ Know how to use the concept of C = [mass]/[air volume] to solve complete evaporation problems.
- ✍ Understand the relationship between the vapor hazard ratio (VHR) and the breathing zone dilution ratio (Dilution Ratio_{BZ}) and their application to industrial hygiene recognition, evaluation, and control.

I. MECHANISMS OF VAPOR GENERATION

While vapor pressure is not the only variable that affects evaporation, it is the only variable that is intrinsic to the chemical. All of the other variables that affect the evaporation rate depend either upon the physical nature of the liquid source (such as size and shape) or upon the speed and turbulence of the passing air.

✍ 1. The liquid and its vapors are always considered to be in equilibrium with each other right at the liquid–air interface. The concentration of a vapor in equilibrium was defined in Chapter 2 as its vapor pressure (P_{vapor}). P_{vapor} was equated to C in mg/m^3 by Equation 2.8 and to ppm by Equation 2.9. The latter equation is rewritten (and "boxed") here as Equation 5.1 because of its practical importance.

$$\boxed{\text{ppm at the source} = \frac{P_{vapor} \times 10^6}{P} = \frac{(P_{vapor}\ \text{mmHg}) \times 10^6}{760}} \tag{5.1}$$

2. As the vapors are swept away from the liquid surface by the passing air, more molecules evaporate from the liquid to maintain that same, very localized equilibrium. Thus, conceptually, the rate of evaporation is determined by how fast vapor molecules can diffuse across a thin, low speed *boundary layer* of air that always exists very close to a liquid surface. The thickness of that boundary layer (typically a few millimeters along the vertical axis in Figure 5.1) is determined by the air velocity, its turbulence, and the geometry of the liquid source or its container. The vapor concentration (the horizontal axis in Figure 5.1) at the liquid surface at the bottom of this boundary layer is always equal to the chemical's vapor pressure (P_{vapor}), and the concentration on the "room" side

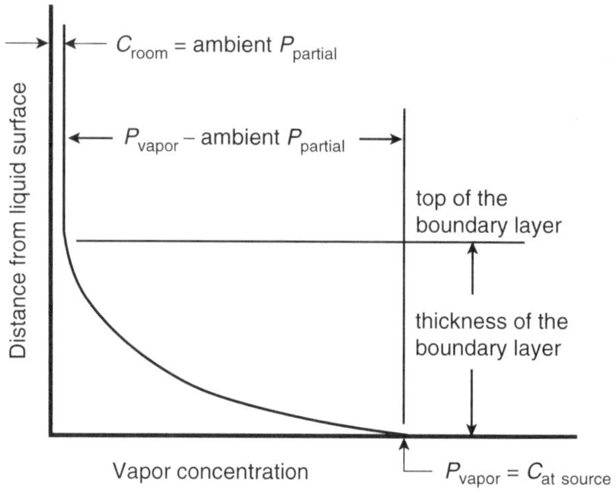

FIGURE 5.1 Depiction of evaporation as diffusion due to a concentration difference across a boundary layer.

of the boundary layer above the boundary layer is the ambient concentration (ambient $P_{partial}$ or C_{room}). These conditions set the parameters for what is modeled as molecular diffusion across that boundary layer. Thus, the evaporation rate is largely determined by a set of environmental conditions and one chemical dependent condition (the chemical's vapor pressure).

3. The chemical engineer's equation to predict evaporation rate is presented as Equation 5.2 primarily to indicate that a conceptual model exists and to point out vapor pressure's role in that model.[1–4] As depicted in Figure 5.1, P_{vapor} at the liquid side of the boundary layer is a major determinant of the evaporation rate. The thickness of the boundary layer depends upon the air's turbulence, which in turn depends upon its velocity [V], its viscosity [ν], the shape of the evaporating source (its width, length, and the shape of its container if it has one), and (in the real-world vs. the research laboratory) obstacles upstream of the source that may have already imparted turbulence into the air before it even reaches the source.

$$G_{moles} = \kappa_{mass\ tx}(\text{Width})(\text{Length})^{0.5}\frac{\mathcal{D}^{0.67}V^{0.6}}{\nu^{0.17}}(P_{vapor} - \text{ambient } P_{partial}) \quad (5.2)$$

Industrial hygiene research based upon Equation 5.2 has been reported in many references.[3–10] However, the explicit application of Equation 5.2 to predict G in real workplaces is hampered by the impact of undefined upstream conditions on the boundary layer and the value of $\kappa_{mass\ tx}$. The use of a simpler, approximate model for evaporation rate in the form of Equation 5.3 is adequate for IH field applications. Its importance is lightly highlighted here, but its slightly simpler variation appearing later as Equation 5.13 will be marked to be remembered! This basic evaporation equation will be encountered many times within this book.

$$\text{evaporation rate } (G_{moles}) = (\text{Geom.Coef.})(\text{Area})V^{0.5}(P_{vapor} - \text{ambient } P_{partial}) \quad (5.3)$$

where

G_{moles} = the evaporation rate in terms of moles of vapor generated per unit of time, e.g., mole or millimole per minute. Mass-based units such as mg/min could easily be created by multiplying this equation by MW (where G [g/min] = MW × G_{moles}). In fact, by the notation used herein, the unsubscripted G means contaminant generation rate in g/min and could apply to any source whether it is volatile or not. G_{moles} applies only to vapors and has virtually no meaning for aerosols; the same could be said for Equation 5.3 because it applies to evaporation. The value of G can be measured in a variety of ways (to be covered in Chapter 20) but is difficult to predict accurately because of the importance of but the imprecision in our ability to estimate values of $\kappa_{\text{mass tx}}$ or the geometric coefficient.

$\kappa_{\text{mass tx}}$ = what engineers would call an evaporative mass transfer coefficient. The $\kappa_{\text{mass tx}}$ is an empirical coefficient that characterizes the effect of the source geometry (its size and shape) on the thickness of the boundary layer right above the liquid (see Figure 5.1). One way to tell that this coefficient is empirical is to look at all of the fractional powers. The units of $\kappa_{\text{mass tx}}$ necessary to make G moles per minute would resemble nothing in the real-world. This $\kappa_{\text{mass tx}}$ was incorporated into the geometry coefficient in Equation 5.3.

ν = the kinematic viscosity of the air. The low 0.17 power of viscosity means that a change in ν has a proportionately smaller effect on G than say the source's width or P_{vapor}. As ν varies only slightly within the temperature range of humanly occupiable workplaces, it too is incorporated into the geometry coefficient in the simpler model.

$\mathcal{D}$ = molecular diffusivity of the contaminant molecule in air. Its effect on evaporation rate is also small both because $\mathcal{D}$ varies only slightly among volatile chemicals and because it too is raised to a power well less than one. As a result, $\mathcal{D}$ was also incorporated into the geometry coefficient in the simpler model.

Geom. Coef. = The geometry coefficient in Equation 5.3 is an empirical "catch all" term that depends primarily upon the geometry of the liquid source (like $\kappa_{\text{mass tx}}$) and the air's turbulence (not easily measured in the field). In addition, the Geom.Coef. also depends slightly upon $\nu^{0.17}$ (which is virtually constant for air across the range of workplace temperatures), $\mathcal{D}^{0.67}$ (whose values do not vary much across organic solvents in IH practice), and $L^{0.5}$ (the source length [and the time or distance the air takes to form a boundary layer over a liquid surface] has a small inverse effect on Geom.Coef. because $W \times L^{0.5}$ in Equation 5.2 is replaced by Area in Equation 5.3). The only geometry for which Geom.Coef. is defined quantitatively is for air flowing smoothly over a nearly circular spill on a flat horizontal surface such as in Equation 5.4. A quantitative value is not available for evaporation from a liquid inside a container. Thus, in practice, the effect of the source geometry can only be predicted qualitatively. For instance, increasing a liquid container's freeboard (the height of the container's walls above the liquid) will decrease the Geom.Coef. by an undetermined amount. Given the qualitative nature of the Geom.Coef. in most circumstances, Equation 5.3 is both conceptually simpler and thus clearer than Equation 5.2 and adequately accurate for semiquantitative IH applications.

Area = the size or surface area of the volatile liquid (ft^2 or m^2). The wetted surface area is usually easy to see and often measurable as the Width × Length of the volatile liquid. The more complicated equation uses $(\text{Width})(\text{Length})^{0.5}$ because the thickness of the boundary layer (perpendicular to the liquid surface) grows as a function of the length of the liquid surface across which the air is blowing or flowing. This effect is reduced in real-world settings because the air flow over the liquid source is usually turbulent.

V = the velocity of the air passing over the evaporating source (e.g., fpm or m/sec). The power of V characterizes the nonlinear way that the thickness of the boundary layer decreases as V increases. Various evaporation models may raise V to a power between 0.5 and 0.67. The power of V to 0.5 in Equation 5.3 is adequately accurate for a fixed geometry coefficient over the range of air velocities encountered in most workplaces, and the behavior of a square root is more intuitive than a 0.6 or 0.67 power.

P_{vapor} = the vapor pressure of the evaporating chemical at its liquid temperature. The units of P_{vapor} are normally mm Hg but can be converted to or from ppm using Equation 2.5 or Equation 2.9. P_{vapor} is the concentration of vapor at the bottom of the boundary layer. Heating a solvent will increase its P_{vapor} and produce more vapors. Similarly, cooling the solvent or changing it to a chemical with a lower P_{vapor} will reduce the evaporation rate and lower the resulting worker exposure.

$P_{partial}$ = the partial pressure (or concentration) of the same vapor in the ambient air passing over the source. The units of ambient $P_{partial}$ must be the same as the units of P_{vapor}. The chemical's partial pressure in the workplace air is usually in the range of its TLV® and/or PEL and very much less than its vapor pressure. Thus, the ambient $P_{partial}$ can usually be disregarded and omitted from this evaporation equation. However, in a poorly ventilated environment, the value of the ambient $P_{partial}$ can approach P_{vapor}. Equation 5.2 and Equation 5.3 both state that if the ambient $P_{partial}$ approaches P_{vapor}, the rate of evaporation will slow, and evaporation will cease when $P_{partial}$ equals P_{vapor} (a vapor in equilibrium with its liquid). Ambient $P_{partial}$ is sometimes called a back pressure because it tends to force molecules back into the liquid.[11]

4. While values for a liquid source's area (Area) and the air velocity over that source (V) can be measured and a value for vapor pressure (P_{vapor}) can be extracted from the literature (e.g., see the list of reference texts in Table 2.1), values for the empirical geometry coefficient are rare. One example of such an evaporation equation was developed for EPA by Caplan[12] for organic solvents evaporating from a smooth surface such as a spill of 0.5 to 3 ft in diameter. He showed that Equation 5.4a can predict the evaporation rate of a fairly wide range of organic chemicals (except for the low vapor pressure alcohols 1-hexanol to 2-octanol) from a smooth surface such as a spill of 0.5 to 3 ft in diameter with an accuracy of a factor of about $\pm 2\times$.

$$G(\text{lb/h}) = 2.37 \times 10^{-4}\text{MW} \times A \times V^{0.625} \times P_{vapor} \tag{5.4a}$$

where

G = the vapor generation rate, lb/h.
MW = the molecular weight, g/mole (added to convert moles/min to g/min).
A = the surface area of the liquid source, ft^2.
V = the air velocity over the liquid surface, fpm.
P_{vapor} = the vapor pressure of the evaporating substance, inches Hg.

Translating the units for the G and P_{vapor} variables in their formula into mostly metric units yields the following variation of their evaporation equation:

$$G(\text{mg/min}) = 0.0706 \times \text{MW} \times A \times V^{0.625} \times P_{vapor} \tag{5.4b}$$

where

MW, A, and V are as above and the difference in G and P_{vapor} units is shown below in bold:

G = the vapor generation rate (**mg/min**).
P_{vapor} = the vapor pressure of the evaporating substance (**mmHg**).

The generic evaporation rate equation (5.3) should be thoroughly understood even though hygienists almost never use this equation as written. One reason for this apparent dichotomy is that, in practice, the value of the geometry coefficient is almost never known. Another reason is that knowing the rate of evaporation gives insight into exposure but does not by itself predict the exposure level. All of the other ways to estimate *G*, to be presented in Chapter 20, suffer the same limitation of not predicting exposure. On the other hand, important ways to control exposure to be presented in Chapter 8 and Chapter 9 will use Equation 5.3 to predict the effect of changing the source or modifying work practices. Sooner or later a hygienist should learn the effect on exposure of modifying each of the variables within this equation.

II. MECHANISMS OF PLUME DISPERSION

Once a chemical becomes airborne (either by the release of a gas, the evaporation of a liquid, or as an aerosol), it more or less goes where the air goes. Exceptions to this pattern include plume buoyancy (described in Section VII of Chapter 2) and the sedimentation and inertial separation of large particles (described in Section VI of Chapter 3). While solid materials lack an intrinsic propensity (such as vapor pressure) to become airborne, the same mechanisms that disperse a vapor plume will also disperse an aerosol plume. As depicted in Figure 5.2, the concentration of the contaminant in a plume decreases slowly along its axis (in its direction of travel) but rapidly across its axis (perpendicular to the plume's direction of travel). The rates of both its axial dilution and lateral spread depend more upon turbulence (eddy currents) than upon molecular diffusion.

1. *Convection* is the bulk movement of air from place to place as in a wind. Convection is usually the most important mechanism transporting contaminants within a workplace. Figure 5.2 depicts the contaminant concentration (in the vertical axis) in a plume being carried downwind from the source along with the bulk air moving horizontally. The plume's direction may be somewhat vertical due to the plume's buoyancy or the gravitational settling of large particles. However, convection in and of itself will not cause a plume to disperse, dissipate, or be diluted.
2. As the plume moves away from the source, it will dissipate, due mostly to *turbulent diffusion* (sometimes called eddy diffusion). How rapidly turbulent diffusion acts to dilute the plume is a function of the moving air's turbulence for much the same reasons that it determined the thickness of the boundary layer right over the source. The apparent angle at which the edge of the plume depicted in Figure 5.2 spreads depends on the ratio between the speeds of convection and eddy diffusion.

   Turbulence can be created in still air by moving objects such as machinery, vehicles, and people either walking past or moving their arms. Turbulence is also created as air moves past or through fixed obstacles such as equipment, furniture, and doorways. Turbulence can be visualized as eddies or moving pockets of circulating air that can carry contaminants from the concentrated plume into the surrounding ambient air. Notice in Figure 5.3 how the turbulent eddies immediately downwind of a person standing upwind of a source can cause some of the plume to move upstream into the person's breathing zone. The eddies are caused by the shedding of vortices that form when the inertial forces

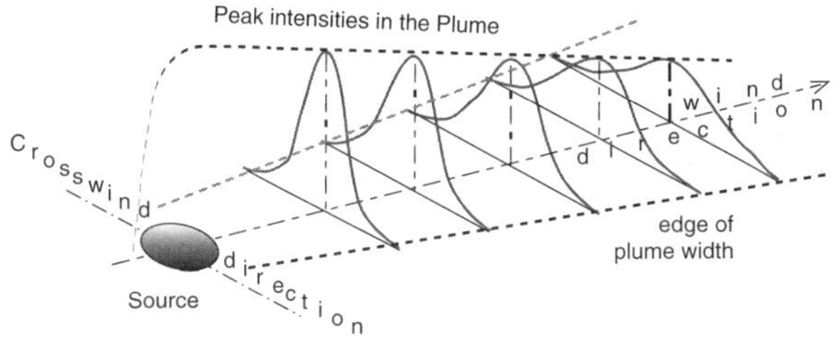

FIGURE 5.2 Depiction of the relative decrease in a plume's concentration axially vs. laterally.

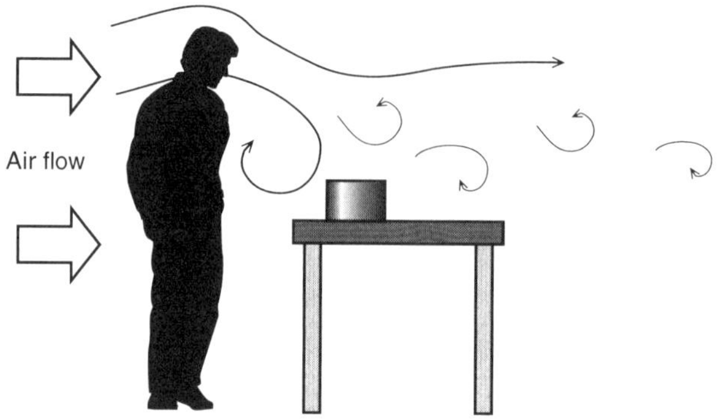

FIGURE 5.3 Depiction of wake turbulence created by air flowing past a person.

pushing the air along exceed the ability of the viscous forces within the air to keep it all moving on one direction. The vortices actually shed off of a body at a predictable frequency.[a] The vortex shedding frequency of a human body, the entrainment of the contaminant within a wake, and the mixing of that contaminant into the upwind breathing zone are all affected by slight changes in the body's position and its movements.[13]

Turbulent diffusion has been modeled in a manner analogous to Fick's law for molecular diffusion as shown in Equation 5.5. In both kinds of diffusion, the flux of molecules through a given cross-sectional area perpendicular to the pathway is proportional to the concentration gradient (the difference in concentrations divided by the path length).

$$\text{Flux (moles/cm}^2\text{/min)} = \mathcal{D}\frac{P_{\text{of } i \text{ at } 1} - P_{\text{of } i \text{ at } 2}}{L_{\text{at } 1} - L_{\text{at } 2}} = \mathcal{D}\frac{\Delta P_{\text{partial}}}{\Delta L} \quad (5.5)$$

where

1 and 2 = any two points selected along the path on which the chemical is diffusing.

P_{partial} = the chemical's concentration or partial pressure at each point or location.

L = the location where P_{partial} is measured; $\Delta L = L_1 - L_2$ is the diffusion path length.

$\mathcal{D}$ = either molecular diffusivity or turbulent diffusivity but rarely a combination of the two because turbulent diffusivity is generally between 100 and 10,000 times greater than molecular diffusivity.[14]

Turbulent diffusivity is also known as eddy diffusivity for probably obvious reasons. While the value of the molecular diffusivity is specific to the particular chemical in the air, the value of turbulent diffusivity depends strictly on the turbulence within the local airstream and not at all upon the particular chemical caught up in that turbulence. The two key parameters to eddy diffusivity are the intensity of the turbulence (characterized by the root mean square of the velocity fluctuations divided by the mean velocity) and the mixing length associated with the size of the obstacles upstream of the source of the emission that generate the turbulence.[1,2,15–17] Since the amount of turbulence is not easily measured in the workplace, the value for turbulent diffusivity in the workplace is usually unknown. Nonetheless, turbulent diffusion remains a useful concept to help us understand contaminant plume dispersion both within the workplace and after exhausted emissions leave the building.

[a] Research has found the frequency (f) at which various shaped bodies will shed vortices is related to a dimensionless coefficient called the Strouhal number (St). For cylinders of diameter D, vortex shedding occurs when $\text{St} = fD/V = 0.21$ where V is the air velocity in units of D per second. Thus, the frequency can be predicted as $f = 0.21\ V/D$. With some rearranging, $V/f = D/0.21$ also predicts that a vortex will form about every 9 ft in the wake of a 2-ft diameter person.

3. In contrast to eddy diffusion, molecular diffusion is *almost never* an important determinant of plume dilution or of personal exposure within the workplace. As implied by Figure 5.4, molecular diffusion can dominate eddy diffusion above and to the left of the line representing the 1.5×10^{-5} m^2/sec molecular viscosity of air. Molecular diffusion is too slow to be an important chemical transport mechanism *except* either over very small distances or over very long times. Eddy diffusion predominates in the times and distances applicable to the workplace (from about 1 min to 8 h and from less than 1 to about 10 m, respectively). The following are examples of just how short the distances or long the times must be for molecular diffusion to be important:
   a. Movement across the 1- to 5-mm boundary layer over an evaporating solvent is a molecular diffusion process. The parameters in Equation 5.2 and Equation 5.3 determine the boundary layer's thickness.
   b. Molecular diffusion across a 1- to 3-mm path length of a passive air sampling device affects its sampling rate.
   c. Gas exchange within the respiratory tract and the deposition of fume particles deep within the lung (as discussed ca. Table 3.2 and Table 3.4) are diffusion processes.
   d. A duration of more than 1 day (8.6×10^4 sec in a day) is necessary for a gaseous contaminant concentration to become homogeneous throughout a 10-ft (3 m) confined space.
4. *Buoyancy* is the force on a body owing to the difference between its density and the density of the fluid the body displaces (the force is actually the difference in density times the displaced volume). A plume's density can be strongly affected by differences between its temperature and that of ambient air but only weakly by its vapor concentration. As discussed in Chapter 2, buoyancy can affect a plume but has minimal influence on either a dilute or an isothermal plume.

   Since molecules in their vapor state have more internal energy than molecules in their liquid state, the process of evaporation absorbs heat from its environment. Evaporation can cool both the evaporating liquid source and the air into which it evaporates. For example, this heat absorption is why the evaporation of perspiration cools our bodies. The more volatile the solvent, the higher will be the rate of cooling. While a highly volatile solvent such as acetone can cause its liquid temperature to drop

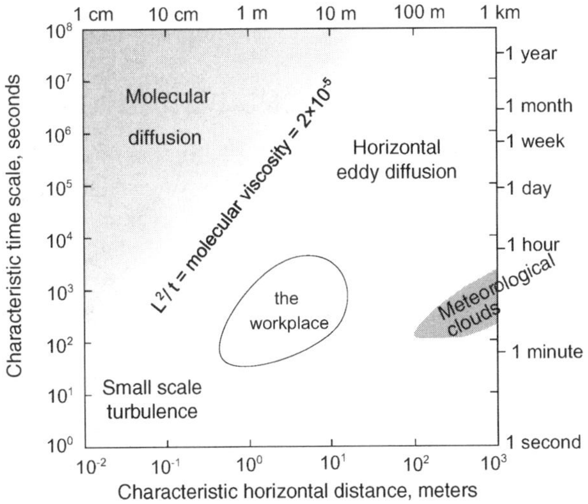

FIGURE 5.4 Adapted from Smagorinsky but originally attributed to Ooyama and Fortag.[18–20]

10°C below the ambient temperature (the consequent decrease in its vapor pressure is a health-protective change), the vapor plume is usually much closer to room temperature because each parcel of air is over the evaporating liquid for only a short time. Thus, a vapor plume's density is usually very close to that of the surrounding air, and the effect of buoyancy on the plume is negligible.

On the other hand, a solvent can be heated above ambient temperature either on purpose (e.g., to increase its solvating power or to decrease its drying time) or by happenstance (as from mechanical friction or from its use in or on a hot machine). Heating increases a solvent's vapor pressure (as per Equation 2.9), its rate of evaporation (as per Equation 5.3), and eventually its density (as per Equation 2.29 and Figure 2.8). The resulting downward motion owing to its negative buoyancy may either increase or decrease worker exposure depending upon their location. As the plume dissipates, this buoyant force will weaken and eventually loose its influence, and the contaminant will disperse throughout a room.

It is also possible for a liquid to be heated to the point of producing a visible plume. A heated vapor will generally cool thermally faster than it can dissipate chemically. If concentrated vapors in a rapidly cooling plume exceed their equilibrium vapor pressure, the supersaturated vapor will condense producing what would appear to be steam if the solvent were water. Technically, such steam is a mist. This visible plume too will dissipate and rapidly reevaporate back into an invisible vapor. This process can be observed by watching water boil or a person's visible breath in cold winter air.

In an open-air (outdoor) setting, the plume is free to dissipate via the above mechanisms as it moves downwind virtually forever, but indoors the plume will be constrained by walls and a ceiling. Local exhaust ventilation can be used to suck most of that plume into a hood and keep it away from workers' breathing zones. That proportion of a plume that is sucked away but is not removed by an air cleaner will be emitted into the outdoor ambient air where it can continue to dissipate. Whatever proportion of the plume that is not exhausted locally will be diluted into the fresh air entering the workroom and will eventually leave with the excess air that must exit the workroom to make room for more fresh air. In this latter case, everyone in the room is exposed to the contaminant. In the worst scenario there is no ventilation, and eventually everyone receives the highest levels of exposure that the source is capable of producing. If work has to be conducted in a setting that cannot be ventilated, then one must limit the time of exposure or rely on personal protective equipment as the last remaining control options.

III. FOUR UNIVERSAL AIRBORNE CHEMICAL EXPOSURE SCENARIOS

The above mechanisms of evaporation, convection, dilution, and accumulation suggest that any occupational exposure to airborne chemicals can be categorized into one of the following four scenarios. While these scenarios are described herein as vapor exposures, the same plume dispersion mechanisms also apply to aerosols. These scenarios comprise threads by which the various parts of this book are linked together.

The first two scenarios involve the plume but no room accumulation.

1. People often work in the "near-field" of the source of a chemical (within about an arm's length), as depicted in Figure 5.5.[b] This figure is based on Popendorf (1984) and unpublished research by Livengood (1985) to be discussed ca. Figure 5.12.[22,23]

[b] Unobstructed vapors in a no-wind condition can disperse horizontally in all directions as reported by Gumaer (1937), somewhat like fog produced by dry-ice when put in hot water; but no-wind conditions are not common in the workplace.[21]

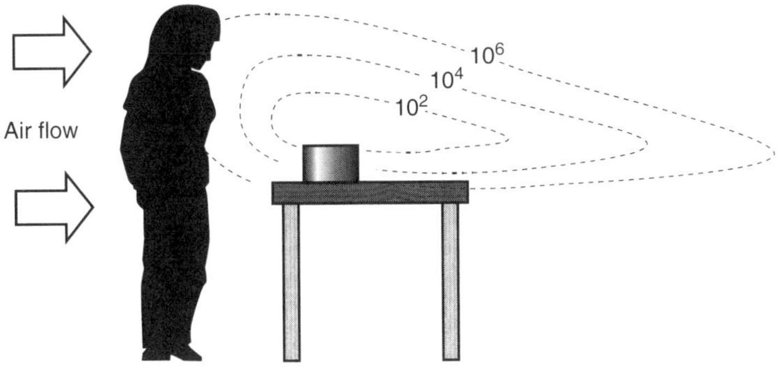

FIGURE 5.5 Depiction of the near-field exposure scenario 1 as seen from the side showing the magnitude by which the vapor concentration at the source might be diluted.

Kim and Flynn[24,25] illustrated that the role of eddy diffusivity in dispersing a plume in the wake of obstacles such as people extends perhaps two body widths downwind or just beyond their arm's length. Recall that the vapor concentration at the source will always equal its vapor pressure. The dotted lines in Figure 5.5 depict the amount by which that vapor concentration might be diluted both upwind and downwind. The amount of dilution between the source and the breathing zone in this "near-field" scenario depends upon several factors. The most obvious factors in this diagram are the employee's proximity to the source and their orientation with respect to its plume (at least upwind vs. downwind). Unfortunately, dilution also depends upon the magnitude of the air's turbulence which is neither visible nor easily measured. Proper training and supervision would be of great benefit to prevent people from working either too close to or downwind from the source before the plume can be adequately diluted. Nonetheless, some exposure will always occur to anyone in the near-field working either upwind or downwind of the plume. The applicability of a near-field plume to worker exposure was introduced qualitatively by Popendorf[22], but much of the quantitative research on exposures in the near-field has been conducted by others.[17,24–29] Near-field controls will be discussed further in the later chapters on "Other source and pathway controls" (Chapter 9) and "Selecting and designing local exhaust hoods" (Chapter 13).

2. If the source depicted in Figure 5.5 is not controlled by local exhaust ventilation, the plume will leave the near-field by convection and continue to dissipate by turbulent diffusion as depicted as the top view in Figure 5.6. Concentrations within a plume decrease much more slowly with distance along the downwind axis of the plume (e.g., towards position D in Figure 5.6) than with lateral distance from the plume's axis (e.g., towards positions B or C in Figure 5.6). Although this plume dispersion pattern differs from the omnidirectional pattern of noise dispersion in a free-field, the same term will be applied to a plume dispersing without substantial effects from the walls of a room, such as a plume generated outdoors. Researchers have even used indoor plume dispersion data to estimate the total source generation rate (G in Equation 5.2, Equation 5.3, or Equation 5.4a,b).[30–34] The structural nature of plumes, as depicted in Figure 5.2, stimulated other researchers to concluded that someone working to the side of the plume (specifically position B in Figure 5.6) will receive about 95% less exposure than working upwind of the plume (in position A).[17,24,25,35] Thus, the magnitude of chemical exposure in this free-field plume scenario will depend more strongly upon whether the person is in the plume or not than upon how far away from the source they are.

The free-field plume exposure scenario depicted in Figure 5.6 can be virtually eliminated or made negligible by local exhaust ventilation as depicted in Figure 5.7.

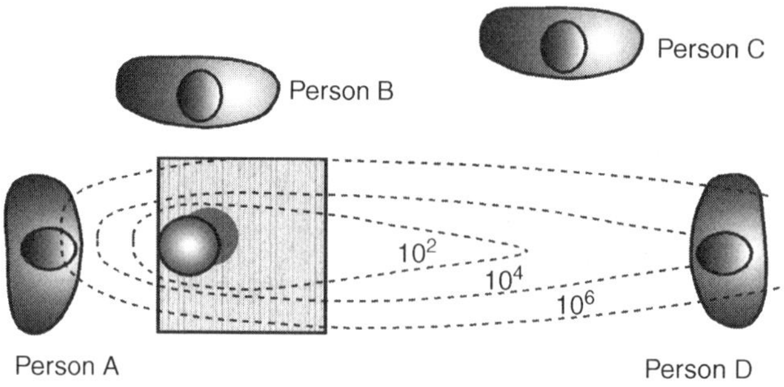

FIGURE 5.6 Depiction of the free-field plume exposure scenario 2 as viewed from above.

However, if a plume is generated indoors and not exhausted locally, it will not travel far before encountering walls and other obstacles that deflect it laterally and eventually cause it to recirculate back towards its source. This recirculation will cause such a setting to transition rapidly into a room accumulation scenario.

The last two scenarios involve room accumulation but not the plume; thus they only apply to indoor work. In the first of these scenarios the room is ventilated; in the last, it is not ventilated. The next section of this chapter (Section IV) will also define two subscenarios within each of these last two accumulation scenarios that differ by how much liquid is present to evaporate.

3. In scenario 3, the room has general ventilation, meaning that some fresh air is always entering the room and an equal amount of room air always leaving the room carries some of the contaminant with it. If the emission (G) is continuous, whatever vapor does not leave the workroom by local exhaust ventilation will recirculate and accumulate within the room until eventually the rate at which it is being removed from the room with the general ventilation air equals the rate at which it is being generated. When these two rates equal each other, the concentration in the room will be at a *steady state*. Steady state means when conditions do not change over time. To glance ahead to Chapters 19 and 20, the more complex dynamic ventilation scenario (where concentration does change over time) is described mathematically by a differential equation such as Equation 19.4, is solved in the form of Equation 19.15, and used in the form of Equation 20.1 in research.[36–41] The $K_{\text{room mixing}}$ and $Q_{\text{effective}}$ in Equation 20.1 relate to how well the air in the room mixes with the contaminant (as will be explained in Chapter 19).

$$C_{\text{eventual or steady state}} = \frac{G}{Q/K_{\text{room mixing}}} = \frac{G}{Q_{\text{effective}}} \approx C_{\text{TWA}} \qquad \text{(to be Equation 20.1b)}$$

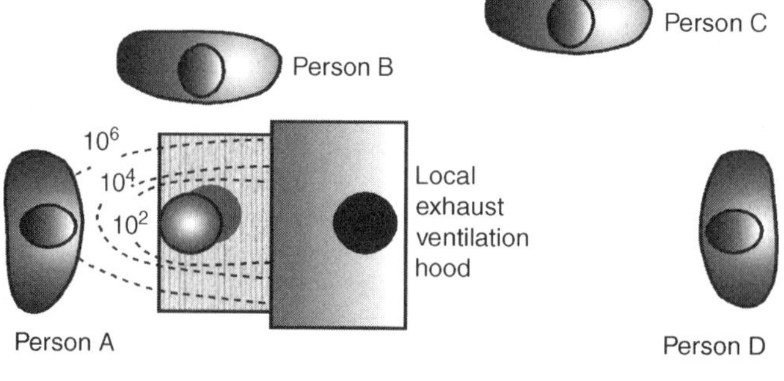

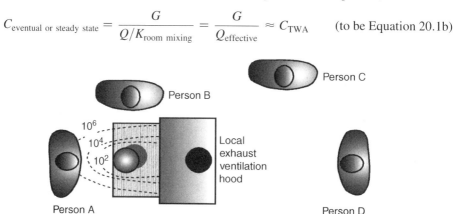

FIGURE 5.7 Depiction of the free-field plume being captured and removed by local exhaust ventilation.

If in this scenario the room air is reasonably well mixed, most of the people in the room will be exposed to some concentration, as depicted in Figure 5.8. To use again an analogy to noise, this accumulation scenario is similar to a reverberant field where exposures within the room are relatively uniform with distance. The control of most airborne chemical hazards via general ventilation usually requires a large air flow rate, meaning that it is only economically feasible when the chemical in use requires relatively little dilution. Thus, employee protection can generally be achieved more cost-effectively from modified work practices, local exhaust ventilation, or even the use of respirators than from general ventilation.

4. Scenario 4 is a closed room or confined space with no ventilation. If vapors cannot escape, they will accumulate but eventually be limited to a maximum vapor concentration defined either by the amount of liquid available to evaporate or by the liquid's vapor pressure, depending upon the quantity of liquid solvent initially present. The vapor concentration resulting from a volume of liquid large enough for its vapors to reach equilibrium (depicted on the left in Figure 5.9) is the liquid's vapor pressure. This subscenario will be discussed further in Section V of this chapter. The vapor concentration resulting from a small amount of liquid (depicted on the right in Figure 5.9) is simply the liquid's mass divided by the airspace's volume. This subscenario will be discussed further in Section VI of this chapter. Equilibrium is always the worst case (highest) of these two vapor concentrations.

   The concentration of an aerosol that is actively being generated within a confined space has no intrinsic upper limit, a foible that can result in the kind of explosive aerosol concentration mentioned in Chapter 4. If an ignitable aerosol or vapor is being generated inside a confined space that cannot be ventilated, then even a respirator would not be a good control option (see Section VI of Chapter 22).

Thus, in almost all of the scenarios described above, vapor exposures depend at least in part upon the source's vapor pressure (P_{vapor}) and in three out of four scenarios upon its associated evaporation rate (G), although in no case can exposure be determined solely by knowing even both of these variables. Anyone within about arm's reach of an evaporating chemical will have a near-field exposure. Good work practices can keep people out of the strongest portions of the near- and free-field plumes, and local exhaust ventilation can virtually remove the free-field plume altogether.

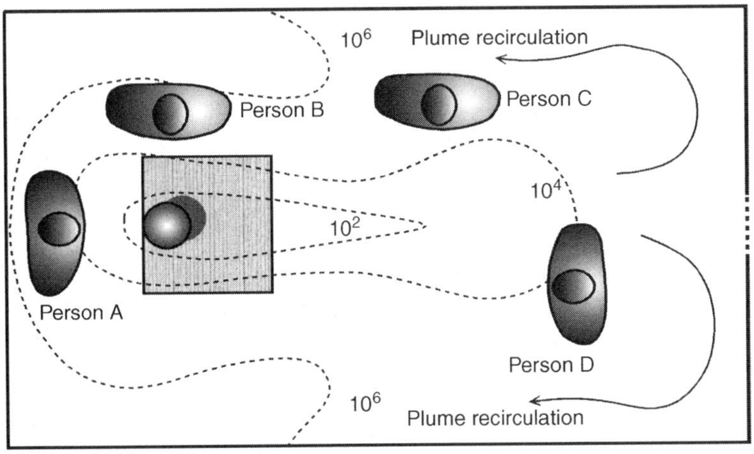

FIGURE 5.8 Conceptual depiction of a contaminant dispersing within a ventilated, occupied room where the plume, initially flowing from left to right, eventually doubles back on itself to form accumulation scenario 3.

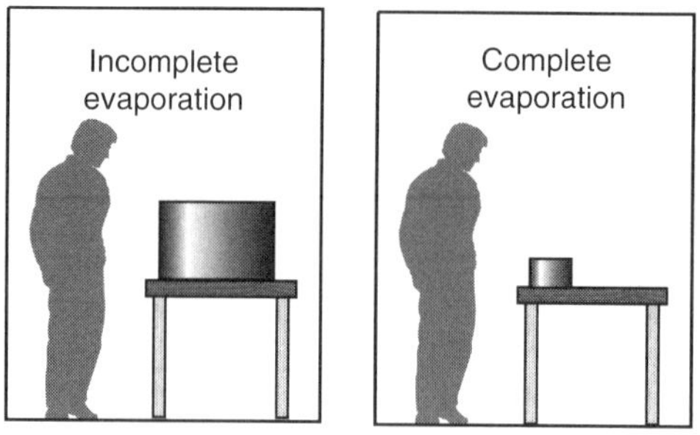

FIGURE 5.9 Depicting two subscenarios of the closed room exposure scenario 4: a large volume of liquid on the left and a small volume on the right.

Thus, not every setting will have either a free-field plume or someone in that plume. A review by Rodes et al. of air samples from various work environments showed that reported personal breathing zone vapor concentrations measured in the near-field exceeded concentrations in general area samples (collectively including both the free-field and accumulation scenarios) by median ratios ranging from about 2:1 to 15:1.[41] More focused research by Cherrie found similar to slightly higher ratios that varied with room size and ventilation.[42] While all workers in some workrooms can be overexposed, the data presented by Rodes et al. and Cherrie supports the expectation that workers in the near-field plume are at the greatest risk of overexposure. Local exhaust ventilation can also minimize (if not virtually eliminate) the accumulation of a contaminant within a room. The level of accumulation without local exhaust will depend upon the presence and effectiveness of the room's general ventilation system. Few work settings are unventilated or considered to be confined, but any work that must be conducted in a confined space requires the utmost management supervision.

IV. FOUR SETTINGS WITH VAPOR ACCUMULATION

The last two of the above universal exposure scenarios each describe a setting with vapor accumulation. The discriminator that separates scenario 3 from scenario 4 is the presence or absence of room ventilation. A second discriminator that can divide each of these two accumulation scenarios into two subscenarios is the amount (volume or mass) of chemical available to evaporate. These two discriminators of ventilation and liquid volume define the four settings or subscenarios shown in Table 5.1.

At least three of these four settings have practical as well as predictable implications for hygienists. Trying to predict a ventilated setting with a small amount of liquid is not too practical because its concentration is "dynamic" (changing in time, first increasing then decreasing, as will be discussed but not dwelt upon in Chapter 19). A ventilated setting with a continuously evaporating liquid is a common setting, although a setting with accumulation in the room (analogous to setting $3_{\text{large liq.}}$) will not be discussed in detail until Chapter 20. Workplaces with no or very little ventilation are not common, but settings $4_{\text{small liq.}}$ and $4_{\text{large liq.}}$ will be discussed at great length within the next three sections of this chapter for reasons that will be explained. The essence of all four settings is summarized below in the order from the simplest to the most complex; the simplest is the worst case.

1. Setting $4_{\text{large liq.}}$ occurs when the amount of volatile liquid available to evaporate into an unventilated workplace is sufficiently large that not all of it can evaporate before the vapor reaches its equilibrium concentration, as depicted in Figure 5.10. The air in the room or workplace may start out initially pure (where ambient $P_{\text{partial}} = 0$), but with no

TABLE 5.1
Scenarios 3 and 4 Subcategorized by the Amount of Chemical Available to Evaporate and the Presence or Absence of Ventilation Present in the Room

	Small Liquid Volume	Large Liquid Volume
Scenario 3 with general ventilation	setting $3_{\text{small liq.}}$ Transient; no steady state C changes with time	setting $3_{\text{large liq.}}$ Continuous evaporation C_{TWA} becomes G/Q
Scenario 4 with little or no ventilation	setting $4_{\text{small liq.}}$ Complete evaporation C_{TWA} becomes mass/Vol	setting $4_{\text{large liq.}}$ Incomplete evaporation C_{TWA} becomes P_{vapor}

ventilation the vapors will accumulate until equilibrium is achieved, which by definition is the liquid's vapor pressure, P_{vapor}. When the concentration in the ambient air P_{partial} equals P_{vapor}, no more evaporation can occur. This incomplete evaporation setting is the worst case (the one that results in the highest concentration) within both the four universal vapor exposure scenarios previously described and the four subscenarios being described here. Incomplete evaporation is typified by entry into a confined space that has a residual volatile liquid still present.

2. The vapor concentration that results after a small amount of liquid evaporates into an unventilated workplace (setting $4_{\text{small liq.}}$ as depicted in Figure 5.11) will be the total mass of the liquid chemical that was initially available to evaporate divided by the air volume within the closed room. The result of dividing mass in milligrams by volume in cubic meters defines C in mg/m^3. While this scenario may not occur often in real workplaces, its prediction is simple, the underlying concept is powerful, and the resulting concentration is a useful point of reference for other accumulation scenarios. The key feature that distinguishes this complete evaporation scenario from the incomplete evaporation scenario (above) is whether or not the liquid initially present is small enough so that all of the liquid can evaporate before the equilibrium concentration is reached.
3. The method to predict the steady-state vapor concentration in setting $3_{\text{large liq.}}$ employs elements used to predict the concentration in both setting $4_{\text{small liq.}}$ and $4_{\text{large liq.}}$. The rate of evaporation as stated by Equation 5.3 is dependent (in part) upon the chemical's vapor pressure, similar to setting $4_{\text{large liq.}}$. After vapors have reached a steady-state concentration (where C is constant with time), the rate at which the chemical evaporates

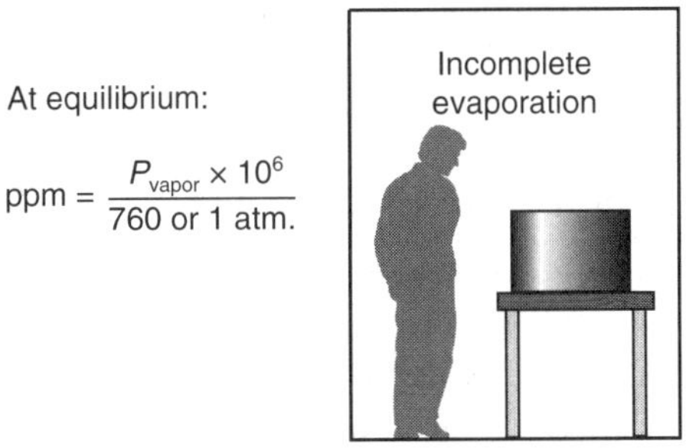

FIGURE 5.10 Concentration from incomplete evaporation as in scenario $4_{\text{large liq.}}$: no ventilation and a large source.

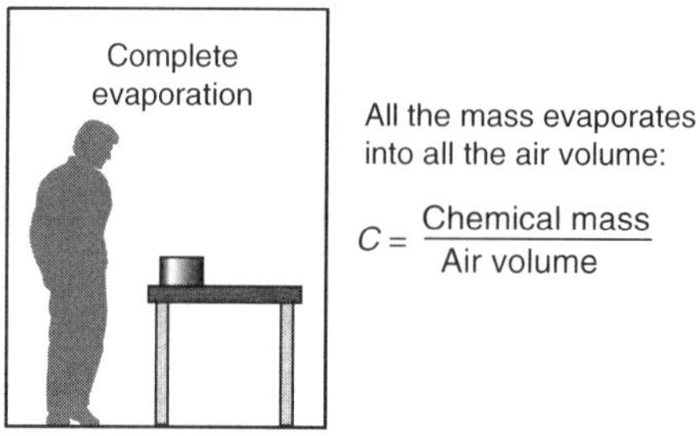

FIGURE 5.11 Concentration from complete evaporation as in scenario $4_{\text{small liq.}}$: no ventilation and a small source.

equals the rate at which these vapors leave the workplace with the room's ventilation air. Of particular interest to industrial hygienists is that when the evaporation rate in mg/min is divided by the ventilation rate in cubic meters per minute, the units are again the familiar mg/m^3 concentration units found in setting $4_{\text{small liq.}}$. This steady-state concentration is also closely related to what hygienists call the room's time-weighted average concentration (the C_{TWA} in the preview of Equation 20.1 shown above). Thus, employee exposures in two of the three predictable accumulation subscenarios either depend upon or are limited to P_{vapor}, the vapor pressure.

4. The vapor concentration resulting from a small volume of solvent evaporating into a ventilated room in scenario $3_{\text{small liq.}}$ are typified by a spill. While such concentrations can be predicted, the concentration in transient settings will be bounded by one of the previous steady-state concentrations.[44,45] A small amount of any airborne contaminant generated over a short time interval in a ventilated room will produce a transient increase to some peak concentration, perhaps remain there for a short time, followed by a decrease back towards zero (or the background ambient concentration). A learning goal of Chapter 19 will be to gain a sense for the typically short duration of the increasing transient and how the duration of the decreasing transient depends upon which exposure criterion is applicable (fire, oxygen, or health). While a transient's peak concentration is rarely predictable quantitatively because it depends upon the net effects of evaporation, dispersion, and removal of vapors from a room, the following rules-of-thumb may help anticipate such transient conditions:
   a. The peak concentration will be less than that determined for the case of complete evaporation with no ventilation (setting $4_{\text{small liq.}}$).
   b. The time-weighted average concentration will approximate that for the steady-state condition that would result if that small liquid volume were to evaporate into the volume of ventilation air entering the room over the duration that it takes for the liquid to evaporate (setting $3_{\text{large liq.}}$).

V. VAPORS FROM INCOMPLETE EVAPORATION WITH NO VENTILATION

The principle used to predict the worst case vapor concentration that will result from the incomplete evaporation of a large volume of liquid within a closed or unventilated setting (subscenario $4_{\text{large liq.}}$) is so fundamental to IH that it should be thoroughly understood. To be incomplete means that some liquid must always be present. Recall from one of the definitions given in Chapter 2 that a closed room means simply a room with no ventilation, not a completely sealed chamber. While most workplace settings are ventilated, examples of closed rooms at work include a storage container, a shipping tank, any poorly ventilated room (where $Q \approx 0$), or many OSHA-defined confined spaces.[c] As the vapors in any unventilated setting have nowhere to go, they will eventually reach an equilibrium with their evaporating liquid, and that equilibrium concentration is (by another definition given in Chapter 2) that liquid's vapor pressure. This definition was stated in quantitative IH terms by Equation 5.1 at the beginning of this chapter.

$$\text{ppm at the source} = \frac{P_{\text{vapor}} \times 10^6}{P} = \frac{(P_{\text{vapor}}\ \text{mmHg}) \times 10^6}{760}\ \text{at NTP} \qquad \text{(repeat of Equation 5.1)}$$

Example 5.1

What concentration of toluene vapors could be expected if the plant's maintenance staff has to conduct unscheduled pump repairs inside a toluene storage tank with residual toluene still present?

The fact that residual solvent is still present in this setting is clear evidence that evaporation is incomplete. The fact that the tank is used to store toluene indicates that sufficient time has passed for the vapors to have reached their equilibrium. Another giveaway that Equation 5.1 is applicable would have been if the question were something like "What is the *maximum* concentration of toluene vapors?"

$$\text{The equilibrium ppm for toluene} = \frac{P_{\text{vapor}} \times 10^6}{P} = \frac{22 \times 10^6}{760} = 28{,}950\ \text{ppm}$$

VI. VAPORS FROM COMPLETE EVAPORATION WITH NO VENTILATION

While hygienists do not often calculate the vapor concentration that will result from a small volume of liquid evaporating into a closed airspace (subscenario $4_{\text{small liq.}}$), this prediction is also so simple and the concept so fundamental to IH that it too should be thoroughly understood. The vapor concentration generated by a small volume of a liquid chemical that completely evaporates into a closed room or airspace will simply equal the mass of the evaporated chemical divided by the closed air volume into which it is diluted. This principle is stated mathematically in Equation 5.6.

$$\text{For complete evaporation,}\ C_i\ (\text{mg/m}^3) = \frac{\text{mass}_i\ (\text{mg})}{\text{air Vol}\ (\text{m}^3)} \qquad (5.6)$$

where

C_i = gaseous concentration (mg/m^3) of contaminant i.
mass_i = mass of contaminant i that evaporates (mg or converted into mg).
air Vol = the air volume within the chamber or room (m^3 or converted into m^3).

Three forms of Equation 5.6 will be used in this section (Equation 5.6a, Equation 5.6b, and Equation 5.6c). The concept underlying this equation will be used several more times throughout this book. Learn the concept. The following points could also prove to be useful to apply this concept.

- If (as is common in the real-world) one is given or wants to know the liquid volume rather than the liquid mass, one needs to know the liquid's density (ρ_ℓ) or its specific gravity. The definition of liquid density given in Equation 5.7a is analogous to that for gas density given in Equation 2.23a except for two differences. One is the subscript notation ℓ that is consistently used herein to differentiate liquid parameters (such as density)

[c] Headspace is the air above the liquid in any closed container. Although the head space in a closed vial cannot be occupied, it meets the same conditions. Analytical methods to analyze headspace air are cited in Chapter 6.

from gaseous parameters that are written without that subscript. The other is that the units used for liquid density are g/mL vs. the g/L units for the much smaller density of gases and vapors. Units of liquid density given in kg/L, g/mL, or mg/μL are all interchangeable. *Specific gravity* is something's density compared with water, and is therefore exactly equal to and interchangeable with liquid density when the latter is given in any of the above metric units.[d]

$$\rho_\ell \ (\text{g/mL}) = \frac{\text{mass of liquid (g)}}{\text{liquid Vol (mL)}} \tag{5.7a}$$

$$\text{mass of liquid (g)} = \text{liquid Vol (mL)} \times \rho_\ell \tag{5.7b}$$

- The concentration of a released gas may also be solved in terms of ppm using the volume-to-volume ratio shown in Equation 5.8 (just make sure the units of both volumes are the same). See Example 5.4.

$$\text{ppm} = \frac{\text{gaseous volume of contaminant released} \times 10^6}{\text{air volume into which the contaminant is dispersed}} \tag{5.8}$$

- If desired, one could also use the same volume/volume technique in Equation 5.8 to solve a liquid evaporation problem by using a variation of Equation 2.23b to find the equivalent volume of the vapor from an evaporating liquid. The volume is called equivalent because the calculated result is equivalent to the theoretical volume as if the vapor could form a pure gas. Thus, the equivalent volume of vapor from any version of Equation 5.9 can be inserted into Equation 5.8 in place of "gaseous volume of contaminant released."

Equation 5.9a converts the mass in grams into vapor in liters (at NTP).

$$\text{Equivalent volume of vapor (L)} = \frac{\text{mass (g)} \times 24.45}{\text{MW}} \text{ at NTP} \tag{5.9a}$$

Equation 5.9b makes the same conversion if the amount of liquid evaporating is given in mL.

$$\text{Equivalent volume of vapor (L)} = \frac{\text{liq. Vol (mL)} \times \rho_\ell \times 24.45}{\text{MW}} \text{ at NTP} \tag{5.9b}$$

Equation 5.9c makes the same conversion on a larger scale from liters of liquid evaporating into the equivalent volume in cubic meters of the vapor.

$$\text{Equivalent volume of vapor (m}^3\text{)} = \frac{\text{liq. Vol (L)} \times \rho_\ell \times 24.45}{\text{MW}} \text{ at NTP} \tag{5.9c}$$

The principle embodied in Equation 5.6 can be used to solve *at least* three kinds of potential IH problems, depending on whether *C*, mass, or air volume is the unknown variable.

[d] Knowing one of the metric equivalents between ft^3 and m^3 can help to solve problems involving volume. In addition to the 1 ft^3 = 28.3 L conversion factor provided near the front of this book, it may help some readers to remember that a Greenburg-Smith midget impinger was designed to sample at a rate of 0.1 ft^3/min for which the pump is calibrated at 2.83 L/min.

1. To find an unknown C, the appropriate form of Equation 5.6 is Equation 5.6a:

   The goal of this problem is to calculate the vapor concentration in mg/m^3 or ppm resulting from the complete evaporation of a known amount of solvent into a closed room or chamber of known volume. In American workplaces, the liquid volume is commonly known in units of gallons and the air volume in units of cubic feet. It is easier if one knows all of these parameters in metric units.

   The appropriate form of Equation 5.6 in this case is

$$C = \frac{\text{mass}}{\text{air Vol}} \tag{5.6a}$$

a. Air volume in m^3 = room volume in $ft^3 \times \frac{28.3\ L}{ft^3} \times \frac{m^3}{10^3\ L}$ (known)

b. Solvent volume in L = Liquid volume in gal $\times \frac{3.785\ L}{gal}$ (known)

c. Solvent mass in mg = Solvent volume in liters $\times\ \rho_\ell \frac{g}{mL} \times \frac{10^3\ mg}{g} \times \frac{10^3\ mL}{L}$

d. Airborne concentration C in $mg/m^3 = \frac{(\text{mass in mg})}{(\text{air volume in } m^3)}$ (the unknown)

e. Optionally, the concentration in ppm = $mg/m^3 \times \frac{24.45}{MW} \times \frac{760 \times (°C + 273)}{P\ (mmHg) \times 298}$

(Equation 2.21b)

EXAMPLE 5.2

What concentration would result if a full liter jug of toluene were to break open, and its entire contents were to evaporate within a small laboratory with dimensions of $8 \times 18 \times 10\ ft^3$ tall with poor ventilation so that the entire liquid volume would evaporate before much vapor is removed?

To proceed, one would first need to look up toluene's liquid density ($\rho_\ell = 0.867$ g/mL) and molecular weight (MW = 92.13). NTP is assumed when T and P are not stated.

a. air volume = $8 \times 18 \times 10 = 1440\ ft^3 \times 0.028\ m^3/ft^3 = 40.3\ m^3$

b. solvent volume = 1 L

c. solvent mass = $1\ L \times 0.867\ g/mL \times 10^3\ mg/g \times 10^3\ mL/L = 8.7 \times 10^5$ mg

d. $C = \frac{8.7 \times 10^5\ mg}{40.3\ m^3} = 21{,}500\ mg/m^3$ (using Equation 5.6a)

e. ppm = 21,500(24.45/92.13) = 5710 ppm

The result of any calculated complete evaporation concentration could be checked against the equilibrium concentration at its vapor pressure to assure that all of the spilled solvent could evaporate before reaching that equilibrium. In Example 5.1, the equilibrium ppm for toluene was found to be 28,950 ppm. This equilibrium ppm is about 5 × more than the above 5710 ppm if the liter completely evaporated. Thus, Example 5.2 is clearly a complete evaporation subscenario.

2. To find an unknown mass (mass) or liquid volume, the appropriate form of Equation 5.6 is Equation 5.6b:

   The goal of this problem is to calculate the volume of solvent needed to generate a known concentration in a known volume of air such as for instrument static calibration, to generate an initial concentration of a tracer gas or vapor to test a dilution ventilation system, or to find the maximum volume of solvent that will not exceed a chosen concentration limit.

The appropriate form of Equation 5.6 is

$$\boxed{\text{mass} = C \times (\text{air Vol})} \qquad (5.6\text{b})$$

a. If needed, C in mg/m^3 = (Known ppm) $\times \dfrac{\text{MW}}{24.45} \times \dfrac{760 \times (°\text{C} + 273)}{\text{P (mmHg) } 298}$ (known)

b. Air volume in m^3 = (Air volume in ft^3)$\dfrac{28.3\ \text{L}}{\text{ft}^3} \times \dfrac{\text{m}^3}{10^3\ \text{L}}$ (known)

c. Mass in mg = (C in mg/m^3)(Air volume in m^3) (the unknown)

d. Optionally, the Mass in g = (Mass in mg)$\dfrac{\text{g}}{10^3\ \text{mg}}$

e. Liquid volume in mL = $\dfrac{\text{(Mass in mg)}}{\rho_\ell\ (\text{g/mL})}$

Example 5.3

Suppose that one chose to control overexposure within a small room by limiting the allowed solvent container size to that which if spilled will not produce enough vapors to exceed the chemical's TLV for the room in use. Apply that policy to the previous example. In other words, what volume of liquid toluene would have to be released into an 8 × 18 × 10 ft tall room to generate its TLV of 50 ppm? Proceed in accordance with the above steps:

a. C = 50 ppm × 92.13/24.45 = 188 mg/m^3
b. Air volume = 8 × 18 × 10 = 1440 ft^3 × 0.028 m^3/ft^3 = 40.3 m^3
c. Mass of toluene = 40.3 m^3 × 188 mg/m^3 = 7576 mg (using Equation 5.6b)
d. Mass of toluene = 7576 mg/10^3 mg/g = 7.6 g
e. Liquid volume = 7.6 g/(0.867 g/mL) = 8.7 m ≈ 9 mL

This 9 mL is a fairly small volume. If containers were limited to this size, it may not be feasible to complete the task for which the solvent is needed. However, keep in mind that unless the TLV is a ceiling limit, the 8-TWA can be exceeded on occasion (hopefully the accidental spill is a rare occasion). For chemicals that do not have an STEL, the TLV Committee recommends that excursions should not exceed the TLV–TWA by more than 5× nor happen more than three times per day. This flexibility would allow a perhaps more practical container size of 45 mL (1.5 oz) which still might not be feasible for some tasks. In a ventilated workplace, the solvent vapors would probably be diluted into about three times the room volumes of fresh air over the time it was evaporating; thus, this same policy could be modified to allow about 150 mL or about one cup. At each step of this analysis, control is more flexible but still would not be feasible in all settings.

Example 5.4

How much gas such as sulfur hexafluoride [SF_6] should be released into a 4000 ft^3 room to create an easily detectable 1 ppm concentration. SF_6 is commonly used as a ventilation tracer gas because its TLV is 1000 ppm but its detection limit using the MIRAN is 0.05 ppm, giving it a wide working margin of over 10^4 between its detectability and its presence as a hazard. Solving this special case can proceed using Equation 5.8:

a. ppm $= \dfrac{\text{released vol.} \times 10^6}{\text{air volume}}$ (using Equation 5.8)

b. released vol. $= \dfrac{\text{ppm} \times \text{air vol.}}{10^6} = \dfrac{1 \times 4000\ \text{ft}^3}{10^6} \times \dfrac{\text{m}^3}{35.5\ \text{ft}^3} = 113 \times 10^{-6}\ \text{m}^3 = 113\ \text{mL}$

3. To find an unknown air volume (air Vol), the appropriate form of Equation 5.6 is Equation 5.6c:

The goal of this problem is to calculate the air volume needed to dilute the vapors from a given or known liquid volume of solvent down to a known concentration (either chosen or set by regulation).
The appropriate form of Equation 5.6 in this case is air

$$\text{air Vol} = \frac{\text{mass}}{C} \tag{5.6c}$$

a. If needed, concentration in $\text{mg/m}^3 = \text{ppm} \times \frac{\text{MW}}{24.45} \times \frac{760 \times (°\text{C} + 273)}{\text{P (mmHg)} \times 298}$ (known)

b. Liquid volume in liters = liquid volume gallons $\times \frac{3.785\ \text{L}}{\text{gal}}$ (known)

c. Mass in mg = Liquid volume in L $\times\ \rho_\ell \frac{\text{g}}{\text{mL}} \times \frac{10^3\ \text{mg}}{\text{g}} \times \frac{10^3\ \text{mL}}{\text{L}}$

d. Air volume in $\text{m}^3 = \frac{(\text{Mass in mg})}{(\text{Allowable mg/m}^3)}$ (the unknown)

e. *Optional* air volume in ft^3 = Air volume in $\text{m}^3 \times \frac{\text{ft}^3}{28.3\ \text{L}} \times \frac{10^3\ \text{L}}{\text{m}^3}$

EXAMPLE 5.5

How much air would be needed to dilute the vapors from 1 L of toluene to its TLV of 50 ppm? Again, first look up the liquid density $\rho_\ell = 0.867$ g/mL and MW = 92.14, then follow the above steps:

a. Allowable $C = 50$ ppm (92.14/24.45) = 188 mg/m^3
b. Liquid volume = 1 L
c. Mass = 1 L $\times$ 0.867 $\times$ 10^6 = 8.7 $\times$ 10^5 mg
d. Air volume = 8.7 $\times$ 10^5 mg/188 mg/m^3 = 4630 m^3 (using Equation 5.6c)
e. Air volume = 4630 m^3 $\times$ 28.3 ft^3/m^3 = 163,500 ft^3

VII. DIFFERENTIATING BETWEEN COMPLETE AND INCOMPLETE EVAPORATION

As stated above, the concepts underlying incomplete evaporation and complete evaporation in an unventilated setting are both fundamental to IH. They can be used in recognition, evaluation, and control. Predicting the concentration in either setting is simple once one can decide which kind of problem it is! Such a differentiation is based on the ratio of the volume of volatile liquid relative to the spatial volume of the room that must be filled by its saturated vapor. The following subsections will describe one qualitative and two quantitative ways to decide when a liquid volume is large enough to reach equilibrium. If the liquid volume is not large enough, then it is small enough to evaporate completely.

1. Almost by definition, incomplete conditions will always apply inside of a partly filled storage tank or any confined space with lots of liquid and enough time for the liquid and vapor in the headspace atmosphere to come into equilibrium. Settings in a poorly ventilated room with either a large open solvent storage container or a large spill (enough liquid to cover the entire floor) also imply incomplete evaporation. For academic purposes, *any problem given to you where the volume is stated as large can always be assumed to be an incomplete evaporation setting*. For practical purposes, any quantity that can be carried, poured, or otherwise manipulated by hand can be assumed to be small.

The following two quantitative ways to differentiate an incomplete from a complete evaporation setting parallel the uses in the previous section of Equation 5.6b and Equation 5.6c. The only way to use Equation 5.6a is so roundabout that it is best skipped.

2. An approach closely related to Equation 5.6b (mass = volume × concentration) can be used to find the maximum mass or liquid volume of the contaminant that if released into an enclosed air volume would generate an equilibrium concentration equal to the liquid's vapor pressure.

EXAMPLE 5.6

What is the maximum volume of toluene that if released into the same 8 × 18 × 10 ft tall room used in Example 5.2 and allowed to completely evaporate, would just reach its saturated equilibrium? The following solution proceeds using Equation 5.6b and toluene's $P_{vapor} = 22$ mmHg and MW = 92.13 looked up from the previous example.

a. the maximum $C = 10^6 \times P_{vapor}/P \times MW/24.45 = 10^6 \times 22/760 \times 92.13/24.45$ $= 109{,}000\ mg/m^3$
b. air volume = $8 \times 18 \times 10 = 1440\ ft^3 \times 0.028\ m^3/ft^3 = 40.3\ m^3$ (as above)
c. mass of toluene = $40.3\ m^3 \times 109{,}000\ mg/m^3 = 4{,}400{,}000$ mg (using Equation 5.6b)
d. mass of toluene = $4{,}400{,}000\ mg \times g/10^3\ mg = 4400$ g
e. liquid volume = $4400\ g/(0.867\ g/mL) = 5070\ mL \approx 5$ L

3. The other quantitative approach uses Equation 5.6c only indirectly but is perhaps the easiest way to classify the liquid volume as large. This third approach (eventually embodied into Equation 5.11) also uses both the equivalent volume of vapor evolved from a liquid stated in Equation 5.9c.

$$\text{Equivalent volume of vapor (m}^3) = \frac{\text{liq.Vol (L)} \times \rho_\ell \times 24.45}{\text{MW}} \tag{5.9c}$$

and the principle from Dalton's law (reformatted as Equation 5.10) that the fraction of the room's total air volume occupied by the vapors at equilibrium is the same as both the molar fraction of the vapor (Y_i) and the ratio of the chemical's vapor pressure to the ambient air's total pressure.

$$\frac{\text{Equivalent volume of vapor (m}^3)}{\text{Room air Vol (m}^3)} = Y_i = \frac{P_{vapor}}{P} \tag{5.10}$$

Equation 5.11 was derived by inserting the right-hand side of Equation 5.9c into Equation 5.10 and solving for the liquid volume necessary for the volume of vapor to equal the chemical's partial pressure at equilibrium. A large volume of liquid would be any volume in excess of that defined by Equation 5.11:

$$\text{"large" liq. Vol (L)} > \frac{\text{Room Vol (m}^3) \times P_{vapor} \times \text{MW}}{P \times \rho_\ell \times 24.45} \tag{5.11}$$

If the room volume is known in terms of cubic feet, the value of 24.45 in the denominator of Equation 5.11 could be replaced with the value of 0.8634 ft^3/mole (derived from $24.45 \times 0.035315\ ft^3/L$). At the ACGIH definition of normal (21°C or 70°F), these values would be 24.13 L/mole and 0.8522 ft^3/mole, respectively. The following example converts the room volume into meters before using Equation 5.11 as written.

EXAMPLE 5.7

What is the minimum liquid volume of toluene that would create an incomplete evaporation setting in a classroom measuring $31 \times 30 \times 9$ ft^3 high?

To use Equation 5.11 one needs to know (perhaps from the *NIOSH Pocket Guide*[46]) that toluene's liquid density is 0.867 kg/L, its vapor pressure is 22 mmHg, and its MW is 92.13. As a practical convenience in most rooms, one may disregard any solid furniture, machinery, or people; thus, the air volume equals 8370 ft^3 = 237 m^3.

$$\text{large liq. Vol} \geq \frac{237\ \text{m}^3 \times 22\ \text{mmHg} \times 92.13\ \text{g/mole}}{760\ \text{mmHg} \times 0.867\ \text{g/mL} \times 24.45\ \text{L/mole}} = 29.8\ \text{L} \approx 7.8\ \text{gallons}$$

✓ At toluene's density of about 7 lb per gallon, the liquid volume in this example would weight about 50 pounds (20 kg), which supports the rule-of-thumb that the minimum liquid volume necessary to qualify a problem as an incomplete evaporation setting is more than one can comfortably carry.

Finally, some semiquantitative insight into complete vs. incomplete evaporation can be gained by knowing the ratio between the volume that a chemical will occupy as a vapor and its volume as a liquid. Equation 5.12a (derived by rearranging either Equation 5.9b or Equation 5.9c) yields a theoretical ratio if the vapor could be pure (but that is really only possible for a gas, not for a vapor).

$$\frac{\text{hypothetical pure vapor volume (L)}}{\text{liquid volume (L)}} = \frac{\rho_\ell \times 24450}{\text{MW}} \approx \frac{24450}{\text{MW}} \tag{5.12a}$$

✓ As the density (ρ_ℓ) of most organic solvents only varies between approximately 0.7 and 1.5, the ratio of the volumes depends mainly on the solvent's molecular weight. For a molecular weight of about 100 g/mole (about midway between the range of 80 and 120 common to many industrial solvents), the hypothetical ratio will be about 250:1. That is, the molecules of a chemical of near-unit density with a MW of approximately 100 g/mole (literally 102) will occupy *at least* 250 times more volume as a vapor than as a liquid.

However, a vapor (by definition) can never comprise all of an air volume. Thus, the ratio of the volumes of contaminated air to evaporated liquid will be greater than the hypothetically pure vapor, depending upon how contaminated the air is. The more practical ratio is given by the first part of Equation 5.12b where Y_i could be the chemical's measured concentration, its lower limit of flammability, or its exposure limit, but Y_i can never exceed the chemical's saturated vapor mole fraction ($P_{\text{vapor}}/P_{\text{atmospheric}}$ in the last part of Equation 5.12b). Using the nominal solvent MW $\approx$ 100 and P_{vapor} of 7.6 mm Hg, this vapor to liquid ratio would be $\geq$24,450!

$$\frac{\text{saturated air volume (L)}}{\text{liquid volume (L)}} = \frac{\rho_\ell \times 24450}{Y_i \times \text{MW}} \approx \frac{24450}{Y_i \times \text{MW}} \geq \frac{\rho_\ell \times 24450 \times P}{P_{\text{vapor}} \times \text{MW}} \tag{5.12b}$$

VIII. VAPORS FROM CONTINUOUS EVAPORATION WITH VENTILATION

The above solutions are all for unventilated settings such as a leak or an open container in a closed room. However, people rarely work in unventilated rooms. Routine work with volatile chemicals is much more likely to involve more-or-less continuous evaporation into a ventilated room. Chapter 19 will show that the concentration outside of the plume from continuous evaporation in a ventilated setting (such as from an open container) will fairly rapidly achieve a steady state as described in subscenario $3_{\text{large liq.}}$ in Table 5.1. The following discussion provides an important transition between the previous unventilated scenarios and the ventilated settings that comprise the bulk of the succeeding chapters. It also explains how the model to predict exposure in the steady-state ventilated subscenario $3_{\text{large liq.}}$ uses principles from both complete and incomplete evaporation in unventilated subscenarios $4_{\text{small liq.}}$ and $4_{\text{large liq.}}$, respectively.

1. Predicting the rate of continuous evaporation is partly analogous to incomplete evaporation in that the rate of evaporation is proportional to the chemical's vapor pressure (P_{vapor}). Recall that in the model for evaporation rate (G_{moles}) previously introduced as Equation 5.3, the liquid and vapor are considered always to be in equilibrium at the bottom of the boundary layer of air flowing over or past the liquid source. Equation 5.13 takes Equation 5.3 that next small step of assuming that ambient $P_{partial}$ is well less than P_{vapor} to make the simplest and hopefully the most memorable form of the evaporation equation, herein called the basic evaporation equation.

$$\text{evaporation rate } G\text{ (moles/min)} = (\text{Geom.Coef.}) \times A \times V^{0.5} \times P_{vapor} \tag{5.13}$$

where

Geometry Coefficient is the catch-all empirical constant that characterizes the thickness of the boundary layer due to the shape of the liquid source and air turbulence and converts the units with the odd powers on the right side of the equation into those on the left.

A is the wetted area or the surface area of the volatile liquid.

V is the air velocity passing the liquid source that creates the plume.

P_{vapor} is the vapor pressure of the particular chemical at its source temperature.

While the role of P_{vapor} in the rate of evaporation given by Equation 5.13 is analogous to incomplete evaporation, the average concentration that will result from a source that is continuously evaporating into a ventilated room given by Equation 5.14 is also analogous to complete evaporation. Equation 5.5 stated that the concentration from complete evaporation is simply the mass evaporating divided by the unventilated air volume. Equation 5.14 uses this same principle except that the concentration is the mass of chemical evaporating *in a given amount of time* diluted into the air volume flowing through the zone of concern *during that same amount of time*. In more quantitative terms, the steady-state concentration (still *C* in mg/m^3) equals the mass rate of evaporation (*G*, given in or converted to "mg per minute" = $G_{moles} \times MW_{i)}$ divided into some apparent volumetric flow rate of fresh air (*Q* in m^3/min or converted from ft^3/min into m^3/min with a subscript to be explained).

$$\text{steady state } C\ (\text{mg/m}^3) = \frac{G\ (\text{mg/min})}{Q_{apparent}\ (\text{m}^3/\text{min})} = \frac{G\ (\text{moles/min})\text{MW}}{Q_{apparent}\ (\text{m}^3/\text{min})} \tag{5.14}$$

$Q_{apparent}$ is the volumetric flow rate of ventilation air into which the contaminant appears that it is being diluted. The term "apparent" is used because the stream of contaminant represented by *G* is not diluted into any real volume flowing per minute. It only appears to be diluted because the concentration (*C*) at any location (whether in the near-field breathing zone (scenario 1) or around the free-field plume (scenario 2) is always less than $C = P_{vapor}$ at the source. In fact, the dilution will depend strongly on one's orientation and distance from the source. Flynn and George and Cherrie both used this concept in separate empirical mathematical models of worker exposures in the near field.[17,43] A similar-looking but more explicit $Q_{effective}$ will be used in Chapter 8 and defined in Chapter 19 in the context of the room accumulation scenario 3.

Equation 5.14 is the first of three times that this important concept will be stated within this book (see also Equation 19.14 and Equation 20.1). Equation 5.14 can also be viewed as simply Equation 5.6 put into the time domain. For instance, if the 1 L of toluene spilled all at once in Example 5.5 were released slowly and allowed to evaporate at a constant rate into a ventilated room over an 8-h workday, the vapors would still need at least 163,500 ft^3 of fresh air to be diluted to their TLV. However, rather than the fixed room volume into which a spill would have to be diluted, the volume for this slow release would be the ventilation air that flows into and out of that room over those 8 h. Ventilation rates are normally expressed as a volume per minute. In this case, dividing that 163,500 ft^3 by the 480 min in 8 h would yield a necessary equivalent ventilation rate of 340 ft^3/min (written as 340 ft^3/min), about the same amount of air flow rate that should be provided to a classroom for 20 to 30 students.

2. At least in theory, Equation 5.13 and Equation 5.14 could be combined into Equation 5.15a to predict the average concentration from continuous evaporation to which someone would be exposed. The concept is theoretical because Q_{apparent} would depend upon the air flow pattern and the person's proximity to the source and the plume in a way that is currently difficult if not impossible to predict. In practice, Geom.Coef. is also difficult to predict for most sources. Nonetheless, this equation can be manipulated to good effect as explained below.

$$\text{vapor } C\ (\text{mg/m}^3) = \left(\frac{(\text{Geom.Coef.}) \times (\text{Area}) \times V^{0.5}}{Q_{\text{apparent}}} \right) \times \text{MW} \times P_{\text{vapor}} \qquad (5.15\text{a})$$

First, recall the two ways of expressing vapor concentration embodied within Equation 2.14b and Equation 2.15b:

$$C_i = P_{\text{partial}} \frac{\text{MW}_i \times 10^6}{24.45 \times P} \qquad (\text{Equation 2.14b})$$

$$C_i = \text{ppm}_i \frac{\text{MW}_i}{24.45} \qquad (\text{Equation 2.15b})$$

These two equations make C (mg/m^3), P_{partial}, and ppm virtually interchangeable. Since the numeric value of Geom.Coef. in Equation 5.15a depends upon the units of the other variables (see the discussion ca. Equation 5.3), it could just as easily be expressed in terms of the partial pressure of the vapor. Thus, Equation 5.15b has pressure terms on both sides of the equation.

$$P_{\text{partial}}\ (\text{mmHg}) = \left(\frac{(\text{Geom.Coef.}) \times (\text{Area}) \times V^{0.5}}{Q_{\text{apparent}}} \right) \times P_{\text{vapor}} \qquad (5.15\text{b})$$

The arrangement of Equation 5.15b suggests that the vapor concentration (P_{partial}) at any location in the vicinity of an evaporating source can be viewed as having two groups of determinants: a group of environmental determinants (those in the parentheses on the right side of Equation 5.15b) and a chemical determinant (the chemical's vapor pressure on the far right side of Equation 5.15b).

- None of the environmental determinants of exposure within the parentheses of Equation 5.15b has anything to do with the chemical *per se*. The numerator in the parentheses comprises all of the physical characteristics of the source that affect its evaporation rate. The denominator encompasses all of the physical mechanisms that dilute the plume in the pathway from the source to the specific location at which the vapor concentration is being described (such as someone's breathing zone). Together, this whole group of environmental variables determines by how much the vapors will be diluted between the P_{vapor} concentration existing at the source and the P_{partial} concentration at the location of interest in any of the first three exposure scenarios. Environmental factors only affect dilution; they do not affect the initial vapor concentration at the source.

- The singular chemical determinant of exposure on the far right side of Equation 5.15b has nothing to do with the environment in which the chemical is being used. The vapor pressure (P_{vapor}) is a characteristic of the particular volatile chemical and its liquid temperature. The chemical component only affects the source of exposure, not the pathway or its dilution.

Most aerosols (except for fumes) have no chemical-specific vapor pressure term, but an equivalent G (mg/min) for aerosols can be thought of as a characteristic of any particular aerosol source. Once an aerosol is generated, its dilution still depends on $Q_{apparent}$ (if it is small enough ($< 10\ \mu$m) to not be affected much by sedimentation).[47] Thus, an aerosol concentration could be predicted using Equation 5.14 in much the same way as for vapors if the aerosol generation rate G and environmental dilution due to $Q_{apparent}$ were known. However again, $Q_{apparent}$ can rarely be predicted quantitatively in practice.

3. To explain the implications of this partitioning, let us first discuss the environmental determinants of exposure. Equation 5.16 isolates the environmental determinants by dividing Equation 5.15b by P_{vapor} and inverting. The resulting ratio is the amount by which the vapor pressure concentration at the source (P_{vapor}) is diluted to create the concentration ($P_{partial}$) at any location affected by that source. Since the ratio in Equation 5.16 is the amount of dilution created by the environment between the source and any point in that environment, this ratio will be called the environmental dilution ratio or EDR. Again, notice that all of the variables that comprise the EDR relate to the environment, not to the chemical being used in that environment. This ratio is unitless as long as the units of concentration in both the numerator and denominator are the same (or made to be the same by using Equation 2.14b or Equation 2.15b).

$$\text{Environmental Dilution Ratio (EDR)} = \frac{P_{vapor}}{P_{partial}} = \left(\frac{Q_{apparent}}{(\text{Geom.Coef.}) \times (\text{Area}) \times V^{0.5}}\right) \tag{5.16}$$

Figure 5.12 and Figure 5.13 present two currently unique examples of actual vapor plume dilution profiles generated downwind of a full-size mannequin standing 10 cm in front of

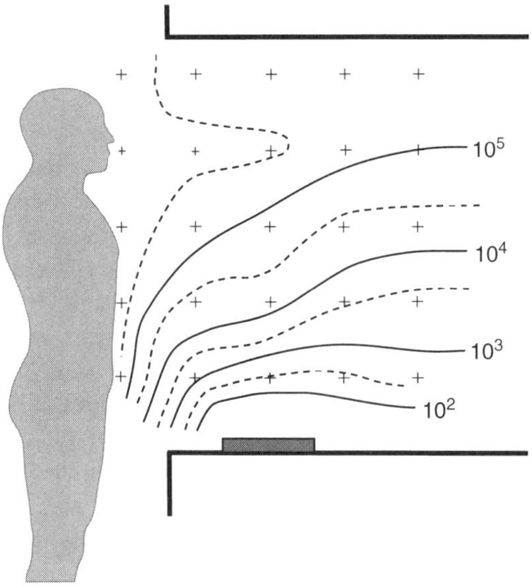

FIGURE 5.12 Patterns of dilution ratios (see Equation 5.16) resulting from a broad pan of acetone inside a laboratory exhaust hood with a face velocity of 50 fpm.

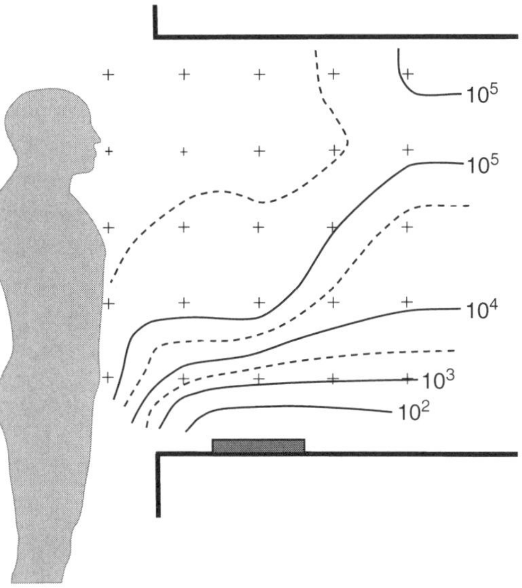

FIGURE 5.13 Patterns of dilution ratio (see Equation 5.16) resulting from a broad pan of acetone inside a laboratory exhaust hood with a face velocity of 100 fpm.

a laboratory chemical hood containing downwind of the mannequin a shallow, 19-cm wide pan of acetone.[23] The vapor concentrations (P_{partial}) were measured with a Miran 1A at 25 points on a 6 by 6-in. grid pattern (marked in the figures by the small + symbols) in a plane centered on the mannequin. The air concentration at each point (P_{partial}) was divided by the calculated vapor pressure (P_{vapor}) to yield the EDR as defined in Equation 5.16. The resulting data were analyzed using a SAS procedure to produce the lines of constant EDR as marked.

One can see from these dilution profiles that EDRs of 10^2 to 10^3 occur by the time the plume is only 6 in. from the source in this particular setting. The dilution ratios of 10^4 and 10^5 near the breathing zone imply that some vapors are migrating upwind due to the high eddy diffusivity in the wake of the mannequin (as depicted schematically in Figure 5.3). Similar near-field plume dilution profiles would apply to any other solvent placed in the same geometry (with slight effects due to differences among plume densities as discussed in Section VI of Chapter 2).

An EDR could be specified at any location (even beyond those shown in the above figures), but industrial hygienists are typically interested in the EDR in the breathing zone of a person exposed to a chemical. For simplicity, this ratio will be abbreviated as EDR_{BZ} and defined for future use in Equation 5.17.

$$\text{Environmental Dilution Ratio}_{BZ}\ (EDR_{BZ}) = \frac{P_{\text{vapor}}\ (\text{or units of } C_{\text{at } P_{\text{vapor}}})}{P_{\text{partial,BZ}}\ (\text{or units of } C_{BZ})} \tag{5.17}$$

The Miran was not quite sensitive enough to be able to measure acetone in the breathing zone of the mannequin in Figure 5.12 and Figure 5.13, but a little extrapolation suggests that the EDR_{BZ} for this particular source geometry at an air velocity of 50 fpm might be near 10^6. The EDR_{BZ} at 100 fpm appears to be somewhat higher but not much. Before trying to apply these data too simply to other source or breathing zone geometries, keep in mind that changing the position of the breathing zone is not quite the same as envisioning the person's head and body closer to the source or within either one of these plumes because the plume would change as the geometry changes.

4. Having separated the environmental from the chemical determinants of exposure allows chemical hazards to be looked at in a new way. An industrial hygienist usually measures a chemical's concentration in someone's breathing zone, then evaluates the acceptability of that exposure by the ratio of the measured breathing zone concentration (denoted herein as either C_{BZ} in units of mg/m^3 or ppm_{BZ} in units of ppm) to that chemical's TLV, PEL, or other exposure limit in the same units. Such a ratio as defined in Equation 5.18 may be called a compliance ratio. To be acceptable, the compliance ratio must be equal to or less than one.

$$\text{Compliance Ratio} = \frac{\text{measured } C_{BZ} \text{ or } ppm_{BZ}}{\text{TLV or PEL}} \text{ must be} \leq 1 \text{ to be acceptable} \quad (5.18)$$

Knowing that P_{vapor} (or its equivalent in either ppm or mg/m^3) is the maximum concentration at which a given chemical can exist as a vapor, one can anticipate that the maximum compliance ratio that a chemical's vapor can potentially create is equivalent to the ratio of how much greater P_{vapor} is than its exposure limit. The vapor hazard ratio or VHR as expressed mathematically in Equation 5.19 is the maximum potential hazard that a given chemical's vapors can generate.

$$\text{Vapor Hazard Ratio} = \frac{P_{vapor} \text{ in units of ppm or mg/m}^3}{\text{exposure limit in the same units}} \quad (5.19)$$

The two properties that comprise the VHR (its vapor pressure and its exposure limit) are both intrinsic to a given chemical (and have nothing to do with the environment *per se*). Together, they specify the minimum amount by which a given chemical needs to be diluted by the environment to reduce its concentration from its vapor pressure at the source to its exposure limit in the breathing zone. If the vapors in the breathing zone are not diluted by as much as the chemical's VHR, then someone will be overexposed. The VHRs will be discussed in Chapter 8 as being very useful when selecting a substitution control for volatile chemicals.

5. Of course nothing is wrong with evaluating compliance as simply the ratio of measured results to an exposure limit, but consider the advantages of viewing compliance in terms of the two concepts of chemical dilution and environmental dilution separately. Think of looking at evaluation as comparing the amount of dilution that some chemical needs to the amount of dilution a given environment actually creates in someone's breathing zone. This alternative view of IH evaluation is the same as asking if the EDR in someone's breathing zone is equal to or greater than the VHR for the chemical being used, as expressed in Equation 5.20.

$$\text{Compliance Ratio} = \frac{C_{\text{measured in the BZ}}/P_{vapor}}{\text{TLV or PEL}/P_{vapor}} = \frac{\text{VHR}}{\text{EDR}_{BZ}} \leq 1 \text{ to be acceptable} \quad (5.20)$$

Conceptually viewing compliance in its separate chemical and environmental components can yield several benefits. For instance, one could anticipate that a given chemical should only be used in an environment that can dilute the vapors reaching the breathing zone (the EDR_{BZ}) by as much as the toxicity of a given chemical requires its vapors to be diluted (its VHR), as stated in Equation 5.21.

✍ To be acceptable, the "EDR_{BZ}" must be $\geq$ a given chemical's "VHR" (5.21)

If an evaluation reveals that an employee is being overexposed, then one can conclude that the environment cannot create sufficient dilution for the chemical being used. One could then compare the advantages of attempting to achieve exposure control either by reducing the chemical's intrinsic VHR (e.g., by substituting an intrinsically safer chemical) or by increasing the environment's ability to dilute a chemical reaching someone's breathing zone (e.g., by modifying the work practice, by adding local exhaust ventilation, or by changing the general ventilation to increase the environment's ability to dilute the plume of whatever chemical is being used).

This conceptual separation can also improve a hygienist's ability to compare similarities and differences among settings and to recognize potentially hazardous conditions more readily. Viewing compliance as dilution might improve our ability to predict the effectiveness of ventilation controls for an otherwise similar source (imagine other versions of Figure 5.12 and Figure 5.13 or tables of dilution ratios for a wider range of settings). This separation also has direct applications to control banding.

6. The following discussion is presented to give the reader a glimpse of dilution and compliance in real settings. Figure 5.14 plots the cumulative frequency (using a probability scale on the *x*-axis) of the assumed vapor dilution ratio (on the *y*-axis) from about 8500 personal air samples collected by OSHA for these particular chemicals between 1989 and 1998. The term assumed is used here because the OSHA database does not specify the source of the sampled vapor; it was assumed to be a pure liquid (rather than a mixture) at NTP (neither heated nor above sea level). (The effect of liquid temperature was discussed in Chapter 2. Emissions from mixtures will be discussed in Chapter 6.) In only 1 to 2% of these settings was the dilution ratio less than 100 (many of these were when xylene was in use). The majority of these settings had dilution ratios between 10^3 and 10^6, the same range as the dilution profiles in Figure 5.12 and Figure 5.13.

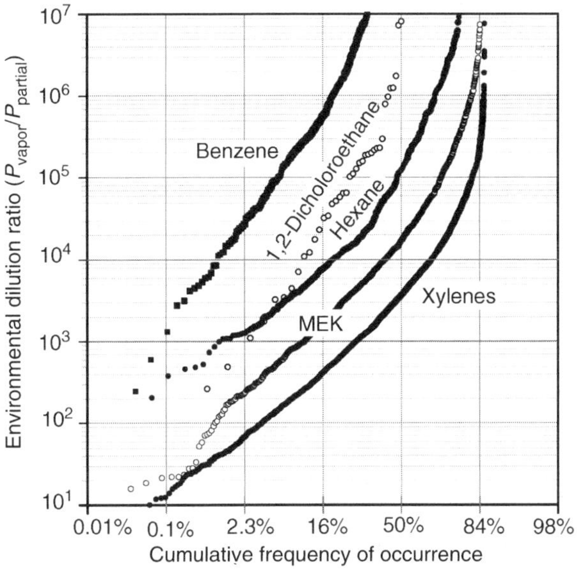

FIGURE 5.14 About 8500 OSHA vapor sample results expressed in terms of EDRs.

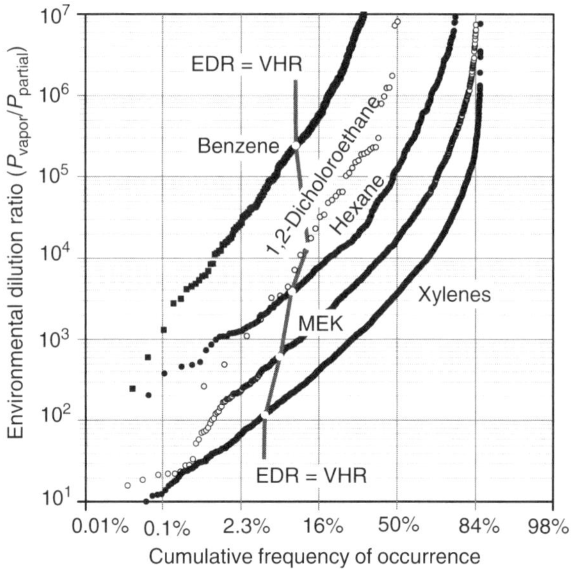

FIGURE 5.15 The same data as in Figure 5.14 to which a line is added that connects the points in each distribution where an EDR equals that chemical's VHR.

Figure 5.15 contains the same data as Figure 5.14 to which a line was added that passes through the point in each chemical's distribution where its EDR equaled its VHR. The VHR for each of these chemicals is listed in Table 5.2. Exposures above the point within each chemical's data where the dilution ratio is greater than the chemical's VHR are in compliance. Because the equal ratio points happen to be aligned nearly vertically in this set of data, compliant exposures appear to the right of the added line, while noncompliant exposures appear to the left of this line. Noncompliant exposures comprise only 5 to 15% of the samples within this particular set of data. In fact, more than 50% of these chemicals were diluted hundreds to thousands of times more than their VHR demanded (the fraction of these samples that are two or more orders of magnitude above the EDR = VHR point for each chemical).

The magnitude of these apparent dilutions can also be compared to the VHRs of a much broader array of volatile chemicals listed in Appendices A and B of this book that range from more that 10^7 to less than 10^{-3}. The data in Figure 5.15 show that more that 50% of the environments in which benzene or 1,2-dichloroethane were used created

TABLE 5.2
The Vapor Hazard Ratios for the Chemicals in Figure 5.14

Chemical Name	CAS No.	TLV (ppm)	P_{vapor} (mmHg)	VHR	VHI
Benzene	71-43-2	0.5	95.2	250,526	5.4
1,2-Dichloroethane	107-06-2	10	78.9	10,382	4.0
n-Hexane	110-54-3	50	151	3974	3.6
Methyl ethyl ketone (MEK)	78-93-3	200	95.3	627	2.8
Xylenes (mixed)	1330-20-7	100	8	105	2.0

VHI = $\log_{10}$(VHR)

dilution ratios of more than the 10^7, but for whatever reason, fewer than 16% of the settings in which MEK or xylene was used created EDRs greater than 10^7. At the other extreme, the lowest EDR observed was on the order of a $10\times$ dilution.

The database from which these results were extracted does not contain much information about the use conditions that might explain some patterns within these data. For instance, was there some reason that the chemicals that need more dilution (i.e., those with high VHRs) had more dilution (i.e., are plotted higher in Figure 5.14) than those chemicals with low VHRs? Were these chemicals all selected by someone with the vapor hazard and setting or compliance in mind (unlikely) or is this pattern merely a coincidence? Similarly, what is the reason for each of these distributions starting out as the straight line expected of log-normally distributed data (as at least the left-hand portion of each chemical's samples appear to be) before deviating upwardly? Could the upward deviation be the result of some very low evaporation rates (e.g., from small sources or because the vapor measured was only one component of a solvent mixture)? Or is the deviation because of abnormally large distances between the source and the sample? How many of these samples were collected just to satisfy an employee complaint or concern when the inspector knew from experience that the exposure would be minimal if even detectable analytically? The answers to these questions are unknown.

In summary, vapor pressure strongly affects a chemical's evaporation rate and its vapor concentration whether in an unventilated setting with a large amount of liquid or in a ventilated setting with a continuously evaporating liquid. In settings with similar source geometries and airflow conditions, the vapor concentration will be proportional to the P_{vapor} at each plume's source. The amount of dilution of an airborne contaminant depends upon the environment and is largely independent of the chemical being diluted. In fact, once airborne, vapor and small aerosol contaminants behave much the same way.

- The vapor (or aerosol) concentration will always be highest in the plume near the source before it has much of a chance to be diluted. Exposure in the near-field is dependent upon the person's proximity to the source and orientation to the plume.
- The concentration within a plume is determined by distance and the air's turbulence, but normally the concentration will change much more slowly with distance along the axis of the plume than with distance across the axis of the plume. Thus, the exposure level in a free-field depends much more upon the person's orientation to the plume than their proximity to the source.
- Local exhaust ventilation can have a strong effect on the near-field plume direction and can practically eliminate the free-field plume and accumulating fields. In contrast, general room ventilation has almost no effect on the near-field plume, but it can affect the direction of the free-field plume and its rate of dispersion.
- Only if the plume can be detected or its direction is reasonably constant can one's orientation to the plume be an effective control. Some exposure will occur even to someone working upwind from the source, such as standing in front of a local exhaust hood.
- To the extent that the plume is not completely removed from the room by local exhaust ventilation, vapors will accumulate in a room to become a relatively uniform, elevated, ambient concentration that is dependent in large part on the vapor pressure at the source and the effective amount of general ventilation reaching any location within the room.

The amount of dilution necessary to reduce the vapor concentration within any of these fields to an acceptable breathing zone concentration is the VHR specific to the chemical. The amount of actual dilution between the source and the breathing zone is the EDR that is specific to the environment and work practice. Getting a sense of the EDR from the EDR profiles in Figure 5.12 and Figure 5.13, realizing that the real EDRs in Figure 5.14 span a broader array of values, and relating these values to the range of dilutions that common chemicals need (such as from the list of VHRs in Appendix A or B of this book) is the fastest way this author knows of to put the perspective of an "old head" into a young body.

PRACTICE PROBLEMS

In each of this series of "accidents," one-half liter (1.06 pint) of toluene is spilled in a closed storage building ($20 \times 20 \times 10$ ft^3, about the room volume of a small classroom) with no ventilation. At least initially assume complete vaporization of this small volume, and remember that the pressure inside a closed room always equals the ambient barometric pressure. The first and last question relates to Chapter 5; the second through fourth to Chapter 2; and the fifth and sixth questions to Chapter 4. The following data pertain to toluene:

TLV: 50 ppm (A4, skin)	MW: 92.1	P_{vapor}: 28.2 mmHg (at 77°F)
OSHA TWA PEL: 200 ppm	Boiling point: 232°F	P_{vapor}: 50.4 mmHg (at 98°F)
Ceiling: 300 ppm	Solub. (74°F) 0.07%	FRZ: − 139°F
For 10 min	Flash point: 40°F	UFL: 7.1%
Max. peak: 500 ppm	Specific gravity: 0.866	LFL: 1.1%

1. a. What would be the average concentration of toluene in mg/m^3 if the spill occurred at 77°F (25°C)?

   ans. = 3823 mg/m^3

   b. What would be the average concentration in ppm at 77°F (25°C) and one atmosphere?

   ans. = 1015 ppm

2. a. What would be the average concentration in mg/m^3 from the spill if the building were already heated in the afternoon sun to a temperature of 98°F (36.7°C) before the spill occurred?

   _____ mg/m^3

   b. What would be the average concentration in ppm if the spill had occurred at a room temperature of 98°F? Show your work.

   _____ ppm

3. a. What would be the average concentration in mg/m^3 if the spill occurred in the morning at 77°F (25°C), the vapors had time to disperse within the room, and then the building was heated in the afternoon sun to a temperature of 98°F (36.7°C)? Show your work.

   _____ mg/m^3

   b. What would be the average concentration in ppm if the spill had occurred in the morning and then the room was heated to a temperature of 98°F?

   _____ ppm

4. Why does the concentration change to 1055 ppm in one case and 3681 mg/m^3 in the other? (Only one value changes in each case.)
5. Was there a fire hazard anywhere in the room during the first few seconds right after the original spill in question 1 above? Why or why not?

   The max. Conc. = P_{vapor} = _____ UFL = _____ LFL = _____

6. Will there be a fire hazard at the final (well mixed) concentration in the room? (Another hint: you should get different answers for questions 5 and 6.) Why or why not?
7. How much solvent would have had to have been spilled in the above problem at NTP (problem 1) to have created an incomplete evaporation problem?

   ans. = 18 L

REFERENCES

1. Bird, R. B., Stewart, W. E., and Lightfoot, E. N., *Transport Phenomena*, Wiley, New York, 1960.
2. Sherwood, T. K., Pigford, R. L., and Wilke, C. R., *Mass Transfer*, McGraw-Hill, New York, 1975.
3. Fiegley, C. E., Ehmke, F. M., Goodson, T. H., and Brown, J. R., Experimental determination of volatile evolution rates from coated surfaces, *Am. Ind. Hyg. Assoc. J.*, 42, 365–372, 1981.
4. Bishop, E. C., Popendorf, W., Hanson, D., and Prausnitz, J., Predicting relative vapor ratios for organic solvent mixtures, *Am. Ind. Hyg. Assoc. J.*, 43, 656–661, 1982.
5. Powell, R. W., Estimating worker exposure to gases and vapors leaking from pumps and valves, *Am. Ind. Hyg. Assoc. J.*, 45(11), A7–A15, 1984.
6. Säämänen, A. J., Niemela, R. I., Blomqvist, T. K., and Nikander, E. M., Emission of styrene during the hand lay-up molding of reinforced polyester, *Appl. Occup. Environ. Hyg.*, 6(9), 790–793, 1991.
7. Braun, K. O. and Caplan, K. J., *Evaporation Rate of Volatile Liquids*, 2nd ed., US Environmental Protection Agency, National Technical Information Agency, Springfield, VA, NTIS pub. no. PB92-232-305, 1992.
8. Nielsen, F., Olsen, E., and Fredenslund, A., Prediction of isothermal evaporation rates of pure volatile organic compounds in occupational environments — a theoretical approach based on laminar boundary layer theory, *Ann. Occup. Hyg.*, 39(4), 497–511, 1995.
9. Nielsen, F. and Olsen, E., On the prediction of evaporation rates — with special emphasis on aqueous solutions, *Ann. Occup. Hyg.*, 39(4), 513–522, 1995.
10. Hummel, A. A., Braun, K. O., and Fehrenbacher, M. C., Evaporation of a liquid in a flowing airstream, *Am. Ind. Hyg. Assoc. J.*, 57(6), 519–525, 1996.
11. Jayjock, M. A., Well-mixed model, back pressure — affected generation rate, In *Mathematical Models for Estimating Occupational Exposure to Chemicals.*, Keil, C. B., Ed., AIHA Press, Fairfax, VA, pp. 41–44, chap. 6, 2000.
12. Caplan K., *Evaporation Rate of Volatile Liquids*, Final Report, EPA Contract Nos. 68-2-4248, 1988 and 68-D8-0112, 1989.
13. Anderson, E. A. and Szewczyk, A. A., A look at a universal parameter for 2 and 3-d bluff body flows, *J. Fluids Struct.*, 10, 543–553, 1996.
14. Keil, C. B., Eddy diffusion modeling, In *Mathematical Models for Estimating Occupational Exposure to Chemicals*, Keil, C. B. Ed., AIHA Press, Fairfax, VA, pp. 57–65, chap. 9, 2000.
15. Csanady, C. T., *Turbulent Diffusion in the Environment*, Reidel Publ. Co., Boston, 1973.
16. Roach, S. A., On the role of turbulent diffusion in ventilation, *Ann. Occup. Hyg.*, 24, 105–132, 1981.
17. Flynn, M. R. and George, D. K., A field evaluation of a mathematical model to predict worker exposure to solvent vapors, *Appl. Occup. Environ. Hyg.*, 11(10), 1212–1216, 1996.
18. Smagorinsky, J., Global atmospheric modeling and the numerical simulation of climate, In *Weather and Climate Modification*, Hess, W. N., Ed., Wiley, New York, pp. 633–686, 1974.
19. J. Smagorinsky, Epilogue: a perspective of dynamical meteorology, In *Dynamical Meteorology: An Introductory Selection*, Atkinson., B.W., Ed., Methuen, pp. 205–219, 1981.
20. Atkinson, B. W., Introduction to the fluid mechanics of meso-scale flow fields. In *Diffusion and Transport of Pollutants in Atmospheric Mesoscale Flow Fields*, Gyr, A., and Rys, F-S, Eds., Kluwer Academic Publishers, Dordrecht, Netherlands, pp. 1–22, 1995.
21. Gumaer, P. W., Ventilation of heavier than air vapors, In *Proceedings of 19th Annual Safety Congress. Plant and Process Ventilation*, Hemeon, W. C. L., Ed., Industrial Press, New York, p. 464–476 from p. 22, 1963.
22. Popendorf, W., Vapor pressure and solvent vapor hazards, *Am. Ind. Hyg. Assoc. J.*, 45(10), 719–726, 1984.
23. Livengood R. E., Vapor Dilution Profiles for Near-Field Hazard Evaluation, M.S. Thesis for the Department of Chemical and Materials Engineering, The University of Iowa, 1985.
24. Kim, T. and Flynn, M. R., Airflow pattern around a worker in a uniform freestream, *Am. Ind. Hyg. Assoc. J.*, 52(7), 287–296, 1991.
25. Kim, T. and Flynn, M. R., Modeling a worker's exposure from a hand-held source in a uniform freestream, *Am. Ind. Hyg. Assoc. J.*, 52(11), 458–463, 1991.

26. Kim, T. and Flynn, M. R., The effect of contaminant source momentum on a worker's breathing zone concentration in a uniform freestream, *Am. Ind. Hyg. Assoc. J.*, 53(12), 757–766, 1992.
27. George, D. K., Flynn, M. R., and Goodman, R., The impact of boundary layer separation on local exhaust design and worker exposure, *Appl. Occup. Environ. Hyg.*, 5(8), 501–509, 1990.
28. Flynn, M. R., Chen, M.-M., Kim, T., and Muthedath, P., Computational simulation of worker exposure using a particle trajectory method, *Ann. Occup. Hyg.*, 39(3), 277–289, 1995.
29. Madsen, U., Fontaine, J. R., Nielsen, P. V., Aubertin, G., and Breum, N. O., A numerical study of dispersion and local exhaust capture of aerosols generated from a variety of sources and airflow conditions, *Am. Ind. Hyg. Assoc. J.*, 57(2), 134–141, 1996.
30. Wadden, R. A., Scheff, P. A., and Franke, J. E., Emission factors for trichloroethylene vapor degreasers, *Am. Ind. Hyg. Assoc. J.*, 50(9), 496–500, 1989.
31. Wadden, R. A., Hawkins, J. L., Scheff, P. A., and Franke, J. E., Characterization of emission factors related to source activity for trichloroethylene degreasing and chrome plating processes, *Am. Ind. Hyg. Assoc. J.*, 52(9), 349–356, 1991.
32. Wadden, R. A., Baird, D. I., Franke, J. E., Scheff, P. A., and Conroy, L. M., Ethanol emission factors for glazing during candy production, *Am. Ind. Hyg. Assoc. J.*, 55(4), 343–351, 1994.
33. Scheff, P. A., Friedman, R. L., Franke, J. E., Conroy, L. M., and Wadden, R. A., Source activity modeling of freon emissions from open-top vapor degreasers, *Appl. Occup. Environ. Hyg.*, 7(2), 127–134, 1992.
34. Keil, C. B., Wadden, R. A., Scheff, P. A., Franke, J. E., and Conroy, L. M., Determination of multiple source volatile organic compound emission factors in offset printing shops, *Appl. Occup. Environ. Hyg.*, 12(2), 111–121, 1997.
35. Flynn, M. R. and George, D. K., Aerodynamics and exposure variability, *Appl. Occup. Environ. Hyg.*, 6(1), 36–39, 1991.
36. Brief, R. S., Simple way to determine air contaminants, *Air Eng.*, 2, 39–41, 1960.
37. Repace, J. L. and Lowrey, A. H., Indoor air pollution, tobacco smoke, and public health, *Science*, 208, 464–472, 1980.
38. Goldfield, J., Contamination concentration reduction: general ventilation versus local exhaust ventilation, *Am. Ind. Hyg. Assoc. J.*, 41, 812–818, 1980.
39. Bennett, J. S., Feigley, C. E., Underhill, D. W., Drane, W., Payne, T. A., Stewart, P. A., Herrick, R. F., Utterback, D. F., and Hays, R. B., Estimating the contribution of individual work tasks to room concentration: method applied to embalming, *Am. Ind. Hyg. Assoc. J.*, 57(7), 599–609, 1996.
40. Burton, D. J., General methods for the control of airborne hazards, In *The Occupational Environment: Its Evaluation and Control*, DiNardi, S. R. Ed., American Industrial Hygiene Association, Fairfax, VA, pp. 829–846, chap. 31, 1997.
41. Reinke, P. H., Box models, In *Mathematical Models for Estimating Occupational Exposure to Chemicals*, Keil, C. B. Ed., AIHA Press, Fairfax, VA, pp. 25–40, 2000.
42. Rodes, C. E., Kamens, R. M., and Wiener, R. W., Experimental considerations for the study of contaminant dispersion near the body, *Am. Ind. Hyg. Assoc. J.*, 56(6), 535–545, 1995.
43. Cherrie, J. W., The effect of room size and general ventilation on the relationship between near and far-field concentrations, *Appl. Occup. Environ. Hyg.*, 14(8), 539–546, 1999.
44. Nicas, M. and Armstrong, T. W., Using a spreadsheet to compute contaminant exposure concentrations given a variable emission rate, *Am. Ind. Hyg. Assoc. J.*, 64(3), 368–375, 2003.
45. Keil, C. B. and Nicas, M., Predicting room vapor concentrations due to spills of organic solvents, *Am. Ind. Hyg. Assoc. J.*, 64(4), 445–454, 2003.
46. NIOSH Pocket Guide to Chemical Hazards, DHHS (NIOSH) 97-140. 1997, (Also on-line at http://www.cdc.gov/niosh/npg/npg.html).
47. Bémer, D., Callé, S., Godinot, S., Régnier, R., and Dessagne, J. M., Measurement of the emission rate of an aerosol source — comparison of aerosol and gas transport coefficients, *Appl. Occup. Environ. Hyg.*, 15(12), 904–910, 2000.

6 Vapor Pressure in Mixtures

The learning goals of this chapter:

- ☐ Be able to calculate the vapor pressure of a component within an ideal mixture using Raoult's law.
- ✍ Understand why Raoult's law often does not work well for mixtures involving organic solvents.
- ✍ Be able to differentiate between a mixture that is likely to be ideal and one that is likely to be nonideal.
- ☐ Be able to use a known activity factor (γ) to estimate a component's vapor pressure and to interpret the magnitude of a component's deviation from an ideal mixture and Raoult's law.
- ✍ Understand how Henry's law and Henry's coefficients are a special case of Raoult's law.

The previous chapter showed how the vapor pressure of each chemical plays a leading role in determining its airborne concentration when evaporation is either incomplete in a closed (unventilated) setting or continuous in an open (ventilated) setting. This chapter will discuss the vapor pressure of chemicals in a liquid mixture. The vapor pressure in mixtures has uses other than predicting evaporation or an equilibrium condition. An "out of equilibrium condition" can be used to cause a gas or vapor to move either from a liquid like water into air or from air into a liquid. Such movements can either be designed to occur beneficially or become a hazard by occurring accidentally. For instance, an aerator can be designed to strip contaminants from water into air as a water purification step, or an air scrubber can be designed to reduce air emissions by absorbing airborne toxins into water. In a parallel fashion, the vapor pressure of an unwanted contaminant coming out of water can create an inhalation hazard, and the absorption of oxygen and toxic vapors into the body depends upon the absorption from air into blood or other lung fluids.

Three basic approaches to predicting vapor pressures in mixtures will be presented: Raoult's law for ideal mixtures, an empirical adjustment to Raoult's law for nonideal mixtures, and Henry's law for dilute mixtures in water. Each approach has its advantages. Raoult's law is simple, but many mixtures are not ideal. Empirical coefficients are illustrative of nonideal effects but are not simple to predict, and Henry's law coefficients exist for many chemicals in water but they have limited applicability. No model can predict all mixtures, but models can help to interpolate or extrapolate experimental measurements.

While classical physical chemistry or industrial hygiene examinations test one's knowledge of Raoult's law, this chapter tries to point out how often Raoult's law is incorrect, and while neither the typical IH student nor prospective CIH candidate is expected to master any of these predictive methods, the student should learn when to anticipate deviations and how one could use an empirical adjustment to Raoult's law if presented with a value. Failure to anticipate that a mixture is not ideal usually results in underestimating the health hazard and/or overexposing people to mixtures of organic vapors.

I. MIXTURES

1. Chemical mixtures are used in industry where they meet a particular production need better or more cheaply than a single pure component used alone. Pure chemicals are called "neat" within the petrochemical industry (the dictionary definition of neat is unmixed or undiluted). Chemical mixtures may be either a blended combination of neat chemicals (sometimes called a formulated mixture), a condensed distillate from a more complex mixture (sometimes called a cut), the result of absorption of an airborne chemical into a liquid, the product of an incomplete reaction, or some combination thereof. Only mixtures of volatile chemicals can be distilled. Mixtures of nonvolatile chemicals can only be prepared as a blended mixture or as an impure product.
2. Certain volatile mixtures can be produced by distillation much more cheaply than either pure solvents or blended mixtures of pure solvents. Table 6.1 gives some examples of common solvents distilled at defined (often proprietary) T and P conditions. Distilled solvents are complex and their composition varies somewhat with the feed stock, making accurate predictions of their individual vapors both more difficult and less precise, respectively. As a result, some of the more common distilled solvents have their own TLV® (including diesel fuel and gasoline not listed in Table 6.1). The range of some properties in Table 6.1 reflects the multiple grades of these solvents.
3. With solid aerosols, what you see in the powder or its source is pretty much what you get in the aerosol (e.g., the minerals in soil, foliar dust, and aerosols[5]). Thus, the nature of an airborne solid is easy to predict if you know the nature of the source of that solid. The same can be said for a nonvolatile liquid. The special case of vapors resulting from the complete evaporation of a mixture in an unventilated setting is again very easy to predict because when all the liquid evaporates, the composition of the vapors (Y_i) is the same as

TABLE 6.1
Some Chemical and Physical Properties of Common Distilled Petroleum Solvents[1–4]

	TLV	Prevailing Molecular Species	Molecular Weight (TLV Value)	Specific Gravity	Vapor Pressure (mmHg)	Flash Point Range	Flammable Limits (%)
Petroleum ether (or ligroin)		C_5–C_6	74–77	0.63–0.66	100	−57 to −46°C −70 to −50°F	1.1–8.0
Rubber solvent (or naphtha)	400 ppm	C_5–C_8	~97 (97 mean)	0.674–0.85	unk.	−46 to −13°C −50 to 9°C	1.1–6.5
Petroleum naphtha		C_6–C_8		0.6–0.8	40		
VM&P naphtha	300 ppm	C_7–C_{11}	87–114 (114.0)	0.727–0.76	2–20	−7 to 13°C 20 to 55°F	0.9–6.0
Coal tar naphtha (aromatic petroleum)		C_8–C_{11}	~140	0.86–0.89	<5	2 to 38°C 36 to 100°F	
Mineral spirits (or petroleum spirits)		C_9–C_{12}	144–169	0.77–0.81	0.8	30 to 40°C 86 to 105°F	1.0–6.0
Stoddard solvent (or white spirits)	100 ppm	C_8–C_{12}	135–145 (140.0)	0.75–0.80	4–4.5	38 to 60°C 100 to 140°F	0.9–6.0
Kerosene (or stove oil)	200 mg/m^3	C_9–C_{16}	~170 (varies)	0.8	unk.	38 to 74°C 100 to 165°F	0.7–5.0

the composition of the liquid (X_i). Equation 6.1 states this equality mathematically.

$$\text{For aerosolization and complete evaporation: } Y_i = X_i \tag{6.1}$$

Predictions do not get any easier than this. Equation 6.1 applies to the composition of most aerosols generated from dry dusts and powders (except that solids are not typically defined in terms of molar X_i fractions). The aerosolization of a nonvolatile liquid would also follow Equation 6.1. Evaporation of the volatile portion of a mist (such as an aqueous mist) will not affect the composition or ratio of the dissolved or suspended but nonvolatile solids. In contrast, Equation 6.1 would not apply to evaporation of volatile components because they depend upon vapor pressure. However, before going on to the main discussion of predicting the vapor pressure of a volatile mixture, a very practical and generic issue for IHs dealing with mixtures must be addressed: how does one know X_i?

II. DEFINING LIQUID MIXTURES

All of the models that predict vapor pressure, vapor composition, or evaporation rates from mixtures base that prediction on the liquid molar fraction of each component (X_i as defined by Equation 2.6). Predicting the vapor pressures of mixtures would be much easier if they were defined or specified in molar fractions. Unfortunately, mixtures are almost never defined as molar fractions. Liquid mixture compositions are typically given either in volume percent (vol% such as l of a component per 100 L of mixture), in weight or mass percent (wt% such as grams of a component per 100 g of mixture), or in parts per million by mass (ppm_ℓ such as milligrams of a component per kilogram of mixture). Thus, in order to calculate the liquid molar fraction from these common units of composition, one needs to always account for each component's MW_i, and when starting from vol% one also needs to account for each component's liquid density ($\rho_{\ell,i}$) unless a simplifying approximation is acceptable.

$$X_i = \frac{(\text{Number of } i \text{ molecules within a liquid mixture})}{(\text{Number of all molecules within a liquid mixture})} \qquad \text{(repeat of Equation 2.6)}$$

The following notations and relationships are used to develop the conversions and the approximations shown below.

- As introduced in Chapter 5 (ca. Equation 5.7), the subscript ℓ is used when referring to the liquid state. Specifically, the term ρ_ℓ refers specifically to the density of a liquid (cf. ρ for the density of a gas or vapor) and ppm_ℓ refers to a liquid weight concentration (cf. an airborne ppm).
- The subscript "solvent" refers to the major solvent in the mixture where the minor constituents comprise no more than about 5% of the entire mixture (this criterion will allow future short cuts).

$$\text{wt\%}_{\ell,i} = \left[\frac{\text{grams}_i}{\text{grams}_{\text{mixture}}}\right]100 \tag{6.2a}$$

or

$$\text{grams}_i = \left[\frac{\text{wt\%}_{\ell,i} \times \text{grams}_{\text{mixture}}}{100}\right] \tag{6.2b}$$

$$\text{ppm}_{\ell,i} = \left[\frac{\text{grams}_i}{\text{grams}_{\text{mixture}}}\right]10^6 \tag{6.2c}$$

or

$$\text{grams}_i = \left[\frac{\text{ppm}_{\ell,i} \times \text{grams}_{\text{mixture}}}{10^6}\right] \tag{6.2d}$$

1. Equation 6.3a was derived using Equation 6.2b to find the liquid molar fraction (X_i) if the liquid mixture is given on the basis of weight percent ($wt\%_{\ell,i}$). Notice how the variables "$grams_{mixture}$" and the "100" above and below the line both cancel out.

$$X_i = \frac{mol_i}{mol_{mixture}} = \frac{mol_i}{\Sigma mol_i} = \frac{grams_i/MW_i}{\Sigma(grams_i/MW_i)}$$
$$= \frac{(wt\%_{\ell,i} \times grams_{mixture})/(100 \times MW_i)}{\Sigma(wt\%_{\ell,i} \times grams_{mixture})/(100 \times MW_i)}$$

$$X_i = \left[\frac{wt\%_{\ell,i}/MW_i}{\Sigma(wt\%_{\ell,i}/MW_i)}\right] \quad (6.3a)$$

The denominator in Equation 6.3a also equals 100/the solvent's average MW. If the component of interest is sufficiently dilute such that $\Sigma(wt\%_{\ell,i}/MW_i)$ in the above denominator can be approximated by $100/MW_{solvent}$ (which will be true when one solvent comprises more than about 90 to 95% of the mixture), then the conversion to X_i could be approximated in one step as follows:[a]

$$X_i = \left[\frac{wt\%_{\ell,i}/MW_i}{\Sigma(wt\%_{\ell,i}/MW_i)}\right] \approx \left[\frac{wt\%_{\ell,i}/MW_i}{100/MW_{solvent}}\right] \quad \text{(using Equation 6.3a)}$$

$$X_i \approx \frac{wt\%_{\ell,i}}{100}\left[\frac{MW_{solvent}}{MW_i}\right] \text{for dilute mixtures (when one } wt\%_{\ell \text{ not } i} \geq 95\%) \quad (6.3b)$$

2. The procedure shown in Equation 6.4a (derived in an analogous way as above except using Equation 6.2d) can be used to calculate X_i values if the mixture is given in units of ppm_ℓ by weight or mass (although for water or any other solvent whose density is approximately 1 g/mL, the ppm_ℓ by weight or mass concentration is also its mg/L or g/m^3 concentration; Howard and Meylan use g/m^3).[6]

$$X_i = \left[\frac{ppm_{\ell,i}/MW_i}{\Sigma(ppm_{\ell,i}/MW_i)}\right] \quad (6.4a)$$

Again, the denominator also equals 10^6/the solvent's average MW and one can use Equation 6.4b if the component concentration of interest is sufficiently dilute. The same somewhat arbitrary 90 to 95% primary solvent criterion can be used, but in this case it must be expressed in terms of ppm. For instance, 5% = 50,000 ppm. The difference between Equation 6.3b and Equation 6.4b is moot as long as the MW of the primary solvent is known.

$$X_i \approx \frac{ppm_{\ell,i}}{10^6}\left[\frac{MW_{solvent}}{MW_i}\right] \text{ for dilute mixtures } (\leq 50{,}000\ ppm_{\ell,i}) \quad (6.4b)$$

3. The composition of manufactured chemicals (such as blended mixtures) and the data on most MSDSs is typically given in units of liquid volume percent (vol% as defined by Equation 6.5a). The conversion from vol% to molar fraction is somewhat similar to those above except that one must also account for differences among the components' liquid

[a] The subscript "solvent" is used here to mean the major component in a liquid mixture. Later the companion term solute will be used to mean a minor constituent in a liquid mixture. While interactions will sometimes occur among solutes within the mixture, generally the strongest interaction is between solutes and the solvent (both defined as above).

densities. Pertinent liquid mixture terms used are defined below.

$$\text{vol\%}_{\ell,i} = \left[\frac{\text{mL}_i}{\text{mL}_{\text{mixture}}}\right] \times 100 \qquad (6.5a)$$

$$\text{and by definition } \rho_{\ell,i} = \left[\frac{\text{grams}_i}{\text{mL}_i}\right] \qquad \text{(using Equation 5.7a)}$$

$$\text{Thus, the mass in terms of grams}_i = \rho_{\ell,i} \times \text{mL}_i$$
$$= \left[\frac{\rho_{\ell,i} \times \text{vol\%}_{\ell,i} \times \text{mL}_{\text{mixture}}}{100}\right] \qquad (6.5b)$$

The conversion from vol% to X would again start with Equation 2.6, but in this case use Equation 6.5b (instead of a version of Equation 6.2). Because the constants "$\text{mL}_{\text{mixture}}$" and "100" appear in both the numerator and denominator, they cancel out. The denominator of the exact form of the conversion shown in Equation 6.6a requires summing all of the mixture components' fractions of weight percent over their respective molecular weights. The term $\rho_{\ell,i} \times \text{vol\%}_{\ell,i}/\text{MW}_i$ is the moles of component i per 100 mL.

$$X_i = \frac{\text{mol}_i}{\text{mol}_{\text{mixture}}} = \frac{\text{grams}_i/\text{MW}_i}{\Sigma(\text{grams}_i/\text{MW}_i)}$$
$$= \frac{(\rho_{\ell,i} \times \text{mL}_{\text{mixture}} \times \text{vol\%}_{\ell,i})/(100 \times \text{MW}_i)}{\Sigma(\rho_{\ell,i} \times \text{mL}_{\text{mixture}} \times \text{vol\%}_{\ell,i})/(100 \times \text{MW}_i)}$$

$$X_i = \left[\frac{\rho_{\ell,i} \times \text{vol\%}_{\ell,i}/\text{MW}_i}{\Sigma(\rho_{\ell,i} \times \text{vol\%}_{\ell,i}/\text{MW}_i)}\right] \qquad (6.6a)$$

If the components of interest are sufficiently dilute such that the denominator $\Sigma(\rho_{\ell,i} \times (\rho_{\ell,i} \times \text{vol\%}_{\ell,i}/\text{MW}_i) \approx \rho_{\ell\text{solvent}} \times 100/\text{MW}_{\text{solvent}}$ (again using the somewhat arbitrary 90 to 95% primary solvent criterion), then the conversion to X_i could be approximated in one step as shown in Equation 6.6b. This approximation is often appropriate for organic pollutants in water.

$$X_i = \left[\frac{\rho_{\ell,i} \times \text{vol\%}_{\ell,i}/\text{MW}_i}{\Sigma(\rho_{\ell,i} \times \text{vol\%}_{\ell,i}/\text{MW}_i)}\right] \approx \left[\frac{\rho_{\ell,i} \times \text{vol\%}_{\ell,i}/\text{MW}_i}{\rho_{\ell\ \text{solvent}} 100/\text{MW}_{\text{solvent}}}\right]$$

$$X_i \approx \frac{\text{vol\%}_{\ell,i}}{100}\left[\frac{\rho_{\ell,i}/\text{MW}_i}{\rho_{\ell\text{solvent}}/\text{MW}_{\text{solvent}}}\right] \quad \text{for dilute mixtures (e.g., when one vol\%}_{\ell\ \text{not}\ i} \geq 95\%) \qquad (6.6b)$$

A still further simplification can be made if the densities of all the components in the mixture are similar enough to each other such that $\rho_{\ell,i}$ and $\rho_{\ell\text{solvent}}$ above and below the line of Equation 6.6b will approximately cancel out. The liquid density (ρ_ℓ) of simple hydrocarbons tends to range between 0.65 and 0.8 g/mL; the density of most acetates, alcohols, ketones, ethers, and nitro compounds tends to range between 0.8 and 1.2 g/mL; but halogens can have a liquid density in excess of 1.5 g/mL. Thus, the following even simpler approximation should only be used either for mixtures of organic solvents with similar liquid densities or where low accuracy is acceptable. Using the range of density values given above, Equation 6.6c could easily yield errors of ±25% for organic solvents in water.

$$X_i \approx \frac{\text{Vol}\%_{\ell,i}}{100}\left[\frac{\text{MW}_{\text{solvent}}}{\text{MW}_i}\right] \begin{array}{l}\text{for dilute mixtures (when one vol\%} \geq 95\%) \\ \text{and where all } \rho_{\ell,i} \text{ are approximately equal}\end{array} \quad (6.6c)$$

4. Finally, notice how similar to each other Equation 6.3b, Equation 6.4b, and Equation 6.6c appear. If the mixture conditions are within the defined constraints and/or one is willing to accept a less accurate answer (what is sometimes called a "first approximation"), then Equation 6.3b, Equation 6.4b, and Equation 6.6c can be considered three variants of the same generic solution for X_i for use in either Raoult's law, the empirical adjustments to it to be explained within Section IV, or Henry's law within Section VII.

$$X_i \approx \frac{\text{wt}\%_{\ell,i}}{100}\left[\frac{\text{MW}_{\text{solvent}}}{\text{MW}_i}\right] \quad \begin{array}{l}\text{Equation 6.3b for dilute mixtures} \\ \text{(when one wt\%}_\ell \geq 95\%)\end{array}$$

$$X_i \approx \frac{\text{ppm}_{\ell,i}}{10^6}\left[\frac{\text{MW}_{\text{solvent}}}{\text{MW}_i}\right] \quad \begin{array}{l}\text{Equation 6.4b for dilute mixtures} \\ (\leq 50,000 \text{ ppm}_\ell)\end{array}$$

$$X_i \approx \frac{\text{vol}\%_{\ell,i}}{100}\left[\frac{\text{MW}_{\text{solvent}}}{\text{MW}_i}\right] \quad \begin{array}{l}\text{Equation 6.6c for dilute mixtures} \\ \text{(one vol\%}_\ell \geq 95\%) \text{ if all } \rho_{\ell,i} \text{ are equal)}\end{array}$$

To put this into another perspective, the error caused by omitting liquid density to calculate X_i is usually smaller and more acceptable than the error that will result from omitting nonideal effects within mixtures of organic molecules (methods to predict ✍ nonideal effects will be described in Section V). Thus, if you are told on an exam to apply Raoult's law or if in desperation (or both), then the simplified one-step approximation of X_i implied by Equation 6.3b, Equation 6.4b, and Equation 6.6c is accurate enough.

III. RAOULT'S LAW FOR IDEAL LIQUID MIXTURES

If molecules within a liquid mixture all act independently of each other, then logic would suggest that the number or density of molecules of each component in the mixture that would be available to evaporate from the liquid's surface would be reduced in proportion to that component's molar fraction within the liquid. If vapor pressure were simply proportionate to the surface density of molecules, then one would predict that each component's vapor pressure would be reduced in proportion to its respective molecular fraction within a given liquid. Indeed, this is exactly what Raoult's law predicts. While Raoult's law only applies to ideal liquid mixtures in which the molecules within the liquid do not interact, its logic and simplicity underlie most of the other models of vapor pressure from liquid mixtures.

1. Raoult's law states that the vapor pressure of each component in a mixture ($P_{\text{vapor},i}$) is reduced from its vapor pressure as a pure chemical (P_{vapor}) in proportion to the molar fraction of that component within the liquid mixture. Recall that the symbol X_i is used to indicate each component's molar concentration in the liquid state, viz., X_i is the moles of component *i* per mole of liquid mixture (see Equation 2.6 repeated above in Section II).

Thus, Raoult's law can be written as Equation 6.7.

$$\boxed{P_{\text{vapor},i} = X_i P_{\text{vapor}}} \qquad \text{Raoult's law} \qquad (6.7a)$$

The notation $P_{\text{vapor},i}$ is used to designate the ith chemical's vapor pressure when it is present as a component of a liquid mixture. To make Raoult's law more applicable to IH work, the component vapor pressure can also be written in concentration units of ppm, as shown below.

$$\text{ppm}_i = X_i \frac{P_{\text{vapor}}\ (\text{mmHg}) \times 10^6}{P \text{ or } 760\ \text{mmHg}} \qquad \text{an IH version of Raoult's law} \qquad (6.7b)$$

✍ Chemical mixtures that obey Raoult's law are called an ideal mixture, and mixtures that deviate from Raoult's law are called nonideal mixtures. The behavior of many chemical mixtures at least approximates Raoult's law, but many other mixtures deviate from it, and some deviate a lot. For Raoult's law to be applicable, all the molecules within the liquid must act independent of each other (just as Dalton's law assumes that molecules in gases and vapors act independently of each other). Most chemicals will behave like a component in an ideal mixture when it is present at high liquid concentrations (when more than 50% pure, the chemical acts like it is pure) and/or when its molecules are structurally similar to the molecules of the other components within a mixture. Unfortunately, the molecular interactions within mixtures of organic solvents (and especially between organic solvents and water) are sufficiently strong so that such mixtures often deviate from Raoult's law. Yet despite its limitations, the physical mechanism underlying Raoult's law should be thoroughly understood.

2. The vapor pressures of most of the molten metals listed in Table 6.2 are sufficient for them to create respiratory hazards. However, rather than remaining as a vapor or even a mist, many hot metal vapors rapidly condense into a fume. The Vapor Hazard Ratio (VHR from Equation 5.20) for a molten metal indicates how much the incipient fume concentration at its liquid surface is in excess of its TLV. The VHI = log(VHR) is listed in Table 6.2 to simplify comparing the huge range of potential fume hazards among metals. Any VHI greater than 0 (corresponding to a VHR of 1) indicates that the equilibrium concentration in that metal's vapors at its source exceeds its exposure limit and must be diluted.

   Because molten metals have a significant vapor pressure, metal fumes become an exception to all other aerosol's lack of an intrinsic means to become airborne. As with a solvent, the rate of generation of fumes from a molten metal should be proportional to the heated metal's vapor pressure (although that prediction has not been demonstrated in an IH setting). The fifth column of Table 6.2 lists the Vapor Hazard Index of each metal at its melting point (VHI = log[VHR] as will be covered in Section VII.5 of Chapter 8).

   A second metal (or more) may be present either as an impurity or as an alloy. The behavior of metal vapor pressures in a mixture has apparently not been studied in an IH context, but chemistry and engineering studies have reported that mixtures of common metals generally follow Raoult's law fairly closely (some negative deviations of 2–3 × were observed). In order to melt a mixture of metals, the temperature of the mixture must be raised to the melting temperature of the component with the highest melting point. The wide diversity among metal melting temperatures and the exponential increase of each metal's vapor pressure with temperature make it possible for the fumes of even a small amount of one metal with a low melting temperature when it is raised to the higher melting temperature of the base metal in a mixture, to exceed the fumes generated either by the pure, low melting metal at its normal melting temperature or by the more

TABLE 6.2
The Vapor Pressure and Potential Vapor Hazard of Molten Metals[a]

	TLV (mg/m^3)	Melting Temperature (°C)	P_{vapor} at Melting T (mmHg)	VHI at the Metal's Melting T	P_{vapor} at 1535°C (mmHg)	VHI at Iron's 1535°C
Aluminum	5	660	2.7×10^{-9}	−6.1	0.67	2.3
Antimony	0.5	631	0.014[b]	2.2	390[b]	6.7
Arsenic[c]	0.01	817	2.7×10^{4}[b]	10.0	2×10^{5}[c]	~11
Barium	0.5	725	0.051	3.9	236	6.5
Beryllium	0.002	1278	0.033	3.9	0.93	5.4
Cadmium[e]	0.002	321	0.11	5.5	1.4×10^{5}[d]	11.6
Chromium[f]	0.01	1857	26	6.9	0.44	5.1
Cobalt	0.02	1495	0.0055	2.9	0.0097	3.2
Copper	0.2	1083	4.2×10^{-4}	0.9	0.45	3.9
Hafnium	0.5	2227	0.0011	1.3	1.5×10^{-8}	−3.6
Indium	0.1	157	4.8×10^{-21}	−15.5	29[d]	6.3
Iron	5	1535	0.025	1.6	0.025	1.6
Lead	0.05	328	4.3×10^{-9}	−3.0	267[d]	7.8
Lithium[g]	0.025	181	1.8×10^{-10}	−5.6	3.1×10^{3}[d]	7.7
Magnesium[h]	10	649	2.8	2.8	2.3×10^{4}[d]	6.7
Manganese	0.2	1244	1.2	4.2	32[d]	5.7
Mercury	0.025	25[i]	0.0020	2.9	1.7×10^{6}[i]	12
Molybdenum	10	2617	0.032	1.2	3.7×10^{-9}	−5.7
Nickel	1.5	1453	0.0033	0.8	0.012	1.4
Osmium[j]	0.002	3045[k]	0.033	5.3	1.8×10^{-12}	−4.9
Platinum	1	1772	1.4×10^{-4}	0.1	2.6×10^{-6}[d]	−1.6
Rhodium	1	1966	0.0052	1.5	7.3×10^{-6}[d]	−1.4
Selenium	0.2	217	0.005[b]	2.0	3.0×10^{5}[b]	9.8
Silver	0.1	962	0.0027	2.2	9.7	5.7
Tantalum	5	2996	5.5×10^{-5}[d]	−1.0	2.2×10^{-14}	−10.4
Tellurium	0.1	449	0.2[b]	4.1	1.0×10^{4}[b]	8.8
Thallium	0.1	304	2.9×10^{-8}	−2.5	1.4×10^{3}[d]	8.2
Tin	2	232	6.1×10^{-23}	−18.7	0.46	3.2
Titanium[k]	10	1660	0.0030	0.1	4.5×10^{-4}[d]	−0.7
Tungsten	5	3410	0.002[b,d]	0.6	7.8×10^{-15}[b,d]	−10.8
Uranium	0.2	1132	1.7×10^{-10}	0.0	2.3×10^{-6}	−0.8
Vanadium[l]	0.05	1890	0.018	3.5	9.5×10^{-5}	1.3
Yttrium	1	1522	0.0022	1.0	2.7×10^{-3}	1.1
Zinc[m]	2	420	0.15	3.3	6.0×10^{4}[d]	8.1
Zirconium	5	1852	2.7×10^{-5}	−1.6	2.0×10^{-5}[b]	−1.7

[a] Vapor pressure was calculated as $\log P = 760(A + B/T + \mathrm{CH} \log T + D/T^3)$ from coefficients in the 83rd Edition of the *Handbook of Chemistry and Physics* (2002) except where noted in footnote b.[7]

[b] Vapor pressure was calculated as $\log P = A + B/T$ from the 60th ed. of the *Handbook of Chemistry and Physics* (1979).[8]

[c] Arsenic begins to sublime at 613°C (where its P_{vapor} as a solid is 760 mmHg); so it is virtually a gas at its melting temperature of 814–817°C, and 1535°C is way outside the accuracy limits to predict its P_{vapor}.

[d] Temperature is outside the range stated to predict this chemical's P_{vapor} within ±5% but is consistent with tabulated values.

[e] The cadmium entry is for respirable aerosols (and all fumes should be respirable).

[f] The chromium entry is for water insoluble Cr^{+6} in fumes.

[g] The Lithium TLV is written for lithium hydride (LiH) but the P_{vapor} in its VHR was calculated for elemental lithium.

[h] The Magnesium TLV is written for magnesium oxide (MgO) but P_{vapor} was calculated for elemental magnesium.

[i] Because mercury melts at −39°C, its P_{vapor} at melting is calculated at a normal 25°C. It boils at 357°C, and 1535°C is both well outside the range stated to predict its P_{vapor} and beyond tabulated values.

[j] The Osmium TLV is written for osmium tetroxide (OsO_4) in ppm. It was converted into mg/m^3 on that basis, but its P_{vapor} was calculated on the basis of elemental osmium. Ref. b lists osmium's melting temperature as 2697°C.

[k] The Titanium TLV is written for titanium dioxide (TiO_2) but the P_{vapor} in its VHR was calculated for elemental titanium.

[l] The Vanadium TLV is written for vanadium pentoxide (V_2O_5), but P_{vapor} in its VHR was calculated for elemental vanadium.

[m] The Zinc TLV is written for zinc oxide (ZnO), but the P_{vapor} in its VHR was calculated for elemental zinc.

concentrated base-metal. Two common examples are the presence of zinc when welding galvanized iron and of lead in a recycled iron foundry.

More generally, a metal fume will be a greater hazard as part of a mixture than as a pure metal any time its molar fraction is greater than $P_{\text{vapor at low }T}/P_{\text{vapor at high }T}$, where "low T" and "high T" refer to the pure metal's low temperature and the mixture's high temperature, respectively. Similarly, the fume of a minor component with a low melting temperature that is present at a base metal's higher melting temperature will be a greater health hazard than the fumes of the base metal any time that the $X_{\text{of the minor metal}}$ is greater than $\text{VHR}_{\text{base at high }T}/\text{VHR}_{\text{minor at high }T}$, where the subscripts "base" and "minor" refer to the high melting temperature base metal and low melting temperature minor metal, respectively, when both VHRs are evaluated at the higher melting temperature of the base metal. The data in the last column of Table 6.2 give an indication of the theoretically maximum potential health risks presented by each metal in its pure form if it were heated to the melting temperature of iron (1535°C). Notice that the VHI for iron indicates its vapors or fumes would only have to be diluted about $10^{1.6}$ or 40 times to be an acceptable airborne concentration. However, the much higher VHR or VHI of many other metals (such as antimony, cadmium, or lead) indicates that their fumes are more likely to exceed their individual exposure limits than will iron's fume if they are present in molar concentrations in the molten iron at concentrations greater than 1 ppm, 1 ppb, or 1 ppm, respectively.

EXAMPLE 6.1

How does the vapor pressure of lead present at 0.01% in a batch of recycled iron compare to both its exposure limit and to the vapors of the base metal at its normal melting temperature? The answer to this question combines several concepts. Equation 2.10 was used in conjunction with the coefficients in Table 6.2 to find lead's vapor pressure at its and iron's melting temperatures (only the latter is shown below). Raoult's law (Equation 6.7a) should apply fairly well to $X_{\text{lead}} = 0.01\% = 0.0001$ at the melting temperature of iron. Equation 2.8 shows that a vapor pressure of 0.0267 mmHg is equivalent to a concentration of 300 mg/m^3.

$$\log(P_{\text{vapor}}) = A - \frac{B}{T} = 7.792 - \frac{9701}{(1535 + 273)} = 2.426 \qquad \text{(Equation 2.10)}$$

$$P_{\text{vapor},i} = X_i \times P_{\text{vapor}}(\text{mmHg}) = 0.0001 \times 10^{2.426} = 0.0267\ \text{mmHg} \qquad \text{(using Equation 6.7a)}$$

$$C\,(\text{mg/m}^3) = 53.82 \times P_{\text{vapor}} \times \text{MW} = 53.82 \times 0.0267 \times 201.2 = 300\ \text{mg/m}^3 \qquad \text{(using Equation 2.8)}$$

This information allows us to see that the 0.01% lead present right at the molten liquid surface would be 6000 times its exposure limit. The lead is theoretically still a vapor right at the molten surface but will rapidly cool and condense into a fume, and a Vapor Hazard Ratio of over 6000 represents a substantial dilution requirement and a significant fume hazard.

$$\text{Vapor Hazard Ratio} = \frac{P_{\text{vapor}}\text{in mg/m}^3}{\text{TLV}} = \frac{300\ \text{mg/m}^3}{0.05\ \text{mg/m}^3} = 6000 \qquad \text{(using equation 5.20)}$$

The data in Table 6.2 show that the vapor pressure of pure lead at its melting temperature is only a fraction of its exposure limit (viz., its VHR = 0.001). Thus, the hazard of even a small amount of lead in molten steel is a much greater respiratory hazard than pure lead at its melting temperature.

3. The old saw "Likes repel and opposites attract" rarely applies to different organic molecules within a mixture. It turns out that almost all organic mixtures deviate from Raoult's law in a positive direction (i.e., they are more volatile than predicted by Raoult's law). The deviation from Raoult's law starts slowly as an initially pure chemical starts to be diluted into a nonideal mixture but can become extremely large. For instance, the deviation is rarely more than a factor of 2× at concentrations of 20% or greater, but certain dilute mixtures can be deviate from Raoult's law by a factor of more than 10^6.

However, rather than rejecting the fundamental logic of Raoult's law, much research has been devoted to adjusting it. Three sections to follow describe three alternatives to Raoult's law for nonideal mixtures:

- Section V describes an empirical adjustment to Raoult's law that is not in current IH use but best explains the principles underlying nonideal deviations from the intuitive nature of Raoult's law.
- Section VII describes Henry's law and its commonly available (although not intuitive) coefficients for dilute, aqueous mixtures.
- Section VIII describes how experimental measurements of vapors can be used to anticipate hazards.

IV. THE NEED FOR AN EMPIRICAL ADJUSTMENT TO RAOULT'S LAW

✍ The more different that the molecules of one chemical in a liquid mixture are from the other molecules around it, the more likely they are to interact and their vapor pressure are to deviate from Raoult's law. Thus, the small differences between the members of a homologous organic series (such as hexane and octane) create small deviations, while the larger differences between molecules of different homologous series (such as between aromatic and paraffinic solvents, between alcohols and chlorinated solvents, or between most nonpolar organic chemicals and water) create larger deviations. The opportunity for even grossly different molecules to interact is small as long as there are still many identical molecules around (i.e., when the mixture is relatively concentrated), but deviations will increase when most of the other molecules are different (i.e., when a component in a mixture is dilute). This section will describe this pattern of molecular interactions and various predictive models and empirical adjustments to Raoult's law that have been proposed.

1. Engineers are pragmatists that revere efficiency. They like simple models, ideally linear equations. However, linear equations do not always match experimental observations. Thus, they often develop or adopt empirical correction factors to allow a historically acceptable, simple model to continue to be used either in more complex or realistic settings or with more precision or accuracy. The Cunningham Slip Correction Factor of Chapter 4 was one example of such an empirical correction factor. An activity coefficient is used in chemical engineering to adjust Raoult's simple, logical law for ideal mixtures to match the behavior of nonideal liquid mixtures. This activity coefficient is given the symbol γ (a lowercase Greek gamma). This γ is sometimes also referred to as a fugacity coefficient. Equation 6.8 both defines γ mathematically and expresses the empirical concept behind it.

$$\gamma_i = \frac{\text{measured or actual } P_{\text{vapor},i}}{\text{Raoult's } P_{\text{vapor},i}} = \frac{Y_i \times P}{X_i \times P_{\text{vapor}}} = \frac{\text{measured or actual } Y_i}{Y_i \text{ given by Raoult's law}} \tag{6.8}$$

In a mixture, each component i will have its own γ_i. Each activity coefficient could be used either in Equation 6.9a to predict the absolute vapor pressure in units such as mmHg, or in Equation 6.9b to predict the partial pressure in units more useful for IH applications (equivalent to Equation 6.7a and Equation 6.7b, respectively, for ideal mixtures). Equation 6.9 will revert to Raoult's law (Equation 6.7) when $\gamma = 1$.

$$\boxed{P_{\text{vapor},i} = \gamma_i \times X_i \times P_{\text{vapor}}} = \gamma_i \times \text{Raoult's predicted } P_{\text{vapor},i} \tag{6.9a}$$

$$\text{ppm}_i = \frac{\gamma_i \times X_i \times P_{\text{vapor}}\,(\text{mmHg}) \times 10^6}{P_{\text{atmospheric}} \text{ or } 760 \text{ mmHg}} \tag{6.9b}$$

2. The following three figures are attached to provide some graphical insight into how γ values vary both among mixtures and within mixtures as a function of X. Each figure presents data from different binary mixtures: the first describes γ for three mixtures of ethanol, benzene, and toluene; the second describes the γ for various organic solutes in a simple alkane solvent; and the third describes the γ for 12 organic solutes in water. The activity coefficients (γ values) for all mixtures, except the benzene, in water were calculated using Equation 6.10 with seven values derived by Hirata et al. (1975)[9] from published mixture data.

The first of these figures (Figure 6.1) plots the γ for benzene in ethanol, for ethanol in benzene, and for benzene in toluene. In the first binary mixture, the molar fraction of ethanol in benzene must equal $1 - X_{\text{benzene}}$. Both the X and γ axes are on traditional linear scales that makes it easy to see how the γ values initially increase slowly from $\gamma_i = 1$ for the pure chemical (where $X_i = 1$), but this scale does not make it easy to see how γ approaches its maximum value (γ^{∞}) as X_i in the mixture gets very dilute (as X_i gets into the ppm range on the very left edge of the figure). It is normal for the two γ's in a binary mixture to be asymmetrical. In this case, γ^{∞} is approximately 5 for benzene in ethanol and 15 for ethanol in benzene. The fact that the maximum γ^{∞} for benzene in toluene is as large as 2 demonstrates that even small structural differences between molecules in a mixture can create measurable nonideal effects.[b]

Figure 6.2 shows γ values for several organic solutes in either hexane or heptane (two closely related alkane solvents). The logarithmic scales on both the X and γ axes compress the concentrated mixtures into the lower right corner of this figure, allowing a better view of how (and at what X_i value) each γ approaches its maximum γ^{∞} value. Notice the similar pattern of γ_i as a function of X_i within each of these mixtures. Notice also how the values of γ^{∞} increase as the solute's molecular polarity gets more dissimilar from the nonpolar alkane solvent. This latter pattern is especially evident in the γ^{∞} values for the alcohols in the alkane solvent.

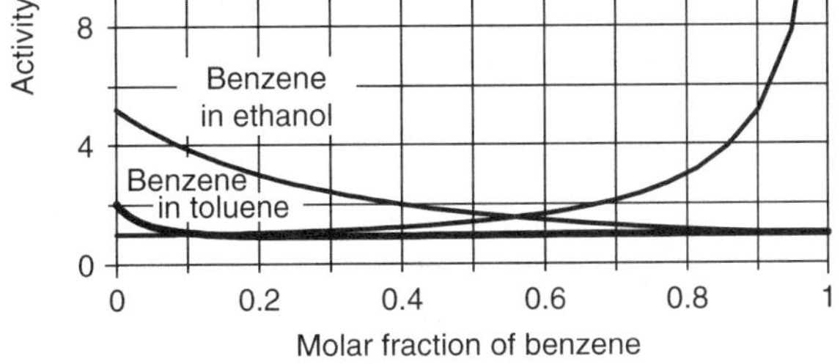

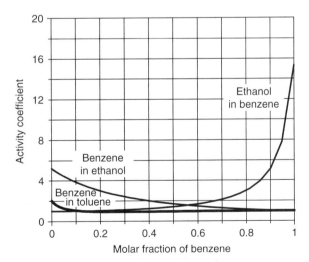

FIGURE 6.1 Activity coefficient (γ_i) values caused by nonideal interactions between ethanol and benzene.

[b] The superscript ∞ notation (as in γ^{∞}) does not mean that gamma is raised to the infinite power (which would not be meaningful) but is just a common notation in this field meaning "highly diluted." See also the Abbreviations section in the front of this book.

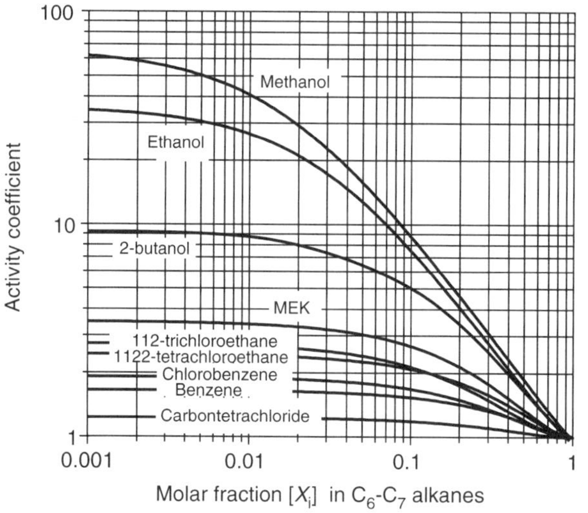

FIGURE 6.2 Activity coefficient (γ_i) values for selected organic solutes in either hexane or heptane.

The last figure of 12 organic solutes in water shows the same pattern of increasing values of γ as the solution gets more dilute and as the molecules become more dissimilar. The γ^{∞} for very symmetric organic molecules (such as hexane) in water can approach 10^6, although the largest γ^{∞} values included in Figure 6.3 are 1730 for benzene and 222 for phenol (both of these γ^{∞} values occur off-scale). The smallest γ^{∞} shown is 1.7 for methanol (most similar to water). Because a γ value greater than one means that the component's vapor pressure is greater than predicted by Raoult's law, most of these organic contaminants when diluted in water will present a surprisingly high vapor hazard if their activity coefficient is not anticipated.

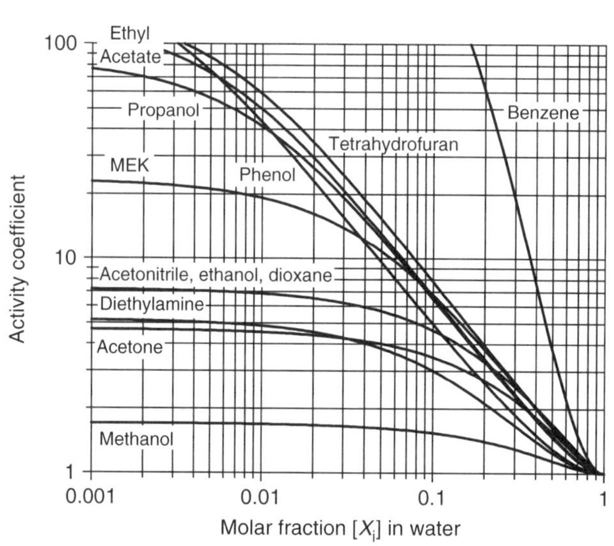

FIGURE 6.3 Activity coefficient (γ_i) values for selected organic solutes in water.

Understanding the effect of an activity coefficient upon Raoult's law is a major learning goal of this chapter. The examples presented in the previous figures should help the reader to better anticipate in what mixtures γ is likely to deviate from one, at about what concentration it will begin to deviate significantly, and about how large that deviation can be. Such knowledge can go a long way toward avoiding the pitfalls of trying to apply Raoult's law to a nonideal mixture or trying to apply Henry's law (which assumes γ is a constant γ^{∞}) to a more concentrated mixture where γ is somewhere between γ^{∞} and one. The next section is presented only to provide the reader with some background into research on activity coefficients. Keep in mind that activity coefficients are not currently a common part of IH practice.

V. METHODS TO PREDICT AN EMPIRICAL ADJUSTMENT TO RAOULT'S LAW

The following four approaches to predict the activity coefficient γ were adapted primarily from Chapter 11 of the *Handbook of Chemical Property Estimation Methods*,[10] a jewel of a resource if one is challenged to estimate an otherwise hard-to-get or nonexistent property value. Other sources of property estimation methods are also available.[11,12] The middle two methods below rely on experimental data for the particular mixture, while the first and last methods can estimate γ without prior mixture-specific measurements.

1. A method first developed by Pierotti *et al.* (1959) and refined by Grain (1982) can be used to predict γ^{∞} for binary mixtures.[13,14] The method shown below in Equation 6.10 was described by Grain as "easy" in that its seven coefficients are based on the overall molecular structure of the solute and solvent (rather than on each component's individual moieties as in UNIFAC). Lyman's book contains four pages of tables and internal correction factors for the seven variables included in Equation 6.10.

$$\log\gamma^{\infty} = A_{1,2} + B_2\left[\frac{N_1}{N_2}\right] + \left[\frac{C_1}{N_1}\right] + D(N_1 - N_2)^2 + \left[\frac{F_2}{N_2}\right] \tag{6.10}$$

2. Wilson's equation is a good descriptor of γ as a function of X_i for binary mixtures.[15] The $\Lambda_{1,2}$ and $\Lambda_{2,1}$ values (uppercase Greek lambdas) in Equation 6.11 were estimated by Hirata *et al.* (1975) using a computerized regression technique based on a few measured pairs of X_i and Y_i values. The examples in Figure 6.1, Figure 6.2, and Figure 6.3 (except for benzene in water) were generated using this method from 5 to 20 data points for each binary mixture.

$$\gamma_1 = \exp\left[-\ln(X_1 + \Lambda_{12}X_2) + X_2\left[\frac{\Lambda_{12}}{X_1 + \Lambda_{12}X_2} - \frac{\Lambda_{21}}{\Lambda_{21}X_1 + X_2}\right]\right] \tag{6.11a}$$

$$\gamma_2 = \exp\left[-\ln(X_2 + \Lambda_{21}X_1) - X_1\left[\frac{\Lambda_{12}}{X_1 + \Lambda_{12}X_2} - \frac{\Lambda_{21}}{\Lambda_{21}X_1 + X_2}\right]\right] \tag{6.11b}$$

The differences between these two equations are subtle but important; for instance, they reflect the asymmetry between the γ^{∞} values for benzene and ethanol seen in Figure 6.1. These equations can be used either to interpolate between measured values of γ_i or to extrapolate from γ_i values measured easily at moderately high values of X_i to γ_i values at very low concentrations that are generally harder to measure. They can also use these easier measurements to predict $\ln\gamma_1^{\infty} = -\ln\Lambda_{12} - \Lambda_{21} - 1$ and $\ln\gamma_2^{\infty} = -\ln\Lambda_{21} - \Lambda_{12} + 1$, where again the "$\infty$" means at infinitely small dilution (meaning below approximately 10^{-4} or 10^{-5}).

3. The van Laar equation (ca. 1910) provides the opposite capability: the ability to estimate γ_i at any concentration based only on γ_1^∞ and γ_2^∞ derived either from measurements, from using Equation 6.10, or from any other approach.[9,12,14] Equation 6.12b was used to generate the activity coefficients for benzene in water plotted in Figure 6.3.

$$\ln\gamma_1 = \ln\gamma_1^\infty\left[1 + \frac{X_1 \ln \gamma_1^\infty}{X_2 \ln \gamma_2^\infty}\right]^{-2} \tag{6.12a}$$

Equation 6.12b (derived by converting Equation 6.12a into a power function) gives a more direct (although still not easily interpretable) view of how γ_i varies with X_i. Keep in mind that for a binary system $X_2 = 1 - X_1$ and all of these subscripts could be reversed to find γ_2.

$$\gamma_1 = \gamma_1^{\infty\left[1 + \frac{X_1 \ln \gamma_1^\infty}{X_2 \ln \gamma_2^\infty}\right]^{-2}} \tag{6.12b}$$

4. The most robust model to predict activity coefficients is via the group contribution method of UNIFAC (fostered by Prausnitz, Gmehling, and Fredenslund).[12,16] "Robust" in this case means that it is able to predict γ for a multicomponent mixture (more than just a binary) without any mixture-specific experimental data. UNIFAC bases its predictions not on whole molecules but on the expected interaction of specific chemically functional groups or moieties that make up organic molecules. For instance, a normal alkane molecule is made up of two end CH_3 moieties and some number of interior CH_2 moieties; alkenes comprise those moieties and one or more $CH_2{=}CH$ or $CH{=}CH$ moieties; benzene is made up of six saturated aromatic CH moieties vs. toluene that is made up of five such moieties and one aromatic $C{-}CH_3$ moiety. Each molecule in the mixture must be defined in terms of its constituent groups that might include alcohols, ketones, aldehydes, esters, ethers, amines, nitriles, chlorides, and other structural entities. To implement UNIFAC, a computer program is needed that solves an algebraic array of interaction parameters that have been developed for approximately 50 functional groups that comprise many (but certainly not all) chemicals of interest to the IH community. However, the lack of a database that defines the structures of even the TLV chemicals or/and a user-friendly interface has kept these programs from being easily accessible to most hygienists. The following are some examples of currently available programs:
   a. S.I. Sandler published a BASIC language version of UNIFAC.[17] The downloadable program is at http://www.che.udel.edu/thermo/basicprograms.htm. To run UNIFAC.BAS, the user would need to have identified the functional groups in each of the molecules in the mixture of interest and (for certain types of outputs) either the Antoine (vapor pressure) constants for those substances or at least their vapor pressures at the temperature of interest. All the results are displayed on the screen.
   b. An Excel version of UNIFAC was written by Randhol and Engelien at the Norwegian University of Science and Technology (NTNU).[c] XLUNIFAC gives the user the ability to add and edit the group volume and area parameters as well as the interaction parameters, to define new components, and to add new subgroups and main groups to the model. Its results include the calculated values of intermediate variables within the

[c] XLUNIFAC (version 1.0, 2000) is accessible at http://www.nt.ntnu.no/users/randhol/xlunifac/. This program is free software usable under the terms of the GNU General Public License as published by the Free Software Foundation, Inc., 675 Mass Ave, Cambridge, MA 02139, USA.

equations from which the user can hopefully get a better understanding of the UNIFAC model.

c. VLECalc Version 1.3 (©2001) by Kovats is also a Freeware program written in Visual Basic to calculate vapor–liquid equilibrium and interface the output with Microsoft Excel 97®. A full description is available at http://www.vlecalc.org, and the program can be downloaded at my.net-link.net/~wdkovats/.

Data in Table 6.3 excerpted from Grain (1982) allows one to compare activity coefficients at infinite dilution (γ^{∞}) for five common organic liquids in water The first column of γ^{∞} values were experimentally measured; the second column of γ^{∞} values were calculated by UNIFAC. The range of the percent errors by which the calculated γ^{∞} differs from the measured γ^{∞} that are presented in the third column of data suggests that UNIFAC does not always predict activity coefficients accurately.

However, even the highest of these errors (131% for toluene in water) is much better than the 339,000% error that would result if no γ^{∞} value were used at all (equivalent to assuming Raoult's law or $\gamma^{\infty} = 1$ vs. the γ^{∞} of 3390 measured experimentally). The ratios in the right-most column are representative of the variable but larger γ^{∞} values for organic solvents in water than for the γ^{∞} of water in the respective solvent. Hexane and water are an extreme example in both the magnitude of γ^{∞} (see also Figure 6.3) and the difference in the interactions of two solvents.

5. Neither Equation 6.10, nor Equation 6.11 nor Equation 6.12 is very intuitive. However, the above examples lead to the following rules-of-thumb for γ and the behavior of vapors emanating from mixtures:

✓ The γ of an ideal component always equals exactly one. The γ of even a nonideal component will approximate one if that component is present in the liquid at a high concentration. For instance, γ_i is almost always less than 2 as long as X_i is greater than 0.5 and often even when X_i is as dilute as 0.3 or 0.2 if the molecules are reasonably similar. Thus, approximately equal mixtures of two or even three chemicals should not deviate far from an ideal mixture.

✓ The more dissimilar the structure of the components' molecules are from each other, the more γ^{∞} will differ from 1. The direction and magnitude of γ^{∞} is related to the components' intersolubility and hence to differences in their polarity. Because the molecules of most organic solvents are significantly different from water, the γ^{∞} for most organic chemicals (and for all hydrophobic chemicals) in water will be >1, while the γ^{∞} of hydrophillic chemicals in water (such as ammonia and formaldehyde) will be <1. For example, the γ^{∞} for ammonia in water is 0.031. Acetone in chloroform is another rare example of two highly compatible organics where γ^{∞} is also <1.

TABLE 6.3
Examples of the Accuracy of Activity Coefficients Calculated by UNIFAC[14]

	Experimental γ_i^{∞} in Water	Calculated γ_i^{∞} in Water	% Error	Experimental γ_{water}^{∞} in Organic	$\gamma_{organic\ solvent}^{\infty}/\gamma_{water}^{\infty}$
Hexane	489,000	402,000	−18	1880	260
Toluene	3390	7820	131	3320	1.0
Benzene	1730	1670	−3.5	226	7.7
Aniline	34.2	50.6	48	4.98	6.9
Acetone	6.80	5.69	−16	5.64	1.2
Average absolute error =			43		

✓ Deviations will increase as a nonideal component becomes more dilute, but γ values do not increase forever. The deviation from an ideal mixture eventually reaches a maximum γ^{∞} value. As a slightly cumbersome rule-of-thumb, γ will be within 5% of its maximum γ^{∞} by the time X_i drops to approximately $1/10(\gamma^{\infty})^{1.1}$. Thus, for $\gamma^{\infty} < 10$, γ_i will be nearly constant by the time X_i drops below 1%; for $\gamma^{\infty} < 100$, γ_i is nearly constant by the time X_i drops below 10^{-3}; and for $\gamma^{\infty} < 500$, γ_i will be constant by the time X_i drops below 10^{-4}.

✓ As suggested by the above rules-of-thumb, but which may not be obvious from the above figures, the γ correction factor almost never rises as fast as X_i falls. Thus, the $P_{\text{vapor},\,i}$ in most mixtures will never exceed the P_{vapor} of the pure liquid. However, a couple of highly symmetric molecules such as hexane, toluene, or benzene in water are exceptions (see the almost vertical $\gamma^{\infty}_{\text{benzene}}$ line in Figure 6.3).

VI. EXAMPLES OF MIXTURES

A series of three examples of mixture component vapor pressures is presented below. The first two examples demonstrate the use of a known activity coefficient for a nonideal mixture. The last example demonstrates the derivation of an activity coefficient. Keep in mind that the tables shown in the first two examples are to aid learning and are not expected to be reproduced or required to be used on a timed exam.

Example 6.2

Find the mixture component vapor pressures assuming that a vol/vol mixture of 99% ethanol and 1% benzene (defined in terms of volume %) is ideal. An ideal mixture obeys Raoult's law, viz. $P_{\text{vapor},\,i} = X_i \times P_{\text{vapor}}$ in Equation 6.7. The more exact formula for the molar X_i fraction (Equation 6.6a) will be used to calculate $P_{\text{vapor},\,i}$ in the right-most column in the following table.[d]

$$X_i = \left[\frac{\rho_{\ell,i} \times \text{vol\%}_{\ell,i}/\text{MW}_i}{\Sigma(\rho_{\ell,i} \times \text{vol\%}_{\ell,i})/\text{MW}_i}\right] \qquad \text{(using Equation 6.6a)}$$

For convenience, the denominator of Equation 6.6a (effectively the number of moles of component i per 100 mL of mixture), is calculated and labeled as $\text{mol}_i/\text{L}_{\text{mix}}$ in the fourth column of data in the table below.

$$\text{mol}_i/\text{L}_{\text{mix}} = \frac{10^3 \times \rho_{\ell,i} \times \text{vol\%}_{\ell,i}}{\text{MW}_i} \qquad \text{(see the fourth column of data below)}$$

	Liquid Phase						Ideal Vapor
	Vol% (%)	$\rho_{\ell,i}$	MW_i	$\text{mol}_i/L_{\text{mix}}$	X_i	Pure P_{vapor} (mmHg)	$P_{\text{vapor},i} = X_i\,P_{\text{vapor}}$ (mmHg)
Ethanol	99	0.79	46	17.0	0.994	59	58.6
Benzene	1	0.88	78	0.11	0.0064	95	0.61
				$\Sigma\text{mol}_i/L_{\text{mix}} = 17.11$			$\Sigma P_{\text{vapor},i} = 59.21$

[d] While in this example neither component's ρ_ℓ is very close to 1, they are close enough to each other that the shorter Equation 6.6c could have been used without generating a significant error in the calculated X_i.

EXAMPLE 6.3

Find the vapor pressures for the same mixture used in Example 6.2 but applying γ values. In general, γ_i values may be either found in the literature, predicted from one of the previous models, or measured experimentally. In this case, the γ_i for benzene in alcohol can be found in Figure 6.1. These γ_i values are inserted into the ideal-mixture solution of Example 6.2, and Equation 6.9a is used instead of Equation 6.7a to generate the $P_{vapor,\ i}$ values in the right-most column of the following table.

$$P_{vapor,i} = \gamma_i \times X_i \times P_{vapor} = \gamma_i \times \text{Raoult's predicted } P_{vapor,i} \qquad \text{(using Equation 6.9a)}$$

	Liquid Phase					Pure	Nonideal Vapor	
	Vol% (%)	$\rho_{\ell,i}$	MW_i	mol_i/L_{mix}	X_i	P_{vapor} (mmHg)	γ_i	$P_{vapor,i} = \gamma_i X_i P_{vapor}$ (mmHg)
Ethanol	99	0.79	46	17.0	0..994	59	1.0	58.6
Benzene	1	0.88	78	0.11	0.0064	95	5.1	3.1
				$\Sigma mol_i/L_{mix} = 17.11$				$\Sigma P_{vapor,i} = 61.7$

As expected, the major (99%) component within the mixture acts as if it were ideal (viz., the γ for the 99% ethanol is 1.0), but the value of γ for the 1% benzene indicates it is five times more volatile than expected for an ideal mixture. This could be a significant correction when predicting health hazards. Just to broaden your perspective, first notice that ethanol is even more nonideal than benzene when mixed with hexane or heptane as shown in Figure 6.2, and second notice how more volatile benzene is when it is diluted in water as shown in Figure 6.3, instead of in alcohol as shown in Figure 6.1.

The last example below uses van Laar's method (Equation 6.12a) to extrapolate using a known γ^{∞} value to find the activity coefficient γ_i and corresponding $P_{vapor,\ i}$ at another more concentrated X_i.

EXAMPLE 6.4

Find the γ of 1% benzene in water from the γ^{∞} data given in Table 6.3. For consistency, the same 1% by volume benzene was used in this example as was used in the previous example (although the major solvent here is water instead of ethanol). Because $\rho_{\ell,solvent} \approx \rho_{water} = 1$, the shorter calculation method of Equation 6.6b will be used here to find $X_i = X_1 = X_{benzene}$.

$$X_1 = \frac{mol_1}{mol_{mix}} \approx \frac{vol\%}{100}\left[\frac{\rho_1/MW_1}{\rho_{\ell,solvent}/MW_{solvent}}\right] = 0.01 \times \frac{0.88/78}{1/18} = 0.0020 \qquad \text{(using Equation 6.6b)}$$

$$X_2 = 1 - X_1 = 1 - 0.0020 = 0.998$$

$$\ln\gamma_1 = \ln\gamma_1^{\infty}\left[1 + \frac{X_1 \ln\gamma_1^{\infty}}{X_2 \ln\gamma_2^{\infty}}\right]^{-2} \qquad \text{(using Equation 6.12a)}$$

$$\ln\gamma_1 = \ln 1730\left[1 + \frac{0.002 \ln 1730}{0.998 \ln 226}\right]^{-2}$$

$$\ln\gamma_1 = 7.456\left[1 + \frac{0.0149}{5.410}\right]^{-2} = \frac{7.456}{1.0055} = 7.415$$

$$\gamma_1 = e^{7.415} = 1660$$

Note that a $\gamma_{benzene}$ of 1660 at $X_{benzene} = 0.002$ is only 4% below the $\gamma^{\infty} = 1730$ at infinite dilution. Thus, the activity coefficient in any mixture less concentrated than 0.2% will lie within the narrow range between 1660 and 1730. This range of a fixed coefficient is where Henry's law is applicable (see the next section). However, trying to apply such a fixed coefficient to any mixture between this 0.2% concentration and $\gamma_{\text{benzene}} = 1$ at $X_{\text{benzene}} = 1$ could grossly over-estimate the vapor hazard. Using Equation 6.12 would allow reasonably accurate predictions across the full range of mixture dilutions.

VII. HENRY'S LAW

1. Henry's law assumes that the vapor pressure of component i ($P_{\text{vapor, }i}$) is proportionate to X_i via a fixed empirical coefficient called either "Henry's law constant" or "Henry's law coefficient" and typically given a symbol such as H_i or HL_i. Henry's law has many applications and variations. In the context of vapor pressure, Henry's law might look like Equation 6.13a.

$$P_{\text{vapor},i} = H_i \times X_i \tag{6.13a}$$

Henry's law is occasionally written in reverse (especially for gases), as shown in Equation 6.13b. For instance, Equation 6.13b predicts how well an inhaled gaseous irritant can be absorbed into the wet mucus membranes of the upper respiratory tract, the fraction of the total amount of an inhaled gas or vapor that can be retained by the lungs, and the ultimate efficiency of an air cleaner to collect an exhausted gaseous contaminant.

$$X_i = S_i \times P_{\text{of gas } i} \tag{6.13b}$$

where $S_i = 1/H_i$ is a solubility coefficient (cc/mL atm). A Henry's law constant can also be viewed as a generic air-to-liquid partition coefficient for dilute mixtures. A partition coefficient is the ratio of the equilibrium concentrations of any chemical between two media, such as between air and a liquid, as shown in Equation 6.13c.

$$H_i = \frac{P_{\text{vapor},i}(\text{in air})}{X_i(\text{in liquid})} = \frac{Y_{\text{vapor},i}(\text{in air})}{X_i(\text{in liquid}) \times P_{\text{atmospheric}}} \tag{6.13c}$$

2. In comparison to Raoult's law with activity coefficients, Henry's law has one big advantage and two limitations: the extensive library of coefficients is its advantage, but the coefficients are only for aqueous mixtures and the law is only applicable to dilute mixtures where γ is the constant γ^{∞} value.

The major advantage of Henry's law is the widespread availability of its constants.[6,10,18–20] At the same time, these diverse sources have been developed with a considerable amount of variability in the units of H_i. For example, viewing H_i from the perspective of Equation 6.13c, if P_{vapor} is in mmHg and X_i is a unitless molar fraction in the liquid, the units of H_i are mmHg. If P_{vapor} is again in mmHg but X_i is moles/m^3 in the liquid, then the units of H_i would be mmHg m^3/mol. H_i can be nondimensional if C_i is used instead of P_{vapor} and C_i and X_i are in the same concentration units such as mg/mL.

The disadvantage of Henry's law lies in its two major limitations. Although the concept embodied in Henry's law could be applied to any combination of solvents, Henry's law constants reported in the literature refer exclusively to aqueous mixtures (fortunately for us, mixtures of organic chemicals in water are a fairly common IH problem).

Its second limitation is that Henry's law only applies to dilute mixtures where H_i is in fact a constant, independent of X_i. This limitation can be viewed in the context of how H_i is related to γ_i. The relationship between these two coefficients can be found by equating a component's vapor pressure predicted by the empirical adjustment to Raoult's law (Equation 6.9a,b) to that predicted by Henry's law (Equation 6.13).[e]

$$P_{\text{vapor},i} = H_i X_i = \gamma_i X_i P_{\text{vapor}} \qquad \text{(repeating Equations 6.9a and 6.13a)}$$

$$H_i = \gamma_i^{\infty} P_{\text{vapor}} \tag{6.14a}$$

$$\gamma_i^{\infty} = \frac{H_i}{P_{\text{vapor}}} \tag{6.14b}$$

Henry's law coefficient (H_i) will be constant as long as the component's concentration is sufficiently dilute that γ stays equal to the constant γ^{∞}. This range is best seen in Figure 6.2 and Figure 6.3 and glimpsed in Example 6.4. The curves of most activity coefficients flatten out to a constant γ^{∞} by the time the liquid molar concentration decreases to approximately 10^{-3}. The γ of a few chemicals with small γ^{∞} values are already constant by the time X_i is only 10^{-2}, while those with large γ values are not constant until X_i is less than 10^{-4}. However, in all cases, continued use of a Henry's constant for more concentrated mixtures will eventually over-estimate the real component's vapor pressure and its resulting airborne exposures.

3. This subsection on converting the units of Henry's law constants focuses on Howard and Meylan's handbook because it lists constants for such a vast number of chemicals.[6] The passing reader can easily skip this subsection (and go directly to Section V.4) until they are actually faced with a real problem. On-line converters are also available at the following sites:

   http://www.mpch-mainz.mpg.de/~sander/res/henry-conv.html

   http://www.epa.gov/athens/learn2model/part-two/onsite/henryslaw.htm

   Howard and Meylan's version of Henry's law is shown as Equation 6.15. They use $Y_{P_{\text{vapor}},i}$ where Equation 6.13a,b,c uses $P_{\text{vapor},\,i}$, and the units of their HL_i are atm·m^3/mole (convertable at the above EPA site).

$$Y_{\text{at the vapor pressure}}(\text{atm}) \text{ in Howard and Meylan} = \text{HL}_i \times X_i(\text{mol}_i/\text{m}^3) \tag{6.15}$$

Some conversion is necessary to match either the liquid mixtures that IHs use or the airborne results IHs generally want. Sections I through III in the Gas and Vapor Physics chapter covered the conversion of airborne units in atmospheres to most any other unit. Thus, the emphasis here will be on converting those units typical of environmental liquids into the units of mol/m^3 used in Howard and Meylan's version of Henry's law.

In the rare event that the mixture has been defined in terms of its unitless X_i liquid molar fractions, the conversion into mol/m^3 would proceed as shown in Equation 6.16a.

$$\frac{\text{mol}}{\text{m}^3} = X_i\left[\frac{\text{moles}_i}{\text{moles}_{\text{mixture}}}\right]\left[\frac{\text{moles}_{\text{mixture}}}{\text{MW}_{\text{mixture}}\ \text{g}_{\text{mixture}}}\right]\left[\rho_{\text{mixture}}\frac{\text{g}_{\text{mixture}}}{\text{mL}_{\text{mixture}}}\right]\left[\frac{10^6\text{mL}_{\text{mixture}}}{\text{m}^3_{\text{mixture}}}\right]$$

$$\frac{\text{mol}}{\text{m}^3} = X_i\left[\frac{\rho_{\text{mixture}} \times 10^6}{\text{MW}_{\text{mixture}}}\right] \tag{6.16a}$$

[e] While the relationship in Equation 6.14 is conceptually true, technically, γ^{∞} can only be unitless as shown if both HI and P_{vapor} have the same units. Unfortunately, neither are their units usually the same, nor are they units commonly used within IH.

In order for Henry's law to apply, X_i must be small (<0.1%), where $\rho_{\ell,\text{mixture}}$ can be assumed to approximately equal $\rho_{\ell,\text{solvent}}$ (the notation here is the same as was used circa Equation 6.2). We will further assume that the solvent is water where $\rho_{\text{solvent}} = 1$ g/mL and MW = 18.015. Therefore, Equation 6.16a can be approximated by Equation 6.16b.

$$\frac{\text{mol}}{\text{m}^3} \text{ in water} \approx X_i \times 55{,}509 \text{ to within } \pm < 1\% \tag{6.16b}$$

If the mixture is defined in units of vol% (as is typically provided by chemical manufacturers and their MSDS), then the conversion to mol/m^3 would proceed as shown in Equation 6.17.

$$\frac{\text{mol}_i}{\text{m}^3} = \text{vol\%}\left[\frac{L_i}{100\,L_{\text{mix}}}\right]\left[\frac{\rho_{\ell,i}\text{kg}_i}{L_i}\right]\left[\frac{10^3\text{ gm}\times 10^3\text{ L}}{\text{kg m}^3}\right]\left[\frac{\text{mol}_i}{\text{MW}_i\text{ gm}}\right]$$

$$\frac{\text{mol}_i}{\text{m}^3} = \text{Vol\%}_i\left[\frac{\rho_{\ell,i}\times 10^4}{\text{MW}_i}\right] \tag{6.17}$$

where $\rho_{\ell,i}$ is the liquid density of the component of interest.

If the mixture is defined in units of ppm$_\ell$ by weight (as is typical from laboratory analyses or is convertible from % by weight [wt%]), the conversion to mol/m^3 would proceed as shown in Equation 6.18.

$$\frac{\text{mol}_i}{\text{m}^3} = \text{ppm}_\ell\left[\frac{\text{gm}_i}{10^3\text{ kg}_{\text{mixture}}}\right]\left[\frac{\text{mol}_i}{\text{MW}_i\text{ gm}_i}\right]\left[\frac{\rho_{\text{solvent}}\text{ kg}_{\text{solvent}}}{L_{\text{solvent}}}\right]\left[\frac{10^3 L_{\text{solvent}}}{\text{m}^3}\right]$$

$$\frac{\text{mol}_i}{\text{m}^3} = \text{ppm}_i\left[\frac{\rho_{\text{solvent}}}{\text{MW}_i}\right] \tag{6.18}$$

If the solvent is water or any other time $\rho_{\text{solvent}} = 1$ g/mL, the above conversion merely means dividing the ppm$_\ell$ in the liquid by the contaminant's MW$_i$. Notice the subtle but important contrast between this conversion from ppm by weight using the density of the solvent (ρ_{solvent}) and the above conversion in Equation 6.17 from vol% using the density of the reported component of interest ($\rho_{\ell,i}$).

In either case, the resulting liquid concentration X_i in units of mol$_i$/m^3 can be multiplied by Howard and Meylan's HL to yield the equilibrium partial pressure P_i in units of atmospheres using Equation 6.15. Conversely, the liquid contamination that must exist to create a known equilibrium airborne concentration can be predicted by a reverse calculation or by measuring the airborne concentration in the head space above that liquid within a sealed container as will be described in the next section of this chapter.

4. **Example 6.5**

Suppose that you were tasked to provide IH supervision for remediation work where ground water was contaminated with up to 500 ppm$_\ell$ methyl ethyl ketone (MEK, CAS 78-93-3, MW = 72.1). Anticipating the highest vapor concentration that your remediation workers might encounter would greatly help decide if respiratory protection is needed and, if so, what form is adequate. The highest vapor concentration is, of course, the chemical's vapor pressure.[f]

a. A simple (but usually wasted) first step is to use Raoult's law (Equation 6.7) to check this mixture component's probable minimum potential vapor pressure. If the solvent presents a hazard using Raoult's law (where by default $\gamma = 1$), it will certainly present a hazard for $\gamma > 1$.

$$\text{Ideal } P_{\text{vapor},i} = X_i \times P_{\text{vapor}} = 0.0005 \times 95.3 \text{ mmHg} = 0.0476 \text{ mmHg} \quad \text{(using Equation 6.7a)}$$

$$\text{Ideal ppm} = \frac{0.0476 \times 10^6}{760} = 62.7 \text{ ppm MEK} \quad \text{(using Equation 6.7b)}$$

This 63 ppm MEK would apply if the mixture were ideal. In this case, MEK appears to be safely less than the 200 ppm 8-TLV; however, the structural differences between MEK and water molecules suggest that MEK is likely to be more volatile than predicted by Raoult's law.

b. One handy source for an activity coefficient in water is to look at the chemicals in Figure 6.3. The γ for MEK in water seems to approach a γ^{∞} of approximately 23 (its γ curve flattens to a nearly horizontal line as X_i approaches the 1000 ppm left-hand margin of that figure). Equation 6.9b shows how to adjust the ideal 63 ppm found by Raoult's law for the nonideal $\gamma = 23$ at 500 ppm_{ℓ}. The resulting 1440 ppm equilibrium vapor concentration is significantly above the 200 ppm 8-TWA.

$$\text{Empirically adjusted ppm} = 23 \times 62.7 = 1440 \text{ ppm MEK} \quad \text{(using Equation 6.9b)}$$

c. If activity coefficient data were not available, one could use Henry's law for MEK. Howard and Meylan list an HL for MEK of 5.69×10^{-5} atm m^3/mole. To use their HL in Equation 6.15 requires converting the 500 ppm_{ℓ} by weight into units of mol_i/m^3 using Equation 6.18, as shown below. Because the solvent in this mixture is water, $\rho_{\text{solvent}} = 1$.

$$\text{mol}_i/\text{m}^3 = \text{ppm}_{\ell}\left[\frac{\rho_{\text{solvent}}}{\text{MW}_i}\right] = 500\frac{1}{72.1} = 6.93 \text{ mol/m}^3 \quad \text{(using Equation 6.18)}$$

and

$$Y_{\text{MEK}} = 5.69 \times 10^{-5}\frac{\text{atm m}^3}{\text{mole}} \times 6.93\frac{\text{mole}}{\text{m}^3} = 39.4 \times 10^{-5} \text{ atm} = 394 \text{ ppm} \quad \text{(using Equation 6.15)}$$

Having used both an empirical adjustment and Henry's law in this case but resulted in different answers, which is the right prediction? This question is often difficult to answer. The value of 1440 ppm using $\gamma_{\text{MEK}}^{\infty} = 23$ predicted by MEK's activity

[f] Checking the solute's solubility is slightly divergent but can be a meaningful point of reference from which to view this level of contamination. One can also note in Howard and Meylan that the water solubility of MEK is 223 g/L = 223,000 ppm_{ℓ}.[6] So right away you can sense that while 500 ppm_{ℓ} sounds like significant water contamination from the environmental perspective, it is nowhere near saturated for MEK.

coefficient in Figure 6.3 is approximately 3× larger than the Y_{MEK} of 400 ppm predicted from Henry's law. Sometimes that is all we know, and sometimes a range is good enough. For instance, from an IH perspective, even this 400 ppm predicted maximum vapor concentration is more than both the 200 ppm TWA–TLV and 300 ppm STEL for MEK; so clearly some form of vapor exposure control will be needed, and the 1440 ppm predicted maximum concentration is less than 10 × the TLV, so organic vapor cartridges on a half-mask air-purifying respirator should suffice in even the worst equilibrium case. If you really want to delve deeper, sometimes one can look back at the data for an answer. In this case, one would find that the original liquid concentrations for the MEK data that went into Wilson's equation that created Figure 6.3, only went as low as $X_{MEK} = 0.03$. Thus, the extrapolation to predict a γ^{∞} at a low concentration of $X_{MEK} = 0.0005$ may be slightly less accurate than a Henry's law constant. If the original data are not available, there is always the experimental method (to be described). And certainly either prediction is better than assuming the vapor is ideal!

5. Henry's law should not be used for concentrated mixtures. The limitation in Henry's law of assuming that the partition ratio between the liquid and its vapor is always constant (independent of X_i) becomes more obvious if we try to extrapolate to a mixture more concentrated than approximately 10^{-3}. For instance, what would be the equilibrium vapor concentration if the water in the above example were saturated with 223 g/L MEK? As shown below, one might be tempted to use the same Henry's law constant at this concentration and simply insert 223,000 ppm into Equation 6.15 instead of the original 500 ppm.

$$Y_{MEK} = 5.69 \times 10^{-5} \frac{\text{atm m}^3}{\text{mole}} \times 3093 \frac{\text{mole}}{\text{m}^3}$$

$$= 0.176 \text{ atm} = 176{,}000 \text{ ppm} \qquad \text{(using Equation 6.15)}$$

However, it turns out that the vapor pressure predicted above by Henry's law for approximately a 0.22 molar faction of MEK is greater than the 125,400 ppm vapor pressure of pure MEK! Remember that mixture components tend to become more ideal and γ values decrease as the liquid concentration increases. Therefore, Henry's law (using a fixed HL value and, in effect, a fixed γ value) will usually over-estimate vapor exposure as the contaminant concentration approaches saturation. While this is a health-conservative error, specifying that employees must wear a respirator when in fact they are unnecessary is both costly and creates artificial hazards of its own. One way to extrapolate to a relatively high concentration would be to use γ^{∞} in Equation 6.12a,b (perhaps using Equation 6.14b to solve for γ^{∞}). Another way is to measure the more concentrated mixture experimentally.

VIII. MEASURING MIXTURES EXPERIMENTALLY

If neither an activity coefficient nor a Henry's law constant for a particular mixture is available, measuring the vapor pressure experimentally is still a potential option. Experimentally, one can measure the vapor concentration in the head space in a closed but not completely sealed vial over a mixture of known X_i at a known if not controlled temperature. The result can be expressed as the activity coefficient as previously defined in Equation 6.8.

$$\gamma_i = \frac{\text{the measured or actual } P_{\text{vapor},i} = Y_i P}{\text{Raoult's } X_i P_{\text{vapor}}} \qquad \text{(repeating Equation 6.8)}$$

Controlling the liquid temperature and pressure while measuring low concentrations in both the liquid and gaseous media is not a trivial procedure. The laboratory methods to measure headspace is a topic for further research,[12,21,22] but Mackay et al. (1992)[19] recommended the following methods:

- Equilibrium batch stripping.[23,24]
- EPICS (Equilibrium Partitioning in Closed Systems).[25,26]
- Head space analysis.[27–29]

Any experimental method may or may not be slower or more costly than a little desk time finding published constants, depending upon the laboratory vs. library resources that one has available. To its advantage, the experimental method can be used for basically any mixture of measurable solvents. Some consideration should be given to how γ_i varies as a function of X_i.

1. The simplest yet least accurate approach would be to analyze two or more samples whose liquid concentrations were created to span the range of the X_i expected in the work environment. Calculate each γ using Equation 6.8, plot γ_i as a function of X_i, and pick the appropriate γ for any particular need for use in Equation 6.9.
2. The other approach would be to analyze only a pair of highly dilute samples (i.e., $X_i < 10^3$), calculate γ_1^∞ and γ_2^∞ for each pair of components using Equation 6.8, and then use them in Equation 6.12 to generate γ_i at any other desired X_i. The same data could also be used to calculate a Henry's law constant using Equation 6.13.

A third, more circuitous approach is possible but not recommended. Hirata *et al.* describe how they calculated their Wilson Λ parameters from experimental data from which γ values were subsequently calculated using Equation 6.11a (and from which Figure 6.1, Figure 6.2, and Figure 6.3 were generated).[9] Their experimental data generally comprised between 6 and 10 samples whose liquid concentrations spanned molar fractions from $X_i \approx 0.9$ down to $X_i \approx 0.001$, but often only to $X_i \approx 0.1$. Using Equation 6.11a to calculate γ_i at any other concentration is an approach that is good applied science but will probably have an unacceptably long learning curve for most field applications.

IX. APPLYING PREDICTIVE METHODS TO MIXTURES

Small deviations (ca. 2–3×) from Raoult's law should still yield decent IH predictions. However, deviations of >10× make poor predictions. The point should be clear from the above discussion that failure to anticipate nonideal molecular interactions within the liquid will almost always result in underestimating vapor exposure, sometimes severely. Severe underestimates of exposure are unacceptable.

The fact that many if not most chemicals used in industry are mixtures creates one of many challenges that make IH work interesting. Learning how to use the above tools to predict $P_{\text{vapor},\,i}$ in mixtures exceeds the more modest learning goals of this chapter. The student is expected to learn the concept underlying Raoult's law, when to expect deviations from this law, and how to use γ values if they are known. Hygienists planning work with mixtures commonly face many questions, such as finding the ratio of one vapor to another vapor or the ratio of the vapor hazards of one component to another. This section discusses these and some other relationships that are useful when working with volatile liquid mixtures.

1. One pair of good questions are "Which vapor component is present in the highest concentration?" or "By how much will the highest concentration exceed the next most common vapor?" The vapor molar fraction of any component *i* relative to the total vapor concentration (herein called Rel.Y_i for short) can be calculated by the ratio of each component's vapor pressure to the total vapor pressure of all components as shown below. Rel.Y_i is basically the vapor molar fraction of component *i* in the mixture (Y_i) as if air were not present (as would be true inside a sealed chemical reactor).

$$\text{Relative } Y_i = \text{Rel.}Y_i = \frac{P_{\text{vapor},i}}{\Sigma P_{\text{vapor},i}} = \frac{\gamma_i X_i P_{\text{vapor of pure } i}}{\Sigma \gamma_i X_i P_{\text{vapor of pure } i}} \qquad (6.19)$$

Example 6.6

The relative molar fractions in the vapor are calculated below using Equation 6.19 for the same 1% benzene in ethanol mixture that was used in Example 6.2 and Example 6.3. The Relative Y_i for an ideal vapor is only shown for completeness (having already found in Example 6.2 that this mixture is not ideal).

	Pure		Ideal Vapor		Nonideal Vapor		
	P_{vapor} (mmHg)	X_i	$P_{\text{vapor},i} = X_i P_{\text{vapor}}$ (mmHg)	Rel.Y_i	γ_i	$P_{\text{vapor},i} = \gamma_i X_i P_{\text{vapor}}$ (mmHg)	Rel.Y_i
Ethanol	59	0.994	58.6	0.990	1.0	58.6	0.950
Benzene	95	0.0064	0.6	0.010	5.1	3.1	0.050
			Σideal $P_{\text{vapor},i} = 59.2$			$\Sigma P_{\text{vapor,i}} = 61.7$	

This example shows that while benzene comprises only 1% of the liquid mixture by volume, the nonideal Rel.Y_i value for benzene (3.1/61.7 = 0.050) shows that it comprises 5% of the vapor. The Rel.Y_i can be useful to plan sampling; however, it still does not give any information regarding either each component's vapor hazard or the collective hazard of vapors from the mixture as a whole.

2. Thus, other good questions are "What is the vapor hazard of a given mixture?" or "Which component contributes the most to the additive exposure limit for the mixture?" In order to answer this question, one can apply the additive effects model. This model (used by OSHA, the ACGIH TLV Committee, and others) states that the concentration of each chemical (C_i) contributes in an additive way to the whole hazard in proportion to its individual exposure limit (EL_i where EL could be either the PEL, TLV, etc.). To be an acceptable exposure, the sum of all C_i/EL_i ratios cannot exceed 1, as stated by Equation 6.20.

$$\frac{C_1}{\text{EL}_1} + \frac{C_2}{\text{EL}_2} + \frac{C_3}{\text{EL}_3} + \cdots + \frac{C_n}{\text{EL}_n} \quad \text{must be} \leq 1 \text{ to be acceptable} \qquad (6.20)$$

If the hazards of all the components in the vapor plume are additive, then the $P_{\text{vapor},i}/\text{TLV}_i$ ratios will also be additive. The fraction that each component's $P_{\text{vapor},i}/\text{TLV}_i$ ratio adds to the sum of all such ratios comprises that component's contribution to the total vapor hazard from that mixture. These "relative hazard" fractions are shortened in Equation 6.21 to Rel.Haz$_i$. The $P_{vapor,i}/TLV_i$ ratios have the mixed units of mmHg/ppm; however, because these units are both in the numerator and denominator of Equation 6.21, they cancel out, allowing Rel.Haz$_i$ to be expressed within the table below as a percent.

$$\text{The Relative Hazard of vapor } i \text{ within the mixture} = \text{Rel.Haz}_i = \frac{P_{vapor,i}/EL_i}{\Sigma P_{vapor,i}/EL_i} \quad (6.21)$$

Example 6.7

The calculations shown below find the relative hazard for each vapor component from the same 1% benzene in ethanol mixture that was used in Example 6.6 using the TLVs for the generic EL$_i$ in Equation 6.21.[g]

			Nonideal Vapor			
	Liquid Vol%	Pure P_{vapor} (mmHg)	$P_{vapor,i} = \gamma_i X_i P_{vapor}$ (mmHg)	TLV$_i$	$P_{vapor,i}/TLV_i$	Rel.Haz$_i$ (%)
Ethanol	99	59	58.6	1000	0.0586	0.94
Benzene	1	95	3.1	0.5	6.2	99.1
				$\Sigma P_{vapor,i}/TLV_i$ =	6.2586	

The component with the highest $P_{vapor,\ i}$/TLV ratio will have the most hazardous vapor. Thus, while the values of the Relative Y_i in the previous example showed that ethanol comprises 95% of the vapor emanating from this mixture, the values of the relative hazard found in this example show that benzene comprises approximately 99% of the total vapor hazard due to its much lower TLV.

3. A not uncommon circumstance is to have the ability to measure one component in a vapor separate from the other components. For instance, a Miran® or a detector tube can often measure one component with no or negligible cross-response from the other components. Measuring a single component can also result from method incompatibilities (e.g., a charcoal does not collect all kinds of vapors). A logical question when faced with this situation is "By how much should the measured component's exposure limit be reduced because of the presence of the other not measured or unmeasurable vapor component(s)?" Setting an adjusted exposure limit is vital if only one component of the vapor can be measured.
To answer this question, we will again apply the additive effects model and the principle that if as expected, each component of vapor from a liquid mixture is diluted by the same ratio, then the concentration of each vapor from a volatile mixture must stay in proportion to its vapor pressure in that mixture. A generic Environmental Dilution Ratio was previously defined by Equation.

$$\text{Environmental Dilution Ratio} = \frac{P_{vapor,i}}{P_{partial,i}} \quad \text{(repeat of Equation 5.16)}$$

[g] Looking both back to Equation 5.20 and forward to Equation 8.4, the ratio of a chemical's P_{vapor} to its EL is called its Vapor Hazard Ratio (VHR). A chemical's unitless VHR is the amount by which the equilibrium concentration right at the liquid surface (P_{vapor}) must be diluted to an acceptable breathing zone concentration (TLV or PEL). Thus, the VHR is a good way to quantify the inherent potential for any solvent's vapors to create an over-exposure condition. In this case, multiplying the $\Sigma P_{vapor,i}/TLV_i$ = 6.2586 mmHg/ppm ratio in Example 6.7 by 10^6 ppm/760 mmHg will create a unitless VHR of 8235.

In the context of this question and the form of Equation 6.20, the term "EDR" (literally in quotes) as defined by Equation 6.22a will be substituted for the Environmental Dilution Ratio. The quotes indicate that an adjustment in units is necessary to convert the concentration (C_i) in the same units as the exposure limit (EL_i) back into its equivalent $P_{partial,i}$ but not shown for clarity.

$$\text{"EDR"} = \frac{P_{vapor,i}}{C_i} \tag{6.22a}$$

$$C_i = \frac{P_{vapor,i}}{\text{"EDR"}} \tag{6.22b}$$

By inserting the right side of Equation 6.22b into Equation 6.20, then an acceptable "EDR" (dilution not forgetting to adjust for units) in which someone can work must be at least as large as the sum of all of the ratios of each component's vapor pressure to its exposure limit. A just barely acceptable "EDR" would equal the sum of these ratios as expressed by Equation 6.23.

$$\frac{P_{vapor,1}}{EL_1} + \frac{P_{vapor,2}}{EL_2} + \cdots + \frac{P_{vapor,n}}{EL_n} = \text{the minimally acceptable "EDR"} \tag{6.23}$$

Because Equation 6.23 is true, then the ratio by which each individual component vapor pressure must be diluted to reach its adjusted exposure limit must also be at least equal to that same Environmental Dilution Ratio as expressed by Equation 6.24.

$$\frac{P_{vapor,i}}{\text{adjusted } EL_i} = \sum \frac{P_{vapor,i}}{EL_i} = \text{the same minimally acceptable "EDR"} \tag{6.24}$$

After solving Equation 6.24 for the adjusted exposure limit as shown in Equation 6.25a, one can see that the appropriate adjustment for the presence of the other toxic airborne components is simply to multiply any given chemical's individual exposure limit by its relative hazard as previously defined in Equation 6.21.

$$\text{adjusted } EL_i = \frac{P_{vapor,i}}{\Sigma P_{vapor,i}/EL_i} = EL_i \frac{P_{vapor,i}/EL_i}{\Sigma P_{vapor,i}/EL_i} = EL_i \times \text{Rel.Haz}_i \tag{6.25a}$$

In Example 6.7, exposures to benzene would have to be reduced below its 0.5 ppm TLV by an insignificant fraction (× 0.991) for the presence of ethanol because the latter has such a low toxicity. However the same logic leads to the conclusion that the 1000 ppm exposure limit for ethanol would have to be reduced by a multiplying factor of 0.0094 to account for the presence of the much more toxic benzene, as calculated below:

$$\text{adjusted TLV}_{ethanol} = 1000 \frac{0.0586}{6.2586} = 9.4 \text{ ppm} \qquad \text{(using Equation 6.25a)}$$

4. Yet another useful ratio (not shown) could be developed to adjust the reading of a nonspecific detector (such as a PID or combustible gas detector) for the presence and simultaneous detection of multiple components. The response of such detectors typically differs among individual chemicals. Thus, this calculation would not only use the same assumptions of additive effects (the sum of the ratio of exposures to allowable exposure limits for all the components must be less than or equal to unity) and that the vapors are present in the previously predicted ratio of component vapor pressures, but would also require knowledge of the chemical specific response factors for each component.

5. The above methods could be used to predict employee exposures to vapors from real mixtures in any of the four universal exposure scenarios introduced in Chapter 5 with one important caveat. If the relative volatilities of the components are not equal, there will be some undefined enhancement or depletion of the previously predicted rates of evaporation that will vary over time.
   a. The relative volatility of any mixture component (α_i) compares the fraction of that component in the vapor phase to its fraction in the liquid state. Without air (e.g., inside a chemical reactor), a component's relative volatility could be calculated using Equation 6.26.

$$\alpha_i = \frac{Y_i/X_i}{(1 - Y_i)/(1 - X_i)} = \frac{Y_i(1 - X_i)}{X_i(1 - Y_i)} \rightarrow \left[\frac{Y_i}{X_i} \text{ for dilute mixtures when } X_i \text{ and } Y_i \text{ are } << 1\right] \quad (6.26)$$

   For vapors in an open-air working environment, the other component's vapor fraction is no longer equal to $1 - Y_i$, but each component's vapor fraction is still proportional to its vapor pressure compared to the sum of all the other components' vapor pressures or to its Relative Y_i as previously defined in Equation 6.19. Thus, an equation for α_i in air can be rewritten as Equation 6.27.

$$\alpha_i \text{ in air} = \frac{Y_i(1 - X_i)}{X_i \Sigma Y_i} \approx \frac{P_{\text{vapor},i}}{X_i \Sigma P_{\text{vapor},i}} = \frac{P_{\text{vapor},i}/\Sigma P_{\text{vapor},i}}{X_i} = \frac{\text{Rel.}Y_i}{X_i} \quad (6.27)$$

   b. You may also recall from chemistry that an azeotrope is a liquid mixture that retains the same composition in the vapor state as in the liquid state. Hygienists working in the petrochemical industry may have access to what are called $X - Y$, vapor liquid equilibrium (VLE), or phase diagrams such as those shown in Figure 6.4. The diagonal 1:1 line represents an $\alpha = 1$ condition (equal volatility).

   Figure 6.4a shows a nearly ideal mixture of hexane in heptane where α is consistently just over three due to hexane's intrinsically higher vapor pressure than heptane; because γ for this mixture never strays far from 1.0, this phase diagram appears quite symmetric.

   The skewing of the phase diagram for benzene in heptane shown in Figure 6.4b reflects an α that varies from one when benzene is concentrated to just over three as the γ for benzene increases from one to a γ^{∞} of 1.7 while the γ for heptane

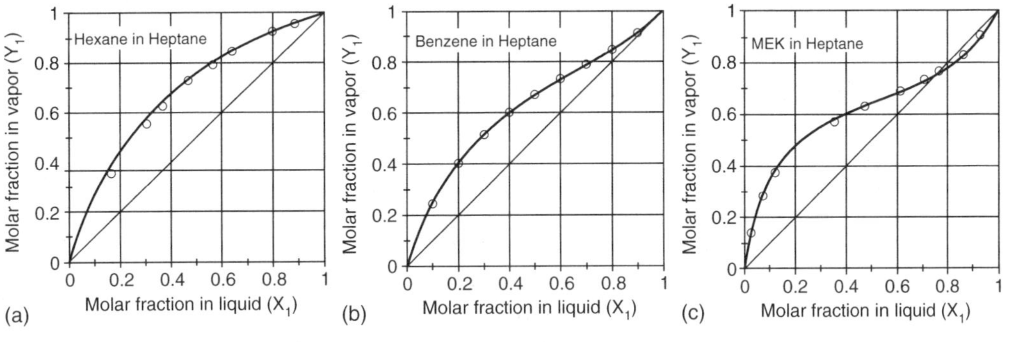

FIGURE 6.4 Examples of X-Y diagrams arrayed in order of increasing peak activity coefficients (γ^{∞}) of 1.1, 1.7, and 3.5, respectively. The small α symbols are experimental data points from Hirata et al. (1975).[9]

decreases from a γ^{∞} of 1.9 toward one as the fraction of benzene decreases (the γ of benzene in heptane is shown in Figure 6.2). The changing γ values cause the curve to be asymmetric, but the fact that the vapor curve is always on the top side of the 1:1 line indicates that benzene is always more volatile than heptane in all mixture proportions.

In Figure 6.4c, the γ of MEK in heptane rises from one to a γ^{∞} of 3.5 (again as shown in Figure 6.2) while the γ for heptane falls from a γ^{∞} of 3.4 toward one, causing the α for MEK to vary from 0.5 when it is concentrated (on the right side of Figure 6.4c) to over 6 when it is dilute (on the left side of this figure). These changes are sufficient for this mixture to form an azeotrope near X_{MEK} of 0.75 where the vapor curve crosses the 1:1 line. In fact, such nonideal interactions within the liquid are necessary (although not sufficient) for an azeotrope to form. For the same reason, an ideal mixture (where γ always equals 1) that follows Raoult's law can never form an azeotrope. The vapor curve of an ideal mixture will be consistently either above or below the 1:1 line because nothing can happen to change the Y_1 to Y_2 ratio.

c. While exposure is always proportional to $P_{\text{vapor},i}$ and $P_{\text{vapor},i}$ is proportional to a given X_i (at least as long as γ_i does not change too much), unfortunately X_i will change in an actively evaporating liquid mixture when its relative volatility is not near unity. If a component's relative volatility (α_i) is much greater than 1.0, that component's molecules will leave the liquid surface more rapidly than the molecules of the other components within that mixture. Those molecules that leave faster than the other molecules will become depleted at the liquid's surface until molecules from the rest of the bulk liquid can either flow or diffuse back into the liquid boundary layer. Any component that becomes depleted in the liquid will appear to have a lower X_i and to be less volatile than its equilibrium $P_{\text{vapor},i}$ predicted using even the best available method presented above. And the reverse would be true if α_i were less than one; that component will effectively accumulate at the surface (while one or more of the other components will become depleted). Over time, the molecules of the more volatile component will become depleted from the whole liquid mixture, and the mixture's composition (its X_i values) will change. The effect of relative volatilities different from unity and changing X_i values can present an almost intractable problem, certainly not one with which most hygienists have the time with which to deal.

d. Some generalizations about the relative volatility of components in mixtures can be made. Fortunately, a good liquid cleaner will be formulated to be azeotropic so its performance properties remain constant over time. This statement applies to cleaners formulated as a mixture of organic solvents, as a mixture of organic solvents in water (to be discussed in Chapter 8), or as a vapor cleaner/degreaser (discussed in Chapter 4). While the vapors from most cleaners are easy to predict, the vapors from paints and surface coatings are the hardest to predict because they are formulated from components chosen purposefully to have diverse relative volatilities (α_i values) and they are spread thinly so that X_i will vary rapidly over the coating's drying time as the more volatile component(s) evaporate first. Even if activity coefficients are known over a range of X_i conditions via any of the above methods, some experiments at the worksite would still be necessary to characterize the length of the drying time and the portion or phase of that drying time during which nearby people are exposed to the most toxic components. There comes a time where one is better off bypassing the models and the experiments and just assess the hazard the old-fashioned way: collect lots of air samples. But even then, the above theory might still be useful to either interpolate or extrapolate those measurements to other times or settings.

In summary, vapor exposures from a liquid mixture can occur in any of the four scenarios discussed in Chapter 5. Exposure levels in three of these four scenarios involve vapor pressure. While none of the above models for vapor pressure from mixtures are perfect, failure to anticipate nonideal effects in a liquid mixture will generally yield worse predictions than a prediction resulting from any of the above models or set of experimental measurements that accounts for such nonideal effects.

PRACTICE PROBLEMS

1. Find the vapor pressures of the following two mixture components assuming that the mixture is ideal and conforms to Raoult's law. Use P_{vapor} data from Appendix A. Remember that X_i is the molar fraction of the ith component in the liquid, and Y_i is the molar fraction of the ith component of the vapor.

	Liquid Phase					Pure	Ideal Vapor
	Vol%	$\rho_{\ell,i}$	MW_i	mol_i/L_{mix}	X_i	P_{vapor} (mmHg)	$P_{vapor,i} = X_i HP_{vapor}$ (mmHg)
Toluene	99	0.87					
Benzene	1	0.88	78.1	0.113		95	
		$\Sigma mol_i/L_{mix} =$					$\Sigma P_{vapor,i} = 29.19$
		or if you don't include the decimal on toluene's P_{vapor},					$\Sigma P_{vapor,i} = 28.79$

2. Would this particular mixture approximate an ideal mixture? Why or why not?
3. Look up the activity coefficient for benzene at its concentration in this mixture in one of the figures in Chapter 6. What is $\gamma_{benzene}$?

$\gamma_{benzene} =$ ________

Why is it just slightly greater than 1?
4. The relative molar fractions $Y_i/Y_1 + Y_2$ of the vapor from a liquid mixture can be assumed to stay in the same proportions anywhere in a workplace as $P_i/P_1 + P_2$ were at the source. Use your answers from above to find the nonideal relative molar fractions in the table below.

$$\text{Relative } Y_i = \text{Rel.}Y_i = \frac{P_{vapor,i}}{\Sigma P_{vapor,i}} = \frac{\gamma_i \times X_i \times P_{vapor\ of\ pure\ i}}{\Sigma \gamma_i \times X_i \times P_{vapor\ of\ pure\ i}} \quad \text{(using Equation 6.19)}$$

		Ideal Vapor			Nonideal Vapor	
	Pure P_{vapor}(mmHg)	$P_{vapor,i} = X_i P_{vapor}$ (mmHg)	Rel Y_i (%)	γ_i	$P_{vapor,i} = \gamma_i X_i P_{vapor}$ (mmHg)	Rel. Y_i (%)
Toluene		28.06		1.0		
Benzene	95	1.13				
	$\Sigma ideal P_{vapor,i} =$		$\Sigma nonideal\ P_{vapor,i} =$		30.10 to 30.32 if $\gamma = 2$	

5. What would be each chemical's contribution to the hazard of the mixture of vapors as a whole? Make this calculation in the table below for TLVs.

Relative Hazard of i within the mixture = Rel.Haz_i

$$= \frac{P_{vapor,i}/\text{TLV}_i}{\Sigma P_{vapor,i}/\text{TLV}_i} \quad \text{(using Equation 6.20)}$$

			Nonideal Vapor			
	Liquid Vol%	Pure P_{vapor} (mmHg)	$P_{vapor,i} = \gamma_i X_i P_{vapor}$ (mmHg)	TLV_i	$P_{vapor,i}/TLV_i$	Rel.Haz_i (%)
Toluene	99	28		50		
Benzene	1	95		0.5		
			Σnonideal $P_{vapor,i}/TLV_i$ =		4.63 to 5.09 if $\gamma = 2$	

6. (Fill in the blanks) Based on the above calculations, benzene comprises approximately ___ % of the vapor but ____ % of the vapor hazard. Why does benzene comprise so much more of the vapor hazard than it does of the actual vapor?
7. A chemical's vapor pressure, its relative molar fraction in the vapor, and its relative vapor hazard are all made worse as the activity coefficient (γ) deviates upwardly from 1.0. How much worse will someone's exposure be to benzene than would be predicted by Raoult's law in this case? (Hint: The answer is simpler than you might try to make it.)

REFERENCES

1. NIOSH: *Criteria for a Recommended Standard ... Occupational Exposure to Alkanes*, HEW Publication No. (NIOSH) 77–151, 1977.
2. IPCS Environmental Health Criteria 20: Petroleum Products, International Programme on Chemical Safety, http://www.inchem.org/documents/ehc/ehc/ech020.htm.
3. NIOSH: *Manual of Analytical Methods (NMAM)*, 4th ed., Method 1550 for Napthas, HEW Publication No. (NIOSH) 94–113, http://www.cdc.gov/niosh/nmam, 1994.
4. Cavender, F., Aromatic hydrocarbons, In *Patty's Industrial Hygiene and Toxicology*, 4th ed., Clayton G. D., and Clayton F. E., Eds., Vol. II, Part B, Wiley, New York, chap. 21, 1994.
5. Popendorf, W., Pryor, A., and Wenk H. R., Mineral dust in manual harvest operations, In *Proceedings of the ACGIH Agricultural Respiratory Hazards Symposium, Ann. Am. Conf. Gov. Ind. Hyg.*, 2, 101–115, 1982.
6. Howard, P. H. and Meylan W. M., In *Handbook of Physical Properties of Organic Materials*, Howard P. H. and Meylan W. M. Eds., CRC Press, Boca Raton, FL, 1997.
7. Lide, D. R., Ed., *CRC Handbook of Chemistry and Physics*, 83rd ed., CRC Press, Boca Raton, FL, 2002.
8. Weast, R. C., Ed., *CRC Handbook of Chemistry and Physics*, 60th ed., CRC Press, Boca Raton, FL, 1979.
9. Hirata, M., Ohe, S., and Nagahama, K., *Computer Aided Data Book of Vapor–Liquid Equilibria*, Kodansha Limited, Elsevier Scientific, Tokyo, 1975.
10. Lyman, W. J., Reehl, W. F., and Rosenblatt, D. H., *Handbook of Chemical Property Estimation Methods*, McGraw-Hill, New York, 1982.
11. Reid, R. C., Prausnitz, J. M., and Sherwood, T. K., *The Properties of Gases and Liquids*, 4th ed., McGraw-Hill, New York, 1987.
12. Poling, B. E., Prausnitz, J. M., and O'Connell, J. P., *The Properties of Gases and Liquids*, McGraw-Hill, New York, 2001.
13. Pierotti, G., Deal, C., and Derr, E., Activity coefficients and molecular structure, *Ind. Eng. Chem.*, 51, 95, 1959.
14. Grain, C. F., Activity coefficient, In *Handbook of Chemical Property Estimation Methods*, Lyman W. J., Reehl W. F., and Rosenblatt D. H., Eds., McGraw-Hill, New York, chap. 11, 1982.
15. Wilson, G. M., Vapor–liquid equilibrium. XI. A new expression for the excess free energy of mixing, *J. Am. Chem. Soc.*, 86(2), 127–130, 1964.

16. Bishop, E. C., Popendorf, W., Hanson, D., and Prausnitz, J., Predicting relative vapor ratios for organic solvent mixtures, *Am. Ind. Hyg. Assoc. J.*, 43, 656–661, 1982.
17. Sandler, S. I., *Chemical and Engineering Thermodynamics*, 3rd ed., Wiley, New York, 1999.
18. Yaws, C. L., *Thermodynamic and Physical Property Data*, Gulf Publ. Co., Houston, TX, 1992.
19. Mackay, D., Shiu, W. Y., and Ma, K. C., *Illustrated Handbook of Physical–Chemical Properties and Environmental Fate for Organic Chemicals*, Vol. I, Lewis Publishers, Chelsea, MI, 1992.
20. Sander, R., *Compilation of Henry's Law Constants (Solubilities) for Inorganic and Organic Species of Potential Importance in Environmental Chemistry*, http://www.mpch-mainz.mpg.de/~sander/res/henry.html#2, 2004.
21. Mackay, D. and Shiu, W. Y., A critical review of Henry's law constants for chemicals of environmental interest, *J. Phys. Chem. Ref. Data*, 10, 1175–1199, 1981.
22. Arbuckle, W. B., Estimating activity coefficients for use in calculating environmental parameters, *Environ. Sci. Technol.*, 17, 537–542, 1983.
23. Mackay, D., Finding fugacity feasible, *Environ. Sci. Technol.*, 13, 1218–1223, 1979.
24. Dunnivant, F. M., Coates, J. T., and Elzerman, A. W., Experimentally determined Henry's law constants for 17 polychlorobiphenols, *Environ. Sci. Technol.*, 22, 448–453, 1988.
25. Gossett, J. M., Measurement of Henry's law constants for C_1 and C_2 chlorinated hydrocarbons, *Environ. Sci. Technol.*, 21, 202–208, 1987.
26. Ashworth, R. A., Howe, G. B., Mullins, M. E., and Rogers, T. N., Air–water partitioning coefficients of organics in dilute aqueous solutions, *J. Hazard Mat.*, 18, 25–36, 1988.
27. Hussam, A. and Carr, P. W., Rapid and precise method for the measurement of vapor/liquid equilibria by headspace gas chromatography, *Anal. Chem.*, 57, 793–801, 1985.
28. Woodrow, J. E. and Seiber, J. N., Vapor pressure measurement of complex hydrocarbon mixtures by headspace gas chromatography, *J. Chromatogr.*, 455, 53–65, 1988.
29. Seto, Y., Determination of volatile substances in biological samples by headspace gas chromatography, *J. Chromatogr.*, 674, 25–62, 1994.

7 Changing the Workplace

The learning goals of this chapter are:

✍ Be familiar with examples of technologic, psychologic, and economic dimensions to "change."

Despite their outstanding qualities, hygienists do not control the workplace. In most workplaces, production managers make the rules and decisions, and production operators make the products or do the things that make the employer money. Unfortunately, their decisions and actions sometimes cause injury or disease. An industrial hygienists goal is to minimize these adverse effects while not stifling the things they do best. Thus, changing either the rules, the chemicals, or the changing production procedures, takes the approval, cooperation, and/or the participation of others, often many others. Obtaining that commitment requires technologic knowledge, psychologic skills, economic awareness, and patience.

I. THE TECHNOLOGIC DIMENSION OF CHANGE

Controlling an existing hazard means changing the way things are currently done. Thus, making a change certainly requires knowledge of current production processes, their hazards, and the people involved. Making a change is aided by knowledge of the past (the history of process development, what has been tried before, and why it failed or was replaced). Selecting a good change is easier with knowledge of what the future might become (what new options are on the horizon). Thus, understanding as much of the production process as possible is a major contributor to controlling a hazard. Many of the following sources of technologic information are much more available on-the-job than as a student.

1. Most industrial production processes are actually a carefully structured sequence of steps that are individually simple. Design engineers refer to these steps as unit operations and unit processes. Every plant has at least a conceptual process flow diagram that describes the sequence of steps they conduct; most plants have them on paper. Process flow affects the way a plant is laid out as the product goes from one step to the next (although the older a plant gets, the more convoluted their plant layout often becomes as segments change and are replaced incrementally). Knowledge of such unit operations, unit processes, and process flows will not seek you out; you have to be curious enough to seek it out. Production engineers should be a good source of knowledge if you can understand their terminology. A good selection of books with terminology and process flow diagrams is available, some tailored to the industrial hygienist[1–7] and others more for the engineer or technician.[8–11] More detailed knowledge exists "on the job" than in the classroom, and when you are out on the job, more knowledge can usually be found in the workplace than in your office.
2. Read the industrial literature. Almost every industrial sector has one or more trade magazine that describe relevant news, history, regulations, processes, and vendors. Detailed information from industrial trade associations is rarely on-line as of this writing

and your company's proprietary information never is (it is business confidential and meant to be kept that way). Some industrial information is on-line such as those described in Section I in Chapter 8 and can be very useful for certain technologies and chemicals, but the source of such information should always be considered.

3. An industrial hygienist does not have an isolated role. While programs are typically managed within specialized departments, some programs and most projects are managed by multi-disciplinary teams.[a] Your participation in such teams benefits both the project, the participants, and their other programs. Participation is an excellent opportunity to learn about other people's and organization's roles and priorities, to become familiar with their processes, capabilities, opportunities, and limitations, and to raise the awareness of health and safety issues among the other participants. Formal opportunities to interact with people in personnel, production, engineering, and/or research and development departments occur in meetings and project teams, but smaller and/or informal meetings are often more productive venues to exchange detailed information and to explore potential solutions to specific health and safety hazards.
4. Participation on technical committees outside the company, both within industry-specific associations and within the professional industrial hygiene organizations, can yield similar results but usually requires some travel. While information regarding health and safety issues is usually exchanged openly among professional colleagues, one must also learn to recognize and protect your employer's proprietary information such as production temperatures, pressures, and durations.
5. Listen to production supervisors and employees describe their processes, tasks, and problems. Finding one person in the company who completely understands the entire process may be difficult if not impossible, and the technical knowledge from that one person is often inadequate for IH purposes. Talking to each operator who may only understand one step in the process takes time but has other advantages. The people who handle the material are the ones who experience the problems. An industrial hygienist can sometimes be told things that employees are unwilling to tell anyone else. Such discussions can often identify problems and issues that are "below the radar" of any existing formal surveillance mechanism. Operators often have a longer personal history than the design engineers, know what has been tried and failed in the past, and thus can be excellent resources to point out old flaws in new ideas. Their participation early on (versus bringing them someone else's solution) will also help ease their acceptance of and transition into any change.
6. Become acquainted with vendors of production oriented machinery and process equipment, of preventive services such as air cleaners, of green chemicals, of personal protective equipment (PPE), and even of specialized IH consulting services. Vendors usually know their product or service, at least some of their competitors' products or services, and what has been tried in similar (or dissimilar) operations. Keep in mind that they are trying to sell their product; so it helps to listen to more than one vendor.
7. Apply sound, basic scientific principles in new ways. Knowledge that comes from a broad and solid science-based education is needed to understand the chemical, physical, and biologic processes used in industry. Knowledge can help see why something works. Curiosity can be the catalyst to ask why not try something new. Thinking creatively outside of the boxes imposed by someone else's process diagram and checklist is the fuel that new professionals can bring by literally walking into an employer's door to help

[a] A program is a function, service, or activity that is performed continuously or repeatedly (either periodically (e.g., monthly or annually) or episodically). A project is a group of related tasks to be performed within a definable time period and/or budget to meet a specific set of objectives. Thus, a project has a finite starting and ending date; programs do not.

them succeed at innovation. Experience must be the leavening to help see what has and has not worked in the past or how to make any good idea become a reality. New industrial hygienists' do not have much experience, so along with knowledge, curiosity, and creativity, bring a little humility and patience.

II. THE PSYCHOLOGIC DIMENSION OF CHANGE

So you have done your homework and have a good idea. Will it fly? Most people do not embrace change until they have to. Perhaps they think like Yogi Berra, "the future ain't what it used to be" or "when we come to that fork in the road, we'll take it." Ideally they should want to change, but at best they have to accept change. Making major changes can require more psychological insight and salesmanship than devising the technical justification for that change. Even more than technology, the workplace is a matrix of groups of people. For hygienists to apply their technical skills (and especially to implement changes), they need to work with many groups of people. Thus, some understanding of group psychology, sociology, and organizational dynamics will help make new ideas become a reality.

1. What are the motivators for change within your organization? Take advantage of other people's or other groups' interests that may not be driven by health, safety, or the environment. Other common drivers include production increases (that often creates the need to purchase more production equipment), productivity increases (that often creates an interest in new technology that can produce the same product more efficiently, quicker, or cheaper but not necessarily safer), or market demand (customers wanting more quality, a better product, or just more, might mean either of the above).
2. Another huge driver beneficial to employee health is the marketable demand for *environmental quality*. A closely related trend is the concept of a sustainable future. Related terms include *industrial ecology* and *green chemistry*. The latter attempts to avoid pollution by utilizing chemicals and processes that are *benign by design*. Typical published resources on green chemistry discuss alternative feedstock and reaction processes, catalysts and biocatalysts, polymer-supported acids and bases, and alternative solvents.[12–14] Many information resources on new alternative technologies for specific processes are new enough to be web-based.
3. Another often unrecognized but important component of corporate culture is the organizational process by which your employer or client (in case you are a consultant) makes and implements its management decisions. How do they systematically identify weaknesses within their current products and operations, limitations to their capabilities, and opportunities for change? Do proposals have to go through channels (established reporting relationships), can a small working group be created to devise a solution? How do groups winnow multiple ideas and reach or assure a consensus? Does one person ultimately have to make the final decision or does that person just need to be assured that the decision went through the proper process? Who enforces those policies? And how well?
4. One goal for any health and safety program is to integrate health and safety into every management decision. Identifying and proposing fixes to potential problems before they are built (while they are still on paper or in a computer file) is always cheaper than changing an existing process after it is built. Thus, a good goal is to integrate health and safety as a standard criterion in every major renovation, expansion, new equipment purchase, or process change. However, when everyone is participating, who gets the credit? The allocation of credit for ideas can be institutional (a reflection of an employer's culture), and it can vary among individual supervisors. Credit can be thought of like computer files: it can be copied and every eligible person can share it.

Whatever your local policy is, keeping your own list of contributions can be very beneficial during annual reviews, reorganizations, etc.

5. The willingness to innovate (or not) is even more of a corporate culture. Organizations that focus on finding a fault vs. fixing a process, or that punish failure more than they reward success, often have more difficulty being the innovator than being the follower of another organization's precedent. The irony is, in business, innovation breeds success. The point is to recognize the management style of your boss and your organization and try to learn to work within that style (or those styles).

Although changes are implemented collectively, we work with people individually. Thus, using some personal psychology is at least as important as group psychology.

6. Even the best idea in the mind of one person can appear to another to be just someone else's opinion. IH ideas are often generated in response to a challenge with which you are dealing as a member of either a project team or program staff. What does the leader of your group want? If the project leader is not your nominal boss, how much involvement does your boss want because it is an IH issue? Does the particular leader want to make all the decisions? Will they carry ideas forward, or will they just sit on them? What does your boss's boss want? Will your collaborators give you credit, or, to get credit, will you have to broach your idea early? The answers to these questions depend on both your organization's policies and your particular setting.
7. While changes usually require money (the topic of the next section of this chapter), cost is often not the strongest reason that people resist change. Psychologic reasons why people resist change can include their comfort with the known vs. discomfort with something new and unknown, because it was not their idea (the "NIH" or not-invented-here syndrome), or because they fear punishment for failure more than they desire the rewards for success. Major changes engender more uncertainty than minor changes; thus, a good major change can be made more acceptable if it can be recast into a series of minor changes.

One can develop one's personal people skills any time and at virtually all times.

8. To implement change, hygienists have to work with people at many organizational levels. The levels differ as you move through the sequence of conceiving, designing, proposing, budgeting, contracting, and training for a change. At times, an industrial hygienist will work with people from virtually an organization's top to its bottom within the same day. Communication skills have many forms from listening to talking, from reading to writing. Few IH classes foster good communication skill development. Students can focus on developing some of these skills in other purposely-selected courses, but becoming a good listener is an important skill to cultivate at any time. A good first step toward good communication is simply to keep an open ear and a closed mouth. A prerequisite to good listening is to keep your mind on the conversation, not on what else is happening right now (such as, "How is my air sampling pump running?") or what needs to happen tonight (such as, "What was it I wanted to get at the hardware or grocery store on the way home?"), tomorrow ("What do I need to do for my project meeting?"), or whenever ("When is the next exam in this class?").
9. When you do hear things from employees about problems or issues that are "below the radar," how you use that information will have a large impact on your future credibility. Using such private information wisely can benefit all concerned, but gaining the personal confidence of employees is a lot harder than losing it by a knee-jerk reaction to place blame or fault. Protecting the source of such private information is harder than purposefully or accidentally identifying a confidant. Proprietary information is another

form of private information with legal implications; understand your organization's policy.

10. Who said, timing is everything? A good idea at the right time is most easily accepted. The same idea at the wrong time needs to be nurtured patiently, worked slowly to build the support one person at a time, or maybe just stored away to await the right time or opportunity for it to be acceptable. Finding the key person or the right time takes people skills, applied psychology, and insight into other people's intentions and limitations. Some problems (especially those that involve individual people) cannot be solved and just have to be worked around. Do not expect success on every attempt.

III. THE ECONOMIC DIMENSION OF CHANGE

1. Occupational health and safety was, and still is, a branch of public health. A challenge common to all branches of public health is to get someone to spend money or take action now to avoid an adverse outcome in the future. In our case, how do we get our employer to fund our budget requests or accept and implement our recommended protective programs? Each of the following approaches can be taken to induce an employer, a manager, an owner, or any other controlling agency to spend money to either initiate, maintain, or strengthen a health and/or safety program. The choice of how to approach your employer will be yours when you get there.
   a. The ethics of being good. One can argue that implementing an employee health protective control is a moral or societal responsibility (e.g., because it is the right thing to do). No one always makes the right decisions in their everyday lives, and there are few incentives for managers to want to do what is right irrespective of costs. However, simply informing some managers of the potential adverse outcomes is sufficient to garner their support; however, I would avoid the pitfall of letting an IH paradigm become a moral imperative.
   b. The legal requirement. One can argue that implementing an employee health protection control will avoid the regulatory policing of the Occupational Safety and Health Administration (OSHA), the Environmental Protection Agency (EPA), or other enforcement agency. In general, the EPA has much stronger enforcement powers than OSHA, but the perception of a potential future financial penalty is rarely a major deterrent to not support a health and safety program (almost a double negative). On the other hand, a publicly announced enforcement action can have even larger negative financial implications with labor, the community, buyers, and stockholders. The threat of enforcement can be a statistical gamble (how often will they visit your plant?).
   c. The "hammer" of a liability lawsuit. One can argue that implementing an employee health protection control will reduce the probability of someone suing their employer for negligence. While such suits (called a tort) may be rare, the cost consequences of a court imposed fine can be much greater than the penalties from regulatory citations. Thus, the probable cost (the product of a suit's cost times its statistical probability) can be in the range of or even much greater than regulatory penalties. This looks like another statistical gamble that is probably harder to quantify.
   d. The "carrot" of increasing profitability. Apart from the increase in efficiency of many new technologies, we can try to persuade others to make decisions based on the anticipated cost of future liabilities. (Recall that anticipation is a part of the definition of IH.) One can argue that funding an employee health protective control now will reduce the liability costs for future adverse health, safety, or environmental outcomes

(disease, injury, or damage, respectively). To quantify the liability costs, one must define one or more adverse outcomes that the current condition is likely to cause, the probable number of those outcomes (per year or over some defined time frame), and the cost to treat, mitigate, remediate, and/or compensate each adverse outcome. By definition, a cost-effective change will cost less to implement now than the money it will save by reducing these future liabilities.

Now imagine yourself in the decision-making role for an enterprise (be it a family, a group, a company, or an agency), whose primary focus is not health and safety. What are your motivators? Which of the above approaches would be your strongest inducement to spend money on health and safety? All other things being equal, a decision is more likely to be made based on economic self-interest than on moralistic grounds or because someone out there may be watching. An opinion supported by quantitative evidence of either increased profit and/or reduced future costs for damage to people, a community, or the environment is likely to be more appealing than an opinion without such benefits.

2. The following approach to quantifying the potential cost (financial liability) of current work practices can be used to set management priorities and/or to justify a change. The first two steps determine the dose. The dose and toxicity determine the probability of or the incidence at which an adverse health effect might occur. The incidence times the number of people at risk determines the expected annual number of cases of that adverse health effect. The nature or severity of that effect determines the cost to treat or compensate for each case of that adverse health effect, and the number of cases times the cost per case determines the financial liability to the employer if the hazard is not controlled.
   a. The setting, use conditions, and work practices determine the dose. The intrinsic ability of the chemical (such as the Vapor Hazard Ratio in Equation 5.19) is just one factor that determines dose. Four categories of other factors will be described in Section VIII.3 in Chapter 8 that act in concert with the VHR to determine the level of the airborne concentration, to allow or inhibit an additional dermal route of exposure, or to vary the length of time that employees are exposed to that environment. Exposure concentration times the time of exposure equals dose.
   b. The dose determines the incidence of the adverse health effect(s). The incidence is the number of people per exposed population who develop or will develop a symptom, disease, or effect per year (e.g., the number of cases per 100 or per 100,000 people per year). Prevalence is the number of people within a population who either exhibit the symptom, have the disease, or suffer from some defined effect at any given time. Both terms are referred to as a "rate." Such rates can be gleaned from toxicity or/and epidemiologic data such as that compiled in the *Documentation of the TLVs*,[15] a NIOSH *Criteria Document*, or an employer's own data if it exists and can be accessed.
   c. The number of people exposed to the chemical or process defines the size of the population at risk. The number of currently exposed employees is fairly easy to identify by counting people in the workplace, but identifying the people and their exposures in the past (retrospectively) is a much more difficult task. The number of cases of that adverse effect equals the product of the incidence of the disease times the number of people exposed, as shown in Equation 7.1.

$$\text{Probable Number of Cases} = \begin{bmatrix}\text{Incidence}\\ (\text{rate or }\%)\end{bmatrix} \times \begin{bmatrix}\text{number of people}\\ \text{exposed}\end{bmatrix} \qquad (7.1)$$

   d. The cost per case of disease or other adverse health effect. The cost of all of the effects related to an occupational disease depends in large part upon the severity of the disease. An employer's costs include both the direct costs per case (determined by the severity of the adverse health effect that must be treated and/or compensated) and the indirect costs (such as for the administrative time to handle the patient, process the paperwork, hire or transfer and train replacement employees, and the lost productivity associated with any particular skills or experience of the affected employee).

1. Medical treatment is the major direct cost (at least to the employer as a whole if not to the employee's specific department or profit center), whether the company is self-insured or their workers compensation insurance rates rise based on the employer's rating. Costs or cost estimates may be available from sources internal to the employer (when you work for them), but costs can also be estimated from external sources. The annual DRG Expert provides guidelines for government reimbursement of medical treatment (although actual costs tend to be greater than these).[16] The following on-line sites can also be useful sources of data.

   Medical-Care Expenditures Attributable to Cigarette Smoking — United States, 1993 at www.oncolink.upenn.edu/cancer_news/1994/cig_1.html.

   Health Futures (information on health/medical costs and hospital services for managed care organizations, human resource managers and consumers) at www.health-futures.org/.

   EPA's *Cost of Illness Handbook* gives an estimate of the cost of medical care for a number of potentially environmentally-related illnesses at www.epa.gov/oppt/coi/index.html.

   OSHA's "$afety Pays" (can estimate both the direct and indirect costs for about six broad categories of chemically induced diseases) at www.osha.gov/dts/osta/oshasoft/safetwb.html.

2. Indirect costs are those expenses to cover activities that must be undertaken by others to respond to the person or persons that are actually affected by the disease or injury. The following indirect costs can easily exceed the direct costs of medical care and are never covered by insurance.
   a. Productivity will decrease (at least temporarily) due to the concerns of other employees, the decreased productivity of the partially disabled employee and/or the newly hired or replacement employee. These costs depend upon the setting, the expertise required, and the pay scale(s). The training time of a replacement employee comprise another cost.
   b. The costs to investigate and/or validate the diagnosis, causation, and/or work-relatedness of a claim may be significant, especially for poorly recognized or easily faked illnesses. These costs depend upon the nature of the workforce, the types of disability claims, and the investigator.
   c. Administrative follow-up of a disabled employee for return to work, retraining, consultation, etc. can be the highest portion of a protracted disability. Programs that provide social support, approve work accommodation, and encourage early return to work are usually very cost effective.
   d. Legal costs can be large should the case go to court. Liability compensation or a court-imposed negligence penalty is rare in occupational settings but could be substantial. Newspapers or legal journals tell a history of at least such exceptionally high costs.

e. Ultimately, an employer should be comparing the benefits from sales of their products or services against the future costs associated with all of the above adverse health effects. These future costs are also called future liabilities. Such future liabilities can be calculated as the product of the total cost of each case or adverse outcome (from step d) times the probable number of cases or adverse outcomes that might be expected per year (from step c), as per Equation 7.2.

$$\text{Probable Liability} = \begin{bmatrix}\text{costs of each}\\ \text{outcome}\end{bmatrix} \times \begin{bmatrix}\text{probable number}\\ \text{of outcomes}\end{bmatrix} \qquad (7.2)$$

3. While you are looking toward the future, understand that management as a whole must live within the economic constraints of the present. In order to negotiate with business managers, one should know some of the language of business economics.
   a. For instance, managers use the concept of financial discounting to compare current expenditures (costs) with future benefits (sales) and risks (liabilities). Discounting to a present value (as in Equation 7.3) reduces the present value of a future dollar transaction by a discount rate for each period of time into the future (n) that the transaction is to be made. A discount rate (an r of typically 4–12%) is similar to interest, but instead of being paid on a past deposit, it is the fraction by which the value of a future transaction is reduced. Discounting is a good compromise between economic and psychologic realism and mathematical convenience.
   b. The basic formula shown below for the present value of some future action, purchase, benefit, or cost can be found in books on economics or accounting, in scientific handbooks such as the *Handbook of Chemistry and Physics*, and in computer spreadsheet functions.

$$\text{Present Value} = \frac{\text{future \$}}{(1+r)^n} \qquad (7.3)$$

   where future \$ is the dollar value of a future action, purchase, sale, benefit, or cost at the time of the transaction. Future costs include liabilities and operating costs for any control; n is the number of years into the future at which each transaction will occur. The location of n as a power term within Equation 7.3 compounds the effect of the discount rate after each time period (like compounded interest); r is the fractional discount rate which is the increment by which a future financial transaction is reduced for each time period into the future. If the discount rate is expressed as a percent, r would equal the percentage divided by 100 to yield a fraction [$r = \%/100$].
   c. A project's net present value (NPV) is probably the best financial analysis tool in common use today. NPV is the sum of the present values of discounted benefits, costs, and liabilities projected for some number of years (n) into the future as defined by a management plan.

$$\begin{aligned}\text{NPV} = &\sum\nolimits_{i=1}^{n} \text{Present Value}_{\text{Benefits},i} - \sum\nolimits_{i=1}^{n} \text{Present Value}_{\text{Costs},i} \\ &- \sum\nolimits_{i=1}^{n} \text{Present Value}_{\text{Liabilities},i}\end{aligned} \qquad (7.4)$$

   To calculate a project's NPV using Equation 7.4, one first estimates the costs, liabilities, and benefits (if any) for each of i time periods (typically years). One then discounts each of these Benefits_i, Costs_i, and Liabilities_i values for their time period using Equation 7.3, and adds them up. For a change to be justified, its net present value must be more (or at least less negative in the case of control costs) than the existing net present value if nothing were changed.
   d. Payback period is the time needed for the benefit or savings to equal the cost of implementing a change. A payback period may be calculated either with or without

discounting. In principle, any value (income or benefit) that occurs after the payback period can be considered to be all profit. Basing financial decisions only on the shortest payback period tends to be short sighted, but comparing the payback period with the projected life span of the hazard being controlled can be useful as a sort of reality-check.

e. Return on investment (ROI) is a calculated profit on the investment expressed in terms equivalent to an interest rate. Managers can compare each project's ROI to the return offered by other business opportunities, to the commercial interest paid on cash reserves for money not spent, or to the cost of capital if the project must be done on borrowed money. One limitation of the ROI is the difficulty in comparing the results of multiple projects with different payback periods.

Such economic tools can be used to justify the value (in reduced costs) of employee protection programs and changing to a less hazardous chemical or process. While subjective benefits such as good health (wellness) or good will (a valued name brand) may be difficult to assign a financial value, the costs of adverse outcomes such as illness, injury, or environmental damage are quantifiable.

4. Ultimately, IH is a form of risk management. Risk management is neither complete risk avoidance nor complete prevention. Risk management is a program to reduce the risks that are necessary to obtain certain economic benefits or to satisfy other defined goals within a framework of limited financial resources. To succeed within this framework, an industrial hygienist should eventually learn how to cast risks into monetary liabilities (such as $) and how to use the tools of business and/or public financial decision-making. It is not the place of this text to teach such tools. However, when the time is right (or get them now and save them until you can use them), the following EPA documents (especially the first) might help a novice learn more about financial planning and accountability:[17–20]

A Primer for Financial Analysis of Pollution Prevention Projects.[17] A 39 pp Image file at www.epa.gov/oppt/library/pdfs/primerfinancialanalysis.pdf.

An Introduction to Environmental Accounting as a Business Management Tool: Key Concepts and Terms.[18] A 39 pp document at www.epa.gov/opptintr/acctg/pubs/busmgt.pdf.

Guide to Industrial Assessments for Pollution Prevention and Energy Efficiency.[19] Access a CD-ROM or 486 pp pdf document via www.epa.gov/ttbnrmrl.

An Organizational Guide to Pollution Prevention.[20] Access a CD-ROM via www.epa.gov/ttbnrmrl.

IV. IMPLEMENTING THE IH CONTROL PARADIGM

Any hazard can be controlled by a variety of approaches. Recall the paradigm of industrial hygiene control priorities from Chapter 1: "Source controls are preferred over Pathway controls are preferred over Receiver controls." Controlling the source of the chemical contaminant includes substitution and modification. Controlling the pathway includes automation, separation, isolation, and ventilation. Local exhaust ventilation is much preferred over general or dilution ventilation, and controlling the receiver via administrative control (schedule adjustments) and PPE (e.g., clothing, gloves, and respirators) is neither cheap nor reliable.

Despite the above control priority paradigm, one should not always try to force the highest IH control priority onto every problem just because the paradigm says so. In fact, one rarely starts the process of control by picking only one approach.

1. First, source control is not always feasible. Some production conditions are not changeable. For instance, substitution is almost never a possible control for the hazards of packaging a toxic chemical. The process of packaging is required for most chemicals, and the chemical being packaged has its own economic benefits of sales and profits. The financial liabilities from the health risks associated with useful toxins will reduce, but will rarely outweigh, the financial benefits of its sales. Our credibility is enhanced by recognizing that reality and focusing on something further down the priority list such as changing the setting or controlling the pathway. Investigating, quantifying, and documenting a reduction in the employer's liability for adverse effects can help to show your services add value, but such data will rarely lead to a decision to discontinue the production and packaging of an existing product.
2. Second, many factors can change what is the best control from one workplace to the next. Although hygienists are taught that substitution always solves the entire hazard, we soon realize that the production schedule, financial resources, market potential, management, and/or production workers cannot always support that dramatic of a change. Recognizing and working within the limitations of the setting is an art for industrial hygienists.
3. Third, we never know everything we need to know. Ventilation may be an obvious option to the IH, but with some discussion and encouragement a production supervisor might suggest a simple modification that we had not thought of that may increase production costs slightly but offers sufficient improvement in the short-term to allow a more cost effective change (such as substitution) to be deferred further into the future.
4. Fourth, things change. We may think that building an enclosure around a source would be best, but then find out that a new technology will soon displace the current product or process. This new technology might either affect the market potential of what is currently being produced, or cause the currently used technology to be abandoned (possibly to be replaced by the new). In either case, the current hazard will not be around for long, justifying the use of personal protection as an acceptable temporary control to yourself, to management, to the employees, and to OSHA (if it applies).

Whether one makes a mental list, a physical list, or an electronic list, the process of selecting a control is evolutionary both within ourselves and through the teamwork of the many colleagues with whom we cooperate to solve problems within the workplace.

An instructor of IH control would be remiss if somewhere they did not point out the irony of committing so much time (and pages in a textbook) to the science of IH when the majority of a professional's daily routine is spent dealing with the issues in this chapter. We envision, we sell, we train, and we manage. To be a competent IH, one needs to know science, but to be an effective IH one has to know how to change the workplace. Industrial hygiene is based on science, but it is a profession of, and for, people.

REFERENCES

1. Goldfarb, A. S., Goldgraben, G. R., Herrick, E. C., Ouellette, R. P., and Cheremisihoff, P. N., *Organic Chemicals Manufacturing Hazards*, Ann Arbor Science, Ann Arbor, MI, 1981.
2. Cralley, L., Cralley, L., Eds., *Industrial Hygiene Aspects of Plant Operations, Process Flows*, Vol. I, Macmillan, New York, 1982.
3. Cralley, L., Cralley, L., Eds., *Industrial Hygiene Aspects of Plant Operations, Unit Operations and Product Fabrication*, Vol. II, Macmillan, New York, 1982.
4. Cralley, L. V., Cralley, L. J., Caplan, K. J., Mutchler, J. E., Woolrich, P. F., Eds., In *Plant Practices for Job Related Health Hazards Control, Production Processes*, Vol. 1, Wiley, New York, 1989.

5. Cralley, L. V., Cralley, L. J., Caplan, K. J., Mutchler, J. E., Woolrich, P. F., Eds., In *Plant Practices for Job Related Health Hazards Control, Engineering Aspects*, Vol. 2, Wiley, New York, 1989.
6. Burgess, W. A., *Recognition of Health Hazards in Industry*, 2nd ed., Wiley, New York, 1995.
7. *Encyclopedia of Occupational Safety and Health*, 4th ed., (2 vols. and CD-ROM), International Labor Office, Geneva, 1997.
8. Shreve, R. N. and Brink, J. A., *Chemical Process Industries*, McGraw-Hill, New York, 1977.
9. Avallone, E. A., Baumeister, T. III, Eds., *Mark's Standard Handbook for Mechanical Engineers*, 10th ed., McGraw-Hill, New York, 1996.
10. Perry, R. H., Green, D. W., Maloney, J. O., Eds., *Perry's Chemical Engineers'; Handbook*, 7th ed., McGraw-Hill, New York, 1997.
11. Krochwitz, J. I., Howe-Grant, M., Eds., *Kirk-Othmer Encyclopedia of Chemical Technology*, 4th ed., Wiley, New York, 1997, See also the "Kirk-Othmer Concise ..." (a one volume abridgement).
12. Anastas, P. T., Williamson, T. C., Eds., *Green Chemistry — Frontiers in Benign Chemical Syntheses and Processes*, Oxford University Press, Oxford, England, 1998.
13. Matlack, A. S., *Introduction to Green Chemistry*, Marcel-Dekker, New York, 2001.
14. Sunol, A. K. and Sunol, S. G., Substitution of solvents by safer products and processes, In *Handbook of Solvents*, George Wypych, Ed., ChemTec Publishing, Toronto, chap. 21, 2001.
15. *Documentation of TLVs for Chemical Substances*, 7th ed., ACGIH Worldwide, Cincinnati, OH, 2001.
16. Ingenix. *DRG Expert: A Comprehensive Guidebook to the DRG Classification System*, Ingenix, Inc., Salt Lake City, Utah, 2005.
17. *A Primer for Financial Analysis of Pollution Prevention Projects*, EPA 600/R-93/059, 1993.
18. *An Introduction to Environmental Accounting as a Business Management Tool: Key Concepts and Terms*, EPA 742/R-95/001, 1995.
19. *Guide to Industrial Assessments for Pollution Prevention and Energy Efficiency*, EPA 625/R-99/003, 1999.
20. *An Organizational Guide to Pollution Prevention*, EPA 625/R-01/003, 2001.

8 Source Control Via Substitution

The learning goals of this chapter:

- ✍ Understand the influence of a chemical's physical nature on its hazard and on control options.
- ☐ Know how to predict the effect of changes in P_{vapor} on exposure concentration.
- ☐ Understand at least one of the various interpretations of the vapor hazard ratio (VHR).
- ☐ Be able to calculate a VHR.
- ☐ Know how to use a new and old VHR to evaluate a potentially acceptable volatile chemical substitute.
- ✍ Be aware of factors in the "hazard rating" other than the VHR that can affect a vapor's hazard.
- ☐ Know some ways in which all exposure limits are not created equal.

The "source" is the origin of the contaminant. The concept of source control is closely related to the following two equations from Chapter 5. All other things being equal (*which if you look too closely they may never be*), the best way to reduce the concentration [C] to which people are exposed is to change the source by either reducing the rate at which the source generates the airborne chemical [G] or reducing the amount of dilution that the airborne chemical needs [such as a volatile chemical's VHR]. As implied in Chapter 1, source control is generally more advantageous than either pathway or receiver controls because it is universal (protecting people both near and far from the source), it is reliable (steady over time), it is reproducible (from site-to-site and from person-to-person), and it will be cost-effective (sooner or later).

✍ $$\text{steady state } C\,[\text{mg/m}^3] = \frac{G\,[\text{mg/min}]}{Q_{\text{apparent}}\,[\text{m}^3/\text{min}]} = \frac{\text{Generation rate}}{\text{Dilution air flow rate}} \quad \text{(repeat of Equation 5.14)}$$

✍ $$\text{Compliance Ratio} = \frac{C_{\text{measured in the BZ}}/P_{\text{vapor}}}{\text{TLV}^{\circledR} \text{ or PEL}/P_{\text{vapor}}} = \frac{\text{VHR}}{\text{Dilution Ratio}_{\text{BZ}}} \quad \text{(repeat of Equation 5.20)}$$

This chapter will focus on substitution as a source control: substituting either a less hazardous manufacturing process for a more hazardous process or a less hazardous chemical for a more hazardous chemical. The first section will present some simple examples of alternative technologies that involve less exposure to hazardous chemicals. The next section will compare the substitution options available for nonvolatile vs. volatile chemicals related to some of the physical differences presented in Chapter 2 and Chapter 3. Section III to Section V will progress from aqueous solvents as potential substitutes for semiaqueous or organic solvents, to semiaqueous solvents as potential substitutes for organic solvents, and on to less *hazardous* volatile organic solvents as potential substitutes for more *hazardous* volatile organic solvents, respectively. The final sections will dwell on substitutions among organic solvents and ways to manage these issues.

I. ALTERNATIVE TECHNOLOGIES

The options to replace a hazardous process with a less hazardous technology are as diverse as the number of processes used in industry and they are a moving (evolving) target. The resources cited in the first subsection below address many large-scale technologies for either specific processes or specific industries. The rest of the subsections describe smaller-scale technologies to which the novice industrial hygienist might relate more easily.

1. The U.S. EPA has been very active in promoting less polluting alternative technologies. The EPA Office of Compliance Assistance has produced a series of "Industry Sector Notebooks" or "Profiles" that may be downloaded as pdf files from the following website: www.epa.gov/compliance/resources/publications/assistance/sectors/notebooks/index.html.

   Previous to that, the EPA Office of Research and Development (ORD), National Risk Management Research Laboratory, Technology Transfer Branch published a series of "Guides to Pollution Prevention."[a] Some of the specific alternative technologies described in this chapter came from these EPA documents.[1–5] Despite their publication dates of 1993 to 1994, many employers are slow to change and the information is still relevant.

   Somewhat similar documents endorsed by but not written by EPA are also available from the Pollution Prevention Resource Exchange (P2Rx.org), a national network of regional centers dedicated to disseminating pollution prevention information. The site at www.p2gems.org lists other resources and links by product or industry, chemical, and process.
2. Many solvents in paint create health hazards to employees and can contribute to smog. The following alternatives to either replace, reduce, or change the solvents, technologies, or processes are discussed in EPA's "Guide to Cleaner Technologies: Organic Coating Replacements."[1]
   a. Waterborne coatings or aqueous emulsions such as acrylic latex, polyurethanes, and epoxy are good substitutes for oil-based paints because they allow many of the same application methods. Even those coatings that still contain substantial organic solvents will reduce VOCs.[6,7]
   b. High-solids paints offer similar VOC reductions but their high viscosity requires that new users modify their paint application equipment.[8,9]
   c. Powder coating with thermoplastic resins offers a more complete reduction in VOC emissions and a longer-lasting and more chip-resistant finish, but they cannot replace all paint applications.[10–12] See also The Powder Coating Institute at www.powdercoating.org.
3. Removing or stripping paint with solvents such as methylene chloride or a glycol or by sand-blasting with silica or other solids also creates health hazards. These processes too can be replaced with a range of safer technologies. The following options are discussed in EPA's "Guide to Cleaner Technologies: Organic Coating Removal."[3] A series of military maintenance depot study reports document field evaluations of several of these alternative technologies (see www.jdmag.wpafb.af.mil/study.htm or 2plibrary.nfesc.navy.mil).
   a. Several alternative solid blasting materials are available. "Poly beads" represent a reasonably mature, often cost-effective technology, although modest concerns persist

[a] The web pathway that seems to yield the most complete list of these guides starts at www.epa.gov/ttbnrmrl. Select "Browse Publications Catalog" (in the left margin; not "Search ... "). Type "Guide" in the Title field, select "Pollution Prevention" in the Subject field, select "Search," then look down the page for Pollution Prevention. An alternative way to get to the same search engine is to start at www.epa.gov/opptintr/library/ppicdist.htm (the Pollution Prevention Information Clearinghouse); look down the list for EPA Guides for Pollution Prevention. Click on it.

with the dust that they generate and the difficulty of recycling used media intermingled with paint chips.[13–15] Baking soda (sodium bicarbonate) is a cheaper and environmentally friendlier material for some defined paint and surface stripping applications, although the baking soda requires more careful metering to be successful and can leave an alkaline residue.[16]

Walnut shells, ground corn cobs, and wheat starch offer a range of biodegradable solid blasting media that require minimal equipment changes. Corn cob products are not only used as a media for finishing, tumbling, and blasting but also as a carrier for fertilizers, insecticides, and feed additives and as an absorbent for hazardous liquids, grease, and oils. Walnut shells are used to remove paint (including graffiti), flash, burrs and other flaws in plastic, rubber molding, and aluminum and zinc die castings; for paint removal and general cleaning in the restoration of buildings, bridges, and outdoor statuary; and for cleaning aircraft engines and steam turbines.

b. Ultrahigh pressure water and xenon-flashlamp stripping can be effective when coupled with a computer automation system. Water applied at 55,000 lb/in.^{2} through a special nozzle has been shown to be effective at stripping paint and other coatings from metal surfaces, and although water has minimal disposal problems (see also a discussion on the disposal of aqueous mixtures in the next section) the commercial system is designed eventually to capture, clean, and recycle much of its water.[17] The high-intensity pulsed light from a xenon flashlamp can ablate a coating into a fine ash that would (by itself) be at least a nuisance if not a hazard; however, the Flash Jet® system (depicted in Figure 8.1) automates the process (to protect the operator's eyes) and combines this technology with a low-pressure stream of dry ice particles that both cools the substrate and blows the dry residue into a capture hood from which it is removed by filtration to create an efficient and benign stripping system.[18]
c. Supercritical fluids comprise a less developed alternative technology. A gas above its critical point (such as above both 31°C and 72.9 atm [1057 lb/in.^{2}] for carbon dioxide) has the wetting power (low surface-tension) of a gas but the high solvent power of a liquid. For a part to be cleaned by a supercritical fluid, it must be placed in a chamber big enough to contain the part (immersion) and strong enough for the pressure. Needless to say, it works best on small parts.
d. The technology to use solid CO_2 pellets for cleaning is fairly well-developed for many applications other than paint stripping (e.g., www.jdmag.wpafb.af.mil/carbon%20dioxide.pdf). Application equipment is specialized but generally available (e.g., www.nortonsandblasting.com/icesys.html). The CO_2 rapidly sublimates and leaves no liquid

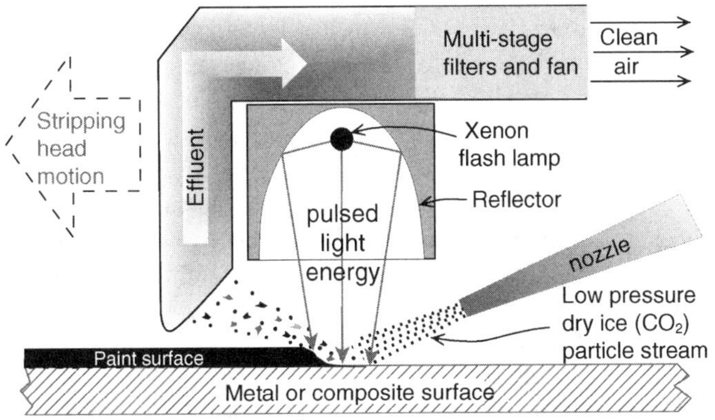

FIGURE 8.1 Depiction of the components of the Flash Jet® stripping head (courtesy of Flash Tech, Inc.).

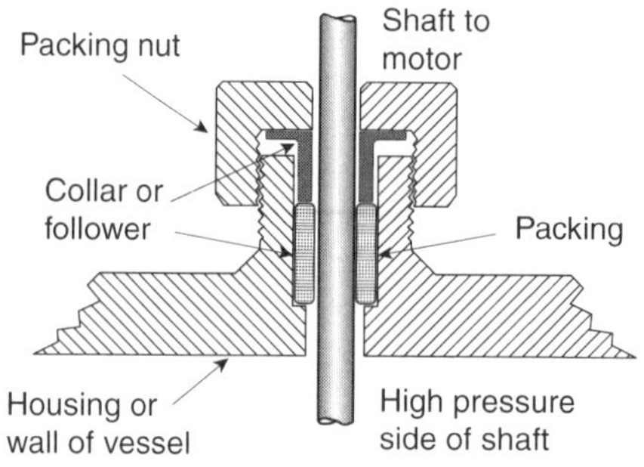

FIGURE 8.2 Packing in a stuffing box around an ordinary pump or valve shaft.

residue. In general, dry ice is too soft to remove hard or well-bonded residues (typical of primers and of finish coats on metals), but its nonabrasive quality makes it advantageous to clean weakly adherent residues or to clean substrates that are susceptible to abrasion such as rubber molds, textured molds (as for polyurethane), molds for blown plastic bottles, foundry core hot-boxes (that may be cleaned *in situ*), baked-on foods in food processing, or grease and oil from building interior surfaces.

4. Alternatives exist for a number of common pieces of industrial equipment; unfortunately, no centralized source(s) of information about alternatives for individual pieces of equipment have been identified. The following example of pumping equipment was chosen because the process of pumping is so common and the pump is often a source of vapor exposure.[19]

   Ordinary pumps and valves have rotating shafts that must pass through the wall or housing of the pump or valve through which toxic liquids can leak. To minimize such leakage, these openings are fitted with a "stuffing box" that is packed with a matting material or (in large systems) string or rope containing a dry lubricant such as graphite. The packing is compressed by a follower ring, gland, or collar that, in turn, is held in place by either a packing nut (as shown in Figure 8.2) or a flanged cap (as shown in Figure 8.3). When compressed tightly around the shaft, the packing minimizes leakage but also increases the force needed to turn or move the shaft. Over time, the packing deteriorates and can leak, creating a health risk to nearby employees. Good maintenance can minimize that leakage, but consistent maintenance is rare and maintenance of a contaminated pump has its own potential for high vapor and dermal exposures. Several substitution control technologies are available.[b]

   One interesting control option is to use a lantern gland, as shown in Figure 8.3. This type of seal comprises two layers of packing separated by a lantern ring. The cross-section of the ring is shaped like an H on its side (shown as the small black piece in Figure 8.3) with a series of holes drilled through the vertical ring. (The name no doubt came from the shape of the ring with holes around the sides similar to a small lantern.) A small pipe is threaded into the side of the housing at the level of the lantern ring to withdraw any chemical (preferably as a vapor) leaking past the first layer of packing. The second layer of packing is more than sufficient to prevent the passage of any vapors that are unaffected by the pressure of the pump. Alternatively, in some cases it is acceptable

[b] One nonsubstitution control is to fit an existing pump with local exhaust ventilation. For instance, the *Manual of Industrial Ventilation*'s VS-15-21 describes a close-fitting enclosure (what will be called a slotted hood in later chapters) that might only need to exhaust 10 to 40 ft^3/min from around such a seal.[20]

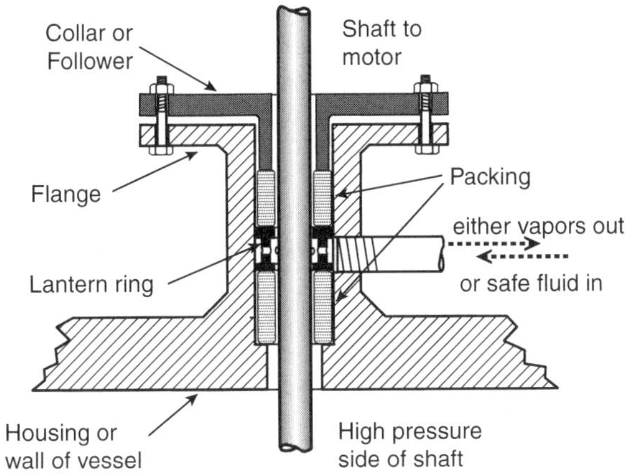

FIGURE 8.3 Packing with a lantern gland on an otherwise ordinary pump or valve.

to pressurize that space with a safe fluid (such as water); the equal pressure prevents any outward leakage of the toxic chemical.

Another substitute technology is to use pumps and values with mechanical seals (sometimes called a rotary seal) instead of stuffing boxes with packed seals. In a mechanical seal, a fixed graphite ring is held tightly against a rotating collar on the shaft by a system of springs. Rotary seals require less maintenance than stuffing boxes, which is an additional advantage when dealing with highly toxic or corrosive fluids.

Sealless or zero-emission pumps avoid the problem of having a rotating shaft penetrate the wall of the high pressure pump chamber. Such pumps include both magnetic-drive and canned-motor pumps as well as pneumatic displacement pumps. In a canned-motor pump, the pump rotor/impeller assembly is driven by the magnetic field of the pump's induction motor and a corrosion-resistant liner or "can" completely seals the stator windings of the motor and prevents its contact with the fluid. With magnetic-drive pumps, the pump impeller is magnetically coupled to the motor at the end of the containment shell housing. Displacement pumps are submerged into the liquid and use compressed air to act like a piston pushing any liquid from its filled pressure chamber; its only moving parts are two check valves (see for example Figure 8.4 or an animated version at www.rmsenviro.com/pitbull_pumping_principle.htm).

An older but still viable option is to use submerged pumps and valves that place the entire rotating portions of the hardware under the surface of the liquid. Any leakage from their submerged mechanisms goes straight back into the liquid being pumped or throttled, respectively. The only health-oriented downside from using submerged hardware is the increased hazard from having to decontaminate both the outside of the equipment and its insides when it is removed for maintenance.

If left to engineers, they would probably select a pump based on the composition of the substance being pumped, its viscosity, the nature of any solids in the stream, and the operating temperature and pressure. The industrial hygienist adds the criteria of toxicity and chemical exposure both in use and during maintenance.

II. VOLATILITY AND ALTERNATIVE CHEMICALS

Liquids are used in industry to either clean, cool, lubricate, propel, and suspend solids or nonvolatile materials. Metalworking fluids epitomize the functional use of liquids to simultaneously clean, cool, lubricate, propel, and suspend solids while machining metal parts. Although much of the following

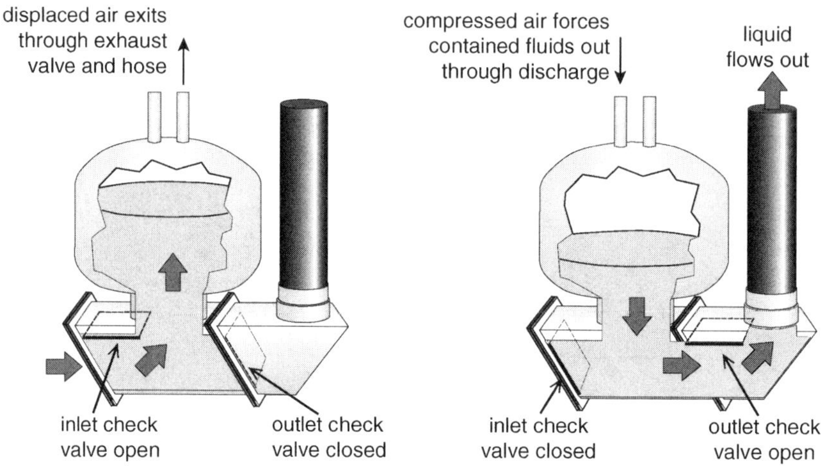

FIGURE 8.4 A pneumatic displacement pump, courtesy of Chicago Industrial Pump Company. The pump chamber is filling on the left and discharging on the right.

discussion was based on the *Guide to Cleaner Technologies: Alternatives for Chlorinated Solvents for Cleaning and Degreasing*,[4] the concepts generally apply to any of the above five functions of liquids. Later sections of this chapter will discuss three groups of liquids: water-based solvents, semiaqueous solvents, and organic solvents. Think of them as three technological dimensions. Reducing VOCs is often a motivator to change either from one organic solvent to another or from an organic to a water-based solvent.[c] Economic factors make aqueous solvents attractive; however, they are not compatible with all products or production processes, and microbial growth in aqueous mixtures can present both production and health hazards, especially to the skin. Semiaqueous solvents tend to be somewhere in between, but some of them also present health risks to the skin, eyes, mucous membranes, or reproductive systems. Without being quantitative, think as you read about aqueous and semiaqueous mixtures about the lessons learned concerning mixtures in Chapter 6.

The physical nature of a chemical hazard has a powerful influence not only on the nature of its hazard but also on the feasibility of certain controls. For the reasons summarized in Table 8.1, what can control a chemical hazard depends strongly upon whether that chemical is volatile or nonvolatile.[d]

1. Volatile materials (discussed in Chapter 2) include all of the liquified gases, most liquids including solvents, most molten metals, and aqueous mixtures with volatile constituents. Some volatile materials can be absorbed through, irritate, or sensitize the skin, but their volatility will limit the time they remain on the skin before they evaporate. Their potential to remain on the skin long enough to present a hazard via the oral route is even less likely. Similarly, volatile materials will not remain long on contaminated surfaces (unless the surface is absorbent). On the other hand, a volatile chemical can always

[c] For practical purposes, the U.S. EPA defines VOCs in 40CFR51.100(s) as any compound of carbon that participates in atmospheric photochemical reactions except specific chemicals that have been determined to have negligible photochemical reactivity. These exceptions include carbon monoxide and dioxide, methane, ethane, acetone, T-butyl acetate, methyl acetate, methyl formate, most methylated siloxanes, and a number of halides including a few simple solvents such as methylene chloride (dichloromethane), perchloroethylene (tetrachloroethylene), and 1,1,1-trichloroethane (methyl chloroform).

[d] Although every material has a finite vapor pressure and no fine line can be defined between nonvolatile and low volatility chemicals, those with vapor pressures of less than 0.01 mmHg will evaporate so slowly that they will remain on any surface for at least several days and thus might as well be considered nonvolatile from an occupational health perspective.

TABLE 8.1
Comparison between the Hazards Presented by Volatile vs. Nonvolatile Chemicals

	Volatile Chemicals	Nonvolatile Chemicals
Ability to become airborne	Evaporation is passive. A vapor's generation rate depends upon the vapor pressure of the source	No passive generation ability. Aerosolization depends upon an external activity
Ability to stay airborne	Vapor molecules will mix with and remain in air indefinitely	Rate of sedimentation depends on aerosol aerodynamic diameter
Ability to be inhaled and location of deposition	Molecules move with air. Absorption depends heavily upon the chemical's water solubility	Inhalation is likely if $d_{\text{aerodynamic}}$ is below 50 μm; deposition in lung only if $d_{\text{aerodynamic}}$ is below 10 μm
Ability to be a dermal hazard	Residence time on the skin is proportional to the chemical's volatility	Absorption depends upon skin contact, octanol-water partitioning, and skin washing
Ability to be an oral hazard	Little time for hand-to-mouth contact during the short skin residence time of a volatile chemical	Hand-to-mouth contact depends upon dermal contact, washing facilities, and eating policies
Ability to be a long-term environmental hazard	Low unless retained by a sorptive surface, soil, or ground water	High; also depends upon clothing usage and housekeeping

evaporate, by which it can create an airborne hazard any time the liquid is exposed to the air. A volatile chemical's intrinsically high vapor pressure complicates the process of selecting a substitute chemical by having to assess both the volatility and the toxicity of two (or more) chemicals at once. This process will be the topic of Section VI, but is generally simplified if the substitute is water based.

2. Nonvolatile materials (discussed in Chapter 3) include solid metals, powders, dusts, pastes, and low volatility liquids. Nonvolatile chemicals have no intrinsic means (such as vapor pressure) to become airborne (fumes from thermally heating a solid material are a major exception). Thus, an external force or activity is required to aerosolize a nonvolatile material. However, nonvolatile materials are also likely to present hazards by routes of exposure other than respiration. Some nonvolatile materials can be absorbed through intact skin or can irritate or sensitize the skin. Nonvolatile materials can also present a hazard via the oral route of exposure where work practices allow eating or even smoking without thorough hand-washing subsequent to hand skin contact. Accumulations on employees' street clothing can also cause the toxic chemical to be carried home. Since nonvolatile materials do not evaporate, they will remain on contaminated surfaces inside the plant for a long time unless removed by good housekeeping. Such surfaces can include tools, the outside surfaces of product mixers or reactors, almost any nearby flat surface, and even structural surfaces. The usual cause of off-site hazards is contaminated ground water, but nonvolatile chemicals can also create off-site hazards by shipping surface-contaminated packages, disposing of wastes improperly, and eventually demolishing the production building.
3. Distinguishing whether a source comprises either a volatile and a nonvolatile chemical will strongly influence the process of selecting a substitute chemical. Section III to Section V of this chapter will discuss a range of water-based solvents, semiaqueous solvents, and organic solvents that parallel but are not synonymous with their range of volatilities. The following comments are a prelude to these sections.
   a. Using the least toxic nonvolatile chemical is always beneficial to employee protection. Substituting a less toxic nonvolatile chemical for a more toxic nonvolatile

chemical with the same physical properties will generally not change the size of the dose, but it will always reduce the fraction of the substitute chemical's toxic dose to which someone will be exposed. Thus, most nonvolatile materials that can serve the same desired function but have a higher TLV or PEL (or other measure of toxicity) would be a good candidate to be used as a substitute for a more toxic nonvolatile material with a lower TLV, PEL, etc.

At the same time, knowing that occupational exposure limits are based on multiple health criteria, a hygienist should thoroughly understand that a lower exposure limit does not necessarily imply a greater occupational hazard. Remember the adage that *all exposure limits are not created equal.* For example, substituting a chemical with reproductive toxicity for a chemical that is only an irritant is generally not a good choice.

b. Substituting a nonvolatile chemical for a volatile chemical is usually beneficial to employee protection; however, such a choice may not always protect the environment or be compatible with the production process. For instance, exchanging a volatile chemical for a nonvolatile water-soluble chemical may just be moving the problem from air into another medium. Exchanging a volatile chemical for a nonvolatile chemical that accumulates in the environment may just be moving the problem into another time frame or into another species. Also, finding a nonvolatile substitute chemical that has the same functional properties is generally not easy. If it were easy, everyone would do it.

c. Selecting one volatile chemical to substitute for another volatile chemical brings into play the dimension of evaporation and vapor pressure, which it turns out is quite unrelated to toxicity. A discussion of the added complexity implicit in substitutions among volatile chemicals will be the topic of Section VI and Section VII of this chapter.

III. AQUEOUS SOLVENTS

Aqueous cleaning and degreasing can now be performed for many applications that were once considered the sole domain of organic solvents. Aqueous mixtures have clear potential benefits such as containing no or very low VOCs, not depleting the ozone, normally being much less toxic (unless they are contaminated with hazardous solutes or microbial growth), and being more likely to be permitted to be discharged into sewers. Most of these potential advantages depend upon the additives and contaminants contained within the mixture. The following information is presented to help understand why such mixtures have so many additives and their various functions.

1. Aqueous cleaners are mixtures of water, surfactants, and other additives that promote the removal of organic and inorganic contaminants from hard surfaces. The other additives include deflocculants, saponifiers, alkalis, and sequestering agents. The deflocculants, sequestering agents, and alkalis (those additives other than saponifiers and stabilizers) are referred to as "builders." Cleaners to be used for an extended period of time are also likely to have an added biocide or antimicrobial agent.

a. Surfactants (also known as surface action agents or wetting agents) lower the surface and interfacial tension of the water so that it can penetrate small spaces better, get below the contaminant, and help lift it from a substrate. Surfactants may be cationic, anionic, or nonionic in their aqueous phase. The anionic and nonionic types are used most often in immersion cleaning. Nonionic surfactants are preferred in applications where agitation is used because they have lower foam-producing characteristics.

Anionic surfactants include sulfosuccinates, long-chain sulfonates, fatty alcohol sulfates, and alkali soaps (also used as an emulsifier). Among the nonionic surfactants, ethoxylated alkylphenols are used as a surfactant and an emulsifier, while glycol ethers (Cellosolves™) and ethylene oxide condensates of alkylphenols are used more as an emulsifier than as a surfactant.[21]

Ethoxylated phenols appear to be skin and respiratory irritants.[22–24] Several of the glycol ethers have recognized reproductive toxicity properties, such as 2-butoxyethanol (or ethylene glycol butyl ether), 2-ethoxyethanol (EGEE or ethylene glycol ethyl ether), and dipropylene glycol methyl ether. Several good reviews of the toxicology of glycols are available.[25–27]

Surfactants are characterized empirically by their hydrophilic-lipophilic balance (HLB) that describes the relationship between their water-soluble (hydrophilic) and oil-soluble (lipophilic) portions. Oil-soluble surfactants have low HLB values, and highly water-soluble surfactants have high HLB values. Typical HLB ranges and applications of surfactants are shown in Table 8.2.

b. Deflocculants are chemicals added to help the dispersion of contaminant particles within the cleaning medium. Deflocculants may also be anionic or nonionic surfactants, or they may be inorganic salts such as alkali phosphates. Because many emulsions remain stable only at elevated temperatures or under alkaline conditions, separation of an oily contaminant from the aqueous cleaner can often be induced in emulsion-type aqueous cleaners by lowering the temperature or, sometimes, by acidifying the mixture.
c. Saponifiers are compounds that react chemically with oils containing fatty acids to form soaps; thus, saponifiers can help to remove heavy or thick contaminants. A classic example is alkali's ability to saponify vegetable oils and animal fats (triglycerides). The organic amines comprise an important class of saponifiers that can react with many common hydrocarbon contaminants. Examples of organic amines used in aqueous cleaners include ethanolamine, diethanolamine, and triethanolamine. The organic amines range in volatility with their molecular weight and (similar to any organic liquid) will evaporate more rapidly when used at high temperatures. Loss from evaporation is one reason aqueous mixtures (such as metalworking fluids) may require routine maintenance to restore their chemical balance.
d. Alkalinity is a property of an aqueous media to neutralize acid not just by its pH alone but also by its ability to buffer itself against an acid. Buffering capacity stabilizes the solution against abrupt pH and other chemical changes and promotes detergency. Aqueous mixtures range in alkalinity from mild alkaline (pH 8 to 10) to highly alkaline (pH 12 and higher). Mild alkaline conditions prevent etching of most metals, especially aluminum and magnesium that are readily etched above pH 11. Mild alkalinity is often maintained by soluble silicates, carbonates, borates, and citrates.

TABLE 8.2
The Range of Hydrophilic–Lipophilic Balance for Surfactants and Their Typical Applications[4,21]

HLB Value	Application
3.5–6	Water-in-oil emulsifier
7–9	Wetting agent
8–18	Oil-in-water emulsifier
13–15	Detergent
15–18	Solubilizer

In some cases, high alkalinity is desired, for example to remove metal oxides and hydroxides from a surface. Highly alkaline (caustic) cleaners are used to prepare metal surfaces prior to plating or to remove heavy grease and oil from corrosion-resistant steels by attacking the saponifiable components of the grease and oil. To achieve a very high pH (>12), strong bases such as sodium hydroxide (lye) or potassium hydroxide may be added.

e. Sequestering agents are added to prevent the mineral content of hard water (e.g., its calcium and magnesium ions) from forming insoluble products within the cleaner, to permit the cleaner to attack only the contaminant, and thus to allow the use of lower concentrations of the cleaner. Common sequestering agents are chelating agents, such as sodium EDTA, NTA, and ODA.

f. Other additives, such as antifoaming agents, corrosion inhibitors, and antimicrobials or biocides may be included to enhance the cleaner's overall performance. Corrosion inhibitors work either by passivating the surface by adsorbing a molecular species onto it that will react with oxygen before the metal can oxidize, or by forming a protective barrier over the surface that excludes oxygen. Antimicrobials and biocides inhibit the growth of microbes initially during storage of a concentrated aqueous product but more importantly during extended use or recirculation of the diluted product. Biocides, especially in metalworking fluids, have long been associated with dermatoses.[28,29]

2. The most important determinant of the viability of aqueous cleaning is whether the product being cleaned and the process doing the cleaning can tolerate water. The compatibility of the product or process with water must be carefully investigated.
   a. Parts should be tested prior to full-scale use. A rust inhibitor may be used along with the cleaner to help prevent some ferrous metals from "flashing" in aqueous environments (flashing is the rapid development of rust). Nonalkaline cleaners help to avoid stress corrosion cracking that can occur in some polymers in contact with alkaline solutions. The manufacturers of chemical cleaners can recommend the appropriate formulations and procedures.
   b. While organic solvents function as a cleaner primarily by dissolving organic contaminants on a molecular level, aqueous cleaners utilize a combination of physical and chemical properties to remove macroscopic amounts of organic contaminants from a substrate. Aqueous cleaning is more effective at higher temperatures and normally is performed above 120°F using suitable immersion, spray, or ultrasonic washing equipment. For this reason, good manufacturing practices and process controls tend to be more important in order to achieve optimum and consistent results in aqueous cleaning than in organic solvent cleaning.
3. When using a water-based cleaner, parts usually need to be rinsed and will remain wet for some time unless action is taken to speed up the drying process. Three common methods for drying parts are evaporation, chemical, and mechanical removal.
   a. Evaporation of rinse water under ambient conditions is slow. Exposing parts to water creates an opportunity for oxidation and for dust to settle onto the part. Consistent with the precursor to the basic evaporation equation (Equation 5.3), the evaporation rate can be improved by moving or circulating the drying air (increasing V), by heating the parts via various means (increasing P_{vapor}), and by dehumidifying the drying air (lowering ambient P_{partial}). An inert gas can also be used to lessen the tendency for oxidation while drying.
   b. Chemical drying methods include displacement and capillary or slow-pull drying. In the displacement technique the water is displaced by some other solvent that is insoluble in water. Oil is commonly used in metal working. The oil can also act as a

rust inhibitor but is inappropriate if the objective of cleaning is to produce an oil-free surface. More expensive alternatives include fluorinated hydrocarbons and methyl siloxanes. With the capillary method, a hot part is extracted slowly from equally hot, deionized water. The surface tension of the water, in effect, peels itself off the part; whatever water is left readily vaporizes.

c. Mechanical removal techniques are also commonly used (sometimes in conjunction with heat). Air knives blow water off the part with high-pressure air. Centrifugal drying can spin the water off.

4. Waste disposal costs can be kept low if the bulk of the used aqueous cleaner may be discharged into a sewer. The cost of performing all of the waste stream pretreatment required to meet municipal or other discharge requirements should be considered and included in estimates of operating costs. When it is necessary, dissolved metals can be precipitated or absorbed onto a substrate using a number of developed technologies. Suspended solids can be removed by small-pore filters (10 μm or less). Emulsified oil can be separated from the aqueous cleaner by means of coalescing equipment or advanced membrane ultrafiltration techniques. Continuous filtering to remove particles and skimming to remove oil will extend a cleaner's lifetime even before disposal.

IV. SEMIAQUEOUS SOLVENTS

Semiaqueous solvents have more additives than aqueous solvents but are still composed mostly of water (ca. 95% water). Semiaqueous cleaners are not as environmentally beneficial as aqueous cleaners but often comprise a more cost-effective substitute for traditional organic solvents than aqueous cleaners. Semiaqueous mixtures will have lower VOCs than organic solvents, will often still be biodegradable, and may be amenable to distillation and membrane filtration to permit recycling and reuse. Semiaqueous mixtures are used in metal fabrication, electronics, and precision parts manufacturing.

The main chemical difference between aqueous and semiaqueous mixtures is in their surfactants. Surfactant added to semiaqueous solvents are sometimes categorized into water-immiscible types (terpenes, high-molecular-weight esters, petroleum hydrocarbons, and glycol ethers [the latter was described as a nonionic surfactant in aqueous mixtures]) and water-miscible types (low-molecular-weight alcohols, ketones, esters, organic amines, and methyl-2-pyrrolidone). Four of these additives are described below.

1. Terpenes are natural hydrocarbons such as D-limonene and α- and β-pinene. Terpene alcohols and *para*-menthadienes are also used as cleaners. Terpenes are derived from plant sources such as citrus (orange, grapefruit, and lemon) and pine oils. Although terpenes are not miscible in water, they form emulsions with water that are stabilized by surfactants and other additives. Their relatively low flash points (about 115 to 120°F) limit their undiluted safe use temperature to about 90°F, but diluting such products with water or other selected higher-molecular-weight hydrocarbons will increase the mixture's flash point. Terpenes are generally not recommended for cleaning polystyrene, PVC, polycarbonate, low-density polyethylene, or polymethylpentene; nor are they compatible with the natural rubber, silicone, and neoprene. Toxicological testing of D-limonene has concluded that there is no risk for human carcinogenicity.[30] While terpenes are reported to have an irritation threshold above 80 ppm,[31] their strong odor may become objectionable to some employees well below that. AIHA has established an 8-h TWA WEEL exposure limit for *d*-limonene of 30 ppm.[32]

2. Esters have good solvent properties for many contaminants and are soluble in most organic compounds. High-molecular-weight esters have limited solubility in water, whereas low-molecular-weight esters are soluble in water.
   a. The types of high-molecular-weight esters most often used in cleaning and degreasing include aliphatic monoesters (primarily alkyl acetates) and dibasic acid esters. A high-molecular-weight ester with a flash point above 200°F may be used either cold or heated to improve its cleaning performance.
   b. Ethyl lactate is one low-molecular-weight ester that is reported to have good cleaning, health, and safety properties.[33] Ethyl lactate is an ethyl ester of lactic acid. Additional information on ethyl lactate is also given in the next section on "Organic solvent cleaners." Other esters studied as cleaning and degreasing solvents include ethylene carbonate and propylene carbonate.
3. Glycol ethers also have good solvent properties for common contaminants. They form emulsions with water that can be separated for recycling. Most glycol ethers also have a flash point above 200°F and can be heated for improved solvency. The ethyl and propyl series of glycol ethers are reported to have some risk for reproductive effects.[25,26] Thus, chemical protective gloves should be selected carefully since a major route of exposure during cleaning is dermal and glycol ethers seem to degrade polystyrene and cause swelling in the elastomers Buna-*N*[e] and silicone rubber (see Chapter 21).
4. *N*-methyl-2-pyrrolidone (NMP) can be used to remove cured paint and hence is a potential substitute for methylene chloride. NMP has been used in the chemical and petrochemical industries as an extraction solvent and as a formulating agent for coatings, paint removers, and cleaners. NMP is completely miscible with water and organic compounds such as esters, ethers, alcohols, ketones, aromatic and chlorinated hydrocarbons, and vegetable oils. Usually, NMP immersion cleaning or paint removing is done either at 155°F in an open tank or up to 180°F if a mineral oil seal is present (just below its flash point of approximately 199°F). NMP presents a slight risk of reproductive and developmental harm.[34,35] Its primary route of entry is also dermal, and again chemical protective gloves should be selected carefully because NMP dissolves or degrades ABS, Kynar™, Lexan™, and PVC and causes swelling in Buna-*N*, Neoprene, and Viton™ (see gloves in Section VI in Chapter 21).

Semiaqueous mixtures may have certain advantages over aqueous mixtures, some of the same limitations, and at least one disadvantage. In general, the semiaqueous cleaners have better solvency for a number of difficult contaminants such as heavy grease, tar, and waxes. They have a lower surface tension that allows them to penetrate small spaces such as crevices, blind holes, and below-surface-mounted electronic components. Semiaqueous cleaners may have a lower corrosion potential with water-sensitive metals. Also, most of the semiaqueous cleaners listed above are available at reasonable prices. Both aqueous and semiaqueous cleaners share some limitations. Both cleaners must be rinsed to avoid leaving a residue on the cleaned parts. If water rinsing is performed, the parts must be dried, and the methods of drying cited for aqueous cleaners also apply here. A notable disadvantage of semiaqueous mixtures is that they require more waste stream management (and perhaps more pretreatment capacity) than cleaning with either solvents or aqueous mixtures, which adds to their costs.

Semiaqueous mixtures recently received a major boost over organic solvents (at least in the U.S.). Effective from January 2003, EPA rules 1171 and 1122 require cleaning materials used in repair and maintenance operations and as cold cleaners to contain no more than 2.5% of VOC by

[e] Buna-*N* is a copolymer of butadiene and acrylonitrile that is sometimes also called nitrile. Acrylonitrile gives the copolymer a high resistance to petroleum products but relies on the butadiene for its low temperature flexibility.

weight. This low VOC requirement generated a new class of solvent called a clean air solvent (CAS) that must meet all of the following criteria:

1. VOC concentration is no more than 25 g of VOC per liter of material, as applied.
2. Composite vapor pressure is no more than 5 mmHg of VOC at 20°C (68°F).
3. Reactivity is not higher than toluene.
4. Contains no compounds classified as hazardous air pollutants (HAPs) by the U.S. Clean Air Act (40 CFR 51.112), Ozone-Depleting Compounds, or Global Warming Compounds.

A web site lists certified CAS products and companies at www.aqmd.gov/cas/prolist.html.

V. ORGANIC SOLVENTS

A substitution to an organic solvent is usually from another organic solvent rather than from an aqueous or semiaqueous solvent to an organic solvent. Most organic solvents are VOCs that evaporate readily, and many contribute to smog formation. In addition, some organic solvents that contain halogens contribute to ozone depletion. On the other hand, most organic cleaners have been used for a long time as general-purpose solvents and their solvent properties are well known (c.f., the newer water additives). They are easily used cold in small quantities for niche applications such as bench-top or spot cleaning.

1. Alcohols are polar solvents that have good solubility for a wide range of inorganic and organic soils. The lighter alcohols are soluble in water and may be useful in spot cleaning or drying operations. Alcohols are also used in fully enclosed spray washers.
   a. Ethyl and isopropyl alcohols are commonly used in spot cleaning and touch-up applications. Since they are slightly polar, they tend to be good, general-purpose solvents for both nonpolar and polar organic compounds, and even for ionic compounds. Ethyl and isopropyl alcohols are fully miscible in water but are also quite flammable (flash points of 55 and 53°F, respectively).
   b. Benzyl alcohol is a solvent for gelatin, casein (when heated), cellulose acetate, and shellac and is a paint softener or stripper for aircraft paints (epoxy primers or polyurethane topcoat). If heating benzyl alcohol, keep in mind that its flash point is 101°C or 213°F. Mixtures of 90% benzyl alcohol and 10% benzoic acid are also used for solvent cleaning applications. Pure benzyl alcohol is 4% soluble in water, but is miscible in lighter alcohols and with ether.
   c. Furfuryl alcohol forms a miscible but unstable solution in water that can be used as a general cleaning solvent and paint softener. Furfuryl alcohol is also miscible in lighter alcohols and in ether. Similar to benzyl alcohol, the solvent properties of furfuryl alcohol can be enhanced by moderate heating, but its flash point is only 75°C or 167°F.
   d. Butyl alcohol is a solvent for fats, waxes, resins, shellac, varnish, and gums. It is 9% soluble in water at 25°C, but forms an azeotrope with water (63% butyl alcohol/37% water) that boils at 92°C. Butyl alcohol is miscible in lighter alcohols, ether, and many other organic substances. Butyl alcohol may cause irritation of mucous membranes, dermatitis, headache, dizziness, and drowsiness, and it has a flash point of 36 to 38°C or 97 to 100°F, which is considered flammable.
2. Ketones have good solvent properties for many polymers and adhesives. Lighter ketones, such as acetone, are soluble in water and may be useful for certain rapid drying operations. Heavier ketones, such as acetophenone, are nearly insoluble in water. Ketones generally evaporate completely without leaving a residue.

a. Acetone is a solvent for fats, oils, waxes, resins, rubber, some plastics, lacquers, varnishes, and rubber cements. It is completely miscible in water and in most organic solvents.
b. Acetophenone (acetylbenzene, $C_6H_5-CO-CH_3$) is only slightly soluble in water but is miscible in alcohol, ether, and other organic substances.
c. Methyl ethyl ketone (MEK) and methyl isobutyl ketone (MIBK) were once widely used. However, they are now considered HAPs and are thus more likely targets for substitution than candidates as solvent substitutes (see www.epa.gov/ttn/atw/hapindex.html).

3. Esters and ethers also have good solvent properties. Their low-molecular-weight compounds also dry readily without leaving a residue.
   a. Butyl acetate is a solvent used in lacquer production. It is less than 1% soluble in water at 25°C. Its solvent activity is enhanced by mixing it with butyl alcohol. For instance, a mixture of 80% butyl acetate and 20% butyl alcohol is used to dissolve oil, fats, waxes, metallic resinates, and many synthetic resins such as vinyl, polystyrene, and acrylates. The mixture also dissolves less highly polymerized alkyd resins and shellac. Butyl acetate has a flash point of 22°C, or 72°F, is irritating to eyes and skin (may cause conjunctivitis).
   b. Ethyl lactate is another ester with solvent properties useful for cleaning and degreasing, and it has good solubility properties for cutting fluids, coolants, mold release compounds, and marking inks.[33] Ethyl lactate has a moderate and somewhat variable flash point of 115 to 139°F. However, ethyl lactate hydrolyzes to form lactic acid upon exposure to water (including contact with moist air) which could have a negative impact on metal parts.
   c. A mixture with equal parts of anisole (methoxybenzene, $C_6H_5-O-CH_3$) and ethanol with 10% sodium ethoxide has been reported to be useful for cleaning and paint removal. Anisole is insoluble in water, but is soluble in alcohol and ether, and is otherwise used in perfumery and in organic syntheses.
4. Methyl siloxanes comprise a relatively new class of organic solvents. Their molecular structure may be either linear or cyclic. The linear type has the general formula $(CH_3)_3$ $SiO-[SiO(CH_3)_3]_n-Si(CH_3)_3$, where $0 \leq n \leq 4$.[36] Thus, $n = 0$ corresponds to a 2-SiO-unit chain, $n = 1$ corresponds to a 3-SiO-unit chain, and so on. The linear methyl siloxanes are nonpolar and are most effective in removing nonpolar and nonionic contaminants, such as in precision metal parts, optics, and electronics; however, they can also remove cutting fluids, greases, and silicone fluids. They have low odor, a vapor pressure around 10 mmHg, and can be used in cleaning equipment designed for use with isopropyl alcohol. The most volatile methyl siloxane ($n = 0$) can function as a drying agent. While the volatile methyl siloxanes are nonirritating to the skin and respiratory tract, they are mild eye irritants.
5. Vegetable oils are finding niche uses such as removing printing inks. They also seem to be compatible with elastomers. Vegetable oils typically contain triglycerides of oleic, linoleic, linolenic, palmitic, and stearic fatty acids with usually $< 1\%$ unsaponifiable matter.

VI. TOXICITY VERSUS VOLATILITY

Substitutions among volatile chemicals are slightly more complex than substitutions among nonvolatile or low volatility chemicals. Less toxic nonvolatile chemicals are almost always a safer substitute. Because aqueous or semiaqueous solvents generally have low volatility, substituting a less toxic aqueous or semiaqueous solvent for a toxic volatile organic is generally advantageous where the process will permit it. However, trying to substitute a less toxic volatile chemical for a more toxic volatile chemical is at best only half of the task. Toxicity and vapor pressure both contribute to a volatile chemical's intrinsic hazard as a vapor, and toxicity and volatility are not related to each other.

1. A not uncommon mistake for someone familiar with TLVs or PELs only as a number is to conclude that conditions will be safer if they only work with chemicals with a high exposure limit. Not only should they know that exposure limits are based on different (and therefore unequal) measures of toxicity, but that a solvent's volatility will affect its level of exposure. A different mistake with the same outcome can be made by someone concerned only about volatility who believes that conditions will be safer if they only work with chemicals that are low in volatile (or in quantitative terms, those that have a low vapor pressure).[f]

   While toxicity and vapor pressure both contribute to a vapor's hazard, they are not correlated with each other. Figure 8.5 displays toxicity (in the form of TLV exposure limits on the *Y* axis) vs. volatility (in the form of vapor pressure [P_{vapor}] on the *X* axis) using log–log scales. A log–log scale spreads out data that span a wide range; however, such scales can make the data look more correlated than they really are. In fact, the linear correlation [R^2] between the TLV and P_{vapor} for the 254 chemicals in Figure 8.5 is only about 0.095, definitely not statistically significant. Thus, using either variable as a measure of hazard is at best incomplete and at worst misguided.

   In order to assess the hazard presented by a given chemical, one must assess both toxicity and vapor pressure. The Vapor Hazard Ratio is a convenient way to link both of these components of hazard into one number. The remainder of this section will first review the logic linking vapor pressure to exposure level then explain why and how Equation 8.3 can be used to predict exposure concentrations.The link between the dilution needed for toxicity and the dilution resulting in exposure will be developed in the next section.
2. Recalling from Chapter 5 that for other than the complete evaporation of small quantities of a volatile chemical into an unventilated room, a vapor's concentration will either be equal to or less than but proportionate to its vapor pressure. A vapor's concentration can never exceed its vapor pressure.[g]
   a. In unventilated, incomplete evaporation conditions (where the volume of a liquid is large relative to the available air volume into which its vapor can disperse), the airborne vapor will be in equilibrium with the remaining liquid, and an air concentration in equilibrium with its vapor is defined to be equivalent to its vapor pressure, P_{vapor} as stated in Equation 2.9 and Equation 5.1.

$$\text{ppm at the source} = \frac{P_{vapor} \times 10^6}{P} = \frac{[P_{vapor}\ \text{mmHg}] \times 10^6}{760} \text{ at NTP}$$

(repeat of Equation 5.1)

[f] The belief that odor is associated with toxicity is another common misconception. Someone's logic might go something like "If I can smell the vapor, I am over-exposed to it" or "Let's just work with chemicals that we can't smell." Helping them to understand that odor and toxicity are not related, might or might not be reassuring. Acknowledging their concern, and explaining that you are also working diligently to monitor and keep them safely below exposure limits for chemicals that they cannot smell (maybe not use the IH jargon that it is below the chemical's odor threshold) might help someone separate the perception of odor from the psychology of risk.

[g] The only way a chemical's airborne concentration can exceed its vapor pressure is as an aerosol (either a mist or fume). If the chemical comprising the mist is volatile, the droplets (a colloquialism for mist particles) will evaporate soon after they reach ambient air that is not already saturated with its vapor (see Section II.5 in Chapter 3).

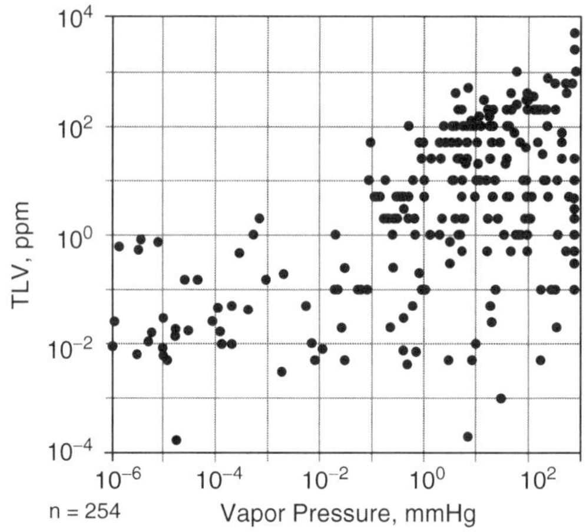

FIGURE 8.5 The correlation between P_{vapor} and TLV for 254 volatile chemicals.

b. In ventilated conditions where the background or ambient vapor concentration is kept well less than P_{vapor}, vapors will be generated from an evaporating liquid at a rate predicted by Equation 5.13.

$$G\,[\text{moles/min}] = \text{Geom. Coef.} \times A \times V^{0.5} \times P_{\text{vapor}} \quad \text{(repeat of Equation 5.13)}$$

The P_{vapor} concentration right at the liquid surface will be diluted into the near-field plume. These vapors will be further diluted in the free-field plume that varies strongly with location. If the source continues to evaporate at a constant rate, resulting molar concentration [P_{partial}/P or ppm$_i$] of any chemical will achieve a steady state or time-weighted average concentration as given for vapors by Equation 5.15, where Q_{apparent} will vary by location in relationship to the plume and to the room, its ventilation rate, and mixing of that general ventilation air.

$$\frac{\text{ppm}_{\text{TWA}} \times P}{10^6} \approx P_{\text{TWA,partial}} = \left(\frac{(\text{Geom. Coef.}) \times A \times V^{0.5}}{Q_{\text{apparent}}}\right) \times P_{\text{vapor}}$$

(repeat of Equation 5.15a)

c. The concept of diluting the P_{vapor} over the volatile liquid source to some partial pressure P_{partial} or equivalent ppm just described above was discussed in Chapter 5. The environmental dilution ratio (EDR) defined in Equation 5.16 can be applied to the dilution anywhere in the workplace. Figure 5.12 gave examples of near-field plume EDRs measured in a laboratory setting.

$$\text{Environmental Dilution Ratio} = \frac{P_{\text{vapor}}}{P_{\text{partial}}} = \left(\frac{Q_{\text{apparent}}}{(\text{Geom. Coef.}) \times (\text{area}) \times V^{0.5}}\right)$$

(Equation 5.16)

The breathing zone dilution ratio defined in Equation 5.17 is the same ratio when applied specifically in the breathing zone. The breathing zone dilution is merely the particular point in the environment that is of most interest to industrial hygienists.

$$\boxed{\text{Breathing Zone Dilution Ratio} = \frac{P_{\text{vapor}}}{P_{\text{partial,BZ}}}} \qquad \text{(Equation 5.17)}$$

In both cases, dilution is dependent solely upon the environment and is independent of the chemical being diluted. The important concept underlying EDRs is that *dilution depends strictly upon the physical conditions of the setting* and not upon the chemical being diluted. Equation 5.16 shows that the EDR depends specifically upon the source's physical geometry and area, the air velocity and turbulence, and the ventilation (both general and local exhaust). Therefore, if nothing is physically changed in the setting (namely, if no changes are made that would affect the source's physical geometry, its evaporating area, the passing air velocity, the total ventilation, or the location, then the EDR in term of either $P_{\text{vapor}}/P_{\text{partial}}$ or $\text{ppm}_{\text{vapor}}/\text{ppm}_{\text{partial}}$ will be the same, no matter what chemical is present!

3. Now apply this concept to control via substituting a new volatile chemical for an old volatile chemical. Equation 8.1 expresses this constant dilution as a dilution ratio with mixed units of mmHg and ppm (a slight variation of Equation 5.16).

$$\text{If nothing in the environment is changed:} \quad \text{then} \quad \frac{P_{\text{vapor}} \text{ at the source}}{\text{ppm at a location}} = \text{a constant ratio} \qquad (8.1)$$

If dilution is constant but the chemical is changed, then the ratio with these mixed terms will still stay constant. Thus, the ratio between the presumedly measured (or at least measurable) concentrations of the original chemical $[P_{\text{vapor,old}}/\text{ppm}_{\text{old}}]$ and the ratio between the anticipated concentrations of the substituted chemical $[P_{\text{vapor, new}}/\text{ppm}_{\text{new}}]$ will be constant.

$$\text{If } \frac{P_{\text{vapor}} \text{ at the source}}{\text{ppm at a location}} = \text{a constant, then } \frac{P_{\text{vapor,old}}}{\text{ppm}_{\text{old}}} = \frac{P_{\text{vapor,new}}}{\text{ppm}_{\text{new}}} \qquad (8.2)$$

Typically, the ppm is measured in the breathing zone, generating a $P_{\text{vapor}}/\text{ppm}_{\text{BZ}}$ ratio. However, in IH practice, the actual value of the $P_{\text{vapor}}/\text{ppm}$ ratio specific to a given setting does not need to be known. The mere fact that it is constant provides a simple yet powerful way to predict the new concentration $[\text{ppm}_{\text{new}}]$ after substituting a new chemical source for an old (or previously existing) chemical source whose concentration was previously measured at that location $[\text{ppm}_{\text{old}}]$.

$$\boxed{\text{ppm}_{\text{new}} = \text{ppm}_{\text{old}} \times \frac{P_{\text{vapor,new}}}{P_{\text{vapor,old}}}} \ \text{in the unchanged setting with the new chemical.} \qquad (8.3)$$

The new concentration predicted using Equation 8.3 can be of interest in and of itself. However, what hygienists generally want to know is, will the vapor exposures from the new substitute chemical be in compliance with its exposure limit or not? To answer this question, one must now test whether ppm_{new} is $\leq \text{TLV}_{\text{new}}$ (or test the compliance ratio using Equation 5.19). Thus, it appears that to answer the question of predicting compliance requires two steps: first, a calculation involving the ratio of two vapor pressures to get ppm_{new} and then a separate step testing ppm_{new} for compliance. Testing the acceptability of one or two specific candidate substitute chemicals using this two-step approach is feasible either in the real IH world or on an examination problem, but searching for an acceptable substitute *de novo* among the wide world of all (or even many) candidates by this two-step trial-and-error procedure can be a long and inefficient process that is not recommended. Fortunately, the vapor hazard ratio provides a workable one-step solution to finding an acceptable substitute chemical.

VII. THE VAPOR HAZARD RATIO

The key to both understanding vapor hazards and choosing good volatile substitutes efficiently lies in assessing both a chemical's vapor pressure and its toxicity simultaneously. The Vapor Hazard Ratio (VHR as defined in Equation 8.4) is a simple ratio of both factors that has many interpretations and uses.[37]

The VHR as expressed in Equation 8.4 and Equation 8.5 is the ratio between the chemical's vapor pressure and its acceptable breathing zone concentration. To be acceptable, the concentration in a person's breathing zone [$P_{\text{partial, BZ}}$] or their time-weighted average exposure [ppm_{TWA}] must be no greater than that chemical's TLV or PEL. Since a chemical's VHR is the amount by which its equilibrium vapors must be diluted to be acceptable, the chemical can only be used acceptably in a setting where the breathing zone dilution ratio is at least as large as the VHR (as originally stated in Equation 5.22).

$$\text{VHR} = \frac{\text{Vapor Pressure } [P_{\text{vapor}} \text{ or } P_{\text{vapor},i}]}{\text{Vapor Exposure Limit, expressed as } P_{\text{partial}}} \tag{8.4a}$$

$$\text{VHR} = \frac{P_{\text{vapor}} \text{ or } P_{\text{vapor},i} \text{ converted into ppm}}{\text{TLV or PEL or STEL as ppm}} \tag{8.4b}$$

$$\text{VHR} = \frac{\text{Equilibrium Concentration at the source}}{\text{Acceptable breathing Zone Concentration}} \tag{8.4c}$$

$$\text{VHR} = \text{the amount of dilution a Saturated vapor needs.} \tag{8.4d}$$

As a stand-alone tool, the VHR is a chemical's own intrinsic ability or propensity to create a vapor concentration in excess of its exposure limit. The higher a chemical's VHR, the more likely it is to create an overexposure (all other things being equal). Similarly, the lower a chemical's VHR, the easier it will be to avoid overexposures (all other things being equal). In fact, any chemical whose VHR is less than unity implies that its vapors can never exceed its exposure limit, even at equilibrium. Figure 5.12 provides a glimpse of a dilution modeling tool that could be used in the future to predict compliance based on an individual chemical's VHR.

As a comparative tool, the ratio of $\text{VHR}_{\text{new}}/\text{VHR}_{\text{old}}$ will be used in Equation 8.9 to compare the acceptably of a new chemical as a substitute for an old chemical to be used in the same setting or plume dilution condition or to predict the actual compliance of a substitute chemical based on a previously measured chemical. The VHR when used in Equation 8.10 is the fastest way to select an acceptable substitute among many candidates. Finally, discussion in the next section will explain how the VHR can be used to help set priorities among vapor hazards from multiple chemicals.

1. The VHR is calculated by dividing the chemical's P_{vapor} by its exposure limit when both are expressed in the same units, such as shown in Equation 8.5. While Equation 8.5a is based on the technically correct way to think of vapor pressure as a physical concentration in units of mg/m^3 as shown in Equation 2.8, it is simpler to calculate the

VHR for vapors using Equation 8.5b by converting P_{vapor} in the numerator (usually expressed in mmHg) into ppm in order to match the ppm units most commonly used to express exposure limits for volatile chemicals in the denominator. At NTP, the P in the denominator is 760 mmHg.

$$\text{VHR} = \frac{\text{MW}_i \times (P_{\text{vapor}} \text{ or } P_{\text{vapor},i} \text{ (in mmHg)}) \times 10^6}{24.45 \times (\text{TLV or PEL or STEL (mg/m}^3)) \times P\,(\text{mmHg})} \tag{8.5a}$$

or

$$\boxed{\text{VHR} = \frac{P_{\text{vapor}} \text{ or } P_{\text{vapor},i}\,(\text{mmHg}) \times 10^6}{\text{TLV or PEL or STEL (ppm)} \times P\,(\text{mmHg})}} \text{ for vapors} \tag{8.5b}$$

2. The acceptability of someone's exposure level can be determined by the balance between the VHR for a given chemical and the EDR in someone's breathing zone [EDR_{BZ}] for a given environment.

$$\text{Compliance Ratio} = \frac{C_{\text{measured in the BZ}}/P_{\text{vapor}}}{\text{TLV or PEL}/P_{\text{vapor}}} = \frac{\text{VHR}}{\text{EDR}_{\text{BZ}}} \leq 1 \text{ to be acceptable}$$

(repeat of Equation 5.20)

The concept underlying Equation 5.20 relies on the principle that *the VHR is an intrinsic property of the chemical while the EDR is an intrinsic property of the environmental setting*. Or stated another way, for someone's exposure to a given chemical in a given workplace to be in compliance with its exposure limit, the environment (including the work practice) must provide at least as much vapor dilution between the source and the breathing zone as is required by the volatility and toxicity of the particular chemical in use. This restatement is expressed mathematically in Equation 8.6.

$$\boxed{\text{EDR}_{\text{BZ}} = \frac{P_{\text{vapor}}}{P_{\text{partial,BZ}}}} \quad \text{should be} \geq \quad \boxed{\frac{P_{\text{vapor}}}{\text{TLV}} = \text{the chemical's VHR}} \tag{8.6}$$

If the measured concentration in the breathing zone [$P_{\text{partial,BZ}}$] is too high, Equation 8.6 suggests that either of two approaches may be taken to control the hazard. Equation 8.7a approaches control from the perspective of the environment: look for ways to increase the dilution of vapors that reach the breathing zone (the breathing zone dilution ratio) such as by increasing the source-to-person separation distance or by adding more ventilation.

$$\text{Acceptable new EDR}_{\text{BZ}} = \frac{P_{\text{vapor}}}{P_{\text{partial,BZ,new}}} \geq \frac{P_{\text{vapor}}}{\text{TLV}} = \text{the chemical's VHR} \tag{8.7a}$$

Equation 8.7b approaches control from the perspective of the chemical: look for ways to reduce the amount of dilution required of a new chemical that has either a lower vapor pressure and/or a higher exposure limit (a lower toxicity), namely, by using a new chemical that has a lower VHR. An acceptable new VHR cannot exceed the existing (or old) breathing zone dilution ratio.

✍ $$\text{Acceptable new VHR} = \frac{P_{\text{vapor,new}}}{\text{TLV}_{\text{new}}} \leq \frac{P_{\text{vapor,old}}}{P_{\text{partial,BZ,old}}} = \text{the existing Dilution Ratio} \quad (8.7b)$$

✓ Even if exposure in the old or existing chemical's breathing zone concentration has not been measured, Equation 8.4 to Equation 8.7 all predict that the chemical with the higher VHR will always require qualitatively more environmental dilution than the chemical with a lower VHR. For any single setting or two settings with the same EDR, users of the chemical with the higher VHR will always have more problems complying with its exposure limit than if they were using the chemical with the lower VHR.

3. Substituting a chemical that intrinsically requires less dilution for its safe use can be a cost-effective alternative to adding more ventilation to increase the dilution ratio provided by the setting. The following discussion explains the logic leading up to Equation 8.10 to predict the acceptability of a substitute new chemical to be used in the same way or in the same setting as an old chemical. If you do not want to bother with the derivation, go straight to Equation 8.10.

   If nothing else in a setting besides the P_{vapor} of the chemical changes, then the EDR should be unchanged. That means that $P_{\text{vapor,new}}/P_{\text{partial,new}}$ should be identical to $P_{\text{vapor,old}}/P_{\text{partial,old}}$ everywhere in the setting including the breathing zone. By dividing both the top and bottom portions of that dilution ratio by the chemical's exposure limit (the TLV is used here although a PEL could just as easily have been used), the numerator becomes the chemical's VHR and the denominator becomes the compliance ratio (much like Equation 5.20). Therefore, the ratio of the old VHR to its compliance ratio should be the same as the ratio of the new VHR to its compliance ratio, just as shown in Equation 8.8a.

$$\frac{P_{\text{vapor}}}{P_{\text{partial,BZ}}} = \text{constant} = \frac{[P_{\text{vapor}}/\text{TLV}]_{\text{old}}}{[\text{ppm}_{\text{BZ}}/\text{TLV}]_{\text{old}}}$$

$$= \frac{\text{VHR}_{\text{old}}}{[\text{ppm}_{\text{BZ}}/\text{TLV}]_{\text{old}}} = \frac{\text{VHR}_{\text{new}}}{[\text{ppm}_{\text{BZ}}/\text{TLV}]_{\text{new}}} \quad (8.8a)$$

By rearranging the diagonal terms in Equation 8.8a, one can see in Equation 8.8b that the ratio of the new to the old levels of compliance equals the ratio of the new to the old VHRs.

$$\frac{[\text{ppm}_{\text{BZ}}/\text{TLV}]_{\text{new}}}{[\text{ppm}_{\text{BZ}}/\text{TLV}]_{\text{old}}} = \frac{\text{Compliance Ratio}_{\text{new}}}{\text{Compliance Ratio}_{\text{old}}} = \frac{\text{VHR}_{\text{new}}}{\text{VHR}_{\text{old}}} \quad (8.8b)$$

If the concentration for the old chemical has been previously measured and its level of compliance determined, it is possible in one step using the ratio of the new VHR to the old VHR to predict the acceptability of any candidate substitute chemical (namely, its new compliance ratio) via Equation 8.9.

$$\boxed{[\text{ppm}_{\text{BZ}}/\text{TLV}]_{\text{new}} = \text{Compliance Ratio}_{\text{old}} \times \frac{\text{VHR}_{\text{new}}}{\text{VHR}_{\text{old}}}} \quad (8.9)$$

Equation 8.9 should be thoroughly understood as a powerful concept. However, the last and probably most useful application of the VHR provided herein comes from realizing that an acceptable substitute chemical must create a new compliance ratio

$[ppm_{BZ}/TLV]_{new}$ that is equal to or less than one. Thus, the VHR of an acceptable new substitute must be as much below the VHR_{old} as the ratio by which the old chemical exposure concentration exceeded its TLV, i.e., $[ppm_{BZ}/TLV]_{old}$. Equation 8.10 is itself almost a restatement of Equation 8.7b, just put in a more useful form.[h]

$$\boxed{\text{acceptable } VHR_{new} \leq \frac{VHR_{old}}{[ppm_{BZ}/TLV]_{old}}} = \frac{VHR_{old}}{\text{compliance ratio}_{old}} \quad (8.10)$$

4. While the concept in Equation 8.9 that the compliance ratio will change in proportion to the two chemical's VHR is important to understand, Equation 8.10 will prove the more useful. Two lists of volatile chemicals are included in the appendix in the back of this book (not in this chapter). Each list includes the chemical's name, CAS number, TLV, its vapor pressure (and reference temperature where it is not 25°C), and its VHR. The first list is ordered alphabetically to make it easy to find an individual chemical. The second list has been rank ordered by the chemicals' VHR to make it easy either to screen candidate substitute chemicals that may be offered to you or to find acceptable candidate substitute chemicals that meet the criteria in Equation 8.10. For instance, if the measured exposures to an old chemical exceeds its exposure limit by a factor of 5 × (namely, if $[ppm_{BZ}/TLV]_{old} = 5$), then using Equation 8.10 indicates that the VHR_{new} must be at least 5 × less than the VHR_{old}. To find acceptable candidate substitute chemicals, it is possible simply to scan down the second list to see all the chemicals with a VHR_{new} equal to or lower than the $VHR_{old}/5$. (One can get to the same end point using Equation 8.7b, but with three times the number of calculations.)

EXAMPLE 8.1

Suppose your environmental measurements showed that laboratory technicians were being overexposed to chloroform by a factor of 50% more than its TLV. What would be an acceptable substitute chemical?

The information given means that the old compliance ratio is 1.5 (1 + 50%/100% = 1.5), meaning that the new VHR must be at least a factor of 1.5 less than the old VHR. As a first step, one would find chloroform on the alphabetical VHR list, in which one could see that the VHR for chloroform is 25,921. Second, use Equation 8.10 to find all the chemicals in the second list rank-ordered by VHR, whose VHR is less than 25,921/1.5 = 17,280. The exposures of employees doing the same task with any chemical with a VHR ≤ 17,280 should be in compliance.

$$\text{acceptable } VHR_{new} \leq \frac{VHR_{old}}{[ppm_{BZ}/TLV]_{old}} = \frac{25,921}{1.5} = 17,280 \qquad \text{(using Equation 8.10)}$$

The user's acceptance of any of these chemicals would depend upon the task being performed, but the large number of candidates from which they might choose makes finding a mutually acceptable substitute more likely. In this case, if the chloroform were being used as a rapidly evaporating cleaner, then methylene chloride (VHR = 11,450), 1-hexene (VHR = 8160), or 1,1-dichloroethane (VHR = 2990) with about equally high vapor pressures might work (except for some of their flammabilities). If the chloroform were being used in an extraction procedure because of its high density ($\rho_\ell = 1.48$), then perhaps 1,2-dichloroethane ($\rho_\ell = 1.24$) or 1,1,2,2-tetrachloroethane ($\rho_\ell = 1.59$) might work. If the laboratory chemist is adamant that chloroform has to be used because it is required by a procedural standard, then substitution may not be successful. However this example should demonstrate VHR's utility to identify multiple acceptable candidates easily and rapidly.

[h] While control by substitution is presented herein as if the old compliance ratio was greater than one, these equations would also apply to a chemical change initiated, for instance, by the production department that might make matters worse.

5. The vapor hazard index (VHI) is a way to shrink the otherwise very wide range of VHR values into a more convenient two digits. (The VHI is mathematically analogous to the Richter scale for earthquakes and similar to the dB scale used in noise but without the 10.)

$$\text{VHI} = \log(\text{VHR}) \tag{8.11}$$

and by analogy

$$\text{an acceptable VHI}_{\text{new}} \leq \text{VHI}_{\text{old}} - \log[\text{ppm}_{BZ}/\text{TLV}]_{\text{old}} \tag{8.12a}$$

$$\Delta\text{VHI} = \text{VHI}_{\text{old}} - \text{acceptable VHI}_{\text{new}} \geq \log[\text{ppm}_{\text{BZ}}/\text{TLV}]_{\text{old}} \tag{8.12b}$$

The main advantage of the VHI is simply in seeing and using fewer digits. The list of chemicals rank-ordered by VHI values is of course the same sequence as the more lengthy VHR values. A simple two-digit number on a chemical label or MSDS would be an easier method by which users could judge each chemical's inherent ability to create a vapor hazard. The VHI value is a good (within less than an order of magnitude) indicator of the additional care with which someone should use one chemical vs. another. Equation 8.13 will use the VHI within a hazard rating index that *could* be used to set management priorities. However, to find an acceptable VHI_{new} using Equation 8.12 requires calculating with logarithms, not a popular topic with many students and not really necessary in practice.[i] While the VHI has some potential hazard communication uses, most people are likely to find the VHR easier to work with to solve problems.

6. It may correctly appear that certain problems can be solved only using P_{vapor} values, while other problems can be solved more easily using VHR values. Being able to recognize when to apply the P_{vapor} ratio vs. the VHR ratio is not essential (because you can always get the right answer sooner or later), but it can save you time when time is critical (as on an examination). One key is not to get bogged down in the equations. They only state with symbols the concepts that you should be learning.
   a. Use the ratio of vapor pressures in Equation 8.3 if you are only comparing concentrations.

$$\boxed{\text{ppm}_{\text{new}} = \text{ppm}_{\text{old}} \times \frac{P_{\text{vapor,new}}}{P_{\text{vapor,old}}}} \qquad \text{(repeat of Equation 8.3)}$$

   The concept underlying Equation 8.3 is that the two chemicals' rates of evaporation and plume concentrations are proportional to vapor pressure. While this calculation is not an efficient way to find an acceptable substitute among many candidates, one can always compare any particular new chemical's concentration $[\text{ppm}_{\text{new}}]$ with its exposure limit $[\text{TLV}_{\text{new}}]$ as a second step.

   b. Use the ratio of the VHRs if you want to compare the compliance or acceptability of a substitution control in one step. Use Equation 8.9 to tell if a given substitution is acceptable.

$$\boxed{[\text{ppm}_{\text{BZ}}/\text{TLV}]_{\text{new}} = [\text{ppm}_{\text{BZ}}/\text{TLV}]_{\text{old}} \frac{\text{VHR}_{\text{new}}}{\text{VHR}_{\text{old}}}} \qquad \text{(repeat of Equation 8.9)}$$

[i] However, if you are comfortable with logarithms, the maximum acceptable VHI_{new} can be found by subtracting the VHI found using Equation 8.12b from the VHI_{old}. For example, if an employee were exposed to an old chemical at two times its TLV, the VHI of an acceptable new substitute must be at least $\Delta\text{VHI} = \log[2] = 0.30$ less than the old chemical's VHI.

Use Equation 8.10 to pick an acceptable substitute from among many candidates.

$$\boxed{\text{acceptable VHR}_{\text{new}} \leq \frac{\text{VHR}_{\text{old}}}{[\text{ppm}_{\text{BZ}}/\text{TLV}]_{\text{old}}}} \qquad \text{(repeat of Equation 8.10)}$$

The VHR ratio can also be used to compare the relative vapor hazard of two or more chemicals when no measurements are available, in which case no compliance data $[\text{ppm}_{\text{BZ}}/\text{TLV}]_{\text{old}}$ can be calculated. The chemical with the higher VHR always has a greater inherent potential to create a hazardous vapor; hence, the name vapor hazard ratio.

VIII. USING VHR AND OTHER INDICES IN RISK MANAGEMENT

While the VHR is informative, it does not by itself comprise a complete management strategy. The following components of risk can (and should) all be used to set risk management priorities or to evaluate the financial liability of a substitute chemical or process.

1. The *VHR* of the chemical is the intrinsic propensity of the chemical to create an unacceptable exposure. While a quantitative risk management model could use the VHR as presented, the mathematics of the VHI = log[VHR] in Equation 8.11 are very compatible with the natural variability of occupational conditions, the imprecision of measuring exposures, the even greater imprecision of predicting dilution ratios, the nonlinear dose-response relationships of most toxins, and the order of magnitude quality of all risk assessment models.
2. A chemical's exposure limit is the other half of the Vapor Hazard Ratio. When comparing exposure limits and hazard, remember that **not all exposure limits are created equal**. For instance, some TLVs are based on irritation, some on neural effects, some on damage to organ systems such as blood, kidney, or liver, some are based on the risk of acute effects, some on chronic disease, and others are based on cancers. While all effects are judged equally in setting the exposure limit, they have no effect upon either the value of the TLV itself or the corresponding VHR.
3. The VHR is only a starting point for an airborne hazard. The actual level of exposure also depends upon *the setting, use conditions, and work practices* that determine the EDR. The following four categories of other factors act in concert with the intrinsic ability of the environment to dilute the chemical to the concentration that reaches the breathing zone, to allow or inhibit an additional dermal route of exposure, to vary the length of time employees are exposed to that environment, and ultimately to determine the dose.
   a. Does the worksite have local exhaust ventilation, and if so, how well does it work? The better the ventilation, the greater the EDR, and the further concentrations will be reduced below P_{vapor} or some other term that would characterize the source. If one is comparing risks based on actual exposure data, then over-exposures in the presence of existing ventilation would suggest a difficult control problem.
   b. Are respirators provided, and if so, are they worn? Respirators should reduce exposures by at least their assigned protection factor (the smallest APF is 10 as will be explained in Chapter 22). Poor fit of a respirator can reduce its protection, but not wearing one's respirator a good proportion of the time will reduce protection even more.
   c. Does the work involve dermal exposure? Are gloves provided and worn? Gloves will decrease dermal exposures and the potential for a dermally absorbed dose. Dermal exposures can range from indirect contact (e.g., handling damp parts after

machining), to direct but incomplete contact (e.g., handling wet parts or being splashed), to immersion or drenching (full coverage of the skin).

d. How frequently does the job involve exposure and for how long? If comparing risks based on actual exposure data, short-term samples taken during intermittent tasks probably overestimate long-term exposures, and 8-TWA samples are generally only collected on days when the chemical is in use. Thus, dose often depends upon more than just sampled airborne concentrations.

4. As pointed out in Section III.2 in Chapter 7, risk management can be viewed as a financial benefit to the employer. A similar motivation can determine the priorities within an occupational health program. One example of a comprehensive risk management priority setting system is the following "hazard rating" system, modeled closely upon but simpler than those proposed by Langner et al. and AIHA.[38,39]

$$\text{Hazard Rating} = [\text{Degree of Exposure}] \times [\text{Severity of Response}] \times [\text{VHI}] \quad (8.13)$$

where

"degree of exposure" are scored categories characteristic of each job that are coded as follows:

0: No contact with agent at work or worksite; no exposure.
1: Either infrequent contact or work in vicinity of a closed system.
2: Either frequent contact at low concentrations (e.g., with exposure controls) or infrequent contact at high concentration(poor control).
3: Work involves regular or frequent contact or worksite has no exposure controls.
4: Frequent contact at very high concentrations (very high exposure).

and "severity of response" are scored categories of the potential adverse health effect expected from exposure near or above the exposure limit that are coded as shown below. Although severity implies cost and therefore liability, the categories as written have no explicit financial qualifier.

0: Reversible effects of little concern or suspected adverse health effects (e.g., irritation without systemic injury such as allergenic or irritating hazards).
1: Reversible health effects without systemic injury.
2: Potentially severe but reversible clinical change due to chronic overexposure (e.g., dermatitis, bronchitis).
3: Irreversible change or tissue damage due to chronic overexposure and potentially life-threatening with acute overexposure.
4: Irreversible and life-threatening to an exposed person or their progeny (e.g., cancer or reproduction, asphyxiation, or an explosion).

[VHI]: The vapor hazard index, as described above (Equation 8.11).

In use, a hazard rating would need to be calculated for each work site in a plant or company. Such a project would require a systematic survey to obtain the setting, use condition, and work practice information needed to assign each site or process a value for its degree of exposure, the severity of the potential response, and the intrinsic VHI for the chemical(s) in use. Numeric values using this particular hazard rating scheme can range from about 0 to extremes approaching 100. This hazard rating when multiplied by the number of employees exposed would yield the employer's relative liability from which monitoring and control priorities can be set. Other predictors of degree of exposure are described in appendices to Chapter 18 as part of a descriptive database to accompany air sample data.

Taken as a whole, the above discussion suggests that a wide range of factors can be used to help set programmatic priorities among potential health hazards within any organization. Some systems such as the hazard rating are relatively complex and require a large investment of human and computerized database resources; others such as the VHR are relatively simple. A common assignment for new hygienists is to compile a list of chemicals (perhaps from or as a part of a hazard communication program). If you are then asked to rank order them by exposure limit (TLV or PEL), do not forget that such a list (or one based on volatility) means nothing regarding the potential hazard of these chemicals. You might need to explain the added value of a list prioritized by the VHR. The data needed to generate such a list for single chemicals is within the appendix of this book; the methods to generate such a list for mixtures is in Chapter 6.

PRACTICE PROBLEMS

1. Answer the following questions for each of the proposed substitute chemicals assuming no other changes are made to the existing workplace. The first pair of chemicals presents an example of the information required and a solution. (You do not really need the old TLV or compliance ratio; it is only for your information.)
   - What concentration would you expect if the proposed chemical were substituted for the existing chemical?
   - Would exposures to the proposed substitute be in compliance? Recall Equation 5.19.

Existing (Old) Chemical	Proposed (New) Chemical	Basis for the Prediction
1,1,2,2-Tetrachloroethane	1,2-Dichloroethane (tried because it has a higher TLV)	
$ppm_{old} = 5$ ppm	$P_{vapor,new} = 78.9$ mmHg	$\frac{P_{vapor,new}}{P_{vapor,old}} = 17.1$
$P_{vapor,old} = 4.6$ mmHg	$ppm_{new} = 85.8$ ppm	
($TLV_{old} = 1$ ppm)	$TLV_{new} = 10$ ppm	
(old compliance ratio = 5.0)	new compliance ratio = 8.6 is not acceptable.	
The increased vapor concentration and higher TLV would actually increase the compliance ratio from 5 to 8.6.		
a. 1,1,2,2-Tetrachloroethane	1,1,2-Trichloroethane (tried for its higher TLV and lower P_{vapor})	
$ppm_{old} = 5$ ppm	$P_{vapor,new} =$ _____ mmHg	$\frac{P_{vapor,new}}{P_{vapor,old}} =$ _____
$P_{vapor,old} =$ _____ mmHg	$ppm_{new} =$ _____ ppm	
($TLV_{old} =$ _____ ppm)	$TLV_{new} =$ _____ ppm	
(old compliance ratio = _____)	new compliance ratio = _____ is or is not acceptable?	
b. 1,1,2,2-Tetrachloroethane	Perchloroethylene (also a higher TLV but still lower P_{vapor})	
$ppm_{old} = 5$ ppm	$P_{vapor,new} =$ _____ mmHg	$\frac{P_{vapor,new}}{P_{vapor,old}} =$ _____
$P_{vapor,old} =$ _____ mmHg	$ppm_{new} =$ _____ ppm	
($TLV_{old} =$ _____ ppm)	$TLV_{new} =$ _____ ppm	
(old compliance ratio = _____)	new compliance ratio = _____ is or is not acceptable?	

2. Again assuming no other changes were made, in essentially one step (as per Equation 8.9) determine the effect on hazard (exposure relative to each chemical's exposure limit) that you would expect if the proposed substitution were made for the initial production situation that you have already evaluated relative to its compliance. These are the same chemicals and settings that were used in problem 1.

Existing (Old) Chemical	Proposed (New) Chemical	Basis for the Prediction
1,1,2,2-Tetrachloroethane	1,2-Dichloroethane	
$\frac{ppm_{old}}{TLV_{old}} = \frac{5\ ppm}{1\ ppm} = \underline{5}$	$VHR_{new} = 10{,}382$	$\frac{VHR_{new}}{VHR_{old}} = 1.71$
$VHR_{old} = 6079$	new compliance ratio $= [ppm/TLV]_{old} \times \frac{VHR_{new}}{VHR_{old}}$ $= 5 \times 1.71$ $= \underline{8.6}$ is not acceptable. (exactly the same result in one step as accomplished in two steps above).	
a. 1,1,2,2-Tetrachloroethane	1,1,2-Trichloroethane	
$\frac{ppm_{old}}{TLV_{old}} = \frac{5\ ppm}{\ \ ppm} =$ _____	$VHR_{new} =$ _____	$\frac{VHR_{new}}{VHR_{old}} =$ _____
$VHR_{old} =$ _____	new compliance ratio $= [ppm/TLV]_{old} \times \frac{VHR_{new}}{VHR_{old}}$ $=$ $=$ _____ is or is not acceptable?	
b. 1,1,2,2-Tetrachloroethane	Perchloroethylene	
$\frac{ppm_{old}}{TLV_{old}} = \frac{5\ ppm}{\ \ ppm} =$ _____	$VHR_{new} =$ _____	$\frac{VHR_{new}}{VHR_{old}} =$ _____
$VHR_{old} =$ _____	new compliance ratio $= [ppm/TLV]_{old} \times \frac{VHR_{new}}{VHR_{old}}$ $=$ $=$ _____ is or is not acceptable?	

3. You should eventually get the same predicted level of compliance with the TLV after the proposed change in problem 1 as you did in problem 2. If all of the chemicals were given to you as proposed substitutes, which of the above methods would be the easiest way to choose among many candidates to find an acceptable substitute for 1,1,2,2-tetrachloroethane? The methods used to answer the next question should give you the answer to this question.
4. The following data is provided for toluene, the initial chemical used in a setting different from above:

NIOSH REL:	100 ppm	MW:	92.1		
TLV:	50 ppm (A4,skin)	P_{vapor}:	28.4 mm (at 77°F)		
OSHA TWA PEL:	200 ppm	BP:	232°F	FRZ:	−139°F
Ceiling:	300 ppm	LFL:	1.1%	UFL:	7.1%
For:	10 min	Density:	0.87		
Max. Peak:	500 ppm	VHR:	747		

Suppose that you sampled and find other workers' 8-h average exposures are 433 mg/m^3 = 115 ppm. Further, suppose that as a matter of policy you and your company try to keep exposures equal to or less than the TLV where it is less than the PEL.

a. How far over the TLV are you? (i.e., what is the ppm_{old}/TLV_{old} compliance ratio?)
b. A compliance ratio >1 means the working environment is not creating enough dilution that would be necessary to use toluene acceptably. Rather than installing

more ventilation to increase the dilution, you decide to look for a substitute chemical that requires less dilution. What is the maximum VHR for a chemical that would be acceptable to use in this setting?

c. After you discussed this problem with the production and engineering departments, they provided you with the following list of candidate substitute solvents that would be acceptable to them. Which of the following substitute solvents would be acceptable to you from a vapor compliance perspective (not considering the nature of the chemical's health hazard)?

	VHR_{new}	Acceptable?
1,1-Dichloroethane	______	Y/N
Methyl chloroform	______	Y/N
Ethyl acetate	______	Y/N
n-Heptane	______	Y/N
o-Xylene	______	Y/N

d. Which of the above chemicals would be your most preferred choice given that not all TLVs are created equal?

e. Aniline is the first numerically acceptable substitute chemical for toluene in the list of chemicals rank-ordered by VHR. (Does that coincide with the answer that you gave for part b above?) Why is it probably an unacceptable substitute toxicologically? (Another reason that not all TLVs are created equal.)

REFERENCES

1. *Guide to Cleaner Technologies: Organic Coating Replacements*, EPA/625/R-94/006, Office of Research and Development, U.S. Environmental Protection Agency, Cincinnati, OH, 1994.
2. *Guide to Cleaner Technologies: Alternative Metal Finishes*, EPA/625/R-94/007, Office of Research and Development, U.S. Environmental Protection Agency, Cincinnati, OH, 1994.
3. *Guide to Cleaner Technologies: Organic Coating Removal*, EPA/625/R-93/015, Office of Research and Development, U.S. Environmental Protection Agency, Cincinnati, OH, 1994.
4. *Guide to Cleaner Technologies: Alternatives to Chlorinated Solvents for Cleaning and Degreasing*, EPA/625/R-93/016, Office of Research and Development, U.S. Environmental Protection Agency, Cincinnati, OH, 1994.
5. *Guide to Cleaner Technologies: Cleaning and Degreasing Process Changes*, EPA/625/R-93/017, Office of Research and Development, U.S. Environmental Protection Agency, Cincinnati, OH, 1994.
6. Günsel, R., Water-based coatings and the environment, *Surf. Coat. Int.*, 1993, 364, 1993.
7. Goldschmidt, G., An analytical approach for reducing workplace health hazards through substitution. *Am. Ind. Hyg. Assoc. J.*, 54(1), 36–43, 1993.
8. Ballway, B., What's new in high solids coatings, *Met. Finish.*, 21, March 1992.
9. Goldberg, D. and Eaton, R. F., Caprolacetone polyols as reactive diluents for high solids, *Mod. Paint Coat.*, 82(11), 36, 1992.
10. Bailey, J., Powder coating: an environmental perspective, *Ind. Finish.*, 68(9), 60, 1992.
11. Major, M. J., Innovation and regulations aid powder coatings, *Mod. Paint Coat.*, 82(13), 6, 1992.
12. Osmond, M., Architectural powder coatings: A review of new advances in exterior durable systems, *Surf. Coat. Int.*, 1993, 402, 1993.
13. Abbott, K. E., Plastic media blasting — State of the technology, *Mater. Performance*, 31(2), 38–39, 1992.
14. Bailey, J., Strip it off, *Ind. Finish.*, 68(2), 24–27, 1992.

15. Nudelman, A. K. and Abbott, K. E., Using plastic media blasting to remove powder coatings from parts, *Powder Coat.*, April 1996.
16. Steiner, R., Carbon dioxide's expanding role, *Chem. Eng.*, 100(3), 114–119, 1993.
17. Howlett, J. J. Jr. and Dupuy, R., Ultrahigh-pressure water jetting for deposit removal and surface preparation, *Mater. Performance*, 32(1), 38–43, 1993.
18. Kozol, J., An environmentally safe and effective paint removal system for aircraft, *J. Miner. Met. Mater. Soc.*, 53(3), 20–21, 2001.
19. Powell, R. W., Estimating worker exposure to gases and vapors leaking from pumps and valves. *Am. Ind. Hyg. Assoc. J.*, 45(11), A-7–A-15, 1984.
20. ACGIH *Industrial Ventilation: A Manual of Recommended Practice*, 24th ed., American Conference of Governmental Industrial Hygienists, Cincinnati, OH, 2001.
21. Ross, S. and Morrison, I. D., *Colloidal Systems and Interfaces. The HLB Scale*, Wiley, New York. p. 274, 1998.
22. Ashworth, J. and White, I. R., Contact allergy to ethoxylated phenol, *Contact Dermatitis*, 24, 133–134, 1991.
23. Cserháti, T., Ethoxylated and alkylphenol ethoxylated nonionic surfactants: Interaction with bioactive compounds and biological effects, *Environ. Health Perspect.*, 103, 358–364, 1995.
24. Popendorf, W., Miller, E. R., Sprince, N. L., Selim, M. I., Thorne, P. S., Davis, C., and Jones, M. L., The utility of preliminary surveys to detect the cause of acute metal working fluid hazards, *Am. J. Ind. Med.*, 30(6), 744–749, 1996.
25. Chapin, R. E. and Sloane, R. A., Reproductive assessment by continuous breeding: evolving study design and summaries of ninety studies, *Environ. Health Perspect.*, 105(Suppl. 1), 199–395, 1997.
26. LaKind, J. S., McKenna, E. A., Hubner, R. P., and Tardiff, R. G., A review of the comparative mammalian toxicity of ethylene glycol and propylene glycol, *Crit. Rev. Toxicol.*, 29(4), 331–365, 1999.
27. Johnson, W., Final report on the safety assessment of ethoxyethanol and ethoxyethanol acetate, *Int. J. Toxicol.*, 21(1), 9–62, 2002.
28. Gruvberger, B. and Bruze, M., Preservatives, *Clin. Dermatol.*, 15(4), 493–497, 1997.
29. Schnuch, A., Geier, J., Uter, W., and Frosch, P. J., Patch testing with preservatives, antimicrobials and industrial biocides. Results from a multicentre study, *Br. J. Dermatol.*, 138(3), 467–476, 1998.
30. Whysner, J. and Williams, G. M., D-limonene mechanistic data and risk assessment: Absolute species-specific cytotoxicity, enhanced cell proliferation, and tumor promotion, *Pharmacol. Ther.*, 71(1–2), 127–136, 1996.
31. Larsen, S. T., Hougaard, K. S., Hammer, M., Alarie, Y., Wolkoff, P., Clausen, P. A., Wilkins, C. K., and Nielsen, G. D., Effects of R-(+)- and S-(−)-limonene on the respiratory tract in mice, *Hum. Exp. Toxicol.*, 19(8), 457–466, 2000.
32. AIHA: *The AIHA 2005 Emergency Response Planning Guidelines and Workplace Environmental Exposure Level Handbook*, American Industrial Hygiene Association, Fairfax, VA, 2005.
33. Hill, E.A. and K.D. Carter, Jr., An alternative to chlorinated solvents for cleaning metal parts, *Proceedings of the 1993 International CFC and Halon Alternatives Conference*, The Alliance for Responsible CFC Policy, Washington, DC, pp. 465–471, 1993.
34. U.S. Environmental Protection Agency, Lifecycle Analysis and Pollution Prevention Assessment for Methyl Pyrrolidone (NMP) in Paint Stripping, Final Assessment, Public RM2 Administrative Record Document, Office of Pollution Prevention and Toxics, Washington, DC, 1993.
35. Solomon, H. M., Burgess, B. A., Kennedy, G. L.Jr., and Staples, R. E., 1-Methyl-2-pyrrolidone (NMP): Reproductive and developmental toxicity study by inhalation in the rat, *Drug Chem. Toxicol,* 18(4), 271–293, 1995.
36. Burow, R.F., Volatile methyl siloxanes (VMS) as replacements for CFCs and methyl chloroform in precision and electronics cleaning. *Proceedings of the 1993 International CFC and Halon Alternatives Conference*, The Alliance for Responsible CFC Policy, Washington, DC, pp. 654–661, 1993.

37. Popendorf, W., Vapor pressure and solvent vapor hazards, *Am. Ind. Hyg. Assoc. J.*, 45(10), 719–726, 1984.
38. Langner, R. R., Norwood, S. K., Socha, G. E., and Hoyle, H. R., Two methods for establishing industrial hygiene priorities, *Am. Ind. Hyg. Assoc. J.*, 40, 1039–1045, 1979.
39. Hawkins, N. C., Norwood, S. K., and Rock, J. C., *A Strategy for Occupation Exposure Assessment*, Farifax, AIHA, VA, 1991.

9 Other Source and Nonventilation Pathway Controls

The learning goals of this chapter:

- ✍ Be familiar with the differences between modifying nonvolatile vs. volatile chemical sources.
- ✍ Be familiar with the effect on exposure of changing each component of the basic evaporation equation.
- ☐ Be able to apply the basic evaporation equation to anticipate the effect of modifying volatile source conditions.
- ✍ Be familiar with, and the limitations of, automation, separation, and isolation as chemical exposure controls.

I. CONCEPTS

The term "Other" in the title of this chapter means to change the physical operating conditions of the source without changing the chemical or the manufacturing process or technology. While the substitution controls discussed in Chapter 8 focused only on the generation of contaminants (the G in Equation 5.14), "other" source controls can both decrease generation and increase dilution in the pathway.

✍ $$\text{steady state } C\ (\text{mg/m}^3) = \frac{G\ (\text{mg/min})}{Q_{\text{apparent}}\ (\text{m}^3/\text{min})}$$ (repeat of Equation 5.14 and a preview of Equation 20.1)

Equation 5.14 is analogous to Equation 5.6a (C = mass/air Vol) for complete evaporation into a fixed volume. In a dynamic environment or when ventilation is present, a toxic chemical's airborne concentration is determined by the rate at which its emissions [G] appear to be diluted into some amount of flowing fresh air [Q_{apparent}]. Equation 9.1 is simply Equation 5.14 solved for the maximum acceptable rate at which a contaminant can be generated (again analogous to Equation 5.6b, mass = $C \times$ air Vol). This equation defines a theoretical goal to which G must be controlled to meet an existing exposure limit in any ventilated setting.

$$\text{Theoretically acceptable } G = C_{\text{in breathing zone}} \times Q_{\text{apparent}} \leq \text{TLV}^{\circledR} \times Q_{\text{apparent}} \tag{9.1}$$

While Equation 9.1 is rarely used by itself because a method to quantify Q_{apparent} and/or G rarely exists, some useful predictions can be made if an existing concentration has already been measured. For instance, Equation 9.2 allows one to compare the old airborne concentration that existed before either a source or pathway modification is implemented (or is even in the design stage) with the new concentration that will (or would) exist after the modification has been (or might be) implemented.

$$C_{\text{new}} = C_{\text{old}} \frac{G_{\text{new}}/Q_{\text{apparent, new}}}{G_{\text{old}}/Q_{\text{apparent, old}}} \tag{9.2}$$

How much change is needed? Equation 9.3 allows one to predict the relative change in the source G or environmental dilution ($Q_{apparent}$) that would be necessary to reduce an initially unacceptably high old concentration to a new concentration that is equal to or less than an acceptable exposure limit such as a TLV. For the same chemical, Equation 9.3 could apply equally well to either mg/m^3 or ppm.

$$\text{acceptable}\frac{G_{new}/Q_{apparent,\ new}}{G_{old}/Q_{apparent,\ old}} \leq \frac{\text{TLV}}{C_{old}} \tag{9.3}$$

This chapter will begin by focusing on ways to modify the source to reduce G (the rate at which a hazard is generated) without changing the chemical. Examples of *other source controls* will first differentiate between modifications applicable to nonvolatile vs. volatile sources. The advantages attributed to substitution in the previous chapter (being universal, being reliable, being reproducible, and being cost-effective) are applicable to many but not quite all of these source modification controls. The last part of this chapter will focus on nonventilation pathway controls. *Nonventilation pathway controls* (such as work practices, automation, separation, and isolation) increase $Q_{apparent}$ or the environmental dilution ratio without actually changing the amount of ventilation air supplied or exhausted. Such nonventilation pathway controls have only a few of the advantages of substitutions.

II. SOME NONVOLATILE CHEMICAL SOURCE CONTROLS

Since the generation of an aerosol depends on some external mechanism, potential controls for nonvolatile and low-volatility chemicals tend to focus on reducing the forces that generate the aerosol. However, aerosols can also be controlled by making the source material less susceptible to being aerosolized or by making the resulting particles either stay airborne for less time or be less respirable.

1. One simple but common example demonstrating both approaches is the use of a dry broom to clean up a toxic powder or fugitive dust (dust for which no specific single source is responsible). Rather than use a dry broom, use either a vacuum (which may not put less force on dust particles but at least directs that force into a vacuum nozzle) or a wet mop (the water makes particles stick together). The Brush-Wellman beryllium employee training videos provide good examples of fugitive dust control.
2. If the work process itself produces or releases the aerosol, using or substituting a process that creates a less physically aggressive force will reduce the rate of aerosol emission, the airborne concentration, and probably the level of any associated fallout dermal exposure. Reducing the speed of a process will generally reduce the contaminant production forces; however, less energetic processes generally mean slower production. High production rates, better resource utilization, and good quality are all *key* to company profitability; thus, slower production rates are rarely an acceptable option.

   Container filling for packaging and/or shipping is a manufacturing step common to both volatile liquids and nonvolatile dust, powders, pellets, etc. The former is likely to produce vapors; the latter, aerosols as the container is filled. A dry nonvolatile product is typically conveyed to the filling site in a conduit such as a pipe or chute and is allowed to flow into the container by gravity. The amount of aerosol generated during a filling operation is directly proportional to the magnitude of the gravitational energy released when the discharged material encounters either the bottom of the container or the material already accumulating within the container (the sequence of drawings in Figure 9.1 depicts that sequence). The magnitude of that gravitational energy depends on the distance the conveyed material is allowed to free-fall before making that impact

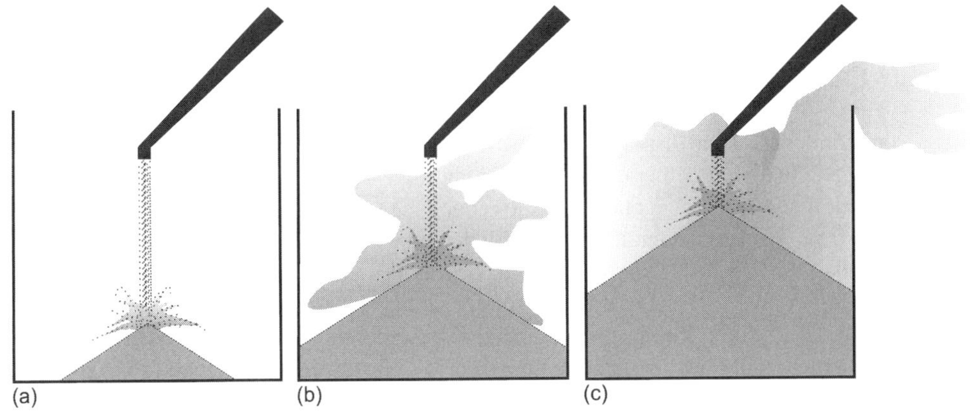

FIGURE 9.1 A large free-fall of a granular product generating a dust aerosol (a), accumulating (b), and releasing the dust aerosol from the top from the container (c).

(basic physics). Minimizing or eliminating that free-fall is one potential means to control airborne dust or the generation of highly concentrated vapors. Dust generation from a material such as agricultural grains can be minimized by keeping the end of the conduit as close as possible to the surface of the grain already in the filling container, as depicted in Figure 9.2. Dust generation could be virtually eliminated if the end of the conduit could be kept at or just below the surface of the grain in the filling container. An option related to both this and the next control is to attach a plastic sleeve to the filling shoot. Although it seems (subjectively) that nonvolatile materials will be even more sensitive to free-fall effects than the volatile liquids to be discussed in Section III.1, no comparative data have been reported on this topic for nonvolatile materials.

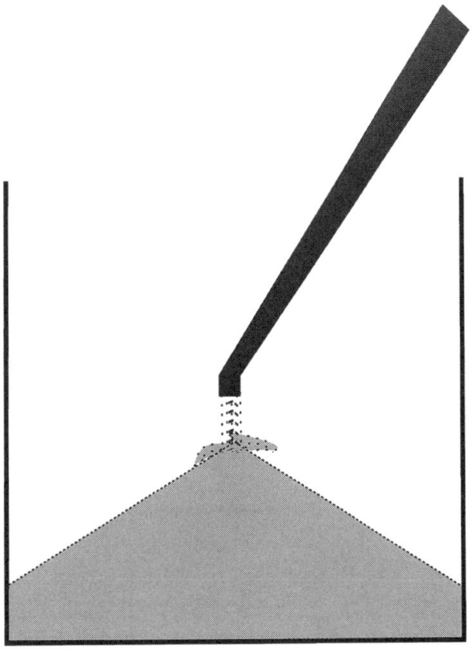

FIGURE 9.2 Shortening the free-fall reduces the amount of aerosol generated and released.

3. The aerosols and vapors generated in the headspace of the container being filled are displaced and driven from the container by the product being packaged (as depicted in Figure 9.1c). Headspace means the volume of air inside a container above its contents. The volume of air displaced will eventually equal the volume of the product that goes into the container. That displaced air will carry whatever airborne dust (or vapor) is generated inside the container during filling. Thus, an alternative or adjunct to reducing the amount of aerosol being generated is to reduce the volume of contaminated air that is displaced from inside the container being filled. One way to reduce this volume is to start with a collapsed bag inside the box, drum, or tote bin container and let the conveyed powder, dust, pellets, etc. expand the collapsed bag as it is poured, as depicted in Figure 9.3. With this filling process, the displaced air is outside of the bag, is uncontaminated, and does not generate a hazard. A plastic sleeve attached to the filling shoot (like an opened-ended bag) is a closely related alternative.
4. Wetting working surfaces that produce powders will decrease a material's susceptibility to become aerosolized and thus is a powerful way to reduce aerosol production. The surface tension of liquids will tend to bind otherwise separate particles either to each other or to solid surfaces. Water is a common dust suppressant for construction work (applied to roadways), during housekeeping (e.g., wet mopping), for asbestos abatement (usually applied to working surfaces in small quantities at low pressure), for mining (applied through hollow tipped tools), and for demolition (applied in any way possible).[1,2] The difficulty in wetting an initially dry dust caused by the surface tension of water can be lessened by adding a surfactant (such as a nonfoaming detergent), but such an additive is usually only feasible where the water volume is small such as in asbestos abatement. Manufacturing a product as a paste, slurry, or suspended in a liquid instead of as a dry powder can be similarly effective.
5. Changing the particle size to one that does not stay airborne for as long or that does not penetrate into the respiratory tract as efficiently is one way to make manufactured products such as granulated or wettable powders less hazardous. Recall from Chapter 3 that larger diameter particles fall much more rapidly than smaller particles, in fact by the square of their diameters. Therefore, increasing an aerosol's mean particle diameter will reduce the time that it can stay airborne, the total number (or concentration) of particles that would be airborne at any given time, and the fraction of that aerosol that is both inhalable and respirable (less able to penetrate into the deep lung). Particles with an $d_{\text{aerodynamic}}$ greater that 100 μm will not stay airborne for long at all. Only about 50% of

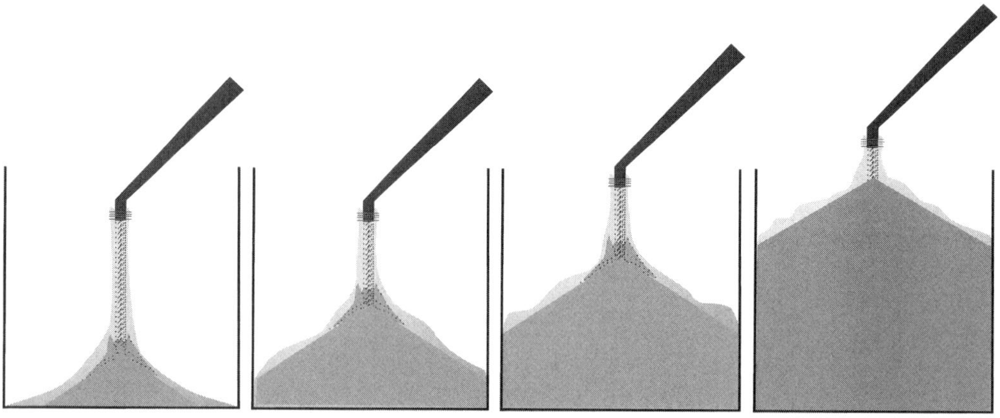

FIGURE 9.3 A container being filled with a bag liner. The bag is unfurling from the dispenser which rises (or the container is lowered) as the container is filled for good bag filling.

particles larger than approximately 50 μm are inhalable; the rest will fall past the mouth and nose. Particles larger than approximately 30 μm will not reach the thoracic region of the respiratory tract. Particles larger than approximately 10 μm will not reach the pulmonary region. The following are notable exceptions:

a. The significant falling velocity of particles greater than approximately 30 to 50 μm allows them to present a potentially significant dermal exposure hazard. Therefore, some consideration to changing a product's skin absorption might also prove cost-effective for such large particles.
b. Particles between approximately 0.1 and 1 μm in aerodynamic diameter are the most difficult to control and have the least pulmonary deposition. Particles smaller than 0.1 μm, such as fumes, are easily deposited in the upper respiratory tract because such small particles diffuse similarly to molecules.

6. Most manufactured particles are polydisperse, so that any given mean size will still contain some (often a predictable fraction) of both smaller and larger particles. Decreasing the variability in the size of manufactured powders (their geometric standard deviation) may also help to reduce the fraction that is inhalable or respirable.

III. SOME VOLATILE CHEMICAL SOURCE CONTROLS

1. The hazards during transfer operations with volatile liquid chemicals largely parallel those for dry nonvolatile chemicals. Similar to dusts, the air displaced by pouring a volatile liquid into a container will be highly contaminated with lots of vapors and some mist; and similar to dusts, the rate at which the contaminated air is emitted from the container will approximate the rate at which the container is being filled. For example, the vapors generated from the bulk loading of petroleum products into tank trucks and tank cars were measured at a variety of sites by MRI (a Kansas City consulting firm) for EPA (1985), who then derived a predictive equation equivalent to the following form that this author translated into IH units:

$$G = \frac{f_{\text{head space}} \times \text{MW} \times \text{Vol} \times P_{\text{vapor}}}{t \times R \times T} = \frac{f_{\text{head space}} \times \text{MW} \times \text{Vol} \times P_{\text{vapor}}}{t \times 24.45 \times 760} \quad \text{at NTP} \tag{9.4}$$

where

G = Vapor generation rate, g/min
$f_{\text{head space}}$ = Head space saturation factor, dimensionless (see Table 9.1)
MW = Molecular weight, g/mole
Vol = Volume of the tank or container being filled, L (gallons × 3.785)
t = Time to fill the tank, minutes
P_{vapor} = Vapor pressure of the chemical, mmHg
R = Universal gas constant, 62.364 mmHg L/mole °K
T = Emitted vapor temperature, °K (used in conjunction with the R value).

Splash filling refers to operating with the dispenser opening above the liquid surface (with concurrent splashing and agitation). Submerged filling refers to operating with the dispenser opening below the more quiescent liquid surface. An $f_{\text{head space}}$ value of 1 would imply that the vapor leaving the tank is saturated at the chemical's vapor pressure. An $f_{\text{head space}}$ value near 0.5 implies that the vapors

TABLE 9.1
Values of $f_{head\ space}$ for Use in Equation 9.4 for Each Mode of Filling

	Splash Filling	Submerged Filling
A previously unused vessel (dry with no preexisting vapors)	1.45	0.5
A previously used vessel (wet and with preexisting vapors)	1.45	0.6

leaving the tank are only about half saturated or that their concentration is about one half of the chemical's P_{vapor}. An $f_{head\ space}$ value greater than 1.0 implies that the vapors being displaced are more than saturated, suggesting that some mist is being carried out with the vapors. Thus, submerged filling (where the outlet is below the liquid surface) seems able to reduce vapor emissions and their potential exposure levels by at least two thirds vs. those generated by typical splash filling. It would not be unreasonable to use Equation 9.4 to estimate G for other sizes of containers being filled with a volatile liquid. Although the generation of a mist is less likely to occur when filling containers much smaller than shipping tankers (Figure 9.4), an $f_{head\ space}$ of around 1 might still be expected at an in-plant, small container filling site.

2. A vapor recovery system is an even better modification than submerged filling for liquid transfer operations. With a vapor recovery system, one end of a second pipe or hose is connected to the top of the tank being filled via a reasonably airtight seal, and the other end is connected back to the headspace of the storage container from which the liquid originates, as depicted schematically in Figure 9.5. This second hose carries vapors from the tank being filled back into the tank being emptied.[a] The effect is to reduce the emission of vapors into the air (and into the breathing zone of someone filling the tank). Vapor recovery is probably better than active local exhaust ventilation because it avoids

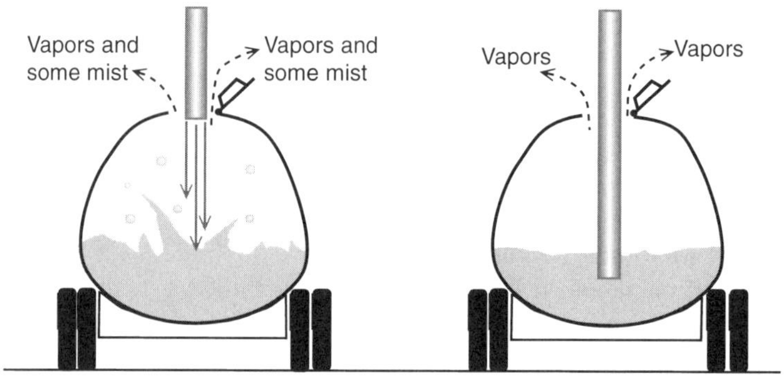

FIGURE 9.4 Depiction of splash filling on the left vs. submerged filling on the right.

[a] If the headspace above both tanks is equally saturated with vapors, the volume of the vapors being displaced from the tank being filled will exactly equal the volume of vapors needed to be added into the tank being emptied. Thus, such vapor recovery systems can operate passively as long as the seals are completely tight and both headspaces are equally saturated.

the need for some form of exhaust air cleaner to reclaim the high vapor concentrations coming from the filling tank's headspace.

3. The basic evaporation equation (Equation 5.13 repeated below) is applicable to virtually any source. Any modification that reduces a variable within that equation will reduce a solvent's evaporation rate.

$$\text{evaporation rate } G \text{ (moles/min)} = \text{Geom. Coef} \times A \times V^{0.5} \times P_{\text{vapor}}$$

(repeat of Equation 5.13)

Equation 9.5 uses Equation 5.13 to define the effect on G of changing each variable from an old condition existing before a modification is implemented to a new condition that will or might exist after a modification is implemented. While some of these variables are more easily quantified than others, virtually any of them can be modified.

$$G_{\text{new}} = G_{\text{old}} \times \left(\frac{\text{Geom. Coef.}_{\text{new}} \times A_{\text{new}} \times V_{\text{new}}^{0.5} \times P_{\text{vapor, new}}}{\text{Geom. Coef.}_{\text{old}} \times A_{\text{old}} \times V_{\text{old}}^{0.5} \times P_{\text{vapor, old}}} \right) \tag{9.5}$$

However, in order to know the effect on airborne concentration, one must evaluate the effect of changing each variable on both the vapor generation rate [G] and the volume of air into which that vapor appears to be diluted [Q_{apparent}], as shown in Equation 9.2.

$$C_{\text{new}} = C_{\text{old}} \times \left(\frac{G_{\text{new}}/Q_{\text{apparent,new}}}{G_{\text{old}}/Q_{\text{apparent,old}}} \right) \quad \text{(repeat of Equation 9.2)}$$

The effect on dilution of changing some variables will turn out to be different in the plume (exposure scenarios 1 and 2 in Chapter 5) than in the room outside of the plume (exposure scenario 3). In other words, each concentration has a different basis for its Q_{apparent}, the apparent volumetric flow rate of fresh air into which the contaminant appears to be diluted. The apparent volume flow rate of air in the plume

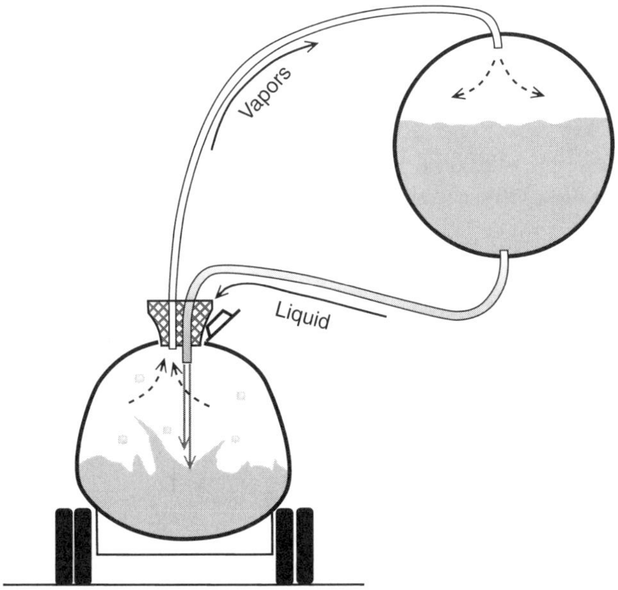

FIGURE 9.5 The upper tank is being gravity fed into the lower tank equipped with a vapor recovery hose.

[$Q_{in\ plume}$] will depend upon both the plume's velocity and its cross-sectional area as depicted in Figure 9.6. The apparent volume flow rate of air into which the entire plume will eventually be diluted will depend upon the general ventilation supplied to the room [Q_{room}] and the amount of mixing between that fresh air and the plume characterized by the term $K_{room\ mixing}$. Section II.3 in Chapter 19 will explain how $K_{room\ mixing}$ is used to make an empirical adjustment to the flow rate of general ventilation air to account for the less than ideal mixing of fresh (outdoor) air with the contaminant generated within a room or work zone, i.e., $Q_{effective} = Q_{room}/K_{room\ mixing}$. For instance, an increase in the general air velocity within a room can increase the mixing within that room, thereby decrease $K_{room\ mixing}$ for the same fresh air volumetric flow rate supplied to the room [Q_{room}], which will increase $Q_{effective}$, and thereby decrease exposure. However, if that increase in air velocity includes the air that passes over the evaporating source, the evaporation rate will increase and potentially increase exposure. However, we are getting ahead of ourselves. The next section will explain these changes one at a time.

IV. EFFECTS OF MODIFYING VOLATILE CHEMICAL SOURCE VARIABLES

The following subsections will describe the effects of changing each variable in the generic evaporation equation first on a volatile source's evaporation rate in the form of Equation 9.5. Since the effect on exposure of modifying a volatile source will depend upon the person's exposure scenario (as stated above and in Section III in Chapter 5), Equation 9.2 will be used to explain the effects on exposure if working within the plume or in a room but outside the plume. These effects will be tabulated in Table 9.2 and summarized graphically in Figure 9.7 and Figure 9.8.

1. The geometry of an evaporating liquid can be changed in various ways to reduce the geometry coefficient in Equation 9.5. The Geom.Coef. will decrease any time that the geometry of an evaporating liquid can be constricted or the thickness of the boundary layer through which the vapors conceptually diffuse can be increased. For instance,

TABLE 9.2
A Summary of the Effect that Changing Each Variable Has on G_{new}, $C_{plume,\ new}$, and $C_{room,\ new}$. In Each Case the New Value Equals the Old Value Times the Ratio Shown

Source Variable	Effect on G	Effect on C_{plume}	Effect on C_{room}
Geom. Coef.	$\left(\frac{\text{Geom.Coef.}_{new}}{\text{Geom.Coef.}_{old}}\right)$	$\left(\frac{\text{Geom.Coef.}_{new}}{\text{Geom.Coef.}_{old}}\right)$	$\left(\frac{\text{Geom.Coef.}_{new}}{\text{Geom.Coef.}_{old}}\right)$
A_{source}	$\left(\frac{A_{new}}{A_{old}}\right)$	$\left(\frac{A_{new}}{A_{old}}\right)^{+0.5}$	$\left(\frac{A_{new}}{A_{old}}\right)$
$V_{over\ source}$	$\left(\frac{V_{new}}{V_{old}}\right)^{+0.5}$	$\left(\frac{V_{new}}{V_{old}}\right)^{-0.5}$	1 to $\left(\frac{V_{new}}{V_{old}}\right)^{+0.5}$
P_{vapor}	$\left(\frac{P_{vapor,new}}{P_{vapor,old}}\right)$	$\left(\frac{P_{vapor,new}}{P_{vapor,old}}\right)$	$\left(\frac{P_{vapor,new}}{P_{vapor,old}}\right)$
Liquid Temperature (Exact effect)	$10^{\left(\frac{B(T_{new}-T_{old})}{(C+T_{new})(C+T_{old})}\right)}$	$10^{\left(\frac{B(T_{new}-T_{old})}{(C+T_{new})(C+T_{old})}\right)}$	$10^{\left(\frac{B(T_{new}-T_{old})}{(C+T_{new})(C+T_{old})}\right)}$
(Approx. effect)	$2^{(T_{new}-T_{old})/12°C}$	$2^{(T_{new}-T_{old})/12°C}$	$2^{(T_{new}-T_{old})/12°C}$

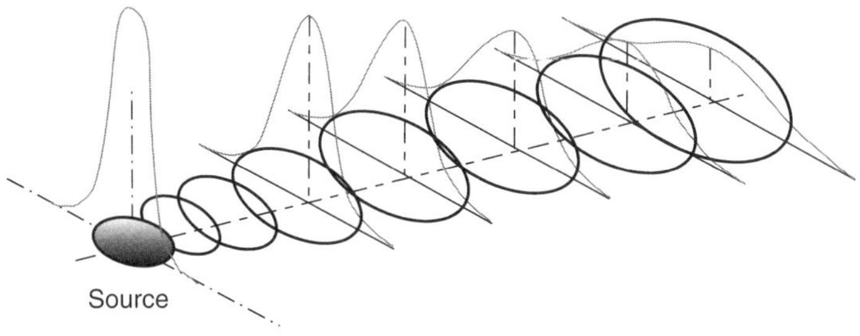

FIGURE 9.6 A modification of Figure 5.2 depicting the plume from a freely suspended source as a series of expanding ellipses.

increasing a container's freeboard (the height of a container's wall above the liquid surface) will shield the liquid source from the passing air, increase the thickness of the layer of air through which the vapors must travel by molecular and turbulent diffusion, and thereby decrease the geometry coefficient and the evaporation rate (the "G" term). The geometry coefficient would decrease if the source were a wet object (not flat) or a cleaning rag (not smooth), but would increase if obstacles were added upwind of the liquid source that would increase the turbulence in the air approaching the source. However, predicting the quantitative decrease in the value of the geometry coefficient is currently not feasible, even for the simple case of changing the freeboard. For the purposes of simplicity, we will assume that changing the source geometry will not change either the velocity of the air flowing over the source or the dilution of the plume (Q_{apparent} in Equation 9.2). Thus, the concentration in both the plume and room will change in direct proportion to the change in Geom. Coef. and in G, as summarized in Table 9.2.

2. The size of the wetted area from which the solvent evaporates (the A term in Equation 9.5) can sometimes be reduced, for instance by using the solvent more conservatively or by reducing the size of its container. (The latter may also change the geometry coefficient, but we will restrict our change here to only changing the source area.) If the liquid container has to be a given size, the wetted surface area available for evaporation can be reduced by floating beads, chips, or other material on the liquid surface, as suggested by ANSI Z9.1 and discussed in Section VII.4 in Chapter 4 of this book. The summary in Table 9.2 shows that the evaporation rate will change in direct proportion to the relative change in wetted surface area (new/old).

   The effect of changing A on airborne concentration is in the same direction as its effect on the G term but not as straightforward. The cross-sectional dimensions of the resulting vapor plume, as depicted by the ellipses in Figure 9.6, are not easily defined, but it is reasonable to assume that the plume's cross-sectional area will change in proportion to the width of the source of liquid solvent but not its length (i.e., the height of the plume is largely unaffected by the downwind length of the source). That being the case and as long as the source is approximately square-to-circular, the plume's cross-sectional area will change in direct proportion to the square-root of the source's wetted surface area, $\sqrt{A}$ or $A^{0.5}$.

   The effect of changes in source area on the concentration within the plume (C_{plume}) is directly proportional to its effect on G but will be inversely proportional to $Q_{\text{in plume}}$ per Equation 9.2. Since G increases linearly with the source area but $Q_{\text{in plume}}$ only increases with the square-root of the source area, C_{plume} will only increase in direct proportion to the square-root of the source area. The mathematical derivation of the source area's effect on C_{plume} in Table 9.2 is provided as a footnote for the curious.[b] On the other hand, the size of the source has no effect on the room air flow, its mixing, or $Q_{effective}$. Thus, C_{room} will increase in direct proportion to the size of the source. This is the first case where the resulting change in exposure depends upon where the person is located, either within the plume (exposure scenarios 1 or 2) or not (exposure scenario 3). However, at least reducing the source area reduces both the concentration within the plume and the concentration in the room outside the plume, just not equally.

3. Air velocity is comparatively the least important of the variables affecting evaporation because of its 0.5 power, and yet its effect on exposure depends even more upon the exposure scenario than did changing the source area. Equation 5.13 and Equation 9.5 clearly show that reducing the air velocity over the source will reduce the rate of evaporation G in proportion to the square-root of V_{new}/V_{old}. However, Equation 9.2 implies that reducing the air velocity over the source will decrease the $Q_{in\ plume}$ linearly with V_{new}/V_{old}. Thus, and perhaps surprisingly, the concentration of the vapor within the plume will increase when the air flow rate over the source decreases. Think of it as if each parcel of air has more time over the source to approach its equilibrium P_{vapor} concentration. The derivation of velocity's effect on C_{plume} in Table 9.2 is also provided as a footnote for the curious.[c]

   This inverse change in plume concentration with change in V over the source should only apply to someone in the downwind plume of scenario 2. Vapor exposures to anyone in the upwind near-field plume of scenario 1 are as yet not well-defined, but it appears from data such as in Figure 5.12 and Figure 5.13 that exposures to someone in the upwind near-field (such as in front of a laboratory hood) will decrease with increasing velocity up to approximately 100 fpm because the air carries away most of the plume. However, at a velocity somewhere above 100 fpm the wake turbulence can start to increase the exposure of someone working upwind of the source.

   Predicting the effect on exposures in the room outside of the plume in scenario 3 is even more ambiguous. The new concentration in the room depends upon what effect (if any) the velocity of the air that passes over the solvent has upon room mixing [$K_{room\ mixing}$]. The maximum effect that the air flow that generates the plume might have on room mixing is unknown but $Q_{efffective,new}/Q_{effective,old}$ might be affected by as much as $\sqrt{V_{new}/V_{old}}$. If V_{new} were decreased in an attempt to decrease G, then this decrease in room mixing and increase in $K_{room\ mixing}$ might be sufficient to balance exactly the decrease in the evaporation rate which would result in no change to C_{room}

---

[b] The new C_{plume} that results from changing A can be found by using Equation 9.2 after inserting the value for $G_{new} = A_{new}/A_{old}$ derived from Equation 9.5 and using $Q_{plume} = V_{plume} \times A_{plume}$ where in this case $A_{plume,new} = A_{plume,old} \times \sqrt{A_{new}/A_{old}}$.

$$C_{plume,\ new} = C_{plume,\ old} \times \frac{G_{old} \times (A_{new}/A_{old})/Q_{in\ plume,\ old} \times \sqrt{\dfrac{A_{new}}{A_{old}}}}{G_{old}/Q_{in\ plume,\ old}} = C_{old} \times \sqrt{\frac{A_{new}}{A_{old}}} \qquad \text{(using Equation 9.2)}$$

[c] The new C_{plume} that results from changing V can again be found by using Equation 9.2 after inserting $G_{new} = G_{old} \times (V_{new}/V_{old})^{0.5}$ derived from Equation 9.5 and using $Q_{plume} = V_{plume} \times A_{plume}$ where in this case $V_{plume,\ new} = V_{plume,\ old}$:

$$C_{plume,\ new} = \frac{G_{old} \times (V_{new}/V_{old})^{0.5}/Q_{in\ plume,\ old} \times (V_{new}/V_{old})}{G_{old}/Q_{in\ plume,\ old}} = C_{old} \times \left(\frac{V_{new}}{V_{old}}\right)^{-0.5} \qquad \text{(using Equation 9.2)}$$

(a coefficient in Table 9.2 of one). If, on the other hand, the air flow that generates the plume is small in volume or for some other reason does not affect room mixing, then $Q_{effective}$ is a constant (unchanged) and C_{room} would decrease in proportion to G. This latter change in C_{room} would be in exactly the opposite direction from the increase in C_{plume} due to a lower velocity over the source. Thus, in general, the effect of changing the evaporating air's speed on exposure to someone *outside* of the plume in a ventilated room might be anywhere between none (no change) to a change in direct proportion to the square root of the change in air velocity over the source. This ambiguity is depicted by the gray area in Figure 9.8.[d]

4. A liquid's evaporation rate will change in proportion to its vapor pressure. This group of controls excludes changing the volatile chemical (that topic was discussed in Chapter 8), but decreasing the liquid temperature will reduce a given chemical's vapor pressure as typified by the Antoine equation given in Equation 2.10.

$$\log P_{vapor} = A - B/(C+T) \quad \text{or} \quad P_{vapor} = 10^{(A-B/(C+T))} \quad \text{(repeat of Equation 2.10)}$$

One could apply Equation 2.10 directly to Equation 5.13 or Equation 9.5 to create the exact coefficients given in Table 9.2; however, algebraically that appears messy. A simpler approximation is to apply the rule-of-thumb that (on average) organic solvent vapor pressures double for every 12°C increase in the liquid's temperature to create the approximate coefficients in Table 9.2. The effect of changing vapor pressure applies equally to both the plume and room concentrations (similar to changing the geometry coefficient). The result is that a solvent temperature change (either decreasing by plan or increasing by accident) of more than about 25°C or 50°F (two doublings) can have a dramatic effect upon exposure concentrations.

The coefficients in Table 9.2 each represent the effect of a change in each source variable on either the evaporation rate per Equation 9.5 or on the plume and room concentrations per Equation 9.2. The ratio of new to old concentrations are plotted on the Y axis in Figure 9.7 and Figure 9.8 for a range of relative changes in each source variable (the ratio of the new to old values of the source variable[e]) plotted on the X axis. To reduce exposures, C_{new}/C_{old} must be less than one. Thus, only the changes in the bottom half of each graph are beneficial. With one major exception, the effects on plume concentration corresponds to the effects on evaporation rate. The one major exception visible in Figure 9.7 is the air velocity (it slopes in the opposite direction from every other source variable). The appropriate way to reduce the exposure of someone within a downwind plume would be to increase the air velocity over the source rather

[d] The effect on C_{room} of changing V can depend upon whether the room's $Q_{effective} = Q \times \sqrt{V_{new}/V_{old}}$ or is constant.

$$C_{room,new} = C_{old} \times \frac{G_{old} \times (V_{new}/V_{old})^{0.5} / Q_{effective,old} \times (V_{new}/V_{old})^{0.5}}{G_{old}/Q_{effective,old}} = C_{old}$$

if V affects room mixing, and

$$C_{room,new} = C_{old} \times \frac{G_{old} \times (V_{new}/V_{old})^{0.5} / Q_{effective,old}}{G_{old}/Q_{effective,old}} = C_{old} \times \left(\frac{V_{new}}{V_{old}}\right)^{0.5}$$

if V does not affect room mixing.

[e] In order to depict the effect of changing the solvent temperature in a manner consistent with the other variables, each temperature doubling (~12°C) is plotted at the number of temperature doublings plus one on the X axis. Thus, no temperature doubling is plotted at 1 resulting in no change in C; one doubling is plotted at $X = 2$ which doubles C, etc.

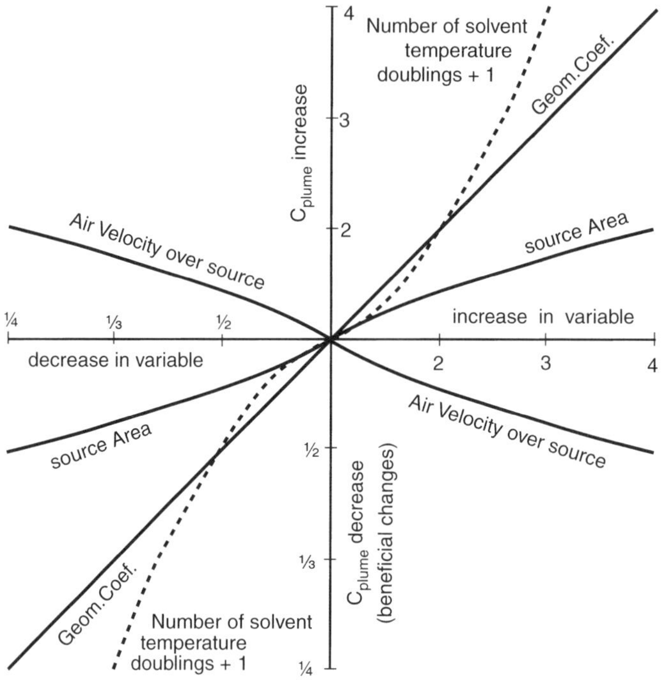

FIGURE 9.7 Graphical depiction of the change in the plume concentration in response to a relative change in a source variable (see footnote e for the change in solvent T).

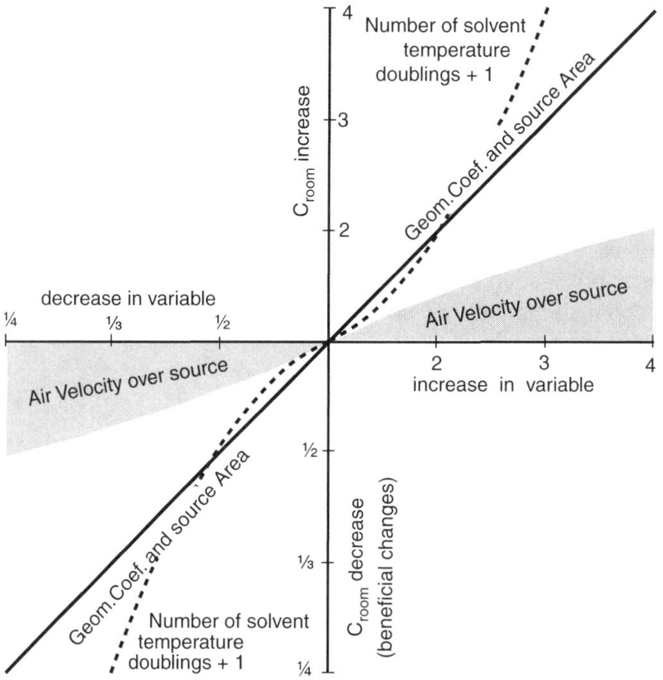

FIGURE 9.8 Graphical depiction of the change in the room concentration in response to a relative change in a source variable (see footnote e for the change in solvent T).

than to decrease it. The effects of reducing the other source variables vary in magnitude, but at least they are all moving in the same direction. Without exception, the effects on room concentration outside of the plume (Figure 9.8) all correspond to the change in the rate of evaporation. (Although the effect of changing the air velocity is slightly ambiguous, at worst, decreasing the velocity over the source may leave the room concentrations unchanged.) The magnitude of the effect of changing these source variables will continue to increase if the variable can be changed more than the range plotted, but the maximum change to any of these source variables in most real settings would probably be limited to about the range that was plotted; in other words, one probably could not expect to reduce one of these source variables by more than about one half to one quarter.

EXAMPLE 9.1

Suppose that someone moved a portable cooling fan near where two to three employees have to hand clean the final product with toluene. Suppose that prior measurements found exposures were about one half of its 50 ppm TLV when air currents through this area were about 100 fpm (1 mph). Now anemometer readings show air speeds are about 300 fpm. Should you bother with more air sampling?

From this description one can surmise some of the employees might be in the plume and other employees will be outside of the plume. We will first solve for the new concentration inside the plume then outside of it in the rest of the room. Both solutions use Equation 9.2 with the coefficient taken from Table 9.2. We will add a prediction for the evaporation rate just for our information.

$$C_{\text{plume, new}} = C_{\text{plume, old}} \times \left(\frac{V_{\text{new}}}{V_{\text{old}}}\right)^{-0.5} = 25\text{ ppm} \times \left(\frac{300\text{ fpm}}{100\text{ fpm}}\right)^{-0.5} = 14\text{ ppm} \qquad \text{(using Equation 9.2)}$$

If the cooling fan increases mixing within the room by at least as much as it increases evaporation, then the effect on the room concentration is one and $C_{\text{room, new}} = C_{\text{room, old}} = 25$ ppm. If the fan had no effect on mixing within the room, then the effect on those outside the plume would equal the effect on G.

$$G_{\text{new}} = G_{\text{old}} \times \left(\frac{V_{\text{new}}}{V_{\text{old}}}\right)^{0.5} = G_{\text{old}} \times \left(\frac{300}{100}\right)^{0.5} = 1.73 \times G_{\text{old}} \qquad \text{(using Equation 9.5)}$$

and

$$C_{\text{room, new}} = C_{\text{room, old}} \times \left(\frac{V_{\text{new}}}{V_{\text{old}}}\right)^{+0.5} = 25\text{ ppm} \times \left(\frac{300\text{ fpm}}{100\text{ fpm}}\right)^{+0.5} = 43\text{ ppm} \qquad \text{(using Equation 9.2)}$$

Thus, despite the fact that the department will have to buy 1.7 times as much toluene (as shown by G_{new}), its vapors will be less within the plume than they were and will be somewhere between unchanged and 43 ppm outside the plume but almost certainly less than 50 ppm. Thus it is possible to save sampling money and time.

V. WORK PRACTICES AS AN EXPOSURE CONTROL

Employees need to be taught good work practices, safe ways to do their job. One of the results of employee hazard communication should be for them to respect toxic chemicals. Employees should be taught how to recognize and avoid the highest concentrations of the chemicals with which they work. They should be taught (and the facilities should be provided) to minimize the time that dermal exposures are allowed to remain on employee's skin and clothing. Good housekeeping should be practiced to minimize the opportunity for employees to be in contact with chemicals that are just lying around.

1. Training employees to avoid working within the very near-field (less than 1 ft) or downwind of the source within a free-field plume (where it can be seen, anticipated, or detected) should be an intrinsic part of hazard communication. Such avoidance can be achieved more effectively by changing one's orientation to the plume than merely by increasing one's distance downwind in the plume. Remember that standing in the lateral position (to the side of a plume) results in even more dilution than standing in the upwind position, i.e., someone in position "*C*" or even "*B*" in Figure 5.6 will receive less exposure than someone in position "*A*."

   Avoiding work in the highest concentration can also be achieved by controlling the employees' *time* of exposure. Employees and their supervisors can both be taught to look for ways to conduct intermittent operations that generate hazardous airborne concentrations when no or few employees are present. For example, if an activity that releases a volatile chemical needs to be conducted only occasionally, conduct that activity then leave that room for a time, or conduct the activity at the end of the work day or on a skeleton shift. These scheduling changes can also be categorized as an administrative control as will be discussed in Section I in Chapter 21.
2. Chemical deposits and residues on the skin do not always look like dirt that needs to be washed off. Increasing the frequency and thoroughness of washing after dermal contact with most chemicals requires a combination of training and *convenient* hand washing facilities. Either element by itself can be ineffective at controlling a dermal or nonvolatile hazard, and the combination of no training and inconvenient washing facilities has almost no chance of control. Other similar examples include requiring showers after work and providing clothing that remains and is laundered at work.

   Eating and smoking at work can create oral exposures. If employees' hands are exposed to chemicals that may remain on their skin, they should not be allowed to eat or smoke without first washing their hands. Providing hand washing facilities at each worksite can be more expensive that prohibiting eating or smoking at the actual worksite. The choice depends on the number of work breaks, the spatial distribution of employees, and the number and location of eating facilities. Both options are really an administrative control, and all administrative controls require supervision (see Section I in Chapter 21). Employees that use even modestly toxic nonvolatile chemicals but have no washing facilities or separate eating facilities at work are at high risk from oral doses.
3. Good housekeeping practices can avoid or at least reduce what are often called secondary routes of exposure. A secondary route means exposure to a chemical by a route not directly the result of the work process. Most secondary routes are the result of either evaporation or aerosolization of surface accumulations. Accumulations on surfaces from leaks, spills, even routine activities or "fugitive" emissions (multiple sources that are not all known) can become the major source of employee exposure via later inhalation, dermal contact, or oral ingestion. Airborne exposures should be detectable by normal air sampling. However, skin contact with such accumulations followed by skin absorption and/or oral ingestion are more insidious routes of exposure that often go undetected.

   Good housekeeping practices can reduce the sources of vapors from volatile chemicals, but they can be even more important in controlling exposures to low and nonvolatile chemicals. If the chemical is toxic and of low volatility, such accumulations will remain a workplace hazard for a long time unless actively cleaned up. Lead is an obvious health hazard that will not go away on its own. Imagine the extent of housekeeping necessary for a manufacturer to keep airborne beryllium below its 0.002 mg/m^3 PEL while machining beryllium metal parts. Good housekeeping can also be based on a safety criterion such as OSHA's grain handling standard 1910.272, which requires good housekeeping to limit the accumulation of surface dust to minimize the potential for the resuspension of such accumulations to create an ignitable or "explosive" airborne dust. The value of good housekeeping must be instilled in both employees and management through education; however, its effectiveness relies on management supervision.

✓ In short, employee training should stress three good work practices. Avoid being exposed to high concentrations. Clean dermal exposures from skin frequently. Do not allow the chemical to lie around to become a means of exposure again and again.

VI. AUTOMATION AS AN EXPOSURE CONTROL

Automation is the use of some form of robotics. Prior to about 1980, the automation of hazardous processes was rarely economically feasible. The advent of the microchip processor, its decreasing cost, and the widespread acceptance of personal computers have caused a virtual revolution in automation. Automation is being instituted on both a small and large scale without IH impetus. Terms such as CAD and CAM (computer-aided design and computer-aided manufacturing) imply a potentially seamless and paperless production process from design to manufacturing with very little human intervention.

1. While some level of automation is generally necessary if separation or isolation controls are to be implemented (as will be described below), most of the advantages of automation are associated with issues other than IH.
   a. From an IH perspective, the most important advantage to automation is to decrease routine production worker exposure to chemicals, physical agents, and ergonomic hazards. However, automation does not alleviate maintenance operations, and may in fact delay them to off-hour shifts (which imposes a not-to-be-forgotten burden on an industrial hygienist's schedule) or increase the frequency of unscheduled maintenance (unscheduled implies a response to breakdowns, malfunctions, or troubleshooting that must be fixed "now").
   b. Automation has several advantages from production management's perspective. Successful automation increases the product's accuracy and precision, can decrease the frequency at which products fail quality checks, and can facilitate operating machines or processes more hours per day (although not all plants want to operate 24 h per day).
   c. Automation should decrease direct labor costs (labor agreements and the availability of appropriately trained employees permitting); however, automation will generate the need to conduct some unique initial and recurring safety training as described below.
2. Automation has its limitations. Automation is limited in the tasks that can be performed (e.g., robots have limited dexterity, tactile or visual feedback, and some are difficult to reprogram in response to production changes). Also, automation increases direct capital costs and entails some often unrecognized indirect labor costs for programming and training. Automation can (and usually does) create new safety hazards that require targeted safety training especially of production programmers and maintenance staff, and at least safety awareness training of all other staff who might have peripheral contact or be too curious about automated equipment for their own good. While automation may be implemented solely for its productivity gains, the industrial hygienist's interests in health and safety can help speed that decision and smooth the implementation process.

Many interesting examples of automation with a clear health benefit can be found. For instance, the Flash Jet stripping head described in Figure 8.1 is positioned automatically by computer control. Another that caught my interest is a robotic submarine by Solex Robotics that inspects the floor of above-ground storage tanks that still contain their normal petroleum or chemical product. It saves not only the labor and health risks of manual inspections but also the cost of transferring the chemicals, the expense of standby tankage, the emission of tons of purge gas, and the flaring of headspace vapors. Of course, it is certified for use in Class I Division 1 environments as covered in Section VI.4 in Chapter 4 (see www.solexrobotics.com/Solex7.html).

VII. SEPARATION AS AN EXPOSURE CONTROL

Increasing the distance between the source and the employee can have a small effect on exposure, especially if the distance is in the direction of the plume.

1. As described in Chapter 5 (Section II), the dispersion of chemical plumes is much slower in the downwind axis than in the lateral axis. Figure 9.9 (a repeat of Figure 5.2) tries to depict a series of cross-sections of the chemical concentration as the plume disperses downwind. The vertical dimension of each cross-section represents the concentration. The shape of each concentration profile in the crosswind or lateral direction exhibits a Gaussian pattern, similar to a normal distribution curve. The peak concentrations decrease almost linearly with distance downwind, while the concentration decreases more rapidly with lateral distance. The angle at which the plume spreads laterally (measured at any multiple of standard deviations similar to a normal curve) depends upon the local air turbulence.[3]

   Since vapors are invisible, the directional orientation between the employees and the plume is not always discernable, but a plume will always exist unless it is exhausted locally. Moreover, the vapor or aerosol within such a plume indoors will accumulate within the work space (as in Figure 5.8). For both of these reasons, an increase in distance between the source and receiver is rarely a reliable control.
2. Separation by distance can be used as an effective control in three exceptional situations:
   a. Separation can be effective if the initial distance from a source is very small. It is hard to avoid the plume if working too close to a source. If someone is working really close to a source (as if they were nearsighted), they should be counseled, retrained, or provided the tools to work at least at a comfortable arm's length from the source.
   b. If the direction of the plume can be seen or sensed, move the source or employee laterally to avoid the plume. The Gaussian plume concentration profile causes even a small lateral move to have a large effect on exposure. For instance, person C in

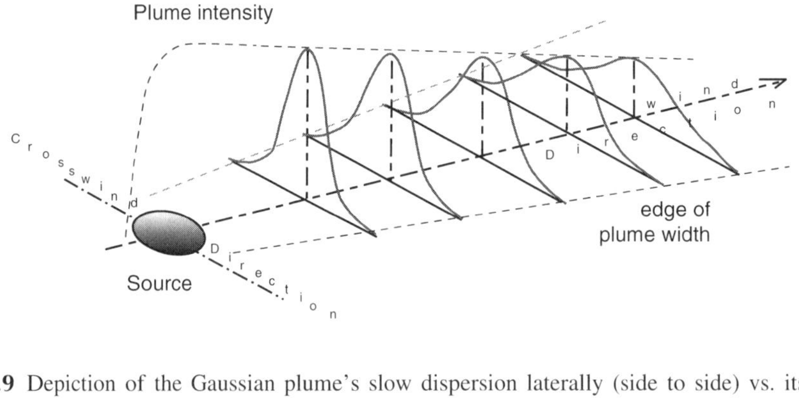

FIGURE 9.9 Depiction of the Gaussian plume's slow dispersion laterally (side to side) vs. its movement longitudinally (in the direction of the wind). The area under each curve will be equal owing to conservation of the mass of the contaminant.

Figure 5.6 will have much less exposure than person D who is standing in the downwind plume of the source. Exposures to person D could be reduced either if they could work nearer to person C, if the source could be moved down the page in the diagram, or if the air movement could be redirected to flow down the page (or up if person C were not already there).

c. Separation can be an effective control if the distance between the source and receiver can become very great. Viewed another way, either completely relocate an identifiable point source to where there are no people, or move the people to where there are no sources. Such a large increase in distance might better be called relocation. Relocation is a relatively easy method of controlling emissions from support machinery or processes that do not require frequent operator interaction. To relocate a machine or process that currently requires operator attention would generally call for automation or at least remote control.

VIII. ISOLATION AS AN EXPOSURE CONTROL

To control via isolation means to create or place a physical barrier between the source and employees.[f] An effective barrier against airborne chemicals generally (but not always) means an enclosure.

1. An enclosure of an evaporating source must generally be ventilated to prevent the buildup of vapors to acutely toxic or perhaps flammable concentrations. An unventilated enclosure would constitute a confined space if it is big enough to be occupied. Some degree of control (reduction in exposure) can also be achieved by a partial enclosure such as a wall, partition, or shield.

   An enclosure that must be opened too frequently is prone to misuse. Closed systems do not always stay closed due either to production expediencies or to poor maintenance. Someone must be assigned to supervise the integrity of removable enclosure access panels.

   Pumps, flanges, and valves (see Figure 8.1 and Figure 8.2) are good examples of point sources that do not require frequent operator access. Assure that seals on enclosed pumps still receive proper maintenance.

   Another example of isolation is the use of either septa and syringes or closed loop valves to collect quality control samples rather than an open valve and sample bottles. Ventilation is again an adjunct control to such enclosures (for an example, see the *Manual of Industrial Ventilation*'s VS-15-30).[4]
2. Isolation can be achieved by enclosing the operator or employee in a ventilated cab or control room. Any occupied enclosure must be ventilated. Contaminants must be cleaned from the air surrounding such an enclosure before it may be safely drawn into the enclosure. Aerosols can be cleaned from the air much more easily than gases and vapors. In either case, a means should be provided to detect saturation or breakthrough of the air cleaner, to respond to a major spill or increase in the outside contaminant concentration, and to deal with either a scheduled or unanticipated shutdown of the external air supply system.

[f] The complete relocation mentioned above is sometimes categorized as isolation although the possible presence of a physical barrier between the relocated people and source is incidental to the extreme increase in distance commensurate with relocation. However, the label put on the move is not as important as the acceptability and effectiveness of the control, so who is to argue with success by whatever name it is given.

IX. FUNDAMENTAL NONVENTILATION CONTROL EQUATIONS

Most of the nonventilation controls rely on the physical principles of gases, vapors, and aerosols that were explained in Chapter 2 and Chapter 3. These principles were presented with the intention that they would be more of a review or recasting of principles from other science courses than as an introduction of totally new material. The four fundamental equations from these chapters that must be remembered are grouped below for the reader's review and future convenience. The student should know how to use more equations from these chapters, but these were the only ones that this author feels are memorable. Three more memorable equations related to the four universal airborne chemical exposure scenarios from Chapter 5 are also listed below for the reader's review and future convenience. Also listed are another two memorable equations from Chapter 6 and three from Chapter 8 (Chapter 8 used Equation 5.13 heavily). A couple more potentially useful but not quite so memorable equations are included. These equations comprise about half of all the memorable equations in this book. A similar summary of memorable ventilation equations is compiled at the end of Chapter 11 (Section VI, page 273).

$$\boxed{P_i = Y_i \times P\ [\text{ambient}]} \quad \text{or} \quad \boxed{Y_i = \frac{P_i\ (\text{of the } i\text{th component})}{P\ [\text{ambient}]}} \qquad \text{(Equation 2.13a and b)}$$

$$\text{at NTP} \quad \boxed{\text{ppm}_i = C_i\ (\text{mg/m}^3)\frac{24.45}{\text{MW}_i}} \quad \text{and} \quad \boxed{C_i\ (\text{mg/m}^3) = \text{ppm}_i \frac{\text{MW}_i}{24.45}} \qquad \text{(Equation 2.15a and b)}$$

$$\text{at NTP} \quad \boxed{\rho_{\text{of a known gas at NTP}} = \frac{\text{mass}}{\text{volume}} = \frac{\text{MW}}{24.45}} \qquad \text{(Equation 2.23b)}$$

$$\boxed{d_{\text{aerodynamic}} = d\sqrt{\rho}} \quad \text{for } \rho \text{ in g/cc or g/mL} \qquad \text{(Equation 3.15b)}$$

$$\boxed{\text{ppm at the source} = \frac{P_{\text{vapor}} \times 10^6}{P} = \frac{[P_{\text{vapor}}\ \text{mmHg}] \times 10^6}{760} \text{ at NTP}} \qquad \text{(Equation 5.1)}$$

$$\boxed{\text{For complete evaporation, } C_i \text{ in } \frac{\text{mg}}{\text{m}^3} = \frac{\text{mass}_i}{\text{air Vol}}} \qquad \text{(Equation 5.6)}$$

$$\boxed{\text{Evaporation rate } G\ (\text{moles/min}) = (\text{Geom.Coef.}) \times (\text{Area}) \times V^{0.5} \times P_{\text{vapor}}} \qquad \text{(Equation 5.13)}$$

$$\boxed{P_{\text{vapor},i} = X_i \times P_{\text{vapor}}} \quad \text{Raoult's law} \qquad \text{(Equation 6.7a)}$$

$$P_{\text{vapor},i} = \gamma_i X_i P_{\text{vapor}} \quad \text{an empirically adjusted Raoult's law} \qquad \text{(Equation 6.9a)}$$

$$\text{ppm}_{\text{BZ,new}} = \text{ppm}_{\text{BZ,old}} \frac{P_{\text{vapor,new}}}{P_{\text{vapor,old}}} \quad \text{a new chemical in an unchanged setting} \qquad \text{(Equation 8.3)}$$

$$\text{VHR} = \frac{\text{Concentration at the source}}{\text{Acceptable BZ Concentration}} = \frac{P_{vapor} \text{ or } P_{vapor,i} \text{ (mmHg)} \times 10^6}{\text{TLV or PEL (ppm)} \times P \text{ (mmHg)}}$$

(Equation 8.4c and 8.5b)

$$[\text{ppm}_{BZ}/\text{TLV}]_{new} = \text{Compliance Ratio}_{old} \times \frac{\text{VHR}_{new}}{\text{VHR}_{old}}$$

(Equation 8.9)

$$\text{acceptable VHR}_{new} \leq \frac{\text{VHR}_{old}}{[\text{ppm}_{BZ}/\text{TLV}]_{old}} = \frac{\text{VHR}_{old}}{\text{Compliance Ratio}_{old}}$$

(Equation 8.10)

PRACTICE PROBLEMS

Suppose that air samples collected during the winter showed that assembly workers were exposed to 75 ppm Stoddard solvent in the vicinity of the plume and 25 ppm when working nearby but not in the plume. The TLV of Stoddard solvent is 100 ppm. Describe the change in the exposures for each group of workers that you would expect following each of the changes in the setting or work practice *if each change were the only change*. Where feasible, be quantitative by using Equation 5.13, Equation 9.5, Equation 9.6, or Equation 9.7 or Table 9.2.

1. What might be the exposures if (to speed production) workers put the solvent in a shorter container of the same diameter and solvent-wetted surface area, resulting in a geometry coefficient twice as large?

In the plume: compared to 75 ppm	In the general room air: compared to 25 ppm

2. What might be the exposures if the surface area wetted by the solvent (and therefore available to evaporate) were to double due to seasonal increases in production?

In the plume: compared to 75 ppm	In the general room air: compared to 25 ppm

3. What might be the exposures if fans that are brought in to help cool workers also double the air velocity passing over the solvent but have no effect on room air mixing?

In the plume: compared to 75 ppm	In the general room air: compared to 25 ppm

4. What might be the exposures if the above doubling in air velocity were caused by fans that are big enough to increase room air mixing by the $\sqrt{2}$?

In the plume: compared to 75 ppm	In the general room air: compared to 25 ppm

5. If the liquid solvent temperature during the winter were 65°F, what might be the exposures in the summer when the liquid temperature is 86°F? (Assume the solvent has average Antoine coefficients.)

In the plume: compared to 75 ppm	In the general room air: compared to 25 ppm

6. For an equally proportionate change in each variable, changing which variable would have the least effect on worker exposure concentrations whether in the plume or not?

In the plume:	In the general room air:

Hints: The effects on the plume concentration are the same for questions 1 and 5 and questions 3 and 4. The effects on the room concentration are the same for questions 1, 2, and 5.

REFERENCES

1. Flynn, M. and Susi, P., Engineering controls for selected silica and dust exposures in the construction industry, *Appl. Occup. Environ. Hyg.*, 18(4), 268–277, 2003.
2. Echt, A., Seiber, K., Jones, E., Schill, D., Lefkowitz, D., Sugar, J., and Hoffner, K., Case studies: control of respirable dust and crystalline silica from breaking concrete with a jackhammer, *Appl. Occup. Environ. Hyg.*, 18(7), 491–495, 2003.
3. Barratt, R., *Atmospheric Dispersion Modelling: An Introduction to Practical Applications*, Earthscan Publications Ltd, London, 2001.
4. Committee on Industrial Ventilation, *ACGIH Industrial Ventilation: A Manual of Recommended Practice*, 24th ed., American Conference of Governmental Industrial Hygienists, Cincinnati, OH, 2001.

10 An Overview of Local Exhaust Ventilation

The learning goals of this chapter:

- ✍ Be able to differentiate a local exhaust ventilation system from a general or dilution ventilation system.
- ✍ Be able to differentiate a containment hood from a collection hood and a simple hood from a compound hood.
- ✍ Know where a control velocity, face velocity, and transport velocity occur and what they are supposed to do.
- ✍ Be familiar with the terms flange, baffle, plenum, and blast gate.
- ✍ Know what are the first three steps in designing an LEV system.

The following Local Exhaust Ventilation (LEV) design topics are only introduced in this chapter and are to be repeated in more testable forms within later chapters. By the end of this chapter the student should:

- ✍ Know the role of air velocity in LEV control and the role of the IH in choosing that velocity (Chapter 13).
- ☐ Know that resistance to flow caused by friction and turbulence is proportional to V^2 and Q^2 (Chapter 14).
- ✍ Know that the fan is selected on the basis of an air flow rate (Q) and FanSP (to be used in Chapter 16).

I. LOCAL EXHAUST VENTILATION DEFINITIONS

1. Local exhaust ventilation systems differ from general ventilation systems. Virtually all occupied commercial and institutional buildings will have a general ventilation system that is designed to bring in and condition outside air for occupant comfort. In contrast, local exhaust ventilation is designed for the purpose of removing an airborne chemical contaminant near its source or point of origin; it is less common than general ventilation, but much more efficient at controlling employee exposures to airborne chemicals.
   a. *General ventilation* is often called dilution ventilation when it is used to disperse a plume of airborne contaminant. Dilution ventilation is simply a particular use of a room's or building's general ventilation system. When chemical emission rates and general ventilation rates are more or less constant, the contaminants generated within a given time are actually diluted into the volume of fresh air entering the room within that time (not into the volume of air in the room at any one time). General ventilation can rely on either natural or mechanical processes to create air flow.
      1. Natural ventilation utilizes wind or thermally induced convection to move the air. Section 1202.4 of the *International Building Code* states that for natural ventilation to be acceptable, the wall area that is openable to the outdoors (windows, doors,

louvers, etc.) must comprise at least 4% of the floor area being ventilated.[1] Most homes rely on natural ventilation.

2. Mechanical ventilation utilizes fans (with or without ducts) to move the air both into and out from a building. Commercial and institutional buildings usually have integrated heating, ventilating, and air conditioning (HVAC) systems.

Although a general ventilation system can cause a particular room's pressure to be either slightly above or below the pressure in neighboring rooms, HVAC systems always blow air into the building as a whole. General ventilation will be discussed in Chapters 19 and 20.

b. *Local exhaust ventilation* (LEV) is a system that extracts contaminated air at or near its source or origin using at least one of each of the following three components, depicted in Figure 10.1:
   1. A hood can be any planned entrance into the exhaust system. Even the plain, open end of an exhaust duct can be considered a hood (not usually a very good one, but it is inexpensive and can (on occasion) suffice).
   2. Ducting comprises the continuous tubes that convey the contaminated air from the hood to the fan and (generally) out of the building.
   3. A fan is a machine that sucks air into its inlet side and blows air out from its outlet side to move air (and the contaminant in this case) through the hood and ducting at some consistent flow rate.

   An LEV system may also contain an air cleaner and an exhaust stack (more ducting downstream from the fan). Figure 10.1 also lists the chapters of this book relevant to sources and pathways of airborne chemical hazards and to each of the above components.

As a very simplified comparison between dilution and exhaust ventilation, dilution ventilation generally blows air into a building while exhaust ventilation always sucks air out (leading some students to apply this phrase to exhaust ventilation in general). Local exhaust ventilation is almost always preferred over dilution ventilation for controlling airborne chemical hazards because of the following reasons:

- LEV protects more people than dilution ventilation by removing the contaminant before it has a chance to spread. Chapter 21 expresses this idea as reducing the total dose and collective risk.
- LEV is more efficient at controlling exposures than dilution ventilation. In other words, it removes less air than would be required of dilution ventilation to achieve

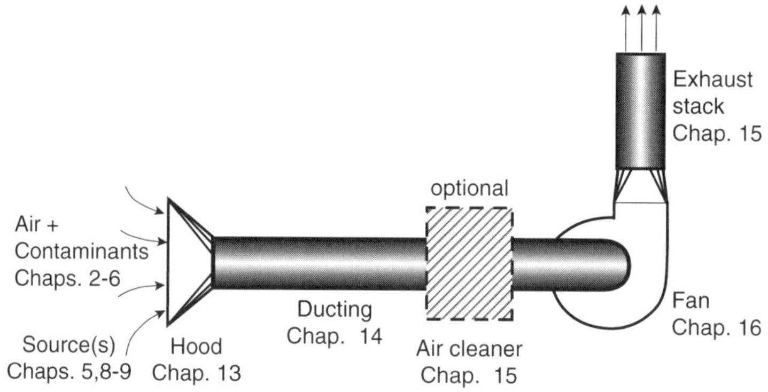

FIGURE 10.1 The components of an LEV system.

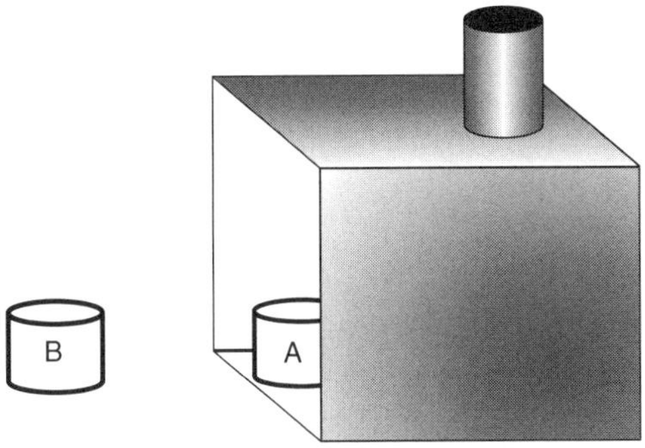

FIGURE 10.2 Depiction of operating as either a containment or a collection hood. Contaminated air exhausts near the rear of the top of this hood.

the same reduction in exposure. Therefore, LEV requires less make-up air which costs less to condition (heat, filter, and often cool).

- The technology to recover the same amount of contaminant in less air is simpler (less expensive) from the initially more concentrated emissions in a local exhaust ventilation system than in the more dilute form intrinsic to emissions diluted by general ventilation.

Any orifice through which air enters into an exhaust system can be considered to be a ventilation "hood". A hood can look like a box with slots or a hole on one side, or it may be the plain, open end of a duct. The following three subsections describe three ways that hoods have been categorized, subdivided, or characterized. Each of these ways will be used herein from time to time.

2. One way that hoods are commonly categorized is by the spatial arrangement between the hood's face and the source of the contaminant being controlled. The source of the contaminant is inside of a containment hood (position A in Figure 10.2), while the source of the contaminant is outside of a collection hood (position B in Figure 10.2). This differentiation is normally based on the hood's design and its intended function rather than how it is actually used. The physical design of some hoods (like those in Figure 10.3, Figure 10.4 and Figure 10.5) is such that they can only function as a collection hood. A hood such as that shown in Figure 10.2 is normally designed as a containment hood, despite the fact that a given user might actually work (inappropriately) with a source emitting an airborne contaminant outside the hood's face.

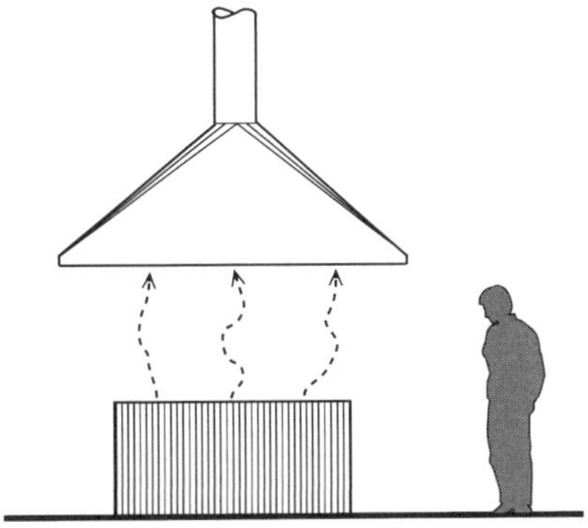

FIGURE 10.3 A canopy hood low over a hot source.

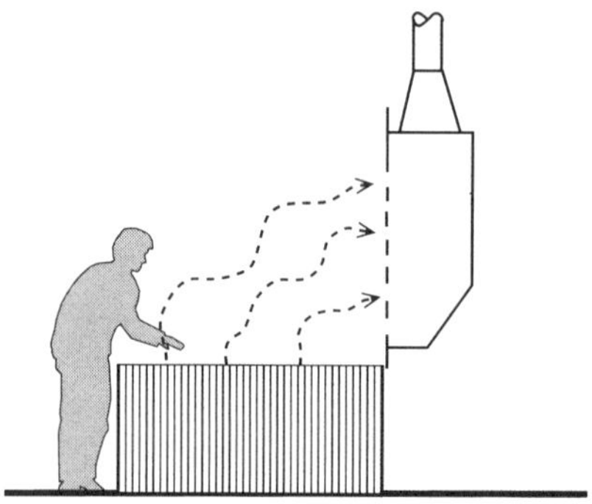

FIGURE 10.4 A side-draft hood next to a hot source.

a. A containment hood is designed so that its walls will virtually surround the source of the contaminant (ideally the source would be further inside location A in Figure 10.2). Any contaminant released or generated inside of a containment hood only needs to be kept from getting out. Examples of containment hoods in Chapter 13 include Figure 13.8, Figure 13.14, Figure 13.15 and Figure 13.22. Examples from the *Manual of Industrial Ventilation* include VS-20-01 (foundry shakeout), VS-35-01 (a laboratory hood), and VS-45-05 (lathe hood, if the cover is closed). Containment hoods are preferred because they are more efficient than collection hoods. That is, a containment hood will collect a larger fraction of a given contaminant than a collection hood exhausting the same amount of air. The *Manual of Industrial Ventilation* calls a containment hood an enclosing hood, although hoods are almost never a complete enclosure.[2]
b. A collection hood is designed to control a source that is outside its face (location B in Figure 10.2). Because the contaminant is generated outside of the hood, it must be drawn into and through the face of a collection hood. Examples of collection hoods in Chapter 13 include Figure 13.6, Figure 13.7, Figure 13.16, Figure 13.17, Figure 13.18, Figure 13.19, Figure 13.20, Figure 13.23, and Figure 13.24. Examples from the *Manual of Industrial Ventilation* include VS-55-10 (pouring station), VS-70-20 (solvent degreasing tank), and VS-75-06 (dip tank). A collection hood will always require a larger air flow rate than a containment hood because the area through which the air flows into the hood expands with roughly the square of the distance outside the hood's face. A predictive model of air velocity with distance (called the DallaValle equation) will be introduced in Chapter 13.IV. The *Manual of Industrial Ventilation* calls a collection hood an exterior hood, even though it is the source that is exterior to the hood (and not the other way around).[2]

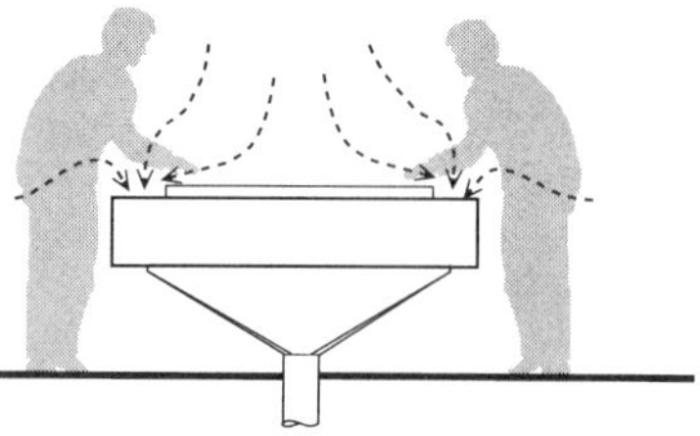

FIGURE 10.5 An example of a down draft hood (see also the VS diagram in Figure 13.17).

   c. A receiving hood can be defined either as a subcategory of collection hoods or as a separate third category. A receiving hood relies in large part on some intrinsic property of the source to move the contaminant into it. Examples of intrinsic properties include the source's inertial trajectory or thermal air currents or buoyancy (e.g., Figure 2.3, Figure 10.3, Figure 13.18, Figure 13.26 and Figure 13.28).
3. A second means of categorizing hoods is to subdivide collection hoods into three groups based on the geometrical relationship between the source and the hood's face. The terminology used describes the direction in which the exhaust air is trying to suck the plume: either up, to the side, or down.
   a. A canopy hood is like an inverted funnel mounted some distance directly (or nearly directly) over the top of a source, as depicted in Figure 10.3. A canopy is often used as a receiving hood for hot sources with buoyant plumes, is often seen in residential and commercial kitchens over a stove, but is incompatible with operations that require unobstructed overhead access.
   b. A side-draft hood (also called a lateral exhaust hood) means that a collection hood is laterally to one side of the source. A lateral exhaust hood draws air sideways across the source and, essentially, horizontally into the hood as depicted in Figure 10.4. The side-draft hood's face may be either one large hole or more commonly a panel with numerous holes or slots cut in it to form a slotted hood.
   c. A downdraft hood is a collection hood that is placed immediately below the source of the contaminant. The distance that a hood is from the source has a strong influence on its ability to collect the contaminant. Placing a hood below a source can result in its face being closer to the source than would be possible by placing the hood either above or to the side of the source. Figure 10.5 depicts workers sorting dusty parts (perhaps with a high quartz content) on a conveyor protected by a down draft hood.
4. A third way to categorize hoods that will be useful later is according to how many locations within the hood generate enough turbulence to cause a significant energy or pressure loss. The number of locations that generate losses is determined by the hood's design. Turbulence always causes an energy loss as the air enters into a duct (e.g., depicted at the top of both hoods in Figure 10.6). The hood on the right (Figure 10.6b) is purposefully designed with panels (called baffles) and spaces (called slots) across its face that create additional losses before the air even reaches the duct.
   a. Simple hoods have only one point or location where a significant energy loss occurs. A hood will always have a significant energy loss due to turbulence where the air enters the duct. Simple examples of "simple hoods" are the plain, flanged, and tapered entrances into a duct depicted in Figure 10.7, Figure 10.8, and Figure 10.9, respectively.[a] The hoods in Figure 10.2, Figure 10.3, and Figure 10.6a (left side) are also simple hoods. Because the velocity of air entering any hood with a large, unobstructed face is usually so much slower than the velocity of air leaving that hood and entering the duct, the air turbulence and energy loss at the face of any large, unobstructed hood is insignificant in comparison to the turbulence and energy loss as the air enters into the duct. Thus, any hood with such a large, unobstructed face is still a simple hood.

[a] Each of Figure 10.7, Figure 10.8 and Figure 10.9 is drawn first two-dimensionally as if it were cut in half (sectioned) to show the inside of the hood, then in a three-dimensional view (called oblique). The air flow pattern can be depicted more clearly in the former. The student should spend some time to be able to visualize the three dimensional structure being depicted by a two-dimensional view. The small twisty shape on the right end of each sectioned view indicates that the duct continues on somewhere in a real system. The arrow indicates the direction of the air flow.

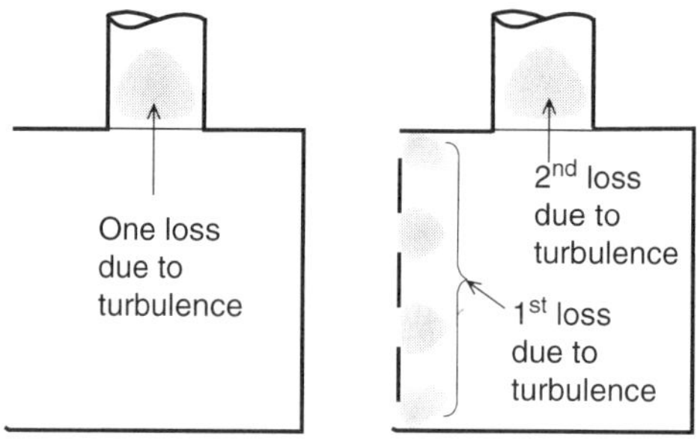

FIGURE 10.6 Depiction of a simple hood on the left (a) and a compound hood on the right (b).

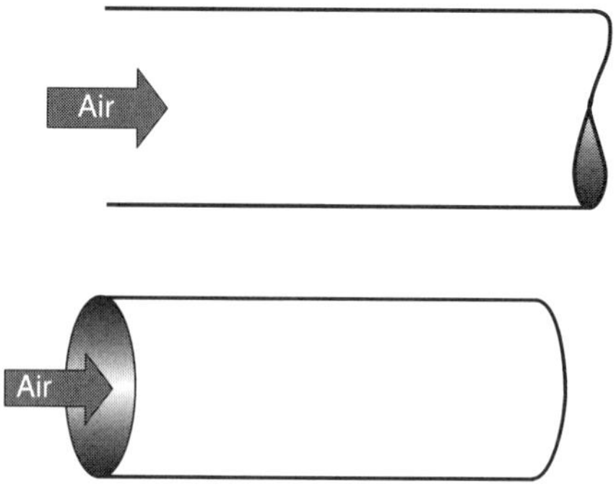

FIGURE 10.7 Plain opening in sectioned and oblique views.

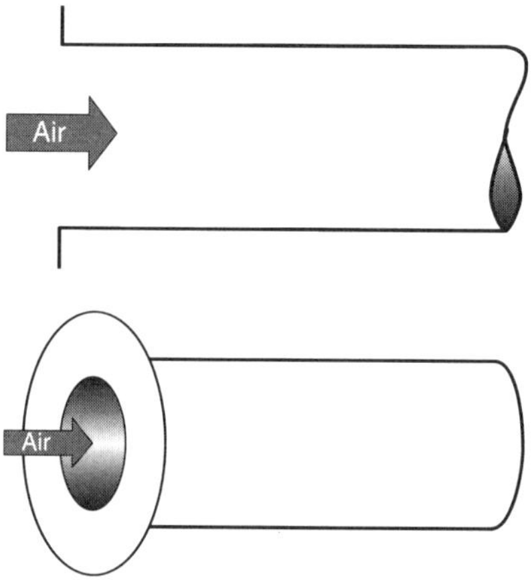

FIGURE 10.8 Flanged opening in sectioned and oblique views.

b. Compound hoods have two or more locations that generate a significant energy loss. Some examples of a compound hood are the slotted hoods in Figure 10.4, Figure 10.6b (right side), and Figure 13.11. Chapter 13 explains that slots are added to a collection hood to create a more uniform air velocity at least one slot width in front of the slots.

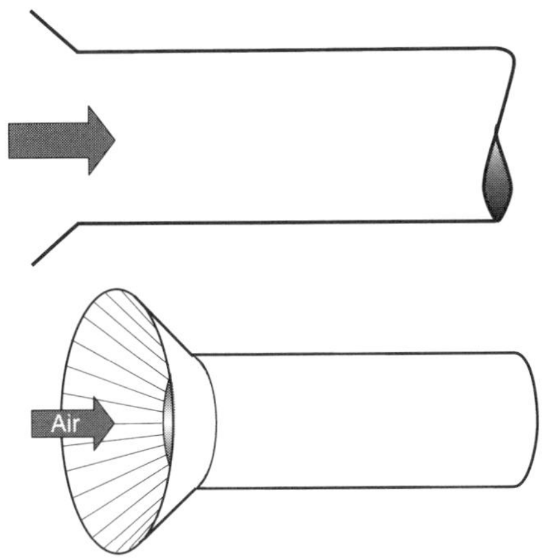

FIGURE 10.9 Tapered opening in sectioned and oblique views.

Another example of a compound hood is the two or three gaps in a baffle in the back of almost all laboratory hoods on campus (Figure 13.14 or Figure 13.15) that serve much the same purpose of creating a more uniform velocity across the face of a hood. Turbulence and energy losses occur both as the air flows through the narrow slots and again as it enters into the duct. These two separate losses (whether calculated or measured) are simply added together to determine the total energy the air loses as it enters through a compound hood.

5. Three or four specific air velocity terms are used in ventilation.
   a. The control velocity ($V_{control}$) is the air velocity at the critical location or zone in the plume's pathway that largely determines the collection efficiency of a particular hood. While the velocity is specified, the critical zone for a collection hood is often more vague. As $V_{control}$ is the major determinant of employee protection, it is one of the first system design parameters to be considered. During design, $V_{control}$ should be chosen (usually by the industrial hygienist) based primarily upon the inertia of the contaminants emitted by the source. The air speed must be sufficient to either start moving an external contaminant toward a collection hood or keep an internal contaminant from getting out of a containment hood. Thus, the term "control velocity" is the generic air velocity term applicable to either a collection or a containment hood.
   b. The face velocity (V_{face}) is the air velocity directly at the entry-plane of a hood, perpendicular to and flowing into its face. For a containment hood, the face of the hood is the zone of control. Thus, a containment hood's V_{face} is the same as (and therefore equal to) its $V_{control}$. For a collection hood, the V_{face} must be greater than the desired $V_{control}$ somewhere outside of the hood because the air velocity decreases rapidly with distance in front of any exhaust hood. (The DallaValle equation to be presented in Chapter 13 states that control velocity decreases with the square of the distance.)
   c. Transport velocity or duct velocity (V_{duct}) is the air velocity within a duct. The air velocity within any duct should be at least sufficient to keep the contaminant airborne, but not so high as to cause excessive friction or turbulence. The *Manual of Industrial Ventilation* recommends specific "minimum duct velocities" in each of its VS diagrams and for some specific operations in its Table 10.99.2. VS diagrams and further guidance to a proper duct velocity is discussed in Chapter 14

of this Book. The duct size should be chosen to provide at least that minimum duct velocity using the $A = Q/V$ version of what will become the basic $Q = V \times A$ equation, Equation 11.7.

d. The ACGIH *Manual* uses the term "capture velocity" in the same sense as the above definition for control velocity. However, believing that the word "capture" applies much more to a collection hood than it applies to a containment hood, the term "capture velocity" is simply avoided herein.

6. A baffle is a barrier, wall, or vane placed inside a hood or duct to redistribute or guide the flow or velocity of air within a hood or ventilation system.[b] Because "baffle" is a generic term, some types of baffles are given a separate name.
   a. One example of a baffle is strips of metal or other material placed across the face of a collection hood to form slots between them (as depicted in Figure 10.4 and Figure 10.6b). Such a hood is called a slotted hood (see definition 8 below and/or more detailed descriptions in Section IV of Chapter 13 on hood design).
   b. A baffle board is a larger plate-like structure placed in the back of a containment hood with only a small slot between the board and the hood's wall (such as the laboratory hood in Figure 13.14 or Figure 13.15). The purpose of baffle boards is, again, to create a more uniform air velocity across the face of the hood to improve control of exposures.
   c. Baffles called turning vanes are formed from thin strips of metal or rigid plastic bent to guide air around tight corners (as shown in Figure 10.10). Turning vanes can make the air flow around corners more smoothly with less resistance.
7. A flange is similar to a baffle except that it extends outside of and parallel to the face of a hood (e.g., Figure 10.4 and Figure 10.8) to block air flow into the hood from uncontaminated areas outside and behind the hood's face that do not need to be controlled. Although a flange blocks air flow from behind the hood, the real function of a flange is to increase the flow of air that comes from in front of a collection hood, thereby increasing the percentage of a contaminant out in front of the hood that is collected into the hood and exhausted. Thus, a flanged hood is said to be a more efficient hood than a plain hood because it will either decrease employee exposure for a given amount of air moved, or require less air to create an equal reduction in exposure compared to a similar hood without the flange. It is also true that removing less air from a room decreases the amount of make-up air that needs to be cleaned and heated or cooled to workroom conditions, thus reducing the costs of operating exhaust ventilation.[c]
8. A slotted hood is any collection hood across whose face several thin baffles are placed separated by narrow gaps to create the slots through which the air must flow (e.g., Figure 10.4 and Figure 10.6b). Such baffles are usually made of sheet metal placed 1 to 2 in. (2 to 5 cm) apart. The open slots are usually horizontal (for no good reason) and run the full width of the hood. Collectively, the slots create a more uniform face velocity than would be achieved without the slots. Chapter 13 will explain that a slotted hood acts and is designed simply as if it were one large opening spanning the width and height of all the slots. Because the small dimension across most slots

[b] The definition in the *Manual of Industrial Ventilation* pp. 3–7 that "A baffle is a surface which provides a barrier to unwanted air flow from the front or sides of the hood" is rather restrictive and at odds with its uses inside hoods and ducts.[2] The *Manual of Industrial Ventilation* contains many examples of baffles internal to a hood (the *Manual of Industrial Ventilation* Figure 3.10, Figure 3.13, or the paint booths in VS-75-01 or VS-75-02) and baffles that extend out perpendicular to the face a hood to block air flowing parallel to the hood's face, called "cross drafts" (in VS-90-01 and VS-99-05).

[c] Examples of hoods with and without flanges in the *Manual of Industrial Ventilation* include their Figure 3.6, Figure 3.7, Figure 3.11, Figure 3.16 and Figure 5.13.[2]

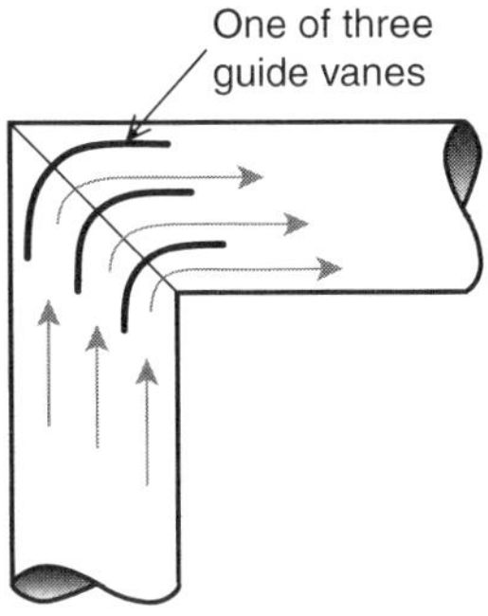

FIGURE 10.10 Turning vanes added to a mitered elbow reduce turbulence and energy losses.

prevents one's hands and/or arms from fitting through the slot opening, a slotted hood cannot be used on a containment hood where a person is expected to access the enclosed source manually.[d]

9. A plenum is simply a chamber that is all at one pressure. Such an equal pressure difference between the plenum and everywhere else can create equal air velocities. Examples of plenums are those created behind properly designed inlet-slots of a slotted hood (Figure 10.4, Figure 10.6b, and Figure 13.11) and behind the baffle plates in Figure 13.14.[e]
10. A blast gate can be any sliding or rotating plate that acts like a damper or valve to regulate the volume flow rate of air through a duct, as depicted in Figure 10.11. Blast gates are commonly used to balance the volume flow rate of air that enters into multiple hoods or portions of a branched duct system (one with multiple hoods and ducts that eventually ties into a single fan). Blast gates can also be used merely to shut off air flow through an unused branch of such a duct system. Blast gates are sometimes included in a ventilation system's design to allow the pressure drop or energy lost from the air passing through the gate to be changed or adjusted after a ventilation system is installed.
11. A bleed-in is an opening in a duct purposefully added to allow additional air flow into a fan, but without having to come through a functioning hood or much of the ducting, and possibly even without coming from the room or interior of the building. A bleed-in can be designed in various ways, including those listed below for a cleanout opening or with a blast gate (see Figure 10.12). A bleed-in can serve several functions depending, in part, on where it is placed. A bleed-in can be used just behind the hood to increase the transport velocity inside the duct without increasing the control velocity and/or to decrease the temperature or dew point of an exhaust stream. A bleed-in can be used just before a branch junction as one way to balance flow in a branch duct system. A bleed-in can be used just before the fan to maintain the fan's flow rate (Q) within its optimal performance range, again without increasing the control velocity, creating too high of a duct velocity (which can cause undesirable noise), or even having to pull conditioned air from inside the building (which can save on heating and/or cooling costs, to be explained in Chapter 17).

[d] The ACGIH *Manual* contains numerous examples of slotted hoods (i.e., the *Manual of Industrial Ventilation*'s Figure 3.1, Figure 3.11, Figure 3.13, Figure 3.18 and Figure 5.13, and VS-15-30, VS-35-40, VS-55-10, VS-75-06, VS-90-01, VS-99-05).[2]

[e] Examples of plenums in the *Manual of Industrial Ventilation* include the space behind a slotted hood's baffles (*Manual of Industrial Ventilation* Figure 3.1, Figure 3.13, or Figure 5.13), a trap or settling chamber (*Manual of Industrial Ventilation* Figure 5.13), and a chamber where multiple ducts meet (*Manual of Industrial Ventilation* Figure 5.4 and Figure 5.5).[2]

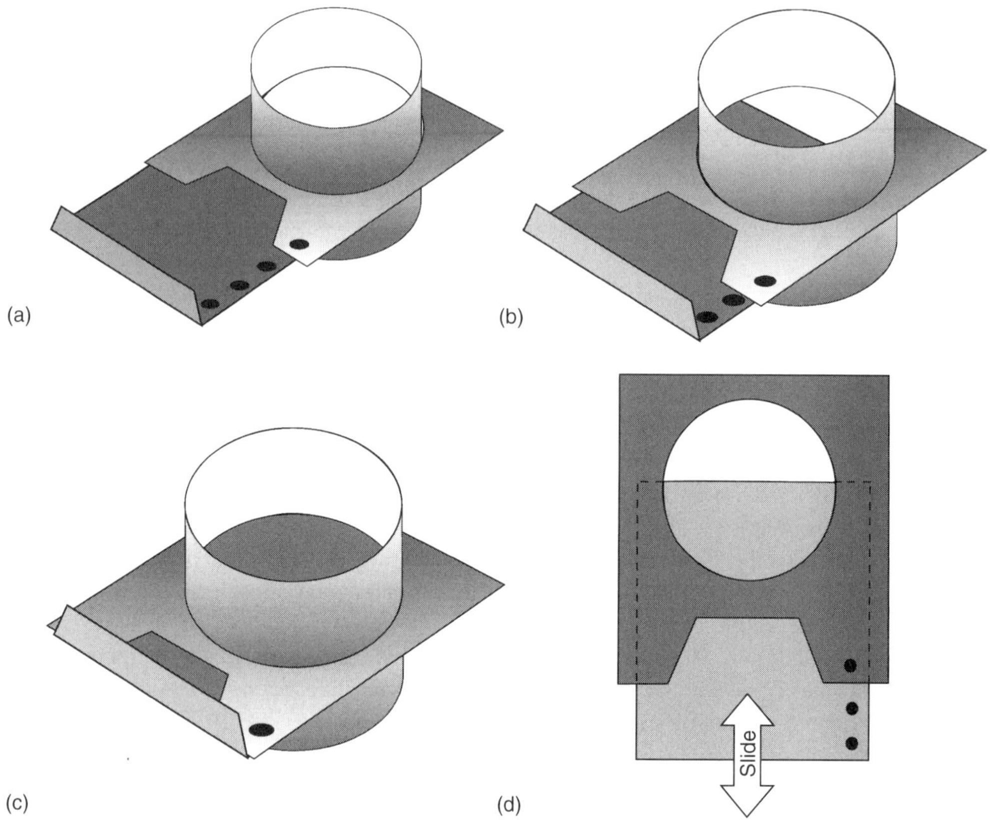

FIGURE 10.11 A blast gate is depicted from left to right in the fully open position (a), the midway position (b), and fully closed position (c). The slider (d) may be locked into position any time the holes on the frame and damper align.

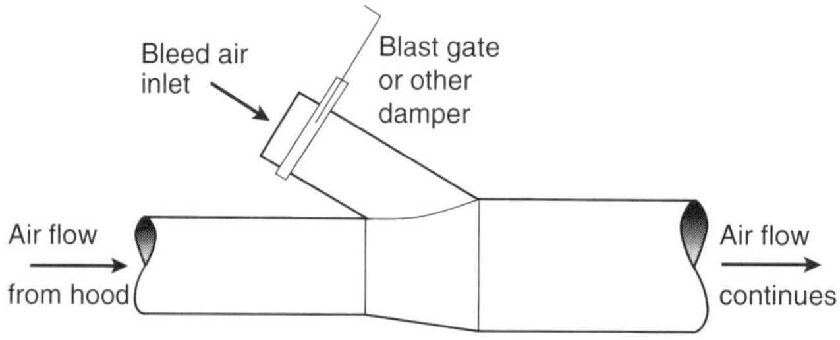

FIGURE 10.12 A bleed-in equipped with a blast gate to adjust the amount of outside air allowed to enter.

12. A cleanout is an opening or port through which the interior of a duct (or hood) can be accessed to remove dust, debris, or corrosion that could otherwise block the air flow. The *Manual of Industrial Ventilation* Figure 5.23 shows several examples of cleanout ports in the side of ducts that either pull out, slide, rotate or involve a preplanned duct disconnect; however, the *Manual of Industrial Ventilation* does not include a removable blind branch cleanout port as shown in Figure 10.13 that is commonly included on an elbow at one or both ends of a long horizontal run of ducting and through which a cleaning tool can be inserted.

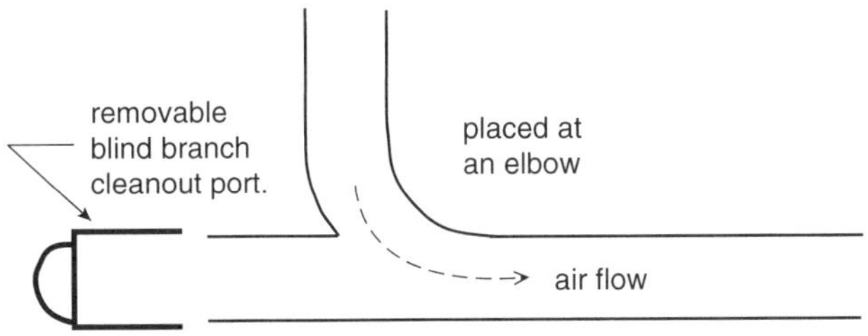

FIGURE 10.13 A blind branch cleanout port may be opened to allow access to the entire straight duct.

II. PRINCIPLES OF LOCAL EXHAUST VENTILATION

The ten principles of LEV expressed here in a nutshell, are elaborated upon in the ten sections to follow:

1. The air flow rate (*Q*) and its velocity (*V*) and direction can control the path of an airborne chemical.
2. A containment hood is preferred over a collection hood, but containment is not always feasible.
3. The exhaust air velocity decreases with the square of the distance out from a hood's face.
4. The ventilation air flow rate (*Q*) equals the air flow velocity (*V*) times the cross-sectional flow area (*A*).
5. Air flows in response to small differences in the surrounding molecular or static pressure (*P*).
6. The pressure difference that moves ventilation air is between the *P* inside a system and the atmospheric *P* outside the system (this difference is abbreviated "SP").
7. Moving air creates a velocity pressure (VP) that is proportional to its velocity squared.
8. The air's total pressure is the sum of its molecular and velocity pressures ($P + \text{VP}$).
9. Each loss from the air's total pressure caused by turbulence and friction is proportional to its VP.
10. The fan must replace the lost energy and create the VP of the air that leaves the ventilation system.

The above principles are used to calculate the following two values needed in design to choose a fan large enough to make the system operate as desired.

- the calculated air flow rate (cubic volume minute or "*Q*") necessary to protect the employee and
- the fan static pressure (FanSP) needed to overcome the resistance to flow through the system

In the real world of industrial hygiene practice, ventilation controls are designed and selected (or not) based both on the initial capital cost to buy and install the system and following two operating cost constraints:

- The cost to clean the exhaust air (remove or reduce the contaminants before they are emitted).
- The cost to condition the make-up air (filter and heat and/or cool the incoming fresh outside air).

1. The important physical principles that affect the generation of gaseous and particulate contaminants were discussed in Chapters 2, 3, 5, 6, and 8. However, once the contaminant is generated and/or leaves the source, the air flow rate, its velocity, and its turbulence all determine its subsequent direction of movement, its dispersion, and its eventual fate. **"Control velocity" is the major design variable of a hood**, not because it is the only variable affecting exposure, but because velocity is so simple to measure. (Recall the definition of control velocity [$V_{control}$] in section I.) Moreover, *the industrial hygienist is generally the most qualified person within an organization to choose, specify, or recommend the desired control velocity* based primarily upon the intrinsic dispersing energy or momentum of the plume, and secondarily upon the acceptable level of exposure. The more energetic or volatile the source and the more toxic the contaminant, the higher the control velocity should be. Chapter 13 will discuss four sources of information to guide a local exhaust ventilation designer in selecting a control velocity.
2. Selecting and positioning the most appropriate local exhaust hood is closely related to selecting the control velocity. One design challenge is to decide whether the source of the contaminant can be contained within a hood or the contaminant must be collected into a nearby hood. For a given amount of air drawn into an exhaust hood, **a contaminant can be contained within a hood more efficiently than it can be collected from outside of the hood** (as will be demonstrated in Chapter 13). The efficiency of a hood to control the contaminant affects the LEV system's operating cost and, ultimately, the degree of employee protection. Thus, the choice between using either a containment hood or a collection hood directly affects a hood's location, its face area, and its face velocity.
3. The primary opening through which air enters into a hood is called its face. Because **the velocity of the air toward the face of the hood decreases with the square of the distance beyond its face**, the spatial relationship between the source of the contaminant and the face of the hood affects the relationship between the control velocity and the hood's face velocity. The mathematical terms of this relationship are stated in the DallaValle equation, also to be presented in Chapter 13.
4. The volumetric rate at which air flows through the control zone (in cubic feet or cubic meters per minute) is not only a major determinant of a hood's performance and employee protection, but it is also the major parameter needed to design the remainder of the LEV system and one of the two major parameters needed to select the fan. The volumetric air flow rate is abbreviated Q as suggested by the pronunciation of its units of cubic feet or cubic meters per minute. **The rate that air flows through any opening is the product of the average air velocity times the area of that opening**. The above principle (stated as $Q = V \times A$ in Equation 11.7) has many uses because the air flow rate that enters a hood is the same air flow rate that moves through the ducts, through the fan, and through the air cleaner. Also, the same air flow rate that is exhausted must be allowed to flow back into the building and/or room. Moving air costs energy, and energy

costs money. Economic considerations favor the minimum air flow rate consistent with protecting the employees.

5. **Differences in air pressure cause air to move**. The potential air velocity between any two locations in a ventilation system is determined by the magnitude of the pressure difference between them, not on the absolute air pressure of the particular setting. Thus, the air flow rate (Q) and the movement of its entrained airborne contaminant is independent of the barometric atmospheric pressure at any location on any given day. Pressure differences can be created by several environmental mechanisms, but a fan is virtually always used to create and maintain the pressure difference that moves air in a local exhaust ventilation system.
6. **The particular pressure difference that is important in ventilation is called "SP"**. SP is the difference between the molecular or static pressure inside the ventilation system and the atmospheric pressure outside the system. A fan creates a pressure of less than one atmosphere on its inlet side and more than one atmosphere on its outlet side. Local exhaust ventilation systems protect people by sucking air into a hood. The two big questions in LEV are (1) how much air (Q) must be sucked from the workplace to protect the employees? (See Principle 4.) And (2) how hard will the fan have to suck to get that amount of air to move through the hood and duct system after it is built? These two questions will be answered in Chapters 13 and 14, respectively.
7. Any moving mass has kinetic energy. Moving air has kinetic energy. In ventilation (as well as in aerodynamics and hydrodynamics), **the kinetic energy of motion is called velocity pressure** (VP). Just as kinetic energy in physics equals one half of an object's mass times its velocity squared, velocity pressure equals one half the air's density times its velocity squared. Velocity pressure is only exerted in the direction in which the air is moving (such as feeling the wind if you put your hand outside of a moving car's window). The equivalence between velocity and velocity pressure will be explained in detail in Chapter 11.
8. While measuring energy is difficult, measuring pressure is relatively simple. Fortunately, **pressure and energy are integrally related to each other, both in concept and in calculations**. The total energy of air is the sum of its molecular energy and its kinetic energy, as reflected by its molecular pressure and velocity pressure, respectively. Quiescent air (static or at rest) has only its static pressure for total energy. Thus, the total energy within the initially quiescent air out in front of a hood is equivalent to the ambient barometric pressure. That is all the energy the air can have until it reaches the fan. As the air flows into the hood and ducts, some of its molecular energy is converted into the kinetic energy of motion called "velocity pressure" and some of its energy is lost due to friction and turbulence.
9. **Friction and turbulence caused by moving air create energy (pressure) losses**. Moving air creates friction in ducting, and it creates turbulence in flow transitions such as hood entrances, elbows, duct contractions and expansions, etc. Friction and turbulence remove energy from the air. The methods to predict these pressure losses that will be discussed in Chapter 14 are based on the principle that **each individual ventilation energy or pressure loss is proportional to the air's kinetic energy as represented by its velocity pressure (VP)** at that segment within the LEV system. The losses in each segment of the LEV system are independent and cumulative (additive).
10. All energy losses must be overcome by the fan. **The LEV fan must be large enough to both move enough air volumetric flow rate (Q) to protect people and create enough pressure difference to overcome the pressure lost from the air due to friction and turbulence**. The size of the fan needed for a given LEV system to perform as desired will eventually be specified by the air flow rate and the fan static pressure. (It turns out

that this "FanSP" is simply the sum of all the energy or pressure losses!) The fan motor delivers the power to move the volume of air per minute (discussed in Principle 4) with enough fan pressure difference (energy per volume) to overcome the flow friction and turbulence losses (discussed in Principle 9). Fans and fan selection will be discussed in Chapter 16.

Economic constraints preclude building a huge ventilation system that can move an unlimited amount of air. **Ventilation costs include the capital needed to build the system and the expenses of operating the system**. Operating costs include cleaning the contaminant from the exhausted air before it is emitted into the environment and conditioning the replacement or make-up air that flows back into the workplace. The necessity for air emission control is determined by the nature of the contaminant, the location of the exhaust stack, and air quality regulations. Air cleaners and the locations of the exhaust stack and inlet for the replacement air (make-up air) are discussed in Chapter 15. The need for conditioning of the make-up air (e.g., heating, cooling, dehumidifying, and/or filtering of dusts) is determined by the building's climate and the activities of its occupants. Such conditioning usually costs much more than the electrical power needed to operate the fan. Both of these ventilation operating costs are covered in Chapter 17.

III. THE LEV DESIGN SEQUENCE

The steps that one goes through to design a ventilation system largely parallel the ten principles of local exhaust ventilation presented in the previous section. While the following eight steps to LEV design can be viewed as a modified reiteration of these principles, it is really provided to give the reader the big picture into which the various steps, methods, and procedures to be presented within the following chapters can be placed. Without the big picture, a student will often get bogged down in some procedure within some topic and forget how each of these pieces fit together. Readers might also consider reviewing this section if they are lost in a later chapter and wonder "Why was it again that I'm doing this?"

A good hygienist should always remember as a sort-of "*zeroeth step*" to **consider better alternatives to ventilation**. In the real world, it is easy to get locked-in to ventilation control as if it were the only solution to every hazard. Local exhaust ventilation is a widely used control but it is really number six on the control priority paradigm introduced in Chapter 1 and discussed in Chapter 7. Use whatever reminders work for you to review some of the other options, such as chemical or process substitution, work or task modification, and source or operator separation or isolation (including automation). These nonventilation controls are often more cost effective than ventilation; they just require a little more creativity.

1. **Determine the appropriate control velocity** ($V_{control}$). Four sources of information are available from which $V_{control}$ can be chosen. The following four sources will be discussed within Chapter 13.
   a. Each VS diagram from the *Ventilation Manual* will list a control velocity (or its equivalent). If a VS diagram is assigned to you as a student or chosen by you in later practice, all the needed information is there on the page.[2]
   b. A value may be found within the list of "miscellaneous specific operations" in the *Manual of Industrial Ventilation*'s Table 10.99.2.[2] This list is not quick to use, but it can be helpful if no other source is obviously applicable and one has the time to peruse the list.
   c. OSHA ventilation standards should be consulted if one applies to a specific hazard (see sections VI and VII of Chapter 4; the latter applies to open surface tanks and ANSI Z9.1 — 1991).[3]

   d. A value can be chosen from within the generic ranges of control velocities from Table 13.2 herein or the *Ventilation Manual*'s Table 3.1.[2] Such a generic $V_{control}$ table can be thought of as a default control velocity when not dealing with a VS diagram or an OSHA standard.
2. Select and design either a containment hood or an efficient collection hood. Decide whether a containment hood is feasible or if the contaminant must be induced to flow into a collection hood. A containment hood is almost always preferred but is not always feasible. Carefully consider both configurations. This decision will affect and interact with the above $V_{control}$ and with A_{face}.

   Determine the hood's face area (A_{face}), the opening through which the exhaust air will flow into the hood. The size and open area of the hood face will depend on the size of the source, what else besides air must be able to move through the face (e.g., people, hands, and material), and (if applicable) the distance that the source is outside the face. If a containment hood is feasible, make its face as small as feasible. If a collection hood is required, locate that hood's face as close to the source as is feasible ($V_{control}$ outside the hood will drop with the square of the distance outside the hood).
3. Estimate the flow rate of air to be moved by the fan. The flow rate of air is abbreviated as Q for either cubic feet per minute or cubic meters per second. Units of cubic flow rate also result from multiplying V (fpm or m/sec) $\times A$ (ft^2 or m^2). For instance, the air flow rate in an LEV system equals the face velocity times the face area, viz., $Q = V_{face} \times A_{face}$ (to be derived as Equation 11.7). The V at the face of the hood will either equal the control velocity (if using a containment hood) or be higher than the control velocity (if using a collection hood and the DallaValle equation). The same Q necessary to protect the employee flows first through the face of the hood, then through the entire LEV system including the fan, any air cleaner, and out the exhaust stack. Q turns out to be one of the two key parameters for fan selection and determines the cost to heat and/or cool the makeup air.

Although system design is always a team activity, the IH typically plays a leading role in these often iterative first three steps. Each of these steps is presented in more detail in Chapter 13.

4. Select the duct's material, route, and size. The following topics are presented in Chapter 14.
   a. Choose the duct material, typically either galvanized sheet metal, plate steel, or PVC. Generally avoid metals if the contaminants are corrosive, and avoid PVC if the contaminants are hot or combustible. Specialty materials include stainless steel and fiber reinforced plastic (FRP). The material can limit the options for the duct's shape.
   b. Lay out the path that the ducting will take from the hood, past or around equipment and building structures, through walls and/or the ceiling to the fan, plus the exhaust stack. Sometimes these obstructions can influence the duct material or shape. Site the exhaust outlet to assure good contaminant dispersal without re-entraining it back into the building with the "fresh" air.
   c. The duct velocity (V_{duct}) must be sufficient to transport the contaminant and any other debris through the duct without clogging. A recommended duct velocity is provided in most ACGIH VS diagrams. Recommended generic ranges of duct velocities are provided in Table 14.2. The duct's size (A_{duct}) must be equal to or less than the flow rate Q from the previous step divided by the minimum design duct velocity (i.e., $A_{duct} \leq Q/V_{minimum}$).
5. Determine any air cleaning requirements. This topic is briefly covered in Chapter 15. Air cleaning may be needed to protect the fan, to avoid local environmental damage, complaints or nuisances, and/or to comply with air emission regulatory requirements.

Select a reliable and energy-efficient air cleaner in consultation with an air pollution manufacturer or engineer. Plan for waste recycling or disposal.

6. Provide for appropriate make-up air to flow back into the workroom, as described in Chapter 17.III.
   a. Minimize the resistance to the free flow of make-up air back into the workplace. Resistance means losing energy, and the more losses that occur means either that one will move less air than desired to protect the employees, or that a bigger fan and more electricity will be needed.
   b. Ensure that the inlet air is not contaminated by the exhaust air (re-entrainment) or by another source of external contamination (e.g., another building, idling engines at a loading dock, or cigarette smoke from a designated smoking or break area).
   c. Make-up air conditioning requirements may include adjusting the temperature (especially heating), dust filtration, and/or humidity controls. If the make-up air must be heated and/or cooled, heating and cooling costs can easily be the largest component of the ventilation control operating costs.
7. Calculate all of the energy (pressure) losses. Although a contractor or plant engineer may be tasked with actually designing the system, calculating pressure losses, and installing the system, the industrial hygienist's role often includes conceptualizing the ventilation design, reviewing someone else's design, and certainly testing the employee-protection performance of installed systems and analyzing any deficiencies. In almost all jobs, the industrial hygienist must be able to talk to the people responsible for the operation of the ventilation system in their language and anticipate the effect of changes, both intentional and unintentional, on system performance.
   a. Pressure losses are calculated in one definable portion of the system at a time. Then, each loss is simply added up to find the total loss. Pressure losses will occur at the following locations or segments of a system (see also 1910.94(d)(7)(ii)):
      - hood entry (loss factor(s) from turbulence × VP)
      - duct (loss factor from friction and/or turbulence × VP)
      - air cleaner (specified by manufacturer that may be proportional to either VP or V)
      - exhaust stack (more loss from duct friction and turbulence × VP as above)
      - make-up air (caused by obstructions to outdoor air flowing back into the building × VP).
   b. Most of these pressure loss are proportionate to the velocity pressure (VP) at that location. Looking ahead, losses will be expressed in terms of a loss factor times the local velocity pressure or change in velocity pressure within an expansion or contraction, as expressed by Equation 10.1.

$$\text{energy or pressure loss} = \text{Loss Factor} \times \text{VP} \tag{10.1}$$

   Velocity pressure (VP) is proportional to the velocity squared, as will be discussed in Chapter 11. The loss factor for most of the segments of a system defined above can be read from tables or figures to be discussed in Chapter 14. Thus, the sum of all of a given LEV system's energy pressure losses will be proportional to V^2 and (by application of $Q = V \times A$) to Q^2.
   c. If a branched duct system is being designed, each branch will have the same molecular energy inside the duct at their junction, whether you design it that way or not. The trick is to design or create the losses in each branch to get the desired Q (not just accept the Q that "happens").
8. Select the air mover (fan) whose performance matches the Q and pressure losses of the system as designed. These two parameters are needed to specify a fan from the manufacturer's literature. The fan must be capable of producing a fan static

pressure (FanSP) equal to the sum of the pressure losses caused by friction and turbulence calculated in step 7 for the system while moving the Q calculated in step 3 that the system requires to protect employees.

These two parameters are just like drinking soda through a straw. Imagine the straw is the ventilation system you just designed, and you are the fan. How fast the soda flows depends upon how hard you suck. You have to suck harder to drink through a small straw or one with more bends or constrictions than through a larger or straighter straw.

Other considerations presented in Chapter 16 that can affect fan selection include (1) capital costs that vary mostly by fan design and secondarily by manufacturer, (2) operating costs (electrical energy) for the fan motor, (3) resistance to clogging, erosion, and/or corrosion, and (4) any physical size, weight, inlet/outlet alignment, or noise and vibration constraints. Chapter 17 will explain that the cost of the energy to run the fan to move the air is small in comparison to the cost to condition the air (and usually less than the air emission cleaning costs).

A hygienist's job is not over even after a system has been designed, is installed, and someone turns on the power switch. There are virtually no existing models that will predict exposures before a system is built. So the first step is to test it by measuring the employee's exposure. A plan for periodic inspection, monitoring, and re-evaluation should be put in place. Both system management and troubleshooting of systems that no longer work well are described in Chapter 18. However, two other chapters are inserted before the chapters on system design. Chapter 11 will explain the important relationships between pressure and energy that are used within both system design and trouble shooting. Chapter 12 will cover the air flow measurement tools (manometers and anemometers) that are used to evaluate, monitor, and trouble shoot systems, and even these tools use some of the principles and concepts in Chapter 11.

REFERENCES

1. *International Building Code,* 2003, publ. by the International Code Council, Inc., Falls Church, VA, 2003.
2. *Industrial Ventilation: A Manual of Recommended Practice*, 24th ed., American Conference of Governmental Industrial Hygienists, Cincinnati, OH, 2001.
3. *AIHA/ANSI Z9.1-1991: Practices for Ventilation and Operation of Open-Surface Tanks*, American Industrial Hygiene Association, Fairfax, VA, 1991.

11 Ventilation Flow Rates and Pressures

The learning goals of this chapter:

- ✍ Understand the principle of Conservation of Mass, and why and when ρ_{air} can be considered a constant.
- ☐ Know $Q = A_1V_1 = A_2V_2$ very well. (Its use saves a lot of time in design, on exams, and in the field.)
- ✍ Understand the principle of Conservation of Energy and the resulting equivalence of pressure and energy.
- ☐ Know that the measured SP = $P_{\text{inside the system}}$ − the ambient $P_{\text{outside the system}}$ and the predicted $SP_{\text{before the fan}} = -[VP + \Sigma \text{ losses}]$.
- ✍ Understand why, in ventilation, we predict and measure SP rather than the static P inside a system.
- ☐ Know that the measured VP = Total P − static P and the predicted VP = $(V/4005)^2$.
- ✍ Understand how the relationship between velocity and velocity pressure changes with air density due to altitude or/and temperature.
- ☐ Understand that Total pressure = P + VP, and why it is an important concept but a rarely calculated value.

I. APPLYING CONSERVATION OF MASS TO VENTILATION

The principle of conservation of mass applies to any system in which the mass of something within any defined boundary can neither be created nor destroyed. If mass is conserved, then a mass balance can be conducted. The following subsections will first explain how a mass balance can be applied to ventilation, then show how, in ventilation, conservation of mass can usually be simplified to conservation of volume.

1. A mass balance means that the sum of the masses moving across any conveniently chosen three-dimensional boundary (e.g., the flow of a mass of air into and out from a room, a hood, or a duct) must equal either zero ("what goes in must sooner or later come out") or the difference in mass must either be generated, consumed, or stored within the bounded region.

$$[\text{mass flowing in}] - [\text{mass flowing out}] = [\text{change in mass within}] \qquad (11.1a)$$

When applying this principle to ventilation, the mass is the rate that air flows into and out of the boundary per unit of time. The term "mass flow rate" refers to the amount of material moving into or out of a boundary per unit of time, typically per minute. Typical boundaries in ventilation could be the walls of a room or the sides and two ends of a duct.

$$[\text{mass flow rate in}] - [\text{mass flow rate out}] = [\text{rate of change in mass within}] \quad (11.1b)$$

Air is compressible but reasonably stable (not easily created nor destroyed). Air does not normally accumulate within a workplace (only if it has a big compressor and an air storage tank, or the room is inflatable; both circumstances are rare). Few workplaces generate new air (an electrolysis plant might be releasing oxygen, but that is even more rare), nor is air destroyed (combustion is a molecular reaction that adds a usually negligibly small amount of carbon and/or hydrogen mass to the air, but let us assume the workplace is not on fire). Thus, in a practical sense, the mass of air within a building does not change.[a]

$$[\text{mass flow rate of air in}] - [\text{mass flow rate of air out}] = 0 \quad (11.2a)$$

$$[\text{mass flow rate of air in}] = [\text{mass flow rate of air out}] \quad (11.2b)$$

2. Ventilation air flow is normally measured as a volume flow rate, not a mass flow rate. Density is the ratio between something's mass and volume. In Chapter 2, air density was mostly written in metric units such as Equation 2.11a, where ρ [g/L] $= n \times \text{MW/Vol}$. American ventilation texts such as the ACGIH *Ventilation Manual* are written primarily using English units where the density is weight density [ρ_w] (although ACGIH also publishes a metric version).[1] The weight density in units such as lb/ft^3 is proportional to mass density (ρ) using Equation 11.3 where g means the acceleration of gravity. Weight density ρ_w is used below and extensively throughout this book; however, the same derivations shown here are also possible in metric units through the simple proportionality within Equation 11.3.

$$\rho_w = \frac{\text{weight}}{\text{volume}} = \frac{(m \times \text{“}g\text{”})}{\text{Vol}} = \text{“}g\text{”}\,\frac{n \times \text{MW}}{\text{Vol}} = \text{“}g\text{”} \times \rho \quad (11.3)$$

Thus, the mass flow rate of air can be calculated by any form of Equation 11.4.

$$\text{lb/min} = (\text{volume of air moved per minute}) \times (\text{density}) \quad (11.4a)$$

$$= Q(\text{ft}^3/\text{min}) \times \rho_w\,(\text{lb/ft}^3) \quad (11.4b)$$

$$= (\text{area}) \times (\text{velocity}) \times \rho_w\,(\text{lb/ft}^3) \quad (11.4c)$$

$$\text{lb/min} = \text{A}(\text{ft}^2) \times V(\text{fpm}) \times \rho_w\,(\text{lb/ft}^3) \quad (11.4d)$$

If mass is conserved, we can conclude that the rate that the mass flows into a walled room or duct system through one portal ("1") is the same as the rate at which it passes out at some other defined portal ("2" or at many portals if the walls have doors and windows or the duct leaks, but we will keep it simple). Equation 11.5 states that the flow rates through portal 1 and portal 2 are mathematically equal.

$$(\text{lb/min})_1 = (\text{lb/min})_2 \qquad (\text{g/min})_1 = (\text{g/min})_2 \quad (11.5a)$$

$$\text{A}_1\text{V}_1\rho_{w1} = \text{A}_2\text{V}_2\rho_{w2} \qquad \text{A}_1\text{V}_1\rho_1 = \text{A}_2\text{V}_2\rho_2 \quad (11.5b)$$

3. Flow area and velocity are much easier to measure thatn the air's mass. The cross-sectional area through which the air could flow into the open face of a hood or through a door, a window, or a duct can easily be measured with a tape measure or other convenient tool, and the air velocity can be measured with methods to be described in

[a] Unfortunately, the same cannot be said for contaminants. Airborne contaminants are often generated within a workroom. In fact, contaminant generation is the source of the hazard in settings of interest to IH. Airborne contaminants from a source within a workroom can accumulate in that workroom for a short time, but eventually they will reach a steady-state where accumulation and absorption are zero. A mass balance on the contaminant is a topic in Chapter 19.

the next chapter. Thus, Equation 11.5 would be much more useful if the fluid density (either ρ_w or ρ) were constant and could be taken out of the equation. That is not a problem for plumbers because water is incompressible; its density is always 1 g/mL (at least within the first three or four significant figures). However, air is compressible, and its density is not always constant. The challenge to be answered below is under what conditions can air density be considered a constant?

The Ideal Gas law was used in Chapter 2 to derive Equation 2.19b. That equation, rewritten here as Equation 11.6a, defines the ratio of the density of air (or any other gas) at two temperature and/or pressure conditions, "1" and "2," respectively.

$$\frac{\rho_2}{\rho_1} = \left(\frac{P_{2,\text{absolute}}}{P_{1,\text{absolute}}}\right) \times \left(\frac{T_{1,\text{absolute}}}{T_{2,\text{absolute}}}\right) \tag{11.6a}$$

In a ventilation system, the initial value of P_1 is actually unimportant. As described in the previous chapter, the pressure of the air outside of a ventilation system starts at ambient conditions; it will decrease slightly within a ventilation system until the air reaches the fan; the fan will boost the pressure to slightly above the ambient pressure causing it to flow back to the same ambient pressure at the exit plane of the exhaust outlet. Thus, it is sufficient and in fact easier to express condition 2 as a change [Δ] from condition 1 (whatever that initial condition might be).

$$\text{If we let } \Delta P = P_{2,\text{absolute}} - P_{1,\text{absolute}} \text{ then } P_{2,\text{absolute}} = P_{1,\text{absolute}} + \Delta P \tag{11.6b}$$

$$\text{and if } \Delta T = T_{2,\text{absolute}} - T_{1,\text{absolute}} \text{ then } T_{2,\text{absolute}} = T_{1,\text{absolute}} + \Delta T \tag{11.6c}$$

Replacing the values of P_2, and T_2 within Equation 11.6a eventually yields Equation 11.6d.

$$\frac{\rho_2}{\rho_1} = \frac{T_1(P_1 + \Delta P)}{P_1(T_1 + \Delta T)} = \frac{1 + \Delta P/P_{1,\text{absolute}}}{1 + \Delta T/T_{1,\text{absolute}}} \tag{11.6d}$$

Equation 11.6d allows one to define either the effect on air density of any expected change in P or T from condition 1 or the maximum allowable change in either P_1 and/or T_1 within which ρ can still be considered to be a constant. Let us now examine the allowable changes in P and T separately.

a. The maximum ΔP throughout most LEV systems is typically approximately 4–6 in. of water, and in the extremes, a simple system might have a ΔP of only 1–2 in., and a system with long and/or narrow ducts and an air cleaner might have a ΔP of as much as 20 in. of water. Thus, the typical $\Delta P/P_1 = 5/407 \approx 1\%$ change, and the maximum $\Delta P/P_1 \leq 20/407 = 5\%$ change. Equation 11.6d predicts that a 1% change in P will produce an equally small 1% change in ρ, which turns out to be well within the expected agreement between a ventilation system's design and its installed performance. Variations between ventilation system design and its performance of $\pm 5\%$ are more typical and are considered acceptable.
b. We will also use this same albeit somewhat arbitrary $\pm 5\%$ limit on ΔP as a benchmark for changes in temperature. Clearly, if T_1 is near room temperature, then there is little incentive for T to change within a ventilation duct. Thus, ρ can always be considered a constant when exhausting isothermal contaminants. On the other hand, as the temperature of the source and the initial exhaust gas increase, then a 5% decrease in the absolute temperature of the exhaust stream becomes more likely. Although predicting the actual ΔT of hot air within a duct is too difficult to consider here, predicting the ΔT necessary to cause a 5% change in ρ is relatively easy to predict. For instance, 5% would correspond to only a 34°F drop from an initial

212°F exhaust gas (viz., 5% × (212 + 460)). The cooling of hot exhaust streams (such as from metal fumes, drying ovens, or smoke) can easily cause ρ to increase by at least 5% and the exhaust air volume to decrease as it moves through a ventilation system. This shrinkage in Q will cause a decrease in V, in VP, and in the corresponding losses calculated in proportion to VP. Adjusting for such changes is an important real-world design consideration; however, for the purposes of the classroom and virtually all IH exam problems, we will assume that the exhaust air is near room temperature (typical of most toxic gases and vapors).

4. Thus, because $P_2 \approx P_1$ in virtually all ventilation systems, and where $\Delta T \leq 5\%$ of T_1 (or for duct air changes of up to approximately 50°F), we can conclude that $\rho_1 \approx \rho_2$ and the volume of a given parcel of air that flows from any point 1 to another point 2 is constant.

$$\boxed{Q = A_1 \times V_1 = A_2 \times V_2} \tag{11.7}$$

This relationship can be envisioned as the volume of air Q moving through a duct of constant diameter and cross-sectional area (A) in one second or minute, as depicted in Figure 11.1. The distance the molecules, initially at the first location (1), will have traveled to reach the second location (2) in that time (using distance = $V \times t$) is "V." Thus, the volume of air (Q) that will have passed location 1 and moved through that duct to reach location 2 is $Q = A \times V$. This same process theoretically happens and at least the first part of Equation 11.7 applies if air moves through any measurable opening whether the opening is connected to a duct or not.

Thus, this incompressible fluid assumption leading to a constant Q (or $Q_1 = Q_2$ if you want to look at it that way) can be applied to isothermal flow through any definable area, into or out of any humanly occupiable room, or through any ventilation duct. A constant Q is one of the most powerful tools in ventilation design or field analysis, as suggested by the following sequence of equalities:

Q = volume flow rate of air into a room
= volume flow rate of air out of that room
= Σ [(each inlet area into that room) × (each inlet's velocity)]
= (total inlet area) × (average inlet velocity)
= Σ [(each outlet area from that room) × (each outlet's velocity)]
= (total outlet area) × (average outlet velocity)

If all of the air that leaves a room passes out through an exhausted ventilation hood, then

Q = (that hood's face area) × (its face velocity) (will become Equation 13.1)

And if all that air flowed through a single duct either entering or leaving the room, then

Q = (duct area) × (average duct velocity) at any point within that duct.

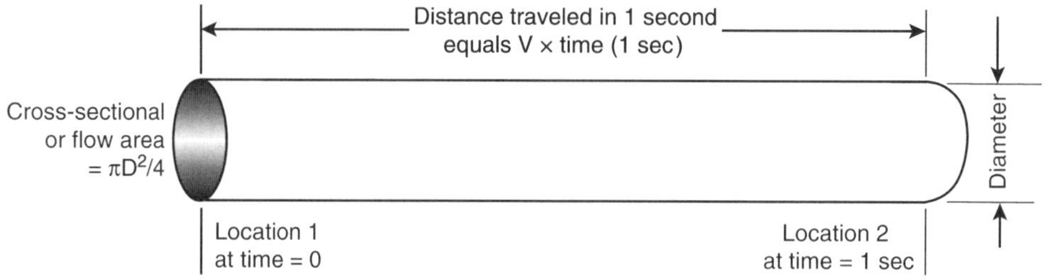

FIGURE 11.1 An example of the volume of air moved in 1 sec.

II. APPLYING CONSERVATION OF ENERGY TO VENTILATION

1. Fluids (ranging from gases such as air in the case of ventilation and aerodynamics to liquids such as water in plumbing, hydrology, and hydrodynamics) can have the following three forms of energy (plus the sum of these three forms that will, logically enough, be called the total energy).
   a. Molecular energy = pressure × volume = $P \times (\text{Vol})$ = the energy inherent in the random motion of individual molecules moving at the speed of sound but going nowhere. The multitude of molecules that continually strike any surface creates a compressive force normal to that surface. Since molecules move in all directions, the molecular pressure is equal in all directions. This pressure pushes on everything including the walls of pressure vessels, storage tanks, and ventilation ducts, which is why (especially in ventilation) this pressure is sometimes called wall pressure. In other fields, the pressure associated with the molecular energy of quiescent air is called static pressure. However, the term "static pressure" has been slightly usurped by the ventilation community, as will be discussed in Section III.
   b. Kinetic energy = $(1/2)\text{mV}^2$ = the energy inherent in the fluid's bulk or convective velocity. Kinetic energy also has an equivalent pressure called velocity pressure (VP).[b] Since velocity pressure is caused by the movement of air, VP is only exerted in the direction of that movement. Thus, VP is a vector. VP cannot be negative; it is always either zero or a positive value. VP will become a central predictor of pressure losses due to the friction and turbulence created as air moves into hoods and through ventilation ducts.
   c. Potential energy = weight × height = $W \times H$ = mass × gravity × H (where H is height, designated in some other texts by Z). At least in theory, the height that a parcel of air is lifted or pushed would be called its head (a term that is used often in ventilation). Buoyancy can affect potential energy if the density of the body differs from the density of the fluid it displaces. In ventilation, a parcel of air has zero weight or potential energy unless its density differs from the air around itself. As shown in Chapter 2, the combination of the molecular weight and toxic concentrations of volatile chemicals do not make the density of vapors much different from air. Thus, potential energy is relevant to most airborne hazards only if the contaminated air is at a different temperature than the ambient air. Practicing IHs regularly encounter hot exhaust gases where the effect of buoyancy can be used beneficially to create a natural updraft or "chimney effect" to aid exhaust flow (to be discussed in Section 13.VIII and Section 15.III). However, most of the learning problems discussed herein will be isothermal (neutral buoyancy as discussed in Chapter 2) where potential energy can be disregarded.
   d. Total energy = Molecular energy + Kinetic energy. Energy can be converted from molecular into kinetic and back again, but the conservation of energy principle dictates that the air's total energy is either constant (if not flowing) or a small amount can be lost due to friction and/or turbulence (if the air is flowing). For instance, when air enters into an exhaust hood or duct, its total energy cannot exceed what it had initially at rest within the room until it reaches the fan that adds energy to the air. This principle will also allow some very useful predictions to be made in ventilation design and system behavior in later chapters.

   No one goes around measuring the energy of air per se, but people do measure pressures. Pressures are central to ventilation design, performance, and system evaluation. Recall Principle 7 in Chapter 10 that a difference in pressure will cause air to move.

[b] Note that "VP" in ventilation is one term, not two. It is not the product of velocity × pressure. Nor is it vapor pressure, as this same notation is apparently sometimes used in chemistry.

Tools for measuring air pressure, air velocity, and air volumetric flow rate are the topic of the next chapter. This chapter will first explain and elaborate on the relationship between the air's energy and pressure.

2. As suggested by the qualitative definitions of molecular, kinetic, and potential energy given above, the terms Pressure, Head, and Energy are qualitatively equivalent. The three forms of fluid energy were expressed mathematically above as they might have been used in a physics course with metric units of Newton-meters or Joules. However, it is not easy to measure the energy of air. On the other hand, air pressures are comparatively easy to measure with a barometer, manometer, or pressure gauge. In ventilation, we are typically interested in small pressure differences that can be measured and expressed as the height of a column of water within a manometer. This difference in height is given a special term called pressure head or usually just "head".

   It turns out that the energy per unit of volume is equivalent to and commonly expressed in terms of pressure, and the energy per unit of weight is equivalent to and common expressed in terms of head or measured height. Energy, pressure, and head are all made equivalent via the fluid's density. Weight density (ρ_{w}, previously defined in Equation 11.3) continues to be used here.

$$\rho_{\mathrm{w}} = \frac{\text{Weight}}{\text{Volume}} = \frac{(m \times \text{“}g\text{”})}{\text{Vol}} \text{“}g\text{”}\rho \qquad \text{(repeat of Equation 11.3)}$$

   First, Equation 11.3 will be solved for a couple of terms that will prove useful in a moment (viz., for fluid volume in Equation 11.8a and fluid weight in Equation 11.8b). Again, "g" means the acceleration of gravity.

$$\text{Vol} = \frac{m\text{“}g\text{”}}{\rho_{\mathrm{w}}} \tag{11.8a}$$

   and

$$\text{weight} = \text{Vol} \times \rho_{\mathrm{w}} \tag{11.8b}$$

   a. One can obtain units of pressure such as psi (pounds per square inch) or Pascals (Newtons per square meter) by dividing either form of energy by a unit of volume. For instance, static pressure is the molecular energy per unit of fluid volume.

$$\frac{\text{Molecular Energy}}{\text{Volume}} = \frac{P \times (\text{Vol})}{\text{Vol}} = P = \text{“Static” or molecular pressure} \tag{11.9}$$

      As in molecular energy, the pressure from molecules is everywhere. Molecular pressure is what inflates a balloon, a basketball, and our lungs. A difference between the molecular pressures in two locations is what causes air to move, for example, from a high to a low meteorological pressure system, from one room to the next, from the hallway into a room (or vice versa), or within a ventilation duct from location 1 to location 2. Similarly, velocity pressure is the kinetic energy per unit of fluid volume.

$$\frac{\text{Kinetic Energy}}{\text{Volume}} = \frac{mV^2}{2(m\text{“}g\text{”}/\rho_{\mathrm{w}})} = \frac{\rho_{\mathrm{w}} V^2}{2\text{“}g\text{”}} = \text{“Velocity pressure”} \tag{11.10}$$

      Consider how Equation 11.8a is used to allow some of the units in Equation 11.10 to cancel and become units of pressure. One can sense velocity pressure as the force exerted on one's hand held out the window of a moving car. If you hold your hand flat, the force is only in the direction of the moving air; however, if you tilt or curve your hand, the force and your hand will be deflected downward or upward. An airplane's wings use a good fraction of the velocity pressure (typically about 10%) to create enough upward lift to fly.

b. The unit that one measures when using a barometer or manometer is the same unit of head that one can obtain by dividing energy by a unit of weight (as the latter is defined in Equation 11.8b).

$$\frac{\text{Molecular Energy}}{\text{Weight}} = \frac{P \times (\text{Vol})}{\text{Vol} \times \rho_w} = \frac{P}{\rho_w} = \text{“Static or molecular head”} \quad (11.11)$$

$$\frac{\text{Kinetic Energy}}{\text{Weight}} = \frac{m\,V^2}{2(m\text{“}g\text{”})} = \frac{V^2}{2\text{“}g\text{”}} = \text{“Velocity head”} \quad (11.12)$$

Again, consider out how some of the units in both of the above equations cancel to become units of height. One can measure a height difference or head using a manometer or equivalent gauge, to be described in Chapter 12. One cannot easily measure the head using air as the fluid (and even if you could, it's density is too small to be practical). Mercury, as used in barometers, is too dense to be deflected by ventilation pressures. However, water is cheap, plentiful and nontoxic, and its density makes it compatible with use in ventilation manometers. Thus, units of head in ventilation design are typically in inches of water [or "wg] in the American ventilation literature or centimeters of water [cm wg] in the metric system. The letter g in these pressures stands for gauge as in gauge pressure discussed in Section I.2.c in Chapter 2 (gauge can also be spelled gage).

3. Within a ventilation system, the total energy in a parcel of air will always be the sum of its kinetic and molecular energies. If energy is conserved, then the sum of these energies in any parcel of air is a constant. If losses occur, then the Total Energy in the air is decreased by the magnitude of that loss. The following list shows these sums in various expressions, terms, symbols, and units:

Kinetic Energy	+	Molecular Energy	=	Total Energy	Eqn. 11.13a
$m\,V^2/2$	+	$P \times \text{Vol}$	=	Total Energy	Eqn. 11.13b
$\rho_w V^2/2g$	+	P	=	Total Pressure	Eqn. 11.13c
Velocity Pressure	+	"Static" Pressure	=	Total Pressure	Eqn. 11.13d
VP[c]	+	P	=	P_{total}	Eqn. 11.13e
Dynamic pressure	+	Wall pressure	=	Stagnation pressure	Eqn. 11.13f
$V^2/2g$	+	P/ρ_w	=	Total head	Eqn. 11.13g
Velocity head	+	Molecular head	=	Total head	Eqn. 11.13h

It will be important to realize that the energy in fluids such as air can be transformed from molecular energy into kinetic energy and back again. Such transformations can theoretically be 100% efficient, but most transformations involve empirically predictable, but generally small, energy losses due to turbulence. This concept can be demonstrated in numerous ways and will later be used in the design of ventilation systems and in diagnosing problems within installed systems.

a. One example is the transformation of some of the initially quiescent room air's ambient pressure P into a certain amount of velocity pressure VP as the air moves into an LEV hood and duct. Since quiescent air (at rest) has no velocity pressure, all of its

[c] Note again that "VP" in ventilation is one term, not two. It is **not** the product of velocity × pressure.

total energy is in the form of molecular energy that equals the ambient atmospheric pressure. When some of that energy is transformed into VP within the duct, the molecular energy or pressure of the air within the duct must become less than one atmosphere. In fact, without any energy losses due to friction or turbulence, the magnitude of the pressure difference between the ambient atmospheric pressure outside the duct and the static pressure inside the duct would exactly equal this VP. This equality will be stated later as Equation 11.24. Of course, it is the fan that creates the pressure difference that causes the air to move in the first place, but the above equality is nonetheless true.

b. Another example is the principle behind a Pitot tube that converts the pressures associated with the kinetic and the molecular energies of air into the potential energy of water within the manometer. The principle behind a Pitot tube will be explained in subsection IV.1.1 of this chapter, and its structure and use as an anemometer will be discussed in Section III.1 in Chapter 12.

The fan will add energy to the air and increase the pressure inside the duct on the exhaust side of the fan to slightly above one atmosphere, sufficient to overcome any resistance to flow from that point to the other end of the exhaust duct. Developing the ability to predict turbulence and friction energy losses is important to make a real ventilation system perform as expected.

III. STATIC PRESSURES (P AND SP)

1. The random molecular motion of air molecules (or any other gas) has energy even when the bulk of the air itself is not moving. The inertia of many, many individual air molecules striking a surface creates the wall or static pressure (*P*) described in Section I of Chapter 2 and in subsection II.1.a of this chapter. Because the typical mean free path of air molecules is only about 0.1 μm, the molecular pressure is equally distributed onto all working and measuring surfaces. The molecular pressure of quiescent air creates the ambient atmospheric pressure, whatever that might be at any location and time.

   As discussed in Chapter 2, atmospheric pressure *P* (as used in Equation 11.9 and Equation 11.11) is an absolute pressure. It turns out to be neither easy nor useful to measure the absolute pressure anywhere within a ventilation system. Because the ambient barometric pressure changes due to passing weather fronts and diurnal effects, measuring the absolute pressure within a ventilation system on one day would not predict what it would be on the next day, and measuring it in one system at a given location would not predict what it would be in a similar system at another location or at another altitude.

2. Recall "Ventilation Principle 1" from Chapter 10: The air flow rate and its velocity can control an airborne contaminant. And "Ventilation Principle 5": Air moves in relation to **differences** in molecular pressure. In contrast to the difficulty and futility of predicting or measuring the absolute molecular pressures (*P*) inside and outside of ventilation hoods and ducts, measuring the difference between the static pressure inside of a ventilation system and the ambient atmospheric pressure outside the system is both easy and appropriate. Exhaust ventilation air initially moves from whatever the ambient atmospheric pressure is outside the hood to a lower pressure inside the hood, to a still lower pressure inside the duct, and to the lowest pressure just before the fan. The fan boosts the pressure to slightly above the ambient atmospheric pressure right after the fan, causing it to flow through the exhaust stack back again to that same ambient atmospheric

pressure upon exiting the exhaust. Thus, the difference between the static pressure inside a ventilation system and the ambient static atmospheric pressure outside the system is a direct measure of the potential for ventilation air to move.

The difference between these two static pressures is central to ventilation performance. SP in Equation 11.14 is basically how hard the fan sucks the air into the fan and pushes it back out from the fan.

$$\boxed{\text{SP} = P_{\text{inside the LEV}} - \text{the ambient atmospheric } P_{\text{outside the LEV}}} \tag{11.14}$$

The use of SP instead of *P* in ventilation is tremendously advantageous because SP does not depend on whether the ambient *P* is one standard atmosphere or not. While the ambient atmospheric *P* varies over time at any location and between locations, the same *P* exists at the hood entrance and exhaust exit of nearly all LEV systems at any time and place. Thus, the SP at any given point within a ventilation system is the same no matter how the ambient *P* outside the system might change.

As will be shown toward the end of this chapter, SP is always a negative value (−) anywhere inside an exhaust ventilation duct before the fan (referred to as the inlet side of the fan). In contrast, the molecular pressure right after a fan (called the outlet side of the fan) must be sufficiently above the ambient atmospheric pressure to push the air through the rest of the ventilation system to the exit plane of the exhaust duct, where *P* is once again that atmospheric pressure and SP is again zero. Thus, $\text{SP}_{\text{outlet}}$ is almost always a positive value (+).

While Equation 11.14 defines SP the way it is measured, Equation 11.15 defines SP before the fan as it must be determined during design.

$$\boxed{-\text{SP}_{\text{before the fan}} = \text{VP}_{\text{before the fan}} + \Sigma\text{Losses}_{\text{since entry}}} \tag{11.15}$$

Of course, real systems always have losses, but if there were no losses, then the drop in SP [or − SP before the fan] would exactly equal the VP to which the air is accelerated.

$$-\text{SP}_{\text{before the fan}}\ (\text{with no losses}_{\text{since entry}}) = \text{VP}_{\text{before the fan}}$$

(Equation 11.15 in an ideal case)

On the outlet side of the fan, the air not only has to have the energy that it will lose due to friction and turbulence in the exhaust stack, but it also has any change in velocity pressure that might happen between the fan outlet and the exit from the exhaust stack. Thus, Equation 11.16 defines SP after the fan.

$$\text{SP}_{\text{after the fan}} = \text{VP}_{\text{at exit}} - \text{VP}_{\text{after the fan}} + \Sigma\,\text{Losses}_{\text{until exit}} \tag{11.16a}$$

When the duct diameter on the outlet side of the fan equals the duct diameter at the exit plane of the exhaust stack (a condition that is common or at least often closely approximated), $\text{VP}_{\text{after the fan}}$ will equal VP_{exit}, leading to the following simplification. Note that SP in Equation 11.16b is always positive (+) after the fan.

$$\boxed{\text{SP}_{\text{after the fan}} = \Sigma\,\text{Losses}_{\text{until exit}}}\ \text{when } V_{\text{in duct after the fan}} \text{ is constant} \tag{11.16b}$$

Figure 11.3 to Figure 11.5 will depict *P* and SP as they might be encountered within a local exhaust ventilation system. Notice that although *P* and SP start out at completely different values, they always change the same amount and in the same direction within an LEV system. At sea level, molecular pressure [*P*] starts out somewhere near one atmosphere in the room, drops below that nominal one atmosphere as the air is accelerated, and gets further below one atmosphere as the air loses energy due to friction and turbulence. SP starts at zero in the quiescent room air, becomes negative as the air is accelerated into a hood or duct, and becomes further negative as it loses more energy flowing through the duct. Where P is above one atmosphere when the air leaves the fan and decreases to one atmosphere at the end of the exhaust stack, SP is positive on the outlet side of the fan and decreases to zero at the stack exit.

3. The only dark side to SP is the potential for some confusion because its initials match both static pressure and suction pressure and it is not quite either one. Outside the field of ventilation, the molecular pressure of quiescent air [P] is called its static pressure. For instance, the side holes on a Pitot tube are called the static pressure ports (in fact, the common Pitot tube anemometer is more correctly called a Pitot-static tube). A similar language is used for the air speed indicator on airplanes. Logically, a suction pressure would be defined by $1\ \text{atm} - P$ (in contrast to the reverse definition above) and would not be a negative value when the pressure inside of a system is below one atmosphere (in fact, a pressure of less than one atmosphere is what defines suction). Moreover, SP in ventilation is not always in the suction direction; it is greater than the ambient atmosphere on the outlet side of the fan. However, somewhere in the murky history of ventilation, people started labeling the pressure difference defined in Equation 11.14 as static pressure and so it has remained.[d]

   A more unique name would avoid this potential confusion. Perhaps a term such as "System Pressure" that retains the same SP abbreviation would be appropriate. In deference to those students who will eventually take a standardized test that is likely to use the common SP notation, the conventional SP notation will be used herein. However, rather than either propagating the static pressure misnomer within the ventilation literature or trying to create a unique term, SP will be used without giving it a name. It may help some readers to keep in mind that the magnitude of SP in ventilation is always much smaller than the atmospheric pressure (rarely more than $\pm$ 10 in. or 25 cm of water versus approximately 407 in. or 1034 cm in one standard atmosphere). Thus, the values of SP are far different from the absolute atmospheric pressures in which we live and work. After a little more time with the topic of ventilation, the difference between SP and P will hopefully become intuitive.

IV. VELOCITY PRESSURE

Velocity pressure (VP) is the pressure associated with the kinetic energy of all the molecules in an airstream moving in one direction. The bulk movement of air creates velocity pressure. If the bulk air velocity is zero, then VP is zero. VP can never be a negative value. Since VP is caused by the bulk movement of air, VP is only exerted in the direction in which the air is moving at that location, like a vector.

1. Velocity pressure can be sensed by putting your hand out the window of a moving car or trying to stand in a strong wind. It is the VP of the wind that lifts a kite or turns a wind mill. But VP cannot be measured independent of the molecular pressure P. Remember

---

[d] The terminology associated with SP has a long and continuing history. Ventilation professionals know so well that duct pressure before the fan is negative that they often mentally and literally drop the minus sign when referring to SP in that portion of the system. They even created the similar looking "SP_h" notation that would not have to be negative (some of its history is summarized below).

SP_h = "static suction at hood throat" [NIOSH (1973) pp. 576 and 616].[2,3]

SP_h = "hood static pressure" [17th ed. of the ACGIH *Manual* pp. 1–3].

SP_h = "hood static suction pressure" [22nd ed. of the *Manual of Industrial Ventilation* pp. 1–6].

$SP_h = -SP$ [22nd ed. of the *Manual of Industrial Ventilation* Equation 1.9].

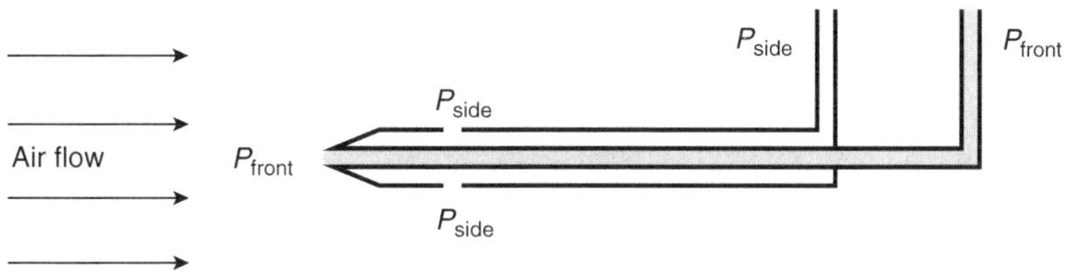

FIGURE 11.2 A section of a Pitot tube (showing its interior features).

that molecular pressure is equally distributed in all directions, so P is sensed everywhere. Figure 11.2 depicts the structure of a Pitot-static tube (pronounced *pē tō*, the second t is silent) that can accurately measure the difference between the total pressure and the local P to find VP. In fact, a Pitot tube represents a nearly-primary standard instrument (its use as an anemometer will be covered within the next chapter, subsection 12.III.1).

a. The hole at the front of a Pitot tube [P_{front}] senses both the wall or molecular pressure P and the velocity pressure VP. Thus, the hole (or port) at the tip measures total pressure (term in Equation 11.13d). Because air does not flow through a Pitot tube but must stop or stagnate right at its tip, the sum of P and VP at the tip is also sometimes called the stagnation pressure (term in Equation 11.13f).
b. However, VP is not sensed by the holes along the side of a Pitot tube because their orientation (the axis of the holes) is perpendicular to the direction of air flow. The pressure sensed by the side port is only the local P, static pressure, or wall pressure (the terms in Equation 11.13c, d, and f, respectively).
c. The VP and P passing the side port of a small and smooth Pitot tube is basically the same as the VP and P that is sensed at the front port. However, the side port only senses and measures P. Thus, when the front and side ports are connected via tubing to a manometer, the pressure difference [ΔP] is VP, as shown below, as implied by Equation 11.13, and as stated mathematically in Equation 11.17. A Pitot tube connected in this way could (and actually often is used to) measure VP in an existing system.

$$
\begin{aligned}
P_{front} &= \text{VP} + P = \text{"total pressure"} \\
P_{side} &= \qquad\quad P = \text{"wall pressure"} \\
\hline
\text{The difference} = P_{front} - P_{side} &= \Delta P = (\text{VP} + P) - P = \text{VP}
\end{aligned}
$$

or

$$\boxed{\text{VP} = \text{Total } P - \text{wall or static } P} \tag{11.17}$$

2. Technically, the VP defined in Equation 11.17 (and Equation 11.10) is a pressure and could be expressed in units such as pounds per square foot or Newtons per square meter.[e] However, the small velocity pressures in ventilation are conveniently expressed as the difference of the height of two columns of water measured in inches or centimeters using a manometer. The resulting Δh is technically a velocity head as stated in Equation 11.12. Equation 11.10 and Equation 11.12 can be mathematically combined to produce Equation 11.18 showing how VP and Δh are related.

$$\text{VP} = \rho_w \times \Delta h \tag{11.18}$$

[e] These units of force per area are actually practical to use in aviation but are too big to use in ventilation.

Measuring velocity head but calling it velocity pressure is also a misnomer but probably more benign than that between SP and P. When velocity pressure is measured or expressed as velocity head, the only conflict is intellectual.

3. What is really needed is a method such as Equation 11.21 that would either predict a Δh corresponding to a V desired from design, or to calculate an actual V corresponding to a Δh measured during a system evaluation. The derivation of the quantitative equivalence between V and Δh or VP is shown below in two parts. Equation 11.21 is important to know, but its derivation is only important to those who might wonder "where did that '4005' come from?"
   a. As part of deriving Equation 11.21, one must adjust for the relative difference between the density of the water in the manometer and the density of the flowing gas (air in the case of ventilation) and for the units of Δh measured in inches of water to its equivalent feet of air. The values of ρ_{water}/ρ_{air} in Equation 11.19 assume that the atmospheric pressure is 760 mmHg, that ordinary water (saturated with air) is in the manometer, and that the temperature is 70°F (21°C) as in the ACGIH *Ventilation Manual.*

$$\frac{\rho_{water}}{\rho_{air}} = \left[\frac{997.9955\ \text{g/L of water}}{1.2008\ \text{g/L of air}}\right]\left[\frac{1\ \text{ft}}{12\ \text{in.}}\right] = 69.259\ \text{ft of air/in. H}_2\text{O} \quad (11.19)$$

   b. For the other part of the derivation, one needs a value for "g" in Equation 11.12 to convert between V in feet per minute (fpm) and velocity head (Δh) in feet of air. It turns out when working at the fourth significant figure, that the Earth's gravitational acceleration varies from approximately 32.130 ft/sec^2 (9.7932 m/sec^2) at 30°N latitude such as along the U.S. Gulf Coast to 32.173 ft/sec^2 (9.8062 m/sec^2) at 45°N such as near the Great Lakes. The ACGIH apparently used a more northerly value of 32.161 ft/sec^2 characteristic of 41–45°N (through the southern Great Lakes or Oregon). Other values of g would cause the result in Equation 11.20b to vary between 4003 and 4006.[f]

$$V = \sqrt{2g\,\Delta h} = \sqrt{\left[2 \times \frac{32.161\ \text{ft}}{\text{sec}^2} \times \frac{69.259\ \text{ft air}}{\text{in. H}_2\text{O}} \times \Delta h\right]}\ \frac{60\ \text{sec}}{\text{min}} \quad (11.20a)$$

$$V = 4005\sqrt{\Delta h} = 4005\sqrt{[\text{VP in}\ ''\ \text{wg}]} \quad (11.20b)$$

4. This "magic number" of 4005 (or the metric equivalent of 1 cm wg = 12.76 m/s) is widely used by virtually all ventilation texts and IH exams. Therefore, the student and prospective certified IH needs to learn one of the following equations (and of course, know how to use them both):

$$\boxed{V\ \text{in fpm} = 4005\sqrt{\text{VP in}\,''\text{wg}}} \qquad V\ \text{in m/sec} = 12.76\sqrt{\textbf{VP in cm wg}} \quad (11.21a)$$

$$\boxed{\text{VP in}\,''\text{wg} = [V\ \text{in fpm}/4005]^2} \qquad \text{VP in cm wg} = [V\ \text{in m/sec}/12.76]^2 \quad (11.21b)$$

The second form of this equation may be easier to remember by visualizing VP as kinetic energy and knowing from physics that kinetic energy is always proportional to velocity squared. In nonmetric settings, the number 4005 will just have to be learned (as you learned 24.45 L/mol at NTP).

5. Developing a sense for the range of values applicable to IH ventilation is also helpful when trying to do a reality check (such as "Is my calculated VP value realistic?").

[f] See the Appendix to this chapter for this and related derivations at 25°C.

TABLE 11.1
Examples of Velocity Pressure (VP) at Some Meaningful Air Velocities

VP or Δh "wg	*V* fpm	*V* mph	*V* in Relation to Typical LEV Velocities	Physical or Personal Effect of the VP
0.0006	100	~1	A low control velocity	Unmeasurable VP
0.01	400	~5	A moderate control velocity	VP ≈ Pitot tube LOD
0.1	1265	~10	A high control velocity	Barely sensible VP
1.0	4005	~40	A modestly high duct *V*	VP can move one's hand
3.0	6935	~75	A very high duct V	Threshold of a hurricane
10.0	12,650	~150	Too high for any LEV	Hold onto your hat!

Remembering that *V* is 4005 fpm when VP is 1 "wg (and vice versa) is one such benchmark. The other examples of VP in Table 11.1 span the range of *V* and VP values applicable to local exhaust ventilation. Velocity pressures of less than 0.0001 or 10^{-4} "wg are unrealistically low, either to be measurable or to provide any employee protection. Velocity pressures approaching 0.1 "wg can be measured comfortably with a manometer and easily sensed on one's hand, and a VP above 3 "wg is literally in the hurricane category (>74 mph).

Notice that ventilation VP is always very small compared to one atmosphere which is approximately 407.6 in. of water. Remembering that 1 mph is 88 fpm can occasionally be useful in the U.S.A. such as when trying to explain ventilation air speeds to anyone unfamiliar with IH or ventilation (which is just about everyone), or to relate VP to the pressure felt when putting one's hand out the moving car window. For instance, the 1 "wg produced at approximately 40 mph is equivalent to 0.036 psi (1 × 14.7 psi/407.6 "wg in one atmosphere) will produce approximately 1 lb of force on an outstretched hand. In contrast, a lab hood face velocity of 100 fpm is just a little more than 1 mph, and its pressure can hardly be felt at all.

6. Because air density occurs in the denominator of Equation 11.19's conversion of inches of water to feet of air, the "magic" number will change from 4005 with the square root of the change in air density. The use of 4005 is fine all of the time in most situations, but low pressure caused by high altitude and high temperature caused by hot exhaust gases can each decrease the air density in some situations enough to justify increasing the referent "magic" number above 4005. The same 5% threshold for changes of ρ within a ventilation system used to justify Equation 11.7 could almost be used here to define a threshold for changing 4005 except for the impact of ρ entering into Equation 11.20 as its square root.
   a. Equation 2.4 was previously given to estimate the normal ambient atmospheric pressure to an accuracy of ±1% at a known altitude in feet above sea level:

$$\text{Normal } P_{\text{at altitude}} = 29.92 \times 2^{(-\text{altitude}/18{,}000)}\ (\text{in. Hg})$$
$$= 760 \times 2^{(-\text{altitude}/18{,}000)}\ (\text{mmHg}) \qquad \text{(Equation 2.4)}$$

The threshold for a change in altitude necessary to create a 5% change in 4005 works out to be approximately 2500 ft or 800 m above sea level. Using 2500 ft in Equation 2.4, one can see that $P_{\text{at altitude}}/P_{\text{normal}} = 2^{(-\text{altitude}/18{,}000)} = 2^{(-2500/18{,}000)} = 0.9082$, meaning there is an almost 10% drop in pressure due to the altitude; but since only the square root of air density enters into Equation 11.20, the 4005 "magic number" will only change by $1 - \sqrt{0.9082} = 4.7\%$.

Remember that if you are not at sea level, the real ambient atmospheric pressure P that could be measured with a mercury-filled barometer will differ from that reported by a weather reporting station because they always adjust their measurements to what the pressure would be at sea level at that location. For example, if the weather station in Logan Utah is reporting a barometric pressure of 29.92 in. Hg, the real P_{ambient} at the campus elevation of approximately 4700 ft is only 24.97 in. as shown below:

$$\text{Normal } P_{\text{at Logan}} = 29.92 \times 2^{(-4700/18{,}000)} = 29.92 \times 0.8344 = 24.97 \text{ in.Hg}$$

This almost 17% decrease in pressure (1−0.8344) due to altitude is sufficient to cause about a 9% increase in the 4005 "magic number" to 4384 that would result from using Equation 11.22a.[g]

b. A similar analysis finds that a ±50°F change in temperature from 70°F is needed to create that same ±5% change in 4005. This ±50°F change is of the same magnitude as the ±33°F change needed to cause a ±5% change in ρ for the purposes of the continuity equation (Equation 11.7). Given the arbitrary but practical basis for using a ±5% error as a threshold of significance, one could not be faulted too severely for using either limit. A ± 50°F (±28°C) temperature allowance around the 70°F (21°C) normal temperature used within the ACGIH *Manual* or the range from 20°F to 120°F (−7°C to 49°C) spans the vast majority of all working conditions. On the other hand, it would only take an exhaust gas temperature of 120°F to match the decrease in air density equivalent to an altitude of 2500 ft. Hot process exhaust gas temperatures can easily differ from normal much more widely than occupied environments can differ from normal. For example, an LEV exhaust gas temperature of 500°F as might be found in a foundry yields a "magic number" of 5355.
c. As a matter of practice, one almost never needs to correct the performance of an anemometer or a ventilation system for daily changes in the local weather. For instance, even large weather systems rarely cause a change in barometer readings of more than ±1 in., i.e., a low pressure might be 28.92 during a rain storm, and a high pressure associated with good weather might be 30.92. A 1 in. change out of 29.92 is equivalent to only a ±2% change in the 4005 magic number.

Therefore, the consistent use of 4005 is justified as long as the ventilated air's temperature does not exceed 120°F (49°C) or the altitude does not exceed 2500 ft (800 m) above sea level. Beyond these limits, the "magic number" will increase above 4005 by more than 5%. In such cases, the appropriate magic number can be derived from the ambient P and the exhausted air's T using either of the following two equations in place of Equation 11.21. Only 4005 needs to be memorized for closed-book testing.

$$V[\text{fpm}] = 4005\left|2^{\frac{\text{altitude}}{18{,}000 \text{ ft}}} \times \left(\frac{°\text{F} + 460}{530}\right)\right|^{1/2} \sqrt{\text{VP}[''\text{wg}]} \quad (11.22\text{a})$$

or its environmentally metric equivalent (although with V still in fpm):

$$V[\text{fpm}] = 4005\left[2^{\frac{\text{altitude}}{5500 \text{ m}}} \times \left(\frac{°\text{C} + 273}{294}\right)\right]^{1/2} \sqrt{\text{VP}[''\text{wg}]} \quad (11.22\text{b})$$

[g] The 9% increase results from $\sqrt{1/0.8344} - 1$ when multiplied by 100 to yield a percent using Equation 2.26.

The same principle described above that causes air density to affect velocity pressure, also affects any flow meter that relies on aerodynamic pressure including rotameters, Pitot tubes, and rotating or swinging vane anemometers (to be discussed in the next chapter on Measuring Ventilation Flow Rates).

V. TOTAL PRESSURE AND PRESSURE LOSSES

The air's total pressure is equivalent to its total energy. As implied by Equation 11.13 and restated below as Equation 11.23, the total pressure can be measured by simultaneously measuring both velocity and molecular pressures at any location. While total pressure is not usually measured for its own sake,[h] the concept that the air's total pressure is its total energy underlies the design and operation of all local exhaust ventilation systems.

$$\boxed{\text{Total Pressure}_{\text{of the air}} = P_{\text{of the air}} + \text{VP}} \qquad (11.23)$$

1. The kinetic energy of quiescent air (at rest or with no velocity) is zero. So, in quiescent air, molecular pressure comprises all of the air's total energy. Thus, the total energy of air at rest before it enters a local exhaust ventilation hood always equals the ambient atmospheric pressure.

$$\text{Total Pressure}_{\text{quiescent air outside the hood}} = P \text{ (the ambient atmospheric pressure)} \qquad (11.24)$$

Because the total pressure comprises all of the air's energy, the only way the exhaust air's total pressure can change is due to either heating or cooling, a loss of energy from friction or turbulence, or a gain in energy imparted by a fan. All isothermal air can do before it reaches the fan is lose energy. Thus, as Equation 11.23 implies and Equation 11.25 concludes, all of the pressure losses that the air experiences inside of an exhaust system before it reaches the fan are subtracted from the total pressure. Or stated another way, the total pressure of the air before it reaches the fan is always less than the ambient atmospheric pressure.

$$\text{Total Pressure}_{\text{before the fan}} = P_{\text{inside the system}} + \text{VP}_{\text{of the air}} \qquad (11.25a)$$

$$\text{Total Pressure}_{\text{before the fan}} = \text{Total Pressure}_{\text{outside the hood}} - \Sigma\ \text{Losses}_{\text{since entry}} \qquad (11.25b)$$

$$\text{Total Pressure}_{\text{before the fan}} = \text{the ambient } P - \Sigma\ \text{Losses}_{\text{since entry}} \qquad (11.25c)$$

When the air reaches the exit end of the exhaust system, its molecular pressure will be back to the same ambient atmospheric pressure where it started, but it will also have whatever additional kinetic energy is associated with its velocity as it leaves the exhaust system ($\text{VP}_{\text{at exit}}$).

$$\text{Total Pressure}_{\text{at exit}} = \text{the ambient } P + \text{VP}_{\text{at exit}} \qquad (11.26)$$

Thus, the total pressure in the air anywhere between the fan and the exhaust stack exit must equal the total pressure it will have when it leaves the exhaust system (as specified in Equation 11.26) plus whatever energy that it will lose due to friction or turbulence

[h] The pressure sensed at the tip of the Pitot tube (Figure 11.2) is total pressure, but the normal purpose of measuring it is to assess VP by the difference between total pressure at the tip and the molecular pressure along the side of the Pitot tube.

between any given location after the fan and the exit plane of the exhaust stack, as described by Equation 11.27.

$$\text{Total Pressure}_{\text{after the fan}} = P_{\text{inside the system}} + \text{VP}_{\text{of the air}} \tag{11.27a}$$

$$\text{Total Pressure}_{\text{after the fan}} = \text{Total Pressure}_{\text{at exit}} + \Sigma\ \text{Losses}_{\text{until exit}} \tag{11.27b}$$

$$\text{Total Pressure}_{\text{after the fan}} = \text{the ambient } P + \text{VP}_{\text{at exit}} + \Sigma\ \text{Losses}_{\text{until exit}} \tag{11.27c}$$

2. Similar to how and why SP is used as a surrogate for P inside a ventilation system, the term "TP" is a surrogate for the total pressure inside a ventilation system. TP is defined in Equation 11.28 (much like the way SP was defined in Equation 11.14, including its own nuances[j]) as the total pressure inside a ventilation system relative to the ambient atmospheric pressure outside the system.

$$\text{TP} = \text{Total Pressure}_{\text{inside the LEV}} - \text{the ambient } P_{\text{outside the LEV}} \tag{11.28}$$

The same either useful, or at least interesting, relationships stated above for total pressure (viz., Equation 11.25 to Equation 11.27) can be derived for TP from this basic definition.

$$\text{TP} = P_{\text{inside the LEV}} + \text{VP}_{\text{inside the LEV}} - \text{the ambient } P_{\text{outside the LEV}} \tag{11.29a}$$

$$\text{TP} = P_{\text{inside the LEV}} - \text{the ambient } P_{\text{outside the LEV}} + \text{VP}_{\text{inside the LEV}} \tag{11.29b}$$

$$\text{TP} = \text{SP}_{\text{inside the LEV}} + \text{VP}_{\text{inside the LEV}} \tag{11.29c}$$

Notice that energy can be exchanged between SP and VP without affecting TP. If there are no losses, the total pressure before the fan remains at one atmosphere, $\text{TP}_{\text{before the fan}}$ always equals 0, and $-\text{SP}_{\text{before the fan}}$ always equals VP. In real systems, $\text{TP}_{\text{before the fan}}$ will decrease due to losses right along with total pressure. The two forms of Equation 11.30 that define TP before and after the fan are exactly analogous to the total pressure defined by Equation 11.25b and Equation 11.27b, respectively. While both sets of equations are important conceptually, they are not the least bit useful.

$$\text{TP}_{\text{before the fan}} = -\Sigma\ \text{Losses}_{\text{since entry}} \tag{11.30a}$$

$$\text{TP}_{\text{after the fan}} = \text{VP}_{\text{at exit}} + \Sigma\ \text{Losses}_{\text{until exit}} \tag{11.30b}$$

The following two subsections will attempt to explain the above relationships diagrammatically, first for ideal flow with no pressure or energy losses and then by including the effects of losses due to friction and turbulence that always occur in real systems.

3. In ideal flow, the air would have no energy losses due to friction or turbulence as it flows from point to point. With no losses, the air's total pressure anywhere inside the ventilation system before reaching the fan will always equal the same ambient atmospheric pressure with which it started, as depicted in Figure 11.3, but the air inside the system has some VP. (VP_{inlet} is shown as a constant throughout this duct but that is not required for these statements to be true.) This kinetic energy has to come from somewhere; and until the air reaches the fan, its only source of kinetic energy is the molecular pressure that it had

[j] The concept of negative TP simply implies a total energy of less than one atmosphere, not a negative total energy. Trying to depict a negative pressure diagrammatically can also be confusing. For example, depicting the direction of the SP arrows as outward in the *Manual of Industrial Ventilation*'s Figure 1-1 is only correct after the fan. While *P* and VP add to yield total pressure, depicting TP as the sum of the SP and VP arrows acting in the same direction (within that figure) is also not correct.

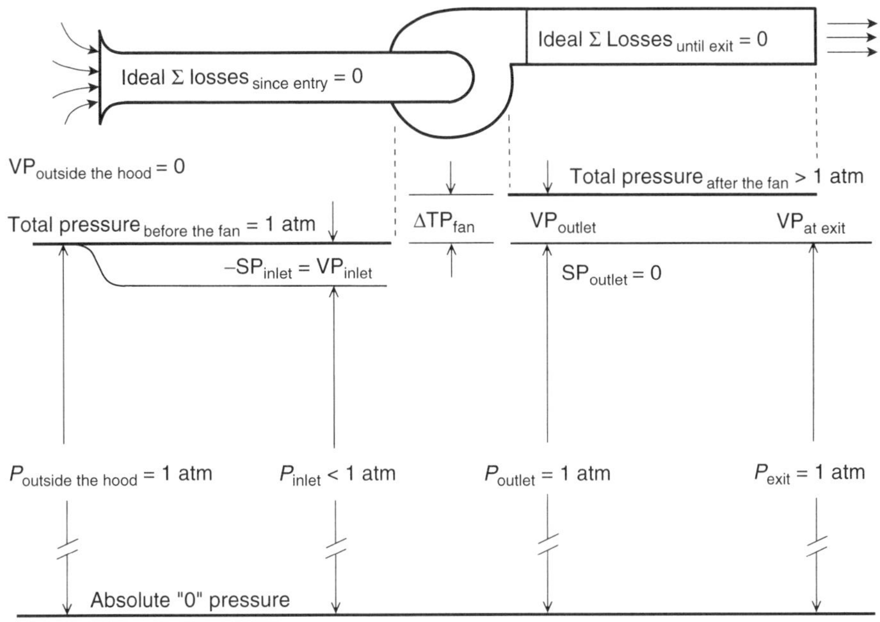

FIGURE 11.3 Depiction of pressures before and after a fan in an ideal local exhaust system with no losses.

initially at rest in the room. Thus, the air converts a little of its static or molecular pressure (P) into kinetic VP as it enters the exhaust hood or duct. With no losses, the conversion between P and VP is exact, and the air's molecular energy before it gets to the fan of an "ideal" ventilation system (shown only as P_{inlet} in Figure 11.3) is less than one atmosphere by exactly the amount of its kinetic energy (VP_{inlet}). That is, $P_{\text{inside the LEV}} -$ ambient $P_{\text{outside the LEV}} = -\text{SP} = \text{VP}$.

At the exit plane of the exhaust duct, where the air leaves the system and reenters the open atmosphere (the right side of Figure 11.3), the molecular pressure is again that same ambient atmosphere and $\text{SP}_{\text{exit}} = 0$. But notice in Figure 11.3 that the air still has VP_{exit} as it leaves the system, and its total pressure at exit is that ambient P plus VP. Indeed, if there are no losses, then the total pressure or energy anywhere between the fan and the exit plane must be the air's kinetic energy plus the ambient atmospheric pressure, again as depicted in Figure 11.3. An energy balance from entry to exit would conclude that with no losses, the fan must put in exactly the VP_{exit} kinetic energy that the air has when it leaves the system.

This seems like a good time to clarify a notation used in this series of figures. The broken dimension arrows for P and one atmosphere means that a portion of the full scale of these variables has been omitted from the diagram. Recall that one normal atmosphere is 407.6 in. of water. If the velocity pressure in Figure 11.3 were drawn to full scale of 1 in. corresponding to 4005 fpm, the bottom of the P scale would be approximately 33 ft down the page! The limitation of page size is only a minor inconvenience compared to the real reason why pressures in ventilation design and system evaluation are measured relative to the ambient atmospheric pressure rather than relative to an absolute vacuum.

4. In contrast to ideal flow, real flow creates friction as molecules move past the walls or other surfaces and turbulence within eddies is created by the air's viscosity. Friction and turbulence in the flow of real fluids create energy losses. Such losses reflect the fluid's resistance to flow. The lost energy is transferred to the environment as noise

(often perceptible) and heat (imperceptible in ventilation). All losses will decrease the air's total pressure, its TP, and its SP. Figure 11.4 depicts the energy loss that occurs as the air enters a hood and in the ducting on its way to the fan (the latter is cumulative as depicted by the gray triangles). The loss created by each section of the system acts independently. Therefore, all such losses are additive (cumulative). The entry loss and the cumulative duct loss shown in Figure 11.4 equals the Σ $\text{Losses}_{\text{since entry}}$.[k]

One can measure SP in an existing system using Equation 11.14. One can also use Equation 11.15 to predict SP on the inlet side of the fan and how hard the fan must suck and/or use Equation 11.16 to predict SP on the outlet side and how hard the fan must blow. Similarly, the VP can either be measured using Equation 11.17 or predicted using Equation 11.21 where V would be calculated from Equation 11.7 ($V = Q/A$ using the air flow rate throughout the system and cross-sectional flow area at each point within the system). We will learn in Chapter 14 how to predict losses.

To make even an ideal exhaust system such as that shown in Figure 11.3 operate, the fan has to put VP_{exit} into the air. To make a real exhaust system such as that shown in Figure 11.4 operate, the fan not only has to put VP_{exit} into the air but it must also put into the air the energy that it loses due to friction and turbulence. Note in Figure 11.4 that all of the positive SP (above one atmosphere) in the air as it leaves the fan [SP_{outlet}] will be lost by the time the air reaches the exhaust exit. In other words, SP after the fan represents how hard the fan must blow.

Again, by working back from the reference point of one atmosphere at the exit plane of the exhaust stack, the total energy at any point on the outlet side of the fan is the kinetic energy in the air when it leaves the system (subscripted "exit") plus whatever energy it

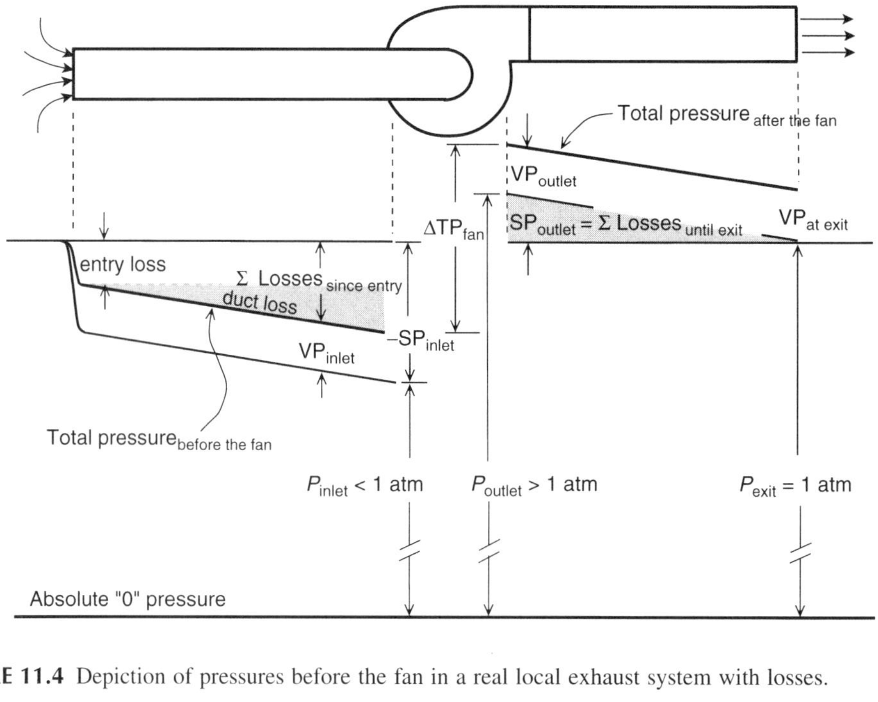

FIGURE 11.4 Depiction of pressures before the fan in a real local exhaust system with losses.

[k] The notation "loss_i" used from time to time means the pressure that is lost as the air flows through the portion of the ventilation system designated as section "i."

will lose between the location of interest and the exit plane of the exhaust stack. As on the inlet side, all of the losses on the outlet side of the fan are independent and additive. Figure 11.4 depicts how duct losses are cumulative on the outlet side just as they are on the inlet side. Thus, when using Equation 11.16, the losses are summed from any given location to the exit plane of the exhaust duct (cf. summing losses on the inlet side of the system from the entrance or hood face to a given location).

5. The pressures in Figure 11.4 exist on a continuum as depicted but are never measured that way in real systems. In real systems, one typically measures the pressure at selected points. Two such points are depicted in Figure 11.5. To access the inside of a duct, one must drill a hole into the side wall of a duct. To measure the wall pressure or molecular pressure (*P*), one can attach a tube onto that hole as long as the inside edge of that hole is smooth (no burrs). Two such smooth holes are depicted in Figure 11.5, one before the fan and one after it. One can achieve the same effect with less work by measuring the pressure at the static or side port of a Pitot tube inserted through a less precisely drilled hole at this location (similar to how the "Total *P*" tubes shown in Figure 11.5 were inserted). Notice that the $P_{\text{inside the duct}}$ before the fan is less than the atmospheric pressure outside the duct, and the water is being pushed up a height equal to SP.

   The tip of the thin, bent "Total *P*" tube (part of the Pitot tube depicted in Figure 11.2) measures the air's total pressure which, according to Equation 11.13, is the sum of its molecular and kinetic energies. The difference between the total pressure and the molecular or wall pressure is the velocity pressure, VP.

   Total pressure starts out equal to the ambient atmospheric pressure in the room, but friction and turbulence cause the loss of some energy. Thus, the TP inside the duct before the fan is less than the ambient atmospheric pressure by exactly the sum of the losses up to that point (Equation 11.27).

   The fan increases both the wall *P* and the TP inside the duct. SP is now positive, meaning that the wall *P* is greater than the ambient atmospheric pressure. Friction and turbulence will cause the loss of all of the SP by the time the air reaches the exit where SP = 0. Thus, the SP inside the duct after the fan is greater than the ambient atmospheric pressure by exactly the sum of the losses from that point to the exit (Equation 11.16). The difference between the total pressure and the molecular pressure is still the velocity pressure, VP, and the total pressure is greater than the ambient atmospheric pressure outside the room.

6. In order to move enough air to control a chemical hazard, someone must specify a fan big enough to supply all of this lost energy. Thus, you (or someone) must predict the magnitude of these losses. Empirical methods to predict energy losses due to friction and

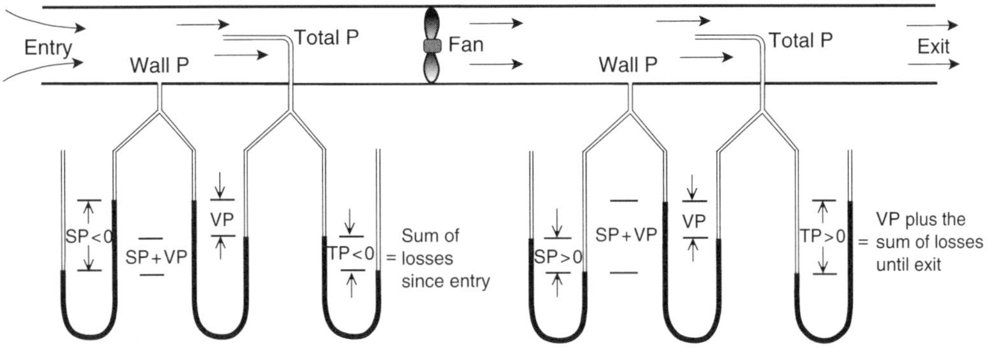

FIGURE 11.5 Measurable pressures on the inlet and outlet sides of a fan.

turbulence as the air flows into a hood and through ducts will be covered in Chapter 14. Fan characteristics and selection will be covered in Chapter 16. The rate at which energy must be supplied to the air (and its cost) will be shown in Chapter 17 to be the product of the air's flow rate Q × the change in the air's total energy between the inlet and outlet sides of the fan [$TP_{fan\ outlet} - TP_{fan\ inlet}$]. The $TP_{fan\ outlet}$ was just stated in Equation 11.30b, and the $TP_{fan\ inlet}$ was stated in Equation 11.30a. The latter's negative value means the fan must provide the change in the air's total energy as will be stated in Equation 16.3:

$$\text{Fan Total Pressure} = VP_{exit} + \Sigma\ \text{Losses}_{since\ entry} + \Sigma\ \text{Losses}_{until\ exit}$$

(preview of Equation 16.3)

Thus, the fan puts in both the kinetic energy and all of the lost energy. Failure to specify a fan big enough to protect the employees' health will result in over-exposure, which in turn can cause such adverse effects as occupational disease, OSHA violations and citations, and the loss of jobs (potentially including that of the "bozo" who did not specify a big enough fan). So there are multiple incentives to learn about ventilation systems.

VI. COMPILATION OF THE FUNDAMENTAL VENTILATION CONTROL EQUATIONS

The seven important ventilation equations pertaining to flow and pressures covered in this chapter are grouped here along with eight other important ventilation equations from future chapters. This list can be used for "one-stop shopping" before an exam (or during if allowed). Remember that equations in a solid box should not only be known by memory but understood almost intuitively. As we cover the remaining memorable ventilation equations in future chapters, you will see that they really only comprise of approximately three or four new concepts. If you understand the equations from this chapter, you are 90% of the way done. If you do not, you will get more chances for insight in the future chapters where they get used. If you have time at this point, you might review how the memorable equations from this chapter fit into the basic ventilation principles presented in Chapter 10. Recall that a similar summary of nonventilation equations was compiled in Section IX of Chapter 9, page 232.

$$\boxed{Q = A_1 \times V_1 = A_2 \times V_2} \qquad (11.7)$$

$$\boxed{SP = P_{inside} - \text{the ambient } P_{outside}} \qquad (11.14)$$

$$\text{Before a fan} \quad \boxed{-SP_{before\ the\ fan} = VP_{before\ the\ fan} + \Sigma\,\text{Losses}_{since\ entry}} \qquad (11.15)$$

$$\text{After a fan} \quad \boxed{SP_{after\ the\ fan} = \Sigma\,\text{Losses}_{until\ exit}} \qquad (11.16)$$

$$\text{If } V_{after\ the\ fan} \text{ is constant} \quad \boxed{VP = \text{Total } P_{of\ the\ air} - P_{of\ the\ air}} \qquad (11.17)$$

$$\boxed{VP \text{ in ''wg} = [V \text{ in fpm}/4005]^2} \quad \text{at NTP} \qquad (11.21)$$

$$\boxed{\text{Total Pressure}_{of\ the\ air} = P_{of\ the\ air} + VP} \qquad (11.23)$$

$$\boxed{Q_{containment\ hood} = A_{face} \times V_{face}} \quad \text{(see Equation 11.7)} \qquad \text{(Equation 13.1)}$$

DallaValle $\boxed{Q_{\text{collection hood}} = V_{\text{control}} \times (10\,X^2 + A_{\text{face}})}$ (Equation 13.2)

$\boxed{\text{Generic Pressure Loss} = \text{LossFactor} \times \text{VP}}$ (Equation 10.1 and 14.4)

$\boxed{\text{Losses} \propto \text{VP and to } V^2 \text{ and to } Q^2}$ (Equation 14.5 and 15.1)

$\boxed{\text{FanSP} = \Sigma\,\text{losses}_{\text{inlet}} + \Sigma\,\text{losses}_{\text{outlet}}}$ (Equation 16.6)

First "Fan Law" $\boxed{\text{RPM}_{\text{new}} = \text{RPM}_{\text{old}} \times Q_{\text{new}}/Q_{\text{old}}}$ (Equation 18.3)

Second "Fan Law" $\boxed{\text{FanSP}_{\text{new}} = \text{FanSP}_{\text{old}} \times (Q_{\text{new}}/Q_{\text{old}})^2} = \text{FanSP}_{\text{old}} \times (\text{RPM}_{\text{new}}/\text{RPM}_{\text{old}})^2$

(Equation 18.4)

Third "Fan Law" $\boxed{\text{kW}_{\text{new}}(\$) = \text{kW}_{\text{old}}(Q_{\text{new}}/Q_{\text{old}})^3} = \text{effect of RPM} \times \text{effect of FanSP}$

(Equation 18.5)

General Ventilation $\boxed{C_\infty = G/(Q/K_{\text{mix}}) \approx C_{\text{TWA}}}$ (Equation 20.1)

see Equation 19.2 for definitions of K_{mix}.

PRACTICE PROBLEMS

1. Suppose that air is exhausted from a typical laboratory hood with a face opening of 5 ft wide by 30 in. (2.5 ft) high at a face velocity of 90 fpm. What is the volume flow rate of air into this hood?

   $Q =$ _____cfm

2. Suppose that a particular laboratory had two such hoods. What is the least amount of air that must flow back into this room to make up for the volume flow rate of air that is exhausted?

   ans. = 2250 cfm

3. Suppose all of this "makeup air" came into this room through four square inlet diffusors each 2 ft by 2 ft. What is the average velocity of the air entering into the room through these diffusors?

   $V =$ _____fpm

4. What is the velocity pressure (VP) of the air coming in through these diffusors? One way to get psi is to equate 406.7 in. of water to 14.70 lb/in.^{2} in one standard atmosphere.

   VP = _____ "wg = 4.4×10^{-5} psi

5. Suppose the area of your open hand is 28 in.^{2} What would be the force of this air pressure on your hand if it were held palm-open in the diffusor? Do you think you could feel this force?

Force = _____ pounds

6. Suppose all of this exhausted air is eventually drawn into one 12-in. diameter round duct. What is the velocity of the air within this duct?

$V =$ _____ fpm

That is about 33 mph.

7. What is the velocity pressure of the air within this duct? And what would be the force on your open hand if you could get it into the duct? Do you think you could feel this force?

VP = _____ "wg = _____ psi Force ans. = 0.517 pounds

8. As the air accelerates into the exhaust duct, it converts some of the atmospheric molecular pressure (P, whatever the ambient pressure in the room is) into kinetic energy (VP). If no energy is lost accelerating this air to this duct velocity, by how much will its molecular pressure be reduced?

ΔP ans. = 0.512 "wg

9. Recall that SP is the difference between the pressure inside the duct minus the one atmosphere of pressure outside the duct. What is SP in this case? (No kidding; the answer to this question is as easy as it should look, except for the plus versus minus sign.)

SP = _____ "wg

10. What would be the resulting absolute molecular or static pressure inside the duct? State your assumptions. To answer this question, you need to assume a pressure outside the duct. If you think too long about this question, the answer will change. It won't be the same tomorrow or 1000 km away. The variability in the answer is why we never really care what the absolute pressure is inside a ventilation system. We only care about the pressure difference, SP.

$P_{\text{inside the duct}} =$ _____ "wg

APPENDIX TO CHAPTER 11

Neither the source for the "407.52 in. of water equals one atmosphere" conversion factor nor the applicable temperatures are defined on pp. 12–25 of the 23rd edition of the ACGIH *Ventilation Manual.*[4]

The 60th edition of the *Handbook of Chemistry and Physics* p. F-243 listed a value of 1 atm = 406.79 in. H_2O also under undocumented conditions.[5] However, a value on p. F-307 of 33.8995 ft of water at 39.2°F (4°C) is 406.79 in., indicating that this value of 406.79 was at 4°C. This 4°C is the temperature at which water (but not mercury) is at its maximum density but is well below normal temperatures.[1]

The 83rd edition of the *CRC Handbook* no longer lists a direct equivalence, but the following equivalences were calculated from conversion factors in its Section 1 for a standard atmosphere of

[1] Ordinary water (expected in a typical laboratory barometer) is saturated with air and is thus always slightly less dense (by approximately 0.00003 g/mL) than pure or "standard" water that contains no air. The calculation of H using the density of ordinary water at 4°C (p. F-5 again from the 60th edition because it is no longer listed in the current edition) yields an equivalent height of water that is only 0.01 in. more than standard water at 4°C, as shown below.

$$\frac{101,325 \text{ Pa in 1 atm}}{(0.9999750 \times 9.80665 \times 25.4)} = 406.792 \text{ in. of ordinary water at 4°C}$$

101,325 Pa (pp. 1–2) and various inch to Pascal conversions (pp. 1–7).[6] The calculated conversions are listed in order of increasing heights, increasing temperature, and (coincidently) decreasing numbers of significant figures:[m]

$$\frac{101{,}325\ \text{Pa}}{1\ \text{atm}} \times \frac{1\ \text{in. water}}{249.0889\ \text{Pa}} = 406.782\ \text{in. of water, ``conventional'' (defined in ISO 31-3)}$$

$$\frac{101{,}325\ \text{Pa}}{1\ \text{atm}} \times \frac{1\ \text{in. water at } 4^{\circ}\text{C}}{249.082\ \text{Pa}} = 406.782\ \text{in. of water at } 4^{\circ}\text{C } (39.2^{\circ}\text{F})$$

$$\frac{101{,}325\ \text{Pa}}{1\ \text{atm}} \times \frac{1\ \text{in. water at } 60^{\circ}\text{F}}{248.84\ \text{Pa}} = 407.19\ \text{in. of water at } 60^{\circ}\text{F}$$

The following single calculation gives the height (H) of a column of water using the density (ρ) of standard water (from the 83rd edition of the *Handbook* pp. 6–5) along with a 9.80665 m/sec^2 acceleration of gravity upon which the standard atmosphere was based.[6] The table that follows uses the Handbook's density of water at other temperatures.

$$P = \rho \times \text{``g''} \times H \tag{11A.1}$$

$$H(\text{in.}) = \frac{P}{\rho \times \text{``g''}} = \frac{101{,}325\ \text{Pa}}{\rho_{\text{water}} \times 9.80665 \times 25.4\ \text{mm/in.}} \tag{11A.2}$$

Temperature	**5°C (41°F)**	**10°C (50°F)**	**15°C (59°F)**	**20°C (68°F)**	**21°C (70°F)**	**25°C (77°F)**	**30°C (86°F)**
ρ Standard water	0.9999668	0.9997021	0.9991016	0.9982063	0.9979948	0.9970480	0.9956511
1 atm in H_2O	406.796	406.904	407.148	407.513	407.600	407.987	408.559

Thus, conversion factors between 407.2 and 408.6 in. would be appropriate over the temperature range of 60–85°F (15–30°C), respectively. The 407.52 in. given in the *Ventilation Manual* is apparently for ordinary water at 68°F, a value that is almost 0.75 in. or 0.25% more than the value of 406.79 at the CRC reference conditions of 4°C. A value of 407.60 in. of water is recommended herein to be consistent with the *Ventilation Manual*'s 70°F (21°C) "normal" temperature used for other ventilation calculations.

In a somewhat related manner, the following data were generated to show how converting Δh measured in "inches of 'ordinary' water" into VP at 25°C (versus the ACGIH use of 21°C) results in a different "magic number" than would result from using Equation 11.22 which assumes that only the air density varies with temperature and the density of water is constant. The first step is equivalent to Equation 11.19:

$$\Delta h = \left[\frac{997.0479\ \text{g/L of water}}{1.1847\ \text{g/L of air}}\right]\left[\frac{1\ \text{ft}}{12\ \text{in.}}\right] = 70.134\ \text{ft of air/in. } H_2O \text{ at } 25^{\circ}\text{C}.$$

[m] The 83rd edition of the *CRC Handbook* specifically cautions the reader that "Additional digits are not justified because the definitions of the units do not take into account the compressibility of mercury [or water] or the change in density caused by the revised practical temperature scale, ITS-90."

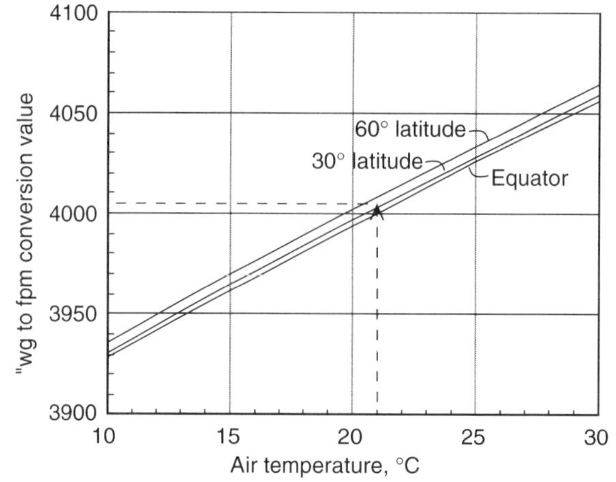

FIGURE 11.6 Depiction of the variation of the 4005 conversion factor with one's location on earth and the temperature of both the air and the manometer's water.

The next step converts Δh in feet of air to velocity as fpm equivalent to Equation 11.20:

$$V = \sqrt{[2g\,\Delta h} = \sqrt{\left[2 \times \frac{32.161\ \text{ft}}{\text{sec}^2} \times \frac{70.134\ \text{ft air}}{\text{in. H}_2\text{O}} \times \Delta h\right]}\frac{60\ \text{sec}}{\text{min}}$$
$$= 4030\sqrt{\Delta h \text{ at } 25°\text{C}}$$

This 4030 differs by only 0.05% from the 4032 that would result from using Equation 11.22, a difference that is typically neglected. Figure 11.6 shows values of this "magic number" at other temperatures (calculated as above) and at three latitudes (using values of gravity that vary due to the Earth's deviation from a perfect sphere). The normal value of 4005 corresponding to a temperature of 21°C and a latitude of between 40 and 45° is indicated by the dashed lines. The point is that temperature can have a much greater effect on the conversion value than can location.

Finally, the above conversion factors for inches of water per atmosphere are also used in Chapter 17 to derive the fan horsepower from a FanTP given in units of "wg (see Equation 17.2b). Equation 11A.3 demonstrates how to derive the conversion factor at 21°C (70°F), the normal temperature used in most of the *Manual of Industrial Ventilation*.[4]

$$\text{hp} = 33,000\ \text{ft} \times \text{lb/min} \times \frac{407.6\ ''\ \text{wg}}{\text{atm.}} \times \frac{\text{atm.}}{14.696\ \text{psi}} \times \frac{\text{ft}^2}{144\ \text{in.}^2}$$
$$= 6356''\ \text{wg} \times \text{fpm} \qquad (11\text{A}.3)$$

A table of such conversion factors at other temperatures is included here for potential future use.

Temperature	**5°C (41°F)**	**10°C (50°F)**	**15°C (59°F)**	**20°C (68°F)**	**21°C (70°F)**	**25°C (77°F)**	**30°C (86°F)**
1 atm in H_2O	406.796	406.904	407.148	407.513	407.600	407.987	408.559
"wg × ft^3/min/hp	6343.5	6345.2	6349.0	6354.7	6356.0	6362.1	6371.0

REFERENCES

1. *Industrial Ventilation: A Manual of Recommended Practice*, 23rd ed., (Metric Version), American Conference of Governmental Industrial Hygienists, Cincinnati, OH, 1998.
2. Mutchler, J. E., Principles of ventilation, In *The Industrial Environment — Its Evaluation and Control*, NIOSH Publications, Cincinnati, OH, chap. 39, 1973.
3. Engineering Staff of George D. Clayton and Associates. Design of ventilation systems, In *The Industrial Environment — Its Evaluation and Control*, NIOSH Publications, Cincinnati, OH, 1973.
4. *Industrial Ventilation: A Manual of Recommended Practice*, 23rd ed., American Conference of Governmental Industrial Hygienists, Cincinnati, OH, 1998.
5. Weast, R. C., Ed., *CRC Handbook of Chemistry and Physics*, 60th ed., CRC Press, Boca Raton, FL, 1979.
6. Lide, D. R., Ed., *CRC Handbook of Chemistry and Physics*, 83rd ed., CRC Press, Boca Raton, FL, 2002.

12 Measuring Ventilation Flow Rates

The learning goals of this chapter:

- ✍ Be familiar with some beneficial uses of qualitative flow indicators.
- ✍ Know the principle of a Pitot tube and how to use it.
- ✍ Be familiar with the major advantages and disadvantages of the Pitot tube, rotating vane, swinging vane, and thermoanemometer.
- ✍ Understand the impact of the boundary layer on ventilation duct flow measurements.
- ✍ Know why and how to conduct a duct velocity traverse.

I. QUALITATIVE AIR FLOW INDICATORS

1. *Smoke* is no doubt the oldest method used to visualize moving air since it naturally forms from virtually any combustion process and any source heated enough to produce a fume. Smoke can now be generated by more controllable methods (several without combustion) that can be used to visualize the movement of an otherwise invisible contaminant. An IH can use smoke to observe a plume's trajectory, the air flow pattern in front of a collection hood, or the kinds of turbulence and back eddies responsible for the breathing zone dilution ratios in Figure 5.12 and Figure 5.13. Smoke is also an excellent method of determining the direction of low speed control air velocities that cannot be felt by hand or differentiated by many quantitative air velocity meters (called anemometers).[1] Several brands of nonthermal smoke tubes are manufactured. One is simply a powder. Several generate an acid mist. The latter brands are sold with a hand-held rubber bulb that can be squeezed to release either a puff of smoke or a small stream of smoke for a few seconds at a time. The exhaust side of an air sampling pump can be used if longer durations of smoke streams are desired.
   a. Inorganic acid mist smoke tubes generate either a hydrochloric or sulfuric acid mist about 0.5 μm in diameter.[2] Two brands (MSA and Sensidyne/Gastech) use stannic chloride impregnated onto an inert granule such as pumice, in which case the following chemical reaction takes place: $SnCl_4 + 2H_2O \rightarrow 4HCl + SnO_2$ (with probably some $SnOCl_2$). Draeger used to make a tube with oleum (a mixture of H_2SO_4 and SO_3) that releases a sulfuric acid mist. The rather acute nasal irritation caused by acid mists makes these tubes useful in qualitative respirator fit tests, but the smoke dissipates rapidly to extremely low or undetectable concentrations.[3] Each glass tube can generate smoke for maybe 10 to 20 squeezes of the bulb. The tube's residual hazards include the acidity of its contents and the sharp broken ends of the tube. Unexpired tubes can be capped for short intervals between usage, but used tubes must be stored carefully. An opened tube can leave a mark on a table surface upon which it might accidently be left, and even the residual acidity in end-caps can etch a hole in a clothes pocket overnight.
   b. A less irritating organic acid smoke tube is also manufactured. Each of these flexible tubes contains two sealed ampules, one with acetic acid adsorbed onto silica gel and one with ethylene diamine adsorbed onto pumace. When the ampules are manually

crushed by purposefully bending the tube, aspirated air mixes the vapors within the tube forming a visible ethylenediamine acetate that smells like vinegar. These tubes will not store long after breaking the ampules.

Smoke can also be generated by conventional pyrolysis. Perhaps the simplest is a smoke bomb used for various celebrations, as a rescue or survival signal, or manufactured specifically for ventilation testing (as from www.SuperiorSignal.com). When ignited, these devices can generate smoke for a duration of about 10 to 30 sec. Their high temperature may make them difficult if not impossible to be hand-held. Their rather high rate of smoke generation precludes their use for measuring air currents but makes them particularly useful to detect leaks to the outdoors, to find and track large volume plumes, and to visualize exhaust air reentrainment. Commercial propane or electrically heated smoke generators (sometimes called a fogger) offer an intermediate option (similar to but larger than smoke tubes yet smaller than a smoke bomb). These devices (developed for either pesticide fumigation or theatrical effects) produce smoke by heating a light oil, and some foggers can be triggered to produce either a thin stream or a burst of smoke.

2. A *streamer* can be any plastic ribbon, tissue paper, tinsel, or other *lightweight* material hung where its deflection from the vertical will show the direction of horizontal air movement. The air's velocity pressure actually causes the deflection. With a little experience, the magnitude of the air velocity can be estimated by the angle at which a given streamer is deflected from the vertical. The negligible cost and intuitive nature of streamers make them an obvious but often overlooked option for either the user or the professional to detect whether any flow is happening or whether the air is flowing in the proper direction (especially when it is too slow to be felt or the meter is nondirectional).
3. Various electromechanical sensors are manufactured to indicate a pressure difference, e.g., between a room and a hood or a duct. The measured pressure difference can be indicated quantitatively on a meter or qualitatively by some sort of warning light (the latter indication is much less dependent upon how and where the pressure sensor is connected). However, a plan needs to be devised, implemented (through training), and supervised regarding who will monitor the sensor, how they will differentiate proper system operation from a malfunction, and how they should respond to a fault indication.

II. AIR VELOCITY AND MANOMETERS

1. The current American industrial hygiene (IH) and general ventilation industry practice is to quantify ventilation velocities in feet per minute (fpm) vs. meters per second (m/sec) in most of the rest of the world. A conversion to or from metric measures of velocity would use two commonly available equalities:

$$1 \text{ m} = 3.2808 \text{ ft}$$
$$1 \text{ min} = 60 \text{ sec}$$

The following conversion of 1 m/sec uses the method of equalities:

$$1\,\frac{\text{m}}{\text{sec}} \times \frac{60 \text{ sec}}{\text{min}} \times \frac{3.2808 \text{ ft}}{\text{m}} = 196.85 \text{ fpm or about 200 fpm}$$

A related conversion for aerosol falling speeds in Figure 3.5 is 1 cm/sec $\approx$ 2 fpm (1/10 of a meter). Yet another occasionally useful ventilation velocity conversion is between units of fpm and miles per hour (mph). For this conversion, it helps to

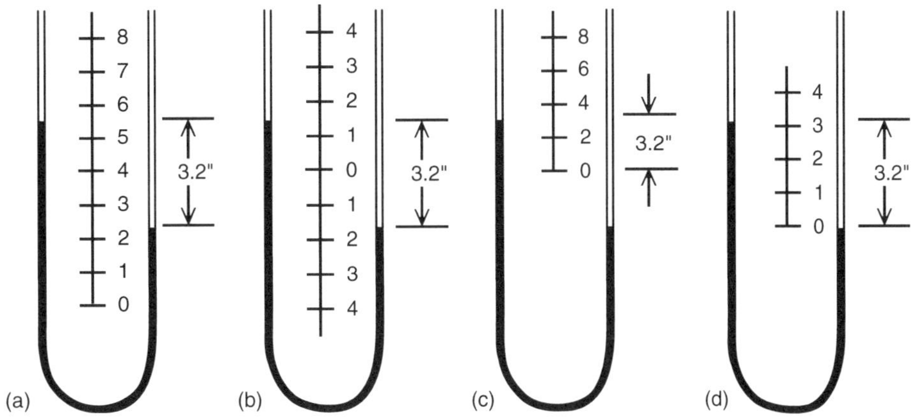

FIGURE 12.1 Examples of four possible scales that could be attached to a manometer.

know that there are 5280 feet in a mile.

$$1\text{ mph} \times \frac{\text{h}}{60\text{ min}} \times \frac{5280\text{ ft}}{\text{mi}} = 88\text{ fpm}\ \text{ or about 10\% less than 100 fpm}$$

2. A manometer is a common tool with which to measure pressure differences in ventilation, including duct pressure, velocity pressure, or total pressure. A manometer is a vertical U-shaped transparent tube containing a liquid such as water, oil, or mercury. In ventilation, the liquid is always water or measured as if it were water. In use, a difference between the pressures applied to each leg of the U-tube (ΔP) will cause a difference in the height or head of the water between each leg of the manometer (abbreviated as Δh). The height difference can be measured by a scale attached to the manometer such as those shown in Figure 12.1. Height differences recorded in units of inches of water are given the abbreviation "wg.[a] The height difference depicted in Figure 12.1 is 3.2 "wg.

   Besides being handy, manometers are fairly cheap to buy or easy to make. Perhaps as a result, they (and their attached scales) can be found in a variety of configurations. The cheapest or homemade manometer might look like the one in Figure 12.1a with an unmovable scale that may not have a zero; the *difference* in the fluid height Δh is calculated by reading the height of the fluid in each leg and subtracting (e.g., $5.6 - 2.4 = 3.2$ "wg). The position of the scale in Figure 12.1b can be adjusted to align the zero with the liquid level when no pressure difference is applied; but the Δh read on either scale has to be doubled (e.g., $2 \times 1.6 = 3.2$ "wg). The scale in Figure 12.1c is also adjusted to align the zero with the liquid level before any pressure difference is applied, but Δh can be read directly from the scale because the scale increments are reduced to account for the fact that the fluid goes down on one side just as much as it goes up on the other. The scale in Figure 12.1d needs to be adjusted to the liquid after the pressure difference is applied to align the zero with the lower liquid level; its Δh can again be read directly using the same scale increments as in scales (a) and (b).
3. The scale over which a manometer's column height (Δh) is read can be expanded if the manometer's high-pressure leg (or at least a portion of it) is inclined toward the

[a] The letter g at the end of "wg stands for gauge which refers to a generic pressure difference vs. an absolute pressure measured relative to a vacuum (see Section I.2 in Chapter 2). Absolute pressure is important when dealing with chemical reactions, while gauge pressure is important when dealing with pressurized tanks, steam vessels, and ventilation.

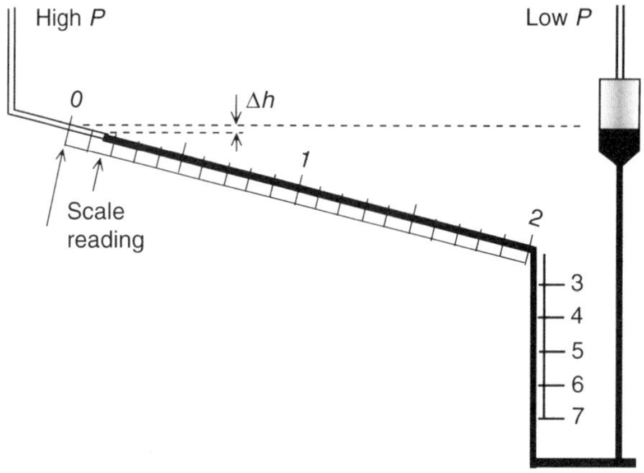

FIGURE 12.2 An inclined manometer reading 0.15 "wg.

horizontal, as depicted in Figure 12.2. The ratio by which a given Δh is expanded on an inclined manometer is determined by the cosecant (1/sin) of its angle from the horizontal. For example, an inclination angle of 5.739° will expand the distance by a factor of 10. The scale attached to an inclined manometer should have been expanded by that same ratio for easier and more precise reading. However, in order for the inclined scale to be accurate the user must set the incline angle accurately (usually by centering a built-in bubble gauge not shown in this depiction) then adjust the liquid level to zero before any pressure difference is applied.

The increments of 0.01 in. greatly improves the precision of reading small deflections. For example, when reading a pressure difference of approximately 1 in. on a vertical scale with 10 increments, each 0.1-in. increment represents a potential reading error of $\pm$ 10%, whereas if reading the same Δh on an inclined scale with 100 increments means each 0.01-in. increment would comprise a potential reading error of $\pm$ 1%. The manometer depicted in Figure 12.2 combines the features of both an inclined and a vertical manometer; the first 2 in. of Δh are read on the inclined portion of the manometer and deflections from 2 to 7 in. are read on the conventional vertical portion of the manometer.

A feature of inclined manometers manufactured by Dwyer® that is transparent to users (both literally and figuratively) is the expanded liquid bore diameter on the low pressure side. By making the diameter of the rising side of the liquid about 4.5 × larger than the falling side, 95% of the liquid level deflection from zero occurs on the falling side (the rising side remains almost static). The increments on their manometer have been adjusted for the ratio of the areas, allowing the scale to look much more like that in Figure 12.1d than that in Figure 12.1b or c.

Reading an inclined manometer filled with gauge oil is particularly easy because the gauge oil's surface tension causes its meniscus to form an almost right angle to the manometer's top wall where the Δh pressure difference is read, as depicted in Figure 12.3.

4. Water is convenient, cheap, nontoxic, but a little volatile. As water will evaporate from permanently fixed manometers, gauge oil is a common alternative to water. Gauge oil's red color also makes it easier to read than clear water. Since oil is less dense than water, the height scale on commercial manometers intended for gauge oil will already have been adjusted (expanded!) when it was manufactured to read correctly in units of inches of water. It is necessary to make sure that anyone refilling a gauge oil manometer uses the appropriate gauge oil (it is dyed red and has a specific gravity of 0.826 g/mL).

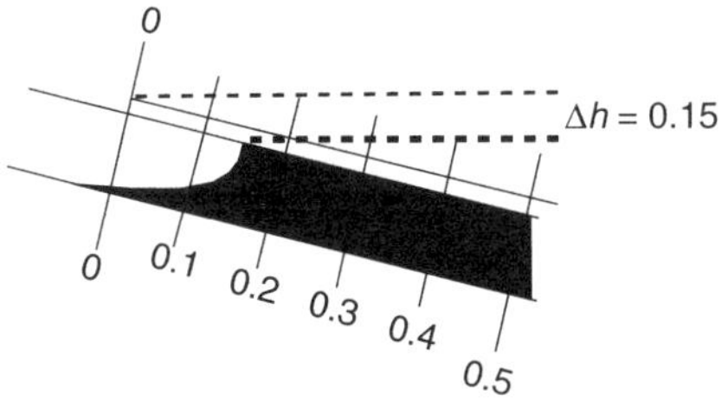

FIGURE 12.3 A typical meniscus within an inclined manometer.

FIGURE 12.4 A Magnehelic™ pressure gauge.

5. A mechanical or electronic pressure gauge is simpler to use in the field than a manometer, sufficiently precise if it reads in increments as small as 0.01 in. (0.2 mm), and accurate if it is calibrated periodically against a manometer. The low internal friction of a Magnehelic™ gauge (Figure 12.4) makes it more precise than other mechanical meters of similar cost. The term aneroid gauge means a meter without water.

III. AERODYNAMIC VELOCITY METERS

By definition, all aerodynamic meters respond to the dynamic or velocity pressure (VP) of the moving air. Most aerodynamic meters use VP to cause an object such as a liquid or a vane to move. Equation 11.22 showed that the air's actual V for any given VP varies with the square root of the ratio of normal air density to the air density at another altitude or/and temperature.

$$V = 4005\left[2^{\frac{\text{altitude}}{18{,}000\ \text{ft}}} \times \left(\frac{°\text{F} + 460}{530}\right)\right]^{1/2}\sqrt{\text{VP}} \quad (11.22\text{a})$$

The same square-root of air density relationship will affect any aerodynamic sensor. Thus, the indicated velocity readings of all aerodynamic meters ($V_{\text{indicated}}$) needs to be corrected to the true airspeed (V_{true}) in accordance with Equation 12.1 when either the altitude or temperature differs significantly from normal.

$$V_{true} = V_{indicated}\left[2^{\frac{\text{altitude}}{18{,}000\ \text{ft}}} \times \left(\frac{°F + 460}{530}\right)\right]^{1/2} \quad (12.1a)$$

or its environmentally metric equivalent:

$$V_{true} = V_{indicated}\left[2^{\frac{\text{altitude}}{5500\ \text{m}}} \times \left(\frac{°C + 273}{294}\right)\right]^{1/2} \quad (12.1b)$$

1. The *Pitot tube* is a fundamental anemometer. A Pitot tube senses VP as the difference between the total pressure and the molecular or wall pressure. Contrary to the similarity of its name to P_{total}, it was actually invented by Henri Pitot in the early 18th century.
   a. The structure of a Pitot tube (depicted in the top part of Figure 12.5) makes it the least mechanical of all the aerodynamic meters. Two separate sets of pressure ports (one hole on the tip and an array of small holes on the sides just downwind from the tip of a Pitot tube) are each connected via two concentric tubes to two fittings near its back end (on the right in Figure 12.5). The tip of a real Pitot tube with both sets of ports is superimposed below the depiction in Figure 12.5. The single port at the tip measures the stagnation or total pressure (see Equation 11.17). This pressure is conveyed via the inner concentric tube to a fitting on the very back of the probe. The ring of ports along the side of the probe measure the wall or molecular pressure. This pressure is conveyed via the outer concentric tube to another fitting on an arm coming out the side of the probe. (As a mental link, think tip connects to tip and side connects to side of a Pitot tube.)

      A manometer or differential pressure gauge is connected between these two fittings to measure the pressure difference, ΔP typically in units of the height of a column of water, referred to as Δh. As explained in Section IV.1 in Chapter 11 of the previous chapter, the atmospheric pressure at the tip and the side are the same, so they cancel out. Thus $\Delta P = (VP_{tip} + P_{static}) - P_{side} = VP$.

      The indicated manometer Δh reading must be converted into velocity via Equation 11.21 (possibly corrected for air density according to Equation 11.22). Units of velocity based on the standard V to Δh conversion scale from Equation 11.21 are imprinted directly onto the scales of some manometers.
   b. Pitot tubes have numerous advantages over other anemometers. They are simple to operate (especially if the manometer or pressure gauge is marked in both velocity and height). A Pitot tube has no batteries or other power source that could cause a spark to

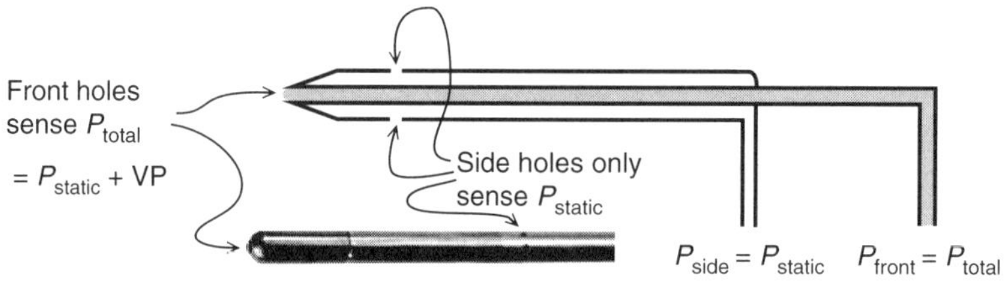

FIGURE 12.5 The tip of a Pitot tube and a depiction of its inner features.

ignite a flammable environment; it is intrinsically safe. Pitot tubes are comparatively cheap to manufacture, without electronic or moving parts such as in the rotating and swinging vane anemometers. Pitot tubes are manufactured in various sizes (especially in various lengths). They have no *practical* upper velocity limit for IH purposes (*local exhaust ventilation velocities are nowhere near the speed of sound!*).

c. Pitot tubes have a couple of disadvantages, most notably the high lower limit of quantification (LOQ) of most manometers. A manometer cannot detect the low velocities typical of control velocities in and around local exhaust ventilation hoods.
   1. If the LOQ for reading Δh is 0.1 in. of water (typical of vertical manometers), the lowest quantifiable V using Equation 11.21 is about 1266 fpm (adequate for measuring velocities typically within a duct but well above the velocity that is typically needed to control a plume).
   2. If the LOQ for reading Δh equals 0.01 "wg (more typical of an inclined manometer), then the lowest quantifiable V is about 400 fpm (still higher than many chemical control velocities).
   3. Magnehelic gauges or electronic manometers are physically smaller and more portable than liquid filled manometers, but even a Δh reading of 0.001 "wg corresponds to 127 fpm, and such mechanical or electronic gauges and meters need periodic calibration.

   Two other characteristics of Pitot tubes are only sometimes disadvantages. While air does not physically flow through a Pitot tube, the high inertia of large particles can carry them into the hole on the tip where they will be deposited (e.g., see stopping distance in Section V in Chapter 3) and can block the small diameter tube. The highly directional characteristics of Pitot tubes can be a disadvantage if the user has no clue to the direction the air is flowing; however, ventilation air inside ducts flows axially in one direction or the other, and it generally flows either into or out from an opening in a ventilation system, such as a hood face.

2. *Swinging vane anemometers* use the force of the VP acting on a vane to deflect an indicator needle. Figure 12.6 depicts the simplest example of a swinging vane anemometer called the Vanometer™. In this device, the portion of the vane inside the meter that can be seen through a window on its side is also the velocity indicator. The portion of the vane that cannot be seen is shown in Figure 12.6 as a dotted line. The Vanometer is not easy to hold level enough to be accurate, but its simplicity beautifully demonstrates the principle behind all aerodynamic sensing anemometers. Air flows

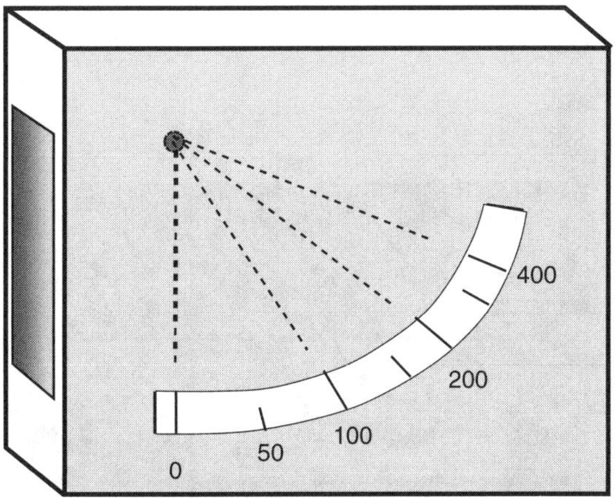

FIGURE 12.6 Depiction of a Vanometer™ showing three different velocity readings.

through the apparatus and deflects the vane. With no velocity entering into the opening on the left, the vane hangs straight down where it indicates 0 fpm. As air flows into the meter, the force exerted on the vane by the velocity pressure deflects the vane from the vertical. The velocity is indicated by values inscribed near the window by the manufacturer. As the velocity of the air increases, the force exerted on the vane deflects it further from the vertical as shown by the third position of the vane at a velocity of about 200 fpm.

The major commercial brand of swinging vane (the Velometer® by Alnor/TSI, depicted in Figure 12.7) uses a sophisticated spring mechanism that allows the meter to be used accurately in any normal position. The Velomcter® is sold with a scrics of adaptors that may be attached to it (shown in Figure 12.8), allowing the meter to measure a wide range of velocities (if one is careful to read the needle deflection on the appropriate scale among several on its face). As with all aerodynamic meters, the user must adjust the indicated readings by the square root of the relative air density (the ratio in Equation 12.1).

Swinging vane anemometers have many of the advantages of a Pitot tube while having alleviated some of their disadvantages. The deflecting force caused by aerodynamic pressure exerted onto the surface of the meter's internal vane has been well-matched to the spring's resistance to make the swinging vane better able to detect low velocities than a Pitot tube. A wide range of air velocities can be measured via

FIGURE 12.7 The Alnor/TSI Velometer® meter with low velocity sensor.

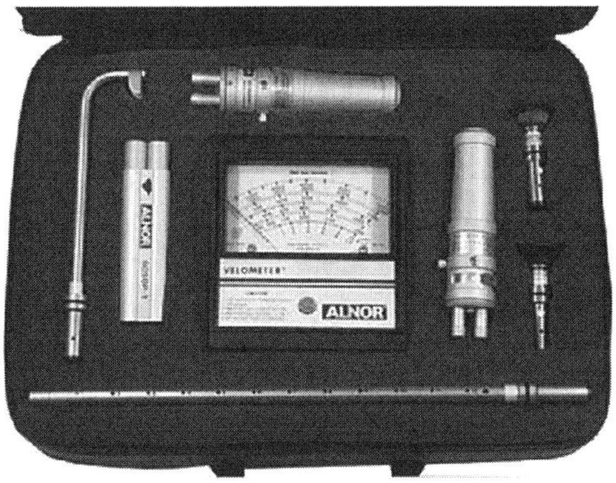

FIGURE 12.8 An Alnor/TSI Velometer® kit with its adaptors.

various adaptors (flow restricting or diverting devices) that can be attached to the meter (including an adaptor to allow it to measure SP through a pressure tap in the wall of a duct). Swinging vane anemometers do not need electric components, and at least the Velometer® is intrinsically safe.

The main disadvantage of the Velometer® is the adaptors that make the instrument bulky. The need to know which of the multiple scales on the face matches the perfomance of the adaptor in use is a nuisance that a little familiarity and care can easily overcome. The Velometer® has filter elements to allow its use in air contaminated by heavy dust; these filters and associated gaskets require periodic cleaning or replacement.

3. In a *Rotating vane anemometer*, the VP is directed against a cluster of inclined fins on a central shaft causing the shaft to turn, similar to a windmill. The shaft of a Davis rotating vane (shown in Figure 12.9) is mechanically linked to a revolution counter. The number of revolutions on the dial after the meter is held in an air stream for 1 min is a direct measure of the air velocity in fpm. A good rotating vane can measure a velocity range from about 30 to over 5000 fpm. Rotating vane anemometers have a couple of advantages. The traditional rotating vane has no electric spark source to ignite a combustible environment, so it is intrinsically safe (and is used, for instance, to monitor air flow in mines). The measurement process provides an integrated (time-weighted) measurement useful in slowly fluctuating flows. In fact, the rotating vane may be the only instrument that will provide a negative reading on the counter to indicate that the flow is going in the wrong direction (useful when V is too low to sense the flow).

   One disadvantage of a rotating vane anemometer is the large size of its sensing element. Thus, its use is inappropriate where the meter would block a large part of the flow area (e.g., inside small ducts) or where the velocity would vary across the vane area (e.g., flow from diffusers, fan outlets, or again inside ducts). Another disadvantage is that the user also has to use a timer. The large dynamic inertia of the meter will prevent the measurement of rapid fluctuations (although this can also be an advantage).

   Other manufacturers connect the shaft of a rotating vane to a DC generator that powers an indicator. Such electronic rotating vanes are smaller, cheaper, and may have a wider velocity range than the traditional all-mechanical rotating vane; but they usually seem to be made too cheaply to be reliable, are still too large for duct work, and are no longer intrinsically safe. A variation on this design (called a bridled vane) is to link vanes similar to those in Figure 12.9 to a spring assembly that does not allow rotation but deflects a needle to indicate the velocity.

FIGURE 12.9 A/2 4″ Davis rotating vane anemometer. Davis Inotek Instruments.

IV. THERMODYNAMIC VELOCITY METERS

As a group, thermodynamic velocity meters are called thermoanemometers. All thermoanemometers respond to the cooling effect of air flowing over a small heated surface. Because some of the earliest thermoanemometers were based on the air's cooling effect on a heated wire, they are often called a hot-wire anemometer despite the evolution of the technology away from a wire. Since the cooling rate is proportional to the mass of air flowing over the heated element, they are also sometimes called a mass flow meter. The importance of mass flow rate to heat transfer means that the indicated velocity readings of all thermoanemometers need to be corrected for the air density by a linear ratio as shown by Equation 12.2 (cf. the square-root relationship for aerodynamic meters in Equation 12.1).

$$V_{\text{true}} = V_{\text{indicated}}\left[2^{\frac{\text{altitude}}{18{,}000\ \text{ft}}} \times \left(\frac{{}^{\circ}\text{F} + 460}{530}\right)\right] \tag{12.2a}$$

or its environmentally metric equivalent:

$$V_{\text{true}} = V_{\text{indicated}}\left[2^{\frac{\text{altitude}}{5500\ \text{m}}} \times \left(\frac{{}^{\circ}\text{C} + 273}{294}\right)\right] \tag{12.2b}$$

1. The classic heated thermocouple anemometer is a clever enhancement to the original hot-wire anemometer that only had a hot wire. These instruments are not as popular as

they once were, but again, their simplicity clearly illustrates the principles underlying all thermoanemometers.

a. A thermocouple is constructed by welding together two wires of different metals. The galvanic potential of any two metals is difference in a predictable millivolts that varies with the temperature of the welded junction. Tables of millivolts vs. temperature for various standard combinations of metals may be found in the *CRC Handbook of Chemistry and Physics*.[4]
b. One thermocouple is attached to a third wire (shown horizontally in Figure 12.10) stretched between two supports (shown vertically in Figure 12.10). Enough current from a separate battery is passed through the third wire to heat it to about 350°F at zero air velocity. This hot junction is cooled by air in proportion to its velocity.
c. A second identical, unheated, thermocouple (called a cold junction) is built into one of wires of the first thermocouple (the left wire in Figure 12.10) and kept in the same airstream to help compensate for the effect of air's temperature on its ability to cool the heated wire.
d. The temperature of the hot wire and its attached thermocouple determines the millivolts that drives the meter's needle scaled to fpm or m/sec.

Heated thermocouple anemometers have several advantages. One of the main advantages is that heated thermocouples can detect very small velocities. The limits to both ends of their velocity sensing range are largely determined by the electronic circuit installed by the manufacturer. Their wide velocity range obviates the need for many attachments, making them quite portable and simple to operate. Their very small sensing element (the smallest of any anemometer) allows measurements to be very precise spatially. Small thermocouples respond to changes in air velocity very rapidly (this responsiveness is necessary if one is trying to measure turbulence and the turbulent

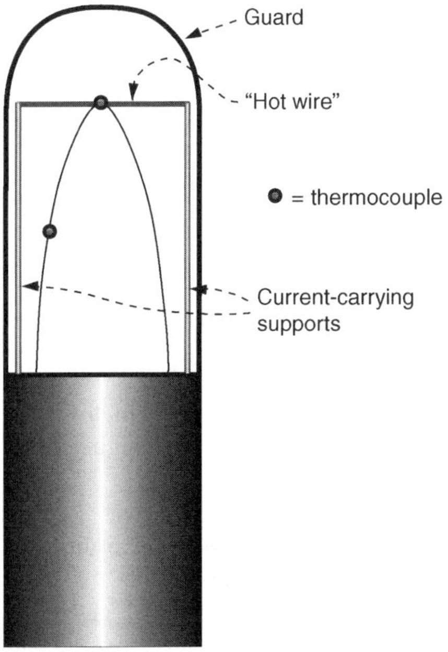

FIGURE 12.10 A hot-wire anemometer sensing element.

eddies discussed in Chapter 5, but routine IH users would prefer rapid meter fluctuations to be damped out electronically).

Hot-wire anemometers also have some disadvantages, at least one of which can be serious. The cooling rate of hot wires (and the velocity indication) will decrease if the angle at which the air passes over the wire deviates from perpendicular by more than about 30°. Alignment within this range is not difficult as long as the air flow can either be sensed or is constrained. Examples of the latter are flow within a duct, flow through an orifice, or if the wire is positioned vertically to measure horizontal wind. However, if one is trying to measure flow into or out of a hood or room that is too low to be felt, the probe cannot indicate which way the air is flowing. The small wires are fragile and difficult to clean if they accumulate dust. The biggest limitation of hot-wire anemometers is that their sensing wire is exposed and conducts enough electrical current that if snapped can cause a spark. It is virtually impossible to make a heated thermocouple anemometer that is intrinsically safe. Thus, they CANNOT be used in potentially combustible or explosive atmospheres!

2. Hot-film or constant temperature anemometers do not have a hot wire *per se* (nor do they use thermocouples), but they utilize most of the same thermodynamic principles as hot-wire anemometers. Perhaps the most important difference for IH uses is that the sensor is larger than a wire (e.g., Figure 12.9), making it more rugged and easier to clean. Most commercial hot-film anemometers are kept at a constant temperature or a constant temperature difference relative to the air (rather than allowed to cool like the thermocouple) which makes it slightly more accurate. However, this latter difference and the electronic differences among manufacturers are largely transparent to users.
   a. The thermal sensor in a hot-film anemometer is built into a small heated probe with no external or fragile wires. The faster the air velocity, the more electrical power it takes to keep the probe at its designed temperature. Alternatively, a probe can be designed to measure the change in the electrical resistance of a resistor owing to a change in its temperature (using what is called a thermistor).
   b. The sensor is packaged with electronic controls and readout in various forms such as shown in Figure 12.12. The electronics measures the change in either current, voltage, or resistance (again, differences among manufacturers is largely immaterial to the user). The hot-film sensor can be temperature compensated in much the same way as a thermocouple anemometer, typically from at least −10 to 60°C (14 to 140°F).

Hot-film anemometers have most of the practical advantages of heated thermocouple anemometers. The small size of the entire meter and its simplicity are equal to that for the heated-thermocouple anemometer. The hot-film detector is not as sensitive to dust as a hot-wire and is more cleanable because of its larger size and less fragile design. Although the larger sensing element does not respond as rapidly to variations in air velocity as does the heated thermocouple (decreasing its use in aerodynamic research), its response is usually still too fast for convenient IH uses without electronic damping (the damping or averaging time can be varied on some meters).

As for their disadvantages, hot-film anemometers have about the same directional characteristics as hot-wire anemometers. The vertical orientation of the sensor within the probe makes orienting them within known flows more convenient, but none can tell if the air is moving forward or backward. The electronics that keeps the film or probe at a constant temperature has a higher energy requirement than a hot-wire anemometer which shortens their battery life (although they all will probably last for nearly a whole workday, and most commercial constant temperature anemometers

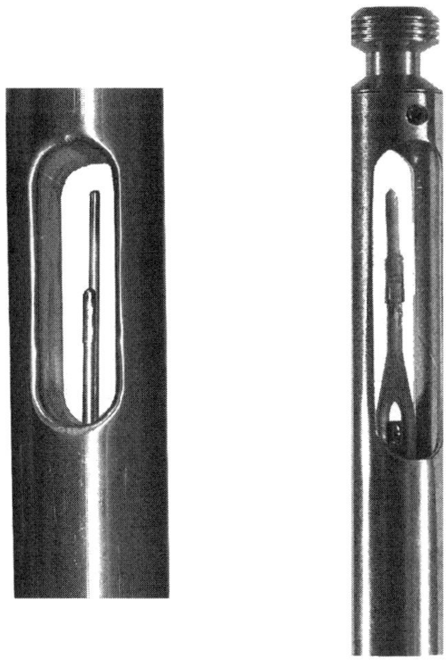

FIGURE 12.11 Two examples of sensing elements for a thermoanemometer. (Courtesy of Kurz Instruments and TSI, Inc., respectively.)

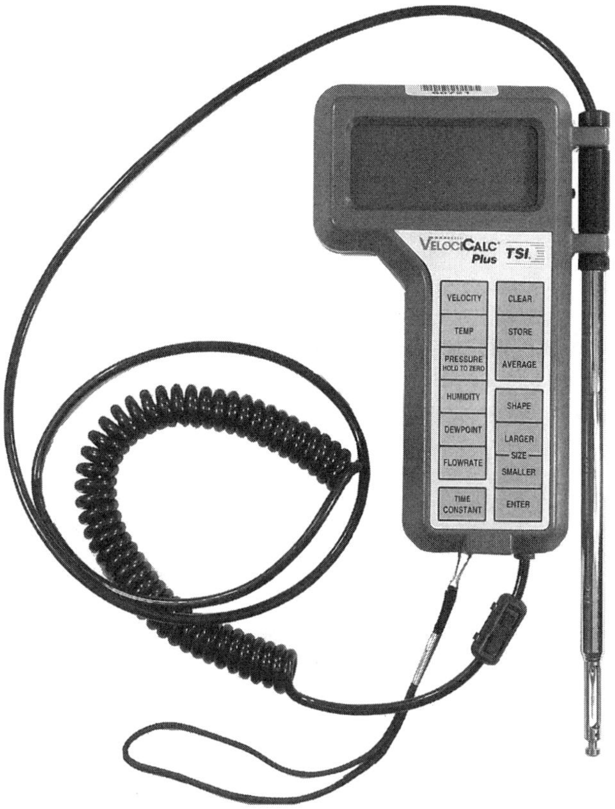

FIGURE 12.12 A TSI VelociCalc®. (Courtesy of TSI, Inc.).

seem to have rechargeable Ni–Cd batteries). The shielded single element detector is more spark resistant than hot-wire anemometers, but still most hot-film anemometers are not intrinsically safe. (Kurz's Series 490IS and TSI's VelociCheck use a resistor for its sensor with a low enough voltage to be intrinsically safe.)

V. THE BOUNDARY LAYER

Ideally, the velocity of air in a duct would be uniform from wall to wall, or the air would move past any solid body at its bulk velocity right up to the surface of that body. Ideally, there would be no friction to stop or slow the molecules right at the wall. Ideally, there would be no viscosity to cause the slow molecules near the wall to interact with the next layer of molecules away from the wall, etc. Unfortunately, real fluids are neither frictionless nor nonviscous.

1. Real fluids are affected by friction and by their own internal viscosity. As air moves through a duct, the molecules of air near the duct walls are slowed as if by friction with the walls. The roughness of the wall affects the magnitude of that friction. Other air molecules that impact the slow molecules near the wall are slowed themselves. The ease with which the fast moving molecules in the bulk air move past the slower moving molecules near the wall is affected by the fluid's viscosity. More technically, *viscosity* is the property of a fluid to transmit forces perpendicular to its flow (what are called shear forces). Thus, the motion of the air molecules near the wall is determined by the balance between the kinetic energy of the air molecules moving in the bulk flow down the duct and the wall friction and the viscous energy or the energy lost to viscosity trying to slow that motion near the wall.

   Conceptually, the velocity changes from a speed of zero at the wall to its bulk or maximum speed some finite distance away from the wall. The depth or distance over which the velocity changes from zero right at the wall or surface of any body to a velocity that approximates the free stream or bulk velocity is called a boundary layer. The depth of the boundary layer depends mostly upon the fluid's *viscosity*, secondarily upon the fluid's velocity and the size of that surface, and tertiarly upon the shape and smoothness of the surface. These parameters together determine whether the boundary layer can be physically characterized as either *laminar* or *turbulent*. The type of boundary layer (whether laminar or turbulent) will predict its size and shape.

   a. A laminar boundary layer has a smooth gradient of velocity with distance with no eddies that could carry energy laterally from a high velocity zone in the bulk flow into the low velocity boundary layer closer to the surface. The lateral transfer of velocity in laminar flow depends upon the molecular interactions that cause viscosity. Thus, a laminar boundary layer is thicker than a turbulent boundary layer. In the extreme, the flow could be laminar all the way across a duct, with the velocity increasing up to a maximum at the duct centerline then decreasing again toward the opposite wall (forming a roughly parabolic curve as depicted in Figure 12.13). Flow tends to be laminar at low velocities where viscous forces predominate.
   b. A turbulent boundary layer has eddy currents that can distribute the bulk velocity laterally across the fluid much more effectively than can the molecular interactions causing viscosity. Such eddies are what cause the turbulent diffusivity discussed in Chapter 5. The turbulent boundary layer depicted in Figure 12.14 would be thinner than the laminar boundary layer shown in Figure 12.13, but the amount of energy that it takes to generate these turbulent eddies is greater than the energy lost due to friction in laminar flow. Thus, the velocity across a duct with a turbulent boundary layer is

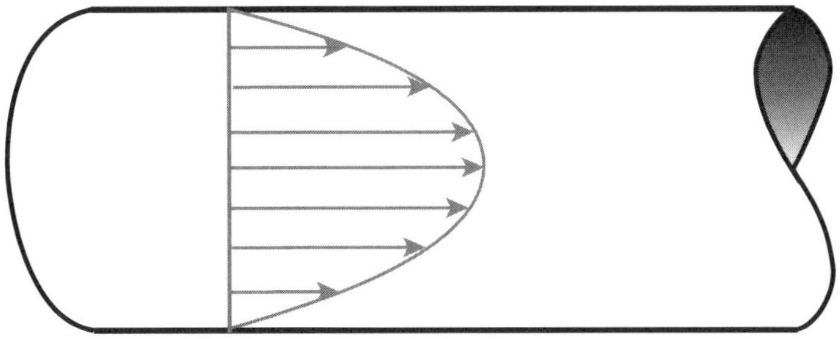

FIGURE 12.13 A depiction of the velocity distribution in laminar duct flow.

more uniform than in laminar flow, but still less uniform than the ideal completely uniform bulk flow. Flow at high velocities tends to be turbulent.

2. A Reynolds number (abbreviated N_{Reynolds} herein may be N_{Re} or as just Re in other texts) is a *dimensionless* empirical coefficient that is widely used to predict fluid flow conditions. The Reynolds number is used to predict the flow both past a solid body such as a building, an airplane wing, or an aerosol particle (see Chapter 3) and through a hollow body such as a blood vessel, a lung passageway, a bulk flow meter, or a ventilation duct. The critical Reynolds number at which the boundary layer transitions from laminar to turbulent varies with the shape of the object. For instance, the flow past a sphere will be laminar if the Reynolds number is less than about 1 (see Chapter 3) but in a duct only if the Reynolds number is less than about 2000.

   The Reynolds number is the ratio between the two major forces affecting fluid flow. The numerator is a characteristic of the kinetic energy of the fluid and is therefore proportional to velocity pressure (ρV^2) times the cross-sectional area of the duct (D^2) or immersed body (such as an aerosol particle in Chapter 3). The denominator is a characteristic of the resisting energy loss internal to the fluid which is related to the fluid's viscosity (η), velocity (V) (similar to Stokes law), and the duct's surface area (which is proportional to D). No matter which system of units is used, the units in the particular combination of variables that make up the numerator and denominator of the Reynolds number all cancel each other out, making the actual Reynolds number unitless and allowing any set of experimental results to be applied directly to another size, velocity, or viscosity that has the same shape and Reynolds number.

   Mathematically, the ratio of these energies simplifies into three or four flow parameters depending upon which version of viscosity is used (as defined just

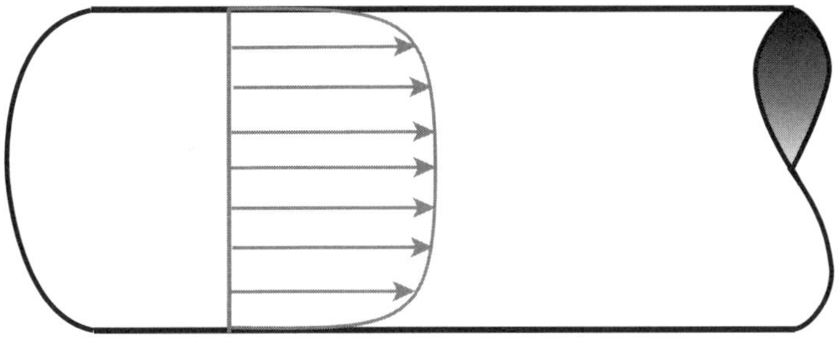

FIGURE 12.14 A depiction of the velocity distribution in turbulent duct flow.

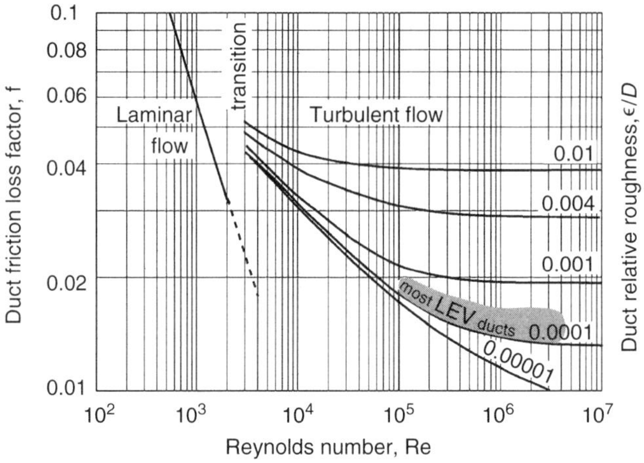

FIGURE 12.15 A simplified Moody diagram.

below Equation 12.3).[b] The most right-hand version of Equation 12.3 with three parameters conveys the simplest essence of the Reynolds number.

$$N_{\text{Reynolds}} = \frac{\text{kinetic energy per unit of length}}{\text{viscous energy per unit of length}} = \frac{\rho V^2 D^2}{\eta V D} = \frac{\rho V D}{\eta} = \frac{VD}{\nu} \quad (12.3)$$

where

D = a characteristic length (duct diameter in the case of ventilation or particle diameter in aerosol dynamics); D^2 represents cross-sectional area, and D represents surface area.

η = dynamic or absolute viscosity (lb sec/ft^2, dyne sec/cm^2, or gm/cm × sec = poise).

ν = kinematic viscosity (ft^2/sec or cm^2/sec = Stokes) $= \eta/\rho$.

ρ = fluid density (ρ for air is 1.2 mg/cm^3 in metric terms and 0.075 lb/ft^3 in English terms).

Technically, ρ, η, and ν all depend upon the fluid and its temperature. The fluid for ventilation and aerosol particles is always air. The values shown below for air at normal temperature are mostly just for your information because they are rarely if ever needed in IH practice.

	English Units	Metric Units
η	3.9×10^{-7} lb$_f$ sec/ft^2	1.9×10^{-4} gm/cm × sec[c]
$\nu = \eta/\rho$	1.6×10^{-4} ft^2/sec	0.15 cm^2/sec

3. Figure 12.15 depicts a Moody diagram that relates the Reynolds number on the bottom axis to the friction factor on the left vertical axis of this diagram and the relative roughness of a

[b] Proportionality factors such as $\frac{1}{2}$ and π used to calculated the surface of some shapes are omitted in the Reynolds number.

[c] The units of viscosity are occasionally given in poise. The viscosity of water at 20°C is 0.01 poise or 0.01 gm/cm × sec = 2.092×10^{-5} lb$_f$ sec/ft^2.[4]

tube, pipe, or duct as a parameter on the right. The experimental research that discovered this pattern was originally conducted in hydrology (water pipes), but the exact same pattern applies to ventilation and ducts. Notice how the friction factor for a laminar boundary layer would continue to drop below the friction factors for a turbulent boundary layer. As the flow velocity and/or the duct diameter increases, the Reynolds number will increase. When the kinetic forces eventually become dominant, the boundary layer will transition from laminar to turbulent. This transition will occur at a Reynolds number between about 2000 and 3500 (the transition zone in Figure 12.15).

Equation 12.3 is used below to show that for the flow to be laminar ($N_{\mathrm{Reynolds}} < 2000$) in a 1-ft diameter duct, for instance, V must be less than 20 fpm, a value that turns out to be unrealistically small for a duct velocity. While friction factors would be less for laminar than for turbulent flow, laminar flow is just not feasible for ventilation ducts.[d]

$$V_{\mathrm{laminar}} = \frac{N_{\mathrm{Reynolds}} \times \nu}{D} = \frac{2000 \times 1.6 \times 10^{-4}\ \mathrm{ft}^2 \times 60\ \mathrm{sec}}{1\ \mathrm{ft} \quad \mathrm{sec} \quad \mathrm{min}} = 20\ \mathrm{fpm}$$

(using Equation 12.3)

The Reynolds number at recommended ventilation duct transport velocities of 2000 to 5000 fpm turns out to range from about 1×10^5 to about 5×10^6 (the shaded area near the lower right side of Figure 12.15). The example below is near the low end of this range.

$$N_{\mathrm{Reynolds}} = \frac{VD}{\nu} = \frac{2000\ \mathrm{ft} \times 1\ \mathrm{ft} \quad \mathrm{sec} \quad \mathrm{min}}{\mathrm{min} \quad 1.6 \times 10^{-4} \quad \mathrm{ft}^2 \quad 60\ \mathrm{sec}} = 208{,}000 \quad \text{(using Equation 12.3)}$$

The Reynolds number at this same velocity in a larger 5-ft diameter duct would be just about 10^6. Thus, the air flow is turbulent in virtually all local exhaust ventilation ducts (and in most flow meters and general ventilation ducts for that matter).

VI. ASSESSING FLOW IN DUCTS

The most common way to assess the volumetric flow rate Q is to measure the velocity at a location where the air is flowing through a known area, then use Equation 11.7 for $Q = V \times A$. However, because of the boundary layer, the velocity across a duct will not be uniform. The standard method to assess the air velocity or volumetric flow rate within a duct with a boundary layer is to conduct a duct velocity traverse.

1. One or more holes need to be drilled through the wall of the duct into which an anemometer will be inserted. In order for flow within a duct to be reasonably symmetric, the location of the traverse should be at least eight straight duct diameters downstream from any major source of air disturbance, such as the hood, an elbow, or a branch duct entry. The hole(s) needs only to be large enough to accommodate the anemometer; it need not be completely burr-free but should not leave material along the inner wall to deflect the air flow. For round ducts, two holes should be drilled 90° apart to test for symmetry of the flow profiles and to average the readings when it is asymmetrical.[5]

[d] The same physical principles also apply to aerosol particle dynamics (Chapter 4) and to aeronautics. Airplane designers have spent a lot of money researching how to keep the air flow over wings laminar.

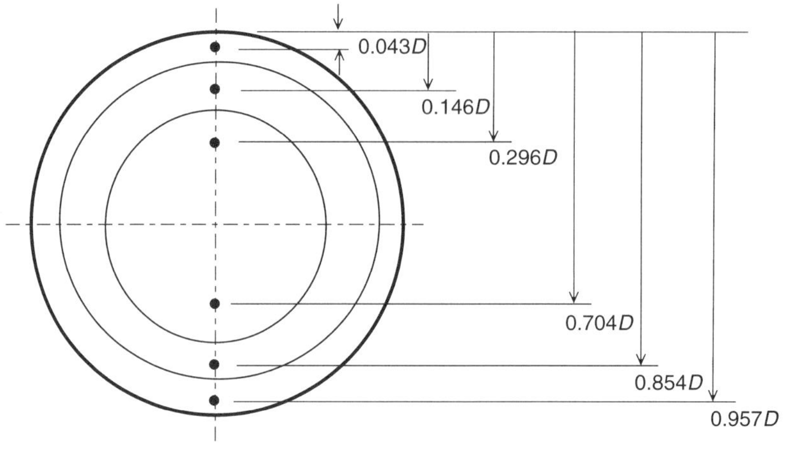

FIGURE 12.16 Depiction of velocity measurements points in the center of six equal annular areas.

TABLE 12.1
Distances from the Wall to the Center of Six Annular Areas Are Given as a Proportion of Any Duct Diameter *D* across the Top and in Decimals and Eighths of an Inch for Three Specific Duct Diameters

	1	2	3	4	5	6
For a Duct						
Diameter of	$0.043 \times D$	$0.146 \times D$	$0.296 \times D$	$0.704 \times D$	$0.854 \times D$	$0.957 \times D$
4 in.	0.17 ($\frac{1}{8}$")	0.58 ($\frac{5}{8}$")	1.18 ($1\frac{1}{8}$")	2.82 ($2\frac{7}{8}$")	3.42 ($3\frac{3}{8}$")	3.83 ($3\frac{7}{8}$")
6 in.	0.26 ($\frac{1}{4}$")	0.88 ($\frac{7}{8}$")	1.78 ($1\frac{3}{4}$")	4.22 ($4\frac{1}{4}$")	5.12 ($5\frac{1}{8}$")	5.74 ($5\frac{3}{4}$")
8 in.	0.34 ($\frac{3}{8}$")	1.17 ($1\frac{1}{8}$")	2.37 ($2\frac{3}{8}$")	5.63 ($5\frac{5}{8}$")	6.83 ($6\frac{7}{8}$")	7.66 ($7\frac{5}{8}$")

2. Measure the velocity at locations and depths using one of the following recommended methods:
   a. For round ducts, experience has shown that it is more accurate and just as easy to measure the velocities in the center of a series of rings of equal areas at unequal distances from the wall rather than to measure the velocities at uniform distances from the wall that would correspond to unequal annular areas. The equal area method takes fewer calculations and also includes measurements deeper into the boundary layer near the wall than would the equal depth method. Figure 12.16 shows the depths as a proportion of the duct diameter for such a six-point traverse. Table 12.1 shows the measurement depths into round ducts of 4 to 8 in. in diameter. More points should be measured in larger ducts to assure that at least one measurement is taken within the boundary layer.[e]
   b. A variable depth method has been developed for rectangles that also allows measurements to be made in the boundary layer. The positions of the measurements listed in Table 12.2 were derived from the Log-Tchebycheff rule.[6] This pattern is touted as an efficient combination of assessing values both inside and outside of the boundary layer. The number of points measured in each direction can be different

[e] The ACGIH *Manual*'s Tables 9.5 to 9.7 show the depth into round ducts for 6, 10, or 20 point traverses.

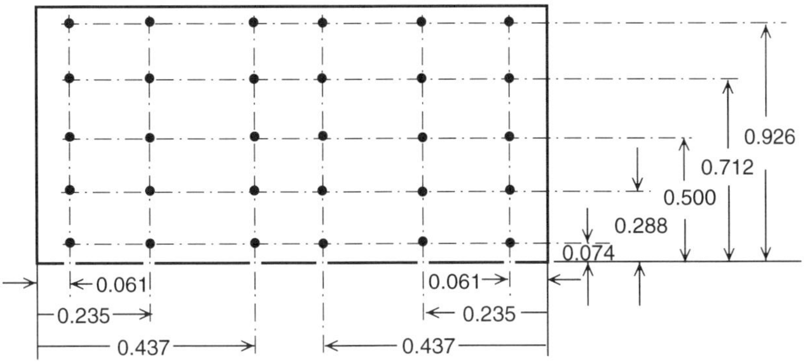

FIGURE 12.17 Example of a variable depth 5 × 6 point velocity traverse in a rectangular duct.

TABLE 12.2
Measuring Point Distances Conforming to the Log-Tchebycheff Rule

Number of Points	1	2	3	4	5	6	7	8
5	0.074	0.288	0.500	0.712	0.926			
6	0.061	0.235	0.437	0.563	0.765	0.939		
8	0.046	0.175	0.342	0.400	0.600	0.658	0.825	0.954

corresponding to the proportions of the rectangular duct, such as the 5 × 6 point traverse shown in Figure 12.17. The distances are symmetric; thus, the same distances from both sides of the duct are shown in Figure 12.17, whereas the holes are only drilled into the bottom of this duct.

c. The *Manual of Industrial Ventilation* still recommends a simple orthogonal X-Y traverse pattern of equal area sectors not more than 6 in. apart for rectangular ducts (see Figure 12.18). This simpler grid pattern is less accurate because it does not place any measuring points within the boundary layer. However, it can be used to assess the face velocity at the entrance to a typical rectangular hood where the air's inertia at $V_{face} < 500$ fpm is negligible and a boundary layer has not yet had time to form. Such a face traverse can be done to assess Q entering the hood, but another useful outcome is to detect an asymmetric velocity profile and to assess its cause such as too large or too many objects within the hood or a baffle that is either missing or out of position.

3. To measure the velocity at each location using one of the above methods, simply insert a small anemometer (such as a Pitot tube) into one of the drilled holes described in step 1 to the appropriate depth as described in step 2. The symmetry of the resulting data can be used to assure that the flow is stabilized at the location and the diameter traversed. In a round duct, two sets of velocities measured across perpendicular traverses at the same location can better assess the average velocity even in the presence of some asymmetry than can measurements across only one traverse. If the distribution is deemed unacceptably asymmetrical, repeat this process at another duct location. One can use $Q = V \times A$ to predict the flow in any desired location from data measured at one convenient location.

It is often helpful to also measure the velocity at the centerline (℄) of the duct. Although $V_{℄}$ is *not* used in the equal annular area calculations, it can be used later as an alternative to velocity traverses. As the centerline is farthest from the wall and its boundary layer, the duct velocity near the centerline is more uniform than at other radial positions, making errors in positioning the probe less critical near the centerline than at any other location.

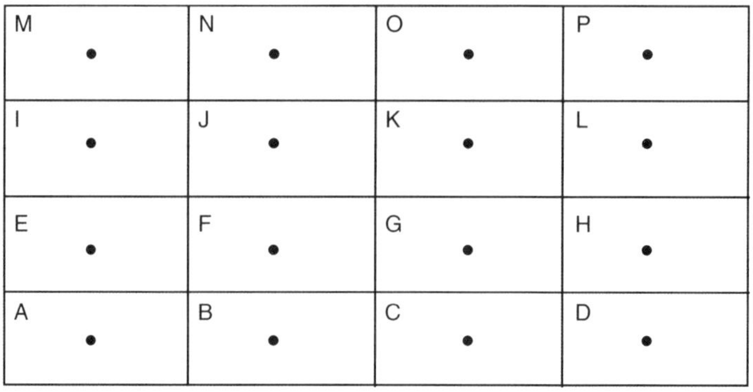

FIGURE 12.18 Velocity traverse across the entrance to a rectangular hood.

For much the same reason, $V_{℄}$ is the least susceptible to changes in wall conditions, making it the best single-measurement baseline for later periodic velocity monitoring. The approximation given in Equation 12.4 can also be used as a time-saving substitute to approximate the results of conducting a full traverse.

$$V_{\text{average}} \approx 0.93 \times V_{℄} \tag{12.4}$$

One of the main reasons this formula is neither as reliable nor as accurate as a full duct traverse is due to unsymmetric flow. Figure 14.13 depicts several examples of unsymmetric flow downstream of an elbow in a round duct. Since duct walls are not transparent, one cannot know what might be happening inside a duct; but one can see and avoid measuring just downstream of an external bend, a duct junction, and an expansion or contraction which are known to create the kind of asymmetric velocity profiles that can adversely affect this approximation.

4. The total Q could be calculated as the sum of the Q_i values flowing through any series of segmented areas within a duct as shown in Equation 12.5.

$$Q_{\text{total}} = [V_i \times A_i] = V_{\text{average}} \times A \tag{12.5}$$

However, a secondary benefit of going to the trouble to measure the V_i in equal areas is to simplify the calculation of V_{average} in Equation 12.6. Notice that if all the ΣA_i terms are equal, they will cancel out in the numerator and denominator to leave n in the denominator needed to calculate the average V.

$$V_{\text{average}} = \frac{Q_{\text{total}}}{A_{\text{total}}} = \frac{\Sigma[A_i \times V_i]}{\Sigma A_i} = \frac{\Sigma V_i}{n} \quad \text{if } A_i \text{ values are equal} \tag{12.6}$$

Should the measurements be taken with a Pitot tube, do not average the Δh or VP values. The velocity corresponding to the average VP is not the same as the average V that you want![f]

[f] The average V based on conservation of mass does not equal the average V^2 based on conservation of energy. Since V varies across the diameter of the duct, the inequality described by Equation 12.13 applies.

$$\text{average velocity} = \frac{\Sigma V_i}{n} \neq \frac{(\Sigma V_i^2)^{1/2}}{n} = \frac{4005(VP_i)^{1/2}}{n} \tag{12.13}$$

For example, the average of velocity measurements of 4 and 8 fpm is 6. However, the average of their squares is $(4^2 + 8^2)/2 = (16 + 64)/2 = 40\ \text{fpm}^2$ whose square root is 6.3, which is not equal to the average V. Close, but no gold ribbon. Therefore, the correct average V cannot be calculated by finding the V of the average VP.

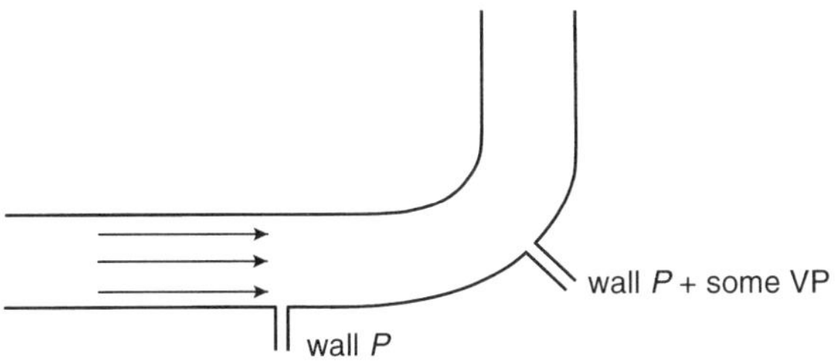

FIGURE 12.19 Elbow pressure tap.

5. Placing a pressure port in the elbow of a duct as shown in Figure 12.19 will yield neither a good measure of the wall *P* (because some VP impinges on the hole) nor a true measure of total pressure (because some of the flow has already changed direction). However, a pair of pressure ports can be used as a low budget duct flow meter by comparing the portion of the total pressure that impinges on the outer wall of an elbow to the wall pressure measured just before the elbow. Such an elbow anemometer with at least a one-point calibration can be a cheap continuous monitor.

VII. CALIBRATING ANEMOMETERS

Anemometers should be calibrated periodically. The calibration frequency must be judged by the stability of the anemometer in use and the potential effect of an error in a given measurement. For instance, a Pitot tube is extremely stable; any change in its accuracy is likely to be due to visible damage such as bending of the tube or cracking or other holes in the hoses. The former problem is obvious, and the latter problem is detectable by a thorough visual inspection. Changes in mechanical and electronic anemometers are less obvious. A commonly recommended recalibration frequency for mechanical and electronic anemometers is annually (probably more for the convenience of scheduling than as the result of any quantitative failure analysis).

An anemometer calibration apparatus can be purchased commercially or constructed. However, most hygienists just send their meters out to its manufacturer or other commercial laboratory for recalibration, in which case knowing how to construct or operate a calibration apparatus is not essential. A calibration wind tunnel creates a stable velocity in the meter's test section whose value is known from the Δh measured downstream in a standard secondary bulk flow meter such as a venturi or orifice.

1. The features and function of the four major components that comprise a typical calibration wind tunnel such as shown in Figure 12.20, are described below. More details are shown in the *Manual of Industrial Ventilation* (Figure 9.5 and Figure 9.6).
   a. Air enters the test section of the wind tunnel through an open bell-mouth. A bell-mouth is a smoothly curved entrance that produces the most uniform face velocity in comparison to a plain or sharp-edged entrance that produces a *vena contracta* and turbulence. A *vena contracta* is an overly constricted flow pattern that would prevent a uniform velocity from existing in the test section (to be discussed further in Section II.3 in Chapter 14).
   b. The anemometer or its sensor is inserted into the test section. Since the usable flow range of any given secondary standard bulk flow meter is limited, some calibration wind tunnels are made to accommodate multiple interchangeable test sections, each of

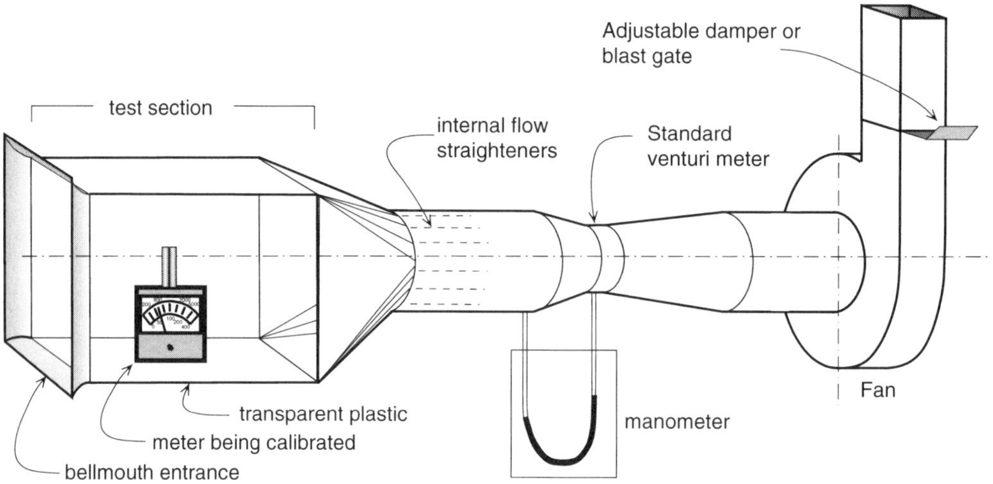

FIGURE 12.20 Depiction of a manually controlled anemometer calibration wind tunnel.

which has a different cross-sectional area so that the velocity obtainable among the multiple test sections spans a wider range. The sensor should be placed as far from the test section's walls as feasible (i.e., near its centerline) and no further than one diameter downstream from the entry face (before the boundary layer has a chance to build up within the test section). Beyond that distance, the boundary layer will cause V to vary across the test section, and the average V will not be equal to the centerline V being sensed and indicated by the anemometer.

c. A reference secondary standard bulk flow meter (sometimes called a full-bore meter) is placed downstream of the test section. Figure 12.21 depicts three common bulk flow meters that could be used in a calibration wind tunnel.[g] The *Manual of Industrial Ventilation* recommends using an orifice meter (Figure 12.21a); it is cheaper, but a venturi meter (Figure 12.21c and depicted in Figure 12.20) would have less resistance to flow and require a smaller fan to achieve the same velocities as an orifice. A manometer connected in-line between the two pressure taps placed as shown by P_1 and P_2, will measure the Δh corresponding to the flow velocity through the meter. Section VII.3 will discuss how to use a bulk flow meter correctly.

d. A calibration wind tunnel must have a fan and motor to power the entire system, and some way to vary the flow. The simplest flow control is a solid state speed control for the motor. Flow can also be controlled by either a bypass inlet (a bleed-in just before the fan) or a damper on the outlet (to vary the total resistance or losses that the fan must overcome). The latter is shown in Figure 12.20.

2. Two readings must be recorded at each flow condition. Measure and record the anemometer's indicated velocity appropriate to the meter (as directed by the manufacturer). Measure and record the Δh across the wind tunnel's standard bulk flow meter. The velocity in and flow through the standard meter (V_{meter} and Q_{meter}, respectively) are calculated using procedures described in the next Section VII.3). The velocity in the test section (V_{test}) is calculated using the simple modification of Equation 11.7 shown in Equation 12.7 (here $Q_{meter} = Q_{test} = V \times A$). This V_{test} is the

[g] A rotameter is not as reliable as a secondary standard nor does it possess as much flow capacity as any of the above meters.

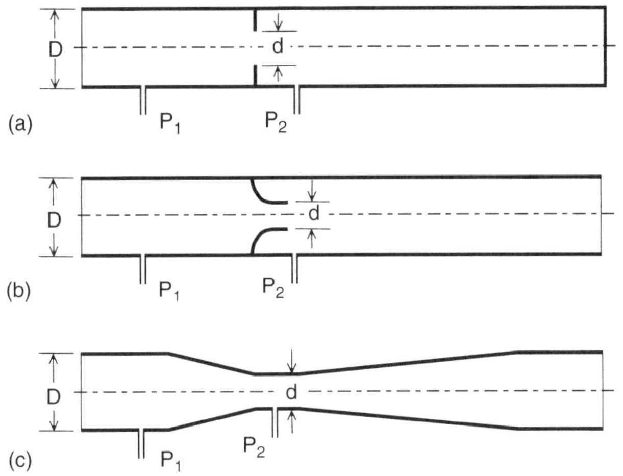

FIGURE 12.21 (a) An orifice bulk flow meter. (b) A nozzle bulk flow meter. (c) A venturi bulk flow meter.

actual velocity to which the anemometer was exposed during calibration and should be the reading on that anemometer if it is properly adjusted.

$$V_{\text{test}} = V_{\text{meter}} \frac{A_{\text{meter}}}{A_{\text{test}}} \tag{12.7}$$

(The remainder of this chapter describes procedures used to determine the actual flow rate through a standard bulk flow meter. The development of the necessary equations leading to the use of the appropriate empirical flow rate coefficient "K_{meter}" [yes, another otherwise nondescript K value] is shown in more detail than is necessary for the average IH unless one were actually to use such a calibration apparatus. Thus, the rest of this chapter is just background information and not required knowledge for all IH students.)

3. As discussed in the previous chapter, air in ventilation and calibration wind tunnels behaves as if it were incompressible. However a boundary layer still exists within bulk flow meters that must be accounted for. The ideal (as if the boundary layer did not exist) velocity V_2 in the constricted portion of the meter can be found using the principles of "conservation of mass" and "conservation of energy" from Chapter 11. The following derivation leading to Equation 12.16 explains why the velocity in the throat of a reference secondary bulk flow meter can be calculated from its Δh.
   a. Equation 12.8 is derived from the principle of conservation of mass with incompressible air Equation 11.7.

$$A_1 \times V_1 = A_2 \times V_2 \qquad \text{(Equation 11.7)}$$

$$V_1 = V_2 \frac{A_2}{A_1} \tag{12.8}$$

Equation 12.9 is derived from the principle of conservation of energy Equation 11.13 and 11.26:

$$V_1^2/2g + P_1/\rho_1 = V_2^2/2g + P_2/\rho_2 + \text{losses} \tag{12.9}$$

b. These two equations can be combined by replacing V_1 in Equation 12.9 with its equality from Equation 12.8, setting $\rho_1 = \rho_2$ (and denoting both as just ρ because the fluid can be considered incompressible), and disregarding for the moment the losses between the approach to the meter (point 1 where P_1 is measured) and the constricted "throat" (where P_2 is measured):

$$\frac{\left(V_2 \times \frac{A_2}{A_1}\right)^2}{2g} + \frac{P_1}{\rho} = \frac{V_2}{2g} + \frac{P_2}{\rho} \tag{12.10}$$

$$V_2^2\left[1 + \left(\frac{A_2}{A_1}\right)^2\right] = 2g\left(\frac{P_1}{\rho} - \frac{P_2}{\rho}\right) \tag{12.11}$$

c. Now taking the square root of Equation 12.11, then letting $\Delta h = \Delta P/\rho = (P_1 - P_2)/\rho$, yields the constricted velocity V_2 for an ideal meter without any boundary layer effects or losses.

$$V_2 = \sqrt{\frac{2g(P_1 - P_2)}{\rho\left(1 - \left(\frac{A_2}{A_1}\right)^2\right)}} = \sqrt{\frac{2g\Delta h}{1 - \left(\frac{A_2}{A_1}\right)^2}} \tag{12.12}$$

4. The empirical bulk flow meter coefficient $K_{\text{bulk meter}}$ converts the Δh reading into the V_2 in the meter's throat by adjusting for the effects of the boundary layer, the *vena contracta*, the pressure losses within the meter (that were disregarded above), and the ratio of the constricted throat area to the area of the duct approaching the throat (A_2/A_1 in Equation 12.12). The first correction is always needed because the radial variations in V affect the measured Q/A or the calculated average V differently than they affect the measured VP or calculated average V^2 (as previously demonstrated in footnote f). The $K_{\text{bulk meter}}$ values in Figure 12.22 are similar to the data in the *Manual of Industrial Ventilation*'s Table 9.4 but cover all of the meters shown in Figure 12.21.

   Values of $K_{\text{bulk meter}}$ are laid out in this figure as a function of the Reynolds number in the throat along the X axis (dV/ν where $d =$ the throat diameter) and the $\sqrt{2g\Delta h}$ as a surrogate for V on the diagonal lines. Note that each slightly diagonal line crosses the horizontal line for $K_{\text{bulk meter}} = 1$ line at the equivalence stated in Equation 12.14.

$$N_{\text{Reynolds}} = dV/\nu = K_{\text{bulk meter}}\, d\sqrt{2g\Delta h}/\nu \tag{12.14}$$

Δh is used as a surrogate for V since one does not know V without knowing $K_{\text{bulk meter}}$. Therefore, one has to enter Figure 12.22 with a value of $d\sqrt{2g\Delta h}/\nu$, find where the corresponding slightly diagonal line crosses the curve for the appropriate bulk meter

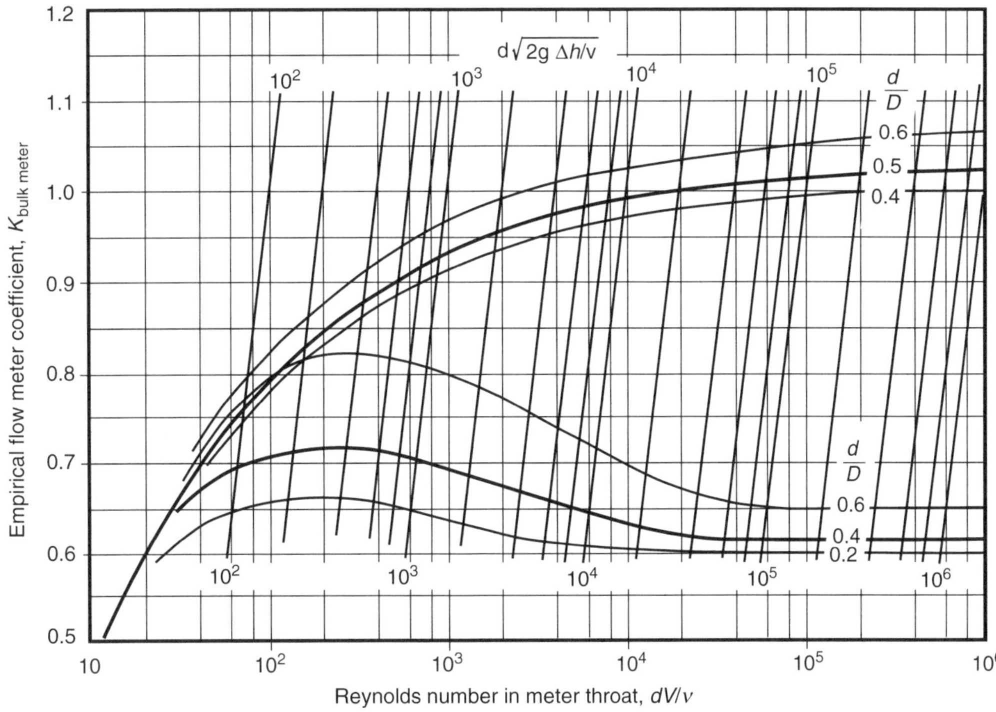

FIGURE 12.22 Empirical flow meter coefficients ($K_{bulk\ meter}$). (From Olson, R.M., *Essentials of Engineering Fluid Mechanics*, International Textbook Company, Scranton, PA, 1962. With permission.)

and d/D ratio, then read $K_{bulk\ meter}$ on the left or Y axis. (The d/D ratio in Figure 12.22 corresponds to the diameters shown in Figure 12.21.)

$$\text{Now solve for } V_2 = K_{bulk\ meter}\sqrt{2g\Delta h} \approx K_{bulk\ meter}\, 4005\sqrt{\Delta h} \tag{12.15}$$

Example 12.1

Find the $K_{bulk\ meter}$ and V_2 for a $\Delta h = 3''$wg in a calibration wind tunnel with a venturi where large $D = 6'' = D_1$ (the diameter of the duct approaching the meter) and small $d = 3'' = D_2$ (the constricted diameter at the meter's throat).

a. To find $K_{bulk\ meter}$, one must first find $d \times \sqrt{2g\Delta h}/\nu$, the temporary equivalent Reynolds number using any set of consistent units. One way to do this is first to use ft/sec in Equation 11.16 to determine $\Delta h \times 69.27 = 208$ ft air, then use a slightly modified form of Equation 11.17 to determine …

$$\sqrt{2g\Delta h} = \sqrt{2 \times 32.2 \times 208} = 115.7 \text{ ft/sec} \quad \text{(using Equation 11.17)}$$

Another way is to modify Equation 11.18 as $\sqrt{2g\Delta h} = 4005 \times \sqrt{(3 \text{ in.})}/60$. In either case, the 115.7 ft/sec can now be used to determine the temporary equivalent Reynolds number.

$$d \times \sqrt{2g\Delta h}/\nu = 0.25(115.7)/1.7 \times 10^{-4} = 1.7 \times 10^5$$

b. The value for $K_{bulk\ meter}$ can now be found on the vertical axis in Figure 12.22 where the appropriate slanted line for the value of $d \times \sqrt{2g\Delta h}/\nu = 1.7 \times 10^5$ crosses $d/D = 0.5$ for the venturi meter.

$$K_{\text{bulk meter}} = 1.01 \text{ or } 1.02$$

While a $K_{\text{bulk meter}}$ so close to one appears to be nearly ideal, note that if one were to use $V_{\text{meter}} = V_2 = \sqrt{2g\Delta h/(1-(A_2/A_1)^2)}$ as shown in Equation 12.12, the denominator would equal $1 - 0.5^4 = 0.9375$. Thus, the true correction from the ideal meter with no boundary layer is 1.02/0.9375 = 1.088 or almost a 10% adjustment!

c. The V_{meter} in the throat is determined using Equation 12.15.

$$V_{\text{meter}} = K_{\text{bulk meter}} 4005\sqrt{\Delta h} = 1.015 \times 4005\sqrt{3} = 7041 \text{ fpm} \qquad \text{(using Equation 12.15)}$$

5. None of these bulk flow meters need to be adjusted for compressibility when they are used in ventilation or a calibration wind tunnel where air speeds are not fast enough to compress the air significantly. Compressibility only becomes important for $V > 13{,}000$ fpm or for a bulk flow meter where Δh is >20 "wg. The errors of neglecting Y altogether will always be $\leq 5\%$ for Δh less than 20 "wg. Figure 12.23 is provided in the unlikely event that you should ever encounter a V beyond those normal in LEV and need to use a compressibility factor. One can find P_2/P_1 using Equation 12.16 where P_1 is approximately the ambient P or 407.6 in. of water at NTP. A linear $Y = a + bX$ can be found for a given d/D. For instance, for $d/D = 0.5$, $Y = 0.375 + 0.625\ P_2/P_1$.

$$P_2/P_1 = (P_1 - \Delta P)/P_1 \approx (407.6 - \Delta h)/407.6 \qquad (12.16)$$

$$\text{And when needed, } V_2 = YK_{\text{bulk meter}}\sqrt{2g\Delta h} \qquad (12.17)$$

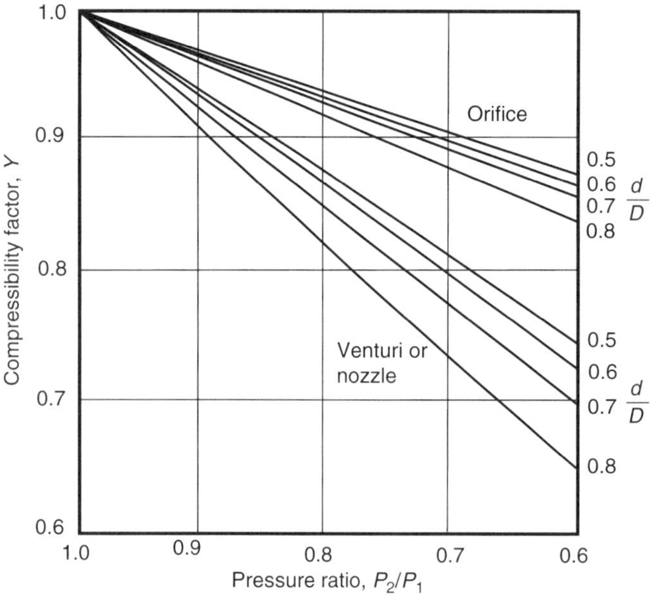

FIGURE 12.23 Expansion factors for air. (From Olson, R.M., *Essentials of Engineering Fluid Mechanics*, International Textbook Company, Scranton, PA, 1962. With permission.)

REFERENCES

1. Maynard, A., Thompson, A. M., Cain, J. R., and Rajan, B., Air movement visualization in the workplace: Current methods and new approaches, *Am. Ind. Hyg. Assoc. J.*, 61(1), 51–55, 2000.
2. Lenhart, S. W. and Burroughs, G. E., Occupational health risks associated with the use of irritant smoke for qualitative fit testing of respirators, *Appl. Occup. Environ. Hyg.*, 8(9), 745–750, 1993.
3. Jensen, P. A., Hayden, C. S., Burroughs, G. E., and Hughes, R. T., Assessment of the health hazard associated with the use of smoke tubes in healthcare facilities, *Appl. Occup. Environ. Hyg.*, 13(3), 172–176, 1998.
4. Lide, D. R., Ed., *CRC Handbook of Chemistry and Physics*, 83rd ed., CRC Press, Boca Raton, FL, 2002.
5. Guffy, S. E. and Booth, D. W., Comparison of Pitot traverses taken at varying distances downstream of obstructions, *Am. Ind. Hyg. Assoc. J.*, 60(2), 165–174, 1999.
6. Goodfellow, H. and Tähti, E., Eds., *Industrial Ventilation Design Guidebook*, Academic Press, London, 2001, Referenced to ISO 3966: 1977, Table 12.8 of Section 12.3.
7. Olson, R. M., *Essentials of Engineering Fluid Mechanics*, International Textbook Company, Scranton, PA, 1962.

13 Designing and Selecting Local Exhaust Hoods

The learning goals of this chapter:

- ✍ Be able to discuss the first three good local exhaust principles and be familiar with the remaining five.
- ☐ Know how to find the control velocity in one of the *Manual of Industrial Ventilation*'s VS diagrams and to pick a control velocity from among the generic values in Table 13.1. Be familiar with the latter's ranges of values.
- ☐ Know that $Q_{\text{containment}}$ always equals $A_{\text{face}} \times V_{\text{face}}$ (Equation 13.1).
- ☐ Know how to use the DallaValle equations to find Q, V_{face}, or V_{control} at a distance X outside the hood's face.
- ☐ Know that for a collection hood, $V_{\text{control}} \approx 10\%$ of V_{face} at one face diameter out in front of a hood.
- ✍ Be able to calculate the effective face area and the slot area of a slotted hood.
- ✍ Be able to recognize or describe a canopy, side draft, down draft, and push–pull hood and at least one condition that would make each of these hoods particularly appropriate or inappropriate.
- ✍ Be able to specify a control velocity for an open surface tank using ANSI Z9.1.

The principles of local exhaust hood design presented in Sections I–IV apply to all the specific hoods that follow in Sections V–XI. These principles include the recommended generic control velocities presented in Section II, the DallaValle equation and use of flanges for collection hoods presented in Section III, the design of slotted hoods in Section IV, as well as the mechanisms that affect source generation and plume behavior presented in previous chapters.

Section V of this chapter describes the format and uses of ACGIH "VS" diagrams. Each VS diagram specifies an exhaust air flow rate (and in effect the control or face velocity), the applicable hood entry pressure loss factor(s), and usually the minimum duct velocity. The OSHA/ANSI criteria for open surface tank ventilation are presented in Section VI. The design procedures for canopy hoods, push–pull hoods, and other specialized hoods presented in the last sections of this chapter are intended to provide the reader with a good starting kit of hood design tools that can grow with experience.

I. LOCAL EXHAUST HOOD DESIGN PRINCIPLES

The effectiveness and acceptability of any local exhaust ventilation (LEV) system are ultimately determined by how well it protects employees. Unfortunately, none of the common design methodologies yield quantitative estimates of employee protection. Thus, a ventilation system has to be built before one can conduct the ultimate test to measure exposures. Existing generic hood design tools are limited to

- Selecting a control velocity: either a process specific velocity (discussed in Section II) or a generic velocity (Table 13.1) whose recommendations are based largely on the source's inertia.

- Using the DallaValle equation: a predictive formula (discussed in Section III) relating a given control velocity and a collection hood's location (distance from the source or plume) to its face velocity.
- Determining the shape and size of a hood: either a specific hood (as exemplified by the *Manual of Industrial Ventilation*'s VS diagrams to be discussed in Section V) or a generic hood (as exemplified by containment, downdraft, lateral exhaust, or canopy hood most of which are discussed in Sections V–VIII).

The following principles can guide the design of any local exhaust hood but are intended especially to guide those designs that go beyond the VS diagrams in the *Manual of Industrial Ventilation*.[1]

1. Air velocity in the critical pathway (viz., in the zone of control) is not the only determinant of the adequacy of exposure control, but it is the most easily measured and the first design parameter to be chosen. The flow rate (Q) at which the contaminated air is exhausted and the turbulence in that air are also important (see LEV Principle 1 in Chapter 10). The more air that is being removed, the lower the control velocity can be. Too much velocity can actually create enough turbulence to increase someone's exposure. A key role of the hygienist during the design phase is to select an appropriate control velocity, and later to assess it in existing LEV systems and possibly to adjust it to achieve the desired level of employee protection (exposure level). The actual process of selecting a control velocity is discussed in the next section of this book. LEV system management is discussed much later in its own chapter.
2. Avoid moving the contaminant from the source toward someone's breathing zone. This principle is violated if someone is allowed to work with their breathing zone under a canopy hood (as in the worker on the left in Figure 13.1) or downwind of a source (as in position D of Figure 5.6 or Figure 5.8). Relying on the employee to avoid the plume is a reliable control only if the plume's trajectory is consistent. A side draft hood (as shown in Figure 13.2) creates both a more consistent plume and one that is easier to avoid than working under a canopy. Again, training someone to work in the lateral position B of Figure 5.6 vs. working in the traditional position upwind of the source (position A of Figure 5.6) will result in even less exposure to the plume but is probably not intuitive to most people.
3. Decide whether a containment hood can be designed to enclose the source or the source must stay outside the hood and its contaminants must be collected into it. This decision (related to Principle 2 in Chapter 10) should be made as early in the design process as feasible because it intimately affects the relationship between the face and control velocities and the size of the open face area of the hood. While two parallel paths of design can be pursued for a while, much of the rest of the design process cannot proceed until a clear choice can be made. By defining "hood efficiency" as the fraction of a source's emitted contaminant that is either collected into or retained within a hood per unit of exhaust air flow rate, a containment hood will always be more efficient than a collection hood; but containment is not always feasible.

   Figure 13.3 presents the professional opinion of the effectiveness of an array of hoods.[2] The "fraction not controlled" represents the fraction of a chemical emission that is not collected by the hood. A value of one indicates that all of the contaminant escapes from the hood, and none is collected (a terrible hood). A value of 0.01 indicates that only 1% escapes and 99% is collected (much better). Thus, the further to the right that a hood's effectiveness is plotted on Figure 13.3, the more the emission is controlled. These data probably reflect more the interests of air pollution emissions than of employee protection

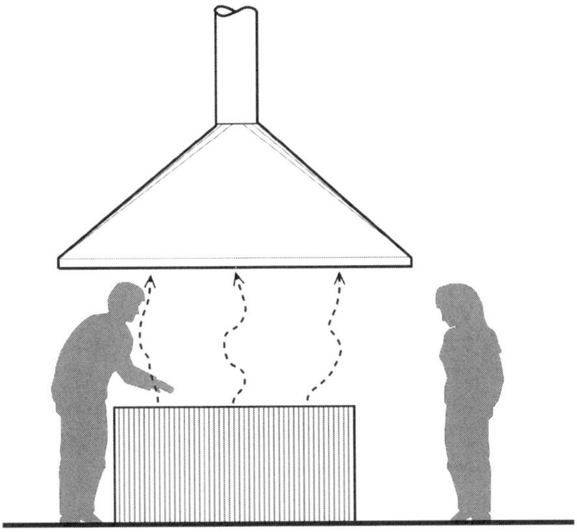

FIGURE 13.1 This canopy draws contaminants into the breathing zone of the person on the left.

or the Environmental Dilution Ratio in the breathing zone. Nonetheless, the pattern of improved control via containment vs. collection is striking.

a. For a containment hood, enclose as much of the source as feasible. Or, viewed from the other perspective, a containment hood should minimize the hood face area while still allowing for:
    - user access as needed including reaching in with ones hands and arms, seeing in visually, and maintenance of any process, tools, or machines that are inside the hood.
    - the passage of production materials entering, and/or processed parts leaving the hood, utility connections for electricity, fluids, air or hydraulic pressure, etc., and excess heat from the enclosed process or machinery.

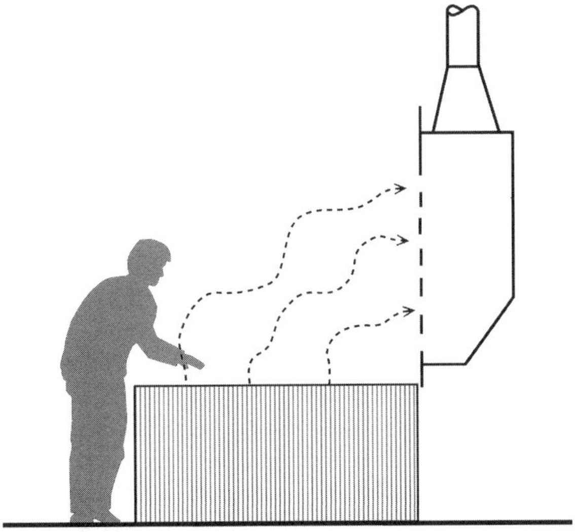

FIGURE 13.2 A side-draft hood avoids moving contaminants into this person's breathing zone.

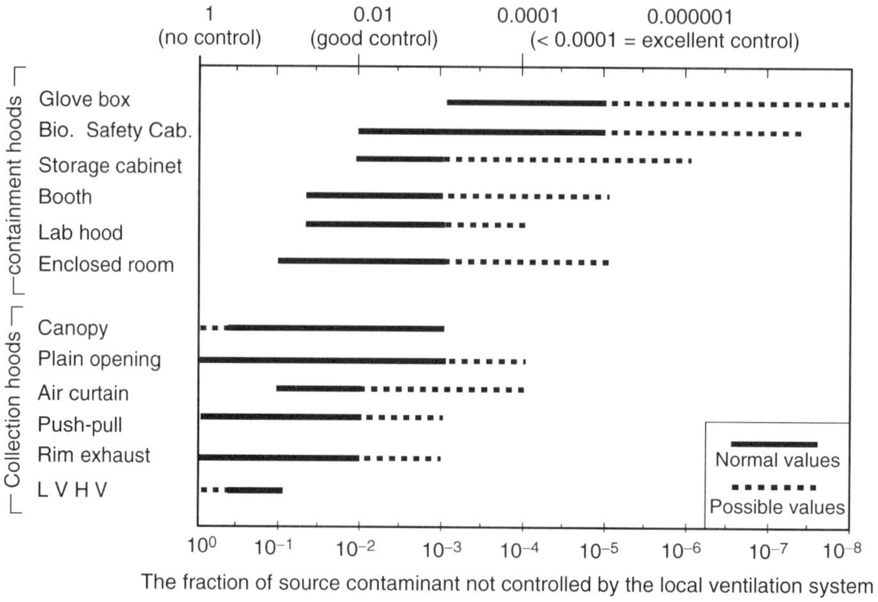

FIGURE 13.3 The approximate effectiveness of emission control via containment vs. collection hoods.[2]

The control velocity in a containment hood is its face velocity, and (as a consequence of Equation 11.7) the exhausted air flow as defined by Equation 13.1 is less than it would be for any collection hood.

$$\boxed{Q_{\text{containment hood}} = A_{\text{face}} \times V_{\text{face}}} \tag{13.1}$$

b. If a collection hood is necessary, its face should be as close to the source as feasible, its flow should be aided by flanges, and its face should approximate the width of the source and its plume. The control velocity outside of a collection hood decreases with the square of the distance outside the face of that hood (to be described by the DallaValle equations in the next section). Therefore, one should minimize the distance from the source to the collection hood's face. One should also use flanges (barriers extending to the sides of the hood's face) to minimize the flow of uncontaminated air from the sides and back of a collection hood. While a theoretically optimum huge face area is not feasible, a combination of observation and/or experience indicates that the width of a good collection hood approximates that of the source and its plume.

4. *Conserve exhaust air.* To "conserve" here means to remove the maximum amount of contaminant for the minimum amount of air exhausted consistent with protecting the employee. Not only does it cost money to move air mechanically, but it costs even more to heat or cool the replacement or "make-up" air. The rest of the principles listed below all contribute to conserving exhaust air.

   a. *Anticipate the natural convective inertia of a source's plume and try to place the hood where it can work with that natural inertia (like a receiving hood) rather than pulling exhaust air against it.* Thermal density is a common cause of natural convection. A hot plume and its contaminants will rise due to its lower density, suggesting the use of a canopy as a "receiving" hood (its design will be discussed in Section VII and Section VIII). A side draft hood designed to collect such a rising plume should be tall (to extend the vertical distance over which the control velocity will act). On the other hand, trying to collect a very cold (and thus thermally dense) vapor with a canopy will be inefficient. The natural convective motion of plumes emanating from moving parts or materials (e.g., chips from a grinding wheel, debris from a conveyor belt, air displaced while filling a container) can be anticipated by the source's shape and speed.

b. *Shape the control velocity profile to accommodate aerosol sedimentation or the plume's thermal buoyancy.* One should try to use baffles and slots to pull most of the air into a containment hood where most of the contaminants would otherwise come out. For instance, if dealing with large particles, more face velocity is likely to be needed toward the bottom of an enclosure where large particles will settle more rapidly and either accumulate or/and spread if not controlled. A lower velocity may be acceptable near the top of such a containment hood where fewer and smaller particles would otherwise come out, but recall from Chapter 3 that sedimentation has a minimal effect on respirable aerosols (diameters $< 10\ \mu m$), and once small aerosol particles leave the vicinity of the source, they will disperse much like gases and vapors.[3]

c. *Avoid placing a hood's face where cross-drafts can disrupt its control.* A cross-draft is any air flow in a direction other than toward the face of the hood. A significant cross-draft (say $> 2 \times$ face velocity) will disrupt either the collection or the containment of any contaminant. Common sources of high cross-draft air velocities in the workplace include cooling fans, the outlets of fresh air supply ducts (called "diffusors"), and open doors and windows. Cross-drafts can disrupt a collection hood more easily than it can disrupt a containment hood. Canopy hoods tend to be especially susceptible to even low speed cross-drafts that can cause the rising plume to miss the hood either partially or completely. Developing policies that prevent the use of cooling fans near local exhaust hoods can help only a portion of these problems, but any policy is only as good as its promulgation and enforcement.

d. *Try to minimize (and avoid creating) turbulence in the plume or in the collection pathway between the source and the hood.* Turbulence will disperse and expand the plume, making the hood less efficient at collecting or containing the contaminant. Turbulent eddies can be generated in front of a hood (e.g., by people passing or forced-air outlets or fans being nearby), at the entrance to the hood (e.g., by the user's body or by sharp or square edges on the hood), or within the hood (e.g., by things moving inside a hood). Some of the research by Flynn et al. and Livengood (cited in Section III.1, Section VIII.3, and Figure 5.12 of Chapter 5) shows that the turbulent "wake" created by an employee's body can carry toxic contaminants upwind into their breathing zone. Working to the side of the source (position B in Figure 5.5) avoids this adverse effect of turbulence.

e. Take advantage of opportunities to *reduce the exhaust rate when emissions are low.* This concept is certainly useful for cyclic sources that emit a majority of their contaminants during a short portion of that cycle. One approach is to operate either two fans or one fan at two speeds where the high flow rate is linked to the high need in some way. For instance, a motion detector can sense when someone is present who could either disrupt the free-flow of air into a hood or be conducting an activity that would release more contaminant than when none is present. On the other end of the cycle, a way is needed (usually just a timer) to drop from a high flow rate back to low. Another approach is to design a "pool hood" in which the spillage from an exhaust canopy is contained within a wider set of baffles. The volume of the larger pool hood must be sufficient to contain the excess emission flow rate for the duration of the transient surge (see Heinsohn, 1999, p. 318).[4]

f. *Generally disregard molecular diffusion or high vapor molecular weight as hood design factors.* Molecular diffusion is too slow of a process to move a significant amount of any contaminant across the dimensions of most rooms (recall Section III.3 of Chapter 5). Molecular diffusion can only be a significant factor over very short distances (e.g., within passive air sampling devices) or in still air and over a relatively long time (e.g., within an unoccupied storage area or confined space). The bulk air velocities that exist in most rooms will move a contaminant much faster than it can be moved by molecular diffusion.

Room air convection will dilute high molecular weight vapors and keep weak buoyant forces of a solvent vapor that is at or near its health exposure limit from deflecting plumes vertically. The only time that a vapor's molecular weight might be important is if concentrated vapors (in the range of their LFL) start out on the floor or ground *in undisturbed air* (with no foot traffic or ventilation). One solution to that problem is to place a low volume exhaust hood near the floor of such a room; another is to disturb (mix) and dilute such vapors with fresh air.

II. SELECTING A CONTROL VELOCITY

This section discusses four sources to guide the selection of an appropriate control velocity. Three of these sources are from the ACGIH *Manual*,[1] and the fourth comes from OSHA standards (scattered from 1910.94 through 1910.252 plus 1926 and 1915 but considered herein to be one source.)[5] However, none of these recommended control velocities (or any other guidelines) are quantitative predictors of exposure control. In each case, the recommended control velocity is based on experience, but they do not guarantee compliance with performance exposure limits, and even if a given velocity decreases exposures to below current exposure guidelines, control criteria change as new information becomes available. Thus, all of these recommendations are just guidelines in a very real sense.

1. Practically every VS diagram depicted in Chapter 10 of the *Ventilation Manual* and discussed in Section V of this chapter lists a control velocity (V_{control}) or flow rate (Q). Most diagrams actually list a Q in either ft^3/min/ft or ft^3/min/ft^2 where the denominator is the dimensions of either the source or the hood. In a few cases, Q is also based on the distance X in front of the hood based on one of the DallaValle equations to be explained in Section III. Analogous to the $V = Q/A$ form of Equation 11.7, any Q given in ft^3/min/ft^2 is equivalent to a V given in fpm. The Q value given in this form allows the designer to match the system to the size of the source or process, and viewing the value as a velocity can help one extrapolate the control velocity for a similar process to another hood design (or on an exam). A summary table at the beginning of the *Manual of Industrial Ventilation*'s Chapter 10 lists hoods by setting and operation, but the applicability of each hood to your setting can only be judged by looking at the diagram and envisioning its use (or not) in your setting. When a VS diagram is applicable, it provides not only the Q or control velocity but also the hood entry energy loss factor(s) and the recommended minimum duct velocity.
2. The *Ventilation Manual*'s Table 10.99.2 lists control velocity information for various miscellaneous specific operations. For most operations, either the air flow rate (Q) or the control velocity (V_{control}) is listed. A few operations list the flow rate per unit area of source or face (Q/A, ft^3/min/ft^2), just as within many VS diagrams. These recommendations are less specific than a VS diagram, but better for the specific operation than the following generic range of control velocity guidelines. The *Manual of Industrial Ventilation*'s table should be perused to see if it contains the process you need to be controlled or a similar process; however, perusing takes time and some familiarity

with the different terms and processes and is therefore not recommended for use during an exam unless you are specifically directed to do so.

3. OSHA ventilation standards applicable to specific hazards were summarized in Chapter 4. The most complex of their ventilation standards is that for open surface tanks. Section VII of that chapter only provided the first two steps of ANSI Z9.1 related to assigning the dichotomus hazard classification values.[6] The remaining steps of using these values in selecting a control velocity will be discussed in Section VI of this chapter. That section will contain two tables: one table specifies control velocities for enclosing, canopy, and lateral exhaust hoods; and the other table specifies the volumetric flow rates (Q) for lateral exhaust hoods used to control various shaped tanks. As in osmosis, gaining further familiarity with other OSHA ventilation standards will be much easier when one is motivated by the pressures of real challenges (and a paycheck) at work. The student should, at this point, be familiar with the kinds of processes covered by OSHA regulations in Chapter 4. However, remember that simply complying with any OSHA ventilation design specification (or a VS diagram for that matter) does not guarantee nor substitute for compliance with the chemical's pertinent exposure limit.
4. Table 13.1 gives ranges for control velocities based primarily on the momentum or inertia of the source's emissions. These generic guidelines are broad rather than specific, but their breadth should cover virtually all of the settings an industrial hygienist might normally expect to encounter in their career. Because of this broad applicability, an industrial hygiene (IH) student should rapidly develop a sense of the magnitude of the control velocities within each of these categories. The examples are intended to provide some concept of the settings or processes envisioned within each range and are certainly not comprehensive. A student might practice finding the recommended control velocity in this table for a setting or process with which they are familiar or for a setting or process described in a VS diagram.

   When using Table 13.1, first choose the most appropriate of the four ranges based upon the momentum of the contaminant in the zone of control. Although this zone is not

TABLE 13.1
Ranges of Recommended Generic Control Velocities (Similar to the *Manual of Industrial Ventilation*'s Table 3.1)[1]

Source Characteristic	Range fpm (m/sec)	Examples of Settings or Processes
Passive vapors with virtually no inertia	50–100 (0.25–0.5)	Passive evaporation from tanks, vats, storage cabinets; normal chemistry laboratory operations; vapor degreasers; slow leaks in quiet spaces
Moderate activity producing low inertia contaminants	100–200 (0.5–1)	Welding, plating or pickling; spray painting well into a large booth hood; vapors from intermittent container filling; dust from a slow speed conveyer
Active generation of the contaminants	200–500 (1–2.5)	Spray painting near the face of a shallow booth hood; rock crushing; continuous packaging, loading, or barrel filling; machining processes
Contaminants with high momentum	500–2000 (2.5–10)	Grinding metal; abrasive blasting; tumbling large objects; active processes involving friable asbestos

explicitly defined, it can be thought of as the region where the exhaust air gains or asserts physical control over the trajectory of the plume. For a containment hood, this zone will be the hood's face. For a collection hood, the zone of control will be somewhere in the plume, generally close to the source, but always before the plume would reach someone nearby or escape into the room as a whole. Subsection V.2 discusses a geometric approach taken by Hemeon to define a zone of control for relatively passive sources, but that zone is not so easily defined for active sources with their own momentum.

After choosing an appropriate range of velocities, choose a value within the range based upon the local environment, the intrinsic hazard of the contaminant, and practical considerations related to production constraints and various costs.

a. The local air flow environment can cause you to choose a value at the high, middle, or low ends of that range. Choose a higher control velocity in proportion to the number and strength of cross-drafts and other conflicting air currents (turbulence). You may choose a lower control velocity in proportion to the additional benefits that will result from either a high total exhausted air volume (its corresponding high flow rate of make-up air will increase the general dilution ventilation within the workroom) and/or a large hood (large hood dimensions will increase the length of the path through which the contaminant must pass while being acted upon by the control velocity).
b. The toxic hazard of the contaminant will affect the amount of dilution that is needed between the source and breathing zone (the Breathing Zone Dilution Ratio in Equation 5.18 or Equation 8.6). The Vapor Hazard Ratio (the VHR in Equation 8.4) is an indicator of the probability or ease of complying with exposure limits for a volatile chemical. However, the VHR does not assess the toxic effect to be avoided such as irritation, systemic effects, delayed or chronic effects, and potential acute fatalities. Toxicological and/or epidemiological information (such as from Refs. [7–9]) is necessary to place a given chemical into one of the ranked categories of "Severity of the potential response" in Equation 8.14.
c. Cost effectiveness is always an IH consideration, including when selecting an appropriate control velocity (not to mention deciding on the best control strategy). The cost of providing make-up air is proportional to the flow rate of the exhaust air, the level of air emission cleaning needed, the amount of conditioning of its make-up air, and the duration of exhaust ventilation use. These components of LEV operating costs are what make conservation of exhaust air an important local exhaust design principle and encourage system efficiency. Moving too much air can decrease productivity if it interferes with the process being controlled (e.g., pulling the shielding gas away from TIG and MIG arc welding or exhausting too large a fraction of an expensive but toxic powder being packaged). While an optimum V_{control} may well exist for each hazard where the LEV control would just barely reduce exposures to the exposure limit, this optimum value is generally unknown. The traditional source control hierarchy in the first chapter of this book was based on generic economic considerations; thus, an economic assessment can sometimes lead to the conclusion that LEV is not a feasible control for a given hazard and that another form of control (either source, another pathway, or receiver control) is more cost-effective.

The student should anticipate that the choice of which source of V_{control} to use will be prescribed on any time-limited examination question, either implicitly or explicitly. If the hood is designed using a specific VS diagram, then use the control velocity in that VS diagram; almost every VS diagram will have one or an equation that substitutes for one. A question about a specific OSHA requirement is likely either to focus on open surface tanks using ANSI Z9.1 or the information within any other specific standard will be provided. If neither a VS diagram nor an OSHA standard is specified, then pick a value from the generic recommendations in Table 13.1 herein (or Table 3.1

in the *Manual of Industrial Ventilation*). The "miscellaneous specific operations" table would generally take too long to find anything in the limited time available for an in-class exam unless the operation was specifically named (the table is alphabetical) and either that table or the specifically applicable data is made available.

III. USING THE DALLAVALLE EQUATION

The DallaValle equation is applicable to virtually any collection hood. The DallaValle equation (developed from experimental studies around 1930) can be used to predict either how the collection velocity decreases with distance in front of a hood, or how the face velocity must be increased as the distance between a hood's face and the zone of control increases. It basically states that the velocity of air flowing toward and into an orifice such as a hood decreases with the square of the distance away from that orifice (see Principle 3 in Chapter 10). Therefore, the face velocity of a collection hood must be greater than the control velocity needed at any distance outside of that hood.

1. The form of the basic DallaValle equation shown in Equation 13.2 can be used in exhaust ventilation design to find the exhaust air flow rate (Q) needed to create a given control velocity (V_{control}) at a distance (X) outside of a freestanding collection hood of face area (A_{face}).

$$\boxed{Q_{\text{collection hood}} = V_{\text{control}}(10X^2 + A_{\text{face}})} \tag{13.2}$$

   X is the distance measured straight out from the hood's face along its centerline. Its units must be the same as both the units that are used to calculate the face area A and the distance units within V, typically feet or meters. Furthermore, the equation is accurate only for X up to approximately $1.5 \times D$ where D is the diameter of the hood's face or approximately $\sqrt{A}$. The square power on X in Equation 13.2 relates to the concept that the air approaches the hood through what approximates a series of concentric spheres as depicted by dashed lines in Figure 13.4. The surface area of actual spheres would increase with the square of their radius, viz., $4\pi X^2$ (see Equation 3.1a). Thus, the surface area of a sphere at a distance X would be $\approx 12.5\, X^2$, cf., $10\, X^2$ in Equation 13.2. The flow of air depicted by solid lines with arrows passes perpendicularly through each pseudo-sphere.

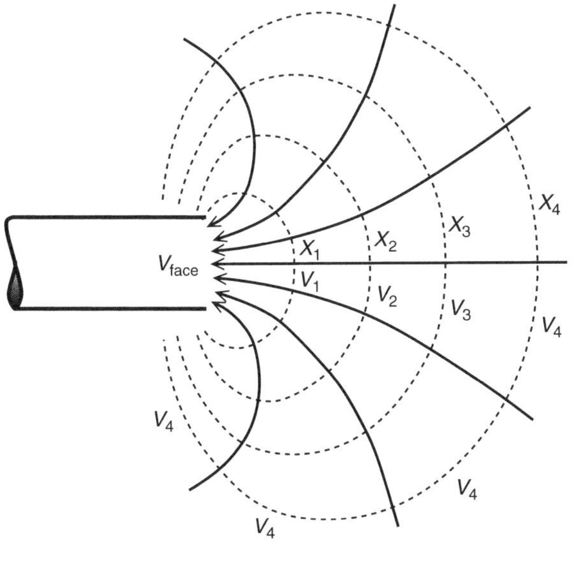

FIGURE 13.4 Depiction of the flow into a plain hood.

2. DallaValle originally had separate equations for round and square hoods and individual coefficients for rectangular openings of W:H ratios up to 10:1. However, he developed Equation 13.2 and Equation 13.3a,b as "a general equation for all ordinary openings".[10] Thus, he recommended approximating D by the $\sqrt{A}$ whether working with a round or square hood. In real terms, the difference between a round and square hood is small (as shown below), although some instructors might be concerned with such details in the classroom and/or on an exam. Check with your instructor to be sure.

$$\sqrt{A_{\text{round}}}/\sqrt{A_{\text{square}}} = \sqrt{\pi \times D^2/4/D^2} = \sqrt{\pi/4} = 0.886 = \text{a } 12\% \text{ difference}$$

(using Equation 2.26)

The approximation that the hood $D = \sqrt{A}$ will be used herein for circular hoods and rectangular hoods whose two sides differ from each other by no more than approximately 2:1. The $X/D \approx X/\sqrt{A}$ ratio is a convenient dimensionless value that will allow the distance from a hood's face to the source or zone of control to be designed, described, or specified in terms of some fraction or small multiple of hood diameter and will help illuminate some useful relationships between V_{control} and V_{face}.

a. Dividing Equation 13.2 by A produces a normalized DallaValle equation that can be used to find the necessary face velocity needed to achieve any desired control velocity at a specified nondimensional distance $X/\sqrt{A}$ (or approximately the X/D ratio) in front of the hood:

$$\frac{Q}{A_{\text{face}}} = V_{\text{face}} = V_{\text{control}}\left[10\frac{X^2}{A} + 1\right] \tag{13.3a}$$

$$V_{\text{face}} \approx V_{\text{control}}\left[10\left(\frac{X}{D}\right)^2 + 1\right] \tag{13.3b}$$

The farther a source is outside of any hood, the larger the $X/\sqrt{A} \approx X/D$ ratio will be, the higher the face velocity must be for the same V_{control}, the more air that must be exhausted, the larger the air cleaner must be, the more make-up air is required, and the higher the operating costs will be to run the fan and condition that make-up air (not to mention the more susceptible that control of the hazard will be to cross-drafts). These problems will get worse by the square of the relative distance, $(X/D)^2$. So, keep X/D as small as possible.

b. The closely related form of the normalized equation shown in Equation 13.4 simply reinforces how the control velocity decreases with the square of the distance outward from the face of a hood.

$$V_{\text{control}} = \frac{V_{\text{face}}}{10(X/\sqrt{A})^2 + 1} \approx \frac{V_{\text{face}}}{10(X/D)^2 + 1} \tag{13.4}$$

For a fixed face velocity, increasing the distance from the hood face to the source means the control velocity in the desired zone of control will rapidly decrease and employee exposure will increase. So, keep X/D as small as possible.

c. Equation 13.5 states a specific, approximate, but memorable result obtained when the source of a contaminant or the zone of control is approximately one diameter away from the face of the hood.

$$\boxed{\text{When } X \approx D, \text{ control } V \approx 10\% \text{ of face } V} \qquad (13.5)$$

This pattern when sucking air into a fan contrasts sharply with that for a jet of air blowing outward where $V_{blowing}$ will not fall to 10% of its initial outward velocity until X is approximately 30 diameters from the outlet nozzle. This much greater blowing distance is the major reason that push–pull exhaust ventilation systems work so well when X is larger than approximately 3 ft or 1 m (to be covered in Section IX).

3. The real challenge during design is not figuring out how small to make X but rather how large to make the hood's face $D \approx \sqrt{A_{face}}$. The fact that X occurs as a "squared" term in Equation 13.2 makes it appear to be the most critical variable for a collection hood. In practice, the collection hood designer should try to make X as small as possible by placing the hood as close to the source as possible, but its proximity is eventually going to be restricted by the production process or setting. Thus, the smallest feasible X value becomes a design limitation rather than an unknown design variable. In design, one must still choose the value for D.

   The most efficient hood would get the most control of the chemical contaminant for the least air exhausted (which is the essence of hood design Principle 4 to conserve exhaust air). The size or X/D of the most efficient hood cannot be found by analyzing or manipulating the DallaValle Equation. A differential analysis of Equation 13.2 (or just looking at $V_{control} = Q/(10X^2 + A)$) shows that the theoretically optimum hood would have a minuscule A or D^2 and suck like crazy. However, a larger A_{face} is not only necessary to keep the DallaValle equation within its accuracy limits of $X/D \leq 1.5$, but it will also prevent V_{face} from getting too large for comfort and allow control over a wider lateral zone. One can get a sense of the potential face velocity for a small hood with a large X/D from the scale on the right side of Figure 13.5. A small hood can necessiate a 40–50 mph (65–80 km/h) face velocity that is high enough to suck up loose parts and cause noise. A larger hood also creates a wider zone of control adequate to accommodate either a large plume or a plume that varies in position due, for instance, to an intermittent cross-draft or other fluctuations that happen in the real world. A 10× velocity ratio is a

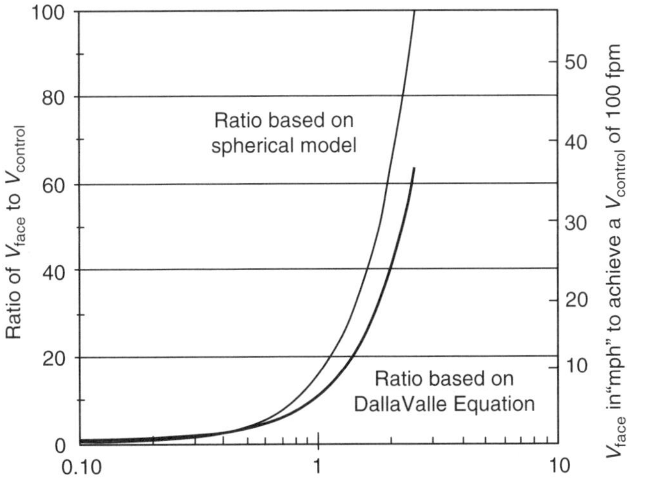

FIGURE 13.5 The increase in ratio of V_{face} to $V_{control}$ as the distance of the hood's face to its diameter increases.

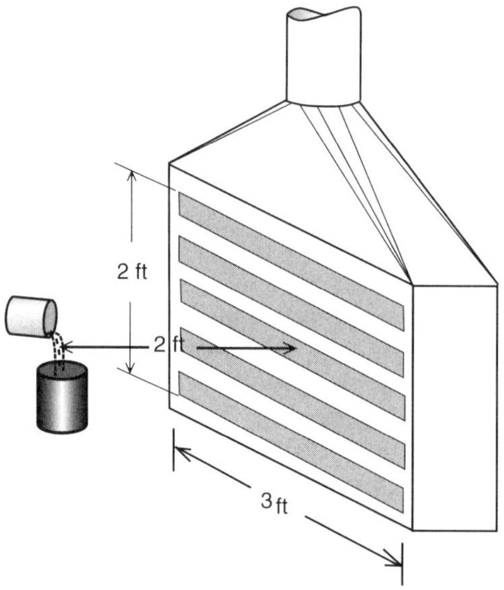

FIGURE 13.6 Depiction of a freestanding, slotted, side draft hood used to control solvent vapors.

good range. In practice, the width of the hood D or $\sqrt{A}$ should be at least as large as either the distance from the source (X) or the width of the source or its plume, whichever is larger. The height of the hood may be less than or equal to its width, but if the plume is buoyant, then the hood might be taller than it is wide, but probably not more than twice as tall as it is wide.

4. A common modification to the plain collection hood shown in Figure 13.5 and Figure 13.6 is to add a flange around its periphery (see examples of flanged hoods in Figure 13.7 and Figure 13.12). A flange, wall, or other barrier in line with a hood's face blocks the flow of air from the back half of the sphere from which air would otherwise approach and enter into a plain hood. Placing a flanged hood on top of a solid surface, such as the bench top in Figure 13.7 or Figure 13.20, blocks the flow of air from three quarters of the sphere.

   Whenever the ineffective flow behind and/or to one side of the face of a hood is blocked, the basic DallaValle equation can be modified by inserting an empirical

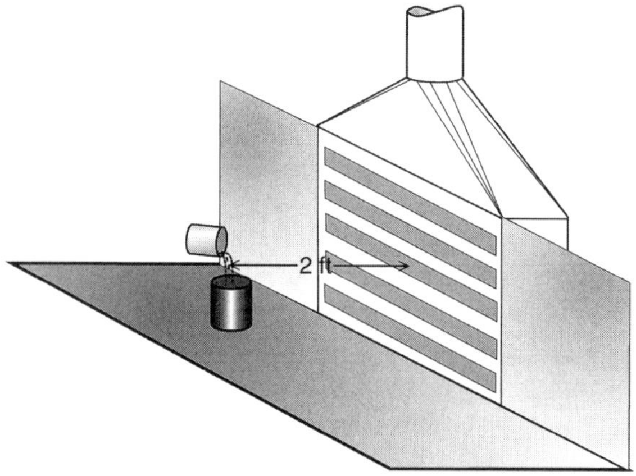

FIGURE 13.7 Depiction of a freestanding, slotted, side-draft hood used to control solvent vapors.

coefficient into Equation 13.2 to Equation 13.4. Rather than using yet another unadorned K notation for this coefficient (as in the *Manual of Industrial Ventilation*), an analogous but more unmistakable K_{hood} is used herein. The effect of using a K_{hood} value of less than one is to reduce the Q per Equation 13.6 and the V_{face} per Equation 13.7 needed to achieve any $V_{control}$ and to increase the control velocity straight out in front of that hood as per Equation 13.8.

$$Q_{collection\ hood} = K_{hood} \times V_{control} \times (10X^2 + A) \qquad \text{(modified Equation 13.2) (13.6)}$$

$$V_{face} = K_{hood} \times V_{control} \times [10(X/\sqrt{A})^2 + 1] \qquad \text{(modified Equation 13.3) (13.7)}$$

$$V_{control} = \frac{V_{face}}{K_{hood}[10(X/\sqrt{A})^2 + 1]} \qquad \text{(modified Equation 13.4) (13.8)}$$

where

$K_{hood} = 1.0$ for no flange around the periphery of a freely suspended hood or plain end of a duct.

$K_{hood} = 0.75$ for either a flange around the periphery of a freely suspended hood (blocking flow from behind the face of the hood) or a plain hood sitting on a bench or table (blocking flow from below the hood's face).

$K_{hood} = 0.50$ for a hood that both has a peripheral flange and is sitting on a bench or table (blocking flow from both behind and below the hood's face).

As a guide, a flange on a small hood might need to be as large as one diameter or side dimension, but little more control can be gained for a flange width of more than 6 to 8 in. outside the face of a large hood. Of course, a flange would have no beneficial effect on a containment hood because its control depends upon its face velocity.

Example 13.1

Suppose that a chemical transfer process such as pouring or bottling solvents were to be conducted in front of a freestanding, slotted, side draft hood such as shown in Figure 13.6. The hood is 3 ft wide by 2 ft tall, and the process (depicted schematically) will be conducted 2 ft straight out in front of the hood (along its centerline). According to Equation 13.2 (used below), if one wanted to have a control velocity of 100 fpm in the vicinity of the pouring operation (nominally its plume), a total of 4600 ft^3/min would have to be drawn into that hood.

$$Q_{plain\ collection\ hood} = 100(10 \times 2^2 + 6) = 4060\ ft^3/min \qquad \text{(using Equation 13.2)}$$

Example 13.2

Now suppose that, rather than a freestanding hood as above, the hood is built with a flange (or into the wall of the room) with a counter top in front of it as shown in Figure 13.7. Assume that the chemical transfer process and all dimensions are the same as above.

This configuration (with a flange and a bench) means that Equation 13.6 would need to be used with K_{hood} set equal to 0.5. The result shown below predicts that this hood can achieve the same $V_{control}$ of 100 fpm with only one half of the exhaust flow rate as the freestanding hood.

$$Q_{collection\ hood} = 0.5 \times 100(10 \times 2^2 + 6) = 2030\ ft^3/min \qquad \text{(using Equation 13.6)}$$

While a flanged hood is better (more efficient) than using a freestanding hood because a flanged hood only requires one-half of the exhaust air flow rate of an unflanged hood to achieve the same control velocity, neither of these collection hoods is as good as using a containment hood such as that depicted in Figure 13.8 (simplified from the laboratory hood shown in Figure 13.14). Even if the top of this hood's face were raised for visibility 1 ft above

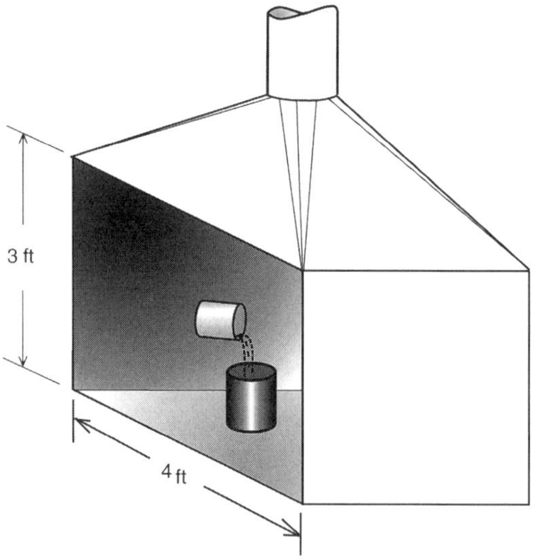

FIGURE 13.8 Conducting the same operation inside a containment hood.

that in Figure 13.6 and Figure 13.7 and the edges of this hood were moved out another foot for more elbow room (to create a 3 ft tall by 4 ft wide opening), such a containment hood with a 100 fpm face velocity (now also its control velocity) would only have to exhaust 1200 ft^3/min, nearly half the flow of even the best collection hood. This difference will translate directly into cost savings for heating and cooling the make-up air.

$$Q_{\text{containment hood}} = V_{\text{face}} \times A_{\text{face}} = 100 \times (3 \times 4) = 1200\ \text{ft}^3/\text{min} \quad \text{(using Equation 13.1)}$$

IV. DESIGNING A SLOTTED COLLECTION HOOD

Slots are often added across the face of a collection hood to achieve a more uniform flow into its face than would otherwise occur without slots. Air will always flow via the path of least resistance. For flow into a plain hood, the path of least resistance is the shortest distance from the hood's face into the exhaust duct. Thus, most of the air would enter the face of a hood without slots in the region nearest to its exhaust duct. For instance, if the exhaust duct exits a hood from the top (such as in Figure 13.8), most of the air would enter the face nearest to the top, as depicted in Figure 13.9. The length of the solid arrows outside the face indicates the face velocity, and the length of the dotted arrows inside the hood indicates the length of the air's path into the duct. Notice that the face velocity is higher and solid arrows are longer closer to the duct entrance, and as the length of the path to the exhaust duct increases near the bottom of the hood, the face velocity decreases and the solid arrows get shorter.

Figure 13.10 depicts the beneficial effect of adding slots to such a hood. Again, the length of the straight, solid arrows outside the face indicates the uniform speed of the air approaching the face of the hood. Notice that the individual slots have no effect on the air flow further out in front of the hood's face than approximately one slot height. The air only senses the slots when it is within about one slot height from the hood's face (as indicated by the path of the curved arrows through each slot). If the pressure drop through the slots is much larger than any pressure difference behind the slots, the pressure difference through each slot will be the same. A uniform pressure difference going into a series of slots of the same size will create the same air velocity through each slot. Most slots are created horizontally, although in principle, the same effect could be created by adding vertical baffles or using a perforated piece of sheet metal with a uniform pattern of closely spaced holes. The space behind the baffles and slots comprises a plenum.

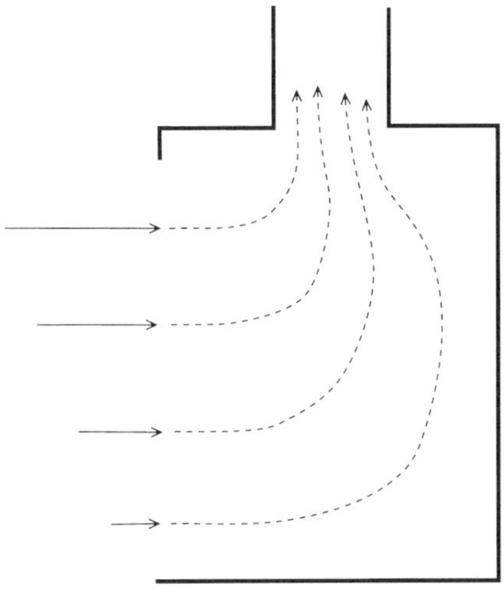

FIGURE 13.9 Depiction of nonuniform air flow into the face of a lateral exhaust hood and out through a duct attached to its top.

Designing a slotted hood can be envisioned as either a two or three step process, as outlined in the next three subsections. The most important step to understand conceptually is step 2, although operationally it can be integrated into the formulas of step 3.

1. A slotted hood is first designed as if it were a simple open-faced collection hood without slots. From the point of view of the air approaching a slotted hood, it does not know the

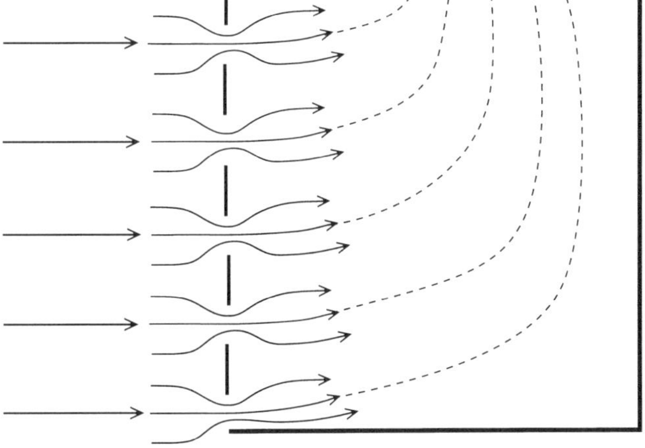

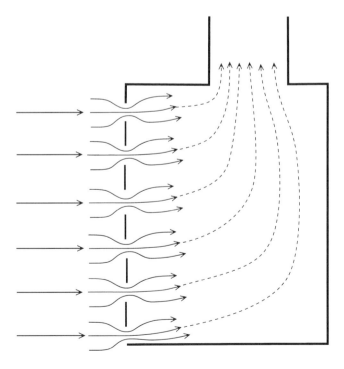

FIGURE 13.10 Depiction of more uniform air flow into a slotted exhaust hood created by adding a series of horizontal baffles across the face of the hood.

hood has slots until it is only approximately one slot width in front of the slots ("width" here means the shorter of a slot's two dimensions). Therefore, the DallaValle equations (Equation 13.2 and Equation 13.6) can be used just as if the slots did not exist (in other words, first imagine the hood without slots). For the purposes of the DallaValle equation, the effective face area of a slotted hood ($A_{\text{effective}}$) is calculated using Equation 13.9.

$$A_{\text{effective}} = \text{Length}_{\text{of slots}} \times \text{Height}_{\text{across all slots}} \tag{13.9}$$

where

$\text{Length}_{\text{of the slots}}$ = the long dimension of the slots. Normally, all slots on any one hood are the same length. This dimension is the width of the hood determined using DalleValle principles. The slot length happens to be the full width of the plenum in Figure 13.11 but is less than the plenum's full width in Figure 13.13 (where the extra space was used to create a flange).

$\text{Height}_{\text{across all slots}}$ = the distance from the top of the top slot to the bottom of the bottom slot, such as in Figure 13.11 or Figure 13.12. The height of the hood is again determined using DallaValle principles.

$A_{\text{effective}}$ can be smaller than or equal to (but never greater than) the size of the original hood opening. In the simplest case, the slots extend to all four edges of the original open hood, as shown in Figure 13.11, and the basic DallaValle equations may be used (Equation 13.2 to Equation 13.4). When a flange is used, as shown in Figure 13.12, the flanged version of the DallaValle equations should be used (Equation 13.6 to Equation 13.8). In Equation 13.10a,b the original variables in Equation 13.6 and Equation 13.7 are modified to use terminology specific to slots.

$$Q_{\text{for a slotted hood}} = K_{\text{hood}} V_{\text{control}} \left(10X^2 + A_{\text{effective}}\right) \tag{13.10a}$$

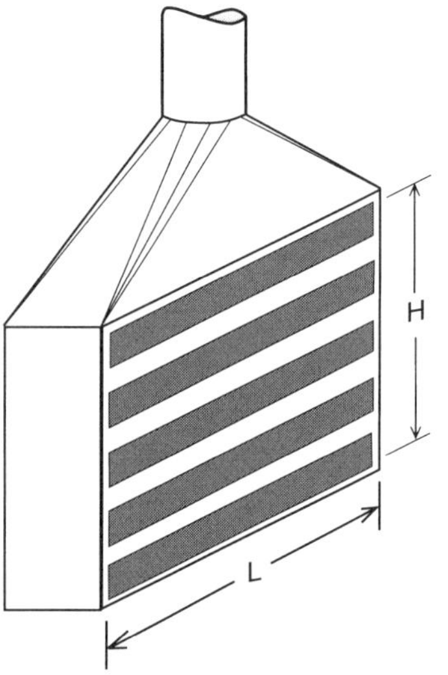

FIGURE 13.11 A lateral slotted hood with no flange.

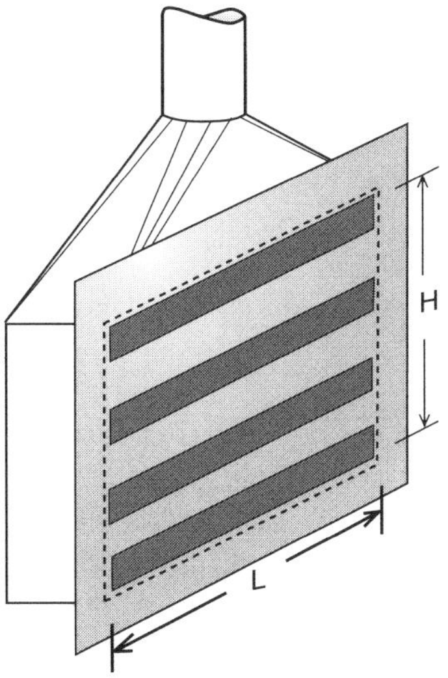

FIGURE 13.12 A lateral slotted hood with an added flange. The dotted lines indicate the face of the original hood.

$$V_{\text{effective face}} = \frac{Q_{\text{for a slotted hood}}}{A_{\text{effective}}} = \frac{Q_{\text{for a slotted hood}}}{\text{Length}_{\text{of the slots}} \times \text{Height}_{\text{across all slots}}} \quad (13.10b)$$

At this stage of design, the size and number of individual slots have nothing to do with the DallaValle equation because the air entering a slotted hood acts as if the hood had a completely open face.

2. The second step in designing a slotted hood (determining the total slot area needed to create a uniform face velocity distribution) is somewhat optional. The flow through an array of equally sized slots will be uniform if the pressure loss of air going through each slot is much greater than the pressure gradient within the plenum behind the slots. Experience has shown that a slot velocity of approximately 2000 fpm is sufficient to create such a uniform flow. Thus, the goal of this second step is simply to find the total slot area that will force the above $Q_{\text{for a slotted hood}}$ to achieve an air velocity through the slots of approximately 2000 fpm. Equation 13.11 is another use of $Q = V \times A$ (Equation 11.7).

$$\Sigma A_{\text{ideal slots}}(\text{ft}^2) = \frac{Q_{\text{for a slotted hood}}}{2000 \text{ fpm } V_{\text{slot}}} \quad (13.11)$$

The criterion that $V_{\text{slot}} \approx 2000$ fpm is critical to making a slotted hood function as desired, but the calculation in Equation 3.11 is optional. Knowing a value for $\Sigma A_{\text{ideal slots}}$ simplifies the equations in the next step, but each of these equations is provided in two forms: one with $\Sigma A_{\text{ideal slots}}$ and one with $Q_{\text{for a slotted hood}}$. A secondary value to knowing $\Sigma A_{\text{ideal slots}}$ is that the ratio of the slot area to the total hood face area gives the designer a sense for the portion of a given hood face that will be open slots.

3. The remaining step is to choose the size and the number (n) of the individual slots that comprise the above total slot area. There is no one right answer to this step in design. Any answer is a balance between performance and cost. The most uniform flow would result from many small slots, but a hood with few slots would be cheaper to build. A slotted hood should have at least three slots. Slots on a large hood are typically between 1 and 2 in. (2 to 5 cm) high, separated by baffles usually between 1 and 3 times as high as the slots.

Slots can be smaller than 1 in. (2 cm) but are susceptible to clogging with debris or corrosion products in harsh environments. All of the slots are full width and usually equally high, yet hood fabrication is simplified if all of the baffles are equally high.

One approach to resolving this ambiguity is to use Equation 13.12 to calculate an initially approximate number of equally sized slots that would result from a somewhat arbitrarily desired slot height ($H_{desired}$) chosen within the above range of 1 to 2 in. (2 to 5 cm). Equation 13.12 is written for $H_{desired}$ to be in inches vs. feet everywhere else. More than one $H_{desired}$ is likely to be tried during this design step.

$$\text{approximate } n = \frac{12 \times \Sigma A_{\text{ideal slots}}(\text{ft}^2)}{H_{\text{desired}}(\text{inches}) \times L_{\text{slot}}(\text{ft})} = \frac{12 Q_{\text{for a slotted hood}}}{2000 H_{\text{desired}}(\text{inches}) \times L_{\text{slot}}(\text{ft})} \tag{13.12}$$

This approximate number of slots would need to be rounded to a nearby integer (rounding could be either up and/or down, as more than one integer is likely to be tried). The resulting integer "n" value can now be inserted into Equation 13.13 to determine the nominal individual slot height for that number of slots. One might try finding the nominal H_n for a couple of combinations of integer "n" values.

$$\text{nominal } H_n \text{ (inches)} = \frac{12 \times \Sigma A_{\text{ideal slots}}(\text{ft}^2)}{\text{“}n\text{”} \times L_{\text{slot}}(\text{ft})} = \frac{12 \times Q_{\text{for a slotted hood}}}{2000 \times \text{“}n\text{”} L_{\text{slot}}(\text{ft})} \tag{13.13}$$

Finally, one should check that the actual V_{slot} in Equation 13.14 is neither much less than the 2000 fpm needed to create a significant and approximately equal pressure drop through each slot, nor so far above 2000 fpm that it would cause either too much energy loss or too much noise.

$$\text{actual } V_{\text{slot}} = \frac{Q}{\Sigma A_{\text{actual slots}}} = \frac{12 \times Q_{\text{for a slotted hood}}}{\text{“}n\text{”} \times H_{\text{desired}}(\text{inches}) \times L_{\text{slot}}(\text{ft})} \tag{13.14}$$

The slot can be built to be any height within the chosen range because it is just space; it is the baffle between the slots that must be constructed. Small variations in the height of individual slots will have little effect on the final air flow pattern.

4. The baffles can be made from individual strips of metal or other thin but stiff and durable material attached across the face of the hood, resulting in a hood such as that shown in Figure 13.11. The "$n - 1$" baffles that separate the n slots are usually equal in height to a fraction or decimal value compatible with the sheet metal contractor or builder purely to simplify construction.
5. When a flange is used as shown in Figure 13.12, the baffle is commonly made from one piece of sheet metal that extends out beyond the slots and may or may not extend out beyond the plenum. Such a flanged hood gains the benefit of a $K_{hood} = 0.75$ instead of $K_{hood} = 1$ for an unflanged hood. While the $L_{\text{each slot}}$ and $H_{\text{across all slots}}$ of the hood depicted in Figure 13.12 are intended to be the same as those depicted in Figure 13.11, the reduction in the face velocity from using the DalleValle equation with a smaller K_{hood} allows a lower $Q_{\text{for a slotted hood}}$ and the number of slots in the flanged hood to be reduced.

   In order to avoid the operational safety hazard of having a thin, metal flange sticking out into pedestrian and/or vehicular traffic, a newly fabricated plenum can be extended out on the back side of the flange to its edge, as shown in Figure 13.13. Again, $L_{\text{each slot}}$ and $H_{\text{across all slots}}$ remain unchanged. This option takes a little more metal but does not change the flow conditions from that in Figure 13.12.

   There is no point in taking the next iterative step of extending the effective face area to encompass the flange. In fact, doing so would be counterproductive, a bigger hood in the

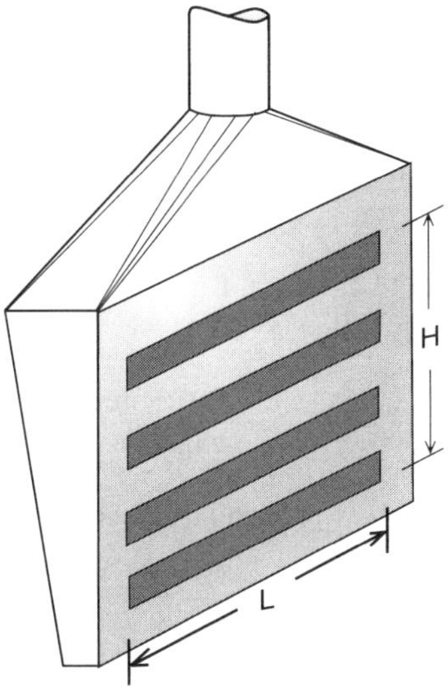

FIGURE 13.13 The same size slots and flange as in Figure 13.12, but with the plenum extended to the edge of the flange along its back side and tapered.

configuration of Figure 13.11 with the same Q but no flange would actually have a lower V_{control} at any distance X.

6. The depth of the plenum is usually not critical. Design the cross-sectional area of the plenum through which the air flows to be at least two times (2 ×) the total area of all the slots ($A_{\text{actual slots}}$). The plenum can either be shaped like the box shown in Figure 13.11 or tapered as shown in Figure 13.13.

Example 13.3

First we will design the slots for Example 13.1. While knowing the total slot area is not required, it is illuminating to know that in order for 4600 ft³/min to move at 2000 fpm, approximately 38% of the 2 ft × 3 ft effective face area will need to be slots.

$$A_{\text{ideal all slots}}(\text{ft}^2) = \frac{Q_{\text{for a slotted hood}}}{2000} = \frac{4600\,\text{ft}^3/\text{min}}{2000\,\text{ft}^3/\text{min}} = 2.3\,\text{ft}^2 \text{ and } \frac{2.3}{2\times 3} = 38\% \quad \text{(using Equation 13.11)}$$

As a first guess, let us assume each slot will be 1 in. high (the bottom of the suggested range).

$$\text{approximate } n = \frac{12 \times A_{\text{slots}}\,(\text{ft}^2)}{H_{\text{desired}}\,(\text{inches}) \times L_{\text{slot}}\,(\text{ft})} = \frac{12 \times 2.30}{1 \text{ in.} \times 3\,(\text{ft})} = 9.2 \quad \text{(using Equation 13.12)}$$

$$\text{nominal } H_9\,(\text{inches}) = \frac{12 \times A_{\text{slots}}\,(\text{ft}^2)}{\text{“}n\text{”} \times L_{\text{slot}}\,(\text{ft})} = \frac{12 \times 2.30}{9 \times 3\,(\text{ft})} = 1.022 \text{ in.} \quad \text{(using Equation 13.13)}$$

While nine slots might work, let us see what eight slots might look like (moving up from the smallest suggested slot height).

$$\text{nominal } H_8\,(\text{inches}) = \frac{12 \times 2.30}{8 \times 3\,(\text{ft})} = 1.15 \text{ in.} \quad \text{(using Equation 13.13 again)}$$

Either option would probably work, but if eight slots each 11/8 = 1.125 were chosen, the actual V_{slot} of

2044 fpm would be just fine.

$$\text{actual } V_{\text{slot}} = \frac{Q}{A_{\text{actual slots}}} = \frac{12 \times 4600 \text{ ft}^3/\text{min}}{8 \times 1.25 \text{ (inches)} \times 3 \text{ (ft)}} = 2044 \text{ fpm} \qquad \text{(using Equation 13.14)}$$

Example 13.4

The design of slots for Example 13.2 with a flange and counter top ($K_{\text{hood}} = 0.5$) will require approximately half of the slot area or that only approximately 19% of the face will need to be slots.

$$A_{\text{ideal all slots}} (\text{ft}^2) = \frac{Q_{\text{for a slotted hood}}}{2000} = \frac{2300 \text{ ft}^3/\text{min}}{2000 \text{ fpm}} = 1.15 \text{ ft}^2 \qquad \text{(using Equation 13.11)}$$

Again, let us assume each slot might be 1 in. high (again the smallest suggested height).

$$\text{approx. } n = \frac{12 \times A_{\text{slots}} (\text{ft}^2)}{H_{\text{desired}} \text{ (inches)} \times L_{\text{slot}} \text{ (ft)}} = \frac{12 \times 1.15}{1 \text{ in.} \times 3 \text{(ft)}} = 4.6 \qquad \text{(using Equation 13.12)}$$

Let us see what both five and four slots might look like.

$$\text{nominal } H_5 \text{ (inches)} = \frac{12 \times A_{\text{slots}} (\text{ft}^2)}{\text{“}n\text{”} \times L_{\text{slot}} \text{ (ft)}} = \frac{12 \times 1.15}{5 \times 3 \text{ (ft)}} = 0.92 \text{ in.} \qquad \text{(using Equation 13.13)}$$

$$\text{nominal } H_4 \text{ (inches)} = \frac{12 \times 1.15}{4 \times 3 \text{ (ft)}} = 1.15 \text{ in.} \qquad \text{(using Equation 13.13 again)}$$

When you think about it, you may have predicted that exactly half the number of slots as in Example 13.3 would be needed to create the same V_{slot} for one half of the flow rate.

$$\text{actual } V_{\text{slot}} = \frac{Q}{A_{\text{actual slots}}} = \frac{12 \times 2300 \text{ ft}^3/\text{min}}{4 \times 1.15 \text{ (inches)} \times 3 \text{ (ft)}} = 2000 \text{ fpm} \qquad \text{(using Equation 13.14)}$$

7. Slotted hoods are "compound hoods," meaning that entry losses occur at two locations: as air passes through the slots and as it passes into the duct. These two losses are merely added together. The pressure loss within a properly sized plenum is small and can be disregarded.[a]
   a. The pressure loss as the air passes through the slots is purposefully designed to be large in comparison to any losses that occur inside the plenum behind the slots. As noted above, the VP_{slot} at the recommended V_{slot} of at least 2000 fpm is 0.25 "wg. Because the slot loss factor is approximately 1.8, the slot loss (at $1.8 \times VP_{\text{slot}}$) will be at least 0.45 "wg.
   b. A pressure loss will always occur as air enters into a duct. The pressure loss as the air flows from the plenum into a duct can range from 0.15 to $0.5 \times VP_{\text{duct}}$ depending upon the taper angle at the entrance to the duct (see Table 14.1). Within the range of recommended V_{duct} values from 2000 to 4000 fpm, the tapered duct entrance loss can range from approximately 0.1 to 0.5 "wg, which is again significant.

It should be clear from the previous figures (and Section I.9 of Chapter 10) that slots are never used on the face of a containment hood! In that sense, the term "slotted collection" hood within the title of this section is somewhat redundant. However, internal baffle boards inside a containment

[a] By designing the plenum cross sectional area to be at least 2 × the slot area, the V_{plenum} will be no more than 1000 fpm and $VP_{\text{plenum}} \leq 0.06$ "wg. This velocity of air passing through a plenum with few obstructions means that very little pressure loss occurs inside the plenum, and what does occur (a small fraction of VP_{plenum}) can be disregarded.

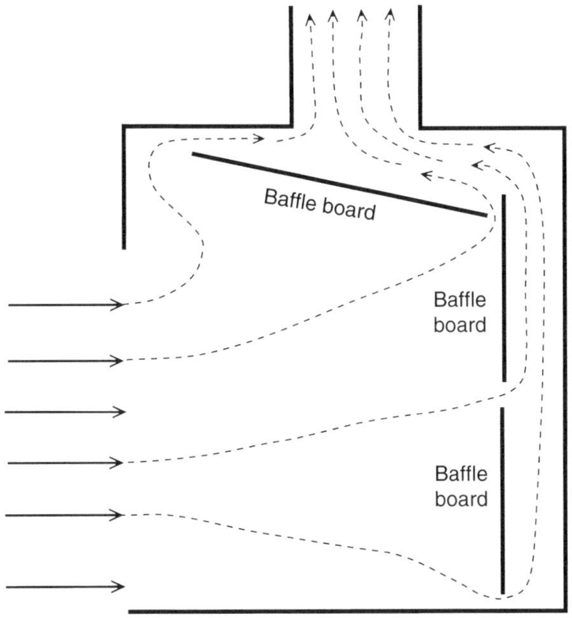

FIGURE 13.14 Cut-away side view of a booth or laboratory hood depicting the placement of baffle boards, used in this case to create four slots: one at the top front and three along its back. In some hoods, the spacing between the baffles may be adjusted in the field to create an even more uniform face velocity distribution.

hood can accomplish much the same effect as slots on a collection hood; they are designed in much the same sequence of steps and, in fact, are common in the construction of laboratory hoods such as shown in Figure 13.14.[b]

V. DESIGNING A HOOD USING AN ACGIH "VS" DIAGRAM

VS diagrams are depicted in the *Ventilation Manual* for over 130 individual operations.[1] Hood design is simplified when you can match one of the ACGIH *Industrial Ventilation Manual*'s VS diagrams to your specific setting or hazard. Each VS diagram typically specifies three pieces of information necessary to design any LEV system (although the latter two values will not be used until Chapter 14):

- the air flow rate (either Q, ft^3/min or V_{face} fpm or their metric equivalents)
- the hood entry pressure loss (the *Manual of Industrial Ventilation* uses "h_e" to signify the hood entry pressure loss), and
- the recommended velocity for the air within the exhaust duct (either the lowest or a range of V_{duct}).

The subsections below introduce the reader to the VS diagram for a common containment hood used in a laboratory and to VS diagrams for an almost generic downdraft hood and a side draft collection hood. The VS diagrams for some less common hood designs will be discussed.

1. Every student in a college chemistry laboratory should have seen and probably used a laboratory hood such as shown in VS-35-01 (copied as Figure 13.15 herein). The common

[b] The use of the term "baffle board" is another vestige in that the baffle can be metal or any other material that is stiff and chemically resistant enough for the environment. In the same vein, the presence of slots in the back of a laboratory hood does not make such a hood a "slotted hood" as that term is reserved for a hood with slots across its face.

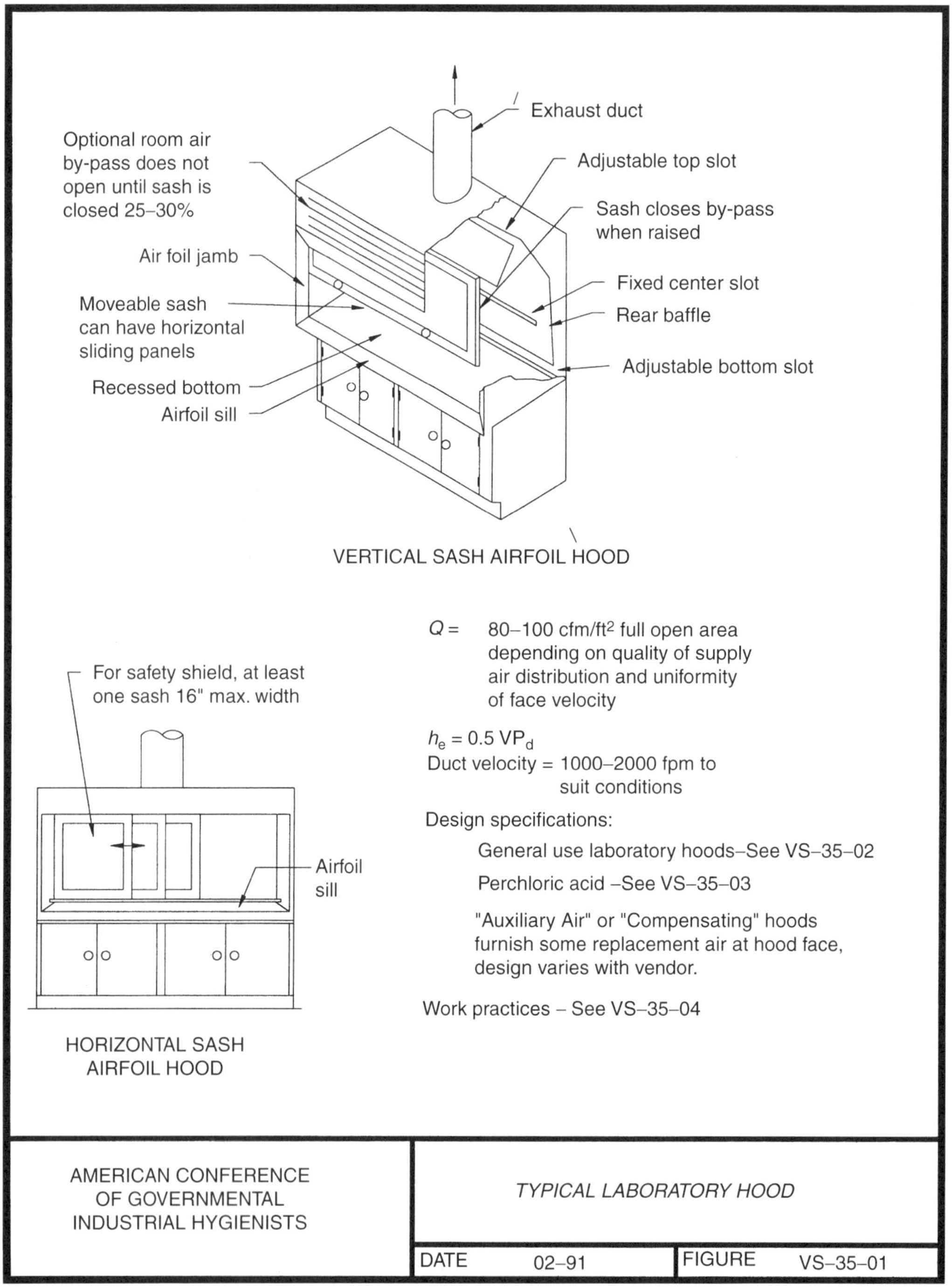

FIGURE 13.15 VS diagram for a laboratory hood from the *Manual of Industrial Ventilation* (with permission from ACGIH).

term "chemical fume hood" is a misnomer to an industrial hygienist since laboratory hoods are probably used to control exposures to gases and vapors much more frequently than they are used to control fumes. The following information provided in the VS diagram for this containment hood is typical of VS diagrams in general.

a. Each VS diagram will indicate the volumetric flow rate (Q) in some form. VS-35-01 indicates "$Q = 80-100$ $ft^3/min/ft^2$ of full open [sash] area." The units of cubic volume per hood face area in this and approximately one third of the other VS diagrams actually specify its face velocity, viz., Q (ft^3/min)/area (ft^2) $= V_{face}$ (fpm). "Area" in this case is the product of the hood's face width × its height.
b. A VS diagram will indicate the hood entry pressure loss as "h_e." As mentioned in Chapter 10, shown in this example, and stressed in the next chapter, the entry pressure loss commonly equals the product of an entry loss factor times the velocity pressure at that entrance. The entry loss factor shown in this VS diagram is 0.5, and the velocity pressure is that within the duct (abbreviated as just VP_d within the *Manual of Industrial Ventilation*).
c. Each VS diagram will also recommend a duct velocity (either a minimum or a range). The duct velocity recommended in VS-35-01 is from 1000 to 2000 fpm. The 1000 fpm value is lower than duct velocities generally recommended for manufacturing settings. A lower duct velocity will minimize the noise that is generated at higher speeds. Audible duct noise in a laboratory setting would be a nuisance that would probably not even be detectable above background noise in a manufacturing setting.

The following features of a good laboratory hood were gleaned from the *Manual of Industrial Ventilation* regarding this hood.[c]

- A good containment hood will not have abrupt or sharp edges at its entrance. A bellmouth or at least a beveled sill and side jams will minimize the *vena contracta* and turbulence that form just behind an abrupt or sharp hood entrance. Turbulence can increase user exposure.
- The recommended average face velocity for a laboratory hood is within the range of 80 to 100 fpm. The best velocity will depend upon the amount of external turbulence that might be caused, for instance, by a nearby room supply air inlet, a doorway, or passing foot traffic. More protection can be expected from an undisturbed 75 fpm than from a disturbed 150 fpm face velocity. See also the discussion on changing air velocity in subsection II.3.c of Chapter 9.
- The maximum variation in the velocity across the face should be not more than ± 15% of the mean. The most common way to achieve a uniform face velocity is to have either fixed or adjustable baffle boards in the back of the hood to form a plenum (shown in Figures 13.14 and VS-35-01).
- Laboratory equipment should not be used closer than 6 in. from the hood's face to minimize the amount of contaminant that can come out through that face and into a user's breathing zone.

Such laboratory hoods are almost always purchased as commercial products (rather than custom designed and built on-site). Other types of specialized commercial hoods (laminar flow and biological safety cabinets) are discussed in the last section of this chapter.

The following three subsections cover of VS diagrams for downdraft, side draft, and canopy hoods.

2. Recall that a downdraft hood (depicted generically in Figure 10.4) comprises a collection hood placed immediately below the source of the contaminant. Examples of downdraft hoods in the *Manual of Industrial Ventilation* include VS-20-03 (foundry shakeout), VS-80-18 (hand grinding bench), and on numerous wood working tools such as (VS-95-10 (drum sander), VS-95-12 (disc sander), and VS-95-20 (jointer).

[c] Sec. 10.35, VS-35-02, and VS-35-04. The *Manual of Industrial Ventilation* VS-35-03 also contains special design features if the laboratory hood is to be used for perchloric acid to prevent its accumulation either within the hood or ducting.[1]

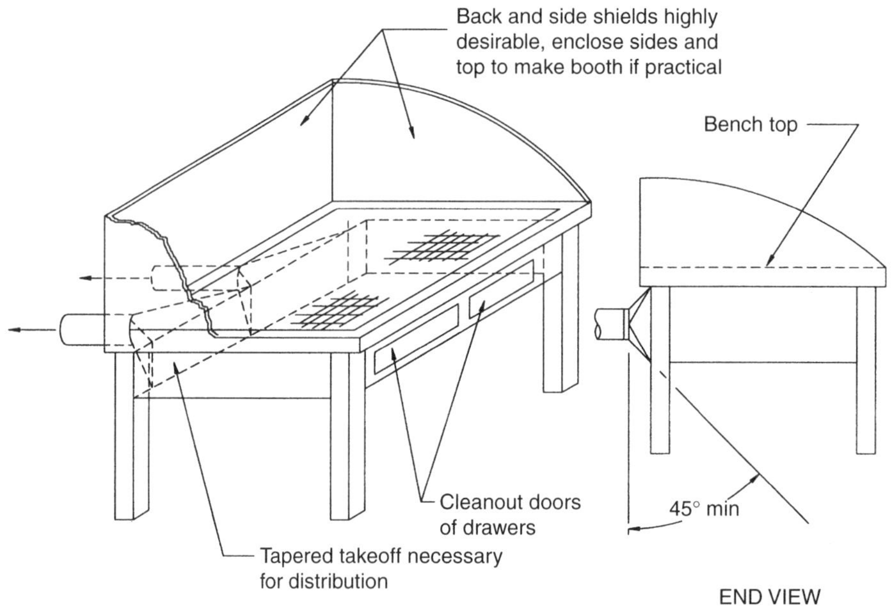

Q = 150–250 cfm/ft^2 of bench area.
Minimum duct velocity = 3500 fpm
h_e = 0.25 VP_d
If slots are used for distribution
h_e = 1.78 Vp_s + 0.25 VP_d

Notes:
1. If grinding in a booth, use 100 fpm face velocity.
2. For downdraft grilles in floor: Q = 100 cfm/ft^2 of working area.
3. Provide equal distribution.
4. Provide for cleanout.

AMERICAN CONFERENCE OF GOVERNMENTAL INDUSTRIAL HYGIENISTS	*HAND GRINDING BENCH*			
	DATE	02–91	FIGURE	VS–80–18

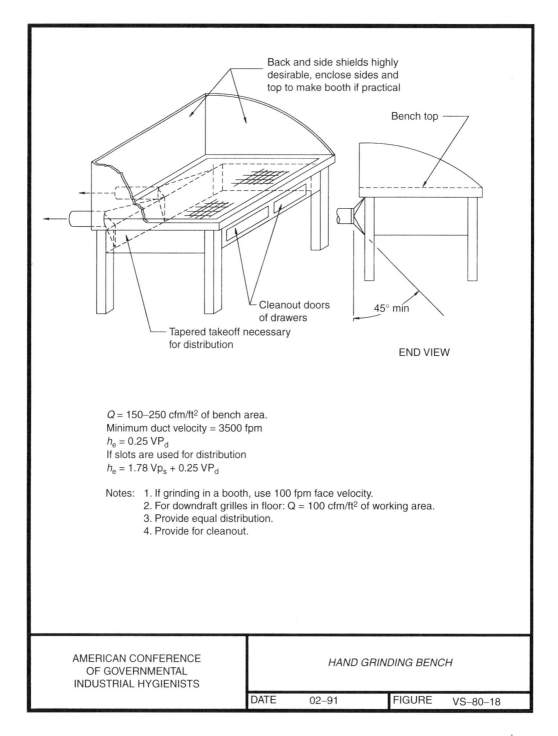

FIGURE 13.16 An example of a VS diagram with a downdraft hood (with permission from ACGIH).[1]

VS-80-18 (shown herein as Figure 13.16) gives the same three pieces of design information typical of all VS diagrams. The units of control velocity are again given in terms of ft^3/min/ft^2 (150–250 ft^3/min/ft^2) but in this case per ft^2 of bench area; however, since the face area in VS-80-18 comprises the whole bench area, this recommendation is equivalent to a V_{face} of 150–250 fpm. The hood entry loss is 0.25 × VP_{duct}. This entry loss is created as the exhaust air enters the duct througha 45° tapered fitting on the back of the hood. A single loss implies that the bench top is a coarse screen or grating that does not cause much resistance to the flow of air, cf., the losses caused by a perforated or slotted face as described in Section IV. The requirement to run the exhaust duct at or below floor level without obstructing the floor or work area can preclude the use of a downdraft hood. One could use the DallaValle equation (Section III) to design a downdraft hood for a task that generates its airborne contaminant some distance above the work bench.

3. A side draft hood (also called a lateral exhaust hood) was depicted generically in Figure 10.4. A side-draft hood is usually less efficient than a containment or downdraft hood, but it is the most feasible option for many settings. Examples of side draft hoods in the *Manual of Industrial Ventilation* include VS-20-02 (foundry shakeout), VS-55-10 (pouring station), VS-75-06 (dip tank included as Figure 13.15), VS-90-01 (welding bench), and VS-99-08 (stripping tank). Only about 10% of the VS diagrams (including VS-75-06 shown herein as Figure 13.17 and the design of a side draft hood for an open surface tank in Section VI of this chapter) indicate a control velocity in terms of $ft^3/min/ft^2$ of some characteristic source area. Less than 5% of VS diagrams (such as VS-55-10) use a form of the DallaValle equation such as $Q = 200\,(10X^2 + A)$ where the number "200" is the control velocity of 200 fpm at a distance X between the source and the face of the collection hood of face area A.
4. A canopy hood mounted directly over a source is shown in Figure 13.18. Canopy hoods can sometimes control an isothermal source; however, canopy hoods work best as a receiving hood for hot sources. A canopy hood is probably not the best choice if any of the following apply:
   a. a person's breathing zone is often in the rising plume under a canopy hood (such as the person on the left in Figure 13.18),
   b. the plume can be deflected by a cross-draft far enough to miss the hood (especially likely if the canopy is rather high above the source), or
   c. the contaminant is generated faster than the hood can contain and remove it resulting in "spillage" from the bottom of the canopy back out into the room (a common problem in cyclic or batch processes with periodic large releases).

   The design procedure from VS-99-03 for a canopy hood over an isothermal source will be deferred to Section VII of this chapter so that it can be compared to another canopy hood design procedure presented in Section VI. Two design procedures for a canopy hood over a hot source will be discussed in Section VIII of this chapter. Examples from the *Manual of Industrial Ventilation* of a canopy hood over specific hot sources include VS-25-12 (ethylene oxide sterilizer), VS-30-10 and VS-30-11 (kitchen hoods), VS-35-41 (oven exhaust), and VS-55-02 (melting furnace tilting).

The last two subsections describe VS diagrams for somewhat specialized kinds of collection hoods.

5. *Movable exhaust hoods*: In principle, movable hoods (such as depicted in Figure 13.19 and Figure 13.20) are a good way to control small individual sources of contaminants that move with the work over a distance that is small but too far to be controlled by any one hood. A moveable hood can be placed either over or alongside the source, qualifying it as either a canopy or a small side draft hood, respectively. Note however, that their success requires that the worker actively place the hood near each source or operation. This requirement for active participation by the user makes moveable exhaust hoods susceptible to some of the same limitations of personal protective equipment (to be discussed in Section IV of Chapter 21, and Sections II and VI of Chapter 22); however, with adequate training and supervision such movable hoods can be placed closer than most fixed hoods and can therefore be very efficient.

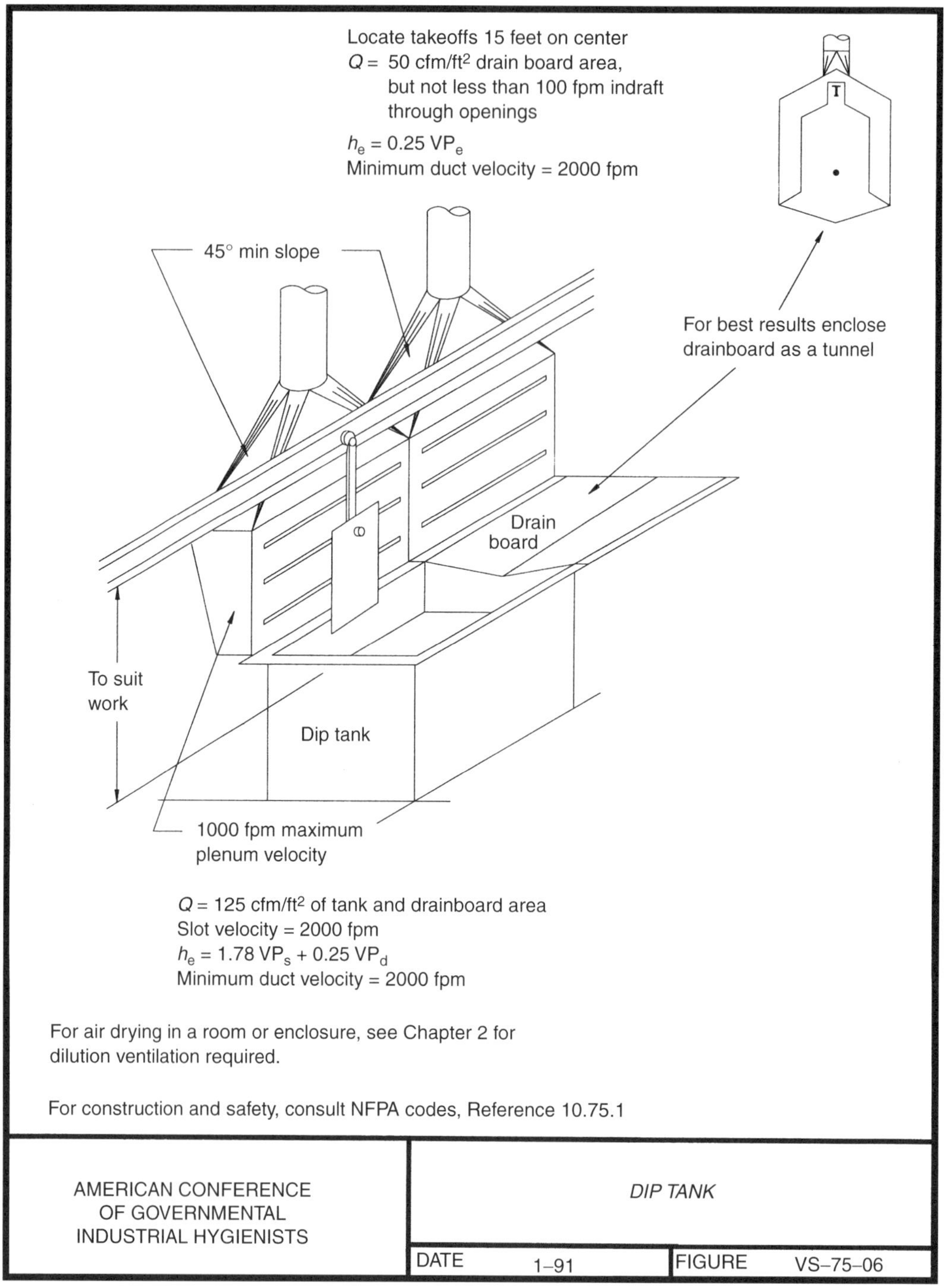

FIGURE 13.17 An example of a VS diagram with a side draft hood (with permission from ACGIH).[1]

Examples of moveable exhaust hoods in the *Manual of Industrial Ventilation* include VS-65-02 (granite cutting and finishing), VS-85-01 (automotive maintenance), VS-90-02 (welding), and VS-90-20 (robotic manipulator arm). The flow rate for these and approximately 20% of all VS diagrams is specified in terms of either fixed ft^3/min values

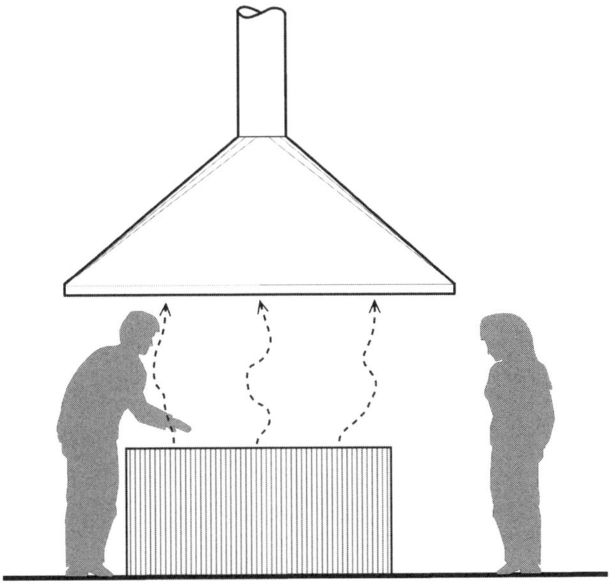

FIGURE 13.18 This hood only protects one of these people.

(that might vary depend on the size or number of sources being controlled) or fixed face velocities. Hood entry losses in these VS diagrams are either similar to those above or not specified (the latter is probably due to large and difficult to predict variations in the pressure loss with the angle of a movable hood's flexible duct).

6. Low-volume/high-velocity exhaust systems are typified by a small containment hood built right onto a small source, such as a hand tool, that generates a

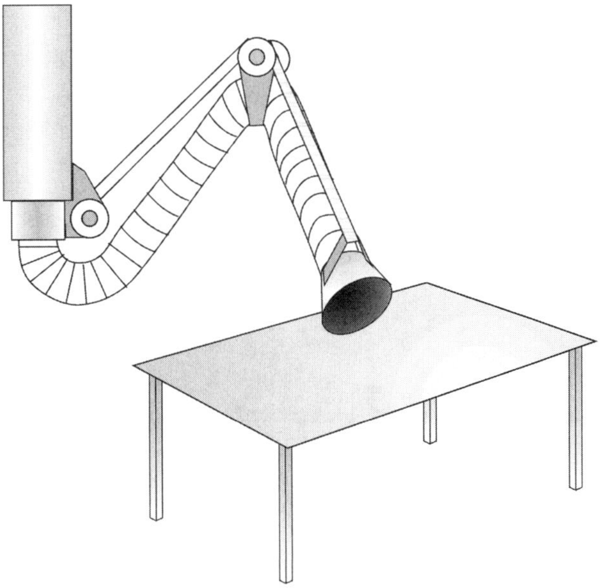

FIGURE 13.19 A self-supported movable tapered hood; see also Figure 15.3.

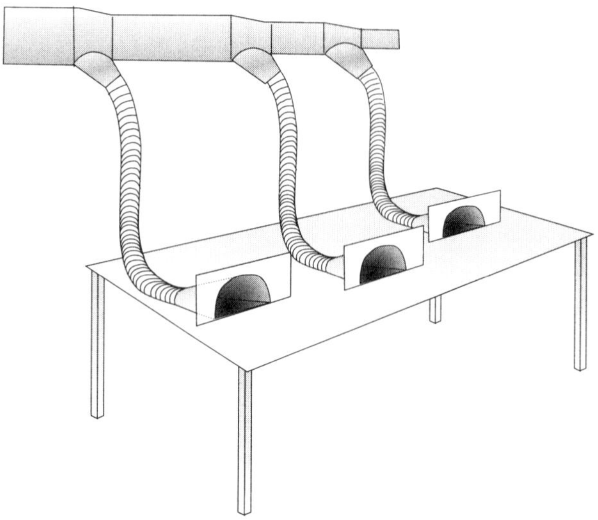

FIGURE 13.20 Three movable tapered hoods with flanges that tie into a branched duct via flexible hoses.

contaminant (e.g., Figure 13.21). Examples in the *Manual of Industrial Ventilation* include grinders (VS-40-01, VS-40-02, and VS-40-04), chippers (VS-40-03), and sanders (VS-40-05 and VS-40-06). These and approximately 25% of all other VS diagrams specify Q in units such as 10 to 50 ft^3/min/in. of tool diameter or other linear dimension appropriate to each system (such as Figure 4.5 that matches VS-80-30). The resulting flow rates for low-volume high-velocity hoods of 100 to 200 ft^3/min are small in comparison to most of the other hoods described herein, but they can create a velocity of from 15,000 to over 35,000 fpm through the rather small hood face area,

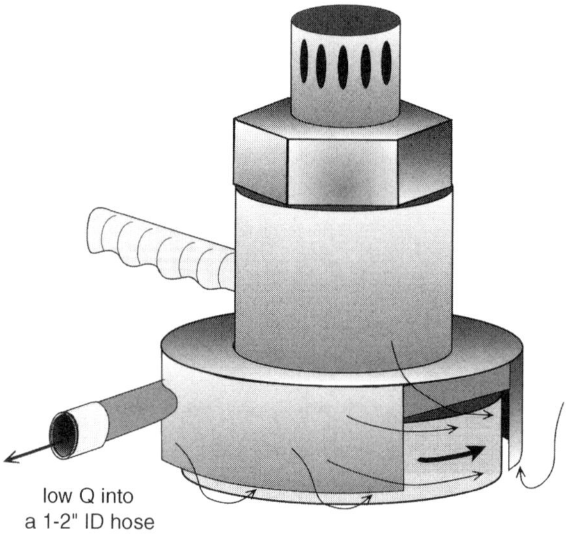

FIGURE 13.21 A low-volume/high-velocity hood on a surface grinder.

usually a gap between the working surface and the attached hood. Hence, the name "low-volume/high-velocity."

The advantages of using a low-volume/high-velocity system include the fact that attaching the exhaust hood to the tool overcomes the limitation of the worker having to manually position the hood, and creating a very high control velocity can control a contaminant released with high inertia. A disadvantage to these systems is the very high pressure losses of 7 to 14"wg (compared to many traditional hoods that might lose 1 to 2"wg) created by the high VP and small, light flexible tubing (in place of ducting).

VI. DESIGNING A HOOD FOR AN OPEN SURFACE TANK

Recall from Chapter 4 that OSHA incorporates ANSI Z9.1 by reference into §1910.6 and §1910.124. The first part of this standard utilizing the dichotomous tank classification system was previously described in Section VII.2 of Chapter 3.[6] The next version of the ANSI Z9.1 standard (after 1991) will probably offer the user two options by which to specify the exhaust air flow rate (Q in ft^3/min or m^3/min): either the "control velocity method" or the "Hemeon method." The former is the traditional ANSI empirical approach to be described below in Subsection 2. The latter method envisions a control velocity passing through a virtual zone of control to be described in Subsection 3.

1. OSHA/ANSI Z9.1 covers three basic types of hoods: viz., enclosing hoods, canopy hoods, and lateral exhaust hoods, depicted in Figure 13.22 to Figure 13.24, respectively.
   a. For a hood to be considered an enclosing hood in Table 13.2, the hood must project over the entire open surface of the tank, and the sides of the hood must be fixed in such a location that the head of the workman, in all his or her normal operating positions while

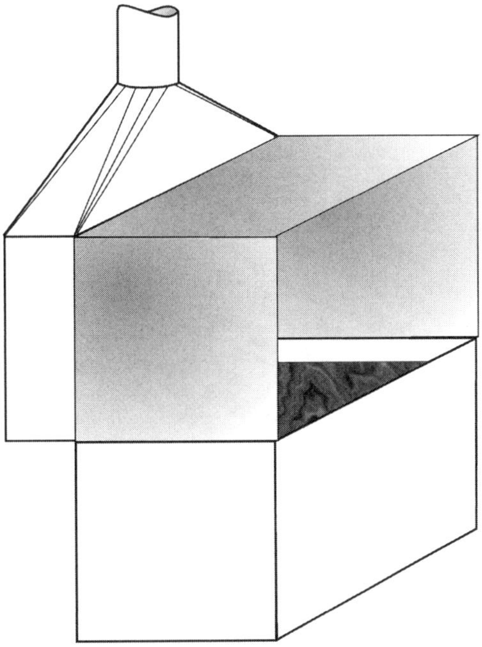

FIGURE 13.22 Depiction of an enclosing hood over a tank.

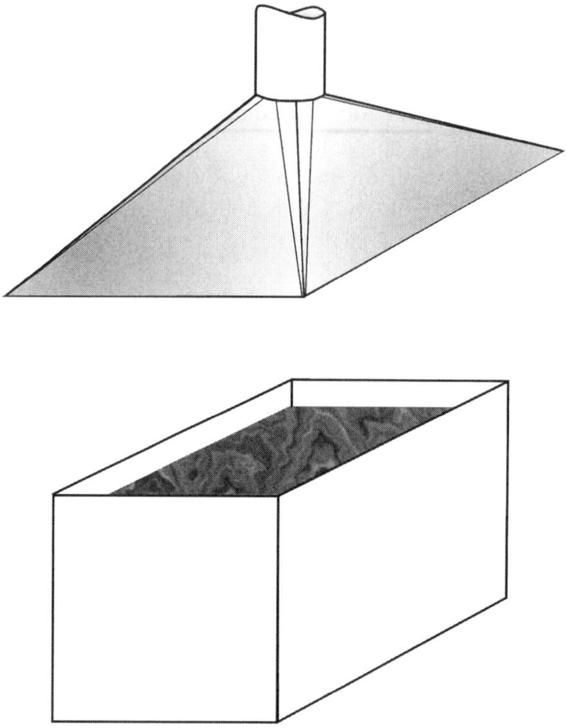

FIGURE 13.23 Depiction of a freely suspended canopy hood.

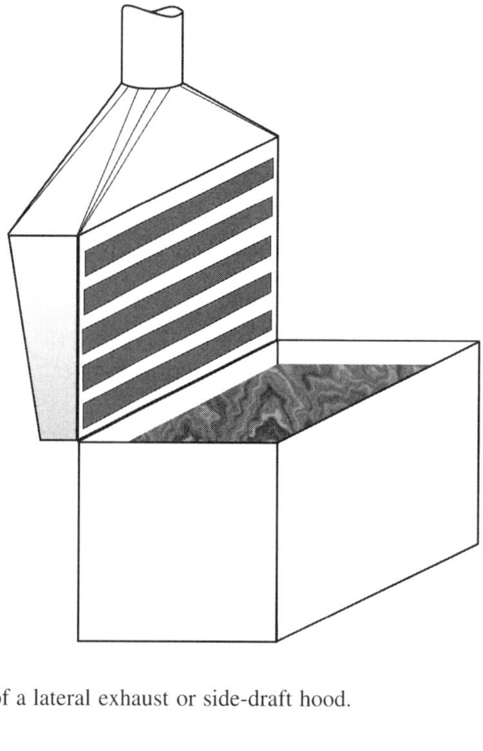

FIGURE 13.24 Depiction of a lateral exhaust or side-draft hood.

TABLE 13.2
Exhaust Ventilation Control Velocities for Undisturbed Locations

	Enclosing Hood[a]				Minimum Lateral Exhaust (see Table 13.3)		Canopy Hood[b]			
	One Open Side		Two Open Sides				Three Open Sides		Four Open Sides	
Risk Class	fpm	m/sec	fpm	m/sec	fpm	m/sec	fpm	m/sec	fpm	m/sec
A-1[b] and A-2[b]	100	0.51	150	0.76	150	0.76	Do not use		Do not use	
A-3[b], B-1, B-2, and C-1	75	0.38	100	0.51	100	0.51	125	0.64	175	0.89
B-3, C-2, and D-1[c]	65	0.33	90	0.46	75	0.38	100	0.51	150	0.76
A-4[b], C-3, and D-2[c]	50	0.25	75	0.38	50	0.25	75	0.38	125	0.64
B-4, C-4, D-3[c], and D-4	General room ventilation is required and may be adequate for this risk class.									

[a] The control velocity of an enclosing (or containment) hood is its face velocity.
[b] The control velocity of a canopy hood is inward between the bottom edge of the canopy and the top edge of the tank. Do not use a canopy hood for Hazard Potential A processes.
[c] Where complete control of hot water is desired, design as next highest class.

working at the tank, is in front of (and outside) all hood openings.[d] A chemistry laboratory exhaust hood such as depicted in Figure 13.13 should be a familiar example of an enclosing hood. The enclosing hood depicted in Figure 13.22 would block the use of an overhead crane to insert and withdraw parts from that tank. The quantity of air (Q) in cubic feet per minute (or cubic meters per minute) to be exhausted through an enclosing hood is the product of the face or control velocity listed in Table 13.2 times the face area, as per Equation 13.1.

b. A canopy hood can be defined as any exhaust hood that projects over the entire tank but does not conform to the definition of an enclosing hood. The canopy hood depicted in Figure 13.23 would also block the use of an overhead crane to insert and withdraw parts from that tank. Canopy hoods can have four, three, or two sides open through which air and/or contaminants can flow (see also Section VII.2).
   1. A traditional canopy hood is freely suspended or freestanding, meaning it is open on all four sides between the bottom edge of the canopy and the top edge of the tank or other source as depicted in Figure 13.23.
   2. Some advantage can be gained by placing a source and its associated canopy hood next to a wall or other barrier that blocks the flow of air from entering the hood from one side of the canopy. Thus, the control velocities in Table 13.2 for a canopy hood that has only three sides open are lower than for a freestanding canopy.
   3. An additional advantage can be gained by placing the source and its associated canopy in a corner, blocking two sides and leaving two sides open. Notice that the control velocities for a canopy with two sides open are equivalent to the lower velocities listed in Table 13.2 for an "enclosing hood with two sides open." Their hood configurations are physically the same.

   The quantity of air in cubic feet per minute (or cubic meters per minute) necessary to be exhausted through a canopy hood should not be less than the product of the

[d] An enclosing hood is normally completely enclosed on at least three sides, but it is possible, and occasionally useful, for portions of all of its sides to have at least partial openings.

TABLE 13.3
Minimum Ventilation Volumetric Flow Rate per Unit of Tank Surface Area Needed for a Lateral Exhaust Hood to Maintain the Control Velocity Derived from Table 13.2 for Various Tank Width to Length Ratios

Minimum Control Velocity From Table 13.2		Tank Width (W) to Tank Length (L) Ratio.[a] Note that "m^3/m" is Cubic Meters per Minute									
		0.0 to 0.09		0.1 to 0.24		0.25 to 0.49		0.5 to 0.99		1.0 to 2.0	
fpm	m/sec	$ft^3/min/ft^2$	$m^3/m/m^2$	$ft^3/min/ft^2$	$m^3/m/m^2$	$ft^3/min/ft^2$	$m^3/m/m^2$	$ft^3/min/ft^2$	$m^3/m/m^2$	$ft^3/min/ft^2$	$m^3/m/m^2$
If the hood is either against a wall, has a flange[b], or is a manifold hood along the tank's centerline[c]											
50	0.25	50	15	60	18	75	23	90	27	100	30
75	0.38	75	23	90	27	110	34	130	40	150	46
100	0.50	100	30	125	38	150	46	175	53	200	61
150	0.76	150	46	190	58	225	69	250	76	250	76
If the hood is without a flange or not against a wall (i.e., free standing)[c]											
50	0.25	75	23	90	27	100	30	110	34	125	38
75	0.38	110	34	130	40	150	46	170	52	190	58
100	0.50	150	46	175	53	200	61	225	69	250	76
150	0.76	225	69	250	76	250	76	250	76	250	76

[a] Trying to capture contaminants across the long dimension of a tank whose W/L ratio exceeds 2 is not practicable. Doing so when W/L exceeds 1 is undesirable. For a lateral exhaust along up to one half the circumference of a circular tank, use $W/L = 1$; for over one half the circumference, use $W/L = 0.5$.

[b] A flange as defined in Section 10.A.8. (The ANSI standard refers to a flange as a "baffle.") A lateral exhaust hood that is against a wall or close to it, can be said to be "perfectly flanged."

[c] Use $W/2$ as the tank width when either the hood is a manifold (exhausting into both sides) along the tank centerline or separate exhaust hoods are used on two parallel sides of a tank.

control velocity times the net area of all openings between the bottom edges of the hood and the top edges of the tank.

c. For the purposes of both Table 13.2 and Table 13.3, any hood that does not project over the entire tank (i.e., that is not an enclosing hood), and that moves the contaminated air in a substantially horizontal direction (cf., in a vertical direction as into a canopy hood), is considered to be a lateral exhaust hood. A lateral exhaust hood's face may be one large opening but is more commonly a slotted hood as described in Section IV. While a lateral exhaust hood such as depicted in Figure 13.24 is the only one of the three options covered by ANSI Z9.1 that does not block the use of an overhead crane or other tall obstructions that might extend above a source, its design entails an extra step described in Table 13.3 to accommodate various widths of the tank.

2. The control velocity method specifies that the exhaust air velocity over an open surface tank shall be equal to or greater than those specified in Table 13.2 and Table 13.3. These values apply where the flow of air past the breathing or working zone of the operator and into the hood is undisturbed by local environmental conditions such as open windows, fans, unit heaters, or moving machinery. Where such disturbances exist and cannot be alleviated, the next higher velocity should be used.

   The control velocities in Table 13.2 are specified by risk classifications and by the type of local exhaust hood. The dichotomous risk classification scheme (letter and numeric category) for the open surface tanks was determined in Section VII in Chapter 4. The type of local exhaust hood is determined by the system designer. The appropriate control velocity (ranging from 50 to 175 fpm) is found by cross-matching the dichotomous risk classification (along the left side of Table 13.2) with the type of hood to be used (across the top of Table 13.2).

   Table 13.3 applies only to lateral exhaust hoods. While the total exhaust volume (Q) for all other hoods is the product of the control velocity found in Table 13.2 times the open face area, Table 13.3 gives the exhaust volume as $ft^3/min/ft^2$ ($m^3/min/m^2$) of tank surface area in various columns specified by the tank's width (W) to length (L) ratio. The tank width (W in Table 13.3) means the effective width of the tank over which the hood pulls air. For example, where the hood face is set back from the edge of the tank, this set-back must be added to the tank's measured width.[e]

   The W/L ratio indicates the width of the tank away from the hood in proportion to the length of the tank, implicitly assuming that the hood is as wide (horizontally) as the tank is long. In effect, the larger total Q for the wider tanks at any given control velocity is a tabular application of the DallaValle equation (or at least its principles). The volumetric flow rate of air to be exhausted from the tank's hood should be not less than the product of the area of tank surface area times the ft^3/min per square foot (or cubic meters per minute per square meter) determined from Table 13.3. Also, the exhaust volume should be increased if the rate of vapor evolution (including steam or products of combustion) from the process to be controlled is equal to or greater than 10% of the exhaust volume calculated using the above methods.

[e] The 1991 ANSI standard included the following additional guidance[6]:

a. for $W \leq 20$ in. or 51 cm, an exhaust along one side of the tank is satisfactory.
b. for W between 20 and 36 in. (51 and 91 cm), an exhaust along both sides is desirable.
c. for W between 36 and 42 in. (91 and 107 cm), an exhaust along both sides is necessary.
d. for $W > 42$ in. (107 cm), an alternative means of control must be considered such as "push–pull" ventilation, a cover that extends over the tank from above the upper slot, or enclosures.

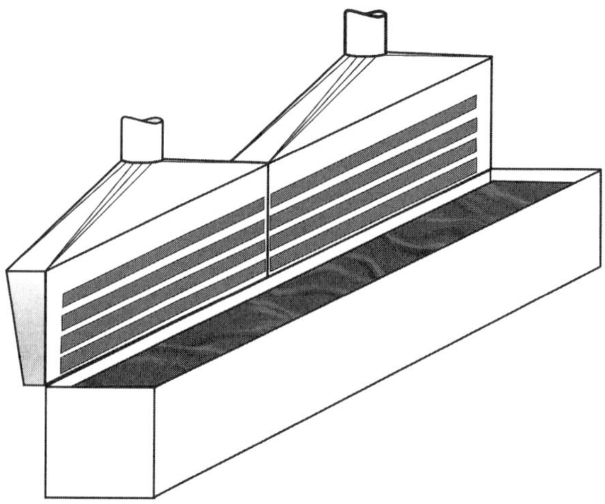

FIGURE 13.25 Example 13.5.

Example 13.5

Design a ventilation control using the ANSI Z9.1 Control Velocity Method for an open surface tank of dilute sulfuric acid with a B-1 risk classification used to anodize aluminum rods. The tank is 3 ft wide and 20 ft long. Baskets of parts are drawn slowly through the long dimension of the tank by an overhead conveyor system, precluding an enclosure or canopy hood. Thus, control will be via a lateral exhaust hood the full length of this tank similar to that shown in Figure 13.25. A common convenience is to split such a long hood into at least two segments.

Table 13.2 shows that the minimum control velocity for a lateral exhaust hood in an undisturbed location would be 100 fpm and refers the reader to Table 13.3. The W/L ratio for this tank is $3/20 = 0.15$. The volumetric flow rate (Q) for such a tank W/L, hood with a flange, and control velocity is 125 $ft^3/min/ft^2$. For the total tank surface area of 60 ft^2, this translates into a Q of 7500 ft^3/min. If the hood height were also 3 ft, this also translates into a face velocity of 375 fpm.

3. In the Hemeon method, the exhaust air volume flow rate Q is calculated as the product of what he called the "critical capture velocity" (basically $V_{control}$) times the control surface area ($A_{control}$).[11,12]

   During design, $V_{control}$ should be at least equal to or greater than the appropriate critical capture velocity for an undisturbed location given in Table 13.4. Notice that the Hemeon method utilizes only one set of control velocities that apply to each of the same groupings of dichotomous risk categories found in Table 13.2.

TABLE 13.4
Typical Critical Capture Velocities for Undisturbed Locations

Risk Class	Critical Capture Velocity	
A-1, A-2	150 fpm	0.76 m/sec
A-3, B-1, B-2, C-1	100 fpm	0.51 m/sec
B-3, C-2, D-1	90 fpm	0.46 m/sec
A-4, C-3, D-2	75 fpm	0.38 m/sec
B-4, C-4, D-3, D-4	50 fpm	0.25 m/sec

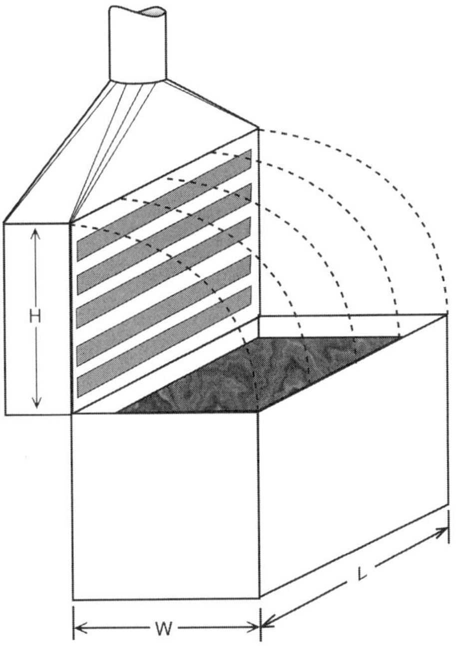

FIGURE 13.26 The virtual surfaces through which the exhaust air must pass at the critical capture velocity are depicted by dotted lines.

$A_{control}$ is the area of a hypothetical contour surface comprising the outer boundary of the control zone through which the $V_{control}$, at least conceptually, would pass. Hemeon's text offers a series of diagrams that each suggests a conceptual three-dimensional zone of control for a specific configuration of source and hood similar to Figure 13.26, and one or more formulas by which to calculate the surface area of its contour surface. The boundary of the zone of control depends not only on the geometry of the tank (as in the control velocity method) but also upon the type, location and shape of the hood and the presence of any flanges, baffles, etc. around that hood.

Figure 13.26 gives an example of the zone of control for a lateral exhaust hood. The boundary of the surface area through which the air must pass (depicted by dotted lines) comprises one quarter of the imaginary cylinder over and parallel to the long side of the tank and its two quarters of the circle or ellipse that would cap each of the two ends of that imaginary cylinder. The first surface area is approximately $\pi\sqrt{H^2 + W^2}L/2\sqrt{2}$ (or $\pi WL/2$ if $H = W$), and that of each of the two caps is $\pi HW/4$ (or $\pi W^2/4$ if $H = W$). The necessary exhaust air volume (Q) is simply the product of the critical capture velocity from Table 13.4 times the calculated control surface area.

$$Q = V_{\text{control (from Table 13.4)}} \times A_{\text{outer boundary of the zone of control}} \tag{13.15}$$

Example 13.6

Design a ventilation control for the previous example using the ANSI Z9.1 Hemeon method assuming that the hood will be 3 ft tall (equal to the width of the tank).

Table 13.4 shows that the appropriate critical capture velocity for a B-1 risk classification in an undisturbed location would be 100 fpm. Using Figure 13.27 as a model, the appropriate contour surface for this shaped source would be one quarter of a cylinder 20 ft long with a radius of 3 ft. The cylindrical surface area in this case is $\pi WL/2 = \pi \times 3 \times 20/2 = 94.2\ \text{ft}^2$ and that of both caps is

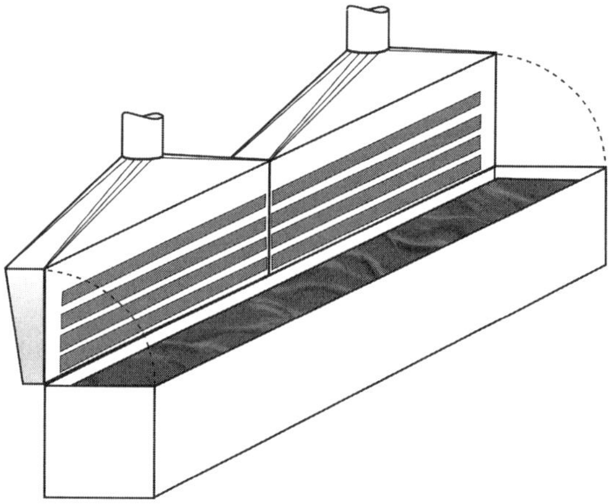

FIGURE 13.27 Example 13.6.

$2 \times \pi W^2/4 = 14$ ft^2. The resulting 10,800 ft^3/min is nearly half-again greater than the 7500 fpm using the control velocity method.

$$Q = 100 \text{ fpm} \times 108 \text{ ft}^2 = 10{,}800 \text{ ft}^3/\text{min} \qquad \text{(using Equation 13.15)}$$

VII. DESIGNING A CANOPY HOOD OVER AN ISOTHERMAL SOURCE

Canopy hoods are probably used in more industrial workplaces than they should be given their significant limitations in terms of employee protection (see also Section V.4):

Most importantly, a canopy is inappropriate if the worker's breathing zone must be "under" the canopy, especially between the source and the canopy.

Canopies are susceptible to cross-drafts that can push the plume laterally past one side of the canopy. Loss of control from cross-drafts is more likely if the hood is placed too high above a source.

A hot plume's buoyancy can cause it to rise into the hood faster than the hood can exhaust the emission. The resulting spillage can either occur routinely or periodically when the production process is cyclic and result in the release of a large volume of contaminant within a short interval.

As stated in Section V.4 of this chapter, a canopy hood can sometimes be used effectively if it is placed low over an isothermal source. Section VI.1.b of this chapter discussed an ANSI method to design a canopy hood over an open surface tank. The following design information is from the *Manual of Industrial Ventilation*'s VS-99-03 for a canopy hood over other than a hot source. (While admittedly the phrase "other than a hot source" is not quite the same as "an isothermal source," the latter phrase will be used herein for simplicity.) A separate design procedure for a canopy hood over a hot source will be discussed in the next section of this chapter.

1. The zone of control for a canopy over an isothermal source is the vertical (or near vertical) plane along the open perimeter between the top of the source and the bottom edge of the canopy. Thus, the control velocities in Table 13.5 are for the practically horizontal flow through that nearly vertical face depicted by the arrows on the right side of Figure 13.28 (not the vertical flow into the horizontal face across the bottom of the canopy!).

TABLE 13.5
A Comparison between Design Formulae and Control Velocities for Canopy Hoods with One, Two, Three, or All Four Sides Open

As per the *Manual of Industrial Ventilation*'s VS-99-03		As per Table 13.2 (from OSHA/ANSI)		
Open type canopy	$Q = 2.8(W + L)HV$ where $50 \leq V \leq 500$ fpm	Canopy hood	Four open sides	125 to 175 fpm or 2.4 × one side open
			Three open sides	75 to 125 fpm or 1.6 × one side open
Two sides enclosed	$Q = (W + L)HV$ where $50 \leq V \leq 500$ fpm	Enclosing hood	Two open sides	75 to 150 fpm or 1.4 × one side open
Three sides enclosed	$Q = WHV$ or LHV where $50 \leq V \leq 500$ fpm		One open side	50 to 100 fpm, the lowest $V_{control}$

2. The left side of Table 13.5 shows three formulae for Q that appear in VS-99-03. In each case, H is the height from the top of the source to the bottom of the hood. W and L are the width and length of the source, respectively. A canopy hood's "face" is the sum of all of the vertical openings through which the control velocity passes.

   It is unusual for a VS diagram to have three formulae for Q, but a canopy can have at least that many different combinations of its sides being either open or closed. A side can be closed by either a panel (typically built of metal), a curtain (such as strips of polyurethane), or simply the result of the canopy being placed adjacent to one or two existing walls.

a. All four sides are open in a classic, freely suspended canopy hood. Although the VS diagram indicates the hood should extend 20% of H beyond the width and length of the

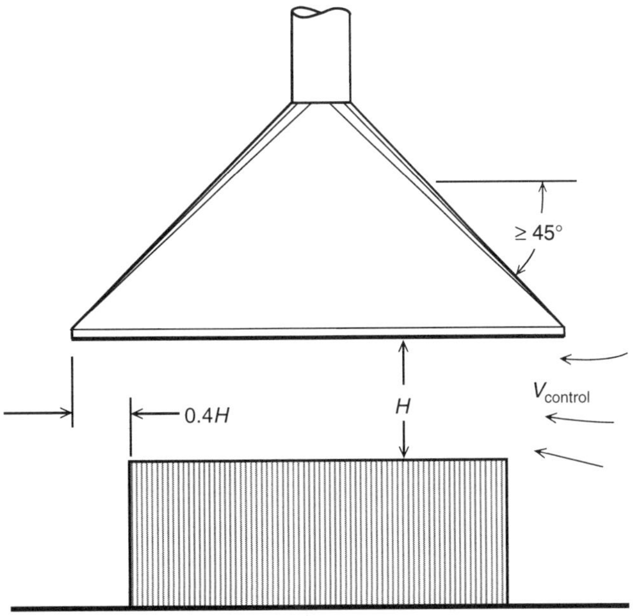

FIGURE 13.28 The design of an isothermal canopy.

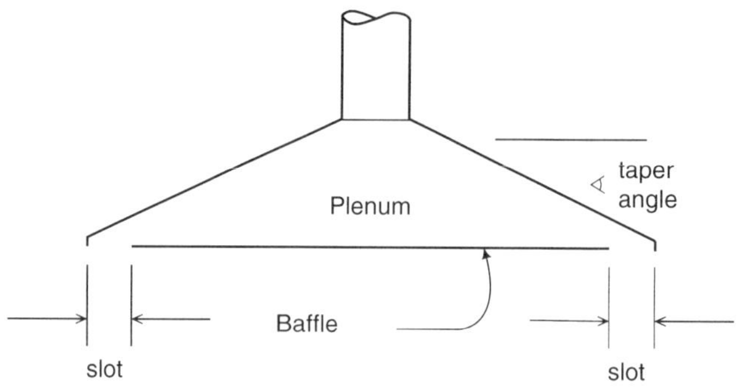

FIGURE 13.29 A sectional view of a large flat canopy hood with a baffle and plenum to increase the uniformity of the control velocity around its periphery.

source, only in the case of the freely suspended canopy does it account for that overhang in the formula. Thus, the total face area of this canopy is $1.4(2W + 2L)H$ or $2.8 \times$ Perimeter of the source $\times H$.

b. Three sides would be open if the hood were installed against one wall. Although this case is not covered by the VS diagram, if one long side were blocked and the width of the hood only extended 20% to the front and two sides, the total face area would be $(1.4L + 2.8W)H$.
c. Only two adjacent sides would be open if the canopy were installed in a corner. The formula not showing any overhang implies that the authors think two walls will block most crosswinds and alleviate the need for an overhang.
d. And if three sides were enclosed, the canopy would, in effect, constitute a booth enclosure.

The guidance for the control velocity in VS-99-03 is less specific than in most other VS diagrams. As reflected in the left side of Table 13.5, the recommended control velocity is only specified within a tenfold range between 50 and 500 fpm. The designer is directed to the generic recommended control velocity information covered in Table 13.1, herein. VS-99-03 also includes a hood entry loss factor ($h_e = 0.25$ VP_d) and recommended range for the duct velocity (1000–3000 fpm). These latter pieces of information will be discussed in the next chapter.

3. The right side of the table copies the control velocity specified by the OSHA ANSI Z9.1 standard for open surface tanks when using the same array of hoods.[f] The control velocities for open surface tanks tend toward the low end of those recommended in VS-99-03, and they increase as the number of open sides increases versus an appearance of being constant in VS-99-03. The lowest velocities in Table 13.5 are also consistent with the generic control velocity recommended in Table 13.1 for tanks.
4. A useful option (especially for a large, low canopy where the taper angle may be restricted to be less than 45° because of a nearby ceiling, overhead deck, ducts, etc.) is to place a baffle just up and inside the face of the canopy to create a slot between the lip of the canopy and the outer edge of the baffle (as shown in Figure 13.29). The slot should be sized to create an

[f] It may also be interesting to some readers that the ANSI authors focused on the number of sides that are open while the *Manual of Industrial Ventilation*'s authors focused on the number of sides that are enclosed. Moreover, a close look at Table 13.2 shows that the ANSI Z9.1 standard authors consider a canopy with two open sides to be an enclosing hood.

air velocity passing through the slot of around 2000 fpm (see also Section IV on slotted hoods). A taper angle of 45° or more makes the need for a separate baffle unnecessary.

VIII. DESIGNING A CANOPY HOOD OVER A HOT SOURCE

A canopy is better suited to capture and exhaust thermally hot emissions that rise due to their buoyancy than to capture isothermal sources as discussed in Section VII. This section will present two design procedures for canopy hoods over a hot source from the *Manual of Industrial Ventilation*: the first is for a canopy to be placed low over a hot source and the second design procedure is for a canopy to be placed high over a hot source.[1] For this purpose, a canopy is considered to be low if it is either within one source diameter or less than 3 ft above the hot source. Conversely, a canopy is considered to be high when the distance from the hot source to the hood is at least 3 ft and more than one source diameter.

1. Design procedures for a low canopy over a hot source are relatively simple. The procedures shown below differentiate between a circular and a rectangular hood. While a low canopy hood's shape usually corresponds to the shape of the source in terms of width and length, a rectangular canopy is often used over a round source because it is easier to construct from sheet metal than to construct a circular canopy. Using the procedures shown below, the area of a round hood $= \pi D^2/4$ is approximately three quarters × the area of a square hood whose side dimensions $= D$. Thus, the air flow exhausted through a square hood is about $4/3 = 1.33$ more than the flow through a round hood over a round source.[g]

a. The design of a circular canopy low over a hot source, as shown in Equation 13.16, is the easier of the two procedures (although a round hood is not as easy to construct as a square hood).

$$Q_{\text{total}}\ (\text{ft}^3/\text{min}) = 4.7(D_{\text{face}})^{2.33}(\Delta T)^{0.41} \tag{13.16}$$

where
D_{face} = circular hood diameter (ft) = the diameter of the source + 1 ft.
ΔT = temperature difference (in °F) between the initial plume and the ambient air.

b. The width (b) and length (L) of a rectangular canopy low over a hot source should each be designed to be 1 ft larger than the source's width and length, respectively. The hood total air flow rate Q_{total} should then match that specified by Equation 13.17.

$$Q_{\text{total}} = 6.2(b)^{1.33}L(\Delta T)^{0.42} \tag{13.17}$$

where
b = source width + 1 ft and L = source length + 1 ft.

The design of a canopy hood high over a hot source is based on a previously published model of the rising plume as a uniformly expanding cone of hot, contaminated air[13]. The dispersion of the plume causes it to expand, to become more dilute, and to cool; the cooling causes the plume's vertical velocity to slow.

a. Figure 13.30 depicts the expansion of a hot plume as if it were an inverted cone whose apex (origin) is a hypothetical point source some predictable distance $Z = (2D_{\text{source}})^{1.138}$ below the surface of the actual hot source. The distance that the cone

[g] The ratio between the "6.2" in Equation 13.17 and the "4.7" in Equation 13.16 is 1.319 or ≈4/3.

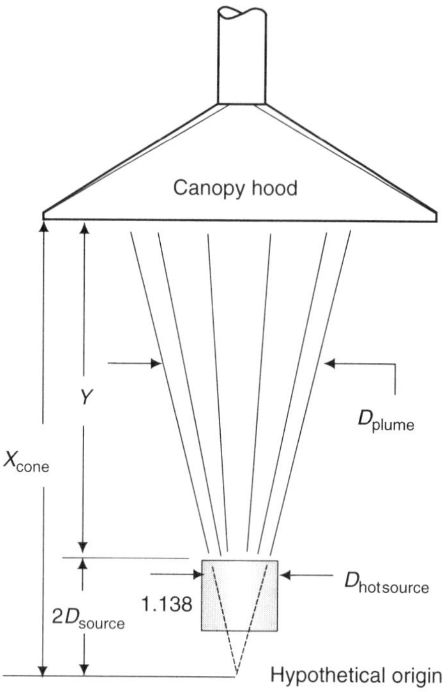

FIGURE 13.30 Model for the design of a canopy high over a hot source.

appears to be above its hypothetical point-source (the virtual plume height $X_{cone\ height}$) is defined by Equation 13.18 as the sum of the actual height above the source ($Y_{above\ source}$ in ft) plus that hypothetical distance.

$$X_{cone\ height} = Y_{above\ source} + (2D_{source})^{1.138} \tag{13.18}$$

b. The plume expands as it rises. Equation 13.19 predicts the diameter of the plume (D_{plume}) at any virtual cone height.

$$D_{plume} = 0.5\,X_{cone\ height}^{0.88} \tag{13.19}$$

c. The velocity of the plume (V_{plume} in Equation 13.20) is based on the horizontal size of the hot source (A_{source} in ft^2) and the temperature difference between it and the ambient air (ΔT in °F). As the plume rises above the source (as the $X_{cone\ height}$ value increases), the plume cools, its buoyant force and its plume velocity decrease.

$$V_{plume} = 8(A_{source})^{0.33}\,\frac{\Delta T^{0.42}}{X_{cone\ height}^{0.25}} \tag{13.20}$$

where A_{source} = either length × width for a rectangular source or $\pi \times D_{source}^2/4$ for a circular source (in units of ft^2 in either case).

d. The diameter of the canopy hood (D_{face} as given in Equation 13.21 and shown in Figure 13.31) should be larger than the diameter of the hot plume given in Equation 13.19 to compensate for the likelihood that the plume trajectory will be deflected from the vertical due to perhaps intermittent or/and variable but undefined

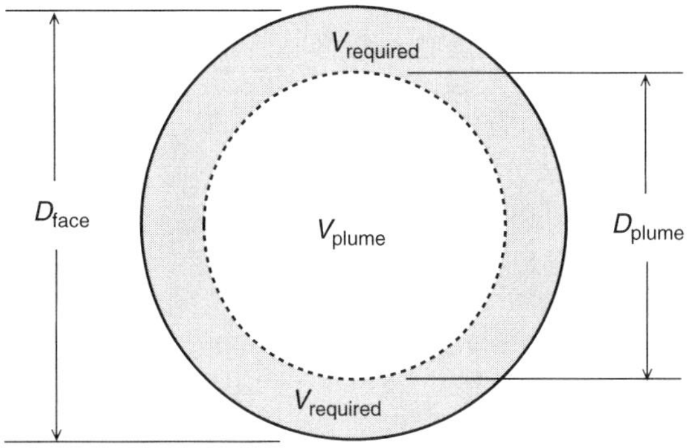

FIGURE 13.31 Depicting both the relationship between D_{plume} and D_{face} and how the face velocity of a canopy hood over a hot source can be designed in two portions.

cross-drafts.[h]

$$D_{face} = D_{plume} + 0.8\,Y \tag{13.21}$$

e. While the face velocity of the entire canopy could be designed to be equal to the plume velocity, designing it in two portions usually conserves more exhaust air. The face velocity in the portion of the canopy through which the plume enters must be at least as fast as the plume velocity defined by Equation 13.20. A lessor velocity would result in spillage when the plume enters the bottom of the canopy faster than air is exhausted out through its top. The face velocity in the rest of the canopy beyond the plume (as depicted beyond the dashed circle in Figure 13.31) could be the lessor of either V_{plume} or a generic velocity that would otherwise be recommended for a containment hood face velocity in that setting; approximately 100 fpm is suggested.

Thus, the total rate at which air should be exhausted from the canopy hood (Q_{total} as shown in Equation 13.22) is the sum of the volume flow rates flowing in from the plume ($V_{plume} \times A_{plume}$) plus that required ($V_{required}$) over the rest of the hood's A_{face} (other than the plume) as described above. The two areas can be calculated using the D_{plume} and D_{face}, respectively.

$$Q_{total} = V_{plume}A_{plume} + V_{required}(A_{face} - A_{plume}) \tag{13.22}$$

The above circular hood would be appropriate for a source that was either nearly circular or square in shape. If the source were considerably longer than it is wide, then a rectangular canopy would be more appropriate. A rectangular canopy high over a hot source would be calculated as above using the width and length of the source independently in place of the diameter.

EXAMPLE 13.7

Design a canopy hood to be placed 4 ft over a brass melting pot used to cast decorative pieces in a furniture shop. The users ladle 1770°F brass out of the top of the melting pot 24 in. (2 ft) above the floor. The solution shown below simply implements the sequence of five equations explained above.

$$X_{cone\ height} = 4 + (2 \times 2)^{1.138} = 8.84\ \text{ft} \qquad \text{(using Equation 13.18)}$$

[h] Equation 13.21 is corrected for several typographical errors in section 3.9 of the *Manual of Industrial Ventilation* including their Equation 3.11.

$$D_{\text{plume}} = 0.5 \times (8.84)^{0.88} = 3.40 \text{ ft} \qquad \text{(using Equation 13.19)}$$

$$V_{\text{plume}} = 8\left(\frac{\pi\, 2^2}{4}\right)^{0.33} \frac{(1770 - 77)^{0.42}}{8.84^{0.25}} = 154 \text{ fpm} \qquad \text{(using Equation 13.20)}$$

$$D_{\text{face}} = 3.40 + (0.8 \times 4) = 6.6 \text{ ft} \qquad \text{(using Equation 13.21)}$$

$$Q_{\text{total}} = 154 \times \left(\frac{\pi\, 3.4^2}{4}\right) + 100 \times \left(\frac{\pi\, 6.6^2}{4} - \frac{\pi\, 3.4^2}{4}\right) = 3900 \text{ ft}^3/\text{min} \qquad \text{(using Equation 13.22)}$$

Thus, the canopy hood should be designed to be 6.6 ft in diameter and exhaust 3900 ft^3/min (or it could be slightly larger).

A horizontal draft or crosscurrent will deflect a plume from its natural vertical rise by an angle determined by the ratio of the crosscurrent velocity to plume velocity predicted by Equation 13.23.

$$\text{degrees deflection from the vertical} = \tan^{-1}\left[\frac{\text{horizontal room air velocity}}{\text{vertical plume velocity}}\right] \qquad (13.23)$$

A glimpse of the problem of trying to collect a nearly isothermal plume with a canopy hood can be seen by comparing the vertical plume velocities in Figure 13.33 (calculated using Equation 13.20 for Y equal to 1 ft) with the air currents that might exist in a typical workplace. The latter tend to be at least 20 fpm, more typically 40 fpm, and can be higher if the room is well ventilated, if the room's occupants are active, or if machinery in the room disturbs the air. These data indicate

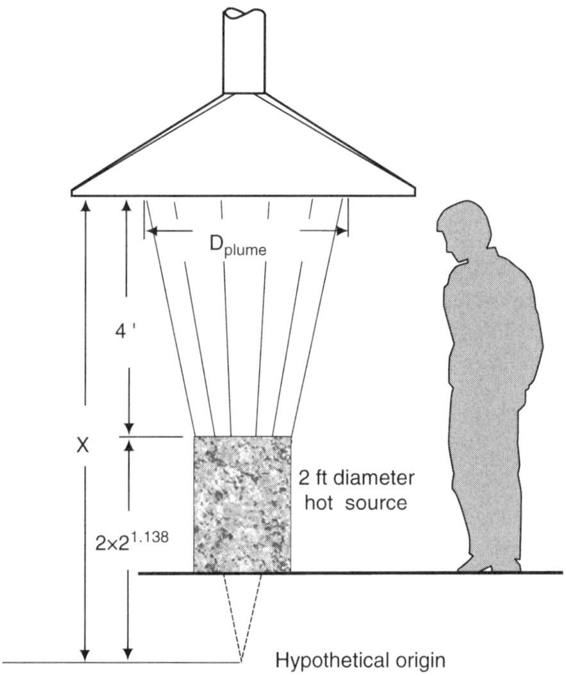

FIGURE 13.32 Example 13.7.

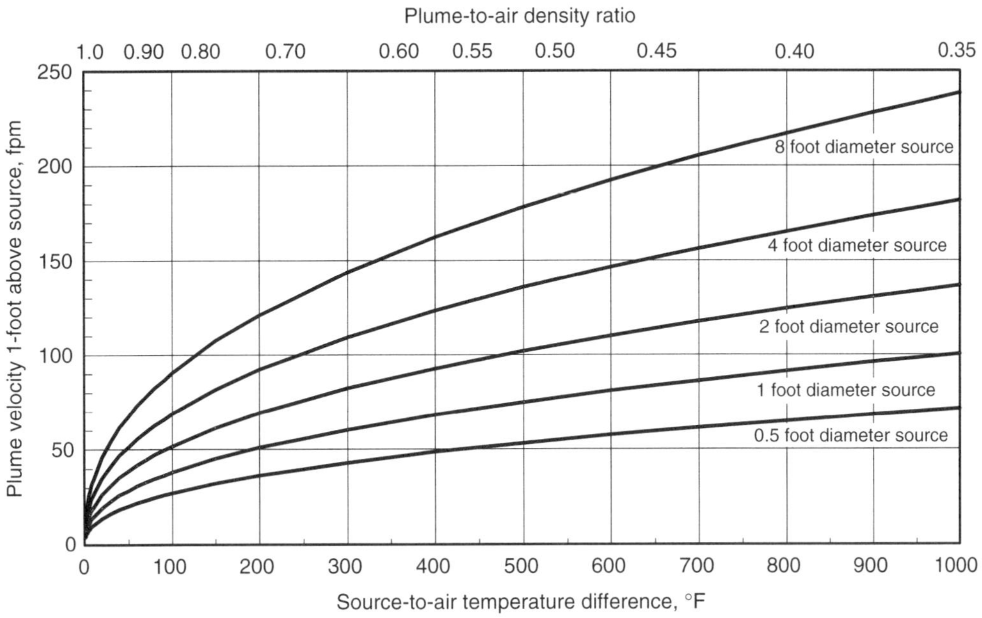

FIGURE 13.33 The effect of the temperature difference between a hot plume and the surrounding air on the plume's vertical velocity 1 ft above a hot source. The corresponding density ratio from Equation 2.25b is across the top.

that virtually any size source that is at least 300°F above room temperature will create a plume velocity in excess of typical room air currents. However, a source has to be both hot and large for its plume velocity to be sufficiently in excess of room air currents for its trajectory to be nearly vertical.

IX. DESIGNING A PUSH–PULL HOOD SYSTEM

When the width of an open surface tank as described above exceeds 36 in. (~ 1 m) the size and air flow rate of a lateral exhaust hood will exceed its practical operating limits. A good solution to such a challenge is to install a "push–pull" exhaust ventilation system (ANSI Z9.1-1991 suggests installing a push–pull system for tanks that are greater than 42 in. or 1.1 m). Such push–pull systems can be used to control the hazard emanating from not only a horizontal surface such as that depicted in Figure 13.34, but also a vertical surface as long as no (or only small) obstacles exist in the path from one side to the other.

Think of push–pull ventilation as a system comprising two parts. The first part (shown on the left side of Figure 13.34) is the portion that pushes or blows a curtain of air that entrains the contaminant as it rises out of a source and pushes it toward a collection hood. The second part of the system (shown on the right side of Figure 13.34) is a hood that more collects rather than pulls the curtain of air and its entrained contaminants into its face and exhausts them before they can disperse into the room. The push side of the system is supplied by a fan that is separate from the typical exhaust fan on the pull side.

The success of push–pull systems relies on two phenomena. The first phenomenon is that the pushed air starts with a high momentum (remember in physics, that momentum is mass times velocity). The distance over which that momentum can be maintained is much further than the short distance that room air initially at rest with no momentum can be pulled or sucked into the face of a

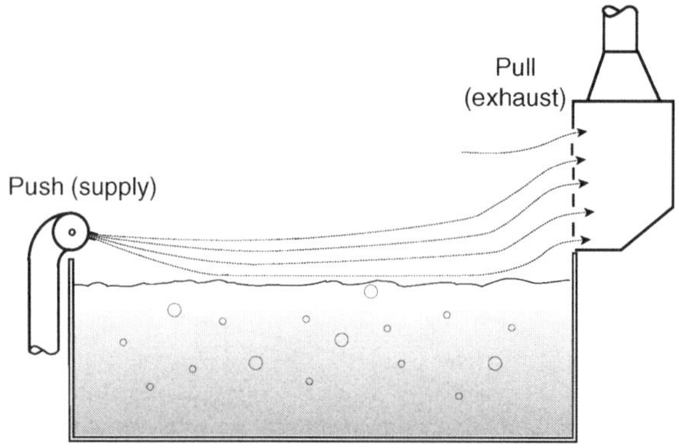

FIGURE 13.34 A depiction of a push–pull ventilation system over an open surface tank.

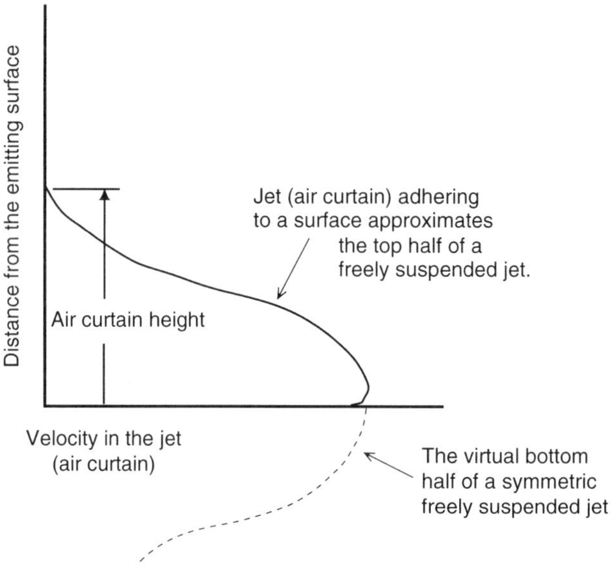

FIGURE 13.35 Detailed depiction of the vertical velocity profile within a jet of air with a surface boundary layer.

collection hood. The DallaValle equation in Section III defined the limited influence of a collection hood acting alone.

The second phenomenon is the tendency for the flow of a jet or curtain of air striking a smooth surface at an incidence angle of less than approximately 20–25° to cling to that surface with minimal dispersion. This latter phenomenon (sometimes called "attached flow") allows a jet of air to act and be modeled just as if it were flowing next to a mirror image of itself that is below the solid or liquid surface, as depicted in Figure 13.34 and in more detail in Figure 13.35. The net effect is that an attached air curtain will expand at a shallower angle than a free-standing jet or air curtain.[i]

Push–pull systems can be used advantageously to control emissions coming off of any flat, wide source up to at least 10 ft (3 m) across. A push–pull system can obtain the same level of control while exhausting less air than a traditional LEV hood that only pulls from the exhaust side.

[i] The effect of the boundary layer that will exist between the adherent flow of an air curtain above a flat surface and that surface (and depicted in Figure 13.35) is negligibly small.

TABLE 13.6
Experimentally Derived Values of "a" and "α" Reported by Baturin (1972)[14] for a Single Circular Jet Outlet or a Series of Jet Outlets Aligned in a Plane

	Single Circular Jet $\tan(\alpha) = 3.4$"a" (13.23a)		Air Curtain Jet $\tan(\alpha) = 2.4$"a" (13.23b)	
Shape of the Jet Outlet	a	α	a	α
Convergent slot or nozzle 	0.068	13°	0.11	15°
Sharp edged slot or drilled holes 			0.12	16°
Tube 	0.078	15°	0.125	17°
The "b" coefficient	half of the slot height or "b" from Equation 13.24		the full slot height or "b" from Equation 13.24	

This advantage will result in significant reductions in the cost of conditioning the room make-up air (heating and/or cooling to be discussed in Chapter 17). On the other hand, obstacles or obstructions in the air's straight pathway from the push supply side to the pull exhaust side (such as suspended parts, support frames, hoists, etc.) will disrupt the air curtain and prevent any push–pull system from being an effective control. In fact, the inappropriate application of a push–pull system to a source with such obstructions can actually increase dispersion of a contaminated plume into the workplace.

The design of a push–pull system can be envisioned in four steps. Equation 13.26 uses Baturin's "a" and "b" coefficients to give the initial jet velocity leaving the outlets on the push-side of the system (V_{outlet}) needed to obtain a desired control velocity at a distance X away ($V_{at\ X}$). Equation 13.27a,b gives the rate that supply air must be provided to the outlets (Q_{supply}). Equation 13.28 gives the exhaust air volume ($Q_{exhaust}$) which must be significantly bigger than Q_{supply} to handle both the supplied air and the entrained air volumes. Equation 13.29 defines the height of the exhaust hood so that it will be at least as tall as the air curtain is high.[j] The following subsections discuss each of these equations in more detail.

1. A key to the first step is the formulas and tables describing the behavior of "jets" of air complied by Baturin (1972)[14]. The expansion angle (α) and Baturin's design coefficient "a" are related to each other via Equation 13.23 embedded within Table 13.6. The value for both α and "a" is related to the construction and shape of the air outlet that together define the coherence of the air curtain. The more coherent the air stream is when it leaves the outlet, the less mixing will occur between the jet and the surrounding air and the lower will be the value of "a" and the expansion angle of the jet (α).

   Table 13.6 provides several descriptive and pictorial examples of air outlets gleaned from Ref. [14] for both a single, free standing, circular jet and a plane jet or "air curtain" made up of multiple outlets.[k] However, notice that the data within each group of columns in Table 13.6 span a relatively narrow range of values. Hughes (1990)[15] recommends using a design coefficient ("a") of 0.13 for an adherent air curtain (corresponding to an expansion angle (α) of approximately 17°). A round supply

[j] The *Manual of Industrial Ventilation* describes some of this information in its Section 3.8, and push nozzle manifold pressure information is provided in VS-70-10 through VS-70-12.

[k] An air curtain jet can be created either by inserting a series nozzles or cylindrical tubes into the side of a supply air manifold or by cutting a continuous, sharp edged slot into the side of such a manifold.

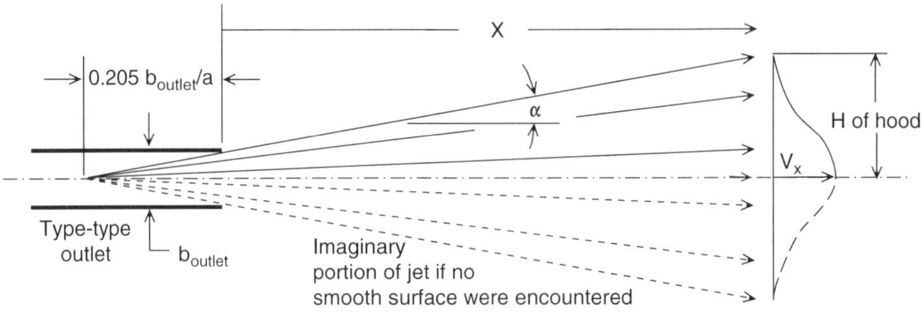

FIGURE 13.36 Schematic description showing how an expanding plane jet in contact with a smooth surface behaves as if it were one-half of a freely expanding symmetric jet. The virtual or nonexistent portion of the jet is shown as dotted lines.

manifold (as shown end-on at the right side of Figure 13.34) that can be easily rotated after installation is recommended to simplify adjusting the angle at which the jets are released and strike the liquid surface to obtain the best control in a given setting. The air flow from a series of 1/4 to 3/8 in. (5 to 9 mm) diameter air outlets spaced 3 to 6 diameters apart (approximately 20 to 50 mm) will merge into one plane jet or "air curtain" within about 15 to 20 outlet diameters (<5 in. for this size holes).

2. Baturin's "b" coefficient is the effective vertical height of the slot from which the jet emanates (perpendicular to its length). As implied in Table 13.6, a freely suspended jet will expand in each direction as if the nozzle were half as wide as it appears; thus divide the "b" as found below by 2. A jet that emanates parallel to an adjacent wall or liquid surface will attach to that surface and expand in only one direction slightly more slowly. Thus, if the air curtain emanates from a continuous slot, its "b" value should equal the height of that whole slot b_{outlet} as shown in Figure 13.36). If the air curtain emanates from a linear series of circular nozzles or holes cut into the side of the manifold, the "b" value should be the average outlet height derived from Equation 13.25. A "b" value of 0.125 to 0.25 in., 0.01 to 0.02 ft, or 3 to 6 mm is reasonable for push lengths of 4 to 8 ft.

$$\text{total outlet area} = \text{outlet area per length} \times \text{length of manifold } [L] = \text{“b”} \times L \quad (13.24)$$

$$\text{“b”} = \frac{\pi (D_{\text{each jet outlet}})^2/4}{\text{separation between outlets}} \quad (13.25)$$

3. The velocity of any jet or air curtain will decrease with distance as it entrains adjacent quiescent air. The initial velocity at the outlet of the jet (V_{outlet}) needs to be high enough to assure that at least the jet's centerline velocity at the far side of the source (the distance X in $V_{at\ X}$) is still sufficient to control the emitted contaminant. The *Manual of Industrial Ventilation* suggests that $V_{at\ X}$ should be at least 75 fpm for an isothermal source and should be increased to $V_{at\ X} = 0.4°F + 15$ fpm as the source temperature rises above 150°F. Alternatively, $V_{at\ X}$ could be set equal to the open tank exhaust ventilation control velocity determined from Table 13.2. Consider increasing the $V_{at\ X}$ value from either source to accommodate cross-drafts. Equation 13.26 can be used to specify an initial velocity at the outlet of the jet ($V_{\text{at jet outlet}}$) in order to achieve any specified centerline velocity at the pull hood ($V_{at\ X}$).

$$V_{\text{at jet outlet}} = V_{\text{at }X} \frac{\sqrt{[(\text{“a”}X/\text{“b”}) + 0.41]}}{1.2} \quad (13.26)$$

where

"a" = a design coefficient chosen from Table 13.6 based on the type of push outlet in the supply manifold; a nominal value of 0.12 is recommended herein.

"b" = the effective outlet height (either real or found using Equation 13.25, as described in Subsection IX.2).

X = the distance from the push outlets along the jet flow's centerline in the same units as "b." Pay attention to the units here since the units of "b" may be much smaller than the units of X. For practical design purposes, X can equal the distance from the push outlet to the face of the collection hood, and the additional distance to the theoretical origin of the expanding jet in Figure 13.36 can be disregarded.

4. The necessary Q_{supply} implied by Equation 13.27a,b equals the above $V_{at\ jet\ outlet}$ times the total outlet area derived from Equation 13.24 (be sure to adjust the units of "b" and L as necessary to match those of V). In the case of an attached air curtain, the outlet area is either the actual slot height or the "b" value (since they are equal) times the length of the push manifold parallel to the liquid surface (the latter usually equals the width of the source). Hughes and the *Manual of Industrial Ventilation* also refer to the Q_{supply} (as well as the $Q_{at\ X}$ in Equation 13.28) per foot or meter of source or tank length (L).

$$Q_{supply} = V_{at\ jet\ outlet} \times \text{total } A_{at\ jet\ outlet} \quad (13.27a)$$

$$Q_{supply} \text{ per unit of } L = V_{outlet} \times \text{b}_{outlet} \text{ in the units of } V \quad (13.27b)$$

5. The momentum of a jet turns out to be well conserved as the jet entrains ambient air. Conservation of the jet's momentum means that $Q_{at\ X} \times V_{at\ X}$ is constant and equal to $Q_{supply} \times V_{at\ jet\ outlet}$. In other words, as the jet moves across the source, its volume flow rate ($Q_{at\ X}$) will increase in inverse proportion to the decrease in its velocity ($V_{at\ X}$).

$$Q_{at\ X} = Q_{supply} \times \frac{\sqrt{[(\text{"a"}X/\text{"b"}) + 0.41]}}{1.2} \quad (13.28)$$

To assure that the exhaust flow rate is adequate, the pull hood should have the capacity to exhaust slightly more than the $Q_{at\ X}$ of the now contaminated air within the expanded air curtain given by Equation 13.28. A 20% increase will conveniently cancel out the 1.2 in the denominator of $Q_{at\ X}$. However, if small obstructions are present in the path of the jet that can deflect the pushed air curtain, then the exhaust air flow $Q_{exhaust}$ should be increased from 1.5 to 2 times that $Q_{at\ X}$.

6. The height of the hood's face on the pull side of the push–pull system (Height_{hood}) should be tall enough to accept the height of the (real) upper half of the air-curtain depicted in Figure 13.36. Equation 13.29 is really a restatement of Equation 13.23b using the definition of the tangent of α as the opposite side (the Height_{hood}) over the adjacent side (the distance X) of the right triangle depicted in Figure 13.36.

$$\text{Height}_{hood} = 2.4[\text{"a"}X + 0.41 \times \text{"b"}] \approx 2.4 \times \text{"a"} \times X = X \times \tan(\alpha) \quad (13.29)$$

EXAMPLE 13.8

Design a push–pull system to control the 3-ft wide open-surface tank used in Example 13.5 and Example 13.6 (as depicted in Figure 13.37). Designing a push–pull system has some trial-and-error or experience aspects similar to designing a slotted hood. For slightly more assurance, the hood will be designed to achieve 150 fpm at the far side of the tank ($V_{at\ X}$) instead of the 125 fpm control velocity in Table 3.3. Suppose that a colleague suggested using a 1/4-in. holes drilled 1-in. (four outlet diameters)

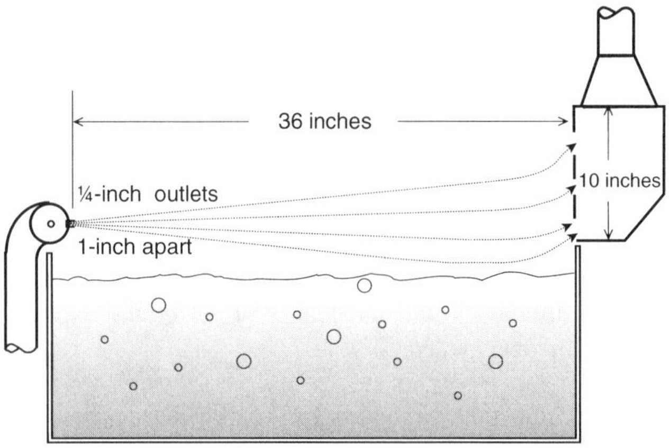

FIGURE 13.37 Example 13.8.

apart as the air outlets.

$$\text{“b”} = \frac{\pi\,(D_{\text{each outlet}})^2/4}{\text{outlet separation}} = \frac{0.049\ \text{in.}^2}{1\ \text{in.}} = 0.049\ \text{in.} = 1.25\ \text{mm} \qquad \text{(using Equation 13.25)}$$

$$V_{\text{outlet}} = 150\ \text{fpm}\frac{\sqrt{[(0.12\times 36/0.049) + 0.41]}}{1.2} = 1180\ \text{fpm} \qquad \text{(using Equation 13.26)}$$

$$Q_{\text{supply}} = V_{\text{outlet}} \times A_{\text{all outlets}} = 1180\ \text{fpm} \times 0.082\ \text{ft}^2 = 80\ \text{ft}^3/\text{min} \qquad \text{(using Equation 13.27)}$$

$$Q_{\text{at}\,X} = Q_{\text{supply}} \times \frac{\sqrt{[(0.12\times 36/0.049)+0.41]}}{1.2} = 7.9 \times Q_{\text{supply}} = 630\,\text{ft}^3/\text{min} \qquad \text{(using Equation 13.28)}$$

$$\text{Height}_{\text{hood}} = 2.4\times[0.12\times 36 + 0.41\times 0.049] = 10.4 \approx 11\,\text{in.} \qquad \text{(using Equation 13.29)}$$

Increasing Q_{exhaust} to approximately 1.25 times $Q_{\text{at}\ X}$ or approximately 785 ft^3/min is still only approximately 10% of the exhaust volume found using the control velocity method in Example 13.5, a significant saving. However, these push–pull design values are below what would be expected for a tank of over 4 ft wide (e.g., a 6 ft wide tank with some other changes would approximately double the flow rate). Marzal *et al.* (2003)[16] showed how smoke can be added to the supply side air to visualize and optimize the performance of a push–pull ventilation system.

X. USING AN AIR SHOWER

Reflecting back on the controls in this chapter, Sections IV–IX each described one or more LEV hoods that keep the contaminant out of someone's breathing zone by containing or collecting the plume and removing it from the workplace. Implicit in all but the last of these exhaust hoods is the expectation that the air in the rest of the work space, which is being drawn through the employee's breathing zone and into the exhaust hood, is of acceptable breathing quality. In most settings, that expectation is met (although one should never get too complacent about making that assumption).

However, when all of the air in the vicinity of someone is contaminated, then simply exhausting a local plume will not protect that person. For example, most of the air that is near the working face of an underground mine is already contaminated with dusts. Another example might be core makers in a foundry surrounded by innumerable pieces of hot-box cores each emitting formaldehyde. Toll booth operators surrounded by a sea of cars emitting CO and other products of combustion are yet another example. In each of these circumstances, the air that would be pulled into someone's breathing zone to replace the air that would be exhausted if a local exhaust hood were installed would already be contaminated, and the person in front of that hood would not be protected.

When faced with such circumstances (and having already considered and rejected source controls), one might either use personal respiratory protection (its limitations will be covered in Chapter 22) or install an air shower as depicted in Figure 13.38. Such a system has also been called a canopy air curtain system, a fresh air island, an "overhead air supply island system" (or its acronym "OASIS"), or simply a local air supply.[17,18] The originator of the OASIS term described an application that directed 6000 ft^3/min of fresh air at 300–400 fpm through a 2 to 2.5 ft wide fresh air outlet placed 1.5 to 2 ft directly over a worker. Air showers can be both larger and smaller and oblong instead of square, but any time the person's head leaves that zone, they are being overexposed. Thus, each system should be sized to protect the breathing zone over most of the workday.

While an air shower has significant advantages over respirators, it also has some of the same limitations. Reductions in exposure of around 90% are typical, similar to a basic respirator with an air purifying cartridge or filter. Control will decrease if the worker must leave the shower zone; thus, some training and employee compliance with that training is needed. For instance, providing visual guidance, such as the floor markings depicted in Figure 13.38, can help to remind the employee of the boundaries of their most protected position. However, employees are more likely to prefer this limitation to the nuisance of having to wear a respirator, and employee acceptance is increased when the air is tempered for comfort.

An air shower also has some of the disadvantages of dilution ventilation to be discussed in Chapter 19. For instance, the added flow of air will disperse an initially localized contaminant, potentially exposing more people nearby. However, that limitation is not applicable if the

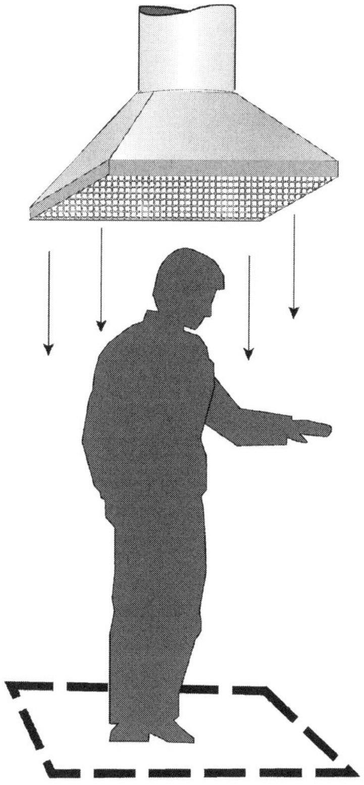

FIGURE 13.38 Depiction of a person working under an air shower.

contaminant is already widespread. An air shower will generally only add an insignificant amount of additional fresh air to a room. For example, the 10 to 15% reduction in the contaminant concentration in the room in which an air shower was installed, as reported by Volkwein (1988a)[17] and others, is an apparent by-product of introducing more dilution air into those rooms. However, this reduction is probably toxicologically insignificant and, in general, other employees outside of the air shower are unprotected.

Nonetheless, this approach has its niche uses. In addition to the circumstances mentioned at the outset of this section, local air supply might be feasible when faced either with too many sources to control via individual local exhaust hoods or systems, with unacceptable cross-drafts that would render a local exhaust hood ineffective, or with any physical constraint (such as building structure) that would preclude an exhaust hood. The canopy air curtain concept was tested successfully in lead reprocessing operations,[19] on a mining machine in an underground coal mine, on a loading machine in a salt mine, in the partially enclosed cab of a load haul dump unit,[20] and to reduce dust exposures during the weighing of chemical powders,[21] powder dye handling operations,[22] and in a bagging operation.[23] An air curtain is a long, narrow air shower that can be designed using the same principle as the push side of the push pull ventilation system described in Section IX to protect the air in some oblong work zones.[24,25]

XI. USING COMMERCIAL AND OTHER SPECIALTY HOODS

All of the previous LEV hoods in this chapter were intended to protect people outside the hood from some toxic contaminant by either drawing it into or containing it within the hood. However, sometimes, the user also wants to protect what is inside the hood from the environment. Examples of this latter problem occur when working with products such as pharmaceuticals, microbiological media or samples, aseptic instruments, or semiconductor substrates that will be damaged by microbes and/or any dust particles. Two forms of commercially manufactured hoods designed specifically to deal with such a task are a laminar flow clean bench (LFCB) and a biological safety cabinet (BSC).

1. The function of an "LFCB", such as depicted in Figure 13.39 to Figure 13.41, is simply to pass filtered air smoothly over the working surface to protect products inside the hood.
   a. Figure 13.39 and Figure 13.40 depict two horizontal flow LFCBs. Models with a bottom-mounted fan have a lower profile than those with a fan on top. (See also the *Manual of Industrial Ventilation*'s VS-35-30.[1]) In each case air is pulled from the room usually through a prefilter (a low efficiency filter that only takes out relatively big particles), pushed through an HEPA filter by an integral fan (an HEPA or high efficiency particulate aerosol filter is at least 99.97% efficient at taking out small particles), and flows horizontally over the enclosed working surface and out through the fully open front of the enclosure. The contents within an LFCB are bathed in highly filtered air, but any contaminant within the LFCB is carried right out at anyone working in front of the bench.
   b. Figure 13.41 depicts a vertical LFCB in which the filtered air flows downward over the enclosed product. Some of the air should continue to flow downward through a perforated work area and be vented either back into the room or outside. By design, some of the downward flowing air must exit the face of the LFCB to keep any particles in the ambient air from contaminating the enclosed product. While this pattern of air flow should result in less exposure to the person in front of the hood than the outward horizontal LFCB, the vertical flow LFCB is still not designed to actually protect that person from the material in the hood.

   Either form of LFCB can actually create health problems. First, the air that flows out through the face of either a horizontal or vertical clean bench can carry any airborne

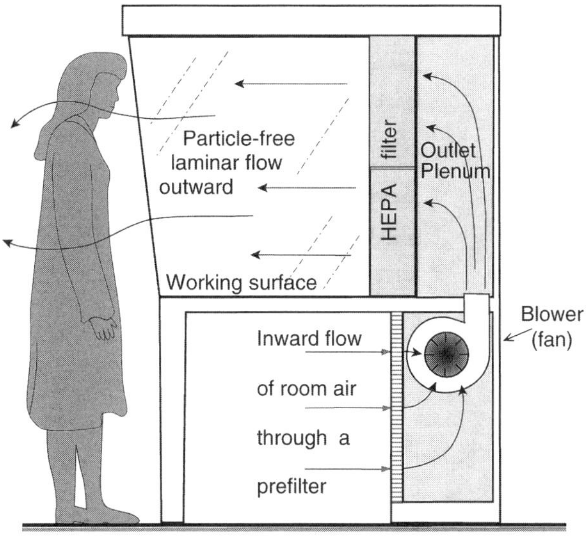

FIGURE 13.39 Air flow within a horizontal laminar flow clean bench with a bottom-mounted fan.

contaminant inside the hood right into a user's breathing zone. A horizontal flow LFCB has no filter on its exhaust air, and a vertical flow LFCB can but does not have to be filtered to protect the environment. An LFCB might easily be mistaken for a laboratory exhaust hood. Thus, an unsuspecting user may not realize the difference and use the hood for an inappropriate purpose, and end up exposing both themselves and potentially other occupants

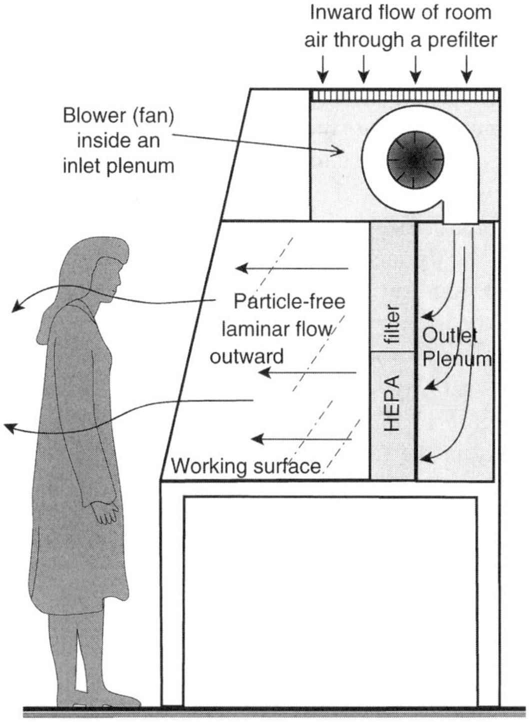

FIGURE 13.40 Air flow within a horizontal laminar flow clean bench with a top-mounted fan.

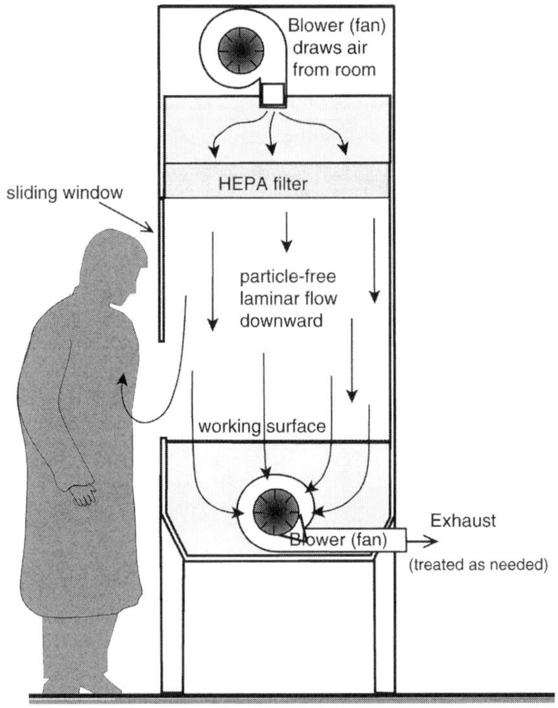

FIGURE 13.41 Air flow through a vertical laminar flow clean bench.

within the room or laboratory to a hazardous chemical or infectious agent. Thus, an IH should be aware of all LFCBs on their site and strive to assure that users of each unit are trained and aware of their limitations. All users should also be made aware of the meaning of each unit's operating light or pressure meter, under what conditions to be concerned, who to call to report a malfunction, and when to stop work or seek appropriate personal protection.

A Biological Safety Cabinet (BSC) is a safer, albeit more expensive, alternative to a LFCB. BSCs come in three classes (I, II, and III), one of which has two types. Air flows into the face of all BSCs, thereby protecting the user out front. All BSCs filter the exhaust air to protect the environment from particulate emissions. However, only a Class II BSC protects the product inside the hood.

2. A Class I BSC (such as the counter top configuration depicted in Figure 13.42) has an inward-flowing face velocity to protect the user much like the traditional laboratory hood depicted in Figure 13.15. However, the inward flowing air also carries any other contaminants already in the room onto the working surface. The filtered exhaust air from a Class I BSC may be vented outside (like a traditional laboratory hood must do), or it may be vented directly back into the room. Users should be made aware that filtration only removes particles; it does not remove gases or vapors that may be released within the hood. In summary, a Class I BSC protects personnel by the inward air flow, protects the environment from particulate contaminants by HEPA filtering the exhaust air stream, but does not protect the product on the working surface.
3. Class II BSCs combine the user protection features of a Class I BSC with the product protection features of a LFCB. Class II BSCs are categorized into two types (type A and type B) that differ by how the exhaust air is handled.

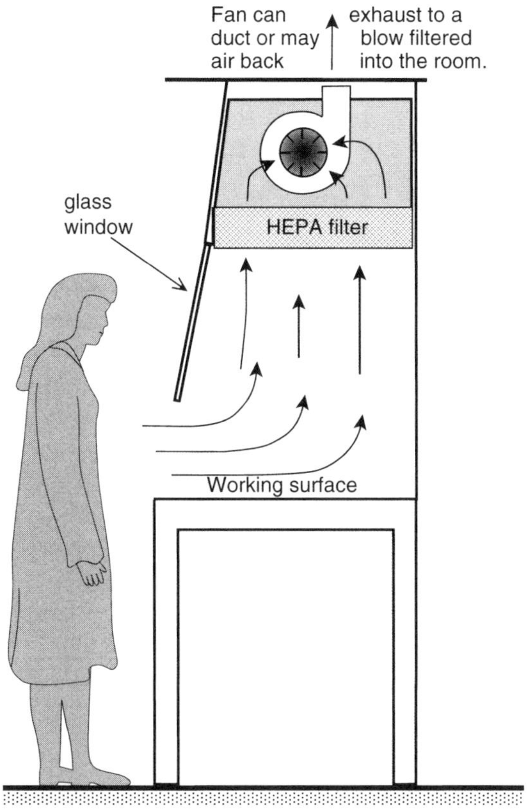

FIGURE 13.42 Depiction of air flow through a Class I biological safety cabinet.

a. A Class II Type A BSC (such as depicted in Figure 13.43 and the *Manual of Industrial Ventilation*'s VS-35-10) recirculates up to 70% of the exhaust air through a supply HEPA filter within the hood and returns the rest of the air back into the work room through an exhaust HEPA filter. The filtered supply air flows down smoothly over the working surface and then through slots both in front of and behind the solid working surface. The room air should only flow into the BSC as far as the front slots where it cannot contaminate the product on the working surface. The correct ratio between the filtered laminar flow and the inward face velocity can be obtained by adjusting a balancing damper. The detailed shape of the front slots differs among commercial manufacturers. Neither a recirculating Class I BSC nor a Class II type A BSC is recommended for work that involves gases or vapors because HEPA filters only remove aerosols.
b. Class II Type B BSCs come in two subtypes: a Class II Type B1 BSC (such as depicted in Figure 13.44) recirculates approximately 70% of the HEPA filtered air internally within the enclosure, while a Class II Type B2 BSC (such as depicted in Figure 13.45) exhausts 100% of the air that enters the hood. Thus, a Type B2 BSC is sometimes called a "total exhaust cabinet." Both subtypes are depicted in VS-35-11 in the *Manual of Industrial Ventilation*. In neither Type B hood does any of the exhaust air go back into the workroom (cf., the way all Class II Type A BSC recirculate exhaust air back into the room and a Class I BSC can recirculate air back into a room). Other than not recirculating air back into the room, both type B hoods share most of the other features

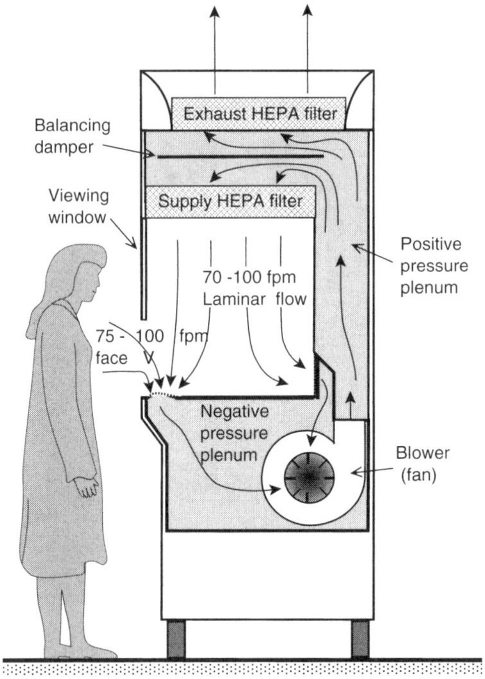

FIGURE 13.43 Depiction of air flow through a Class II Type A biological safety cabinet.

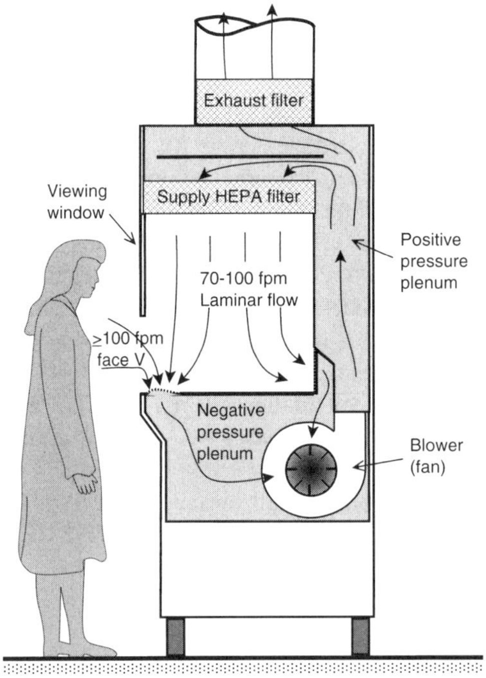

FIGURE 13.44 Class II Type B1 biological safety cabinets recirculate some filtered air within the hood (typically approximately 30%).

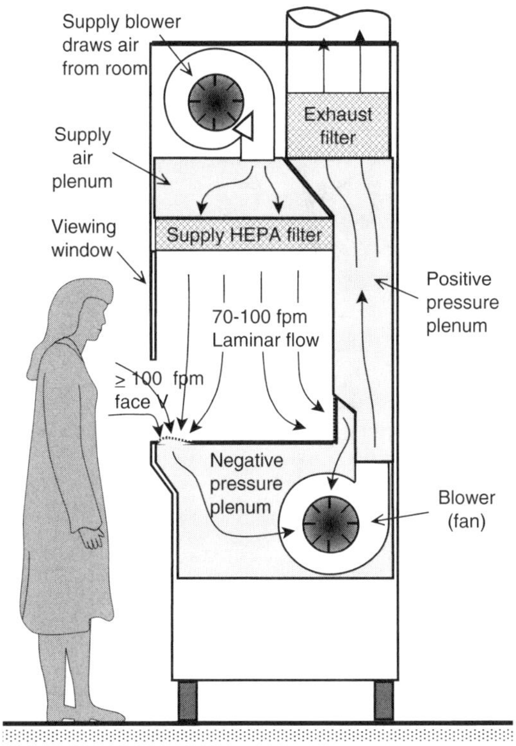

FIGURE 13.45 Class II Type B2 biological safety cabinets do not recycle any air. All laminar air is fresh.

of their type A cousin. In both types of Class II BSCs, only filtered air reaches the work zone to protect the product, while room air flows inward only near the front of the BSC to protect users, and all of the exhaust air is HEPA filtered to protect the environment.

4. Class III BSCs (such as depicted in Figure 13.46) comprise a total enclosure. Access to the working surface is only via gloves that are sealed into the closed face of the cabinet. The entire enclosure is sometimes called a glove box (although not all glove boxes are Class III BSCs). The exhaust air is double HEPA filtered to increase protection (or it may be incinerated). The exhaust fan is generally remote to keep the entire duct under negative pressure. Because a Class III BSC is fully enclosed, the air flow rate is very small (depicted by lightly dashed lines inside the cabinet) which also means that dust or contaminants accumulate on the filters at a slower rate than with the open circulation through the other BSCs. Some further details of a Class III BSC are shown in the *Manual of Industrial Ventilation*'s VS-35-20.

All of the *Manual of Industrial Ventilation*'s VS-35 series of diagrams suggest that BSCs should have the same 80 to 100 fpm face velocity as for laboratory hoods; however because of their complexity, BSCs are usually purchased as complete units from commercial manufacturers who will provide a complete hood along with the pressure loss and duct velocity data needed to design the rest of the LEV system as described in Chapter 10. The use and accumulation of pathogens or toxins within a BSC will create hazards for maintenance workers who replace used HEPA filters and repair or service unfiltered (thus contaminated) exhaust system components such as plenums, fans, motors, etc. Specialized training is available for those needing to learn procedures to conduct such maintenance operations safely.

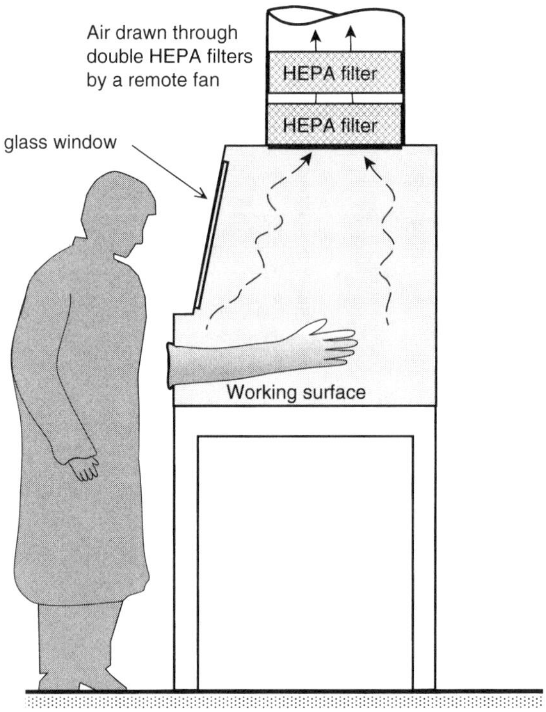

FIGURE 13.46 Depiction of a Class III biological safety cabinet (glove box).

The intent of the first three sections of this chapter (I–III) was to provide readers with the basic principles of local exhaust hood selection and design including control velocities and the DallaValle equation. The succeeding six sections (IV–IX) were intended to provide the reader with a good starting kit of specific hood design tools that will improve over time with more experience. Section XI was to familiarize industrial hygienists with commercial hoods that are designed by specialized professionals. Several other hoods and advanced design tools were not included herein. For instance, a gas storage cabinet is another specialized hood that can be bought commercially or constructed on site.[26,27] Aaberg exhaust hoods, developed in the 1990s, are a novel design that reinforce the normal distance over which a collection hood can be effective. In the axisymmetrical version of an Aaberg hood, an almost normal-looking flange is fitted with a plenum with a narrow slot around its periphery through which fresh air is blown.[28] Researchers at the National Taiwan University have shown some success at controlling exposures in front of a collection hood with a cross-draft by generating a wake from a flat plate placed in that cross-draft.[29] Also, nothing was covered herein about computational fluid dynamics (CFD), an advanced design technique that has been used by design engineers to predict actual hood efficiency although more for emission control than for employee protection.[30,31] As with many aspects of industrial hygiene (and of any science for that matter): the closer you look at a topic, the more you find that you need to learn.

PRACTICE PROBLEMS

1. Use the generic control velocity table (Table 13.1) to recommend a mid-range control velocity for each of the following settings.

Setting	$V_{control}$ range	$V_{control}$
a. Sheet metal cut and bent for use as desk-top computer cases that are hung to air dry after being dip-cleaned with an organic solvent prior to being painted	____ to ____	_____ fpm
b. Manually soldering components onto custom electronic circuit boards on a small scale production line processing approximately 20 boards per hour at a work bench as depicted in Figure 13.47	____ to ____	_____ fpm
c. Scooping 25 lb batches of silica-based clay from a weighing platform into a mulling machine	____ to ____	_____ fpm
d. Operating a 14-in. diameter disc sander in the wood working department of a furniture manufacturing plant	____ to ____	_____ fpm

2. If the dip tank in problem 1.a above were 3-ft long (side-to-side) and 2-ft wide (fore-to-aft) and the adjacent drip board were another 3-ft long (as in Figure 13.17), how much air does VS-75-06 indicate should be exhausted?

   Q = ______ ft^3/min

3. Suppose the organic solvent in the previous problem was methylene chloride (CAS 75-09-2) for which the following data is provided:

NIOSH REL:	Possible carcinogen	MW:	84.9	
TLV:	50 ppm (A3)	P_{vapor}:	435 mmHg	
OSHA 1910.1052 (standard)		BP:	104°F	FRZ: − 139°F
TWA PEL	25 ppm	LEL:	13%	UEL: nonflam
STEL:	125 ppm	Fl.Pt.:	approx. 18°F	
for	5 min. in any 2 h	Sp. Gr.:	1.33	

In Practice Set 4 Problem 4 you found (or at least should have found) that if this chemical were used in an open surface tank at normal temperature, that ANSI Z9.1-1991 rated this chemical's Hazard Potential a "B" and its vapor evolution category would be "2". Although a canopy is usually not a good IH choice for a room temperature ("isothermal") source, as a self-check of your reading of Table 13-2, it specifies a control velocity of 125 fpm for a canopy hood with three open sides over an open surface tank.

a. What is the minimum control velocity specified by ANSI Z9.1 for a lateral exhaust hood used to control this chemical in any shape of open-surface tank?

   minimum $V_{control}$ = _________ fpm

b. Think of the drain board in Problem 2 as being just a continuation of the tank. What would Table 13.3 from ANSI Z9.1 indicate to be the "cfm per square foot" and the "total cfm" to achieve the above control velocity for this shape of tank? Assume that this side draft hood has a flange.

   $W/L =$ _____ ______ $ft^3/min/ft^2$
   $A =$ _____ ft^2 Q ans. = 900 ft^3/min

4. Now design a slotted side-draft hood (similar to that in VS-75-06) a third time using the DallaValle Equation and a generic control velocity for mostly passive vapors.

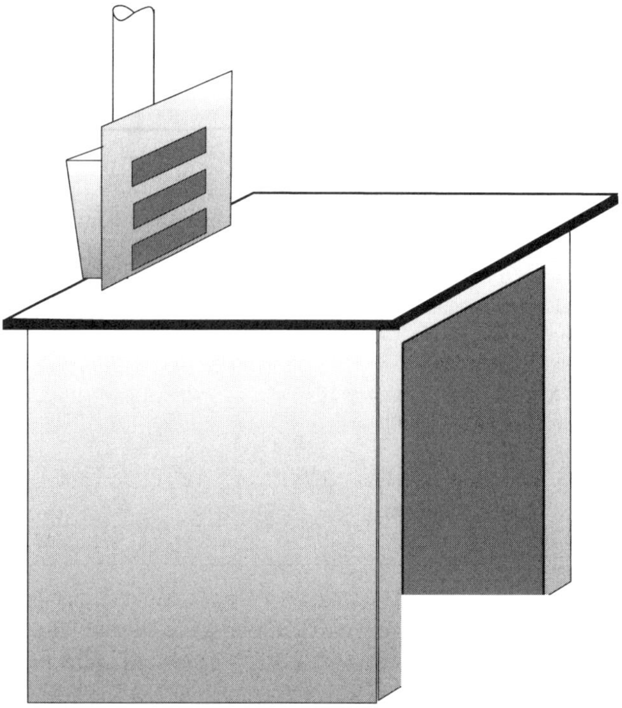

FIGURE 13.47 Work bench.

Make each half of the hood face square, so that the entire face (both halves of the hood) is 6 feet wide by 3 feet vertically, and let the plenum extend beyond the slots to form a flange. Remember that the DallaValle is applied as if the slots are not really there.

a. What would be an appropriate generic control velocity from Table 13.1 at the front edge of the tank?

$V_{control}$ = _____

b. What would be the appropriate K_{hood} factor to use in the DallaValle equation for a flange and a blocking horizontal surface? (Consider the liquid surface to be a blocking horizontal surface.)

for K_{hood} = _____

c. What would the DallaValle equation predict would be the necessary flow rate [Q] to achieve that control velocity on the front edge of the tank?

$Q = 2175$ cfm
using $V_{control} = 75$ fpm

One can see that the recommendations of 1500, 1800, and 2175 cfm in problems 2, 3, and 4, respectively, differ by almost 50% (high to low). This range is typical of the precision of local exhaust ventilation design.

5. Design a small slotted hood to control soldering fumes produced during the manual assembly of custom electronic circuit boards described in Problem 1b. Each worker assembles approximately 20 boards per hour at work benches similar to that shown on the right. Because these circuit boards are approximately 8 in. by 5 in., design the hood to be 1 ft wide and 1 ft tall, with full width slots, and a flange that extends another 6 in. above and to both sides of the hood, similar to that in Figure 13.48.

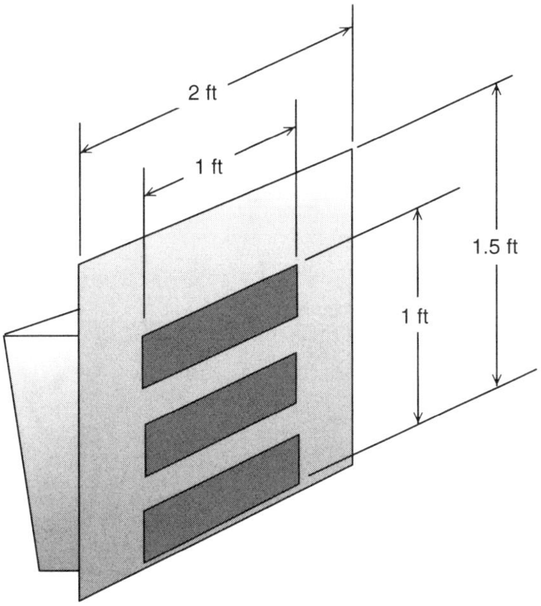

FIGURE 13.48 Hood with 6-in flanges above and on the sides.

a. A slotted hood is first designed as if it were a single open faced collection hood. What is the effective face area of this slotted hood ($A_{\text{effective}}$)?

$A_{\text{effective}} =$ _____ ft^2

b. Based on Table 13.1, what control velocity would be appropriate to collect this warm fume with the boards being moved periodically?

$V_{\text{control}} =$ ______ fpm

c. What is the Q needed to achieve a control velocity of 135 fpm at a distance of 1 ft in front of the hood? Include a K_{hood}.

$Q =$ _____ ft^3/min

d. What is the slot area that will produce a slot velocity of at least 2000 fpm?

$A_{\text{slot}} =$ _____ ft^2

e. Assuming each slot is the width of the hood (L) but only 1 in. high (H), how many equally sized slots n will be needed in this hood?

$n =$ _____

6. Design a high canopy for use over a 2 ft diameter melting pot of aluminum at its melting temperature of 1220°F (660°C) similar to that which might be used in a small metal fabricating shop. The hood needs to be above head height, so design it to be 7 ft above the floor. Assume the melting pot stands 2 ft tall above the floor itself. Thus, the Y value in the ACGIH model is 5 ft.

a. What is the distance from the canopy face to the hot source ($Y_{\text{above source}}$ in Figure 13.49)?

$Y_{\text{above source}} =$ ____ ft

b. What is the distance from the canopy face to the hypothetical point-source origin of the plume?

$X_{\text{cone height}} =$ ____ ft

c. What is the diameter of the plume by the time it reaches the face of the canopy?

D_{plume} ans. $= 3.74$ ft

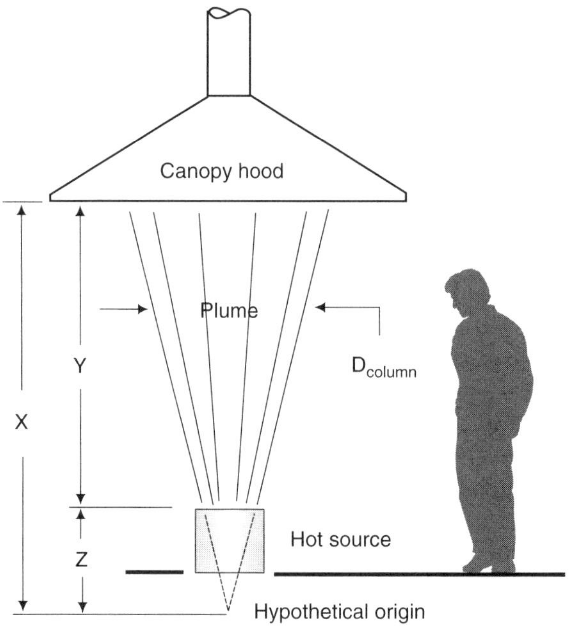

FIGURE 13.49 Canopy above melting pot.

d. What should the diameter of the hood be?

D_{face} = ____ ft

e. What is the velocity of the plume by the time it reaches the canopy?

V_{plume} = ____ fpm

f. If the hood is designed to have 100 fpm in its face area outside (around) the plume, for what flow rate (Q) should the fan be selected?

Q_{total} = _____ ft^3/minQ_{total} ans. ≈ 5000 ft^3/min

REFERENCES

1. *Industrial Ventilation: A Manual of Recommended Practice*, 24th ed., American Conference of Governmental Industrial Hygienists, Cincinnati, OH, 2001.
2. Olander, L., General, in industrial ventilation, In *Industrial Ventilation Design Guidebook*, Goodfellow, H. D., Täthi, E. Eds., Academic Press, London, chap. 10.1, 2001.
3. Bémer, D., Callé, S., Godinot, S., Régnier, R., and Dessagne, J. M., Measurement of the emission rate of an aerosol source — comparison of aerosol and gas transport coefficients, *Appl. Occup. Environ. Hyg.*, 15(12), 904–910, 2000.
4. Heinsohn, R. J., *Industrial Ventilation, Engineering Principles*, Wiley, New York, 1991.
5. Chapter 29 of the U.S. Code of Federal Regulations, Part Number 1910, Occupational Safety and Health Standards.
6. AIHA/ANSI *Z9.1-1991: Practices for Ventilation and Operation of Open-Surface Tanks*, American Industrial Hygiene Association, Fairfax, VA, 1991.
7. Clayton, G. D., Clayton, F. E., Eds., *Examples of Toxicity Resources*, 4th ed., *Patty's Industrial Hygiene and Toxicology*, Vol. II, Wiley, New York, 1994.
8. *Documentation of the Threshold Limit Values for Chemical Substances*, 7th ed., ACGIH Worldwide, Cincinnati, OH, 2001.

9. *Registry of Toxic Effects of Chemical Substances* (RTECS) was originally a NIOSH database now available via commercial vendors such as CCOHS at www.ccohs.ca/products/databases/rtecs.html.
10. DallaValle, J. M., *The Industrial Environment and Its Control*, Pitman Publ. Co., New York pp. 116–118, 1948.
11. Hemeon, W. C. L., *Plant and Process Ventilation*, Industrial Press, New York, 1963.
12. Burton, D. J., ed., *Hemeon's Plant and Process Ventilation*, 3rd ed., Lewis Press, Boca Raton, FL, 1999.
13. U.S. Public Health Service, *Air Pollution Engineering Manual*, Publication No. 999 AP-40, 1973.
14. Baturin, V. V., *Fundamentals of Industrial Ventilation*, 3rd ed., Pergamon Press, New York, 1972.
15. Hughes, R. T., An overview of push–pull ventilation characteristics, *Appl. Occup. Environ. Hyg.*, 5(3), 156–161 1990.
16. Marzal, F., Gonzalez, E., Minana, A., and Baeza, A., Visualization of airflows in push–pull ventilation systems applied to surface treatment tanks, *Am. Ind. Hyg. Assoc. J.*, 64(4), 455–460 2003.
17. Volkwein, C. J., Engle, M. R., and Raether, T. D., Dust control with clean air from an overhead air supply island, *Appl. Ind. Hyg.*, 3(8), 236–239, 1988.
18. Säämänen, A. J., Kulmala, I. K., and Enbom, S. A., Control of exposure caused by a contaminant source in the near wake region, *Appl. Occup. Environ. Hyg.*, 13(10), 719–726, 1998.
19. Burton, D. J., Simonson, A. V., Emmel, B. B., and Hunt, D. B., *Demonstrations of Control Technology for Secondary Lead Reprocessing*, NIOSH Publications, Cincinnati, OH, Contract No. 210-81-7106, 1983.
20. Volkwein, C. J., Research to replace respirators in mining, *Appl. Ind. Hyg.*, 3(11), F8–F10, 1988.
21. Gressel, M. G. and Fischbach, T. J., Workstation eesign improvements for the reduction of dust exposures during weighing of chemical powders, *Appl. Ind. Hyg.*, 4(9), 227–233, 1989.
22. Anonymous, *Control of Dust from Powder Dye Handling Operations*, DHHS (NIOSH) Publication No. 97–107, 1997, Also available at www.cdc.gov/niosh/hc13.html.
23. Cecala, A. B., Zimmer, J. A., Smith, B., and Viles, S., Improved dust control for bag handlers., *Rock Prod.*, 103(4), 46–49, 2000, Also available at www.cdc.gov/niosh/hc31.html.
24. Etkin, B. and McKinney, W. D., An air-curtain fume cabinet, *Am. Ind. Hyg. Assoc. J.*, 53(10), 625–631, 1992.
25. Jayaraman, N. I., Erhard, L., and Goodman, G., Optimizing air curtains for dust control on continuous miner faces: a full-scale model study, *Appl. Occup. Environ. Hyg.*, 11(7), 838–844, 1996.
26. Conroy, L., Gas storage cabinets. In *Industrial Ventilation Design Guidebook*, Goodfellow, H. D., Tähi, E. Eds., Academic Press, London, chap. 10.2.3.4., 2001.
27. Burgess, W. A., Ellenbecker, M. J., and Treitman, R. D., *Ventilation for Control of the Work Environment*, Wiley, Hoboken, NJ, pp. 187–190, 2004.
28. Ingham, D. B., Lommel, A., and Wen, X., Aaberg exhaust hoods, In *Industrial Ventilation Design Guidebook*, Goodfellow, H. D., Tähi, E., Eds., Academic Press, London, chap. 10.4.4, 2001.
29. Huang, R. F., Liu, G. S., Lin, S. Y., Chen, Y. K., Wang, S. C., Peng, C. Y., Yeh, W. Y., Chen, C. W., and Chang, C. P., Development and characterization of a wake-controlled exterior hood, *J. Occup. Environ. Hyg.*, 1(12), 769–778, 2004.
30. Davidson, L. and Schälin, A., Computational fluid dynamics in industrial ventilation, In *Industrial Ventilation Design Guidebook*, Goodfellow, H. D., Tähi, E., Eds., Academic Press, London, chap. 11.2, 2001.
31. Burgess, W. A., Ellenbecker, M. J., and Treitman, R. D., *Ventilation for Control of the Work Environment*, Wiley, Hoboken, NJ, chap. 14, 2004.

14 Predicting Pressure Losses in Ventilation Systems

The learning goals of this chapter:

- ☐ Know that moving air causes friction. Expanding airflow causes turbulence. Both cause pressure losses.
- ☐ Know that most pressure losses are proportional to velocity pressure (and therefore to V^2 and Q^2).
- ☐ Know that individual losses add cumulatively to the resulting total losses on either side of the fan.
- ✍ Understand what a *vena contracta* is and how it relates to pressure losses.
- ☐ Know how to find the recommended duct transport velocity in one of the *Manual of Industrial Ventilation*'s VS diagrams and to pick one from among the generic values in Table 14.2. Be familiar with the latter's ranges of values.
- ✍ Be able to find a hood entry loss factor (especially the loss factors in virtually any VS diagram and in tabulated data for generic hoods and duct adaptors) and to use it to calculate hood entry losses.
- ☐ Be able to calculate straight duct friction losses using either a tabular method or a formula method.
- ✍ Be able to find and use loss factors for elbows, contractions, and expansions.
- ✍ Be broadly familiar with the blast gate and design methods to balance a branched duct system (but realize that calculating such a balance is too time-consuming for an in-class examination).

The first section of this chapter is intended to give the reader an overview of the principles underlying energy and pressure losses that occur inside ventilation systems and to introduce the reader to the concepts used to predict those losses in systems they might design or/and investigate (troubleshoot). Subsequent sections deal sequentially with predicting specific energy losses in hoods, straight ducting, curved ducting, various duct fittings, and the ventilation system as a whole.

I. A PRIMER ON ENERGY AND PRESSURE LOSSES

1. Recall from Chapter 11 that pressure is a form of energy. Therefore *in ventilation, the air's energy and its pressure are synonymous terms.* The former is physically more correct, but the latter is what we measure. Consistent with Principle 8 (that total pressure equals P + VP), *all of the energy of quiescent (static) air in a room is in the form of its local atmospheric pressure* (P). Recall Principle 5 from Chapter 10 that *airflows in response to small differences in molecular or static pressure.* The negative pressure *difference* that a fan creates on its inlet side ($-\text{SP}_{\text{before the fan}}$ in Section III of Chapter 11) causes the quiescent air in the room (where its SP is zero) to move into the hood and duct system, and its movement causes some of its initial static or molecular energy to become kinetic energy. Thus, *the P of the air inside of an ideal exhaust duct (one with no friction or turbulence) will drop below the local atmospheric pressure by an amount exactly equal to its VP inside that duct and both of these will exactly equal whatever* $-\text{SP}_{\text{before the fan}}$

that the fan is capable of generating. This basic concept was previously implied by the ideal version of Equation 11.15.

✍ $$\text{Ideal VP}_{\text{before the fan}} = -\text{SP}_{\text{before the fan}}(\text{with no losses}_{\text{since entry}})$$
(Equation 11.15 in an ideal case)

2. Air moving past or over a surface causes friction. The faster the air moves, the more friction it generates. Expanding airflow causes turbulence. The more free the air is to expand (generally corresponding to the more abruptly the walls that surround, constrain, or guide the air's expansion diverge from straight flow), the more turbulence it generates. The energy lost due to friction and turbulence is nothing more than the air's resistance to flow.[a] Equation 14.1a states that the energy lost in any real system will cause the real $\text{VP}_{\text{before the fan}}$ to be less than the ideal VP described in Equation 11.15.

$$\text{Real VP}_{\text{before the fan}} = -\text{SP}_{\text{before the fan}} - \Sigma\text{Losses}_{\text{since entry}} \quad (14.1\text{a})$$

Equation 14.1b states the slightly more useful version of the same concept that *the sum of the losses within any given system account quantitatively for the difference between what the ideal* VP *with no losses would be and what the real* VP *with losses in that same system is or could be.*

✍ $$\text{Real VP}_{\text{before the fan}} = \text{Ideal VP}_{\text{before the fan}} - \Sigma\text{Losses}_{\text{since entry}} \quad (14.1\text{b})$$

These equations reflect the even more important concept that the air velocity in real systems is always less than it could be in an ideal system because losses always happen. As important as these two concepts are, the challenge in design is to predict these losses so that a local exhaust ventilation system can move air fast enough to protect an employee or to control a toxic air emission. If someone does not predict these losses correctly, the system will not protect or perform as well as desired.

Imagine a soda straw as an analog for the inlet side of a ventilation system. Sucking on the straw is like the fan creating the − SP pressure difference on its inlet side. The soda flow rate you need to quench your thirst is similar to the airflow rate you estimated in design (Chapter 13) to protect the employee or to control an air emission. If there were no resistance in the straw, you would just have to suck hard enough to draw the soda up against gravity. However, friction and turbulence will resist the flow of soda. If the straw is too small in diameter, the flow may not be enough to quench your thirst. If the straw becomes bent or constricted, the flow rate will decrease unless you suck harder. Thus, the flow rate (or VP as it turns out) will vary directly with how hard the fan sucks and inversely with the resistance to flow through the system, as restated in Equation 14.2. A garden hose would be a better analog for the outlet side of a ventilation system, but the same concept applies.

✍ $$\text{VP} = \left[\frac{V[\text{fpm}]}{4005}\right]^2 \propto \frac{\Delta P \text{ or SP}}{\text{resistance to flow}} \quad (14.2)$$

[a] While friction and turbulence are the major forms of energy exchange in ventilation, thermal energy may also be exchanged with the environment through the walls of a duct. Since the exhaust air's T and P are related via the ideal gas law, the cooling of a hot exhaust gas will cause a decrease in the pressure inside of a duct just like that created by a fan; however, heat loss (or gain) from a hot exhaust gas is left as an advanced topic not covered herein.

Again this equation (similar to Equation 14.1a,b) is conceptually important (and will be used in Chapter 18), but it is useless in design. In design, one knows what velocity and VP are needed to protect the employee or to control an emission but not the SP needed to overcome the resistance. In order to find out how hard the fan needs to suck (the SP), one needs to predict the losses, almost the reverse of Equation 14.2.

3. In design, the desired VP at each location is known (determined by the flow rate and the duct size). Research has shown that most pressure losses can be calculated as a *loss factor* times the local VP (or change in VP in some cases). Equation 14.3 defines a generic "loss factor" as the proportionality constant between the loss and the velocity pressure.[b]

$$\text{Loss Factor} = \left[\frac{\text{any energy or pressure loss due to friction or turbulence}}{\text{velocity pressure or change in velocity pressure}}\right] \quad (14.3)$$

This was Principle 9 from Chapter 10, originally stated in Equation 10.1, implied by Equation 14.2, and repeated here as Equation 14.4 only boxed! Understand the concept in Equation 14.2 as a cause-and-effect relationship but learn to use Equation 14.4 as a predictive tool.

$$\boxed{\text{energy or pressure loss} = \text{Loss Factor}\,(\text{VP or } \Delta\text{VP})} \quad \text{(repeat of Equation 10.1 as 14.4)}$$

The pressure loss is not subtracted from the VP either when used in design or when actually measured in an existing system. Similarly, VP (or V) does not decrease as the air flows through a constant size duct with losses. During design (before the system is built), the VP used in Equation 14.4 is calculated from (and therefore corresponds to) the air velocity needed to protect an employee or to control a chemical emission. During evaluation (after the system has been built), the VP in Equation 14.4 corresponds to the air velocity that one measures. Thus, VP is always known or can be measured. Moreover, since $\text{VP} = (V/4005)^2$ or some other magic number at other than NTP (as stated in Equation 11.18), *each energy loss is also proportional to* V^2 *and to* Q^2 at the location where the loss occurs, as summarized in Equation 14.5.

$$\boxed{\text{each energy or pressure loss} \propto \text{VP} \propto V^2 \propto Q^2} \quad (14.5)$$

Equation 14.2, Equation 14.4 and Equation 14.5 are interrelated; the latter two are boxed. Equation 14.4 will apply to all the losses within this chapter. Equation 14.5 will be used to explain fan selection in Chapter 16. Equation 14.2 will apply to troubleshooting and adjusting existing ventilation systems in Chapter 18.

4. A significant energy and pressure loss occurs at some hood entrances, at all duct entrances, in straight ducts, and in duct fittings.
   a. Losses as air enters a hood are attributed to turbulence. Turbulence always occurs as the air enters a duct and can occur in other parts of the hood where the air speed is high. As a rule-of-thumb, (significant turbulence losses can be created only if V is greater than 1000 fpm. Since the face velocity of large hoods is usually well less than 1000 fpm, the energy loss as air flows into a large hood is rarely significant (except for a slotted hood where $V_{\text{slot}} \approx 2000$ fpm). In contrast, because duct velocities are almost always greater than 1000 fpm, the energy loss as the air enters into a duct is always significant. Both of these "entry losses" will be discussed in Section II.

[b] Rather than use different letters to denote different kinds of loss factors (as the *Manual of Industrial Ventilation* does), this book will denote all specific loss factors by "LossFactor" with some hopefully obvious subscripts such as $\text{LossFactor}_{\text{hood}}$.

b. Since friction is generated as air flows over or past any straight surface, energy losses in straight ducts are attributed to friction.[c] The friction loss per foot (or even per meter) of duct is small, but the ducting in most LEV systems is long enough to make duct friction a significant contributor to a system's overall pressure loss. As will be explained in Section IV, losses in a straight duct are calculated as a *loss factor* times VP^{κ} where $0.9 < \kappa < 1$, but κ is close enough to one to justify the general statement that losses are proportional to VP.
c. Turbulence tends to be high as the air flows through fittings such as elbows (curved ducts), expansions and contractions (changes in duct diameter), branch junctions (where multiple ducts join together), and within most air cleaners.[d] Turbulence losses will be discussed in Section VI and Section VII. The constricted flow of make-up air into "tight" buildings or rooms also results in turbulence losses (pressure losses due to tight buildings are just more difficult to anticipate).

The common notation for a pressure loss in the ventilation literature is a subscripted letter "h" standing for "head loss" because pressure losses and pressure differences in ventilation are measured as "head". Chapter 11 explained how head is measured by the inches or centimeters of water in a manometer. Rather than use h_e for hood entry losses, h_f for straight duct friction losses, and h_L for other generic loss (as in the *Manual of Industrial Ventilation*),[1] this book will use the slightly longer but clearer notation h_{entry}, $h_{\text{duct friction}}$, and h_{Loss}, respectively, or another subscripted full term such as h_{elbow} for a loss through an elbow.

5. In general, each individual loss is independent of the other losses and contributes additively to the total loss of pressure or energy that is subtracted from the air's static or molecular pressure as it moves through the ventilation system. Equation 11.14 and Equation 11.15 are combined below to form Equation 14.6a; similarly, Equation 11.14 and Equation 11.16 are combined to form Equation 14.6b and Equation 14.6c.[e]

$$P_{\text{outside the system}} - P_{\text{inside the system}} = \boxed{-\text{SP}_{\text{before the fan}} = \text{VP}_{\text{before the fan}} + \sum \text{Losses}_{\text{since entry}}} \quad (14.6a)$$

$$P_{\text{outside the system}} - P_{\text{inside the system}} = \text{SP}_{\text{after the fan}} = \text{VP}_{\text{at exit}} - \text{VP}_{\text{after the fan}} + \Sigma\text{Losses}_{\text{until exit}} \quad (14.6b)$$

$$P_{\text{outside the system}} - P_{\text{inside the system}} = \boxed{\text{SP}_{\text{after the fan}} = \sum \text{Losses}_{\text{until exit}}} \ \text{if } V_{\text{after fan}} \text{ is constant} \quad (14.6c)$$

[c] Do not be confused by recalling the discussion from Section V of Chapter 12 that the boundary layer within straight ventilation ducts is turbulent (and not laminar). The turbulent boundary layer within straight ducts is caused by friction, but the turbulence within a boundary layer is on a much smaller scale than that within the whole duct caused by obstructions and fittings.

[d] The laminar flow conditions encountered within filters and spray towers comprise the most important exception to the general rule in Equation 14.4 that pressure loss is proportional to velocity pressure or in Equation 14.5 to velocity squared.

[e] This book focuses on the pressure loss rather than on SP as in the *Manual of Industrial Ventilation*. In practice (and as long the total Σ Losses does not exceed the 20" wg (50 cm) accuracy limit for the assumption of constant air density discussed in Section I.3.a of Chapter 11), there are only two occasions that one really needs to know SP: at the junction between two ducts in a branched system (covered in Section VII of this chapter) and when one wants to compare a system's measured performance to its predicted performance (such as in Chapter 18, Troubleshooting). Just focusing on losses makes calculations and fan selection much easier.

The greater the losses before the fan, the harder that fan must suck at its inlet side to generate a desired VP. Similarly, the greater the losses after the fan, the harder that fan must blow on its outlet side. Duct losses are calculated and added the same way on either side of the fan. Determining all the losses within a ventilation system is simply a bookkeeping process. Bookkeeping is facilitated by visualizing the system as a series of definable segments, calculating the loss for each segment, and summing all of these losses. A design calculation worksheet is supplied both within this book and within the *Manual of Industrial Ventilation* as a bookkeeping tool to help organize this calculation process. Computer software for ventilation calculations is also available and justifiable if one designs many systems. Both the worksheet and software can be helpful tools, but neither one supplants an understanding of the underlying concepts.

6. Eventually *all of the energy lost from the air within a ventilation system must be transferred back into the air by the fan*. In order to specify a fan that is big enough to protect the person out in front of the hood, one needs to know the volume of air per minute that must be moved and how hard the fan must suck and blow. The former is determined by how much control is needed. The latter is determined by how much pressure (energy) that air loses as it is pulled and pushed through a ventilation system.

II. ENERGY LOSSES FROM HOOD OR DUCT ENTRY

Most of the energy loss in hoods is caused by turbulence beyond a *vena contracta* that forms just downwind of any inwardly contracting airflow. A *vena contracta* is the region of excessive convergence that results from the inertia of air flowing radically inward towards the axis of the hood's face farther than it needs to be. For instance, a *vena contracta* occurs just inside the entrance of a duct (such as depicted in Figure 14.1, Figure 14.2, Figure 14.3, Figure 14.4 and Figure 14.5). A similar *vena contracta* occurs in air flowing through a duct contraction (to be shown in Figure 14.8). The inertia of the inward flowing air causes it to constrict momentarily into a flow area that is smaller than the area of the duct or hood face. The resulting velocity and velocity pressure in the *vena contracta* are higher than they need to be if the air had no inertia, exactly as predicted by the $V_{\text{vena contracta}} \times A_{\text{vena contracta}} = V_{\text{ideal}} \times A_{\text{duct or hood face}}$ relationship of Equation 11.7.

The recovery of the excessive velocity pressure within the *vena contracta* back into static pressure as the air speed slows back down behind the *vena contracta* is almost never 100% efficient owing to turbulence. In fact, a more general statement can be made that *the expansion of flow that becomes separated from its guiding walls (called free expansion) always has turbulence losses.* Theoretically, turbulence losses are proportional to the VP within the *vena contracta*. However, trying to find the exact location of the *vena contracta* and measure its excess velocity or its VP is neither easy nor productive. The practical alternative is to assume that the hood entry loss is simply proportional to the VP of the average air velocity across the face of the hood or duct.

An exactly analogous flow pattern occurs within a duct expansion and along the inside wall of an elbow (also to be shown in Figure 14.8), except that the expansion in these fittings is not from a real *vena contracta* but from what can be thought of as a "virtual" *vena contracta*. In a virtual *vena contracta* the air is not artificially constricted by its own inertia but is free to expand when one or more walls bends away from the previously straight airflow before that fitting. Turbulence losses occur as the flow separates from the diverging walls within an elbow or an expansion fitting. Such losses are still proportional to either VP or the change in VP. Losses from fittings will be discussed in more detail within Section VI.[f]

---

[f] The concept of a virtual *vena contracta* is not the only explanation of turbulence, but it is useful to provide a consistent model for understanding the generation of turbulence and comparing losses among different duct fittings or components.

1. Turbulence always occurs where air enters into the duct because of the turbulent expansion of the *vena contracta* that will form just beyond the duct entry. Because all hoods have a duct entry, one hood entry loss can always be calculated as shown in Equation 14.7 using a $\text{LossFactor}_{\text{duct entry}}$ and the VP of the average velocity inside the duct either as designed or as measured using one of the velocity profile methods discussed ca. Equation 12.13. Do not try to use the velocity right at the entrance to a duct because the *vena contracta* will have already started to form by that location and the face velocity will be greater than the average velocity inside the duct.

$$\text{simple hood entry loss} = h_{\text{entry loss}} = \text{LossFactor}_{\text{duct entry}} \times \text{VP}_{\text{duct}} \quad (14.7)$$

2. Entry loss factors for the following four classic *simple hoods* provide useful benchmarks that are worth learning. These hoods are both simple as defined in Section I.4 of Chapter 10 by having only one point of loss and simple in that they are merely the end of a duct (the first three hoods were also pictured three-dimensionally in Figure 10.7, Figure 10.8, and Figure 10.9, respectively). Turbulent eddies are indicated in these diagrams by the small circular arrows. Notice that the turbulence in each case occurs slightly downstream of the *vena contracta* where it expands. Technically, this is also where the loss occurs.
   a. The simplest of simple hoods is the plain, open end of a duct, as shown in Figure 14.1. Since air approaches such a plain hood from all directions (including some from behind its face), the air develops a high inward inertia that creates the largest *vena contracta* of these four hoods, the most turbulence (reflected by the number of eddies), and the highest hood entry loss factor (almost one VP) of any simple hood.
   b. By adding a flange to the plain hood (Figure 14.2), the approaching air has less inward inertia, produces less of a *vena contracta*, creates less turbulence, and only loses about one half VP. A flange is a common addition to a collection hood. Note that these same flow conditions and loss factors will occur at an untapered duct entrance from a hood (Figure 10.2 or Figure 13.15a) or at an abrupt contraction in a duct's diameter (Figure 14.18).
   c. The beveled lip of a tapered hood as depicted in Figure 14.3 is similar to a flange bent forward. A tapered hood will further reduce the *vena contracta*, the subsequent turbulence, and the energy loss. Table 14.1 shows loss factors for various taper angles. A taper angle of 20 to 25° (a major angle of 45°) has the smallest loss, but a taper angle of 45° (as shown in Figure 14.3) is more convenient. The losses for a major angle of less than 45° (not shown) will again increase due to friction from the resulting longer taper length.

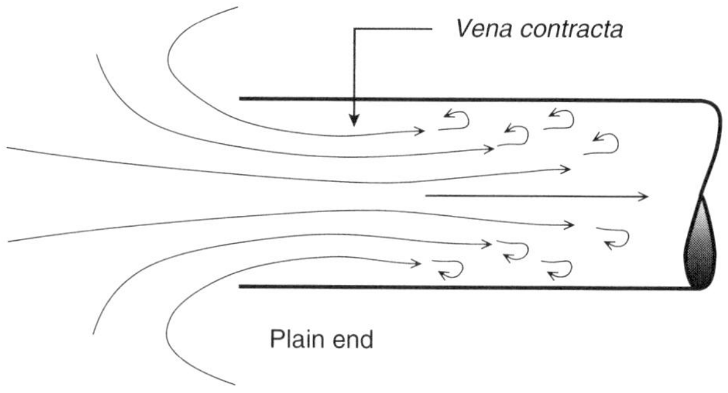

FIGURE 14.1 Plain duct hood $\text{LossFactor}_{\text{duct entry}} = 0.93$.

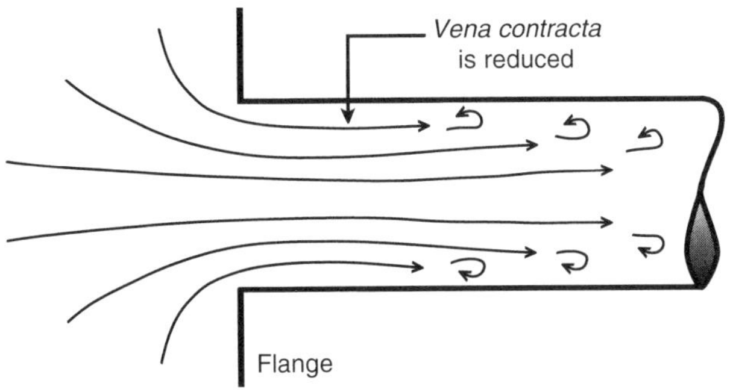

FIGURE 14.2 Flanged duct hood $\text{LossFactor}_{\text{duct entry}} = 0.49$.

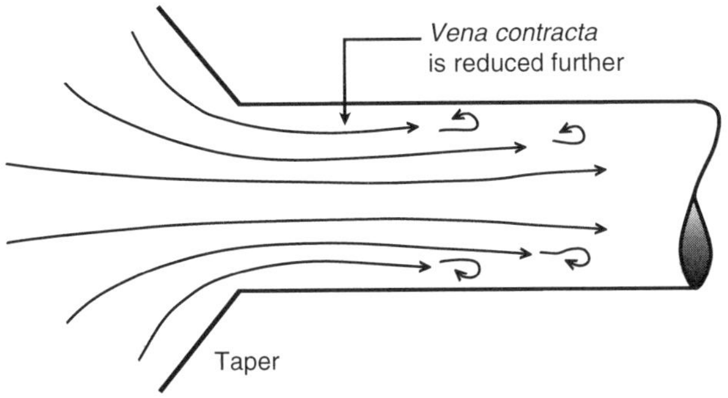

FIGURE 14.3 A 45° tapered hood $\text{LossFactor}_{\text{duct entry}} = 0.15$ for a round taper and 0.25 if the taper starts rectangular.

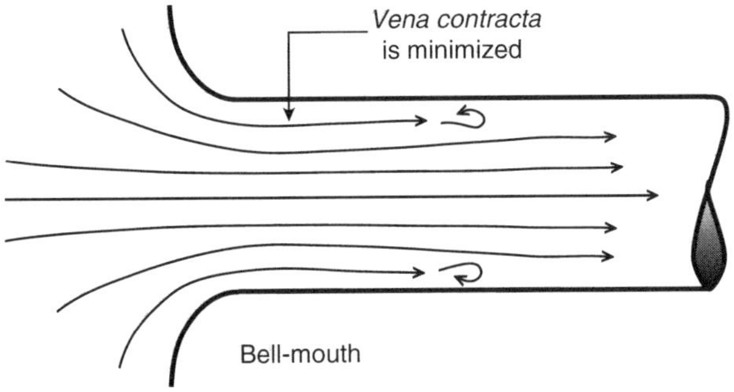

FIGURE 14.4 Bell-mouth hood $\text{LossFactor}_{\text{duct entry}} = 0.03$.

d. A bell mouth (a duct entrance shaped like the outlet to a tuba or sousaphone as shown in Figure 14.1) produces almost no *vena contracta*, turbulence, or entry losses. The air entering the hood from the side adheres to the bell's curved wall much like the air in a push–pull hood clings to a flat surface, as depicted in Figure 13.35. However, the cost to fabricate a bell mouth can easily exceed its operational advantages.

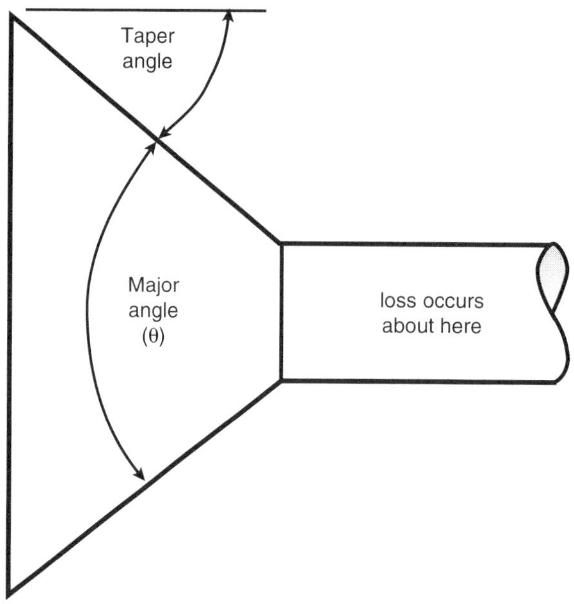

FIGURE 14.5 Two ways to characterize a tapered hood.

TABLE 14.1
Examples of Tapered Hood or Duct Entry Loss Factors to Be Multiplied by VP_{duct}

Major Angle (θ)	Taper Angle	Hood Type	Rectangular Face $LossFactor_{duct\ entry}$	Circular Face $LossFactor_{duct\ entry}$
—	—	Rounded lip (bell mouth in Figure 14.4)	0.03	0.03
45°	22.5°	Highly tapered lip or tapered duct	0.15	0.06
90°	45°	Moderately tapered lip (Figure 14.3) or duct	0.25	0.15
120°	60°	Slightly tapered lip or tapered duct	0.35	0.26
180°	90°	Flanged hood (Figure 14.2) or untapered duct	0.50	0.50
360°	0°	Plain end (unflanged, as in Figure 14.1)	0.93	0.93

As a comparison, the entry loss factor for a slotted hood is 1.8 although it is $\times\ VP_{slot}$ instead of VP_{duct}.

3. The entry loss factors in Table 14.1 can be applied either to a freestanding hood on the end of a duct (e.g., Figure 14.1, Figure 14.2, Figure 14.3 and Figure 14.4) or to the connection between the interior wall of a larger hood and a duct (e.g., Figure 13.15, Figure 13.17, and Figure 14.6). The data in Table 14.1 (taken from the *Manual of Industrial Ventilation*'s Figure 5.13 and other sources) are arranged in order of increasing loss factors.[g,h]

[g] The *Manual of Industrial Ventilation*'s Figure 5.13 is a veritable treasure trove of generic hood entry loss information.[1] This figure is also available in the *ABIH Candidate Handbook* and at http://www.abih.org/application.htm (where it can be downloaded).

[h] An old term called the coefficient of entry (C_e) is the ratio of the actual air flow over the ideal or *potential* flow if there were no losses, namely, $C_e = \sqrt{\frac{\text{actual flow with losses}}{\text{ideal flow with no hood losses}}}$. To convert a C_e into a $LossFactor_{hood} = (1/C_e^2) - 1 = (1 - C_e^2)/C_e^2$.

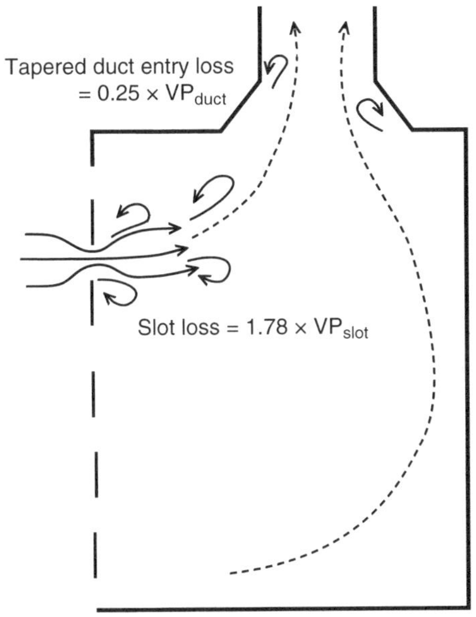

FIGURE 14.6 Depiction of compound losses in a slot and at duct entry.

Note that the *Manual of Industrial Ventilation* characterizes a hood by its major angle but characterizes duct expansions and contractions by their taper angle. Both of these characterizations are shown in Figure 14.5 and tabulated in Table 14.1.

The values shown in Table 14.1 are for tapers that start as either a rectangular or a circular opening at their face but always transition to mate with a round duct. The losses for round-faced tapers are about 0.1 × VP less than for tapers that start as a rectangle, but tapers that start rectangular (as will be explained regarding Figure 14.7) are easier to fabricate and attach to a hood's flat wall.

4. Compound hoods always have a duct entry loss as defined by Equation 14.7, plus they have at least one other significant loss at another location. These two losses are simply added together to determine the total loss of the compound hood as shown in Equation 14.8 for a slotted hood (similar to the way all of the other pressure losses that occur within a ventilation system will be added to determine the total loss of the system). Slotted hoods are a particularly common example of a compound hood (see Figure 13.10, Figure 13.12, Figure 13.17, or Figure 13.24 and Figure 14.6). As described in Section IV.7 of Chapter 13 and depicted in Figure 14.6, the first location at which a loss occurs is behind the *vena contracta* caused as the air passes through the slots, and the second location is the loss that always occurs as the air enters into a duct. There should be no significant loss within a properly sized plenum, but slots generate a huge amount of turbulence because there are no guiding walls behind the baffles. The magnitude of the slot loss (approximately $1.8 \times VP_{slot} = 0.45$ "wg for a $V_{slot} = 2000$ fpm) is about twice the magnitude of a typical duct entry loss (approximately 0.25 "wg for a 45° rectangular tapered entry and a duct velocity of 4000 fpm).

$$\text{slotted hood entry loss} = (\text{LossFactor}_{slot} \times VP_{slot}) + (\text{LossFactor}_{duct\ entry} \times VP_{duct}) \quad (14.8)$$

At this point, the reader might peruse Chapter 10 of the *Manual of Industrial Ventilation* to see examples of loss factors for other hoods.[1] Each VS diagram will contain either one or two hood entry loss factors depending on whether the hood is simple or compound. The vast majority of these loss factors are attributed to duct entry for which the *Manual of Industrial Ventilation*'s authors recommend a 45° tapered duct entry resulting in $\text{LossFactor}_{\text{duct entry}} = 0.25$ (see Table 14.1). The only reason for not using some form of tapered duct entry would be if space were not available.

III. SELECTING A DUCT VELOCITY

Since the losses both at the duct entry and within the duct depend upon VP and V^2, the duct velocity must be known before one can predict either of these pressure losses. The duct velocity depends upon the flow rate (Q) needed to protect the employee or prevent air emissions and upon the duct area (A_{duct}) selected during design, and it should at least equal or be greater than the minimum design duct velocity.

The minimum design duct velocity is that transport velocity that will not only prevent the accumulation of settled particulate contaminants within ducts (especially in long horizontal ducts) but is also sufficient to reentrain settled particles following an abrupt malfunction or shutdown (as from an electrical outage).

The air velocity within the duct (defined in Chapter 10 as the transport velocity or just V_{duct}) should be chosen in much the same way as the control velocity was chosen in Chapter 13. Each of the four sources of recommended control velocities previously described in Section II of Chapter 13 of this book also provides guidance on the minimum duct transport velocity. Since the following list of sources is virtually identical to the sequence of sources for control velocity recommendations, one can often get both the recommended control velocity and the minimum duct velocity from the same source of information.

1. Essentially every ventilation system (VS) diagram in Chapter 10 of the *Manual of Industrial Ventilation* recommends a minimum duct velocity for that specific application.[1] The minimum design duct velocities recommended in the ACGIH *Manual* are based on the further assumption that resistance within a new duct will increase over time and that the fan performance will decrease over time due to corrosion or erosion, even if both receive normal maintenance. If using a VS diagram for its specific hazard, select the duct size to achieve at least the minimum duct velocity that is recommended on the VS diagram. If one is using the hood design from a particular VS diagram to control another hazard, consider using a different minimum duct velocity selected from one of the following sources.
2. The "Miscellaneous Specific Operation Standards" in Table 10.99.2 of the *Manual of Industrial Ventilation* recommends minimum design duct velocities for other settings. Where such settings are applicable to your hazard, these specific recommendations would take precedence over generic recommendations. However, trying to find an appropriate application somewhere within the Miscellaneous table is likely to take too long for an in-class examination. Thus, *unless you are asked to implement a particular Miscellaneous Operation or to use a particular VS diagram on an in-class examination, use the generic table, namely,* Table 14.2.
3. A few OSHA standards specify duct velocities.[2] For instance, 1910.94(b) specifies a duct velocity of 3500 to 4500 fpm for grinding, polishing, and buffing operations, as summarized in Section VI.2 of Chapter 4 herein. As with control velocities, such OSHA specification standards are secondary to their performance standards for the airborne

TABLE 14.2
Generic Ranges of Recommended Minimum Design Duct Velocity Based on the Size and Nature of the Contaminant (Adapted from Table 3.2 of the *Manual of Industrial Ventilation*)[1]

1000–2000	fpm	For gases, vapors, and smoke (1500 fpm is an overall minimum V_{duct} that should be recommended for any industrial exhaust hood)
2000–2500	fpm	For very fine aerosols such as metal fumes, as from welding
2500–3000	fpm	For fine, low density (small d_{aero}) aerosols such as cotton lint, flour, or printing powders. This is also a rule-of-thumb optimum industrial exhaust duct velocity that will keep a typical dry dust moving but not create too much noise or pressure loss
3000–3500	fpm	For dry dusts and powders like finished rubber dust, wood or leather shavings (v. their dusts), or cotton dust (v. its lint)
3500–4000	fpm	For average size dusts such as grinding, buffing lint, granite dust, limestone, silica flour, processed asbestos
4000–4500	fpm	For heavy dusts like sawdust, metal turnings, foundry shakeout, sand blast dust, wood chips, or metal powders or dust
>4500	fpm	For large size and wet dusts such as metal chips, moist cement, chunks of transite pipe,[i] quick-lime (CaO) dust

chemical hazard being controlled and are rarely in conflict with the other recommendations given herein.

4. Table 14.2 contains generic guidance for a duct velocity when other specific guidance (such as that above) is not applicable. These generic guidelines are not as broad as those in Table 13.1 for selecting a control velocity, but in other ways the selection process is much the same. One first chooses the most appropriate of the most appropriate velocity range based upon the size and nature of the contaminant. After choosing an appropriate range, choose a velocity within the range based upon the intrinsic hazard of the contaminant, practical considerations related to costs, and the effect of adding noise to the local environment.
   a. The toxic hazard of the contaminant can be considered to the degree that having to do maintenance on a duct clogged with a highly toxic contaminant is a task worth the cost of preventing by choosing a velocity in the high end of a given range.
   b. Cost-effectiveness is always an IH consideration, including when selecting an appropriate transport velocity. An economically optimum duct velocity can be calculated based on balancing the increasing purchase costs required for large, low speed ducts against the increasing operating costs caused by higher friction and turbulence losses from small, high speed ducts. However, the time and cost to conduct such an optimization calculation is generally only justified for large systems, and even then is best done by ventilation engineering specialists (not a typical industrial hygienist).

[i] Transite is board-like material made from portland cement and selected mineral fibers that is neither combustible nor electrically conductive, has low thermal conductivity and high corrosion and chemical resistance, and will not chip, delaminate, or mold when exposed to prolonged dampness.

c. The final duct velocity is typically slightly above the minimum design duct velocity, but not so high that it can create excessive noise. The velocity in the duct at the fan should also not exceed the fan entry speed recommended by the fan manufacturer. The latter problem can be easily solved by reducing V_{duct} as needed with a duct expansion just before the fan. What determines excessive noise depends upon the setting. While ASHRAE has developed an approximate formula for octave-band sound power levels produced by airflow as a function of duct velocity, duct dimensions, the duct component pressure loss, and other factors (with a proclivity toward supply duct components),[3,4] the following guidance and generalizations about airflow making noise in ducts seems more appropriate for the learning goals of this textbook:

1. The design of ductwork for quiet environments such as offices and classrooms should attempt to meet a noise criteria rating such as NC-30 or NC-35. These criteria would limit the maximum duct velocities to between 1000 and 2000 fpm depending upon the nature of the fittings or components within the duct that generate the most turbulence and consequent noise.
2. Ducts with a higher duct transport velocity will produce more noise ranging from about 60 dB or 45 dBA at 4000 fpm to about 75 dB or 60 dBA at 6000 fpm. The noise from such higher velocities would be quite noticeable (if not annoying) in an otherwise quiet environment; however, it will not contribute significantly (if it is noticeable at all) in a setting where the noise produced by other machinery or processes might already be 75 dBA or greater.
3. The noise level generated by flowing air increases in proportion to between the fifth and sixth power of the velocity (namely, the dBA increases somewhere between 15 and 18 dB for each doubling of V). However, a duct velocity of above 10,000 fpm (50 m/sec) that would be necessary to produce noise above 80 dBA is beyond any recommendation and unlikely to be encountered in a LEV system.

IV. SELECTING THE DUCT SIZE AND MATERIAL

1. In the design stage, the duct size is determined by selecting a duct that will satisfy Equation 14.9 (derived from the $Q = V \times A$ conservation of mass and volume equations in Chapter 11). In practice, ducts are not available in every size. The actual duct V will generally correspond to the next smaller duct area of a commercially available or easily manufactured size of ducting to assure that V_{duct} exceeds the minimum design duct velocity. Round sheet metal ducts are usually fabricated in 0.5-in. steps upto 6-in. diameter, 1-in. steps upto 20-in. diameter, and 2-in. steps beyond that. The selected duct size can be found by starting with the maximum duct size that will provide the minimum design duct velocity according to Equation 14.9, then choosing the next smaller standard size. These steps are laid out below for the novice.

a. $$\text{The maximum duct Area} = \frac{Q\ \text{(previously designed)}}{\text{minimum design duct velocity}} \tag{14.9}$$

b. $$\text{The maximum duct diameter} = \sqrt{\frac{4 \times \text{maximum duct Area}}{\pi}} \tag{14.10a}$$

c. Select the next smallest available prefabricated duct diameter, D

d. The selected duct Area $= \dfrac{\pi(\text{selected duct diameter})^2}{4}$ (14.10b)

e. The calculated duct $V = \dfrac{Q(\text{previously designed})}{\text{The selected duct Area}}$ (14.10c)

2. Selecting the duct material is rarely an IH function, but an industrial hygienist should be knowledgeable enough not to let an inexperienced contractor or design engineer select an inappropriate duct material and mature enough to educate them about why their initial choice was not a wise one.
   a. Polyvinyl chloride (PVC) is corrosion resistant and the cheapest duct material but is limited to preformed sizes (diameters), shapes (round), and operating temperatures (<130 to 175°F). The International Mechanical Code defines limitations on the use of nonmetallic ducts, and local fire codes may prohibit the use of PVC in ducting that conveys combustible materials.[5] Where used, such ducts must be listed and labeled. Vatavuk (1988) derived two correlation formulas for estimating the price for PVC ducting depending upon its diameter (D, in inches)[6,7]:

      PVC ducting price for $6 \leq D \leq 12$ in.: (\$/ft) $= 1.0 \times D^{1.05}$
      PVC ducting price for $14 \leq D \leq 24$ in.: (\$/ft) $= 0.9 \times D^{1.98}$

   b. Galvanized sheet metal can be custom fabricated or built on the site, is not flammable, and is only somewhat more expensive than PVC. It is resistant but not immune to corrosion and limited to gas temperatures $\leq$400°F. Vatavuk estimated the price (\$/ft) for galvanized steel ducting $6 \leq D \leq 40$ in. to be $2.7 \times D^{0.8}$. The International Mechanical Code specifies the minimum thickness of galvanized sheet metal ducting utilized to convey hazardous exhausts, as shown in Table 14.3.
   c. Carbon steel is much more expensive than either PVC or sheet metal, but it may be required for abrasive contaminants. Prices for carbon and stainless steel ducting are based on a cost per pound of steel and the labor of welding sections into position. Vatavuk estimated the weight per foot (lb/ft) of steel duct materials to be $2.7 \times D$ for D in inches. Fabricated carbon steel can cost \$1.20/lb; therefore, the price (\$/ft) for most diameter carbon steel ducting is approximately $3.2 \times D$.
   d. Stainless steel may be needed for corrosive contaminants and/or high temperature gases (up to 1500°F) but at a price (\$/ft) of about $11 \times D$, is considerably more expensive than anything else.

TABLE 14.3
Minimum Galvanized Steel Duct Wall Thickness When Conveying Hazardous Exhaust[5]

Duct Diameter or Its Maximum Side Dimension	Minimum Nominal Duct Wall Thickness					
	Nonabrasive Materials		Nonabrasive/ Abrasive Materials		Abrasive Materials	
0–8 in.	0.028 in.	24 gage	0.034 in.	22 gage	0.040 in.	20 gage
8–18 in.	0.034 in.	22 gage	0.040 in.	20 gage	0.052 in.	18 gage
19–30 in.	0.040 in.	20 gage	0.052 in.	18 gage	0.064 in.	16 gage
Over 30 in.	0.052 in.	18 gage	0.064 in.	16 gage	0.079 in.	14 gage

e. Other duct materials are listed in the Table 5.4 of the *Manual of Industrial Ventilation*.[1] For example, fibrous glass reinforced polystyrene (FRP) has many of the same properties of PVC but comes in larger diameters because it is molded rather than extruded. Vatavuk's price ($/ft) for FRP between $24 \leq D \leq 60$ in. is about $2 \times D$.

V. ENERGY LOSSES FROM FRICTION IN STRAIGHT DUCTS

While turbulence plays a role in air losing energy while traveling through a straight duct, friction is considered the primary cause. This section will describe four optional ways to predict the friction losses ($h_{\text{duct friction}}$) in round ducts that all use the Loeffler equation. Section V will explain how to predict the loss in other-than-round ducts by calculating an equivalent duct diameter.

The origin of all of these prediction methods is the classic D'Arcy-Weisbach equation, Equation 14.11a. This classic equation uses a dimensionless friction factor that was determined either graphically or by a formula using the Reynolds number and the roughness of the duct material.

$$h_{\text{duct friction}} = \text{“friction factor”} \frac{\text{Duct Length}}{\text{Duct Diameter}} \times \text{VP} \tag{14.11a}$$

where

$h_{\text{duct friction}}$ = the friction loss in straight ducting ($''$wg).

“friction factor” = an “f” value from a Moody diagram as in Figure 12.15. The ε in the ε/D ratio equals the surface roughness to be shown in Table 14.4.

Duct Length = straight duct length in feet (or 100 ft depending on the source of the friction factor).

Duct Diameter = duct diameter in inches or feet (again depending upon the source of the friction factor).

VP = velocity pressure, inches of water.

A modified version of the D'Arcy-Weisbach equation called the Wright equation (shown below as Equation 14.11b) was used to determine straight duct losses in editions of the *ACGIH Manual* prior to 1995, but it is no longer in favor, apparently because it provided neither a loss factor that was linear with VP nor a simple method to account for differences in duct material roughness.

$$h_{\text{duct friction}} = 2.74 \frac{\text{Length}}{100} \frac{(V/1000)^{1.9}}{\text{Diameter}^{1.22}} \tag{14.11b}$$

where V = duct velocity, fpm. All other variables remain the same as above.

*Be forewarned if you ever do use the old version of the Wright equation (*Equation 1.16 *in as recent as the* 23*rd edition of the Manual of Industrial Ventilation* (1998) *or corresponding graphical solutions in the Manual of Industrial Ventilation up to the* 21*st edition in* 1995), *that it gives the friction loss in terms of "inches of water per* 100 *feet" of round ducting. Equation* 14.11b *was modified by dividing the equation in the old Manual of Industrial Ventilation by* 100 *to calculate the loss correctly when l is in feet as it is consistently used herein. Watch that* 100 *ft!*

All four of the following optional ways to predict pressure losses in straight ducts are based on the Loeffler equation (1980). They each utilize the same database for the duct materials listed in Table 14.4 and yield the same results, which are claimed to be accurate within $\pm 5\%$ for the duct materials shown.

TABLE 14.4
Constants for the Duct Friction Loss Factor (Taken from the Table 1.2 of the *Manual of Industrial Ventilation*)[1]

Duct Material	"ε" (ft)	*a*	*b*	*c*
Aluminum, black iron, stainless steel, PVC	0.0002	0.0425	0.465	0.602
Galvanized sheet metal	0.0005	0.0307	0.533	0.612
Flexible duct, fabric covered wires	0.003	0.0311	0.604	0.639

1. Equation 14.12 presents the Loeffler equation as it is expressed within the *Manual of Industrial Ventilation*. The first term in this equation uses three coefficients (a, b, c) listed in Table 14.4 to create the same type of dimensionless duct loss friction factor that appears in the D'Arcy-Weisbach equation. Despite the appearance in Equation 14.12 of duct friction losses being linear with VP, the presence in the loss factor of both V and Q to fractional powers actually negates that advantage. In fact, having to enter three variables (VP, V, and Q) that are all functions of each other and three coefficients from Table 14.4 can be construed to be a disadvantage of this calculation option.

$$h_{\text{duct friction}} = \frac{aV^b}{Q^c} \times L \times \text{VP} \tag{14.12}$$

where

L = duct length in feet.
VP = air velocity pressure, inches of water.
V = air velocity within the duct, fpm or fpm.
Q = air volumetric flow rate in cubic feet per minute or ft^3/min.
a, b, and c = coefficients to be read from Table 14.4.

2. The second option is to use a tabulated value of a $\text{LossFactor}_{\text{friction}}$ in Equation 14.13. Table 14.5 (similar to Table 5.6 of the *Manual of Industrial Ventilation*) lists $\text{LossFactor}_{\text{friction}}$ values for relatively smooth ducts made of PVC or welded metal, and Table 14.6 (similar to Table 5.5 of the *Manual of Industrial Ventilation*) lists $\text{LossFactor}_{\text{friction}}$ values for rougher galvanized sheet metal ducts. Values for both tables were calculated using Equation 14.12 with VP and L each set to one. To use either of these tables, follow the four steps listed below:
   a. Enter the appropriate table with the duct diameter D (in inches) in the first column on the left.
   b. Move across the row to the appropriate column of V_{duct}. Interpolation may be needed if V_{duct} falls between two columns and the corresponding $\text{LossFactors}_{\text{friction}}$ differ by more than about 5% (rarely necessary for ducts greater than 18 in. in diameter).
   c. Read the $\text{LossFactor}_{\text{friction}}$ as the fraction of the duct VP per foot of straight duct length.[j]

[j] The label of the *Manual of Industrial Ventilation*'s tables "No. VP per foot" really means that the units are the "fraction of the VP lost per foot of duct." The *Manual* uses the abbreviation H_f for $\text{LossFactor}_{\text{friction}}$.

TABLE 14.5
Loss Factors for Black Iron, Aluminum, Stainless Steel, or PVC Ducts Tabulated by Duct Diameter and Duct Velocity. See Equation 14.13

		Duct Velocity in Feet per Minute (fpm)					
Duct Diameter (in.)	Round Duct Area (ft^2)	1000	2000	3000	4000	5000	6000
4	0.0873	0.0716	0.0651	0.0616	0.0592	0.0574	0.0560
4.5	0.110	0.0621	0.0565	0.0535	0.0514	0.0498	0.0486
5	0.136	0.0547	0.0498	0.0471	0.0453	0.0439	0.0428
5.5	0.165	0.0488	0.0444	0.0420	0.0404	0.0391	0.0382
6	0.196	0.0440	0.0400	0.0378	0.0364	0.0353	0.0344
7	0.267	0.0365	0.0332	0.0314	0.0302	0.0293	0.0286
8	0.349	0.0311	0.0283	0.0267	0.0257	0.0249	0.0243
9	0.442	0.0270	0.0245	0.0232	0.0223	0.0216	0.0211
10	0.545	0.0238	0.0216	0.0204	0.0197	0.0191	0.0186
11	0.660	0.0212	0.0193	0.0182	0.0175	0.0170	0.0166
12	0.785	0.0191	0.0174	0.0164	0.0158	0.0153	0.0149
13	0.922	0.0173	0.0158	0.0149	0.0143	0.0139	0.0136
14	1.069	0.0158	0.0144	0.0136	0.0131	0.0127	0.0124
15	1.23	0.0146	0.0133	0.0125	0.0121	0.0117	0.0114
16	1.40	0.0135	0.0123	0.0116	0.0112	0.0108	0.0106
17	1.58	0.0125	0.0114	0.0108	0.0104	0.0101	0.0098
18	1.77	0.0117	0.0106	0.0101	0.0097	0.0094	0.0092
19	1.97	0.0110	0.0100	0.0094	0.0091	0.0088	0.0086
20	2.18	0.0103	0.0094	0.0089	0.0085	0.0083	0.0081
22	2.64	0.0092	0.0084	0.0079	0.0076	0.0074	0.0072
24	3.14	0.0083	0.0075	0.0071	0.0068	0.0066	0.0065
26	3.69	0.0075	0.0068	0.0065	0.0062	0.0060	0.0059
28	4.28	0.0069	0.0063	0.0059	0.0057	0.0055	0.0054
30	4.91	0.0063	0.0058	0.0054	0.0052	0.0051	0.0050
32	5.59	0.0059	0.0053	0.0050	0.0048	0.0047	0.0046
34	6.31	0.0054	0.0050	0.0047	0.0045	0.0044	0.0043
36	7.07	0.0051	0.0046	0.0044	0.0042	0.0041	0.0040
38	7.88	0.0048	0.0043	0.0041	0.0039	0.0038	0.0037
40	8.73	0.0045	0.0041	0.0039	0.0037	0.0036	0.0035
44	10.56	0.0040	0.0036	0.0034	0.0033	0.0032	0.0031
48	12.57	0.0036	0.0033	0.0031	0.0030	0.0029	0.0028
52	14.75	0.0033	0.0030	0.0028	0.0027	0.0026	0.0026
56	17.10	0.0030	0.0027	0.0026	0.0025	0.0024	0.0023
60	19.64	0.0027	0.0025	0.0024	0.0023	0.0022	0.0022

This table is similar to Table 5.6 of the *Manual of Industrial Ventilation.*[1]

d. Calculate the friction loss ($h_{\text{duct friction}}$ in $''$wg) by multiplying the loss factor read in the table by the duct's straight length (L) in feet and its VP in inches of water using Equation 14.13.

$$h_{\text{duct friction}} = \text{LossFactor}_{\text{friction}} \times L \times \text{VP} \quad (14.13)$$

TABLE 14.6
Loss Factors for Galvanized Sheet Metal Duct Tabulated by Duct Diameter and Duct Velocity. See Equation 14.13

Duct Diameter (in.)	Round Duct Area (ft^2)	Duct Velocity in Feet per Minute (fpm)					
		1000	2000	3000	4000	5000	6000
4	0.0873	0.0791	0.0749	0.0726	0.0709	0.0697	0.0687
4.5	0.110	0.0685	0.0649	0.0628	0.0614	0.0603	0.0595
5	0.136	0.0602	0.0570	0.0552	0.0540	0.0530	0.0523
5.5	0.165	0.0536	0.0507	0.0491	0.0480	0.0472	0.0465
6	0.196	0.0482	0.0456	0.0442	0.0432	0.0424	0.0418
7	0.267	0.0399	0.0378	0.0366	0.0358	0.0351	0.0346
8	0.349	0.0339	0.0321	0.0311	0.0304	0.0298	0.0294
9	0.442	0.0293	0.0278	0.0269	0.0263	0.0258	0.0255
10	0.545	0.0258	0.0244	0.0236	0.0231	0.0227	0.0224
11	0.660	0.0229	0.0217	0.0210	0.0206	0.0202	0.0199
12	0.785	0.0206	0.0195	0.0189	0.0185	0.0182	0.0179
13	0.922	0.0187	0.0177	0.0171	0.0168	0.0165	0.0162
14	1.069	0.0171	0.0162	0.0157	0.0153	0.0150	0.0148
15	1.23	0.0157	0.0149	0.0144	0.0141	0.0138	0.0136
16	1.40	0.0145	0.0137	0.0133	0.0130	0.0128	0.0126
17	1.58	0.0135	0.0127	0.0123	0.0121	0.0119	0.0117
18	1.77	0.0126	0.0119	0.0115	0.0113	0.0111	0.0109
19	1.97	0.0118	0.0111	0.0108	0.0105	0.0103	0.0102
20	2.18	0.0110	0.0104	0.0101	0.0099	0.0097	0.0096
22	2.64	0.0098	0.0093	0.0090	0.0088	0.0086	0.0085
24	3.14	0.0088	0.0084	0.0081	0.0079	0.0078	0.0077
26	3.69	0.0080	0.0076	0.0073	0.0072	0.0070	0.0069
28	4.28	0.0073	0.0069	0.0067	0.0066	0.0064	0.0063
30	4.91	0.0067	0.0064	0.0062	0.0060	0.0059	0.0058
32	5.59	0.0062	0.0059	0.0057	0.0056	0.0055	0.0054
34	6.31	0.0058	0.0055	0.0053	0.0052	0.0051	0.0050
36	7.07	0.0054	0.0051	0.0049	0.0048	0.0047	0.0047
38	7.88	0.0050	0.0048	0.0046	0.0045	0.0044	0.0044
40	8.73	0.0047	0.0045	0.0043	0.0042	0.0042	0.0041
44	10.56	0.0042	0.0040	0.0039	0.0038	0.0037	0.0036
48	12.57	0.0038	0.0036	0.0035	0.0034	0.0033	0.0033
52	14.75	0.0034	0.0032	0.0031	0.0031	0.0030	0.0030
56	17.10	0.0031	0.0030	0.0029	0.0028	0.0028	0.0027
60	19.64	0.0029	0.0027	0.0026	0.0026	0.0025	0.0025

This table is similar to Table 5.5 of the *Manual of Industrial Ventilation.*[1]

3. In the third option, the flow conditions are only entered once. The following three versions of Equation 14.14 were each derived from a set of coefficients in Table 14.4 and the area of a circular duct of diameter D. The use of D reduced the three variables in Equation 14.12 (V, Q, and VP) into the two variables shown in Equation 14.14. The duct diameter D is in inches and L is in feet. The powers of VP from 0.93 to 0.98 in these equations indicate how much (or how little, depending on your point of view) duct losses deviate from being linear with VP (linear would be a power of one).

$$h_{\text{duct friction}} = \frac{0.314}{D(\text{inch})^{1.204}} L(\text{ft})\, \text{VP}^{0.9315} \quad \text{(for PVC and other smooth ducts)} \qquad (14.14a)$$

$$h_{\text{duct friction}} = \frac{0.387}{D(\text{inch})^{1.224}} L(\text{ft})\, \text{VP}^{0\ 9605} \quad \text{(for galvanized sheet metal ducts)} \qquad (14.14b)$$

$$h_{\text{duct friction}} = \frac{0.650}{D(\text{inch})^{1.278}} L(\text{ft})\, \text{VP}^{0.9825} \quad \text{(for covered-wire flexible ducts)} \qquad (14.14c)$$

4. A fourth option is presented not because it is practical but merely because it provides a different insight. The fact that pressure losses in Equation 14.15 vary approximately with the fifth power of duct diameter illuminates the importance of the duct diameter specified during design (or the sensitivity of changes in the duct diameter that might be made to balance the losses in a branched duct system, as will be discussed in Section XII). Equation 14.15 also illuminates how duct losses vary approximately with Q^2 (similar to the way Equation 14.14 showed how friction losses vary almost linearly with VP).

$$h_{\text{duct friction}} = 0.00100 \frac{Q(\text{cfm})^{1.863}}{D(\text{inch})^{4.930}} L(\text{ft}) \quad \text{(for PVC/smooth ducts)} \qquad (14.15a)$$

$$h_{\text{duct friction}} = 0.00103 \frac{Q(\text{cfm})^{1.921}}{D(\text{inch})^{5.066}} L(\text{ft}) \quad \text{(for galvanized ducts)} \qquad (14.15b)$$

5. Round ducting is usually chosen for LEV applications because it requires less material than any other shape (a circle has the smallest perimeter per unit of area) and it can withstand a greater negative pressure than can rectangular or other duct shapes with flat sides. However, rectangular ducts are used at times and places because they have other advantages (such as increasing the ceiling clearance).

   In order to use the above methods to determine losses in noncircular ducts, one must determine the equivalent duct diameter ($D_{\text{equivalent}}$), which is the diameter of an imaginary round duct that would have the same friction loss as a duct whose shape is other than round. Equation 14.16a provides a generic equivalence, while Equation 14.16b provides $D_{\text{equivalent}}$ if the duct is rectangular of height H and width W.[k] In either case, $D_{\text{equivalent}}$ will be in the same dimensional units as the sides and may be used as D (with the original V or VP) in either the second or third of the previous methods to derive the duct friction loss.

$$D_{\text{equivalent}} = 1.546 \frac{(\text{Area})^{0.625}}{(\text{Perimeter})^{0.25}} \qquad (14.16a)$$

$$D_{\text{equivalent}} = 1.3 \frac{(HW)^{0.625}}{(H+W)^{0.25}} \quad \text{for rectangular ducts} \qquad (14.16b)$$

[k] $D_{\text{equivalent}}$ can also be read in Table 5.9 of the *Manual of Industrial Ventilation* for the larger dimension A (down the side of the table) and smaller dimension B (across the top).[1]

TABLE 14.7
Comparison of Losses in a 13.54 in. Diameter Round Galvanized Sheet Metal Duct with Losses in Various Rectangular Ducts of Equal Flow Areas

Aspect Ratio	L	W	$D_{equivalent}$	$D_{equivalent}$/Round D	% Loss per Foot vs. Round Duct
1:1	12	12	13.1	0.9689	+3.9
2.25:1	8	18	12.9	0.9497	+6.5
3:1	6.928	20.78	12.7	0.9345	+8.6
4:1	6	24	12.4	0.9163	+11
6:1	4.899	29.39	12.0	0.8861	+16.0
9:1	4	36	11.5	0.8527	+22

EXAMPLE 14.1
Find the equivalent duct diameter for a rectangular duct 8 × 18 in. (the ratio of something's long to short dimensions is called an aspect ratio, which in this case is 2.25:1).

Using Table 5.9 of the *Manual of Industrial Ventilation*: $D_{equivalent} = 12.9$ in.

Using Equation 14.16b: $D_{equivalent} = (1.546 \times 144^{0.625})/52^{0.25} = 1.546 \times 22.33/2.685 = 12.9$ in.

Square ducts act very much like round ducts, but friction losses increase as a rectangular duct's aspect ratio increases (namely, as the duct gets less and less square). The data in Table 14.7 compare six rectangular ducts with a round duct. Each duct has a cross-sectional area (A) of 144 in.^{2} and therefore has the same V for any Q. The diameter of a round duct with an area of 144 in.^{2} $12\sqrt{4/\pi} = 13.54$ in. The last two columns compare the rectangular duct's equivalent duct diameter and pressure loss per foot, respectively, to that of a 13.54 in. diameter round duct. One can see that frictional loss increases as the aspect increases. However, the loss in a rectangular duct with an aspect ratio of up to 3:1 is less than 10% more than in round ducts.

As will be pointed out in Section VII.4, elbows in rectangular ducts will have lower turbulent losses than comparable elbows in round ducts; however, the savings in lower operating costs due to lower losses in elbows made from rectangular ducts is generally not large enough to offset either the higher frictional energy loss in the straight portions of nearly square ducts or the differences in capital costs between round and rectangular ducts.

VI. AN EXAMPLE OF HOOD ENTRY AND DUCT FRICTION LOSSES

EXAMPLE 14.2
Design a local exhaust ventilation system to keep titanium dioxide powder (TiO_2) used as a pigment during paint manufacturing from becoming airborne as bags are opened and manually emptied into a hopper near floor level. This company manufactures its paint in batches; thus, this operation is intermittent (several times per day). The bold text below denotes a cross-reference to the steps in "The LEV Design Sequence" from Chapter 10.III of this book that starts with "Step 0".

Consider better alternatives to ventilation, the unnumbered step that should always be a starting point. Section V of Chapter 7 tried to suggest that nonventilation controls are not just considered once at the beginning and then either pursued or discarded. One's perception of "the best control" evolves and can change as a ventilation design matures and as new ideas are created. One nonventilation option here is to try to change how the pigment is packaged; that might take working with the pigment vendor or perhaps changing vendors to one that uses soluble bags or that ships in a "tote-bin" that can connect directly to the hopper without an air gap through which dust can escape into the workplace. Another option might be to develop a semiautomatic bag opening operation within an enclosure.

1. *Determine the appropriate control velocity*. Recall that sometimes the first two steps can be iterative processes, where selecting the hood can cause one to change the chosen control velocity. Recall too that the selection options suggested in Section III of Chapter 13 included VS diagrams, the Miscellaneous Specific Operations table, OSHA standards, and the recommended generic control velocities. The VS diagram eventually selected in the next step, recommends a face velocity of 150 fpm. In the real-world, one would have the time to peruse Table 10.99.2 of the *Manual of Industrial Ventilation*; in this case, that table's closest operational match is 250 fpm for "asbestos dumping", but its much greater inherent toxicity than TiO_2 suggests that a lower velocity could be used in this case. No OSHA specification standard applies to either this operation or this chemical. Within the range of generic control velocities in Table 13.1, this bag emptying process could be categorized as "active" generation, suggesting a control velocity of 200 to 500 fpm, but its intermittent nature could allow one to choose the low end of that range.
2. *Design and locate either a containment hood or an efficient collection hood*. Complete containment is not feasible because of the size of the bags and their manual handling. The chosen option (in the bottom of VS-50-10) is a simple open-front booth hood that semiencloses the open end of the bag being emptied. A face size of 2.25 ft wide × 4 ft high was chosen to accommodate the bag opening operator. The booth should enclose at least half of the hopper, but the depth of the hood is of no consequence to pressure losses because of the low speed before the air enters the duct. Baffles could be placed near the rear of the hood to improve the uniformity of the face velocity (although none are shown in this VS diagram), but they would incur an additional pressure loss that would need to be accounted for in design.

   Running the exhaust duct out of the back of a hood takes up more floor space, which is often a concern to plant or department managers, but connecting the duct to the top of this booth hood would probably result in a much higher face velocity at the top of the hood than at the bottom (a baffle would be difficult to incorporate across the top of such a shallow hood). Connecting the duct closer to the bottom of the hood might help by putting the highest control velocity closer to where the dust is aerosolized. The duct was left in the middle for simplicity in this example.

   As discussed in Section II, the transition from the hood to the duct is an important determinant of pressure loss. While this hood could be designed with any transition, a 45° tapered fitting was chosen, as shown in the side view of Figure 14.7 (on the right). The 45° tapered fitting produces about one half as much loss as would occur if the duct were attached with no taper (as reflected by the ratio of the loss factors in Table 14.1 of 0.25 and 0.50, respectively). By looking at the front view in Figure 14.7 (on the left), one can see that the taper starts out being rectangular where it leaves the hood but transitions to mate with a round duct. The fine lines drawn on the walls of the taper in both views indicate that a rectangular-to-round transition is occurring. In fact, those lines can actually be seen on fabricated metal transitions where small but visible bends are made in the sheet metal to make that transition possible.

   The "$Q = 150$ cfm/ft^2 face" near the bottom of VS-50-10 (and Figure 14.7) constitutes a *de facto* control velocity. When that Q is multiplied times a face area in ft^2, the resulting value is 150 fpm. Notice how this value is less than the 200 fpm low-end of the generic range towards which we were inclined in step 1. Given ventilation's

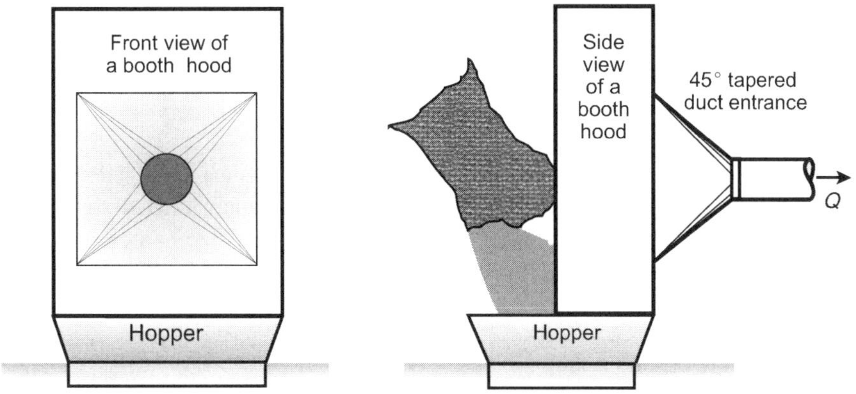

FIGURE 14.7 Depiction of a booth hood with a rectangular tapered duct transition to control manual loading. The design information was obtained from the bottom half of the *Manual of Industrial Ventilation*'s VS-50-10.

less-than-quantitative prediction models, there would be nothing intrinsically wrong with either 150 or 200 fpm in this case, but remember that the turbulence produced by a higher control velocity passing around the employee and the bag can actually increase the employee's exposure. Additionally, titanium dioxide's TLV of 10 mg/m^3 in this case is fairly high as aerosol exposure limits go. Finally (but of no true concern to employee protection), the exhaust air will simply remove excess pigment which the company must buy and is effectively "money up the stack."

3. *Estimate the flow rate of air to be moved by the fan.* Had we chosen 200 fpm or any other actual control velocity, Equation 11.7 would apply explicitly. The $Q = 150$ cfm/ft^2 face specified in VS-50-10 means we have to modify Equation 11.7 just slightly to use its Q per unit face area.

$$Q = V \times A = \frac{Q \text{ cfm}}{\text{ft}^2} \times A \text{ ft}^2 = 150 \text{ fpm } (2.25 \times 4 \text{ ft}^2) = 1350 \text{ cfm}$$

(basically Equation 11.7)

4. *Select the duct's material, route, and size.* Since the contaminant (TiO_2) is not corrosive, galvanized sheet metal ducting is an acceptable option. As it is not combustible (and assuming building codes allow), PVC ducting is also an acceptable option. The capital costs for these two options can be compared for an 8-in. diameter duct (although the size (D) is only known in hindsight):

$$\left.\begin{aligned} &\text{cost of sheet metal} = 2.7\,D^{0.8} = 2.7 \times 5.3 = \$14.25 \text{ per ft.} \\ &\text{cost of PVC} = 1.0\,D^{1.05} = 1 \times 8.4 = \$8.40 \text{ per ft.} \end{aligned}\right\} \text{ from data in Section III.4}$$

The PVC appears to be cheaper and will therefore be used unless thwarted by obstacles along the way or local building codes. The smoother PVC ducting will also translate into lower frictional pressure losses (to be determined in Step 7) and slightly lower dollars per year of electrical bills for the fan (to be determined in Chapter 17).

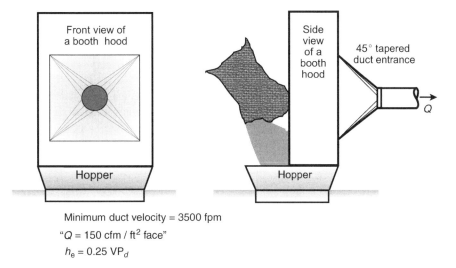

Using PVC means the duct will be round, but beyond that the route is determined largely by convenience (and again in compliance with local building codes). A typical LEV layout requires elbows or other fittings that are yet to be discussed. However, assume that the duct is to run 4 ft horizontally out of the back of the hood, have a 90° elbow to allow it to run vertically upward 18 ft through the ceiling, through another 90° elbow, and run 3 ft horizontally across the roof to a fan, as depicted in Figure 14.8. Another 8 ft of ducting is added to the outlet side of the fan to serve as an exhaust stack (to be discussed in Section VI of Chapter 15).

A duct (especially one exhausting large particles) must be sized to maintain a sufficient duct velocity. VS-50-10 recommends a minimum duct design velocity of 3500 fpm. Recall from Section III.3 that selecting a duct size is a circuitous process using Equation 14.9 and the various forms of Equation 14.10.

a. $A_{max} = \dfrac{Q_{\text{to protect employee}}}{V_{\text{minimum duct design}}} = \dfrac{1350\,\text{cfm}}{3500\,\text{fpm}} = 0.3857\,\text{ft}^2 = \pi D^2/4$ (Equation 14.9)

b. $\text{max. duct } D = \sqrt{\left[\dfrac{4A_{max}}{\pi}\right]} = \sqrt{\left[\dfrac{1.543}{\pi}\right]} = 0.700\,\text{ft} = 8.4\,\text{inch}$ (Equation 14.10a)

c. Rather than have someone custom build ducts exactly 8.4 in. in diameter, one would select the next smaller prefabricated duct size, which means the actual transport velocity will be slightly greater than the 3500 fpm minimum V_{duct} (a larger size would yield a velocity below the minimum recommended). Thus, in this case, the actual duct diameter is 8 in.

d. $\text{The selected duct } A = \dfrac{\pi D^2}{4} = \dfrac{\pi 8^2}{4} = 50.27\,\text{inch}^2 = 0.3491\,\text{ft}^2$ (Equation 14.10b)

Table 14.5 or Table 14.6 can be a handy way to find areas in square feet for any nominal duct diameter, especially *if you tab the table page for quick access.*

e. $\text{The calculated duct } V = \dfrac{\text{calculated } Q}{\text{selected duct } A} = \dfrac{1350\,\text{cfm}}{0.3491\,\text{ft}^2} = 3867\,\text{fpm}$ (Equation 14.10c)

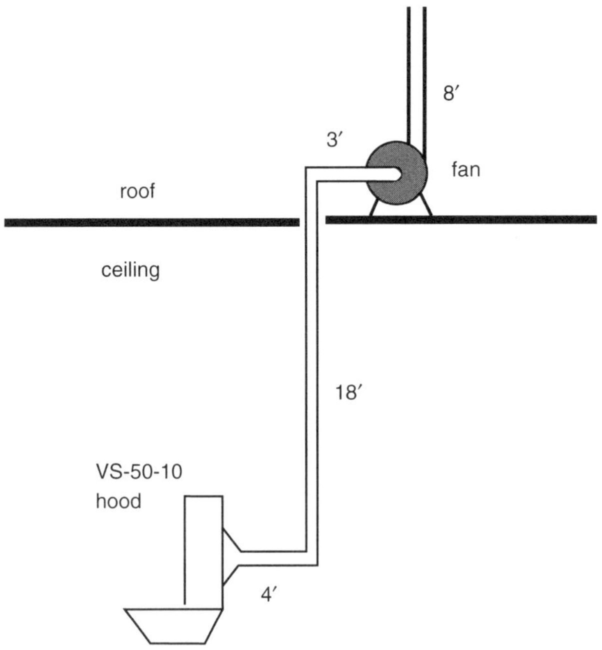

FIGURE 14.8 An elevation or horizontal view of the route for the LEV system in Example 14.2.

5. *Determine any air cleaning requirements*. Emission control of this dust would depend upon local regulations, but assume for this example that no air cleaner is required (air cleaners will be covered in Chapter 15).
6. Also assume for convenience that adequate and appropriate *make-up air is provided* back into the workroom with no additional pressure losses for the exhaust fan to overcome (make-up air will be covered in Chapter 17).
7. *Calculate all of the energy (pressure) losses*. The first step in calculating the energy losses for a ventilation system is to define the system as a sequence of component parts such that the loss in each individual portion can be defined by one of the calculation methods presented herein. Then the sequence of losses will simply be added together. In this case, losses will include those from the duct entry from the hood, the four segments of straight ducting, and two elbows.
   a. Since pressure losses always occur as the air both enters into a duct and within it and each energy loss is proportional to the velocity pressure at the point of the loss, the V_{duct} and VP_{duct} must be known in order to calculate either loss. If V_{duct} was not calculated in the previous step, it would have to be calculated now. The VP in the duct at NTP is calculated according to Equation 11.21.

$$VP_{duct} = \left(\frac{3867 \text{ fpm}}{4005}\right)^2 = 0.932 \text{ ''wg} \qquad \text{(using Equation11.21)}$$

   b. Theoretically, a hood entry loss can occur both as the air enters the hood and as it transitions from the hood into the duct. However, experience will show that any hood with a large face area (at least 2 × the size of the duct) is a simple hood. A simple hood only has one source of pressure loss. Just looking at the large face of this booth hood should suggest it is a simple hood. (The fact that VS-50-10 shows only one loss term at the duct entrance is another clue.)

   The duct entry loss shown in VS-50-10 as $h_{entry} = 0.25\ VP_{duct}$ is an application of Equation 14.7. The 0.25 is the entry loss factor from the rectangular column of Table 14.1 for the 45° tapered hood or a "major angle" θ of 90°.

$$h_{\text{duct entry loss}} = \text{LossFactor}_{\text{duct entry}} \times VP_{duct} = 0.25 \times 0.932 = 0.23 \text{ ''wg} \qquad \text{(using Equation 14.7)}$$

   The insignificant loss at the hood's face will be confirmed as a teaching lesson. The face of the booth is physically equivalent to a plain duct end (as in Figure 14.1). The $\text{LossFactor}_{\text{duct entry}} = 0.93$ for a plain hood (or plain duct end) can be obtained either from Figure 14.1 or from Table 14.1. The energy loss at the face of the booth hood for a face velocity of 150 fpm (VP = 0.0014 ''wg) is calculated in exactly the same way as above:

$$h_{\text{face entry loss}} = 0.93(0.0014) = 0.0013 \text{ ''wg} \qquad \text{(using Equation 14.7)}$$

   Obviously, the 0.0013 ''wg lost entering the booth is very much less than the 0.23 ''wg lost entering the duct and can be neglected (as you too will come to expect).
   c. The next energy loss is from friction in the straight ducting ($h_{\text{duct friction}}$). This calculation is usually carried out for each segment of straight ducting, with losses for the intervening elbows calculated in between, and all of these losses added up (as will be described in Section XIII). For simplicity, the calculation will be made once

for the total length of straight ducting that has the same diameter and airflow (the same D, Q, V, and VP). The total length of straight ducting in Figure 14.8 is 33 ft $(4 + 18 + 3 + 8)$. For instructional purposes, all three ways to find $h_{\text{duct friction}}$ that were presented in Section IV will be shown below.

1. Using Equation 14.12 (the ACGIH Loeffler version) with the PVC constants from Table 14.4 (the constants for PVC are the same as those for aluminum, iron, or stainless steel):

$$h_{\text{duct friction}} = \frac{aV^b}{Q^c} L \times \text{VP} = \frac{0.0425\,(3867)^{0.465}}{1350^{0.602}} 33\,(0.932)$$
$$= 0.794''\text{wg} \qquad \text{(Equation 14.12)}$$

2. Using the tabular method, one can find the following two $\text{LossFactor}_{\text{friction}}$ in Table 14.5 for an 8-in. duct in the two velocity columns that bracket our duct velocity of 3867 fpm.

$$\text{For } V = 3000 \text{ fpm}, \ \text{LossFactor}_{\text{friction}} = 0.0267$$
$$\text{For } V = 4000 \text{ fpm}, \ \text{LossFactor}_{\text{friction}} = 0.0257$$

The difference between the loss factors $(0.0010 = 0.0010/0.0257 \approx 4\%)$ may not be enough to warrant interpolating, but to interpolate between the two velocity columns one must subtract the proportional difference of 0.0010 that 3867 fpm is between 3000 and 4000 fpm from the 0.0267 loss factor at 3000 fpm.

$$0.0267 - 0.0010\,[3867 - 3000]/[4000 - 3000] = 0.0267 - 0.00087 = 0.0258$$

The resulting $\text{LossFactor}_{\text{friction}}$ is 0.0258 in. of loss per inch of VP per foot of duct. Stating the units this way should remind you that to find the loss the value of this loss factor must be multiplied by both L and VP , using Equation 14.13.

$$h_{\text{duct friction}} = 0.0258 \times 33 \times 0.932 = 0.794''\text{wg} \qquad \text{(again using Equation 14.13)}$$

3. Using the third method presented as Equation 14.14a results in the same exact loss:[1]

$$h_{\text{duct friction}} = \frac{0.314}{D^{1.204}} L \times \text{VP}^{0.9315} = \frac{0.314}{8^{1.204}} 33 \times 0.932^{0.9315}$$
$$= 0.794\ ''\text{wg}^{1} \qquad \text{(Equation 14.14a)}$$

Any difference among the results from these three methods is only due to rounding. Which is the easiest for you? You might try each method to figure out beforehand which you want to use on an examination.

---

[1] Calculations indicate that galvanized sheet metal would have produced 18% more loss than the smoother PVC:

$$h_{\text{duct friction}} = \frac{0.387}{D^{1.224}} L \times \text{VP}^{0.9605} = \frac{0.387}{8^{1.224}} 33 \times 0.932^{0.9605} = 0.937\ ''\text{wg} \qquad \text{(using Equation 14.14b)}$$

d. The losses in the two elbows (to be presented shortly) total 0.28 "wg between them (0.14 each).
e. The losses in each segment are now simply added to obtain the loss in the entire LEV system. Thus, the total predicted pressure loss to get 1350 ft^3/min to flow through the system is 1.30 "wg.

0.23 "wg	hood entry loss
+0.79 "wg	loss in the straight duct
+0.28 "wg	loss in the two elbows
1.30 "wg	predicted system loss

8. Now one would select an air mover (fan) that is capable of moving 1350 ft^3/min (Q) against a pressure loss (resistance) of 1.3 "wg. Fan selection will be covered in Chapter 16.

VII. TURBULENCE IN DUCT FITTINGS

The three subsections to follow will help set the stage for the calculation of losses due to turbulence in elbows (Section VIII), contractions (Section IX), expansions (Section X), and branch junctions (Section XII).

1. A *vena contracta* much like that at a duct or hood entrance is also generated by a duct contraction (depicted in Figure 14.9a). A "virtual" *vena contracta* occurs along the inside wall of an elbow (Figure 14.9b) and within a duct expansion (Figure 14.9c) in the sense that the airflow separates from these walls as they diverge from the previously straight flow. No matter whether the *vena contracta* is real or virtual, turbulence occurs as the air expands and turbulence generates energy losses.[m]
2. The turbulence losses are proportional to VP in one way or another. The turbulence losses are directly proportional to the VP in the air as it enters an elbow, an expansion at the end of a duct, an abrupt contraction, and a branch junction. The turbulence losses are proportional to the change in VP ($\Delta VP = |VP_2 - VP_1|$) within a tapered expansion or tapered contraction within a duct. Loss factors are available for each of the following sources of turbulence:

	In This Book	In the *Manual of Industrial Ventilation*
Elbows in round ducts	Figure 14.13	Figure 5.14
Elbows in rectangular ducts	Figure 14.15	Figure 5.14
Tapered contractions	Figure 14.16	Figure 5.16
Abrupt contractions	Figure 14.17	Figure 5.16
Expansions	Figure 14.18	Figure 5.16
Branched duct junctions	Figure 14.25	Figure 5.15

In each fitting, the more and more rapidly that the air can expand, the more turbulence is produced. In other words, more turbulence will be produced, more energy will be lost,

[m] The more eddies depicted in Figure 14.9c vs. Figure 14.9a accurately suggest (as will be shown later) that the turbulence and losses created by a virtual *vena contracta* in a duct expansion will be greater than those created by the smaller but real *vena contracta* created in a duct contraction of the same geometry (the same taper angle and ratio of duct diameters).

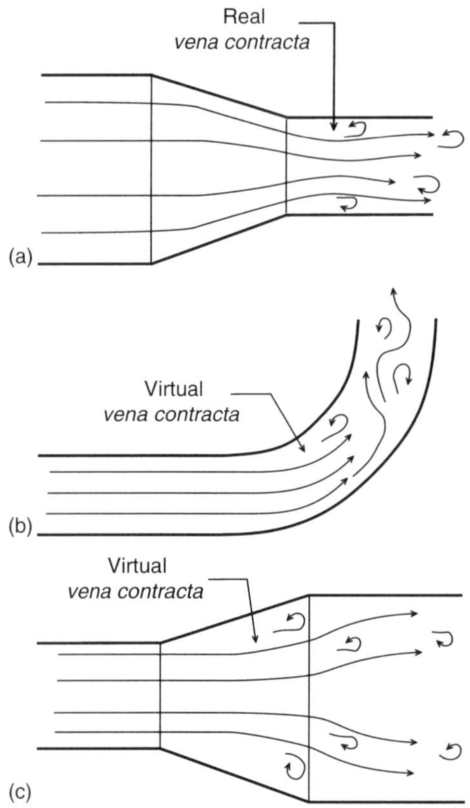

FIGURE 14.9 Depiction of a real *vena contracta* that results from a contraction within a duct and the virtual *vena contracta* that is formed by the diverging walls of an elbow and expansion within a duct.

and the loss factor will increase as the turning radius in an elbow becomes tighter, as an expansion or contraction's diameter ratio becomes larger, as its taper angle becomes wider, and as the branch junction entering angle increases.[n]

3. Before embarking on contractions and expansions, it may be useful to review a little trigonometry applicable to a taper angle.[o] If one views a tapered duct from the side (as in Figure 14.10), it appears to form a triangle with the horizontal side being the length of the taper along the axis of the duct, the diagonal side being the longer length along the tapering wall, and the vertical side being one half of the difference between the two duct diameters. Thus, the taper angle can be calculated using one of the two versions of Equation 14.17a,b, depending upon which length one is specifying or measuring. It proves to make little difference until either length approximately equals the difference in duct diameters.

[n] This book presents the tabulated loss factor data from the *Manual of Industrial Ventilation* in a graphical format.[1] The graphs have the advantage of picturing trends within a fitting, simplifying comparisons among fittings, and avoiding the need to interpolate.

[o] Note that the taper angle used in the *Manual of Industrial Ventilation*'s and these figures is the exterior angle of the tapering cone, or 1/2 of the major or interior angle θ used in Figure 5.13 of the *Manual of Industrial Ventilation* to define tapered hoods.

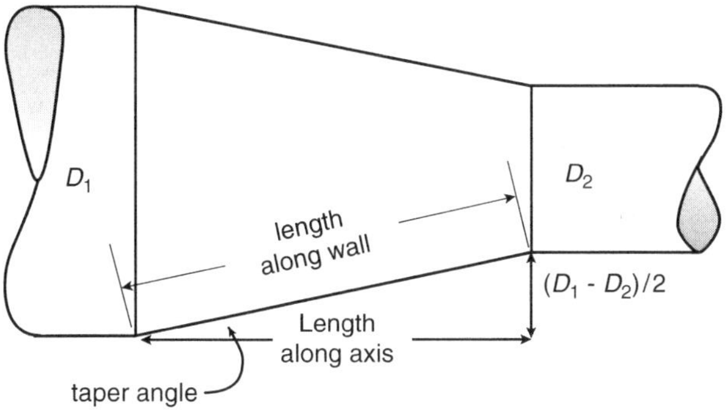

FIGURE 14.10 A depiction of how the taper angle in duct contractions and expansions is defined and can be measured.

During design, the taper angle is usually specified to achieve the desired diameter change within a chosen distance along (parallel to) the axis of the duct where Equation 14.17a applies.

$$\text{taper angle} = \tan^{-1}\left(\frac{(D_1 - D_2)/2}{\text{taper axis length}}\right) \tag{14.17a}$$

During a field evaluation of an installed system, it is usually easier to measure the length of the taper along its wall where Equation 14.17b applies.

$$\text{taper angle} = \sin^{-1}\left(\frac{(D_1 - D_2)/2}{\text{taper wall length}}\right) \tag{14.17b}$$

VIII. LOSSES FROM ELBOWS

Most of the energy loss in an elbow is due to turbulence (not wall friction). The outward curvature of the inside wall of an elbow deviates from the previously straight flow creating a virtual *vena contracta* in that half of the duct. The subsequent turbulent expansion of air from this virtual *vena contracta* creates an energy loss. The turbulent loss in an elbow is proportional to the VP within the duct, as in Equation 14.18.

$$\boxed{\text{loss for an elbow} = h_{\text{elbow}} = \text{LossFactor}_{\text{elbow}} \times \text{VP}_{\text{elbow}}} \tag{14.18}$$

where $\text{LossFactor}_{\text{elbow}}$ is the "elbow loss factor" as shown in Figure 14.13 and Figure 14.15.

1. The detailed shape of an elbow in a round duct depends heavily on the material from which it is fabricated. An elbow fabricated from PVC or another polymer can be molded or stamped as a one-piece continuous curve referred to as a "smooth elbow", as shown at the upper left in Figure 14.11. An elbow fabricated from round metal ducting is made from a series of short straight pieces of that ducting. Each piece is cut diagonally, and their ends are welded or otherwise joined together to form a "segmented elbow". Segmented elbows are further referred to by the number of segments from which they are made, such as a five-piece, four-piece, or three-piece elbow (shown as b through d in Figure 14.11, respectively). The fabrication cost is less for fewer pieces but the pressure loss is more. The four-piece elbow has a good combination of low losses and modest manufacturing costs.

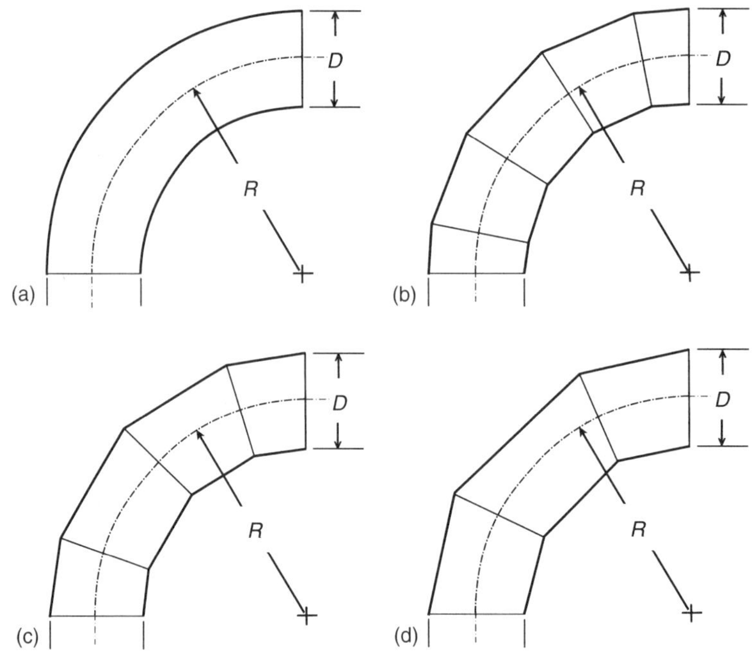

FIGURE 14.11 Depictions of a smooth elbow and ones made from five, four, or three short pieces of straight ducting (or two pieces as in Figure 14.12). Loss factors among these elbows will increase in the order shown.

The use of a two-piece or "mitered" elbow, as shown in Figure 14.12, is rare but sometimes unavoidable when space is not available for a curved elbow owing to nearby physical obstructions such as walls, structural beams, or other equipment. A mitered elbow will have the largest loss by far of any elbow, with a $\text{LossFactor}_{\text{elbow}}$ of 1.2 vs. 0.15 to 0.3 for other elbows. Turning vanes can be installed within a large mitered elbow that can reduce this loss factor by about one half. You may recall from Section I.6 of Chapter 10 and Figure 10.10 that turning vanes comprise several light pieces of a thin

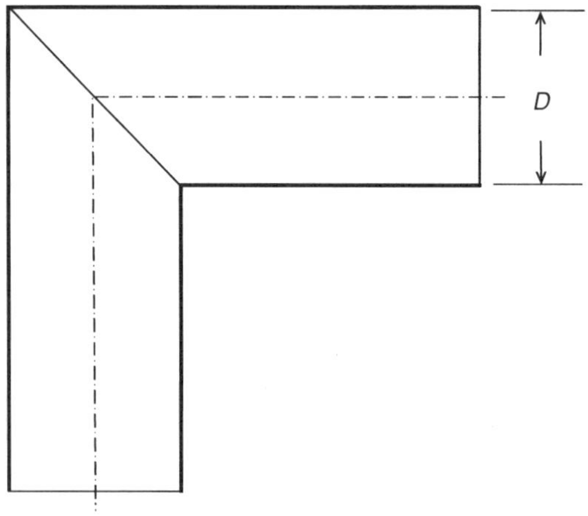

FIGURE 14.12 A two-piece or what is called a mitered elbow.

material bent to guide the airflow around a tight corner. Turning vanes add to the installation costs (especially in a small and/or round duct) and may clog with impacted debris.

2. The loss factors for smooth and segmented elbows, shown in Figure 14.13, vary as a function of the ratio of the elbow's radius of curvature (*R*) to the duct's diameter (*D*) or simply the *R/D* ratio. The radius of curvature is the radius of an imaginary circle drawn through the centerline of the elbow as depicted in each example in Figure 14.11. During design, the *R/D* ratio of a custom-made elbow may be almost any value desired, but the *R/D* ratio of molded or stamped elbows is limited to those values available from their manufacturer. Measuring the *R* value for an elbow in an installed system without a historic record, requires finding its center of curvature using a tape measure and good hand-eye coordination.

   The $LossFactor_{elbow}$ values plotted in Figure 14.13 increase rapidly for *R/D* ratios less than one. Such a tight elbow is cheap and saves space but has high turbulence losses. Turbulence decreases in an elbow with a larger *R/D*, but friction increases with its greater length. Thus, loss factors decrease very little between an *R/D* of 1.5 and 2.5 but will eventually increase for an *R/D* greater than 3. A typical four-piece segmented elbow with an *R/D* of 2 will lose about 0.25 of a VP, much like a 45° tapered hood entrance.

Example 14.3

Find the loss in each of the two 90° elbows in the LEV system for pouring TiO_2 into a hopper (Example 14.2). PVC elbows (similar to all PVC parts) are molded and smooth. The system designer can specify an *R/D* ratio for metal ducts, but the *R/D* of nonmetallic elbows such as PVC are predetermined by the manufacturer. In this example, an 8 in. *D* smooth elbow is available in a radius to

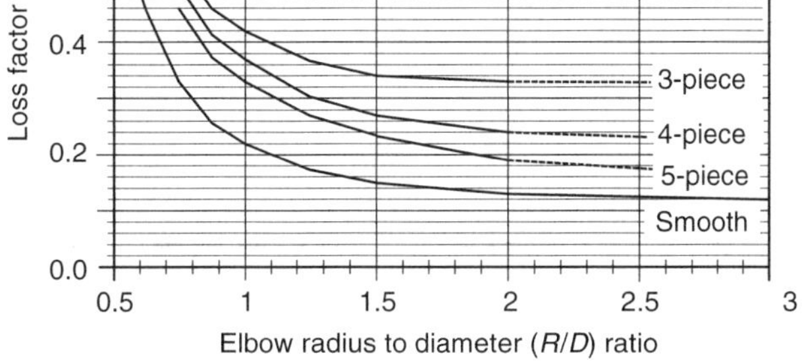

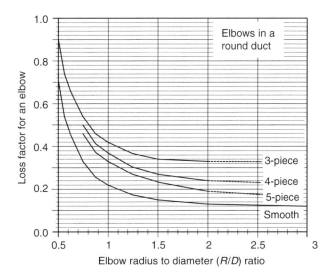

FIGURE 14.13 Loss factors ($LossFactor_{elbow}$) for elbows in round ducts.[P]

[P] The lines in this figure are linear interpolations between data points tabulated in the top portion of Figure 5.14 of the *Manual of Industrial Ventilation*.

diameter ratio of 1.5 (meaning in this case that $R = 1.5 \times 8 = 12$ in.). The loss factor $\text{LossFactor}_{\text{elbow}}$ from Figure 14.13 is 0.15 which is used in Equation 14.18 as shown below.

$$h_{\text{elbow}} = \text{LossFactor}_{\text{elbow}} \times \text{VP} = 0.15(0.932) = 0.14\ ''\text{wg in each elbow} \quad \text{(using Equation 14.18)}$$

3. An elbow in a round duct invariably creates a swirling motion in the downstream air that can cause problems for the unwary designer or system investigator. The swirl is probably caused by the tendency for the air that compresses along the outer edge of an elbow to spill over along one of the round walls towards the low pressure air initially expanding along the elbow's inner wall. This spilling causes the air to swirl as it moves down the duct, as depicted in the upper part of Figure 14.14. While neither the direction of the rotation nor the location of the high velocity proportion of the swirl at any time or place can be predicted, the lower proportion of that figure depicts the hypothetical velocity profile across the duct at three points downstream of the elbow.

   The length of the arrows in these velocity profiles depicts the air velocity at various distances across the duct. The first profile shows a highly unsymmetric flow with high velocity along one wall and highly turbulent flow along the opposite wall, perhaps even with a back-eddy (the backward flowing air initially occurs along the inner wall but could be anywhere as it spirals down the duct). As the air flows down the duct, the asymmetric air velocity distribution starts to return toward a normal symmetric profile, depicted in Figure 10.12. The air must typically flow 8 to 10 diameters down the duct before a reasonably normal (symmetric) velocity profile is reestablished.

   The swirling pattern depicted in Figure 14.14 has several implications of significance to hygienists, both in the design and in field investigations of ventilation systems.

   a. First, a good deal of kinetic energy is involved in the rotational turbulence created by an elbow in a round duct. The loss associated with this turbulence actually occurs over several diameters downstream of the elbow. This is just one example of a more general statement: although we attribute each turbulent pressure loss to just one component or location within a duct, that loss actually occurs over some finite distance downstream of the origin of that turbulence. See another discussion of this effect following Example 14.5.
   b. Second, the initially unsymmetric air velocity profile means that a velocity traverse conducted in the near-vicinity (downstream) of an elbow in a round duct is likely to be in error. Thus, try to avoid conducting a traverse within 8 to 10 duct diameters of such an elbow. Research has shown that conducting two traverses along axes that are 90° apart (as previously described in Section V of Chapter 12 of this book) will significantly moderate this error but not eliminate it.[8]

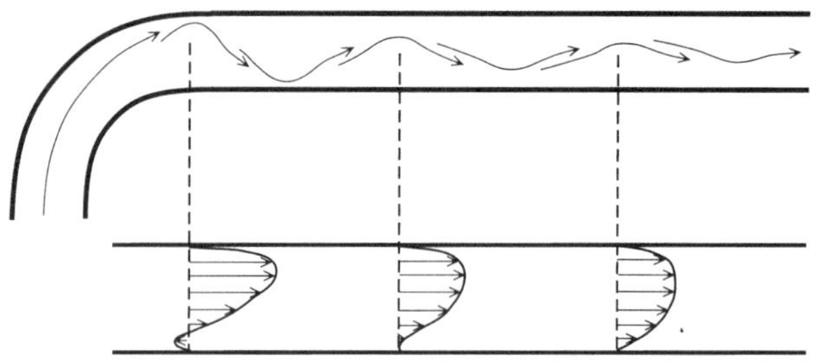

FIGURE 14.14 Depiction of the swirling pattern of high velocity air formed by an elbow in a round duct. The swirl dissipates after flow through 8 to 10 duct diameters.

c. Fans do not perform as well as expected if the air entering them is asymmetric. Not only will an asymmetric airflow pattern decrease the fan's performance, but the variations in the stress put on both the fan blades and the rotating fan shaft's bearings will cause premature wear and shorten the fan's operating life. The same allowance of at least 8 to 10 diameters of straight duct after an elbow before a fan will avoid these fan problems. If room is not available for that much straight duct, the asymmetry might be reduced by adding turning or/and straightening vanes or more losses should be added to the calculations as will be described in Section VI.2 of Chapter 16. The implications of having only about 4 duct diameters of straight duct before the fan in Example 14.2 will not be dealt with until Example 16.2.

4. Smoothly curved elbows, such as shown in Figure 14.11a, can be fabricated out of metal for rectangular ducts comparatively easily. The loss factors for elbows in rectangular ducts, shown in Figure 14.15, depend not only upon the *R/D* ratio (as for round ducts) but also upon both the duct's aspect ratio (the ratio of its width-to-depth or *W/D*) and the orientation of the turn with respect to that aspect ratio. The elbow will have a lower $\text{LossFactor}_{\text{elbow}}$ if it is oriented to make a turn in the direction that is parallel to the narrow side of the rectangle where *W/D* is greater than 1, such as depicted within Figure 14.15.

   By comparing the loss factors in Figure 14.13 and Figure 14.15 for any given *R/D* ratio, one can see that the losses for elbows in rectangular ducts are similar to those for stamped elbows in round ducts and both are about half as much as for segmented elbows in round ducts.[q] This energy savings is a side-benefit of rectangular ducts but not a sufficient advantage by itself to cause one to choose a rectangular duct over a round duct (see the discussion in Section IV.4).

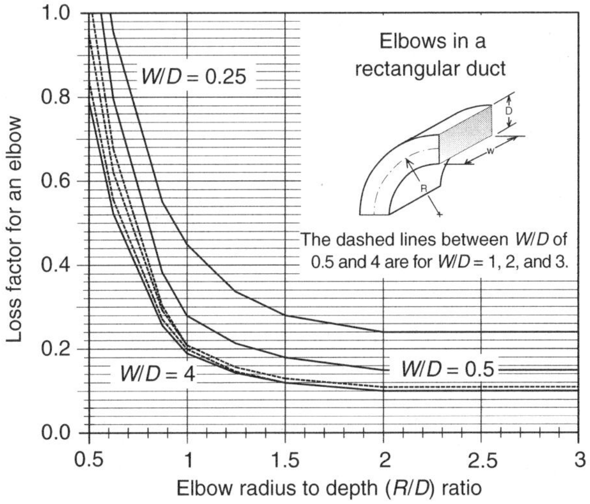

FIGURE 14.15 Loss factors ($\text{LossFactor}_{\text{elbow}}$) for elbows in rectangular ducts. The elbow shown in this figure is turning with a high *W/D* ratio to have the lowest loss.[r]

[q] One can speculate that the lower losses from elbows in rectangular ducts may be due to the rectangular shape blocking the development of the swirling turbulence described above that occurs after round ducts.

[r] The lines in this figure are linear interpolations between data points tabulated in the bottom portion of Figure 5.14 of the *Manual of Industrial Ventilation.*[1]

5. The loss factor for all elbows that only turn part way between 90 and 30° can be assumed to be proportional to the value given in Figure 14.12 or Figure 14.15 times the ratio of the smaller angle to 90°. For example, the loss for a 45° turn would be one half of that for a 90° turn.
6. Flexible ducting (or "flex duct") is used for movable exhaust hoods (Figure 13.18 and Figure 13.19). Loss factors in flexible ducting vary with the number and tightness of its bends (the constants in Table 14.4 are for straight flex duct). Flex duct is not recommended in a permanent elbow that carries particulates because of the erosion and possible corrosion to which elbows in any duct are subjected. A proportion of the air's velocity pressure and a larger proportion of any suspended aerosol will impact on an elbow's back or outer wall owing to its inertia. Thus, abrasive contaminants can wear a hole in the elbow's outer wall, allowing suspended contaminant to leak out and/or unwanted air to leak into the duct. Figure 5.22 of the *Manual of Industrial Ventilation* even provides advice for reinforcing the back wall of rigid elbows when moving abrasive contaminants.[1]

IX. LOSSES FROM DUCT CONTRACTIONS

The walls of a tapered contraction initially guide the air inward with little turbulence. However, the air's inertia will cause it to continue to contract past the end of the taper and to form a real *vena contracta*. The walls of a tapered contraction initially guide the air inward with little turbulence. However, the air's inertia will cause it to continue to contract past the end of the taper and to form a real *vena contracta* as depicted in Figure 14.9c. The magnitude of this *vena contracta* depends upon the taper angle but, in general, a taper's *vena contracta* only comprises a small proportion of the downstream duct's width. The subsequent expansion of that relatively small *vena contracta* will create some turbulence and turbulent losses.

1. This text abbreviates the loss factor for tapered contractions as $\text{LossFactor}_{\text{tapered contraction}}$, whereas the *Manual of Industrial Ventilation* uses "L". Values for $\text{LossFactor}_{\text{tapered contraction}}$ are plotted in Figure 14.16 as a function of the taper angle. Section VI previously described how the taper angle can be defined during system design or measured during a field investigation.

$$\text{loss for a tapered contraction} = h_{\text{tapered contr.}} = \text{LossFactor}_{\text{tapered contr.}} \times (\text{VP}_2 - \text{VP}_1) \quad (14.19)$$

The loss factors for tapered contractions shown in Figure 14.16 increase fairly slowly with taper angle up to approximately 30°. Where space is not a limitation, a contraction's taper angle should be designed to be approximately 20° where the loss will be equivalent to only about 10% of the change in VP. However, even the loss of 20% of the change in VP from a 45° tapered contraction will probably not add too much to the total of a typical system's losses. These loss factors for tapered contractions are lower than those for comparable expansions to be seen in Figure 14.18.

There are no loss factor data for contractions between 60 and 90°. The dotted lines in Figure 14.16 are linear interpolations between the loss factor for a tapered contraction of 60° and the loss factors for the abrupt contractions shown in Figure 14.17.

2. Any taper angle greater than 60° is considered an abrupt contraction and is handled somewhat differently from those above. While the *Manual of Industrial Ventilation* shifts to another symbol "K" for abrupt contractions, this text simply uses $\text{LossFactor}_{\text{abrupt contraction}}$. The loss factors for the abrupt contractions plotted in Figure 14.17 depend on the relative change in duct size, and the actual loss defined in

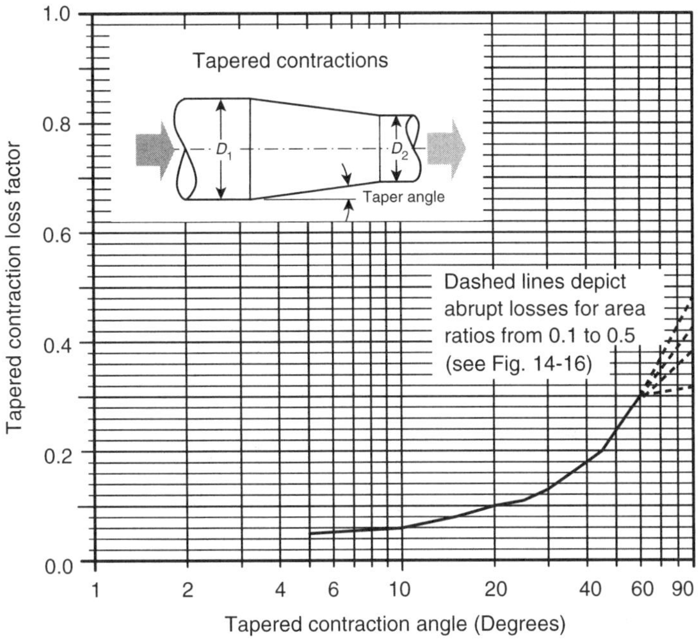

FIGURE 14.16 Loss factors for tapered contractions in round ducts. Taper angles >60° are considered abrupt contractions; see Figure 14.16.[s]

Equation 14.20 is based only on the VP in the smaller diameter duct (VP_2 in D_2), not on the change in VP.

$$\text{loss for an abrupt contraction} = h_{\text{abrupt contr.}} = \text{LossFactor}_{\text{abrupt contr.}} \times VP_2 \quad (14.20)$$

Values for $\text{LossFactor}_{\text{abrupt contraction}}$ range from 0.20 to 0.50, meaning that the equivalent of 20 to 50% of the VP energy is lost in an abrupt contraction, considerably more loss than for a 20 to 30° tapered contraction. $\text{LossFactor}_{\text{abrupt contraction}}$ values appear to increase relatively uniformly with decreasing contraction *area ratio* (A_2/A_1) as shown in Figure 14.17a.[t] However, when viewed as a function of *diameter ratio* (D_2/D_1) as shown in Figure 14.17b, the loss factors appear to increase rapidly to a nearly constant value by the time the downstream diameter is about one half the diameter of the upstream duct.

The loss factors for a large change in duct diameter are extrapolated with confidence from a loss factor of 0.48 at $D_2/D_1 = 0.31$ or $A_2/A_1 = 0.1$ to a loss factor of 0.49 in Table 14.1 for air entering a flanged hood or duct entrance. The near equality is more than coincidental. Note the similarity in their flow geometries! The air approaching an abrupt contraction from an upstream diameter that is more than twice as large as the downstream diameter (a D_2/D_1 of less than 0.5) acts very similarly to the air approaching the flanged entrance to a duct. The fact that most of the air upstream of such a contraction is already flowing in one direction (vs. a freestanding flanged duct entrance) has a negligible effect on the magnitude of

[s] The lines for this figure are linear interpolations of data points tabulated in the bottom portion of Figure 5.16 of the *Manual of Industrial Ventilation*.[1]

[t] For more rapid calculations, some people add another column of diameter ratios corresponding to the area ratios shown in the Figure 5.16 of the *Manual of Industrial Ventilation* (which for many years has a second (erroneously repeated) area ratio of 0.4 that should be 0.5).

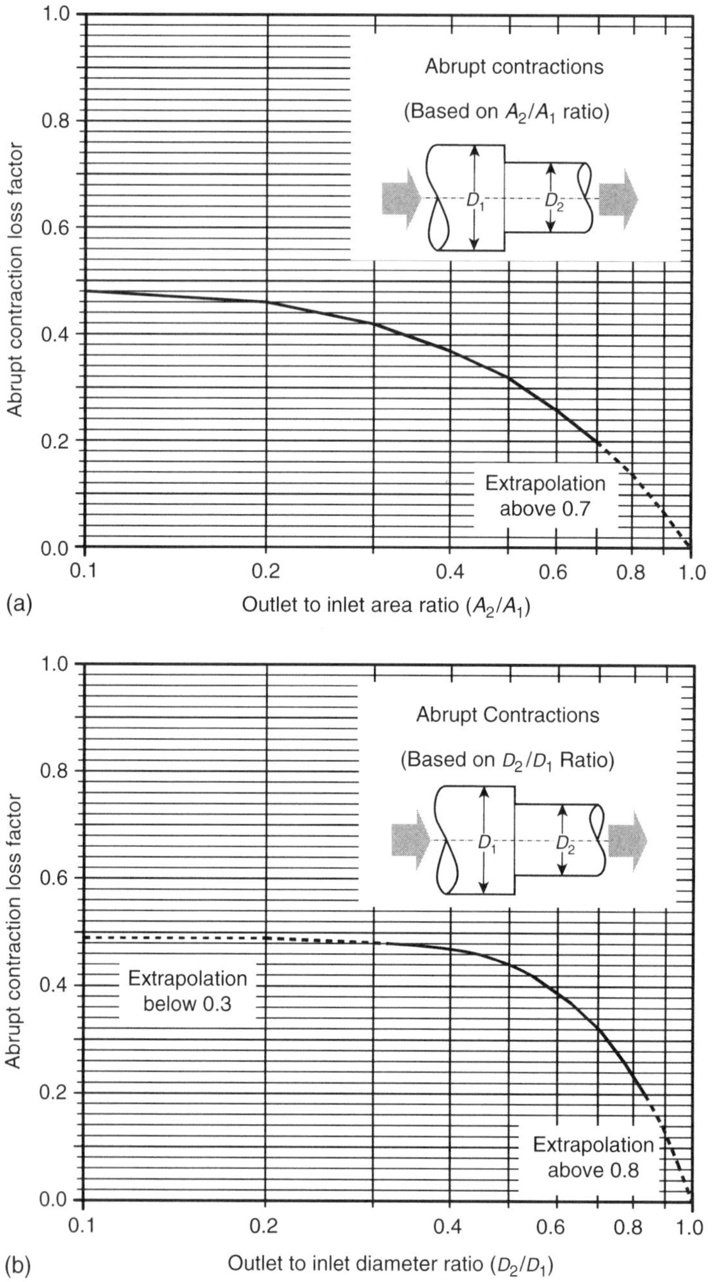

FIGURE 14.17 (a) Loss factors for an abrupt contraction based on its area ratio,[u] (b) Loss factors for an abrupt contraction based on its diameter ratio.

[u] The lines for this figure and Figure 14.17b are linear interpolations of data points tabulated in the bottom right portion of Figure 5.16 of the *Manual of Industrial Ventilation*.

the two *vena contracta* and their subsequent downstream turbulence. The similarity of their converging inertias causes the formation of virtually identical *vena contractas* and the creation of the same loss factors.

X. LOSSES FROM DUCT EXPANSIONS

As the walls diverge from the axis of the duct in an expansion such as that depicted in Figure 14.9c, the previously straight-flowing air forms a virtual *vena contracta*. Energy loss occurs from the turbulent expansion of this virtual *vena contracta*. The magnitude of the virtual *vena contracta* of the expansion in Figure 14.9c was shown realistically to be larger than the real *vena contracta* that forms in a contraction. The greater turbulence parallels the larger loss factor for an expansion than for a contraction of the same geometry. The loss factors for both tapered and abrupt expansions within a duct ($\text{LossFactor}_{\text{expansion}}$) are plotted in Figure 14.18; however, a separate set of data and another figure apply to the very special case of an expansion at the end of a duct.

1. The equation for energy loss from flow through an expansion within a duct parallels the format of losses from flow through a contraction, i.e., a loss factor times the change in velocity pressure.

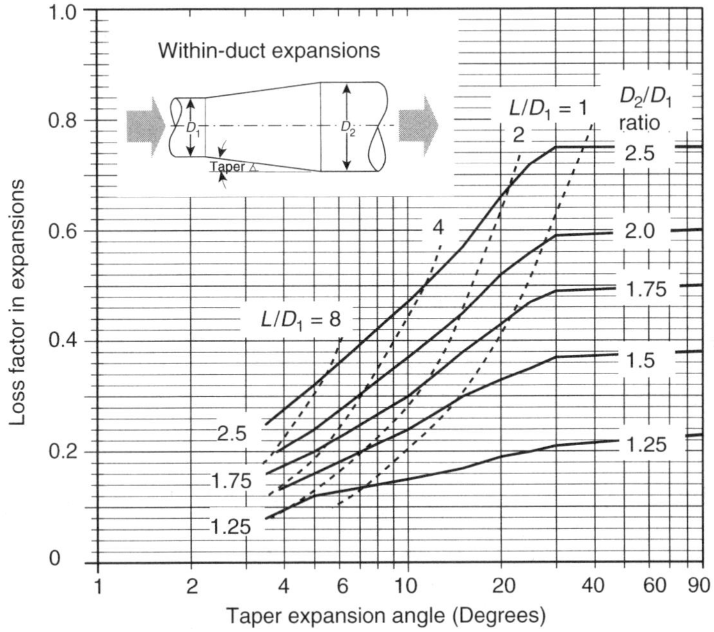

FIGURE 14.18 Loss factors for expansions within a duct.[v]

[v] The data for expansions within a duct was extracted from left side of the upper portion of the *Manual of Industrial Ventilation*'s Figure 5-16. If you were to use the *Manual*'s loss data, it is important to note (and perhaps highlight in your copy of the *Manual*) that those data are a static pressure region factor "R". The region factor is a measure of expansion efficiency rather than inefficiency or the loss factor eventually needed. The use of regain is consistent with the *Manual*'s focus on SP and will be used in Figure 14-19 herein. The energy lost in an expansion is proportional to 1 − R. Thus, the following is equivalent to Equation 14.21 if using "R" from the *Manual*: loss for expansion within a duct $= (1 - R)(VP_1 - VP_2)$.

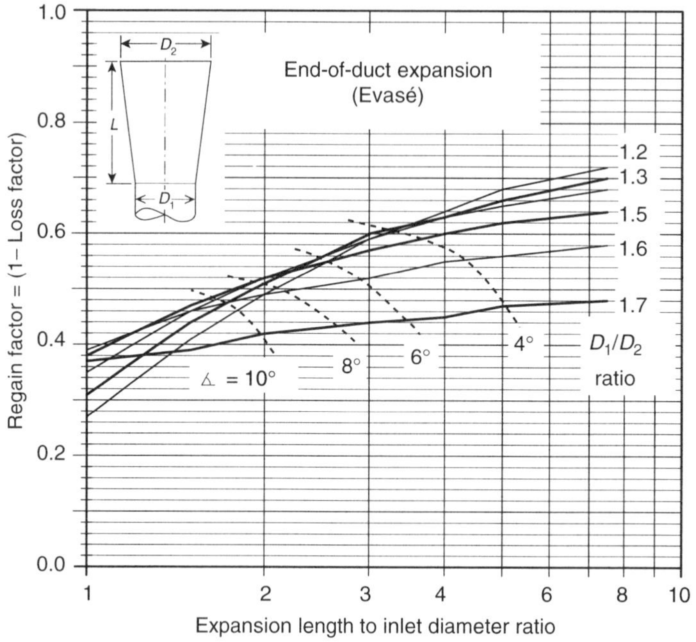

FIGURE 14.19 Regain factors for an *évasé*. The product of the regain factor × the entering duct velocity pressure is to be subtracted from all other losses.

$$\text{loss for expansion within a duct} = h_{\text{expansion}} = \text{LossFactor}_{\text{expansion}} \times (\text{VP}_1 - \text{VP}_2) \tag{14.21}$$

The loss factors in Figure 14.18 are shown as a function of the taper angle and the duct diameter ratio D_2/D_1 (sometimes abbreviated as $\beta = D_2/D_1$) as they are in the *Manual of Industrial Ventilation*. The expansion length-to-inlet diameter ratios (L/D_1) were added as dashed lines to allow future comparisons with the data in Figure 14.19. A comparison between the loss factors in Figure 14.16 and Figure 14.18 will show that the loss will almost always be higher in an expansion than in a contraction of the same taper angle. Moreover, whereas a contraction is not considered to be abrupt until the taper angle exceeds 60°, an expansion is already considered abrupt when its taper angle exceeds 30°. Above 30°, an expansion's loss factor depends only upon its diameter ratio. At the other extreme, loss factors will begin to increase for taper angles somewhere below approximately 4° because friction will start to exceed expansion turbulence losses; however, such shallow taper angles are rarely practical in ventilation systems.

2. If the velocity going out of the exhaust stack is higher than it needs to be, a tapered expansion can be placed at the end of the exhaust duct to recover some of that excess kinetic energy before it is all released into the atmosphere. The name *évasé* is given to such a tapered opening (from the French *évaser* meaning to open out or to flare). The use of an *évasé* is justified by the fact that whatever kinetic energy is in the air as it leaves the exit plane of an exhaust stack is lost from the ventilation system whether an *évasé* is present or not. If some of that excess kinetic energy can be recovered within the system before it leaves the stack, the fan has to expend just a little less energy to move the air. Thus, recovered energy has the effect of having less losses within the system. The *évasé* is depicted vertically in Figure 14.19 because the exhaust stack is almost always

vertical in order to disperse the contaminant safely (as will be discussed in Section VI of Chapter 15). While a tapering expansion will still create some turbulence losses, the designer's interest in an *évasé* is the fraction of that initially high VP that can be recovered or regained before it is released into the atmosphere as it crosses the exit plane of the exhaust stack. Equation 14.20 looks, at first glance, similar to the typical form of a loss factor times the initial VP_1 in the duct (VP_2 is disregarded because none of it is recoverable under any circumstance). However, that is where typical ends for an *évasé*.

$$\text{“Regain” for an expansion at the end of a duct} = -\text{Regain factor} \times VP_1 \tag{14.22}$$

Loss factors for an *évasé* are unique in several respects. Since the main reason to use an *évasé* is to regain a proportion of the kinetic energy before it all leaves the exhaust stack, the proportionality factor is called a regain factor. As the regained energy is added back into the air, an *évasé*'s regain is, in effect, a negative loss and is preceded in Equation 14.22 by a minus sign.[w] The minus value means that the proportion of the excess energy that is regained within an *évasé* is to be subtracted from the sum of all of the previous losses. The values of the regain factors in Figure 14.19 are shown as a function of the duct diameter ratio (D_2/D_1) and the taper length to inlet diameter ratio (L/D_1), as they are in the *Manual of Industrial Ventilation*. The parameter of taper angle was added (as dashed lines) to aid comparisons with the data in Figure 14.18. Notice how the best expansion angles in an *évasé* are at the very low end of the in-duct expansion angles.

An *évasé* should rarely be needed or appropriate in a new LEV system. Chapter 15 will recommend that the air velocity at the exit plane of an exhaust stack should be at least 3000 and 3500 fpm to get good dispersion of the exhaust plume and to keep rain from falling into an open stack. This velocity is also the minimum design duct velocity recommended in Table 14.2 to convey typical or average size aerosols, while the highest minimum transport velocity in Table 14.2 is about 4500 fpm. If the duct velocity in the exhaust outlet is below 5000 fpm, not enough excess VP energy is probably available to justify the cost of building and installing an *évasé*. However, if more hoods have been added to an existing system or their flow rates have been increased after installation, its exhaust duct velocity could easily be much greater than 5000 fpm, at which point the cost to add an *évasé* to recover 50% or more of that excess kinetic energy might be justified. But keep in mind how shallow the expansion angle should be to avoid more turbulence losses and to get back a good fraction of that excess.

XI. COMPARING CONTRACTION AND EXPANSION LOSSES

The above loss factors for contractions range from 5 to 30% of the ΔVP vs. 10 to 75% of the ΔVP for expansions, suggesting that the virtual *vena contractas* of expansions create more loss than the real *vena contractas* of contractions. The following two somewhat contrived examples demonstrate this generality. These examples could conceivably be encountered should the initial 8-in. duct in Example 14.2 not fit between the structural members of a wall through which the exhaust duct must pass. One solution is to reduce the duct diameter for a short distance while it passes through the wall then expand the duct back to its original diameter. Consider the following two questions:

How much more loss would occur in abrupt vs. tapered fittings?

How much more loss would be attributed to the expansion vs. the contraction?

[w] As in the previous footnote, if using regain R values from the *Manual of Industrial Ventilation*, the energy loss for an expansion at the end of a duct = $(R - 1)(VP_1)$, and the result of Equation 14.22 still has to be subtracted from the sum of all other losses.

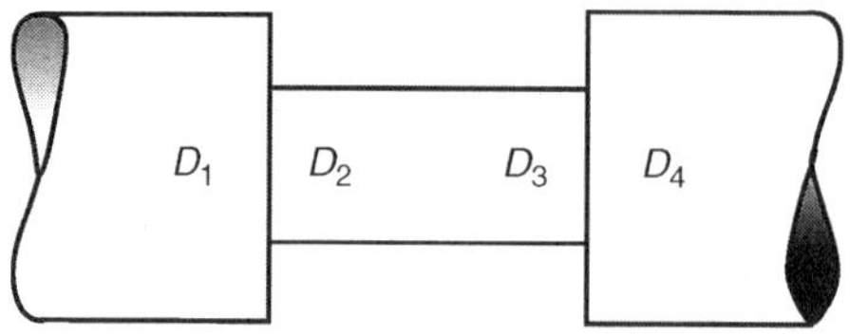

FIGURE 14.20 Depiction of a short segment of ducting between two abrupt fittings.

1. **Example 14.4**

Find the sum of the losses for an abrupt contraction from an 8 to a 4-in. duct, through 1 ft of a 4-in. duct then an abrupt return back to an 8-in. duct, as depicted in Figure 14.20.

$$D_1 = 8 \text{ in.} = D_4$$
$$D_2 = 4 \text{ in.} = D_3$$
$$\text{Length of the 4 in. duct} = 12 \text{ in.}$$

Always begin predicting system losses by dividing the system (or proportion of the system in this case) into a series of individual segments for which the loss in each segment can be defined. The ducting in Figure 14.20 is divided into three segments: the segment from 1 to 2 is an abrupt duct contraction, from 2 to 3 is a short section of straight ducting, and the segment from 3 to 4 is an abrupt duct expansion. Your challenge is to solve for the loss in each segment and to add them.

Recall that losses are always proportional to the local VP or change in VP. Thus, this example could use the $VP_{duct} = VP_1 = 0.932$ "wg from the previous example in an 8-in. duct, but in order for the solution to be generalizable, the losses in these examples will be expressed in terms of the high velocity pressure in the small diameter segment where $VP_2 = VP_3$. The following solution will also use the term β defined as the ratio between the downstream diameter over the upstream diameter. Thus,

$\beta = D_2/D_1 = D_4/D_3 = 0.5$ for contraction

($\beta < 1$ for contractions; $\beta > 1$ for expansions)

$\beta^2 = A_2/A_1 = 0.25$

(the "area ratio" used in some figures and *Manual of Industrial Ventilation* tables)

$Q = V_1 A_1 = V_2 A_2$ (a constant throughout this [and most] duct system)

$V_2 = V_1 A_1/A_2 = V_1/\beta^2 = 4\, V_1$ (also applicable to V_3)

$VP_2 = (A_1/A_2)^2 VP_1 = VP_1/\beta^4 = 16\, VP_1$ (because VP is $\propto$ to V^2)

$VP_4 = (A_3/A_4)^2 VP_3 = (A_2/A_1)^2 VP_3 = \beta^4 VP_3 = 0.0625\, VP_3$ (because VP is $\propto$ to V^2)

a. The turbulence loss from the abrupt contraction of segment 1 to 2 is found using the loss factor from Figure 14.17 in Equation 14.20:

$$h_{\text{abrupt contraction}} = \text{LossFactor}_{\text{abrupt contraction}} VP_2 = 0.44\, VP_2$$

(using Equation 14.20)

b. The friction loss through the 1 ft of the straight 4-in. diameter galvanized sheet metal ducting from segment 2 to 3 through the wall is calculated using Equation 14.14b:

$$h_{\text{duct friction}} = 0.387\,L(\text{VP}_2)^{0.96}/D^{1.22} = 0.39\ 1(\text{VP}_2)^{0.96}/4^{1.22}$$

(using Equation 14.14b)

$$h_{\text{duct friction}} = 0.072(\text{VP}_2)^{0.96} \approx 0.072\ (\text{VP}_2)$$

c. The turbulence loss from the abrupt expansion of segment 3 to 4 is found using the loss factor found in Figure 14.17 in Equation 14.21, where $\text{VP}_4 = \beta^4\text{VP}_3$ and $\beta = 2$ in the expansion:

$$h_{\text{expansion}} = \text{LossFactor}_{\text{expansion}}(\text{VP}_3 - \text{VP}_4) \qquad \text{(using Equation 14.21)}$$

$$h_{\text{expansion}} = \text{LossFactor}_{\text{expansion}}(1 - \beta^4)\text{VP}_3 = 0.60(1 - 0.0625)\text{VP}_3 = 0.56\ \text{VP}_3$$

d. The sum of these losses $= (0.44 + 0.072 + 0.56)\text{VP}_2 = 1.07\ \text{VP}_2$. If this abrupt contraction and expansion arrangement were part of the TiO_2 hood and duct system previously designed in Example 14.2 (with $\text{VP}_1 = 0.932$ and $\text{VP}_2 = \beta^4\text{VP}_1 = 14.9$ "wg), the loss of another $1.07\ \text{VP}_2 = 15.9 \approx 16$ "wg would be a huge addition to the previously calculated loss of only 1.3 "wg.

2. **EXAMPLE 14.5**

Find the sum of the losses for a tapered contraction from an 8 to a 4-in. duct, through 1 ft of that 4-in. duct, and from an equally tapered expansion back to an 8-in. duct, as depicted in Figure 14.21. Each taper is 6 in. long on its centerline.

$$D_1 = 8\text{ in.} = D_4$$

$$D_2 = 4\text{ in.} = D_3$$

Finding the sum again involves finding the loss for each definable segment of the duct system. There are still three segments, and the same β values defined above still apply.

a. The turbulence loss in the tapered contraction from segment 1 to 2 is calculated using the loss factor found in Figure 14.16 in Equation 14.19 for which the taper angle is found using Equation 14.17a.

$$\text{taper angle} = \tan^{-1}\left[\frac{(8-4)/2}{6}\right] = \tan^{-1}[0.3333] = 18°$$

(using Equation 14.17a)

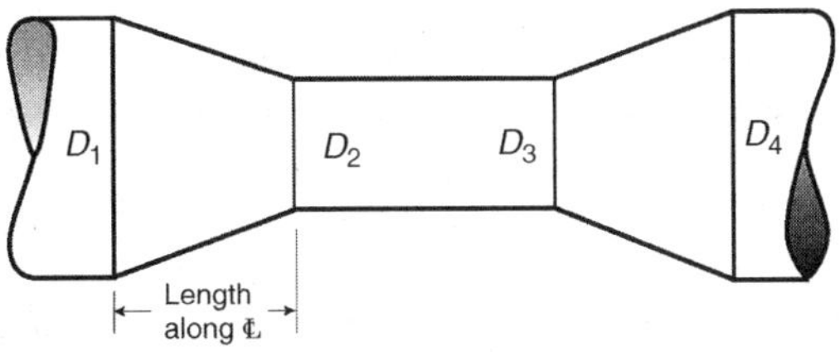

FIGURE 14.21 Depiction of a short segment of ducting between two tapered fittings.

$$h_{\text{tapered contraction}} = \text{LossFactor}_{\text{tapered contr.}}(\text{VP}_2 - \text{VP}_1)$$
$$= \text{LossFactor}_{\text{tapered contr.}}(1 - \beta^4)\text{VP}_2$$
$$= 0.10(1 - 0.0625)\text{VP}_2 = 0.094\ \text{VP}_2$$

(using Equation 14.19)

b. The friction loss in the straight duct segment from 2 to 3 using Equation 14.14b is the same as in the previous example.

$$h_{\text{duct friction}} = 0.072\ \text{VP}_2 \qquad \text{(using Equation 14.14b)}$$

c. The turbulence loss in the tapered expansion segment from 3 to 4 again uses Figure 14.18 to find the loss factor for use in Equation 14.21.

$$\text{taper angle} = 18°\text{(the same as during the contraction)}$$

$$h_{\text{expansion}} = \text{LossFactor}_{\text{expansion}}(\text{VP}_3 - \text{VP}_4) \qquad \text{(using Equation 14.21)}$$

$$h_{\text{expansion}} = \text{LossFactor}_{\text{expansion}}(1 - \beta^4)\text{VP}_3 = 0.50(1 - 0.0625)\text{VP}_3 = 0.469\ \text{VP}_3$$

d. The sum of these losses = $(0.094 + 0.072 + 0.469)\text{VP}_2 = 0.635\ \text{VP}_2$. Again, if this contraction and expansion arrangement were part of the TiO_2 hood and duct system previously designed in Example 14.2 (with $\text{VP}_1 = 0.932$ and $\text{VP}_2 = 14.9$ "wg), the loss of another $0.635 \times 14.9 = 9.5$ "wg would still be a major addition to the previous sum of 1.3 "wg.

3. A comparison of the losses in these two examples demonstrates the previous generalizations.

	Losses Based on VP_2		
	Abrupt	**Tapered**	
Contraction	0.44 VP_2	0.09 VP_2	
Expansion	0.56 VP_2	0.47 VP_2	
Sum of all losses	1.07 VP_2	0.635 VP_2	Including 0.07 VP_2 friction loss

By now, it should not surprise you that the abrupt geometry has more losses than the tapered geometry. What may be surprising is the difference between the tapered expansion and the tapered contraction. This latter difference is consistent with the larger magnitude of turbulence in the expansion from the virtual *vena contracta* in Figure 14.9c than in the real *vena contracta* in Figure 14.9a (again, the two fittings in this figure are intended to have the same geometry). The inertia of the contracting air will not create a real *vena contracta* as large as the virtual *vena contracta* that exists intrinsically in any expansion. The latter is essentially equal to the whole difference between the smaller upstream duct diameter and the larger downstream duct diameter. Thus, in general, losses in the turbulence from a virtual *vena contracta* of an expansion will always be greater than the losses from the smaller, real *vena contracta* that is created just downstream of a contraction of the same geometry.

The similarity between the two tapers in Figure 14.21 and the venturi in Figure 12.15c offers another comparison that might be important on some future project. If you were to compare the predictions made for either of these examples to real measurements, you would find the measured losses would be less than the predicted losses. For instance, the *Chemical Engineer's Handbook*[9] indicates that head losses through a venturi meter can be expected to equal only 0.1 to 0.15 times the ΔVP vs. about 0.35 $\times$ ΔVP that would be predicted using the same techniques as used above for the much more moderate expansion angle characteristic of a standard venturi meter. To generalize, the actual losses from *more than one source of turbulence located in close succession* will almost always be less than the losses predicted by their sum as if they were spread out and acting independently of each other. In other words, one obstruction immediately behind a prior obstruction cannot add much turbulence to air that is already turbulent by the first obstruction. The above example only had three diameters of small ducting between their contraction and expansions. In comparison, the rule-of-thumb suggests that 8 to 10 straight, smooth duct diameters are required to reestablish a normal, reasonably uniform bulk flow.

XII. ENERGY LOSSES IN BRANCHED DUCT SYSTEMS

Many industrial production processes that require LEV are conducted on several machines in clustered locations where it is economically more efficient to join the ducts from multiple hoods into one main duct and fan than to operate each hood and fan independently. A system where the air from multiple hoods flows into one duct is called a branched duct system. Figure 14.22 is a simplified depiction of a branched duct system with three hoods (not shown).[x] Any branched system will have one less junction than the number of hoods in the system. Thus, the system depicted in Figure 14.22 has two branch junctions: where branch number 3 joins branch number 2 and where these two branches join branch number 1.

1. A branched duct system must be balanced to achieve the desired flow rate in each branch. Balancing means to adjust the losses in one or more branches. Balancing the flow is important both during design and later while managing an LEV system. If a system is not balanced, it is unlikely to perform as desired. Although few industrial hygienists might actually design or personally balance such branched systems, it is important for hygienists to know how branched systems should operate and what would happen if they are not designed or operated correctly. There are two keys to understanding branched duct systems.

   The first key is to realize that the *P* (and SP) within each branch will always be equal at any branch junction where two ducts join together. For instance, the $SP_{in\ branch\ 3}$ will always equal the $SP_{in\ branch\ 2}$ at the junction between these two branches (above hood number 2 in Figure 14.22). While the SP values will always be equal, the flow rates (Q) in each branch may not be what you want or need to protect people working at the two hoods or to control the emissions. Recall from Equation 11.15 that before the fan, $-SP = VP + \Sigma$ of the losses up to that point; recall from Equation 14.4 that each loss can be approximated as some generic "LossFactor" $\times$ the local VP; and recall from Equation 14.5 that VP is proportional to Q^2. Thus, the SP in each branch is proportional to $Q^2 \times (1 + \Sigma\text{"LossFactor}_i\text{"})$. Therefore, Q will probably not be the flow you want or need unless the sum of the losses in each branch

[x] The horizontal duct coming into view from the left is not yet functional but was designed and built that way to allow for future expansion to a fourth hood. (With an end-cap attached, it can also serve in the interim as a cleanout port.)

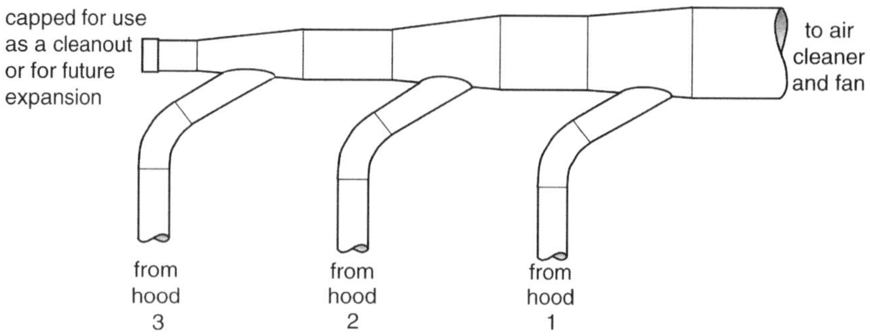

FIGURE 14.22 A generic branched duct system.

are purposefully controlled. The following equations express the above logic mathematically.

$$\mathrm{SP}_{\text{in branch 1}} = \mathrm{SP}_{\text{in branch 2}} \quad (14.23)$$

$$\mathrm{VP}_1 + \Sigma(\text{“LossFactor}_i\text{”}\mathrm{VP}_i)_{\text{in branch 1}} = \mathrm{VP}_2 + \Sigma(\text{“LossFactor}_i\text{”}\mathrm{VP}_i)_{\text{in branch 2}} \quad (14.24)$$

$$Q_1^2(1 + \Sigma\,\text{“LossFactor}_i\text{”})_{\text{in branch 1}} \approx Q_2^2(1 + \Sigma\,\text{“LossFactor}_i\text{”})_{\text{in branch 2}} \quad (14.25)$$

The desired protection is unlikely to be achieved at all the hoods on a branched system by letting nature take its course (without proper design or manipulation). Since the SP's at the junction of any two branches are always equal, Equation 14.25 shows that the Q^2's and Σ"LossFactor$_i$" in each branch are inversely related to each other. The flow in the duct with higher than expected losses will decrease below its design goal, and the flow in the duct with lower than expected losses will increase until the SP at the junction of each branch matches. Visualize what would happen if one wanted the flow rates into all the hoods in Figure 14.22 to be the same. If the duct size and elbow radius in all three branches are the same sizes, then the flow in branch number 2 will be less than in branch number 1 because the duct length in branch number 2 is longer. The flow into the hood in branch number 3 would be less than either branch numbers 2 or 1 for the same reason.

The second key to understanding branched systems is to realize that the losses and loss factors in each lateral branch occur in parallel with (but are independent of) the losses in each other branch. This independence allows either the designer, the installer, or the system manager to manipulate the loss factors (the Σ"LossFactor$_i$") in one branch at a time to achieve the desired Q in that branch and eventually to manipulate the losses in all branches to achieve the desired flow through each branch (or at least to achieve the correct proportion of the total Q flowing into each branch, after which the total flow through the fan may need to be adjusted).

2. There are basically two options to balance a system: the balanced design method (sometimes called the static pressure balance method) and the blast gate method. A less common third method called the plenum system may also be encountered but is generally not recommended. The balanced design method achieves the desired flow rates at the time a system is designed by manipulating the losses in each branch either upward or downward via such design parameters as the size (diameter) of their ducts and/or the tightness of turns. The only way one can balance an unbalanced branched system after it is built is to add more losses via dampers or blast gates to the branch whose flow is too high, provided that dampers have been installed. The advantages and disadvantages of

these two branch balancing options are summarised in Table 14.8. The following descriptions of their operation and design start with the blast gate method because it seems more intuitive.

a. Recall that blast gates are adjustable (either sliding or pivoting) dampers such as shown in Figure 10.10. By adjusting the position of a blast gate, one can easily change the Σ"LossFactor$_i$" on a branch after a system is installed. To adjust blast gates, they must be installed at least in the expected lower resistance branch if not in all branches. Blast gates are often installed near the hood(s) for convenience, but easily accessible dampers should be lockable to prevent unauthorized adjustments from upsetting a good balance. Since adjustments in one branch will change the flow in all the other branches, a particular sequence of systematic adjustments is recommended (similar to the sequence to be used in "balance by design" to follow). Open all blast gates before starting to balance. Start the balancing process at the hood or branch furthest from the fan. Increase the resistance in the branch that has a higher than desired Q until achieving the desired flow rate or proportion of the total rate in that branch. Repeat that process of making adjustments to blast gates at each junction moving closer to the fan at each cycle.
b. A system that is balanced by design has no blast gates. Its design requires an iterative calculation of the SP at each branch junction based on the hood and duct losses at the desired Q in each branch after laying out the overall route of the system. As with the blast gate method, these calculations begin at the junction farthest from the fan. In the first iteration, the designer initially decreases the duct diameter or increases another restriction in the low loss branch to match its SP to the more restricted branch (starting with this step is least likely to violate a minimum design duct velocity). On the next iteration, the designer increases the size and decreases other restrictions in the high loss branch to match its loss to the less restricted branch *while avoiding a transport V that is too low*. The primary design variables that can be used to adjust the SP via Σ"LossFactor$_i$" include:
   1. The duct diameter (in available increments) and its corresponding duct velocity,
   2. The duct entry taper angle that affects the entry loss,
   3. The elbow radius (usually less of an effect than changing the duct diameter).

   The designer can repeat the above cycle until each flow is as close to the desired flow as one chooses (but at least within $\pm 20\%$). This process is repeated at each branch

TABLE 14.8
The Advantages and Disadvantages of the Two Basic Branch Balancing Options

	Advantages	Disadvantages
Blast gate method	Flow rates can be changed more easily after installation (as frequently as per minute or as rare as per year)	Blast gates may corrode or plug (requiring higher maintenance)
	Potentially saving energy if some systems can be shut way down or off	More pressure losses mean a larger fan and higher fan operating costs
Balance by design method	More reliable because flow rates are not changeable by accident or whim	Less tolerant of variations in construction than are blast gates
	Less maintenance is required (*important for highly toxic, radioactive, and explosive materials*)	More difficult to modify the system after installation

duct junction moving closer to the fan at each step. Computer programs are available to facilitate this design process.

c. A plenum exhaust system (Figure 5.4 and Figure 5.5 of the *Manual of Industrial Ventilation*) is a rarely used third alternative or adjunct to balancing branched systems. Recall from Chapter 10 that a plenum is simply a chamber that is all at one pressure. For branched duct systems, the plenum can be a very large duct in which the air is moving slowly enough (between 2000 and 1000 fpm) that a negligible pressure loss occurs within it. Such a plenum simplifies the application of either of the above balancing methods, but its slow duct velocity creates the potential for deposition of particulate matter via sedimentation to occur within the plenum (making it as if it were a crude air cleaner). Therefore, if a plenum is used, provision must be made to allow periodic solid waste removal (see Section V.7 of the *Manual of Industrial Ventilation*).

3. The recommended design parameters of a generic branch junction are shown in Figure 14.23. Most branch junctions involve a primary branch that by convention is a duct in which the flow continues straight; the entering airflow branch is the one that makes the turn. Two ducts can also approach the junction from equal angles, forming the symmetric wye duct junction shown in Figure 14.24.
4. A branch entry creates turbulence and losses immediately downstream from the junction. The branch entry loss factors shown in Figure 14.25 depend only upon the angle at which the two airstreams merge. Although the turbulence does not affect either branch before the junction, by convention the branch loss is calculated as the loss factor multiplied

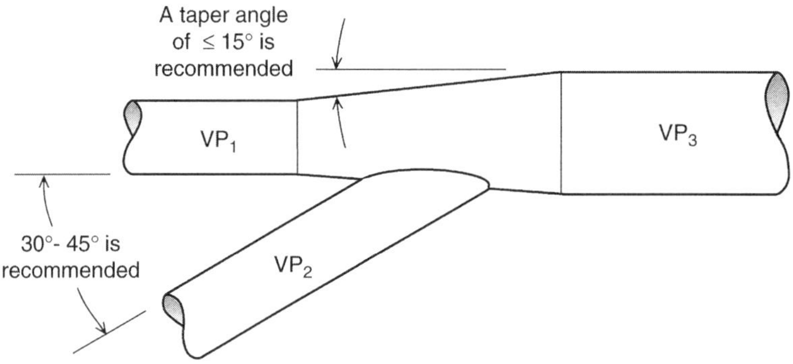

FIGURE 14.23 A well-designed branch duct junction.

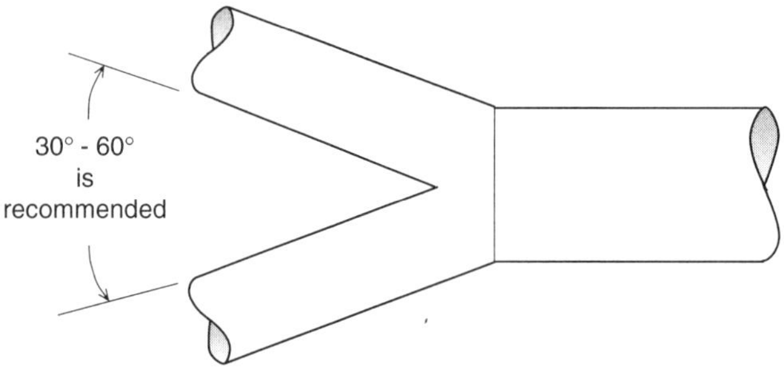

FIGURE 14.24 A symmetric wye duct junction. The branch entry angle for a wye is one half of the major angle shown herein.

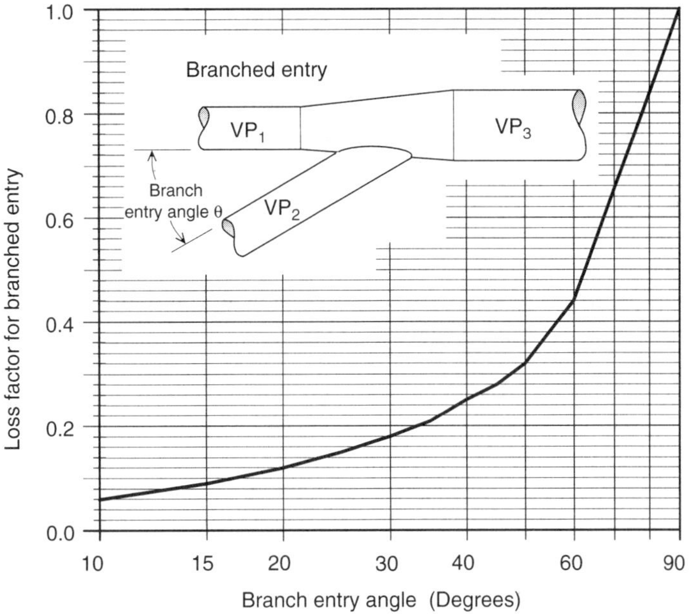

FIGURE 14.25 Loss factors for a branched duct entry. No additional loss factor for the expansion is needed.

times the VP in the entering branch (VP_2 as shown in Equation 14.26 and Figure 14.27). If the branch junction is a symmetric wye, such as shown in Figure 14.24, the decision about which is the entering branch is arbitrary; and if the entering VPs are intended to be equal, the question is moot.

$$\text{“Loss” for a branched duct entry} = \text{LossFactor}_{\text{branch junction}} \text{VP}_{2\ \text{(the entering branch)}} \quad (14.26)$$

The downstream VP_3 can always be derived from Equation 14.27. While an expansion in conjunction with a branch junction is not mandatory, the downstream duct size (represented in Equation 14.27 by $A_{\text{downstream}}$) is normally increased over that in either upstream duct in order to maintain approximately the same duct transport velocity before and after the junction.

$$V_3 = (Q_1 + Q_2)/A_{\text{downstream}} \quad (14.27)$$

$$\text{VP}_{\text{resultant}} = \left(\frac{Q_1}{Q_3}\right)\text{VP}_1 + \left(\frac{Q_2}{Q_3}\right)\text{VP}_2 \quad (14.28)$$

A properly sized downstream duct will result in a VP_3 that is within ± 0.1 "wg of what is called the resultant VP or $VP_{\text{resultant}}$ given by Equation 14.28 (the theoretical average downstream kinetic energy). The energy loss in a properly sized expansion occurs simultaneously with the turbulent branch entry loss and requires no additional loss factor. However, any positive difference between the new VP_3 and the $VP_{\text{resultant}}$ is considered an acceleration loss that should be added to the loss calculated by the branch entry loss factors in Equation 14.26; a negative difference is a slight regain.

$$\text{“Acceleration loss”} = \text{VP}_3 - \text{VP}_{\text{resultant}} \quad (14.29)$$

XIII. THE LEV DESIGN WORKSHEET

It may have become apparent by this point that the calculation side of ventilation design is a small bookkeeping challenge. To deal with this challenge, various design worksheets have been developed to help organize the information and calculations. The LEV Design Worksheet shown as Table 14.9 was modeled after that in the ACGIH *Manual*.[y] In both worksheets, not all cells will be used in any row or column. Several large shaded zones are shown in the USU worksheet to clarify when certain sectors of the worksheet are never used (e.g., losses never occur outside the hood in column A). The following description of the general function of these worksheets will also serve as a review of the design sequence presented above and in Section III of Chapter 10.

1. At the outset, the LEV system to be designed must be divided into as many segments as necessary so that the losses in each segment can be calculated using one of the above methods. One column is then allocated to each such segment. For instance, the USU worksheet allocates three columns to the hood. The first segment of ducting is then assigned to the next available column, the second segment to the next column, and so on until each segment has been assigned to one column. Worksheets typically allocate row 1 of each column for the user to identify each segment. Some people use a single letter or number to identify each segment. Other people prefer to use two numbers or letters, one corresponding to each end of each segment (e.g., 1-2, 2-3, etc.) as being less ambiguous. The choice is personal; whatever works for you.

   Most of the rows are organized into groups of data and equations needed to calculate a particular kind of loss (e.g., hood entry, straight duct, or duct fitting). The bottom rows are reserved to sum the losses as the air progresses along the LEV system. This sum equals the fan static pressure (FanSP) needed to overcome the resistance to pulling and pushing the desired amount of air through the system.
2. In the USU worksheet, the first three columns are preassigned to the hood. The first column (A) is for the zone of control outside of a collection hood, if applicable. The second column (B) is for the hood's face conditions (which may or may not have significant losses). The third column (C) is for the air's entry into the duct (applicable to all hoods). Enter the chosen control velocity in line 2 and all of the chosen or otherwise available dimensions in line 3.
   a. Column A is only used for a collection hood. The desired velocity (fpm) in line 2 is the control velocity and the dimension in line 3 is the distance (X) of the source outside the face. These data will be used in the DallaValle equation (footnote *b in the form) to calculate collection hood's V_{face}, VP, and Q in column B. Enter those results into column B lines 5, 6, and 7, respectively.
   b. The hood's face velocity and dimensions are entered into column B. The hood's dimensions should include both its face diameter and area in lines 3 and 4, respectively. A containment hood's total airflow rate Q (cfm) is easy to calculate from A_{face} in line 4 × V_{face} in line 5. The resulting Q value will often not change throughout the rest of the ventilation system (unless either the exhausted temperature changes appreciably as by the cooling of a hot exhaust gas, if a branch junction is encountered, or if a "bleed-in" is installed).

[y] The rightmost column includes the *Manual of Industrial Ventilation*'s numerical row identifiers where there was a match (a " ~ " indicates a modified line). Unnumbered cells in that column indicate a row unique to this worksheet, and a few of their rows were excluded.

Recall that the hood face entry loss will be insignificant for V_{face} below 1000 fpm. Recall also that the slot area (in line 4) should be small enough to create a V_{slot} of about 2000 fpm. Thus, the most common hood entry loss that needs to be calculated and entered into column B lines 8 to 9 is from flow through slots. The number, size, and total area of the actual slots, the resulting actual V_{slot}, VP_{slot}, and slot loss factor should be entered into their assigned rows.

c. Column C is reserved for the loss created as the air enters into the duct. Use the design sequence given in Section III to select a duct small enough (in column C line 3) to meet the minimum duct transport velocity (in line 2). Enter the duct velocity pressure in line 6. Lines 4 to 6 will stay the same throughout the rest of the ventilation system unless and until the duct size changes or a branch junction is encountered. The entry loss in line 9 equals the duct entry loss factor (e.g., from the VS diagram or Table 14.1) in line 8 times the segment VP in line 6.

3. Beyond this point in the worksheet, the losses will be either from friction in a straight duct or from turbulence in an elbow, a duct diameter transition, or a branch junction. Numerous rows are grouped to help differentiate and guide the user as to which VP to use for each of these types of loss.
   a. If the segment for that column is a straight duct, lines 10 to 12 allow the use of any of the three versions of the Loeffler equation, outlined in Section IV, to calculate the duct loss. Line 11 may be used to list the $\text{LossFactor}_{friction}$ determined either via the coefficients in Equation 14.12 or from its tabular equivalent. The footnote in line 12 describes three calculated loss methods.
   b. If the segment for that column is an elbow, use the description of the elbow in line 13 to find the loss factor entered into line 14 (e.g., from Figure 14.13 or Figure 14.15). Multiply the $\text{LossFactor}_{elbow}$ times the current VP to calculate the turbulent loss to be entered into line 15.
   c. If the segment for that column is an expansion or contraction, insert either the diameter or area ratio (D_2/D_1 or A_2/A_1) into line 16. Calculate the new V and VP (based on $V_1A_1 = V_2A_2$) and insert them into the next column. Calculate the change in velocity ΔVP in line 17 and the angle of the transition in line 18. This information is usually needed to find the loss factor and to calculate the loss based on either ΔVP or the new VP depending upon the type of transition.
   d. If a branched duct junction is encountered, a loss factor is needed but the following calculations should be made first if a balanced design system is desired.
      1. Calculate the cumulative sum of the pressure losses from each hood up to and including the duct segment that reaches the junction by adding the entries in line 28 into line 29 for each branch individually.
      2. Determine the duct static pressure (SP) as each branch enters the junction by adding the VP in that branch (line 6) to the sum of the losses up to that junction (line 29), and enter this value into line 30.
      3. If the SPs in each branch are not equal to within at least $\pm 20\%$, change the dimensions affecting the losses (and the corresponding VPs as needed) and recalculate the SP using the methods described in Section XII.2.

   Allocate a column on the worksheet to the branch entry loss. Calculate that loss as the product of the branch entry loss factor from Figure 14.25 (entered in line 25) times the branch VP (line 6 from the column with the entering branch). The new combined airflow rate (the sum of the two branch flow rates) can be entered into line 7 in the next column, along with any new duct dimensions and subsequent $V_{downstream}$ and $VP_{downstream}$ values. An acceleration loss from Equation 14.27 can be entered as an "Other Loss" in line 27.

TABLE 14.9
Utah State University: Industrial Hygiene: Local Exhaust Ventilation Design Worksheet[a]

				A	B	C	D	E	F	G	H	I	J	K	L	M	*Manual*
1		**Segment Identification** ↓		outside hood	hood face *b	duct entry											1
2		Desired Velocity (control or minimum transport)	fpm														3
3		Dimensions (distance, face, slot, duct, *etc.*)															~ 4
4		Air Flow Area = [width × height] or [$\pi D^2 / 4$]	ft²														~ 6
5		Air Velocity = [Flow Rate] / [Flow Area]	fpm														~ 7
6		VP = ([Velocity] / 4005) ^{2} × [ρ adjustment see *c]	"wg														8
7		Air Flow Rate = Q (adjust for ρ if |SP|>25 "wg)	cfm														2
8	HOOD	Hood entry loss factor, $LossFactor_{hood}$	- -														13 / 17
9	HOOD	Hood entry Loss = [$LossFactor_{hood}$] × [VP]	"wg														20
10	FRICTN.	Straight duct length [L]	ft														23
11	FRICTN.	(opt.) Duct Friction Loss Factor (see *d)	*b														24
12	FRICTN.	Friction Loss using one of the formulas in *e	"wg														25/34

14	TURBULENCE	ELBOW	Elbow R/D ratio and structure															
15			Elbow loss factor [$LossFactor_{elbow}$]	- -														27
16			Loss = [$LossFactor_{elbow}$] × [VP_{elbow}]	"wg														34a
17			Transition D_2/D_1 (or A_2/A_1 for abrupt contraction)															
18			$\Delta VP = \lvert VP_1 - VP_2 \rvert$ for tapered expan. or contr.	"wg														
19			Angle of expansion, contraction, or branch jnct.	°														
20		TRANSITION	Expansion "loss factor" [$LossFactor_{expansion}$]	- -														
21			Loss = ($LossFactor_{expansion}$)×ΔVP or (1-R)×ΔVP	"wg														34b
22			Tapered Contr. loss fac. = $LossFactor_{tapered\ contr.}$	- -														32a
23			Loss = $LossFactor_{tapered\ contraction}$ × (ΔVP)	"wg														34c
24			Abrupt Contr. loss factor = $LossFactor_{abrupt\ contr.}$	- -														32b
25			Loss = $LossFactor_{abrupt\ contr.}$ × (VP_2)	"wg														34d
26			Branch Entry loss factor [$LossFactor_{branch}$]	- -														32c
27			Loss = $LossFactor_{branch}$ × (VP_{branch})	"wg														34e
28			Other losses [see *f]	"wg														36
29			Loss in Segment (repeat Loss entry for column)	"wg														35
30			Cumulative sum of losses = Σ[line#29]	"wg														
31			(optional) -SP = (see *g)	"wg														37

[a] The numbered rows on the far right correspond to a similar form in the 23rd edition of the ACGIH *Manual* (many of which are either missing or repeated herein).

[b] If applicable, use a form of the DallaValle equation: $Q_{collection\ hood} = K_{hood}V_{control}(10X^2 + A_{face}) = V_{face} \times A_{face}$ where $K_{hood} = 0.75$ for a flange or 0.5 for a flange and a bench.

[c] Air density adjustment = (the altitude adjustment $= 2^{(-\,elevation(ft)/18{,}000)}$) × (the temperature adjustment = 535/(°F + 460)).

[d] Optional $LossFactor_{friction}$ as given in Table 14.5 or Table 14.6 (*Manual of Industrial Ventilation* Table 5.5 or Table 5.6) or as the first term in Equation 14.12 as loss per inch of VP per foot of duct.

[e] Loss either $= LossFactor_{from\ table} \times L \times VP$, for galvanized $= 0.387 \times L \times (VP^{0.9605})/(diam.in.^{1.224})$, or for PVC and smooth ducts $= 0.314 \times L \times (VP^{0.9315})/(diam.in.^{1.204})$.

[f] For example, losses due to make-up air resistance, a branch junction perhaps with acceleration, air cleaner, or fan system effect.

[g] Either − SP = VP + cumulative sum of losses before the fan = row 6 + row 30 or SP = cumulative sum of losses in the rest of the duct after the fan.

4. The "Other Losses" category (line 27) has several uses. The most common losses have already been covered, but the following three circumstances can also create "Other Losses". The only one of these other losses for which explicit design or calculation methods are available is for poor fan entry conditions (item c below) which will be covered in Chapter 16.
   a. An air cleaner will produce pressure losses. From the perspective of this book, the air cleaning pressure losses will be provided by the manufacturer, the engineering department, or a consultant (see the later discussion on Exhaust Air Cleaners and Stacks in Chapter 15 of this book and/or Chapter 4 of the *Manual of Industrial Ventilation*). The air cleaner loss can be entered in line 27 in the next available column.
   b. The resistance to the flow of make-up air into the building or room being exhausted is often an unrecognized loss. In a leaky factory or manufacturing building, such resistance should be negligibly small; however, institutional buildings and high-tech industries that are sensitive to heat loss or gain or especially to airborne dust tend to be "tighter" in their construction. Tight buildings do not leak easily. If the air being exhausted from a tight building exceeds the capacity of the general ventilation (HVAC) system to supply make-up air, an additional resistance should be added on line 27 that the fan must overcome.
   c. A turbulent or uneven flow pattern as the air enters a fan or an obstruction immediately after a fan's outlet reduces a fan's ability to produce as much fan pressure as expected from the manufacturer's specifications. Predictable reductions in performance called fan system effect factors (SEF) (to be described in Section F of Chapter 16) are treated as an "Other Loss".
5. Throughout the above design process, the loss in each column should be copied into line 28 simply to help calculate the cumulative sum of the losses across the entire system. Chapter 16 will explain why the sum of these losses is identical to the fan static pressure (FanSP) required of the fan to be chosen. The cumulative sum of the pressure losses from the hood up to the fan and from the fan to where the exhaust exits the system is calculated by cumulatively adding the entries in line 28 into line 29 in the previous column. Thus, the first loss (in whatever column) is entered into both lines 28 and line 29. The loss in line 28 of the next column is added to the sum in line 29 of the previous column, and this sum is entered into line 29 in the next column. That summing process is continued until all the losses have been added together.

The appropriate fan can now be specified by the two parameters just determined within the worksheet, namely:

- Q, the cubic feet per minute you determined in line 7 that was needed to protect the employee.
- FanSP, the fan static pressure you determined as the last entry in line 29, i.e., the cumulative sum of all of the losses (all the resistance) that the fan must overcome as it tries to suck and blow that much air through the system.

6. The calculation of SP is left as an optional calculation in the USU LEV Design Worksheet.[z] SP can be calculated at any segment, but there are practically only two situations where one *must* calculate SP:
   a. At a branch junction (where the predicted SP in each branch should be designed to at least approximately equal that of the adjoining branch as discussed in Section XII of this chapter).

[z] The ACGIH method instead calculates SP for each segment. Such a focus on SP means that one has to repeatedly add and subtract the VP each time a loss occurs, and then subtract it one more time to find the fan static pressure, one of the two parameters needed to select a fan.

b. Anywhere you may want to actually measure SP (e.g., at selected locations after system start-up for periodic monitoring and trouble shooting as will be discussed in Chapter 18). To measure SP, a hole must be created in the wall of the duct either for a static pressure port (with a smooth inner surface) or for a Pitot tube or other anemometer (as discussed in Chapter 11).

A third, rarely encountered case was discussed in Chapter 11 Sections I and IV when the SP exceeds more than about ± 20 in. at which point the change in air density inside the LEV system can significantly change both the volumetric flow rate (Q) according to Equation 11.6 and the VP "4005 magic number" according to Equation 11.22. One source of such a high absolute SP value might be an air cleaner that is too small for the Q that someone is trying to force through it.

PRACTICE PROBLEMS

These questions relate to the 1500 ft^3/min of exhaust air from previous problems and to VS-75-06 (Figure 13.17). Assume NTP throughout this problem.

1. The air passing through the slots of a slotted hood creates the first of two pressure losses of a compound hood.
   a. When designing a slotted hood, the slots are sized to achieve what air velocity through the slots?
   $V_{\text{slot}} =$ _____ fpm
   b. What would be the velocity pressure (VP) at the above slot velocity?
   VP_{slot} ans. = 0.249 "wg
   c. What is the loss factor for air passing through the slots of a slotted hood?
   $\text{LossFactor}_{\text{slot}} =$ _____
   d. What would be the pressure loss as this air passes through the slots?
   $h_{\text{slot loss}} =$ _____ "wg
2. The duct needs to be sized to assure that air passes through this duct at or above a minimum duct velocity.
   a. What is the minimum duct velocity recommended by this VS diagram?
   $V_{\text{minimum duct}} =$ _____ fpm
   b. What is the maximum diameter of a round duct corresponding to this velocity?
   max. $D_{\text{duct}} \leq$ _____ in.
   c. What is the next smaller diameter of a common commercially available round duct? Table 14.5 and Table 14.6 list duct diameters commonly available commercially.
   D_{actual} ans. = 11 in.
   d. What would the velocity be in the duct that you would actually recommend?
   $V_{\text{actual duct}} =$ _____ fpm
3. The air entering the duct creates the second of two pressure losses. All hoods must include the pressure loss as the air enters into the duct. What is the loss for the 45° tapered entry shown in the VS diagram? Notice that the taper starts as a rectangle and transitions into a round duct.
   $\text{LossFactor}_{\text{duct entry}} = 0.25$ $h_{\text{Loss duct entry}} =$ _____ "wg
4. What is the pressure loss as this $Q = 1500$ ft^3/min is next drawn vertically through 20 ft of straight 11-in. diameter round galvanized sheet metal ducting ? There is more than one way to solve this problem (not all of which actually generate or need a $\text{LossFactor}_{\text{friction}}$).
   $\text{LossFactor}_{\text{friction}} =$ _____ $h_{\text{Loss duct friction}} =$ _____ "wg

5. What is the pressure loss if this Q has to go through a 90° four-piece elbow with a 22-in. radius of curvature? If you are doing things correctly, your *R/D* diameter should be on Figure 14.13.

   $\text{LossFactor}_{\text{elbow}} =$ _____ $\qquad h_{\text{Loss elbow}} =$ _____ "wg

6. Now suppose that you decide that 2275 fpm is too slow a velocity for a duct that has to run horizontally about 20 ft to exit onto an intermediate roof on which the fan is located. What would be the pressure loss if you decide to install a contraction 6 in. *long (along the axis of the duct)* to go from an 11- to an 8-in. diameter round duct shortly beyond the elbow?

   What is the taper angle? Taper angle = _____°

   What is the loss? $\text{LossFactor}_{\text{tapered contraction}} = 0.08 \quad h_{\text{Loss tapered contraction}} =$ _____ "wg

7. What is the pressure loss in the succeeding 20 ft of 8-in. diameter ducting? (If you did not try more than one way to solve question 4, you might want to try one of the other ways here to solve this question.)

   $\text{LossFactor}_{\text{friction}} =$ _____ $\qquad h_{\text{Loss duct friction}} =$ _____ "wg

8. What is the total pressure loss from this system up to the fan? Recall that each pressure loss is normally independent of all other losses, and the total is merely the sum of the individual losses. A list like this will not always be provided for you.

   slot loss _____
   duct entry loss _____ $\qquad \Sigma\ \text{Losses}_{\text{inlet}}$ ans. = 1.50 "wg
   20 ft of 11-in. duct _____
   elbow loss _____
   contraction loss _____
   20 ft of 8-in. duct _____

9. What would the static pressure be just before the fan? (Did you get your − or + right?) Again, recall how you found the SP in question 9 in Chapter 11 with no losses was exactly equal to -VP; thus, absolute the value here should be that VP plus the above losses.

   SP = _____ "wg

REFERENCES

1. *Industrial Ventilation: A Manual of Recommended Practice*, 24th ed, American Conference of Governmental Industrial Hygienists, Cincinnati, OH, 2001.
2. Chapter 29 of the U.S. Code of Federal Regulations, Part Number 1910, Occupational Safety and Health Standards.
3. ASHRAE Handbook, *Heating, Ventilating, and Air Conditioning Applications*, Inch-Pound ed. ASHRAE, Inc., Atlanta, GA, pp. 42.16, 1991 and pp. 46.8, 1999.
4. Iqbal, M. A., Wilson, T. K., and Thomas, R. J., *The Control of Noise in Ventilation Systems*, E.&F.N., Spon, London, 1977.
5. International Mechanical Code, International Code Council, Falls Church, VA, 2003, Table 510.8.
6. Vatavuk, W. M., *Estimating Costs of Air Pollution Control*, Lewis Publ., Chelsea, MI, 1990.
7. U.S. EPA: Hoods, Ductwork, and Stacks, In *OAQPS Control Cost Manual*, 5th ed., United States Environmental Protection Agency, Office of Air Quality Planning and Standards. EPA 453/B-96-001, chap. 10, February 1996, Available at http://www.p2pays.org/ref/10/09849.pdf.
8. Guffy, S. E. and Booth, D. W., Comparison of pitot traverses taken at varying distances downstream of obstructions, *Am. Ind. Hyg. Assoc. J*, 60(2), 165–174, 1999.
9. Perry, R. H. and Chilton, C. H., *Chemical Engineer's Handbook*, 7th ed., McGraw-Hill Book, New York, 1997.

15 Exhaust Air Cleaners and Stacks

The learning goals of this chapter:

- ✍ Be familiar with the generic types of air cleaners.
- ✍ Know why each type either is or is not applicable to gaseous vs. particulate contaminants.
- ☐ The head loss in most air cleaners is still proportional to VP or V^2.
- ✍ Understand reentrainment.
- ✍ Be familiar with typically acceptable exhaust discharge velocities.
- ✍ Know why a rain cap is in conflict with the goals of a LEV exhaust stack.

If the chemical that was controlled by exhaust ventilation to protect employees goes out the end of the ventilation duct (called the exhaust stack), it not only remains a health hazard and an environmental pollutant, it is also a potential lost resource that had value when it was purchased and may still have value if it were collected and recycled. More than one resource can be lost out of the exhaust stack:

- Chemicals that were either purchased, manufactured, or refined on the site. The monetary value of chemicals varies with the nature of the air contaminant. Collecting and recycling chemicals means using one or more of the air cleaners that are discussed in this chapter.
- Heat that was probably purchased as electricity, coal, etc. Thermal energy can have value if either the heat or cooling capacity can be reliably separated from the contaminant and reused. The cost of thermal energy will be discussed in Chapter 17.
- Excess velocity (kinetic energy) generated by the fan. In comparison to chemicals and heat, air velocity usually has the least value but it is the most easily recovered via an evasé, as was discussed in Chapter 14 and will be touched on again later in this chapter under "Exhaust stacks."

I. AIR CLEANING NEEDS AND OPTIONS

1. The following industrial processes are typical of those whose air emissions need to be cleaned[1–3]:

Electricity:	Some dust control is needed during coal mining, but most controls are directed toward combustion products (primarily SO_2) and by-products such as fly ash.
Food processing:	Dust from grain elevators, feed mills, odors from livestock production (hogs or dairy cattle), food products (e.g., odors from cheese factories), food by-products (e.g., odors and dusts from rendering plants).
Metals:	Dusts during mining, fumes during smelting, binder gases, nonmetal dusts and metal fumes in foundries, metal dusts during machining, and gases, vapors, and mists during plating and finishing.

Organic chemicals: Petrochemical emission controls vary heavily with the chemical compound, the level of control required by the "air conservation district" into which they are released, and the magnitude of the use.

Pharmaceuticals: Mostly particulate aerosols that need to be held to low emission rates due to their tailored health effects.

Polymers: Vapors during polymerization and fumes during molding. Also complex organic vapors from resins, composites, inks, graphic arts, and printing.

Rubber: Smoke, fumes, and polycyclic aromatics as during curing or other high temperature manufacturing processes.

Wood and paper: Highly visible dusts from woodworking or invisible gases (primarily odoriferous mercaptans) and vapors from paper mills.

No single air cleaner can remove all of the different kinds of contaminants that can be generated. The type of air cleaner that is best suited to each process must be compatible with the physical nature of airborne emissions in general (including those specifically listed above). The following generic types of air cleaners are grouped into those that can only remove particulate contaminants, those that can only remove gaseous contaminants, and those that perform fairly well at removing both particulate and gaseous contaminants.

2. The following air cleaners remove only particulate contaminants because their removal mechanisms all rely in one way or another upon physical forces, primarily (except for ESPs) upon inertial separation.
   a. Filters are efficient and can be sized to match a wide range of dust concentration, chemical property, and air flow conditions. The focus of discussion herein will be on fabric filters arrayed within a structure called a "bag house," but cartridge, rigid ceramic, and granular bed filters are also used.[3,4]
   b. Cyclones (a very recognizable example of an inertial dust collector) work well for high concentrations of large particles ($>10\ \mu m$) but do not remove small particles ($<1\ \mu m$) very well.[3]
   c. Electrostatic precipitators (ESP) are very efficient for small particles, have a low pressure drop, scale efficiently to large air volumes, but are costly to install and to operate. These features make them good for exhaust stacks from combustion fueled electric power plants.[3,5]
3. The following air cleaners remove only gaseous contaminants. Note how each gas cleaner has a different reason for not being able to handle particulates.
   a. Absorbers (notably the "packed tower" or "packed scrubber") circulate a liquid (such as water) over a matrix of solid bodies to absorb the contaminant from the air passing through. The deposition of particulate matter onto the packing material would eventually clog the passageways around the solid bodies.[3,6]
   b. Adsorbers (typically activated carbon or a molecular sieve) are expensive but sometimes used where a water-based absorber is not feasible (e.g., due to temperature extremes or a lack of solubility). Adsorption relies on diffusion to carry gaseous molecules into micropores on the surface of a solid matrix, but particles $>0.1\ \mu m$ do not diffuse well.[3,7,8]
   c. Oxidizers (catalytic, thermal, or direct combustion) break down the chemical into safe or at least safer emissions. Oxidizers are not applicable to many particulate contaminants because they are either already oxides, chemicals that oxidize too slowly or not at all, and/or chemicals whose oxidation would not change their toxicity or visibility.[3,9]
   d. Refrigerated condensers lower the temperature of the contaminated airstream below the vapor's saturation temperature to condense and remove them as a liquid.

Condensation will do nothing for aerosols. Condensation is used as at least a partial control in several industries; however, it will not be discussed further herein.[10–12]

4. The following wet collectors (or "scrubbers") are able to remove both particulate and gaseous contaminants. Although they will generally not clean any one contaminant as well as one of the above air cleaners, they are more resistant to clogging than absorbers or adsorbers.
   a. Spray towers are similar to the packed scrubbers mentioned above but do not have their solid matrix packing, making them less susceptible to clogging by the deposition of particulate matter.[1,3,6,10]
   b. Wet centrifugal collectors and wet dynamic precipitators accelerate the absorbing liquid and/or the air to increase the momentum with which they make contact which increases the efficiency of absorption of both gaseous and particulate contaminants without risking clogging.[1,3,6,10]
   c. Orifice and venturi collectors increase the momentum of contact by accelerating the air through restricted passages instead of accelerating the liquid. They operate at the highest air velocities.[1,3,6,10]

Various biological filtration, absorption, and oxidation schemes for gas purification are relatively recent developments that show promise for their low cost but are not included within this overview.

II. AIR CLEANER SELECTION CRITERIA

Air cleaner selection criteria typically include the physical qualities of both the air and its contaminant, the contaminant's chemical qualities, the efficiency of contaminant removal as dictated by the level of emissions allowed, the resistance to air flow or pressure drop through the air cleaner, and of course costs.

1. The following four physical qualities are important criteria in air cleaner selection:
   a. The volume flow rate of the contaminated air (usually in thousands of ft^3/min or m^3/sec).
   b. The concentration of the contaminant (in the usual units of ppm for gaseous and mg/m^3 for particulate contaminants).[a]
   c. The air's temperature and humidity, and the presence of other condensable vapors that can interfere with the normally dry operation of some air cleaners especially in cold winter climates.
   d. Whether the contaminant is particulate, gaseous, or both; and if the contaminant is particulate, the particle size, shape (e.g., fibers such as lint), stickiness, and abrasiveness. (Recall the particle sizes shown in Figure 3.5.)
2. The chemical qualities of the contaminant that affect air cleaner selection include its flammability, reactivity and possible explosiveness, its corrosivity (affecting the air cleaner's construction materials), and its toxicity (affecting the handling and disposal of the collected contaminant as well as any internal system maintenance).
3. The efficiency of removal is dictated by the air quality and emission criteria written into legal regulations. Whatever the air emission standards are now, history shows that they are likely to become more stringent in the future. The contaminant concentration times the volume of the air flow and the efficiency of removal determine the mass of collected contaminant that must eventually be handled.

[a] The term grain/ft^3 is an odd but once common measure of airborne concentration especially in air emissions. One grain is 1/7000 part of one pound; thus, 1 grain/ft^3 = 2288 mg/m^3.

TABLE 15.1
Flow and Compatible Contaminant Characteristics of Air Cleaners

		Typical h_{Loss} ("wg)[1,13]	Contaminant(s) Each Is Well-Suited to Remove
	(1) Electrostatic precipitator	<1	Solids
	(2) Fabric filter (bag house)	2–5	Solids
	(3) Cyclone[a]	1–4	Solids + liquids
Wet scrubbers	(4) Spray tower	1–2	Solids + liquids + gases
Wet scrubbers	(5) Wet centrifugal collector	2–6	Solids + liquids + gases
Wet scrubbers	(6) Venturi scrubber	5–10	Solids + liquids + gases
	(7) Packed tower	1–4	Liquids + gases
	(8) Carbon bed absorber	1–10	Gases
	(9) Oxidizers[b]		

[a] Add about 2 "wg to this range for a high efficiency cyclone (designed to remove smaller particles). Particles and liquids at the same time (or even sticky particles by themselves) can easily clog an otherwise dry cyclone.
[b] Catalytic oxidizers, thermal oxidizers, and direct combustors can remove any *combustible* contaminant.

4. Resistance to air flow is a significant determinant of operating energy costs. Pressure losses differ among types of air cleaners and between manufacturers of any one type as summarized in Table 15.1, but the loss for most air cleaners is proportional to VP and therefore to Q^2 (as introduced in Chapter 11, first stated quantitatively as Equation 14.5, and restated below in Equation 15.1).

$$\boxed{\text{head}_{\text{Loss}} = \text{constant} \times \text{VP} = \text{constant} \times V^2 = \text{constant} \times Q^2} \quad (15.1)$$

   Exceptions to Equation 15.1 include low velocity air cleaners such as spray towers that operate internally at or near laminar flow conditions where losses are more nearly proportional to velocity.
5. Cost is always a selection criterion at some point. A cost analysis should include both purchase costs (commonly called capital costs) and operating costs. The latter costs include electricity (that tend to be proportional to the pressure drop through the air cleaner), maintenance (broadly a function of system complexity), and the handling of the collected contaminant (recycling can be a positive economic benefit, but disposal costs can dominate a system's operating costs). Not only can some collected chemicals be recycled, but thermal energy (the heating or cooling value of the airstream) can also be recovered and used either within a process or to reduce the cost to heat or cool outside make-up air. The *Manual of Industrial Ventilation*'s Section 4.8 provides some general guidance on costs.[1] Vatavuk (1990) provides more quantitative cost data developed to support EPA regulatory decisions.[14,15] Vendors and contractors are the ultimate sources of costs.

III. PARTICULATE AEROSOL COLLECTORS

1. Fabric Filter Collectors

A fabric filter acts to separate and retain the particulate contaminant as the air passes through the filter.[1,3,4,13] Common filter fabrics listed in the *Manual of Industrial Ventilation*'s Table 4.1 include cotton (for emission temperatures up to 180°F), acrylics and Teflon (for their chemical resistance), glass fiber (for temperatures up to about 500°F), polyester, and various specialty fabrics. Fabric filters can be efficient at removing particles between 10 μm and 0.1 μm. The majority of the particles that are removed by air cleaning filters are deposited onto the surface of the filter and onto particles that have preceded them to that surface. The built-up dust layer on the surface of a filter (commonly called a cake or filter cake) increases the resistance to air flow and must be removed periodically from the filter either by shaking, by a reverse air jet, or by a reverse pressure pulse (see *Manual of Industrial Ventilation* Figure 4.7 and Figure 4.9).[1]

Three configurations of air cleaning filters are described below: tube type and envelope type fabric collectors and rigid HEPA and ULPA filters.

a. Fabric filters can be made in the shape of a tube. The direction of flow through tube-type collectors (shown schematically in Figure 15.1) is from the contaminated air inside the tube to the clean air outside, similar to a home vacuum cleaner. Each tube is typically 4 to 12 in. (0.1 to 0.3 m) in diameter, 3 to 30 ft (1 to 10 m) long, and open at the bottom end like an upside down sock. The open end of each tube is attached to a cell plate that blocks the flow around the tubes and separates the dirty from the clean air. A bag house will contain an array of such tubes. The bottoms (entrance) of two rows are shown in Figure 15.1, but for clarity only one row of actual bags is shown.

   The resistance to air flow through the filter causes the pressure inside the tube to be greater than the pressure outside the tube (but still inside the housing) and keeps each tube inflated. Tube filters are usually cleaned by shaking the mechanical supports on which the tubes are suspended (a reverse air jet can be used if some sort of frame is inserted inside each tube to keep the jet from collapsing it inward). Tube-type collectors tend to be most cost-effective for small and intermittent operations.

b. A fabric filter can also be made in the shape of an envelope (as for postal letters). The direction of flow through most envelope-type collectors is from the contaminated air outside the envelope to the clean air inside. Resistance to inward flow creates a lower or negative air pressure inside the tube, requiring an internal support to keep the filter from

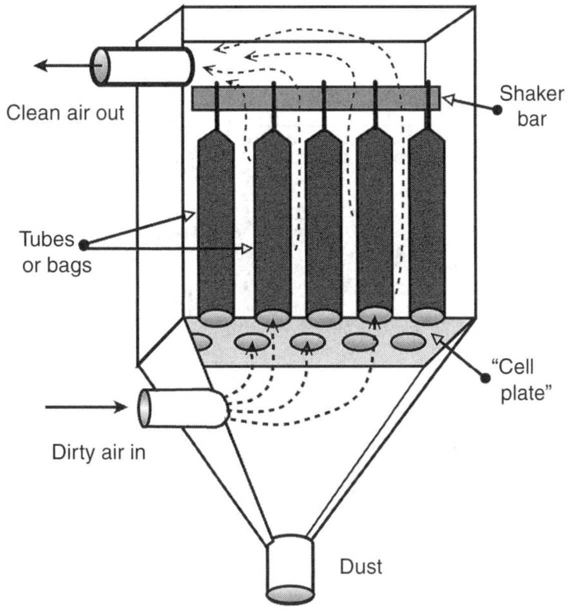

FIGURE 15.1 Schematic depiction of the inside of a bag house with tube collectors.

collapsing (shown schematically in the cut-away filter on the left in Figure 15.2). Envelopes can be round but are typically elongated (similar to an envelope), open at the top (again similar to an envelope), perhaps only a few inches wide in their narrow dimension (shown "across" the page in Figure 15.2), up to 30 in. or 1 m long (going into the page), and perhaps twice as tall. Envelopes that are used for intermittent operations can be cleaned by a shaker or a pulse of reverse air flow, but envelopes that are used for continuous operations work best with a localized jet of air flowing in reverse.

c. HEPA and ULPA are rigid filters that are built into a frame or cartridge through which air can be ducted. HEPA (high efficiency particulate aerosol) filters are "at least 99.97% efficient against a 0.3 μm diameter aerosol." ULPA or "Ultra filters" are even more efficient, with a minimum particle collection efficiency of 99.999% for particulate diameters between 0.1 and 0.2 μm. Neither type of rigid filter is cleanable; thus, when they become too heavily loaded, they must be replaced. Such high efficiency filters are expensive but vital to keep highly toxic or pathogenic aerosols from leaving an exhaust system or most aerosols from entering a clean room. To make their use more economically feasible, such filters are usually used as the final air cleaner in series behind a prefilter (cheaper, perhaps cleanable, but less efficient) or a high-efficiency cyclone as will be shown in Figure 15.5. HEPA filters are also used in small, self-contained units called a unit collector (see Figure 15.3). In many cases, removing more than 99.9% of the contaminant will allow the exhaust from such a unit collector to be recirculated back into the workroom.

Other filter terminology:

d. Filtration velocity is a fundamental parameter for designing filter air cleaners. The filtration velocity is calculated as the ratio of the total air flow (Q) passing through the air cleaner divided by the air cleaner's total filter area (A). Dividing units of Q in ft^3/min by A in ft^2 yields units of velocity in fpm (or m^3/sec divided by m^2 yields m/sec). A manufacturer will recommend a specific range of filtration velocities for their filter material. Collectively, most filtration velocities range from 1 to 12 fpm (0.3

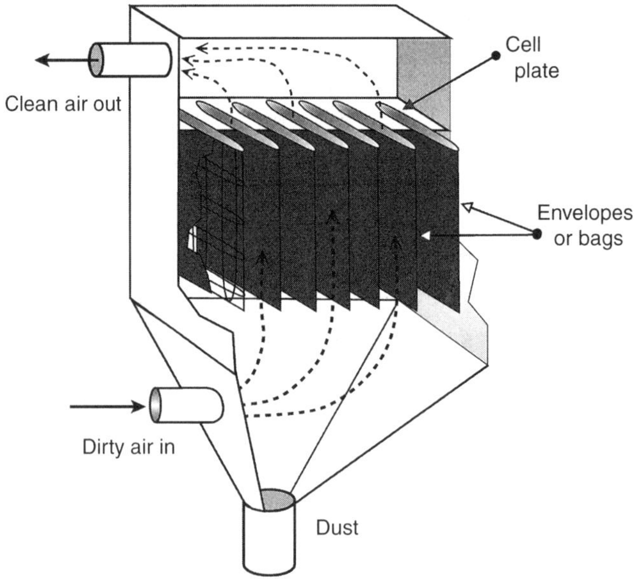

FIGURE 15.2 Schematic depiction of the inside of a bag house with envelope collectors.

FIGURE 15.3 A mobile, commercial dust collector is shown collecting welding fumes. Such unit collectors contain both a cartridge filter and a fan. The filtered air from such mobile units is released back into the room air. Courtesy of United Air Specialists, Inc.

to 3.7 m/min) but are typically about 3 fpm for woven media and about 8 fpm for felted media. The size of the air cleaner is determined by matching the user's needs to the filtration velocity recommended for a given filter material by the manufacturer.

e. Permeability in a filter is the flow rate of air that will pass through a unit area of filter at a specified pressure drop, such as ft^3/min per square foot of filter at 0.5 "wg. Since ft^3/min per ft^2 = fpm is a velocity, the term permeability is in effect a particular filtration velocity. Differences in permeability can affect the selection of the filter material. However, the cake of dust that accumulates on a filter with use will cause the permeability to go down while collection efficiency goes up (barring any failure of the filter itself), thus decreasing the importance of permeability in filter selection. The term filter drag is the pressure drop per ft^3/min per square foot of filter ("wg/fpm), in a way, the inverse of permeability.

2. Cyclone Collectors

A cyclone, such as depicted in Figure 15.4, operates on the principle that air molecules have less inertia than the suspended particles.[1,3,13] Thus, particles respond more slowly to the centrifugal acceleration created by the circular flow of air within a cyclone than does the air, causing them to move outward until they contact the wall that stops and separates them from the air flow. Cyclones are basically self-cleaning. Since the circulating particles strike the wall at a relatively flat angle, they are relatively free to either settle or agglomerate until they are large enough to settle by gravity to the bottom of the cyclone, from which they can be removed periodically.

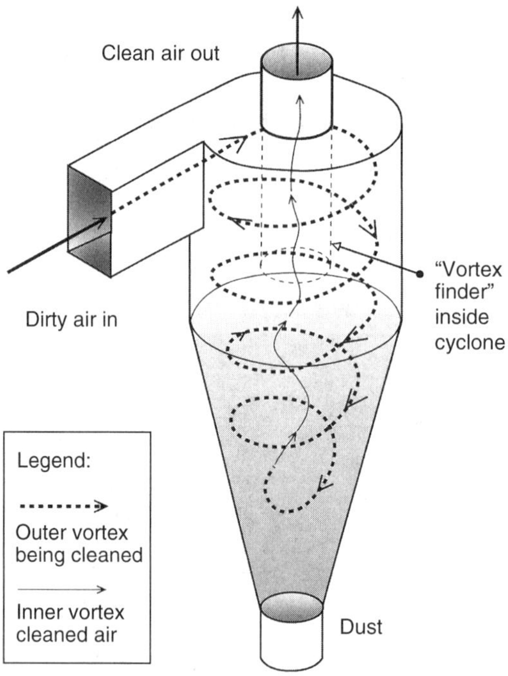

FIGURE 15.4 Depiction of a cyclone air cleaner.

While Equation 15.2 cannot be used by itself to predict the particle diameter that will be removed with a 50% efficiency, it can be used to anticipate the effects of changing a cyclone's operating conditions on that $d_{\text{cut-off}}$. A cyclone's physical size, its shape (taper angle and length of its vortex finder), and the air flow rate all determine the characteristic number of revolutions of the air within the cyclone (the N within Equation 15.2).[16] The group of parameters $\rho d^2/18\eta$ (somewhat hidden within Equation 15.2) is related to Stokes law as explained ca. Equation 3.18.

$$d_{\text{cut-off}} = \sqrt{9\eta D_{\text{inlet}}/2\pi N V_{\text{inlet}}\rho} \tag{15.2}$$

where

$d_{\text{cut-off}}$ = the particle aerodynamic diameter at which 50% of the particles are collected by the cyclone. The particle collection efficiency is $\geq$50% for $d_{\text{aero}} \geq d_{\text{cut-off}}$.

η = the air's dynamic on absolute viscosity.

D_{inlet} = the cyclone's inlet diameter.

V_{inlet} = the cyclone's inlet air velocity.

N = the apparent number of revolutions the air makes within the cyclone before it exits. N might be around 3 to 5 for a large air cleaning cyclone.

ρ = the aerosol particle's density.

The $d_{\text{cut-off}}$ for most cyclones is about 10 to 20 μm. Thus, they are only efficient at removing visible particles and those that could otherwise fall out of the air quickly enough to cause local damage or at least be a local nuisance. They do not collect the small diameter (less than 10 μm) respirable particles. Special high-efficiency cyclones are designed to have a smaller cut-off diameter further into the respirable range described in Section IV in Chapter 3, but even high-efficiency cyclones will not remove respirable-sized particles as well as fabric filters can. On the other hand, a cyclone when placed in-line preceding a fabric or HEPA filter, such as shown in Figure 15.5, will remove the course particles and extend the service life of those filters.

FIGURE 15.5 A cyclone (near center) preceding a dust collector with horizontal cartridge air filters (on the left). Courtesy of United Air Specialists, Inc.

3. Electrostatic Precipitators

The principle of an electrostatic precipitator (ESP) can be thought of in three steps.[1,5,13] First, the ESP must generate small air ions that will attach themselves to contaminant dust particles.[b] Second, the dust particles that become electrostatically charged are in turn attracted to and collected on a plate or grid of the opposite charge. Third, the collected particles must be periodically removed and either recycled or disposed. This flow of charged particles constitutes an electric current. The product of current times voltage comprises the electric power consumed by the ESP. The cost of electricity is a major operating expense of an ESP. (It helps if the user is a power plant that makes its own electricity.)

ESPs are very good at collecting submicron-sized particles. ESPs also have very low pressure losses, making them advantageous for use on combustion or other hot gas exhaust stacks that rely on buoyancy to force the air to flow. A natural draft chimney is the second major exception to this book's focus on isothermal plumes. As stated in Chapter 11, potential energy is not usually important; however, it is relied upon to create the pressure or "head" difference that drives the flow of hot air up a chimney. The potential head of a column of hot exhaust gas rising up a chimney is stated in Equation 15.3a in terms of the height H in *feet of air*.

$$h_{\text{buoyancy}} = \frac{\rho_{\text{hot}} - \rho_{\text{ambient}}}{\rho_{\text{hot}}} \times H(\text{feet}) = \left(1 - \frac{\rho_{\text{ambient}}}{\rho_{\text{hot}}}\right) \times H(\text{feet}) \tag{15.3a}$$

Two other equations are needed to convert Equation 15.3a into something more useful. One is Equation 2.24a to find the ratio of the density of hot air to ambient air related to buoyancy.

$$\frac{\rho_{\text{ambient}}}{\rho_{\text{hot}}} = \frac{(^{\circ}C_{\text{hot}} + 273)}{(^{\circ}C_{\text{ambient}} + 273)} = \frac{(^{\circ}F_{\text{hot}} + 460)}{(^{\circ}F_{\text{ambient}} + 460)} \qquad \text{(Equation 2.24a)}$$

The other is the equivalence of 69.27 ft of air to 1 in. of water, found in Equation 11.19. Together, they yield Equation 15.3b, giving the potential pressure difference in terms of "wg that can be

[b] In general, nonmetallic particles become positively (+) charged more easily, and metallic particles tend to become negatively (−) charged more easily.

created by a natural draft chimney equivalent to what in Chapter 19 will be called a fan static pressure.

$$h_{\text{buoyancy}}('' \text{ wg}) = \left(\frac{°F_{\text{hot}} + 460}{°F_{\text{amb.}} + 460} - 1\right) \times \frac{H(\text{feet})}{69.271} = \left(\frac{°C_{\text{hot}} + 273}{°C_{\text{amb.}} + 273} - 1\right) \times \frac{H(\text{feet})}{69.271} \quad (15.3b)$$

Equation 15.3b predicts that a substantial natural draft (pressure difference) can be created in a tall chimney to move hot air or exhaust gases. In practice, the actual stack temperature is not easy to predict. Moreover, relying on such a natural draft in short chimneys can be dangerous if mechanical fans elsewhere inside the same building are exhausting air without an easy route for adequate make-up air to flow back into the building. Under such circumstances, the mechanical fans can create a larger negative pressure within the building than the natural draft up the chimney, and ambient air will be drawn down the chimney not only preventing the combustion products from escaping but in fact dispersing them within the building. Such problems can occur in gas-fired water heaters, autoclave steam generators, or building space heaters with short and/or poorly insulated chimneys.

IV. AEROSOL AND GAS COLLECTORS

A "wet scrubber" is a generic term for an air cleaner that relies primarily on high-velocity contact between the air and a liquid for rapid absorption (cf., packed scrubbers and adsorbers that depend upon a large surface area). They encompass cleaver designs that allow them to remove both gaseous contaminants and aerosols without clogging. The efficiency of wet scrubbers also depends upon the time (duration) of that contact. Water is the most common scrubbing liquid. Water is cheap, a good solute for many gaseous compounds (especially with some pH manipulation), and it reduces explosion and fire hazards, but it can be corrosive and it has a relatively high freezing temperature. The following wet scrubbers are listed in order of increasing cleaning efficiency, pressure losses, and electrical energy operating costs.[1,6,13]

1. A spray tower is generally a tall, vertical cylinder with one or more tiers of sparging bars (pipes with spray nozzles), as depicted in Figure 15.6 (a similar spray chamber is made to be horizontal). The primary contact and locus of contaminant removal occurs between the sprayed water and the air. Once the spray turns into downward falling water, it has less kinetic energy and absorption power than does the higher velocity spray. The relatively unrestricted air flow pathway creates only a low 0.5 to 1.5'' wg pressure drop.
2. Tray scrubbers are a functional variation of the spray tower where the liquid flows down from tray to tray stacked within a tower that looks on the outside much the same as a spray tower. Tray scrubbers themselves have many variations. The definition of the "number of transfer units (NTU in Equation 15.4) is derived from chemical reaction mass-rate theory and is used in the design of this and other absorbers.

$$\text{NTU} = \text{natural log}\left[\frac{\text{the concentration IN (entering) an air cleaner}}{\text{the concentration OUT (leaving) an air cleaner}}\right] \quad (15.4)$$

3. Vane scrubbers are somewhat like tray scrubbers but use a combination of impaction and centrifugal forces produced by venturi and wheel elements to gain good absorption.
4. Wet centrifugal collectors use a variety of rotating schemes to induce higher speed contact. Each device has a spinning part to create the centrifugal forces on the liquid, typically creating a spray or at least a higher speed than liquid falling.

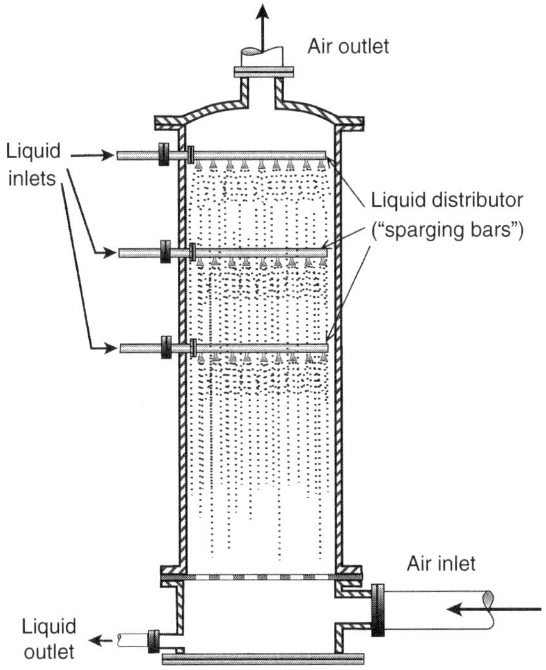

FIGURE 15.6 A spray tower (the simplest wet scrubber).

The liquid spray or droplets absorb the contaminants before being deposited onto the walls of the cyclone, and the contaminated liquid is subsequently removed through a drain at the bottom. The air flowing vertically in most (but not all) of these devices generates a pressure drop similar to a conventional cyclone (and requires the same type of fan).

5. Wet dynamic precipitators use designs without moving parts to impinge the airstream upon a liquid that is either quiescent or flowing slowly. In the scrubber depicted in Figure 15.7, a sheet of liquid flows down a wall that forms a venturi where the main contact occurs with much turbulence. The scrubber depicted in Figure 15.8 is analogous to a large liquid impinger. Such collectors require a larger mechanical fan to create a comparatively high air speed.
6. A venturi scrubber or collector injects a high velocity spray at or into the throat of a venturi. Two versions are depicted in Figure 15.9 and Figure 15.10. The high velocity air passing through the throat maximizes its kinetic energy at the point of contact with the liquid, although the short contact time in the throat limits its use to very soluble gases. A venturi scrubber often must be followed by a centrifugal collector to recover the resulting contaminated mist suspended in the airstream before it is released into the atmosphere.

A key element of any collection scheme involving water and particulates is a provision to deal with the liquid contaminated sludge both within the plant operation and later as a potentially hazardous waste.

V. GAS AND VAPOR COLLECTORS

1. A "packed scrubber" relies primarily upon a large surface area rather than high momentum to absorb gaseous contaminants efficiently. A large surface area is created by filling a tall chamber with a bed of oddly shaped inert solids (the packing material),

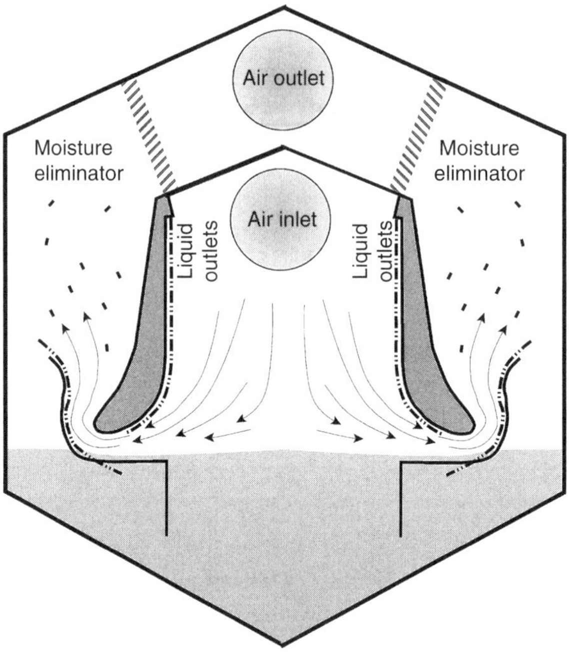

FIGURE 15.7 One version of a wet dynamic precipitator.

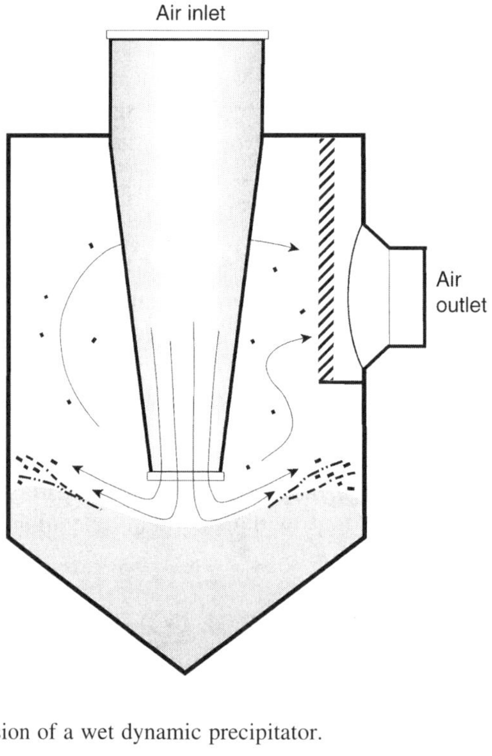

FIGURE 15.8 Another version of a wet dynamic precipitator.

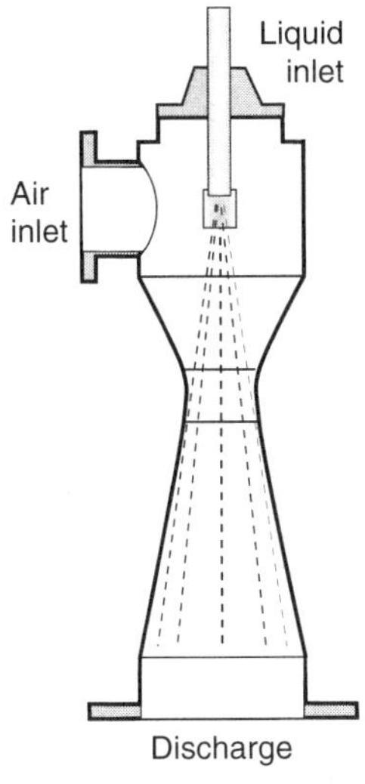

FIGURE 15.9 One version of a venturi scrubber.

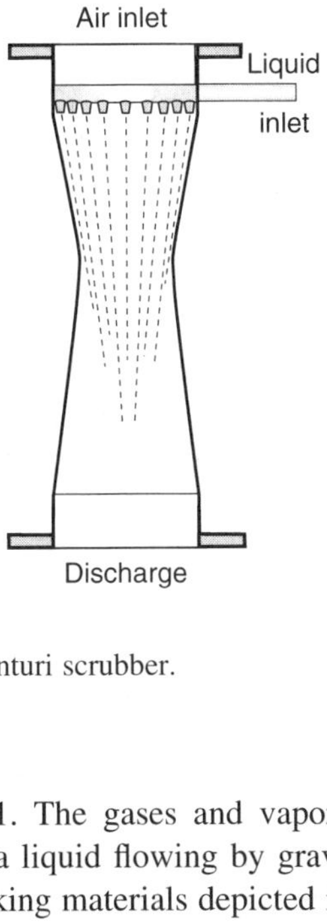

FIGURE 15.10 Another version of a venturi scrubber.

as depicted in Figure 15.11. The gases and vapors flowing upward in the packed scrubber are absorbed into a liquid flowing by gravity over and through the packing material. The shapes of packing materials depicted in Figure 15.12 have evolved over the years, progressing from crude crushed stone or spheres to Raschig rings (a), Berl saddles (b), Pall rings (c), and Intalox saddles (d).[17] Each shape increases the wetted surface area within a tower without creating a high resistance to air flow. The packing material might be a plastic, ceramic, or metal that is chemically inert to the pollutant. The reduced air spaces between the packing do increase the air's velocity, and if the velocity gets too high, the liquid trying to flow past, over, and through the packing can be pushed back up the tower and out of the top. Such a tower is then said to have flooded. High particulate concentrations can also clog a packed tower, reducing the

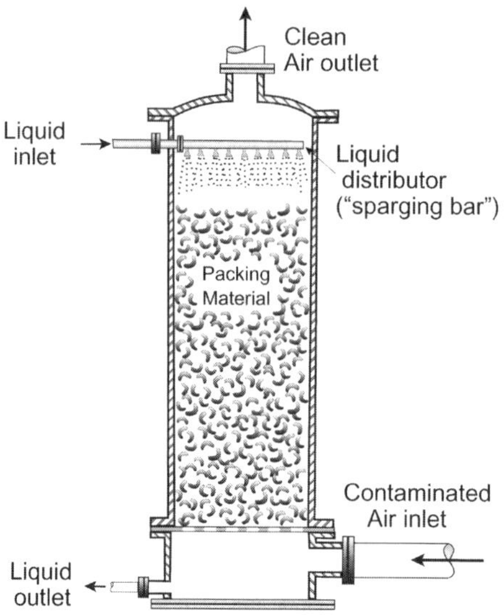

FIGURE 15.11 Depiction of a packed scrubber.

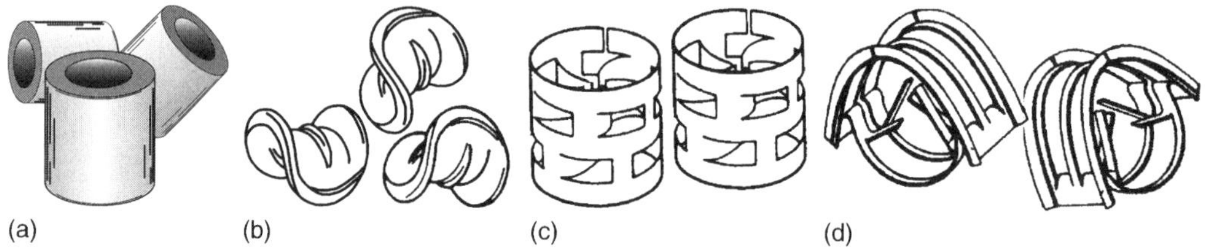

FIGURE 15.12 Four examples of packing media shapes (two rings and two saddles) arranged in roughly chronological order and in order of increasing absorption efficiency. (*Source:* From McCabe, W.L. et al., *Unit Operations of Chemical Engineering*, McGraw-Hill, New York, 1993. With permission.)

open space through which both the air and fluid must pass, decreasing absorption efficiency, and increasing the scrubber's susceptibility to flooding. Nonetheless, packed scrubbers are a common gas emission control because of their comparatively high efficiency and low operating costs.

2. Adsorbers collect gaseous contaminants onto a solid surface or, more precisely, into micropores within solid surfaces. Adsorption by activated charcoal, a molecular sieve (e.g., synthetic hydrophobic zeolite), or a polymer is particularly well suited to control the emission of low concentrations of volatile organic chemicals. A small amount of adsorbent can be single-use (disposable), but for adsorption to be economically feasible on a large scale it must be reusable. Adsorbents are typically regenerated thermally, although vacuum and pressure regeneration are also possible. Particulate contaminants can plug an adsorbent's micropores and thereby reduce its adsorptive capacity. Moreover, high loadings of particles can potentially clog a bed of adsorbers (stop the air flow). Owing to their high capital and operating costs, adsorbers are generally considered to be specialized collectors of volatile organic chemicals from high value processes.[7,8]

3. Oxidizers do not remove the mass of contaminants but reduce their toxicity by some thermal and/or oxidative chemical reaction. Organic compounds are by far the major contaminant treated by oxidizers. Three subgroups of commercially manufactured oxidizers are listed below in order of decreasing operating temperatures. The operating expense of oxidizers can be as high as adsorbers, although adding an ancillary heat recovery component will reduce an oxidizer's operating cost. However, they can be justified to control highly toxic emissions that are difficult to deal in any other way.
   a. Direct combustors simply burn contaminants. The keys to good contaminant reduction by direct combustion are a high temperature and a long retention time at that temperature. An exhaust stream initially above its LFL will sustain combustion, but an auxiliary fuel must typically be added to most emissions, which tends to make a direct combustor the most expensive oxidizer to operate.[9]

      Do not confuse a direct combustor with a flare used to control waste volatile organic chemical vapor emissions from settings such as oil refineries. Such flares were primarily designed to deal with extraordinary events (e.g., processes in transition or upsets such as a loss of cooling water) rather than to be an efficient air cleaner combustor.[10,18]
   b. Recall from Chapter 4 that a chemical's LFL (the concentration at which it will burn) decreases with increasing temperature. By using an external fuel to heat a potentially combustible contaminant stream initially below its LFL to 1000–1500°F, a thermal oxidizer or afterburner can reduce a contaminant's LFL to the point where its existing concentration can sustain combustion.[9]
   c. Catalytic oxidizers use a catalyst such as platinum to allow oxidation to occur at temperatures as low as 700–900°F to remove potentially combustible contaminants initially well below their LFL. The catalyst can be arrayed in either a fixed bed (monolithic sheets or a packed-bed of pellets) or a fluidized bed (in which the air flow suspends and mixes the pellets of catalyst; a liquid can also suspend such a bed in other settings). Fluidized beds are more tolerant of particles in the airstream than are fixed beds, but catalysts can also be poisoned by halogens, sulfur, phosphorous, or heavy metal fumes. Recently developed poison-tolerant catalysts include chromia/a-lumina, cobalt oxide, and copper oxide/manganese oxide.[9]

VI. EXHAUST STACKS AND REENTRAINMENT

An exhaust stack can comprise the same (or similar) duct components as those used elsewhere in the system. Thus, the pressure losses from friction and turbulence in the exhaust duct *after* a fan are calculated using the exact same methods as those previously discussed in Chapter 14. Duct losses in the outlet stack are generally smaller than duct losses before the fan because the exhaust duct is usually shorter and has fewer elbows than the ducting before the fan.

Good exhaust dispersion is needed both to avoid reentraining the exhaust plume back into the building's fresh air inlet and to avoid unacceptable health or environmental effects where the plume does (or might) reach the ground somewhere downwind. The fresh air coming into a building (called make-up air) will be discussed in Section III in Chapter 17. Adequate dispersion of exhaust emissions depends upon the exhaust outlet's location (placement), its stack height, and its discharge velocity. The following four subsections summarize design guidelines for exhaust placement, its height, discharge velocity, and rain protection.

1. Figure 15.13 identifies an attached wake and three recirculation zones around a building in which the exhaust outlet should not be placed. A recirculation zone is similar to a big eddy that tends to stay in one place. A proportion of any contaminant that is released into

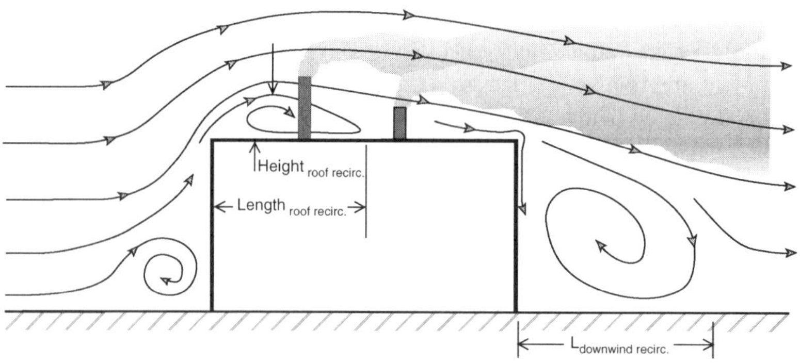

FIGURE 15.13 Depiction of airflow near a building. The depiction of smoke coming out of the stacks is for learning purposes only; such visual emissions would not be permitted on a routine basis. (*Source*: From Figure 5.28, *Industrial Ventilation: A Manual of Recommended Practice*, 24th ed., American Conference of Governmental Industrial Hygienists, Cincinnati, 2001. With permission.)

a recirculation zone will be conveyed laterally along a building and can even be conveyed upwind.

a. A small recirculation cavity is produced low on the upwind side of a building (the side from which the wind is coming). Any exhaust emitted within this zone would have a high likelihood of being reentrained into a building's fresh air inlet, door, or an openable window that might be located within this zone.
b. A modest but important recirculation zone usually exists beginning just behind the leading edge of a flat roof and extending downwind a distance predictable via Equation 15.6. Avoid locating an exhaust outlet within the roof recirculation zone. In principle, the second stack in Figure 15.13 is outside of that recirculation zone; however, what happens when the wind changes direction? One should design any roof-mounted exhaust stack as if it were in a roof recirculation zone.
c. A larger recirculation zone affects most of the downwind side of a building comprising the initial portion of a building's wake. In addition to this recirculating flow, the air flowing downwind off a flat roof will tend to cling to the side of the building (much like the attached flow of the air pushed across a flat surface towards the pull hood described in Section IX in Chapter 13). This attached flow can carry any contaminants that are emitted low onto the roof, back down towards the ground or the base of a building and cause reentrainment into any downwind air inlet along the way. Individual fan coil air conditioning units are particularly susceptible to this problem because they can be installed on every floor and every side of a building (to be described in Chapter 20).

2. One should avoid locating an exhaust outlet in either the upwind or downwind recirculation zones. A wind-rose will indicate the prevailing wind direction and speed; however, because wind is generally so variable, an exhaust outlet cannot be located on any side of a building without compromising dispersion and risking reentrainment. That leaves locating the outlet on the roof but designing to avoid the roof recirculation zone. In general, the higher an exhaust outlet, the better; but higher means more construction costs and a visually less attractive building. ACGIH *Manuals* before 1995 said simply to make the height of the outlet 1.3 to 2 times the height of the building. The current *Manual of Industrial Ventilation* (Section 5.16) advocates using a scaling coefficient ($R_{scaling}$) developed by Wilson.[19,20] The scaling coefficient calculated via Equation 15.5 puts an emphasis on the smaller of the building's height (H) and width (W) dimensions, with an added caveat that the larger dimension used in Equation 15.5 should never be more than

eight times the smaller dimension even when the ratio of the building's two dimensions exceeds that value. If one cannot or chooses not to use the prevailing wind direction for a building with a rectangular floor plan, one should use the widest side of the building in Equation 15.5.

$$R_{scaling} = (\text{Building Dimension}_{smaller})^{0.67} + (\text{Building Dimension}_{larger})^{0.33} \quad (15.5)$$

The resulting scaling coefficient can then be used to predict the following design parameters for a building with a flat roof. Notice in the last design parameter that the scaling coefficient is also the length of the downwind wake.

$$\text{Height}_{roof\ recirc.} = 0.22\ R_{scaling} \quad (15.6a)$$

= the maximum height of the roof turbulent recirculation region as indicated in Figure 15.13. The stack outlet should be either above or outside this zone. (Designated H_c in the *Manual of Industrial Ventilation*.)

$$\text{Length}_{roof\ recirc.} = 0.9\ R_{scaling} \quad (15.6b)$$

= the length of the roof recirculation region as indicated in Figure 15.13. (Designated L_c in the *Manual of Industrial Ventilation*.)

$$\text{Length}_{downwind\ recirc.} = R_{scaling} \quad (15.6c)$$

= the length of the downwind wake recirculation region as indicated in Figure 15.13. (Designated L_r in the *Manual of Industrial Ventilation*.)

These design parameters, along with the stack's placement and its effective height, can be used to assure that the plume (expanding at a 1:5 ratio) does not intersect one of the above recirculation regions.

Figure 15.14 was developed to help visualize the implications of a building's proportions on the value of $R_{scaling}$ and indirectly on the minimum height of an exhaust stack. The vertical axis in Figure 15.14 is the ratio of the scaling coefficient to the building's height ($R_{scaling}/H$). One can see that as the building gets taller, the exhaust

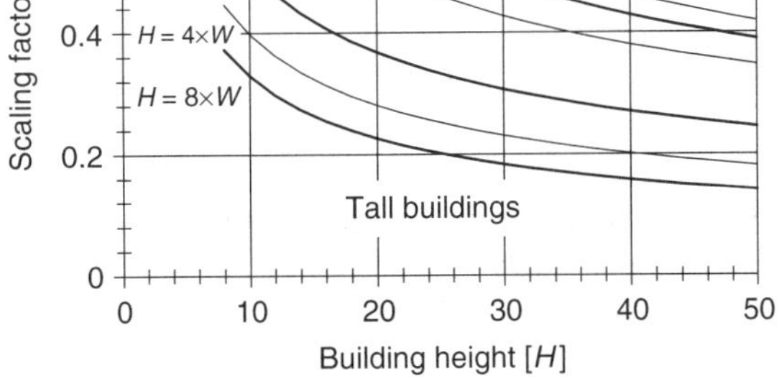

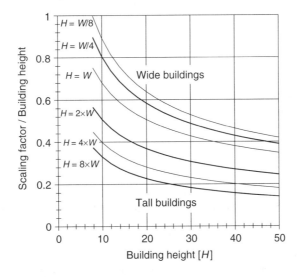

FIGURE 15.14 The ratio of the scaling coefficient ($R_{scaling}$) to a building's height (H) derived from Equation 15.5.

stack needs to be a proportionately smaller fraction of the building's height. For any given building height, the exhaust stack needs to be a larger fraction of a wide building's height than of a narrow or tall building's height. In other words, $R_{scaling}/H$ is larger when W is more than H. The scaling coefficient almost equals the height of a low, wide building (upper left of Figure 15.14), while it is only 20 to 40% the height of a tall, narrow building (lower right of Figure 15.14). Because the effective stack height can be as short as $0.22 \times R_{scaling}$ and (according to Equation 15.5 and Figure 15.14) $R_{scaling}$ is only a fraction of the building height H, one can conclude that a roof-mounted exhaust stack can sometimes be quite a bit shorter than previously recommended.

3. As the potential to reentrain roof emissions into a rooftop fresh air inlet is fairly well known, the roof has in the past not been a common source of fresh air (although it has happened). However, increased awareness of the vulnerability of a building to someone purposefully contaminating a fresh air inlet will no doubt lead to an increased desire to make the inlet less accessible.[21] When one desires to place both a fresh air inlet and an exhaust outlet on the same roof, a tall exhaust stack, a high stack exhaust speed, or both, are needed.
4. The exhaust plume's vertical velocity will follow the same equations that were presented for the single, free-standing circular jet in Section IX in Chapter 13. As practical points of reference, a jet will take over 10 outlet diameters to lose 50% of its initial velocity and about 30 diameters to lose 90% of that velocity (cf., the same 90% reduction in control velocity of a collection hood as described by the DallaValle equation occurs in about one inlet diameter). This exhaust velocity will cause the plume to rise, and this plume rise can be added to the exhaust stack's physical height to calculate its effective height. Briggs developed models for the rise of high and low buoyancy plumes in various atmospheric conditions.[22] Models for buoyant plumes have numerous variables and involve atmospheric conditions, but Equation 15.6 can be used for plumes that are the same temperature as the ambient air.[23]

$$\text{iso-thermal plume rise} = \frac{3 \times \text{stack diameter} \times \text{Stack Discharge Velocity}}{\text{ambient wind speed}} \qquad (15.6)$$

An exhaust stack discharge velocity of 3000 to 3500 fpm is recommended to prevent the plume from contacting the ground in low winds (<20 mph) and to provide adequate dilution before it reaches the ground in high winds. A velocity of 3000 fpm is the minimum to keep the contaminant from being drawn down into the recirculation vortex formed by the stack itself (not to mention to keep rain out, as will be discussed in the next subsection). For a nominal 15 mph wind (1320 fpm), the above discharge velocities, and a 1-ft diameter duct (typical of a small exhaust system designed in previous chapters), the resulting additional plume rise will be 7 to 8 ft that would be added to the physical stack height to determine the effective stack height. Strobic air fans use this principle to their advantage to increase the discharge velocity at the last moment by introducing over 150% outside (roof) air into the exhaust stream (see www.strobicair.com).

While there is no absolute upper limit to the exhaust stack discharge velocity, where a velocity above 5000 fpm would otherwise exist, an évasé exhaust discharge can provide a simple yet advantageous means to recover at least some of the excess velocity energy that would otherwise be lost. You may recall that an évasé is a shallow expansion (about a 4 to 8° taper angle) in the duct diameter at the end of the exhaust stack discussed in Section IX 2 in Chapter 14. The regain factors from Figure 14.18 and the regain formula (Equation 14.22) can be used to show that the P in the exhaust stack before an évasé is actually below one atmosphere by the amount of the regain (whereas the P on the outlet side of the fan would be greater than one atmosphere all the way to the end of an untapered exhaust stack).

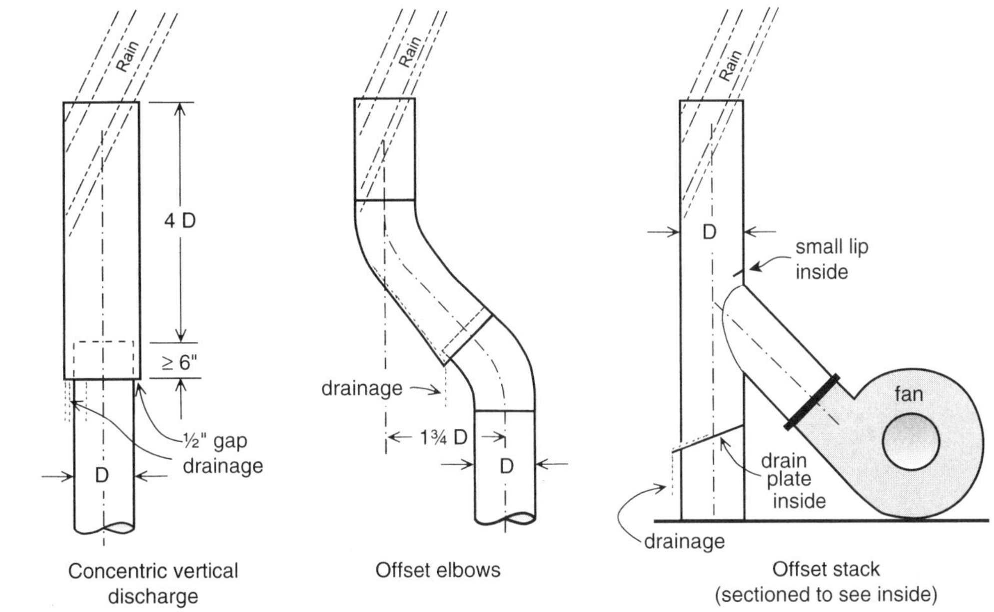

FIGURE 15.15 Three ways to prevent rain from entering the exhaust stack. (*Source*: From Figure 5.30, *Industrial Ventilation: A Manual of Recommended Practice*, 24th ed., American Conference of Governmental Industrial Hygienists, Cincinnati, 2001. With permission.)

Although an évasé will actually save fan power ($) for the same Q, the energy that can be regained from an initial exhaust stack velocity of 4000 fpm or less is not enough to make an évasé cost-effective. Recall that the outlet velocity should never be reduced below the 3000 fpm recommended above for dispersion and below for rain protection.

5. Some form of rain protection is optional for a continuously operated vertical exhaust stack, but rain protection is virtually (but not legally) required for any exhaust stack that is used intermittently. The normal 3000 to 3500 fpm exhaust velocity is sufficient to overcome the 2000 fpm terminal falling velocity of rain (see Figure 3.5). Where such an adequate discharge velocity cannot be either met or sustained (e.g., where the fan runs intermittently), any of the three designs in Figure 15.15 would prevent rainwater from reaching either the fan or running further back into the exhaust system.
   a. A concentric vertical discharge adds an extension at least four original stack diameters long to the original exhaust stack, but about 1 in. wider than that of the original diameter and overlapping it about 6 in. Rain entering the discharge at any angle greater than 15° will impact on the outer duct and flow down its walls to drain outside the actual exhaust stack.[c] This option has the least pressure loss.
   b. An offset elbow comprises two matched elbows that shift the stack opening centerline by more than 1.5 times the exhaust duct diameter. Rain will impact on the outer wall of the top elbow, flow down the diagonal wall, and drain out through a gap or slot in the bottom wall. The upward inertia of the air will actually induce an insignificant upward flow of outside air to come in through this gap or slot and join the exhaust flow (as it would through the annular gap in the vertical discharge). This option is common with molded duct materials.

[c] The falling angle of a raindrop (or other spherical particle) is the inverse tangent of the wind speed (which is dependent upon the weather) and the particle's terminal falling velocity (which is dependent upon the droplet size as in Figure 3.5).

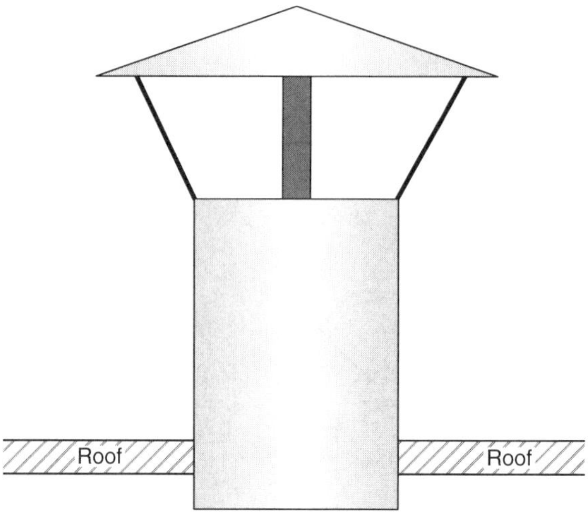

FIGURE 15.16 A weather cap should never be recommended for an exhaust stack that discharges chemical contaminants.

c. An offset stack adds a branch junction to the exhaust stack with a short, blind duct that has a drain hole in the bottom. Rain entering the top will fall or flow past the offset stack into the lower portion of the blind duct and drain out of the bottom. Again, an insignificant volume of exhaust air may enter through this small drain hole because of the upward inertia on all the air induced at the junction. This option is structurally solid and simple to construct, but has the highest pressure loss.

A "cap" such as shown in Figure 15.16 (which may be called a weather cap, rain cap, vent cap, or other descriptive name) should never be used as rain protection for an exhaust stack that discharges toxic chemicals because the cap blocks the coherent upward flow of the plume needed to project the plume well above the end of the stack. Unless the exhaust stack is already very tall, a weather cap will prevent the plume flow from reaching the undisturbed air flowing outside of the roof recirculation zone and will actually force some of the contaminant down onto the roof where it has the potential to be reentrained and/or to expose employees or anyone else on the roof (e.g., for maintenance, building renovations, or sightseeing). If such a rain cap were placed 0.5 to 1 duct diameters about the stack outlet, it would add an additional pressure loss of 0.7 to 0.1 VP, respectively; however, that loss is a less important detriment to using a weather cap than are the health hazards listed above.

PRACTICE PROBLEMS

1. Explain (in the space provided) why each of the suggested air cleaners would be either not applicable, acceptable, or your choice of the three options to control the emissions in the stated settings. In each case, consider the chemical reactivity of the emission, the presence or absence of an aerosol, and the aerosol's particle size (refer to Figure 3.5 as needed).
   a. Foundry fumes:
      cyclone
      electrostatic
      catalytic oxidizer

b. Flour mill dust:
   cyclone
   absorber
   adsorber
c. Odors from cheese or dairy product wastage:
   filter
   spray tower
   combustion scrubber
d. Body shop paint vapors and grinding dust:
   filter
   spray tower
   packed tower

2. Suppose that emissions of airborne wood dust with a 10 μm geometric mean aerodynamic diameter and a geometric standard deviation of 2 from a furniture manufacturer are being controlled by a cyclone operating so that its particle cut-off diameter was right at the aerosol's geometric mean aerodynamic diameter. Unbeknownst to you, the plant has reduced the exhaust air flow rate by about 50% to accommodate changes in production and to decrease ventilation operating costs.
   a. By how much would that change increase the particle cut-off diameter?

$$\frac{d_{\text{cut-off,new}}}{d_{\text{cut-off,old}}} = \underline{\qquad\qquad} =$$

   b. A quantitative version of Figure 3.3 can be used to predict that such an increase in $d_{\text{cut-off}}$ of 0.7 geometric standard deviations will increase the total aerosol emission rate by 26%. Explain how a bleed-in (Section I.11 of Chapter 10) would be a useful response to improve the air emission situation without increasing operating costs significantly?
3. Suppose that you are asked to advise on the placement and design of an exhaust stack for the titanium dioxide dust handling LEV system in Example 14.2. The building in which the system is used is 150 feet wide by 300 feet long with a 24 foot high, flat roof.
   a. What is the value of the "scaling coefficient" for this building? This value is also the length of the downwind wake recirculation region as indicated in Figure 15.13.
   b. What is the maximum height of the roof turbulent recirculation region?
   c. What is the length of the roof recirculation region?
   d. What would you predict would be the additional isothermal plume rise for this emission in a 20 mph (1760 fpm) wind? Recall (from Section VI of Chapter 14) that we designed the duct velocity to be 3867 fpm.

REFERENCES

1. *Industrial Ventilation: A Manual of Recommended Practice*, 24th ed., American Conference of Governmental Industrial Hygienists, Cincinnati, OH, 2001.
2. Burgess, W. A., *Recognition of Health Hazards in Industry*, 2nd ed., Wiley, New York, 1995.
3. Hocking, M. B., *Handbook of Chemical Technology and Pollution Control*, Academic Press, (Harcourt Brace & Co.), San Diego, CA, 1998.
4. U.S. EPA, Fabric filters, In *OAQPS Control Cost Manual*, 5th ed., United States Environmental Protection Agency, Office of Air Quality Planning and Standards, EPA 453/B-96-001, chap. 5, February 1996, Available at www.p2pays.org/ref/10/09848.pdf.
5. U.S. EPA, Electrostatic precipitators, In *OAQPS Control Cost Manual*, 5th ed., United States Environmental Protection Agency, Office of Air Quality Planning and Standards, EPA 453/B-96-001, chap. 6, February 1996, Available at www.p2pays.org/ref/10/09854.pdf.

6. U.S. EPA, Gas absorbers, In *OAQPS Control Cost Manual*, 5th ed., United States Environmental Protection Agency, Office of Air Quality Planning and Standards, EPA 453/B-96-001, chap. 6, February 1996, Available at www.p2pays.org/ref/10/09858.pdf.
7. U.S. EPA, Carbon adsorbers, In *OAQPS Control Cost Manual*, 5th ed., United States Environmental Protection Agency, Office of Air Quality Planning and Standards, EPA 453/B-96-001, chap. 4, February 1996, Available at www.p2pays.org/ref/10/09853.pdf.
8. U.S. EPA, Choosing an adsorption system for VOC: Carbon, zeolite, or polymers?, United States Environmental Protection Agency, Clean Air Technology Center Technical Bulletin, EPA-456/F-99-004, May 1999, Available at www.p2pays.org/ref/10/09863.pdf.
9. U.S. EPA, Thermal and catalytic incinerators, In *OAQPS Control Cost Manual*, 5th ed., United States Environmental Protection Agency, Office of Air Quality Planning and Standards, EPA 453/B-96-001, chap. 3, February 1996, Available at www.p2pays.org/ref/10/09852.pdf.
10. Hunter, P. and Oyama, S. T., *Control of Volatile Organic Compound Emissions*, Wiley, New York, 2000.
11. Seville, J. P. K., *Gas Cleaning in Demanding Applications*, Blackie Academic & Professional (Chapman & Hall), London, 1997.
12. U.S. EPA, Refrigerated condensers, In *OAQPS Control Cost Manual*, 5th ed., United States Environmental Protection Agency, Office of Air Quality Planning and Standards, EPA 453/B-96-001, chap. 8, February 1996, Available at www.p2pays.org/ref/10/09857.pdf.
13. Burgess, W. A., Ellenbecker, M. J., and Treitman, R. D., *Ventilation for Control of the Work Environment*, Wiley, Hoboken, NJ, 2004.
14. Vatavuk, W. M., *Estimating Costs of Air Pollution Control*, Lewis Publishers, Chelsea, MI, 1990.
15. U.S. EPA, Cost estimating methodology, In *OAQPS Control Cost Manual*, 5th ed., United States Environmental Protection Agency, Office of Air Quality Planning and Standards, EPA 453/B-96-001, chap. 2, February 1996, Available at www.p2pays.org/ref/10/09851.pdf.
16. Engineering Staff of George D. Claytone & Associates, Control of industrial stack emissions. In *The Industrial Environment Its Evaluation and Control*, NIOSH Publishers, chap. 43, 1973.
17. McCabe, W. L., Smith, J. C., and Harriott, P., *Unit Operations of Chemical Engineering*, McGraw-Hill, New York, 1993.
18. U.S. EPA, Flares, In *OAQPS Control Cost Manual*, 5th ed., United States Environmental Protection Agency, Office of Air Quality Planning and Standards, EPA 453/B-96-001, chap. 7, February 1996, Available at www.p2pays.org/ref/10/09856.pdf.
19. Wilson, D. J., *Contamination of Air Intakes from Roof Exhaust Vents*, ASHRAE Transactions, 82-1024-1038, 1976.
20. Wilson, D. J., *Flow Patterns over Flat Roof Buildings and Applications to Exhaust Stack Design*, ASHRAE Transactions, 85-284-295, 1979.
21. *Guidance for Protecting Building Environments from Airborne Chemical, Biological, or Radiological Attacks*, NIOSH, DHHS (NIOSH), Pub No. 2002-139, www.cdc.gov/niosh/bldvent/2002-139.html, 2002.
22. Briggs, G. A., *Plume Rise*, USAEC Critical Review Series, TID-25075, Clearinghouse for Federal Scientific and Technical Information, 1969.
23. Barratt, R., *Atmospheric Dispersion Modelling: An Introduction to Practical Applications*, Earthscan Publications Ltd, London, 2001.

FURTHER READING

U.S. EPA, Introduction, In *OAQPS Control Cost Manual*, 5th ed., United States Environmental Protection Agency, Office of Air Quality Planning and Standards, EPA 453/B-96-001, chap. 1, February 1996, Available at www.p2pays.org/ref/10/09850.pdf.

U.S. EPA, Hoods, Ductwork, and Stacks, In *OAQPS Control Cost Manual*, 5th ed., United States Environmental Protection Agency, Office of Air Quality Planning and Standards, EPA 453/B-96-001, chap. 10, February 1996, Available at www.p2pays.org/ref/10/09849.pdf.

16 Ventilation Fans

The learning goals of this chapter:

- ✍ Know when (why) to use the FanTP vs. the FanSP, and vice versa.
- ☐ Know that FanSP $= \Sigma$ inlet losses $+ \ \Sigma$ outlet losses.
- ✍ Be able to differentiate among the major LEV fan types (viz., axial, radial, and backward inclined fans) and what happens to the flow rate if each were to run backwards.
- ☐ Know that $Q \propto \text{RPM}^{+1}$ and FanSP $\propto \text{RPM}^{+2}$ (the most useful of the fan laws).
- ✍ Understand the cause and effect of system effect factors.
- ✍ Be able to select a fan operating condition (RPM) for a given LEV system (Q and FanSP).
- ✍ Be familiar with the nonperformance factors to consider when selecting a fan.

Fans create the pressure difference to make the air move. The fan must be powerful enough to move enough air to protect the employee against the resistance to move that much air through the system. The resistance is the sum of the pressure losses created by friction and turbulence. In order to understand fans as something more than a box with an inlet and an outlet, this chapter will familiarize users of fans with two ways to define that pressure difference, several kinds or categories of fans, and the criteria that one might use to select a fan.

I. FANS HAVE TWO PRESSURES

The title of this section can be interpreted in two ways. Certainly all fans have an inlet and outlet pressure. But the real "two pressures" to which the title refers are two ways to express the pressure difference between a fan's outlet and inlet.

1. Fan total pressure (FanTP or FTP) is the difference between the high total pressure of the air on the outlet side of the fan and the lower total pressure on the inlet side of the fan, just as its name (fan total pressure) implies.

$$\text{FanTP} = \text{Total Pressure}_{\text{outlet}} - \text{Total Pressure}_{\text{inlet}} \qquad (16.1)$$

a. Recall from Chapter 11 that total pressure is the sum of the air's static pressure (P) and velocity pressure (VP). The following derivation starts with the above definition of FanTP, utilizes the definition of SP $= P - 1$ atm from Equation 11.14, and concludes with a working definition of FanTP in terms of variables that can be measured in a ventilation system, such as depicted in Figure 16.1.

$$\text{FanTP} = (P_{\text{outlet}} + \text{VP}_{\text{outlet}}) - (P_{\text{inlet}} + \text{VP}_{\text{inlet}}) \qquad \text{(restating Equation 16.1)}$$

$$= (1 \text{ atm.} + \text{SP}_{\text{outlet}} + \text{VP}_{\text{outlet}}) - (1 \text{ atm.} + \text{SP}_{\text{inlet}} + \text{VP}_{\text{inlet}})$$

$$= (\text{SP}_{\text{outlet}} + \text{VP}_{\text{outlet}}) - (\text{SP}_{\text{inlet}} + \text{VP}_{\text{inlet}})$$

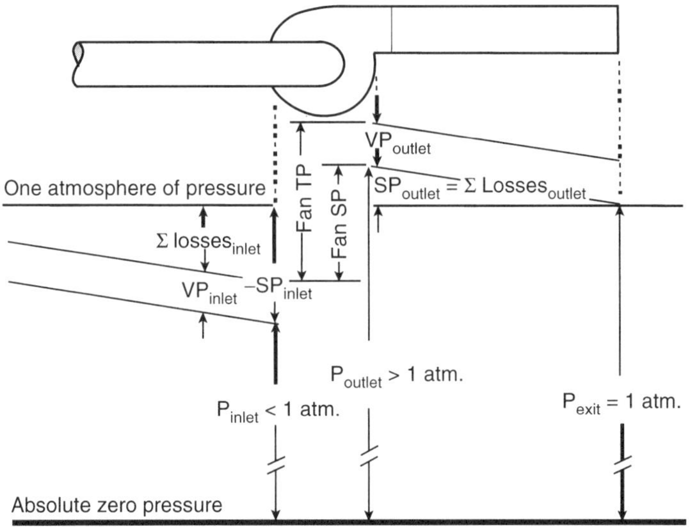

FIGURE 16.1 Depiction of pressures as the air passes through a fan. Notice the difference between FanTP and FanSP and especially that FanSP involves no velocities.

$$\text{FanTP} = \text{SP}_{\text{outlet}} - \text{SP}_{\text{inlet}} + \text{VP}_{\text{outlet}} - \text{VP}_{\text{inlet}} \quad (16.2)$$

This version of fan total pressure (the definition featured in the *Manual of Industrial Ventilation*) is useful because each of the variables in Equation 16.2 can be measured in an existing system.[1] Because SP_{inlet} is always negative, the difference between $\text{SP}_{\text{outlet}} - \text{SP}_{\text{inlet}} = \Delta\text{SP}$ is also the sum of their absolute values.

b. Another view of FanTP can be derived by using the relationships that $-\text{SP}_{\text{inlet}} = \text{VP}_{\text{inlet}} + \Sigma\text{losses}_{\text{inlet}}$ from Equation 11.15 and $\text{SP}_{\text{outlet}} = \Sigma\text{losses}_{\text{outlet}}$ from Equation 11.16 as long as VP is constant in the outlet duct. The resulting definition in Equation 16.3 is easier to use to calculate FanTP during the design stage than Equation 16.2 because the designer has already predicted all of these variables. These variables are also depicted in Figure 16.1.

$$\text{FanTP} = \Sigma\,\text{losses}_{\text{inlet}} + \Sigma\,\text{losses}_{\text{outlet}} + \text{VP}_{\text{outlet}} \quad (16.3)$$

Equation 16.3 says, in effect, that the fan must provide all the energy that is lost from the air as it travels through both the inlet and the outlet portions of the ducting and the energy associated with the velocity pressure in the air as it leaves the fan's outlet. This view is basically an energy balance.

c. Because pressures (including the FanTP) are equivalent to the energy per unit of volume (from Chapter 11), FanTP also equals the energy that the fan adds to the air per unit of volume (see also Principle 10 in Chapter 10), and when FanTP is multiplied by the volume of air flowing per minute (in units such as ft^3/min or L/sec), the product is power such as watts or horsepower. In fact, the next chapter will show how FanTP × Q is (or could be) used to calculate:
   1. the horsepower of the fan motor that is needed to run the system,
   2. the electrical amperage (and indirectly the size of the electrical wiring) of that motor, and
   3. the annual cost for electricity (amps × volts × hours it runs × \$/kWhr = \$).

While FanTP is a good concept that applies directly to energy consumption, it is not convenient to use to specify, select, or buy a fan. Notice that VP_{outlet} is in Equation 16.3; thus, any difference between the outlet diameters of two fans moving the same Q will cause a difference in their outlet areas, their outlet velocities, their outlet velocity pressures, and thus their FanTP values. It is hard to select a fan if one has to continually recalculate the selection variable.

2. Fan static pressure (FanSP or FSP) turns out to be a much more useful variable with which to select a fan than FanTP would be. Despite its title, FanSP is not the difference between the static pressures in the outlet vs. the inlet sides of the fan. However, FanSP is the pressure difference without any direct effect of velocity, as can be seen by looking carefully into Figure 16.1. The *Manual of Industrial Ventilation* presents two equations for FanSP (shown here as Equation 16.4a and Equation 16.5) with no real explanation for why FanSP is not simply SP_{outlet} minus SP_{inlet}. With this in mind, the justifications for three ways to calculate FanSP are provided below.
   a. One can define FanSP in terms of FanTP by thinking that FanTP is the difference in total pressures, total pressure is P plus VP, and VP_{inlet} is a part of the SP_{inlet} but not a part of SP_{outlet}. Therefore, to calculate a difference in static pressures, VP_{outlet} has to be subtracted from the FanTP. Hence:

$$\text{FanSP} = \text{FanTP} - \text{VP}_{\text{outlet}} \tag{16.4a}$$

   And of course the reverse is also true.

$$\text{FanTP} = \text{FanSP} + \text{VP}_{\text{outlet}} \tag{16.4b}$$

   b. Equation 16.5 calculates FanSP from pressures that can be measured on an installed and operating fan. Both a measured and predicted SP_{inlet} include VP_{inlet} that must be subtracted out to exclude all effects of VP on FanSP.

$$\boxed{\text{FanSP} = \text{SP}_{\text{outlet}} - \text{SP}_{\text{inlet}} - \text{VP}_{\text{inlet}}} \tag{16.5}$$

   c. A philosophically easier definition of FanSP can be written by making use of much earlier relationships for static pressures on the inlet and outlet sides of a fan, viz.,

$$-\text{SP}_{\text{before the fan}} = \text{VP}_{\text{before the fan}} + \Sigma\ \text{Losses}_{\text{since entry}} \tag{11.15}$$

   and

$$\text{SP}_{\text{after the fan}} = \Sigma\ \text{Losses}_{\text{until exit}} \tag{11.16}$$

   When one inserts these definitions into Equation 16.5, the only term left over is the sum of all the system losses, as shown in Equation 16.6. This definition reflects the approach taken herein that the most direct way to design an LEV system and to select a fan whose FanSP simply equals the sum of all of the system's losses. All of these definitions can be discerned in Figure 16.1.

$$\boxed{\text{FanSP} = \Sigma\ \text{losses}_{\text{inlet}} + \Sigma\ \text{losses}_{\text{outlet}}} \tag{16.6}$$

No matter how one defines FanSP, its big advantage is allowing one to select a fan independent of the size (diameter or/and area) of its inlet and outlet. As is easily seen in Equation 16.6, FanSP is independent of the VP either entering or leaving the fan. Therefore, FanSP is independent of the dimensions of any particular fan. The desired FanSP is the sum of the pressure losses of the particular system that can be predicted using the methods described in Chapters 13 and 14.

In contrast, FanTP involves VP_{inlet} and/or VP_{outlet} which, for a given Q, depends on each fan's inlet and outlet areas (per Equation 11.7, $Q = V \times A$). Thus, to satisfy any particular system's flow rate and pressure loss operating requirements, the FanTP will differ among fan sizes and designs both within and among manufacturers. Each time that one looks at a fan with a different outlet diameter, one would have to recalculate a new FanTP. Using one fixed FanSP makes it much easier to match a given fan's performance with a system's needs, to compare the performance among different fan sizes, to compare among different fan designs or different manufacturers, and to make the final fan selection.

II. FAN PERFORMANCE

1. A fan's air flow rate (Q) depends upon the resistance to flow against which it must suck and push the air. The resistance is determined as the sum of the losses (Σlosses). Imagine being able to change the resistance within an LEV system, and then being able to measure the Q that a given fan can move against each level of resistance (each FanSP). This might be done by adjusting a blast gate in increments from zero losses (allowing the maximum Q on the lower right of Figure 16.2), past the maximum FanSP the fan is capable of producing at some lower Q, and on down to $Q = 0$ when the flow is completely blocked. The data across the full range of Q comprises what is called a fan performance curve or sometimes just a fan curve. Fan performance curves display the following characteristics[2]:
   a. The highest Q occurs against no flow resistance (i.e., when losses and FanSP = 0).
   b. A high (but rarely the highest) FanSP occurs when the flow is completely blocked (i.e., when $Q = 0$).

   The shape of the Q vs. FanSP curve between these two points of maximum Q and zero Q varies significantly among the different categories of fans to be discussed in Section III and Section IV of this chapter and varies somewhat within a category among the fans made by different manufacturers.

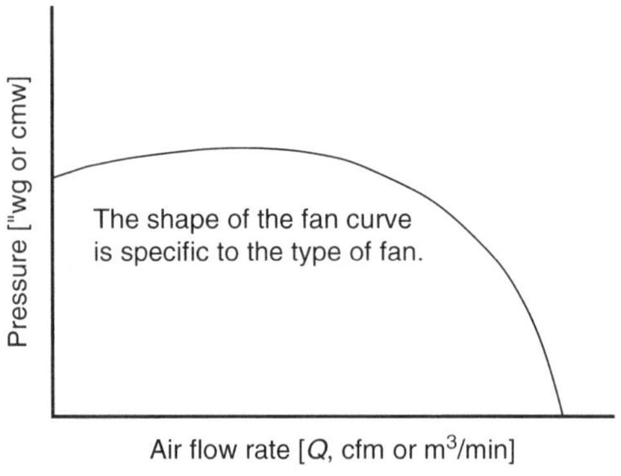

FIGURE 16.2 A fan performance curve.

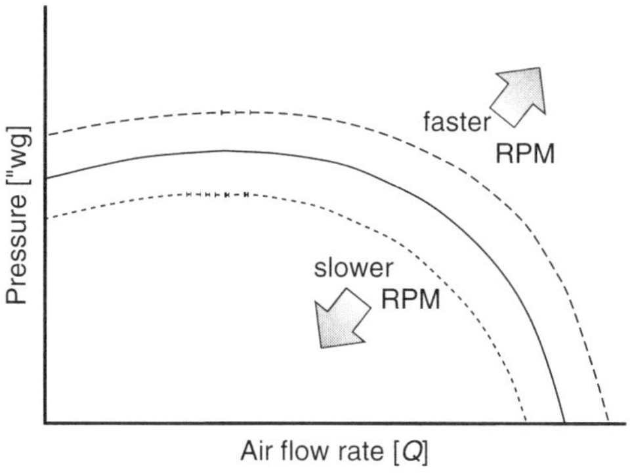

FIGURE 16.3 A family of fan performance curves.

2. Moreover, the position of a fan's performance curve in a plot such as Figure 16.2 depends upon the fan's speed. Fan rotation speed is quantified in revolutions per minute or RPM. If the RPM of the fan is increased, the fan performance curve will move up and to the right, as depicted in Figure 16.3, so that it is capable of producing a higher flow rate against the same resistance, moving the same flow rate against more resistance, or some combination of both. A decrease in RPM will have the opposite effect. Thus, each fan's relationship between *Q* vs. FanSP is actually a family of fan performance curves at different RPMs.
3. Manufacturers measure each of their fan's ability to move air (*Q*) against a range of resistances (FanSP). Voluntary ANSI standards for fan testing have been set by the Air Movement and Control Association International to help manufacturers make their data more accurate, uniform, and comparable (available at www.AMCA.org).[3] A manufacturer's catalogue should indicate that their data conforms to AMCA standardized test conditions (unless they do not follow those testing methods which, of course, would also be nice to know). Manufacturers usually present to the buyer only the portions of such fan performance curves in which they recommend operating their fan. And they typically present such data as a fan rating table rather than as the kind of plot shown above. Examples of such a fan rating tables will be shown in Figure 16.24 to Figure 16.26 in conjunction with examples of fan selection. Think of a fan rating table as a tabular form of the fan curve rotated 90° clockwise on the page.

III. MATCHING FAN PERFORMANCE TO SYSTEM REQUIREMENTS

Two parameters define the operating requirements of any LEV system: the volumetric rate of air flow (*Q*) to protect employees and the static pressure losses (Σlosses) to get that *Q* to flow through the system. An LEV system designer selects a fan and its specific RPM where the fan's performance (the *Q* and FanSP data supplied by the fan manufacturer) matches the *Q* and losses predicted for or generated by a given system.

1. Recall that *Q* started with a $V_{control}$ that was initially selected to achieve the desired level of employee protection based on a chemical's hazard and its momentum upon release from the source, as discussed in Chapter 13. The $V_{control}$ along with the hood design (its face area and sometimes the distance "X" that the source is in front of a hood) determined the hood's face velocity. Together, the V_{face} times the A_{face} specified the air

flow rate (Q) that must pass through the system. Thus, one fan selection parameter is the Q that must be moved through the system.

2. Energy is lost due to friction and turbulence as this Q flows through a given LEV system. This energy must be put back into the air by the fan. The types of losses discussed in Chapter 14 included:
   a. Resistance to the flow of make-up air into the room,
   b. Hood entry loss(es),
   c. Duct losses (from friction and turbulence) both before and after the fan,
   d. Fan installation losses, and
   e. Air cleaner resistance.

   The sum of all of these losses is the FanSP, the second fan selection parameter just defined in Equation 16.6. The proper fan must have a FanSP that is capable of overcoming the flow resistance contributed by all of the losses within the LEV system. What an LEV system designer calls "the sum of the losses" for their system, a fan manufacturer calls the FanSP capability of their fan. Or as stated in ANSI Z9.1-1991, the fan "shall have sufficient capacity to produce the flow of air required in each of the hoods and openings of the system ... [when] operating against the [above] pressure losses, the sum of which is the [fan] static pressure."[4]

These two parameters (Q and FanSP $= \Sigma$losses) are necessary and sufficient to specify a fan big enough to be able to supply the flow needed to protect the employee and to overcome the resistance to flow (pressure losses) expected for the ventilation system you (or someone else) designed.[a] When trying to match fan performance to the system's requirements using Equation 16.7, it is essential to remember that the losses in most of the system's individual components are directly proportional to the VP at each location (as previously stated and boxed in Equation 10.1 and Equation 14.3),[b] and of course that VP is proportional to V^2 (as previously stated and boxed in Equation 11.21).

$$\boxed{\Sigma \text{ losses} = \Sigma\text{"LossFactor}_i\text{"} \times \text{VP}_i} \qquad (10.1 \text{ and } 14.3)$$

$$\boxed{\text{VP in "wg} = [V \text{ in fpm}/4005]^2} \quad \text{at NTP} \qquad (11.21\text{b})$$

Because the sum of all of the losses within a system is proportional to the VP at any location, the losses within a system are also proportional to the Q^2 going through the entire system. This latter principle (previously stated and boxed in Equation 14.4 and Equation 15.1) will underlie fan selection and be used heavily in fan management.

$$\boxed{\Sigma \text{ losses} = \text{a constant} \times Q^2} \qquad (16.7)$$

Now let us imagine again. Imagine being able to vary the fan's flow rate through a given LEV system across a wide range and to measure the sum of the pressure losses at each flow rate. This might be done by adjusting a fan's RPM incrementally. The resulting data would look like that depicted in Figure 16.4. The shape of almost all system requirement curves is

[a] The Q and the outlet diameter of the selected fan could be used to determine its FanTP (per Equation 16.4b), and the FanTP can be used to specify an electric motor large enough to drive that fan and the cost to operate that fan motor. But do not worry: the fan's manufacturer's data (such as Figure 16.24) will tell you the motor power needed without any calculations.

[b] Exceptions to this rule include low velocity filter media (in which losses are linear with Q) and low velocity wet scrubbers (in which losses are relatively proportional to Q; see page 424).

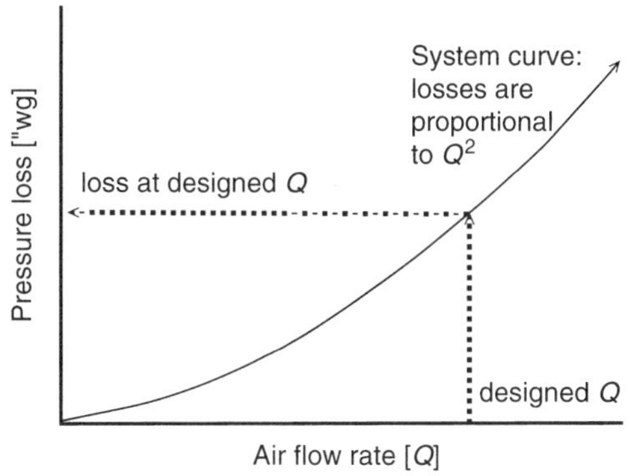

FIGURE 16.4 A system's requirements curve.

parabolic because of the dependence of most losses upon VP and its dependence on V^2 and Q^2 as predicted by Equation 16.7. From this perspective, any specific Q and pressure loss (such as those in Figure 16.4) is in fact only one point on that system's requirements curve.

Thus, the appropriate operating point for a fan is really the intersection of two curves: a parabolic system requirements curve and the fan performance curve, as depicted in Figure 16.5. The shape of the fan performance curve will depend upon the fan category, type, and manufacturer, but for the flow rate to be temporally stable, the fan should be operating where the curve is negatively sloped, again as depicted in Figure 16.5; and the more steeply it is sloped, the better. A steep slope means that the flow rate (Q) will not change much if the system losses increase.

The right fan will be capable of moving enough air against the resistance through a given system to protect the employee(s). The wrong fan will not be able suck hard enough to protect the employee(s). Someone must either select the right fan to be installed or find the right RPM for a fan that has already been installed. But before getting into fan selection any further, let us first cover some basic fan — make that "air mover" — terminology.

IV. SOME AIR MOVER TERMINOLOGY

1. Air mover is a generic term for any device that moves air. A fan or blower is the most common air mover but other types exist. For instance, an air ejector (to be described below) is a specialized air mover without any moving parts.

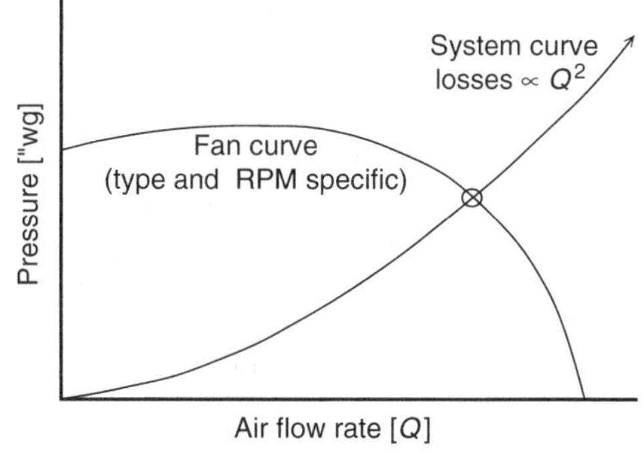

FIGURE 16.5 The fan's performance intersects the system requirements curve at its operating point.

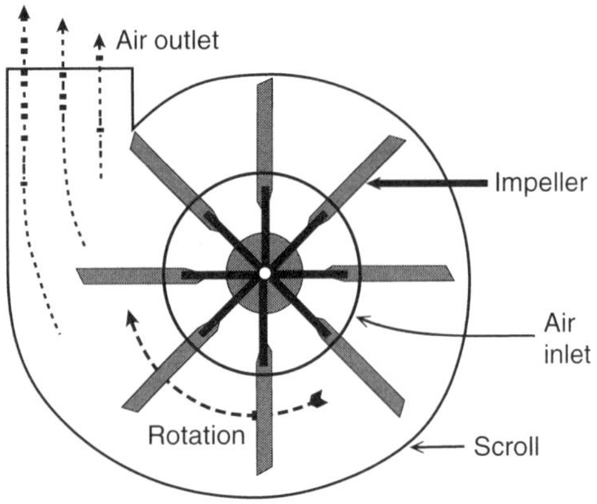

FIGURE 16.6 A fan's impeller and scroll.

2. An impeller is the component of the fan with rotating blades that pushes on the air. Impellers either move air down the axis of their rotation (in what is called an axial fan, as will be shown in Figure 16.9) or move it radially outward away from their axis of rotation (in what is called a centrifugal fan, as shown in Figure 16.6 and Figure 16.12).
3. A scroll is the curved housing that surrounds the periphery of a centrifugal fan's impeller, as shown in Figure 16.6. A centrifugal fan's scroll plays a major role in the efficiency with which it moves the air.
4. Fans are characterized as either axial fans or centrifugal fans. The distinction is made by the direction in which the impeller moves the air.
   a. The impeller of all axial fans moves the air in the same direction as the axis of its rotation. That is, the air enters and exits the fan along its axis. If an axial fan's impeller rotates in the wrong direction, the air will flow backwards through the fan (out of the fan's designated inlet).
   b. The impeller of centrifugal fans moves the air radially outward from its axis due to centrifugal forces. Thus, the air enters near this fan's central axis but exits tangentially along its periphery. Because the air always moves radially outward, a centrifugal fan will move air forward (into the fan's designated inlet, albeit not very well) even when the impeller rotates backward (which just happens to be counter-clockwise in Figure 16.6). Backwards fan rotation will also be discussed in Section VI.5 in Chapter 18.
5. Centrifugal fans are categorized into classes. The class designators (I, II, and III) indicate progressively stronger mechanical strengths. These classes allow manufacturers to fabricate similar fans by design and performance (within the capability of the lower class fan) but range in price and upper flow rate, RPM, and temperature limits. Manufacturers designate the different class fans within their fan tables such as Figure 16.27.
6. An air ejector is a specialized air moving device that has no impeller. Flow through an air ejector is induced via the flow of a usually larger primary airstream (see Figure 16.7). The primary air may be provided by a separate conventional fan, by steam, or by a compressor (the latter primarily for a small ejector). Ejectors are inefficient from an energy perspective (i.e., they have a low mechanical efficiency[c]), but their lack of an impeller makes them uniquely suited to moving corrosive, erosive, or/and flammable

[c] Mechanical efficiency is the fraction of the energy consumed by the electrical motor that actually goes into the air as fan total pressure and cubic feet of air moved (or the fraction of power used); see further discussion in Chapter 17.

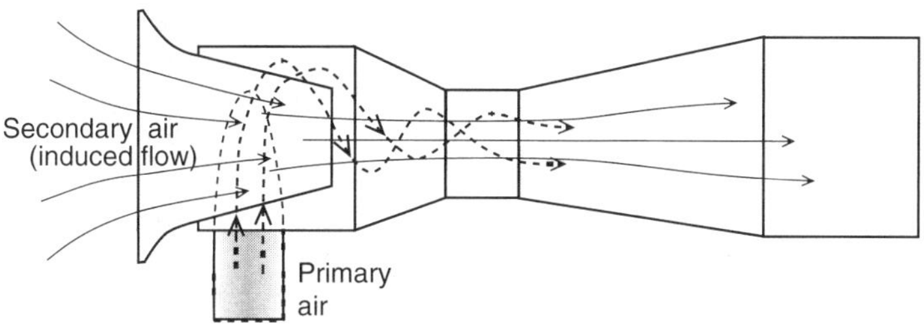

FIGURE 16.7 An air ejector.

contaminants (gases, vapors, or aerosols) in the secondary airstream that cannot go through the moving parts of a conventional fan. A related device sometimes called an educter (shown in Figure 16.8) is designed to move granular materials through the secondary airstream.

The next two sections of this chapter (Section V and Section VI) cover the two major categories of fans: axial and centrifugal. Each category comprises approximately three major types of fans. While the physical differences between the categories of axial and centrifugal fans are obvious and distinct, the differences among the various types within each category are mostly internal (and thus not visible) but still operationally important.

V. AXIAL FANS

Air both enters and exits an axial fan parallel to the axis of the impeller's rotation. The flow into the inlet of an axial fan is aligned with the flow from its outlet (cf. the flow changes direction 90° going through a centrifugal fan). The rotation of an axial fan's blades causes the air velocity near their tips to be much greater than near their center. The following three basic types of axial fans can be purchased.

1. A propeller fan consists of little more than a motor and blades. A wall-mounted fan, as shown in Figure 16.9, also has a short, solid, annular ring around the tips of the fan blades. The blades may look long and thin (much like on an airplane propeller) or be wide, almost disc-like. A propeller fan can be used just to circulate air within a room, and they are commonly used in an opening through a wall or roof where little to no resistance to flow is expected. A propeller fan can only produce approximately 1 to 2"wg of FanSP (and typically produces less). Such a small FanSP is incapable of moving air against much of a restriction, such as if trying to ventilate a "tight building" with few open portals through which make-up air can easily enter the building, if trying to oppose the exhaust suction pressure produced by a more capable centrifugal fan on another exhaust hood, or if trying to blow into a wind impacting onto the fan's exhaust (or onto the wall in which the fan might be mounted). There will be no flow when the wind's VP exceeds the fan's FanTP capability (as described by Equation 16.3 with no losses).

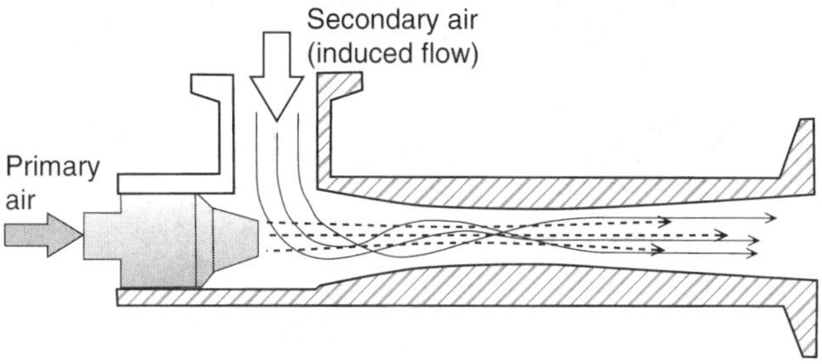

FIGURE 16.8 An air educter.

FIGURE 16.9 A propeller fan. (Courtesy of Chicago Blower, Corp.)

2. Tube axial fans have a close-fitting cylindrical "casing" around the tips of the blades that extends downwind approximately one duct diameter (see Figure 16.10 and Figure 16.11). The casing (or tube) allows an axial fan to be placed within a duct, constrains and directs the air flow more than the ring around a propeller fan can do, and slightly increases its potential FanSP. Tube axial fans are capable of creating a peak FanSP of from 1 to perhaps 2.5 "wg, still pretty low for most LEV purposes.
3. Vane axial fans have radial guide vanes immediately downstream of the impeller, as depicted in Figure 16.12. These guide vanes redirect much of the swirling airflow and

FIGURE 16.10 A tube axial fan. (Courtesy of Chicago Blower, Corp.)

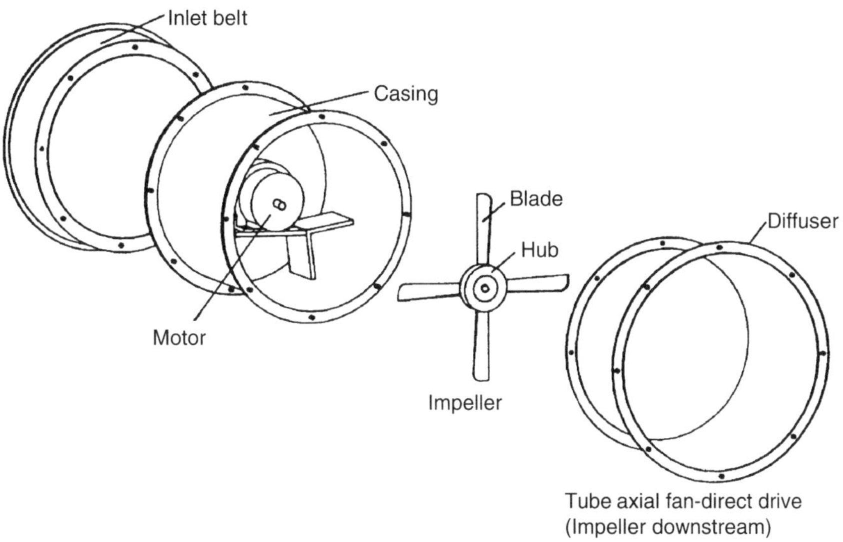

FIGURE 16.11 Exploded view of a tube axial fan. (Courtesy of AMCA.)

thereby recover much of the rotational energy normally imparted to the air by an axial impeller. Some vane axial fans also have upstream guide vanes. Vane axial fans placed within a duct are capable of creating a peak FanSP of from 4 to 8 ″wg, making them competitive with some centrifugal fan applications.

Axial fans have the following general advantages and disadvantages. Because most of the disadvantages of axial fans are important for industrial LEV applications, they are less common in LEV systems than centrifugal fans but they (especially vane axial fans) are often used in institutional LEV applications (e.g., laboratory hoods).

Advantages of typical axial fans:

a. Propeller and tube axial fans can be constructed cheaply, although the more complex vane axial fans are approximately equal to the cost of some centrifugal fans.

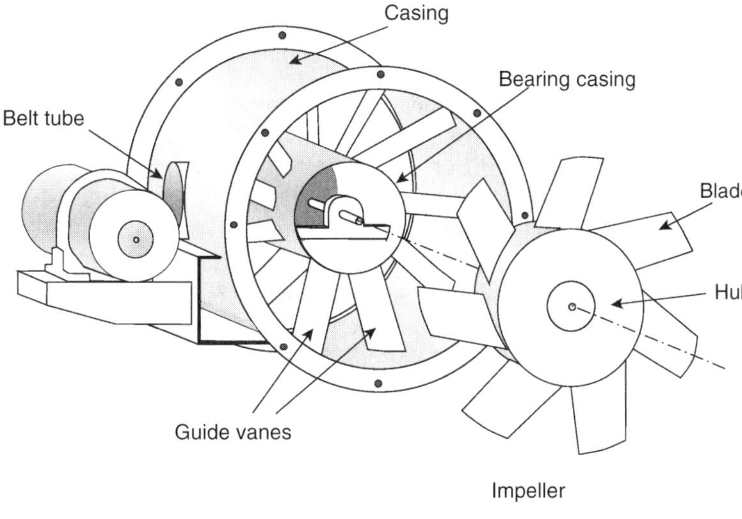

FIGURE 16.12 Exploded view of a vane axial fan. (Courtesy of AMCA.)

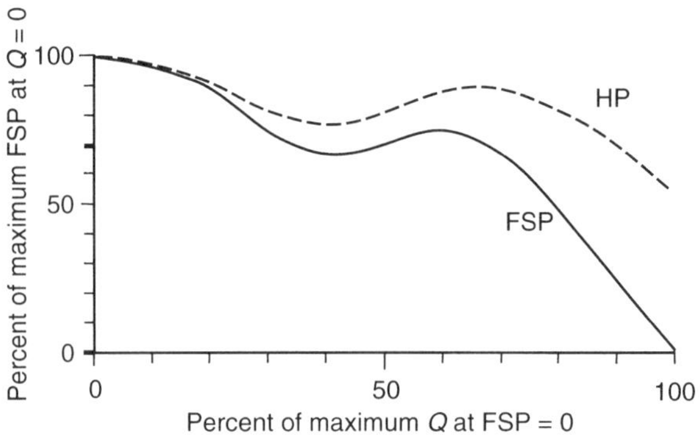

FIGURE 16.13 Depicting a typical axial fan performance curve with scales relative to 100% of Q and FanSP.

b. Axial fans are compact in size, and the fact that the air inlet and outlet are in-line (vs. the 90° change in direction for centrifugal fans) helps them fit into smaller spaces than centrifugal fans.
c. Axial fans can be more efficient in high volume, small FanSP applications such as dilution ventilation or hood-duct systems with few losses.

Disadvantages of typical axial fans:

a. Axial fans have a lower maximum FanSP than most centrifugal fans, have a low tolerance to stalling (not moving much air even though the blades are turning), and create high noise levels.
b. Axial fans typically have a distinct valley or dip in its Q vs. FanSP performance curve, as depicted in Figure 16.13. A fan will not produce a uniform flow if it were to be operated where its performance curve is flat or has a positive slope. The resulting variation in flow rate is detrimental to the control of airborne hazards and employee protection. This fan performance characteristic is similar to a forward curved centrifugal fan (but unlike a forward curved fan, axial fans have a good horsepower versus Q curve that drops at high flow rates).

VI. CENTRIFUGAL FANS

Air enters a centrifugal fan parallel to the impeller's axis of rotation similar to an axial fan, but there the similarity ends. As shown in Figure 16.14, air enters a centrifugal fan only near its axial center (none enters near the blades' tips). Most centrifugal impellers (viz., those in Figure 16.16, Figure 16.18, and Figure 16.20) look more like a wheel than the propeller shape of an axial impeller. The impeller of a centrifugal fan directs the air outward (in the radial direction) where the fan housing (scroll) constrains and redirects the flow tangentially toward the outlet. Because the air exits a centrifugal fan tangentially to its impeller, all centrifugal fans effectively create a 90° turn in the ducting (their universal disadvantage), as depicted in both Figure 16.14 and Figure 16.15. Each of the following three main types of centrifugal fans has distinct performance characteristics that each give separate advantages and disadvantages.[d]

[d] A tubular centrifugal fan could be considered a specialized fourth category. Its impeller is centrifugal, but the entire fan resides within the duct and the air does not make the 90° bend. In fact, from the outside, it is hard to distinguish between a tubular centrifugal and tube or vane axial fan. Its performance is somewhat similar to a backward inclined fan except that it has less volume and pressure capability, lower efficiency, and it may have a dip to the left of peak pressure.

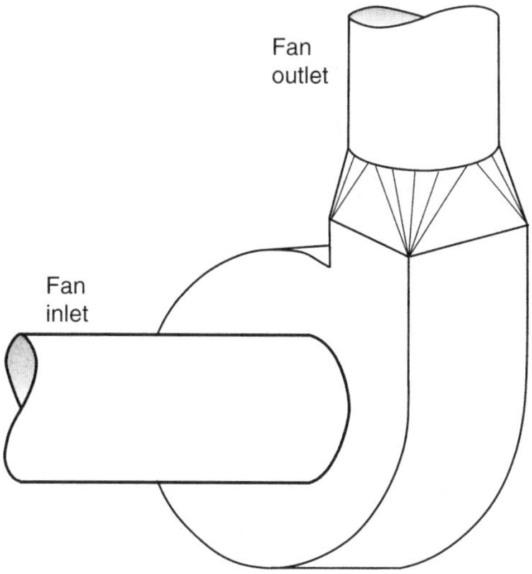

FIGURE 16.14 The external appearance and configuration of a typical centrifugal fan with exhaust pointed upwards.

1. *Forward-curved fans*. The impeller of a forward curved fan has approximately 24 to perhaps 48 shallow blades, usually made of sheet metal, that curve forward at their outer tips (Figure 16.16 and Figure 16.17). The array of many small blades has given forward-curved fans the common name of "squirrel cage" fans. Because the air actually leaves the tip of such blades faster than the blade is moving, forward-curved fans typically operate at a slightly lower RPM than other centrifugal fans. Probably because the air's trajectory in a forward curved fan is due more to the shape of the blade than to the shape of the

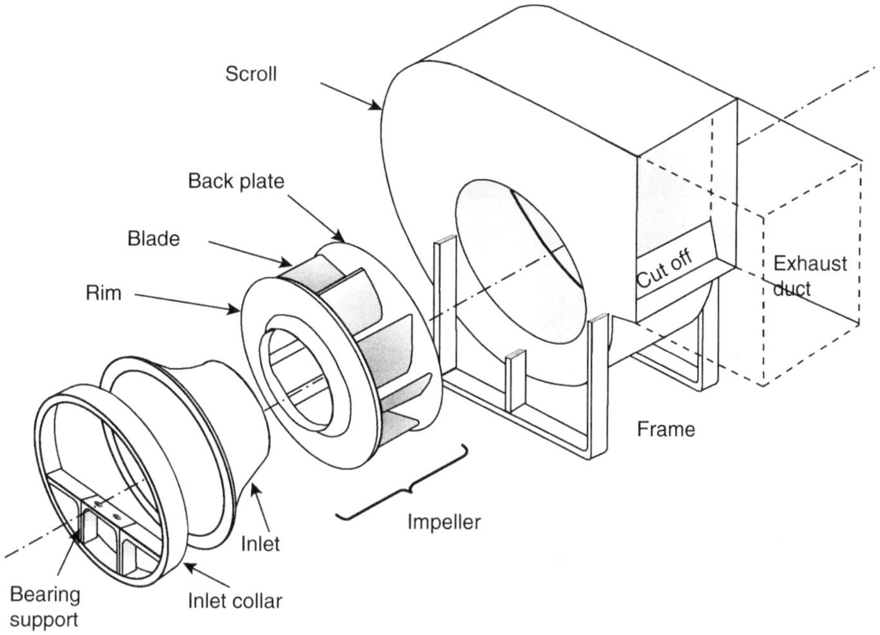

FIGURE 16.15 Exploded view of a backward inclined centrifugal fan with exhaust pointed horizontally. (Courtesy of AMCA.)

FIGURE 16.16 A forward curved impeller.

scroll, its mechanical efficiency is a large function of its fan speed and can vary from approximately 50 to 80%.

Forward curved fans can be made from thin, cheap materials and create low noise. These advantages are the result of both the low RPM at which forward-curved fans can operate, the small blade size, and because the tips of the forward curved blades do not need to pass close to the fan housing or scroll to achieve good flow performance. However, the following disadvantages of forward curved fans pretty much preclude their use in LEV applications.

a. Small blades will clog easily in dirty air or with large debris.
b. Their peak FanSP is typically only 1–2 "wg (some up to 4 "wg).
c. A valley in the FanSP vs. Q performance curve of forward-curved fans (visible in Figure 16.17) can cause large oscillations in Q if the system's pressure losses are underestimated or vary.
d. The forward curved fan's power requirement (and electrical current) rises rapidly at high Q (the dashed line in Figure 16.17). Thus, if the system pressure loss requirement

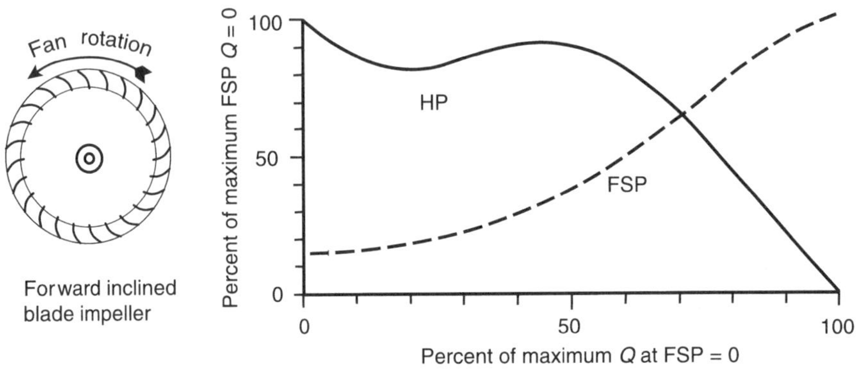

FIGURE 16.17 Depiction of a forward curved fan disk and its typical performance curve with relative scales.

FIGURE 16.18 A radial blade impeller.

is overestimated and the flow rate is much greater than expected, the fan's electric motor capacity may be exceeded, resulting in motor burnout.

The above combination of characteristics limits forward inclined fans to applications where the FanSP is low and narrowly defined (does not vary widely) and where low noise has a high priority.

2. *Radial blade fans*. The impeller of a radial fan has between 6 and 12 straight blades that look kind of like canoe or kayak paddles, typically made of cast iron, plate steel, or heavy gauge sheet metal, and may or may not be attached to a backplate (not shown in Figure 16.18 but depicted by the circle in Figure 16.19). The blades provide both radial and tangential motion to the air, but the scroll (the outer housing of the fan) is crucial to guiding the air's forward momentum efficiently. The mechanical efficiency of radial

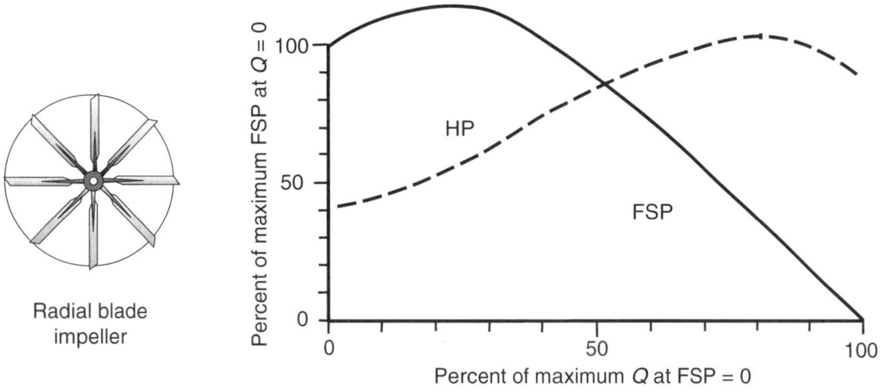

FIGURE 16.19 Depiction of a radial blade fan disk and its typical fan performance curve with relative scales.

blade fans tends to be 60–70%. The following advantages of radial blade fans make them very popular for LEV applications.

a. Radial blades are strong, largely self-cleaning, and therefore resistant to clogging.
b. The simple geometrical shape of radial blade fans allows them to be fabricated at a relatively low cost from cast iron or simple welded parts. Vatavuk's 1988 price formula was $22.1D^{1.55}$ ($\pm$ 10%) where D = fan wheel diameter from 20 to 36 in.[5] For the classroom, use price = $30Q^{0.5}$.
c. The generic performance curve of a radial fan (Figure 16.19) is generally well behaved (steep downward slope and no valley), and its peak FanSP can easily exceed 20"wg (some go as high as 50"wg).
d. Due to the large volume of air moved between adjacent fan blades, a radial blade fan can move more air for a given size and RPM than other centrifugal fans. This feature can also make it quieter than other fans, although that depends greatly on the particular blade and scroll design.

The major disadvantage of radial blade fans is their low efficiency due to the turbulence created as the air flow abruptly changes direction from axial (upon entry) to radial (along the blades) and to circumferentially (as the air spills sharply off the blade tips). This high turbulence limits a radial fan's mechanical efficiency to the moderate 60–70% range.

3. *Backward incline fans.* The impeller of a backward inclined fan typically has 12–24 straight blades whose tips are inclined backward away from the direction of rotation (Figure 16.20 and Figure 16.21). The blades are typically welded either onto one or between two end plates, increasing its manufacturing costs. Entering air is engaged by the leading edge of a backward inclined blade at a less acute angle than by a radial blade and can slide off the tip at a similarly less abrupt angle, thereby increasing its mechanical efficiency to approximately 70–80%. Backward inclined fans typically have the following advantages:
   a. The performance curve of a backward inclined fan (Figure 16.21) tends to be even more stable but generally has a lower peak FanSP (<20"wg) than a radial blade fan.

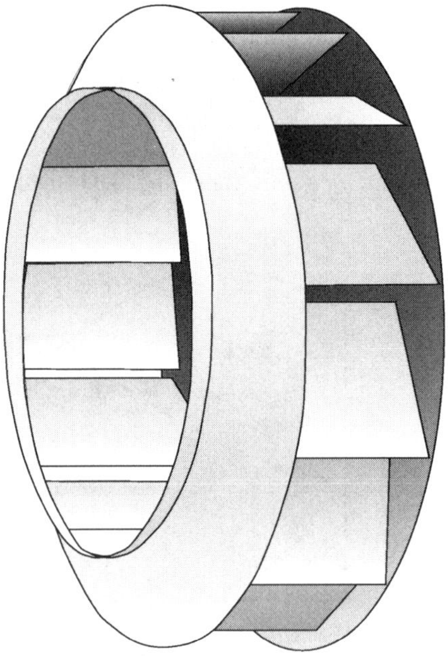

FIGURE 16.20 A backward inclined impeller.

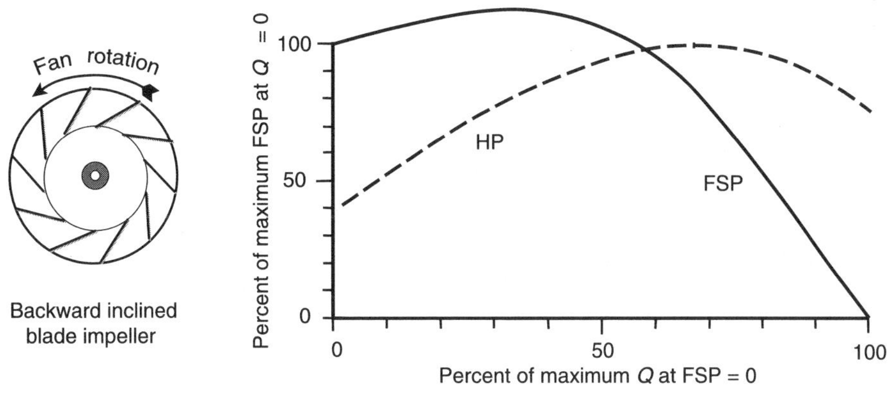

FIGURE 16.21 Depiction of backward inclined blades and its typical fan performance curve with relative scales.

b. Backward inclined fans are more efficient (by ca. 10%) than comparable radial blade fans.
c. The electrical power requirement of a backward inclined fan is flatter and peaks at a lower fraction of its maximum Q, permitting any given backward inclined fan to operate over a wider range of FanSP and Q conditions than most other fans.

The major disadvantage of backward inclined fans is that their welded construction means a higher purchase cost. Vatavuk's 1988 price formula was $42.3D^{1.2}$ ($\pm$ 10%) where again D = fan wheel diameter in inches.[5] For the classroom, use price = $50Q^{0.4}$. This higher purchase cost will eventually be paid back by lower electrical costs due to its higher mechanical efficiency. The performance of backward inclined blades is slightly less resistant to dust build-ups than the radial blade, but that is not a problem for a fan moving primarily gases and vapors or downstream of an aerosol air cleaner.

4. The airfoil blade fan is sometimes considered to be a specialized subcategory of the backward inclined fan. The blades on an airfoil impeller are backward inclined and aerodynamically sculptured for smoother air flow (Figure 16.22). The airfoil shape makes these fans up to 80–90% efficient, slightly less noisy, and able to operate over an even wider range of FanSP and Q conditions than straight backward inclined fan blades (cf. Figure 16.21 and Figure 16.23). Again, their lower operating cost will eventually offset their even higher manufacturing costs.

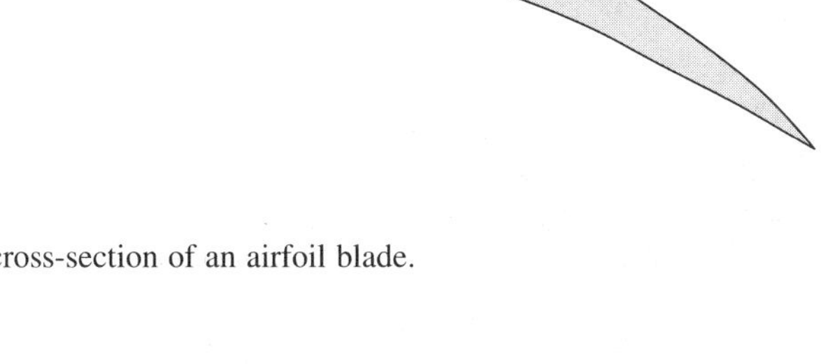

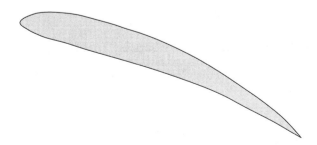

FIGURE 16.22 The cross-section of an airfoil blade.

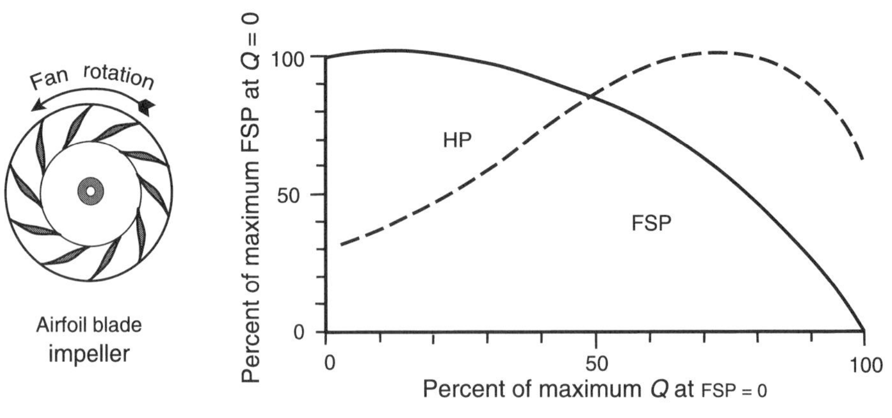

FIGURE 16.23 Depiction of airfoil shaped blades and a typical performance curve of an airfoil blade fan.

VII. THE FAN LAWS

A proper fan must meet the Q and FanSP requirements of the system. However, even after having chosen the category and type of fan, a manufacturer, and that manufacturer's model or series of fans, one must still choose the size and the speed or RPM at which to run a chosen fan. The fan laws allow certain predictions to be made about the effect of fan size, air density, and fan speed on Q and FanSP.

Fan Size D: Most manufacturers offer a range of fan sizes for each model of fan that they make. Each size is typically specified by one dimension referred to herein as "*D*." However, changing one dimension of a given fan design typically means changing all dimensions. Thus, changing the fan's size is analogous to changing the fan's internal volume by D^3. Of course, after a fan is selected, its size cannot be changed.

Air density ρ: A manufacturer tests or reports fan performance data for normal temperature and pressure conditions because fan performance is affected by air density (ρ_{air}). Thus, both the altitude at a given location and the exhaust air's temperature must be known or at least anticipated, but they rarely change for a given use-setting.

Fan speed RPM: Fan speed is measured and reported by its revolutions per minute (RPM). RPM is the only one of these three variables that is controllable and/or can be changed by the user after the fan is selected and even after it is installed.

Table 16.1 describes the fan laws in terms of the effect of changes in fan speed, air density, and fan size on the air flow rate and the FanSP that a given model of fan can generate and the power that it will consume. Fan size, the air's density, and the fan's RPM are all variables before a fan is

TABLE 16.1
The Fan Laws, Emphasizing the Only Variable That Is Controllable On-Site

	Fan Speed	Air Density	Fan Size
Air flow rate, Q	RPM^{+1}	constant	D^{+3}
Fan pressure, FanSP or FanTP	RPM^{+2}	ρ^{+1}	D^{+2}
[a]Power required $= Q \times$ FanTP	RPM^{+3}	ρ^{+1}	D^{+5}

[a] The power required law could easily be included within the memorable box but is excluded because the change in power is always derivable as the product of the change in $Q \times$ FanTP.

selected, purchased, and installed. However, the altitude of a given location (and its effect on air density) will not change. One can assume that neither will the exhaust air temperature (which also affects air density). Fan size is determined when the fan is chosen and purchased with a given air density in mind. After that time, the only parameter that can still vary or be varied on-site is the fan speed. Hence, the effect of fan RPM on Q and FanSP is emphasized in Table 16.1.

The following paragraphs provide a conceptual rationale that might help you to understand, or at best to remember, how each variable in Table 16.1 affects fan performance and appears within the fan laws.

1. Air is moved through the fan by the fan blades. Anticipating the effect of fan RPM on Q is aided by picturing the fan as moving a series of pockets of air between adjacent fan blades as it revolves around its axis. The volume of each pocket is fixed by the number of the blades, their diameter, and their width. The more pockets of air that are moved past the outlet per minute, the greater the volume of air that will be moved per minute. Thus, the flow rate Q changes in direct proportion to the RPM^{+1} of the fan. The effect of RPM on FanSP can be thought of as the indirect result of losses being proportional to VP, to V^2, and to Q^2. Since FanSP is always equal to losses and Q changes linearly with RPM, FanSP must change in proportion to RPM^{+2} (squared).
2. Air density does not affect the volume of air moved between adjacent blades as they go around inside the fan. Thus, Q is unaffected by changes in air density. In contrast, the fan's ability to produce suction (its FanSP) decreases linearly as the density of the air decreases because the fan's blades come into contact with fewer air molecules. Thus, FanSP changes in proportion to ρ^{+1}.
3. The manufacturer's performance literature is always specific to a given fan size, so the effect of D is probably the least important to remember. However, if you see that one size of a given manufacturer's family of fans is too small or too large, that portion of the fan laws can be used to guess what other size fan in that family should be looked at next. While a family of impellers can be made in multiple diameters independent of its width (or vice versa), the fan laws assume all dimensions increase together. In this case, the volume of air between adjacent blades and the volume flow rate (Q) both increase with the cube of its size or D^{+3}. In contrast, the air velocity only increases linearly with fan size because the inlet and outlet flow areas should increase with the square of D. The net result is that fan size has the same effect on FanSP as does RPM, viz., FanSP changes with D^{+2}.
4. The power required of the fan motor varies as the mathematical product of any given parameter's effect on Q and FanSP (as should be expected based on the discussion in Section I.1.c that power is proportional to $Q \times \text{FanTP}$). Therefore, one really only needs to remember the effect of each parameter on air flow rate (Q) and fan static pressure (FanSP) individually and that the fan's motor size and power requirement (electricity) is the product of those changes.

Because the person selecting the fan is working with tables or figures of fan performance data provided by the manufacturer, these fan laws will only have some use in fan selection. They will have more use in Chapter 18 when trouble shooting or making post-installation adjustments to an existing system.

VIII. FAN SELECTION AND INSTALLATION

The desired fan must at least meet, if not exceed, the two basic ventilation system performance requirements:

Air flow rate (Q): Chosen by the LEV designer (maybe the IH) to protect the employee(s).
Fan static pressure (FanSP): Enough pressure difference to overcome the energy losses caused by resistance from trying to move that much air through the system.

1. According to the fan laws (Table 16.1), a low air density will reduce the FanSP that a given fan can produce at any RPM. Since the data in the manufacturer's fan performance tables are based on NTP conditions, one is much more likely to encounter a low air density due to either a higher temperature and/or a higher altitude than a higher air density due to exhausting cold air or being below sea level. In practice, if a lower air density is expected in a particular use setting, the fan simply needs to be sized using the manufacturer's data as if it were to move the designed Q at a higher FanSP.

   Air density ρ was discussed in Chapter 2 (ca. Equation 2.13) and used in Chapter 9 to adjust the 4005 magic number. The ratio given in Equation 16.8 provides a more flexible (and programmable) alternative to using the *Manual of Industrial Ventilation*'s Table 5.10 to find a density correction factor that they call "df."[1]

$$\text{“density correction factor”} = \frac{\rho_{\text{air}}}{\rho_{\text{NTP}}} = \frac{T_{\text{normal}}}{T_{\text{air}}} \times 2^{(-\text{altitude in feet}/18{,}000)} \qquad (16.8)$$

   where

   ρ_{air} = gas density, g/L = MW/24.45 = 1.184 for air at normal T and P,[e]
   T = absolute temperature (either °R = °F + 460 or °K = °C + 273).

   In order to select a fan that will perform as expected in less dense air, the fan laws indicate that the original FanSP that equaled the Σlosses must be divided by the above density correction factor. The selected fan must then be operated at an RPM (determined from the manufacturer's published fan performance data) to match the original Q and this adjusted FanSP. Keep in mind that since the fan will only produce the original Q and losses determined by the original system's design, the actual fan power (horsepower in American fans) will be the horsepower at the new RPM adjusted back down for the low air density as shown in Equation 16.9 The net result is that the new horsepower will be close to (but not exactly equal to) what it would have been at the RPM for the original Q and FanSP.

$$\text{Adjusted Fan power} = [\text{Fan power at new RPM}] \times [\text{“density correction factor”}] \qquad (16.9)$$

EXAMPLE 16.1

What fan performance parameters should be specified to pull a Q of 1350 ft³/min against a resistance of 1.30 ″wg (as designed in Example 14.2) at 70°F (21°C) in Denver, Colorado?

Denver's claim as the "Mile High City" means that its altitude is 5280 ft ASL (or look it up).

$$\text{“density correction factor”} \frac{(70+460)}{(70+460)} \times 2^{(-5280/18{,}000)} = 0.816 \qquad \text{(using Equation 16.8)}$$

Thus, the fan should be selected for a Q of 1350 ft³/min and a FanSP = 1.30/0.816 = 1.59 ″wg.

2. Even after adjusting for air density, a fan will still not perform as well in the field as it is listed in the manufacturer's literature if the physical layout around the fan's inlet or outlet are less optimal than the standard conditions used in the AMCA test.[3] The most common conditions causing reduced fan performance are a disturbed inlet flow (swirling or a nonuniform velocity as described in Figure 14.13) or something other than several diameters of straight ducting on the outlet (even no ducting will reduce the flow). The effects of less-than-optimal fan inlet or outlet conditions are called "system effects." An array of system effect factors has been developed by the AMCA for certain fan inlet and

[e] Density corrections for humidity are covered in the *Manual of Industrial Ventilation* Section 5.13. A special condition results when the air density within a duct system decreases due to cooling and even more so if condensation occurs (see the *Manual* Section 5.13.1).

outlet conditions.[2] Each system effect factor is to be multiplied by the local VP, just as loss factors were used in Chapter 14. The designer needs to recognize such less-than-optimal conditions, either eliminate the condition or calculate the loss(es), and add these losses to all the other losses. (System effects are one of the reasons to use line 28 in the LEV Design Worksheet, Table 14.9.) The net effect of a system effect will be to specify a higher FanSP and to select an appropriately more capable fan. Two types of system effect losses are described below, but the reader is referred to either AMCA Publication 201 or the *Manual of Industrial Ventilation* Section 6.4 for the detailed descriptions and data needed to calculate such losses.[1,2]

a. The air velocity entering a fan should be reasonably uniform and smooth. A fan will not perform as well as expected if the inlet air flow is not uniform and smooth. Recall from Chapter 14 that an elbow in a round duct induces a swirl or spin into the air. The *Manual of Industrial Ventilation*'s Figure 6.21 and Figure 6.23 link various geometries of elbows up to five duct diameters before a fan either to SEF values outright or to SEF letter codes linked to SEF values in their Figure 6.26.[f] Such SEF values range from 0.2 to nearly 3. The *Manual of Industrial Ventilation*'s Figure 6.22 and Figure 6.24 describe the use and effect of turning vanes within rectangular ducts to reduce such SEF values. And the *Manual of Industrial Ventilation*'s Figure 6.25 provides SEF values for mismatched duct-to-fan collar diameters. Each inlet SEF value, when multiplied by the inlet duct VP, will yield an additional loss as shown in Equation 16.10a. That loss, when entered into the design worksheet (Table 14.9) as "other losses," will increase the FanSP that must be specified.

$$\text{"Other Loss"} = \text{System Effect LossFactor} \times VP_{\text{inlet}} \tag{16.10a}$$

b. A section of straight exhaust stack should be present at the fan outlet before the air turns an elbow or exits the system to recover the initially turbulent portion of the energy coming out of a fan. The minimum effective duct length needed to obtain 100% of the manufacturer's stated fan performance is 2.5 duct diameters plus one duct diameter per 1000 fpm of exhaust velocity above 2500 fpm. If an elbow must be placed within 100% of an effective duct length of the fan's outlet, use the *Manual of Industrial Ventilation*'s Figure 6.20 to find the letter code for the particular geometry of that installation. The letter code is again cross-referenced to SEF values in their Figure 6.26, ranging from about 0.2 to 1.2. Their Figure 6.18 and Figure 6.19 provide similar loss data if no or a short exhaust stack is present on either a centrifugal or an axial fan, respectively. This "other loss" can be calculated using Equation 16.10b and added to all the previous losses.

$$\text{Another "Other Loss"} = \text{System Effect LossFactor} \times VP_{\text{outlet}} \tag{16.10b}$$

The basic guidance is that no loss factors are needed if the system has at least eight diameters of straight ducting before the inlet and at least approximately one duct diameter for every 1000 fpm on the outlet side (for the 3000 to 5000 fpm exhaust stack velocities typical of LEV systems).

EXAMPLE 16.2

Now is the time to notice and deal with the short (3 ft) length of straight 8 in. ducting between the elbow and the fan in Example 14.2 (see the close-up in Figure 16.24). This 36-in. of 8-in. ducting represents only 4.5 duct diameters, less than the nominal eight diameters necessary for the swirling turbulence downstream of an elbow in a round duct to subside. Although this subsection does not contain the

[f] System effect loss factors are listed from place to place within the *Manual of Industrial Ventilation* as "SEF," "Fsys," or "loss factor equivalents."

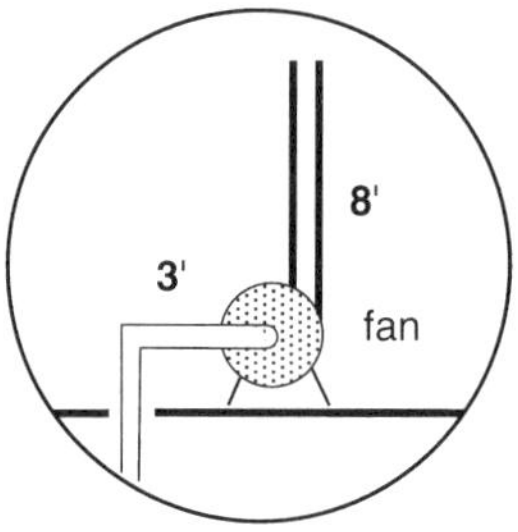

FIGURE 16.24 Close-up of the elbow before a fan in Example 14.2.

system effect data, the *Manual of Industrial Ventilation*'s Figure 6.21 indicates that this inlet condition would result in a system effect factor just below curve *U*, and their Figure 6.26 indicates that curve *U* corresponds to a F_{sys} of 0.4 and curve *V* to a F_{sys} of 0.26. Equation 16.10 with a nominal system effect loss factor of 0.35 and a VP of 0.932 "wg results in a system effect loss of approximately 0.33 "wg that would need to be added to the 1.30 "wg loss previously found in Example 14.2 (unless the system could be redesigned to have 6 ft or more of straight ducting between the elbow and the fan inlet), and applying the 0.816 density correction factor for altitude from Example 16.1 to a total loss of 1.63 means that the fan operating conditions would need to be 1350 ft^3/min and a FanSP of 2.0 "wg.

3. Numerous criteria are (or can be) used when selecting a particular fan to meet the above *Q* and FanSP requirements. While the IH rarely has to make the final selection of the fan, whoever does select the fan has a lot of help, especially from fan manufacturer sales representatives or vendors. The following are a few of the selection criteria.
   a. Debris larger than dust can strongly influence the choice of the fan. The potential for solid material to accumulate on the fan impeller varies with the concentration of the debris, its size, and the stickiness of suspended solids or nonvolatile liquids to be controlled. The recommended practice of placing the fan after the air cleaner (rather than before it) will minimize this consideration.
   b. The chemical and physical nature of the hazard being handled influences the materials from which the fan and its motor should be constructed. The fan should be spark resistant if the exhaust air stream is near, or has the potential to be, a combustible concentration or combustible material (especially solids) that can accumulate within the ducting. The motor of most fans is outside that air stream but may still need to be intrinsically safe in its own environment. The fan should resist the corrosiveness of the contaminant. The fan should withstand the exhaust air temperature; the use of a bleed-in can reduce the exhaust temperature before it reaches the fan. Fans are made of various materials (steel, aluminum, and plastics) or can be custom coated to meet these needs.
   c. The need to be able to vary the flow rate *Q* can influence the type of fan selected. In order to avoid the valley in most axial fans' performance curve (such as depicted in Figure 16.13), they are often run near their maximum flow rate; thus, they are limited in their range of variability. Do not even think about varying the operating range of a forward curved centrifugal fan. In contrast, radial and backward inclined fans (the two work-horses of LEV systems) are stable over a relatively wide range of flow rates that can be adjusted by various means. A belt drive (connecting the motor to the fan) offers a cheap means to change the fan speed for a fixed motor speed by manually changing the diameter of the pulleys (or sheaves) on the motor and fan shafts, but this control is not automatic. Adjustable or variable inlet guide vanes or dampers can control the flow rate, and they can be controlled automatically by pneumatic or electric actuators. Adjustable inlet dampers are common for large radial and backward inclined fans. The newest and probably best option is a variable speed drive, an electronic device

that varies the frequency of the alternating electrical current reaching the fan's motor. Its expense is offset by its ability to control the speed of either belt driven or direct drive fans automatically and remotely more efficiently than inlet guide vanes.

d. A fan's capital cost (purchase price) and installation costs are only a portion of a fan's total or life-cycle cost. Future operating costs should also be included in the fan selection decision.[g] Operating costs include the costs for power consumed (the mechanical efficiency of fans varies among fan categories, types, and manufacturers), maintenance (a fan's tolerance to clogging and the frequency at which a given fan may need to be cleaned), and the time to its eventual replacement (its projected useful life based on such factors as corrosion, abrasion, bearing life, and the fan's loading). An inefficient fan will use more electricity and cost more to run than an efficient fan. Chapter 17 will include a further discussion on the mechanical efficiency of fans.

e. Noise can be an important selection parameter, especially if the fan is in a noise-sensitive area. Installing an LEV fan outside the building is always recommended not only to minimize noise exposure to employees inside the building, but also to maintain a negative static pressure difference inside of all the ducting that is inside the building (if SP were positive, any leakage would be outward creating an unexpected source of the hazardous chemical being exhausted). However, placing a fan outside can create a source of community noise complaints. The following options are possible if the fan noise is excessive:
   1. If selecting a fan, a large-slower fan is usually quieter than a smaller-faster fan. Keeping any fan's speed (RPM) within the manufacturer's recommended range will minimize its noise.
   2. Relocate the fan. If it is indoors, move it outdoors but beware of community noise.
   3. If an indoor fan cannot be relocated, consider acoustically enclosing the fan and usually also the fan motor. An acoustical enclosure can be open on one or more sides that are not direct pathways for noise to reach potential receivers (either employees or community neighbors). If the opening must be pointed at a reflective wall, add noise absorption material to that wall.
   4. Consider adding an acoustical lining or a silencer to the fan's inlet and/or outlet (whichever is left unducted); a snubber (without a straight noise pathway) would be even better.
   5. If a high system effect factor is contributing to the high noise level, consider improving a fan's poor inlet conditions.[6]
   6. Consider whether the duct is a secondary noise source. If it is, physically isolate the fan from its ducting (installing a flexible collar between the duct and the fan's inlet is normal practice).

   In the event that the air passing through the ducts is the source of the noise, try to reduce the duct velocity (especially if it is above 6000 fpm). Recall from Section III in Chapter 14 that noise from turbulence should decrease with approximately the 5th power of its velocity (viz., $\Delta dB = 50 \log[V_2 / V_1]$). If decreasing the flow is incompatible with maintaining employee protection, one solution is a larger duct. Another is to add lagging exterior to the duct to reduce breakout noise.

f. Some fan candidates can be eliminated simply by their physical size, weight, or ducting configuration. A convenient (although not a mandatory) choice is a fan whose inlet diameter is approximately the same size as the ducting. A centrifugal fan's motor on the back side of the fan inlet will contribute to the size of its foot print. Weight can be a problem if the fan is to be installed on an upper floor, platform, roof, or within a

[g] Chapter 7 mentioned possibly discounting such future costs in a "net present value" (NPV) calculation.

duct. The 90°-change in alignment between the inlet and outlet of all centrifugal fans means that a duct rising through a roof will have to make a 90° turn into a horizontal duct that enters the fan. An installation without much straight duct between the elbow and the fan will have a poor system effect factor (as in Example 16.2), which also decreases the fan efficiency and increases its operating costs. Adding turning vanes within a nearby elbow can help reduce a system effect factor (see Figure 10.10).

4. Once a fan is selected, its operating RPM must be determined. Manufacturers commonly display their fan's performance envelope in fan tables such as shown in Figure 16.25 to Figure 16.27. Recommended operating conditions are either marked by obvious notations within a fan table or are the only data presented by a manufacturer. If one can anticipate that the fan's performance requirements (Q and FanSP) will probably never change, then choose a fan that meets all of the above criteria and will run near its peak efficiency at those Q and FanSP conditions. If conditions are likely to change in the future, then choose as large a size fan as feasible that can meet not only the above criteria but also the potential future criteria. Reasons that the exhaust air flow rate may need to be increased include to add more hoods, to comply with tighter exposure limits, or (given the largely generic nature of the design guidance for the control velocity) to increase V_{control} just to meet current employee exposure limits. One reason that the resistance to flow may increase in the future (for the same air flow rate) might be to add a higher performance air cleaner. Fortunately, most radial and backward inclined fans are capable of operating over at least a 5 × range of flow rates (Q) by varying its operating r/min.[h] Improving the performance of an existing fan will be discussed in Section V in Chapter 18.

EXAMPLE 16.3

At what fan RPM would you operate a small radial blade fan such as the Chicago Blower Corp. SQI 11LS whose performance table is provided in Figure 16.25 to run the 1350 ft^3/min system in Example 14.2 at an altitude of Denver in Example 16.1 (where the density correction factor is 0.816) at a total system loss of 2.0 "wg from Example 16.2?

The system loss equals the FanSP. Thus, the relevant data from Figure 16.25 are shown below:

CFM	Outlet V (fpm)	r/min	bhp
1320	2000	1056	0.72
1452	2200	1084	0.83

Interpolate between the above two RPMs in order to determine the RPM for 1350 ft^3/min.

$$\text{Interpolated RPM} = 1056 + \frac{(1350 - 1320)}{(1452 - 1320)} \times (1084 - 1056) = 1056 + 6 = 1062\ \text{RPM}$$

$$\text{Interpolated hp} = 0.75\ \text{hp}$$

Recall that the actual horsepower to run the fan in this less dense air is less than that listed by a manufacturer at sea level (NTP). Thus, the actual hp is reduced by the same density factor of 0.816 from 0.75 to 0.61 hp. The manufacturer can also provide fan efficiency data, but in this case we will use a technique to be described *circa* Equation 17.4 to estimate the fan efficiency by comparing the listed horsepower to the ideal horsepower required to move this air (or $hp_{\text{imparted to the air}}$ 0.401 hp to be found in Example 17.1).

Implied efficiency is 0.401 ideal hp/0.61 actual hp = 60%.

[h] If for some reason the fan's RPM cannot be varied, consider picking a larger than needed fan and either designing in a damper (blast gate) after the hood to increase the FanSP and bring Q back down closer to its originally designed flow rate or/and providing a "bleed-in" somewhere shortly before the fan to keep the Q that goes into the fan high while keeping the Q that goes into the hood close to its originally designed condition.

CHICAGO

SQI

Wheel Diameter: 19-1/8"
Outlet Area: 0.660 sq. ft.
Tip Speed (fpm) = 5.00 x rpm

SIZE 11 LS

RPM	Wheel	Shaft
	3853	2980

CFM	OV FPM	2" SP RPM	2" SP BHP	3" SP RPM	3" SP BHP	4" SP RPM	4" SP BHP	5" SP RPM	5" SP BHP	6" SP RPM	6" SP BHP	8" SP RPM	8" SP BHP	10" SP RPM	10" SP BHP	12" SP RPM	12" SP BHP	14" SP RPM	14" SP BHP	16" SP RPM	16" SP BHP	18" SP RPM	18" SP BHP
792	1200	979	0.39	1185	0.59	1362	0.81	1519	1.03	1662	1.26												
924	1400	993	0.46	1195	0.68	1369	0.92	1525	1.17	1666	1.42	1919	1.96										
1056	1600	1011	0.54	1208	0.78	1379	1.04	1532	1.31	1672	1.59	1924	2.17	2146	2.78								
1188	1800	1032	0.63	1224	0.89	1391	1.17	1542	1.46	1680	1.77	1929	2.40	2151	3.05	2352	3.73						
1320	2000	1056	0.72	1242	1.01	1406	1.31	1555	1.62	1691	1.95	1937	2.63	2157	3.33	2357	4.06	2542	4.81	2715	5.58		
1452	2200	1084	0.83	1263	1.15	1424	1.47	1569	1.80	1704	2.14	1946	2.86	2164	3.62	2363	4.39	2547	5.19	2719	6.01	2881	6.85
1584	2400	1114	0.95	1287	1.29	1443	1.64	1586	1.99	1718	2.35	1958	3.12	2173	3.91	2370	4.74	2553	5.58	2724	6.45	2885	7.33
1716	2600	1146	1.08	1314	1.45	1465	1.82	1605	2.20	1735	2.58	1971	3.38	2184	4.22	2379	5.09	2560	5.98	2730	6.90	2891	7.83
1848	2800	1179	1.23	1343	1.62	1490	2.02	1626	2.42	1754	2.83	1986	3.66	2197	4.54	2390	5.46	2569	6.40	2738	7.36	2897	8.34
1980	3000	1212	1.38	1374	1.81	1517	2.23	1650	2.66	1774	3.09	2003	3.97	2211	4.89	2402	5.84	2580	6.82	2747	7.83	2905	8.86
2112	3200	1247	1.55	1406	2.01	1546	2.46	1675	2.91	1797	3.37	2022	4.30	2227	5.25	2415	6.24	2592	7.27	2757	8.32	2914	9.39
2244	3400	1283	1.73	1439	2.22	1576	2.70	1703	3.18	1821	3.67	2042	4.64	2244	5.64	2431	6.67	2605	7.73	2769	8.83	2925	9.94
2376	3600	1320	1.93	1472	2.45	1608	2.96	1732	3.47	1848	3.98	2064	5.01	2262	6.05	2447	7.12	2620	8.22	2782	9.35	2937	10.51
2508	3800	1358	2.14	1507	2.69	1640	3.24	1762	3.77	1876	4.31	2087	5.39	2283	6.48	2465	7.60	2636	8.73	2797	9.90	2950	11.10
2640	4000	1397	2.37	1542	2.95	1674	3.53	1794	4.10	1905	4.66	2112	5.80	2304	6.94	2484	8.10	2653	9.28	2812	10.48	2964	11.72
2772	4200	1437	2.61	1578	3.23	1707	3.84	1826	4.44	1936	5.03	2139	6.22	2328	7.42	2504	8.62	2671	9.85	2829	11.09	2980	12.37
2904	4400	1478	2.88	1615	3.52	1742	4.16	1859	4.80	1968	5.42	2168	6.67	2352	7.92	2526	9.17	2691	10.44	2847	11.73		
3036	4600	1521	3.18	1653	3.84	1777	4.51	1893	5.17	2000	5.83	2197	7.13	2379	8.44	2550	9.75	2712	11.07	2866	12.40		
3168	4800	1564	3.49	1691	4.17	1813	4.88	1927	5.57	2033	6.26	2228	7.62	2407	8.98	2574	10.34	2734	11.72	2887	13.10		
3300	5000	1608	3.84	1731	4.53	1850	5.26	1962	5.99	2067	6.71	2259	8.13	2436	9.55	2601	10.97	2758	12.39	2908	13.82		
3432	5200	1653	4.21	1772	4.91	1887	5.67	1997	6.43	2101	7.18	2291	8.67	2465	10.14	2628	11.61	2783	13.09	2931	14.57		
3564	5400	1698	4.61	1813	5.32	1925	6.10	2033	6.89	2135	7.68	2324	9.23	2496	10.76	2657	12.29	2809	13.82	2955	15.35		
3696	5600	1744	5.03	1856	5.76	1964	6.56	2070	7.38	2170	8.20	2357	9.81	2528	11.40	2686	12.99	2836	14.57				
3828	5800	1791	5.49	1898	6.23	2004	7.04	2107	7.89	2206	8.74	2391	10.42	2560	12.07	2717	13.71	2865	15.35				

Performance shown is for installation type B: Free inlet, Ducted outlet. Power ratings (BHP) do not include drive losses.
Performance ratings do not include the effects of appurtenances in the air stream.
All performances in the shaded areas are within two percent of peak efficiency.

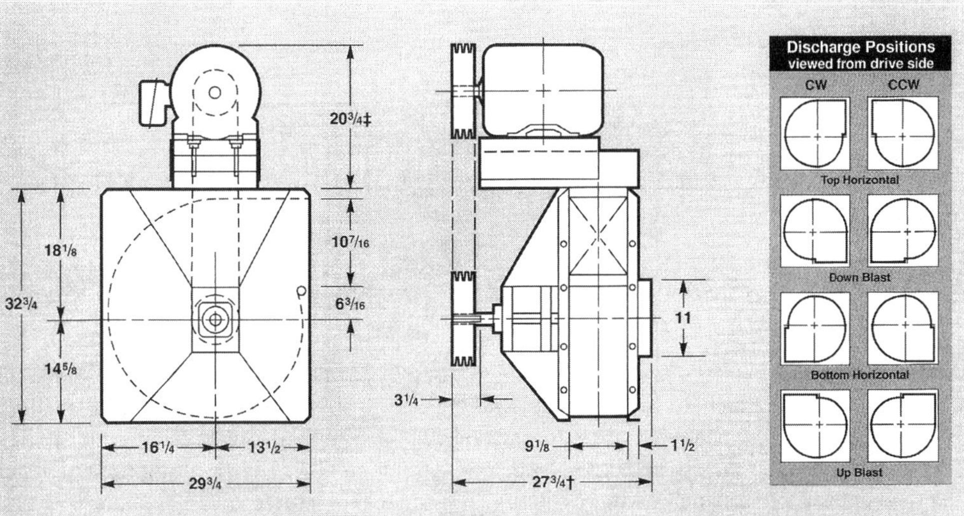

	Shaft Dia.	Keyway	Weight*
Size 11	1⁷⁄₁₆	⅜ x ³⁄₁₆	162

Notes: *Arrangement 9T shown. Arrangement 1 does not include motor or base.*
† Add 3/8" when spark resistant const. (200°F & above) is required.
‡ Dimension applies to maximum motor frame size.
** Bare fan weight in pounds, Arrangement 1.*

DIMENSIONS ARE SHOWN IN INCHES. DO NOT USE FOR CONSTRUCTION UNLESS CERTIFIED.

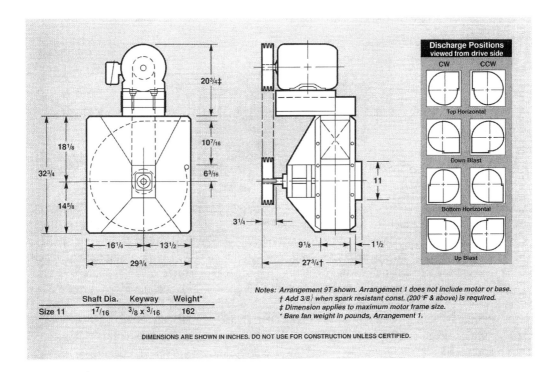

FIGURE 16.25 The fan table for a small radial blade centrifugal fan. (Courtesy of Chicago Blower Corp.)

EXAMPLE 16.4

At what fan RPM would you operate a small backward inclined blade fan such as the Chicago Blower Corp. SQB $12\frac{1}{4}$ whose performance table is provided in Figure 16.26 to run the system in the previous example?

The relevant data from Figure 16.26, the interpolations, and efficiency are shown below:

CFM	Outlet *V* (fpm)	r/min	bhp
1320	1500	1836	0.63
1408	1600	1889	0.69

$$\text{Interpolated RPM} = 1836 + \frac{(1350 - 1320)}{(1408 - 1320)} \times (1889 - 1836) = 1836 + 18 = 1854 \text{ RPM}$$

Interpolated hp = 0.65, although the actual hp is reduced by the density factor to 0.53 hp.

Implied efficiency is 0.401 ideal hp/0.53 actual hp = 76%.

EXAMPLE 16.5

At what RPM would you operate a small airfoil blade fan such as the Chicago Blower Corp. SQA $12\frac{1}{4}$ whose performance table is provided in Figure 16.27 to run the system in the previous example?

The relevant data from Figure 16.27, the interpolations, and efficiency are shown below:

CFM	Outlet *V* (fpm)	r/min	bhp
1320	2000	1982	0.60
1408	2200	2041	0.65

$$\text{Interpolated RPM} = 1982 + \frac{(1350 - 1320)}{(1408 - 1320)} \times (2041 - 1982) = 1982 + 20 = 2002 \text{ RPM}$$

interpolated hp = 0.62, although the actual hp is reduced by the density factor to 0.51 hp. Implied efficiency is 0.401 ideal hp/0.506 actual hp = 79%.

The above system and fans represent a small-scale example. Each fan is a small member of a family of fans that can handle much higher flow rates and pressures (50,000 ft^3/min at a FanSP as high as 15 "wg for the backward inclined and airfoil fans or 40,000 ft^3/min and over 20 "wg for the radial fan). The three example fans were chosen to be comparable in size. The resulting fan efficiencies are not the best in each case but are representative. Of the manufacturer's predicted noise levels of 36, 56, and 64 dBA, respectively, the low noise of the radial fan reflects the much lower RPM at which it needs to operate to run this system.

In summary, the selected fan must provide the Q and FanSP calculated for your system, while operating somewhere to the right of its peak FanSP to be stable and within the efficiency range recommended by the manufacturer. After deciding on the fan type and manufacturer, there are only two variables left in the fan laws: fan size and RPM. After installing a fan, there is only one variable left: its RPM. Each fan actually has a family of Q and FanSP fan performance curves as a function of its RPM (as in Figure 16.3). With all the calculations in hand and the information resources available, selecting the right fan is one of the easiest steps in this whole design process.

CHICAGO **SQB**

Class IIS
RPM 4036

Outlet Area: .88 sq. ft.
Maximum BHP = .104 (rpm ÷ 1000)^3
Tip Speed (fpm) = 3.46 x rpm

SIZE **12 1/4**

CFM	OV FPM	1/4" SP RPM	1/4" SP BHP	1/2" SP RPM	1/2" SP BHP	3/4" SP RPM	3/4" SP BHP	1" SP RPM	1" SP BHP	1-1/2" SP RPM	1-1/2" SP BHP	2" SP RPM	2" SP BHP	2-1/2" SP RPM	2-1/2" SP BHP	3" SP RPM	3" SP BHP	3-1/2" SP RPM	3-1/2" SP BHP	4" SP RPM	4" SP BHP
528	600	688	.03	862	.06																
616	700	747	.04	896	.07	1048	.11														
704	800	811	.05	945	.09	1075	.12	1208	.17												
792	900	877	.07	1003	.10	1116	.14	1233	.18												
880	1000	948	.09	1063	.12	1168	.16	1271	.21	1483	.31										
968	1100	1021	.11	1126	.15	1227	.19	1319	.23	1509	.34	1703	.46								
1056	1200	1096	.13	1191	.17	1287	.22	1375	.27	1546	.37	1723	.49								
1144	1300	1173	.16	1259	.20	1349	.25	1434	.31	1590	.41	1753	.53	1917	.67						
1232	1400	1250	.19	1330	.24	1413	.29	1494	.35	1644	.46	1792	.58	1942	.71	2095	.87				
1320	1500	1328	.23	1403	.28	1479	.33	1557	.39	1702	.51	1836	.63	1976	.76	2118	.92	2261	1.09		
1408	1600	1408	.27	1476	.32	1547	.38	1621	.44	1761	.57	1889	.69	2018	.83	2149	.98	2283	1.14	2416	1.33
1496	1700	1487	.32	1552	.37	1619	.43	1687	.49	1821	.63	1947	.76	2065	.90	2188	1.05	2312	1.21	2438	1.39
1584	1800	1567	.37	1629	.43	1691	.49	1753	.55	1883	.69	2005	.84	2118	.98	2231	1.12	2348	1.29	2466	1.46
1672	1900	1647	.42	1706	.49	1764	.55	1824	.62	1947	.76	2065	.91	2176	1.06	2282	1.21	2391	1.38	2500	1.55
1760	2000	1728	.49	1784	.55	1838	.62	1896	.69	2012	.84	2126	1.00	2235	1.16	2337	1.31	2437	1.47	2542	1.65
1936	2200	1890	.63	1941	.71	1991	.78	2041	.85	2144	1.01	2252	1.18	2355	1.35	2453	1.53	2547	1.70	2638	1.87
2112	2400	2052	.81	2100	.89	2146	.97	2192	1.05	2286	1.21	2383	1.39	2479	1.58	2573	1.77	2664	1.96	2750	2.15
2288	2600	2216	1.01	2260	1.10	2302	1.19	2345	1.27	2431	1.45	2517	1.63	2609	1.83	2697	2.03	2784	2.24	2868	2.44
2464	2800	2379	1.25	2421	1.34	2461	1.44	2500	1.53	2578	1.71	2661	1.91	2741	2.11	2827	2.33	2908	2.55	2989	2.77
2640	3000	2543	1.52	2582	1.62	2620	1.72	2657	1.82	2731	2.02	2806	2.22	2881	2.44	2958	2.66	3037	2.89	3113	3.12

CFM	OV FPM	4-1/2" SP RPM	4-1/2" SP BHP	5" SP RPM	5" SP BHP	5-1/2" SP RPM	5-1/2" SP BHP	6" SP RPM	6" SP BHP	6-1/2" SP RPM	6-1/2" SP BHP	7" SP RPM	7" SP BHP	7-1/2" SP RPM	7-1/2" SP BHP	8" SP RPM	8" SP BHP	9" SP RPM	9" SP BHP	10" SP RPM	10" SP BHP
1496	1700	2563	1.59																		
1584	1800	2585	1.66	2703	1.86																
1672	1900	2612	1.74	2725	1.94	2838	2.16														
1760	2000	2646	1.83	2752	2.03	2860	2.25	2967	2.47	3073	2.71										
1936	2200	2730	2.06	2827	2.26	2922	2.47	3018	2.69	3116	2.92	3212	3.16	3310	3.42	3405	3.68				
2112	2400	2834	2.33	2918	2.53	3005	2.73	3092	2.95	3179	3.18	3268	3.42	3357	3.67	3446	3.93	3624	4.48		
2288	2600	2949	2.65	3027	2.85	3105	3.06	3181	3.27	3263	3.50	3343	3.73	3423	3.98	3506	4.23	3669	4.77	3833	5.35
2464	2800	3068	2.99	3143	3.22	3217	3.43	3288	3.65	3361	3.87	3431	4.10	3508	4.35	3583	4.60	3733	5.13	3883	5.70
2640	3000	3189	3.36	3263	3.60	3334	3.84	3404	4.08	3472	4.31	3538	4.54	3606	4.78	3672	5.03	3813	5.56	3951	6.12
2816	3200	3314	3.77	3384	4.02	3454	4.27	3522	4.53	3588	4.79	3653	5.03	3716	5.28	3779	5.53	3904	6.05	4036	6.61
2992	3400	3443	4.21	3510	4.47	3567	4.74	3643	5.01	3708	5.28	3771	5.55	3832	5.83	3893	6.09	4011	6.62		
3168	3600	3574	4.69	3640	4.97	3703	5.25	3765	5.53	3829	5.82	3891	6.10	3951	6.39	4010	6.68				
3344	3800	3707	5.22	3771	5.51	3833	5.81	3894	6.10	3953	6.40	4012	6.70								
3520	4000	3847	5.81	3904	6.10	3965	6.40	4024	6.71												
3696	4200	3991	6.45																		

If V-Belt driven, do not operate above 3600 RPM

Performance shown is for installation type B: Free inlet, Ducted outlet.
Power ratings (BHP) do not include drive losses.

Performance ratings do not include the effects of appurtenances in the air stream.

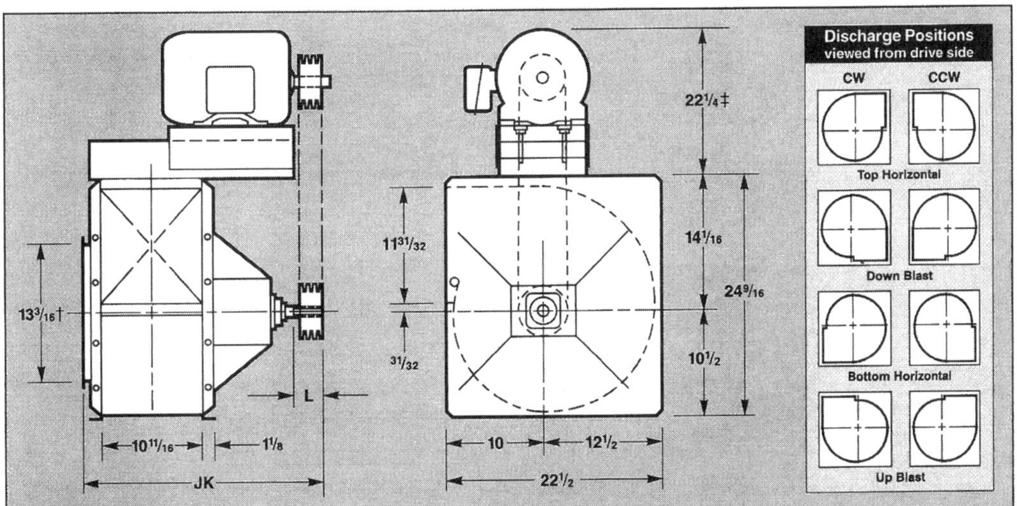

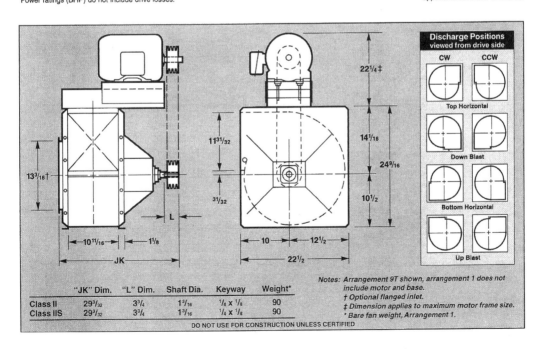

	"JK" Dim.	"L" Dim.	Shaft Dia.	Keyway	Weight*
Class II	29 3/32	3 3/4	1 3/16	1/4 x 1/8	90
Class IIS	29 3/32	3 3/4	1 3/16	1/4 x 1/8	90

Notes: Arrangement 9T shown, arrangement 1 does not include motor and base.
† Optional flanged inlet.
‡ Dimension applies to maximum motor frame size.
** Bare fan weight, Arrangement 1.*

DO NOT USE FOR CONSTRUCTION UNLESS CERTIFIED

FIGURE 16.26 The fan table for a small backward inclined centrifugal fan. (Courtesy of Chicago Blower Corp.)

CHICAGO

SQA

Class I Class II
RPM 4046 4280

Outlet Area: .88 sq. ft.
Maximum BHP = .077 (rpm ÷ 1000)^3
Tip Speed (fpm) = 3.46 x rpm

SIZE 12¼

CFM	OV FPM	1/4" SP		1/2" SP		3/4" SP		1" SP		1-1/4" SP		1-1/2" SP		1-3/4" SP		2" SP		2-1/2" SP		3" SP	
		RPM	BHP	RPM	BHP	RPM	BHP	RPM	BHP	RPM	BHP	RPM	BHP	RPM	BHP	RPM	BHP	RPM	BHP	RPM	BHP
440	500	684	.02																		
528	600	743	.03	921	.05																
616	700	810	.04	964	.06																
704	800	880	.05	1021	.08	1154	.11														
792	900	953	.06	1086	.09	1204	.13	1328	.17												
880	1000	1027	.07	1153	.11	1263	.15	1369	.19	1476	.24	1580	.29								
968	1100	1104	.09	1222	.13	1328	.18	1425	.22	1521	.27	1618	.32	1714	.37						
1056	1200	1183	.11	1294	.16	1395	.20	1487	.25	1576	.30	1664	.35	1753	.41	1841	.46				
1144	1300	1263	.14	1368	.19	1464	.23	1553	.28	1636	.33	1717	.39	1799	.44	1881	.50	2042	.63		
1232	1400	1344	.16	1442	.22	1534	.27	1620	.32	1701	.37	1777	.43	1853	.49	1928	.55	2081	.68		
1320	1500	1426	.20	1518	.25	1607	.31	1689	.36	1768	.42	1841	.48	1912	.54	1982	.60	2125	.73	2267	.87
1408	1600	1511	.23	1597	.29	1681	.35	1760	.41	1836	.47	1908	.53	1975	.59	2041	.65	2173	.79	2308	.93
1496	1700	1594	.27	1676	.33	1755	.39	1832	.46	1905	.52	1975	.58	2042	.65	2104	.71	2229	.85	2355	1.00
1584	1800	1678	.31	1756	.38	1831	.44	1906	.51	1975	.58	2044	.64	2109	.71	2171	.78	2289	.92	2407	1.07
1672	1900	1762	.36	1837	.43	1909	.50	1980	.57	2048	.64	2112	.71	2177	.78	2238	.85	2352	1.00	2464	1.15
1760	2000	1846	.41	1918	.49	1987	.56	2054	.63	2122	.71	2183	.78	2246	.85	2305	.93	2419	1.08	2525	1.24
1936	2200	2017	.54	2084	.62	2146	.70	2208	.78	2269	.86	2331	.94	2388	1.02	2444	1.10	2552	1.27	2656	1.44
2112	2400	2188	.68	2252	.77	2308	.86	2365	.94	2422	1.03	2478	1.12	2535	1.21	2588	1.30	2690	1.47	2789	1.65
2288	2600	2359	.85	2418	.95	2473	1.04	2526	1.13	2579	1.23	2630	1.32	2682	1.42	2734	1.52	2832	1.71	2927	1.90
2464	2800	2536	1.05	2590	1.15	2640	1.25	2688	1.35	2738	1.45	2786	1.56	2786	1.66	2883	1.76	2978	1.97	3067	2.17

CFM	OV FPM	3-1/2" SP		4" SP		4-1/2" SP		5" SP		5-1/2" SP		6" SP		7" SP		8" SP		9" SP		10" SP	
		RPM	BHP	RPM	BHP	RPM	BHP	RPM	BHP	RPM	BHP	RPM	BHP	RPM	BHP	RPM	BHP	RPM	BHP	RPM	BHP
1320	1500			2534	1.19																
1408	1600	2441	1.09	2566	1.26	2688	1.42														
1496	1700	2480	1.16	2605	1.33	2722	1.50	2839	1.68												
1584	1800	2526	1.23	2644	1.40	2762	1.58	2873	1.76	2982	1.95										
1672	1900	2576	1.31	2689	1.49	2801	1.66	2912	1.85	3016	2.04	3121	2.24	3321	2.66						
1760	2000	2632	1.41	2737	1.58	2845	1.76	2951	1.95	3057	2.15	3157	2.35	3351	2.76	3538	3.21				
1936	2200	2753	1.61	2849	1.79	2946	1.97	3042	2.16	3140	2.37	3237	2.57	3427	3.01	3605	3.46	3777	3.94		
2112	2400	2885	1.84	2974	2.03	3063	2.22	3150	2.42	3239	2.62	3328	2.83	3505	3.27	3682	3.75	3849	4.24	4008	4.74
2288	2600	3019	2.10	3107	2.30	3190	2.50	3271	2.70	3353	2.91	3434	3.13	3598	3.58	3762	4.06	3927	4.57	4083	5.09
2464	2800	3156	2.38	3241	2.60	3324	2.81	3402	3.03	3478	3.24	3553	3.47	3705	3.93	3858	4.42	4010	4.93	4162	5.47
2640	3000	3297	2.71	3378	2.93	3459	3.15	3535	3.38	3611	3.62	3682	3.84	3823	4.32	3964	4.82	4106	5.33		
2816	3200	3439	3.06	3519	3.29	3596	3.53	3671	3.77	3745	4.01	3815	4.26	3950	4.75	4083	5.26				
2992	3400	3585	3.44	3664	3.69	3737	3.94	3809	4.19	3880	4.45	3949	4.70	4083	5.23						
3168	3600	3734	3.86	3812	4.13	3882	4.39	3949	4.65	4019	4.92	4087	5.19								
3344	3800	3884	4.32	3960	4.60	4030	4.88	4095	5.16	4160	5.43										
3520	4000	4038	4.81	4108	5.10	4178	5.41														

If V-Belt driven, do not operate above 3600 RPM

Performance shown is for installation type B - Free inlet, Ducted outlet.
Power ratings (BHP) do not include drive losses.

Performance ratings do not include the effects of appurtenances in the air stream.

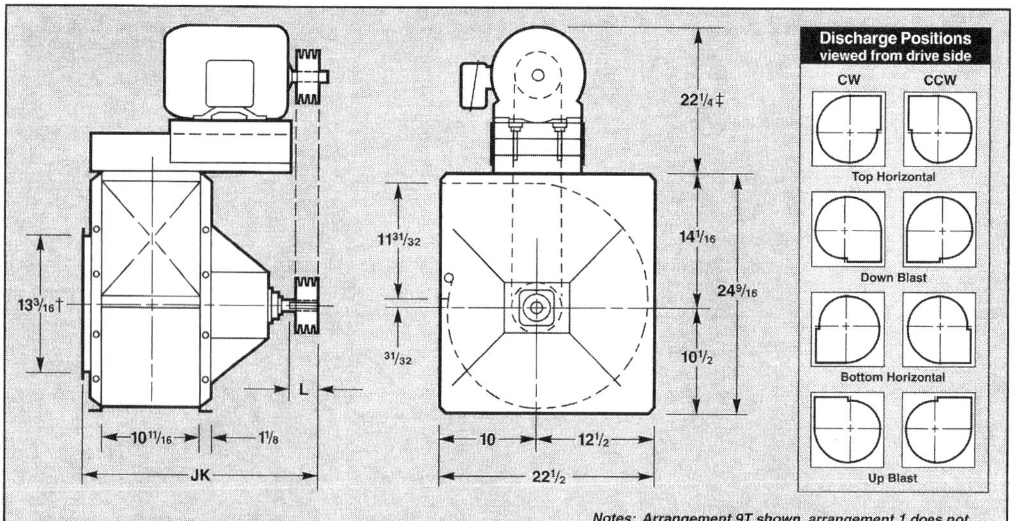

	"JK" Dim.	"L" Dim.	Shaft Dia.	Keyway	Weight*
Class I	31	3½	1 3/16	¼ x ⅛	90
Class II	31	3½	1 3/16	¼ x ⅛	90

DIMENSIONS ARE IN INCHES. DO NOT USE FOR CONSTRUCTION UNLESS CERTIFIED.

Notes: Arrangement 9T shown, arrangement 1 does not include motor and base.
† Optional flanged inlet.
‡ Dimension applies to maximum motor frame size
** Bare fan weight in pounds, Arrangement 1.*

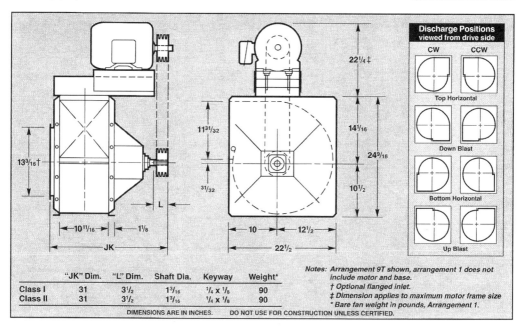

FIGURE 16.27 The fan table for a small airfoil blade centrifugal fan. (Courtesy of Chicago Blower Corp.)

PRACTICE PROBLEMS

The following losses were (or should have been) determined in Chapter 14's Practice Problems for 1500 ft^3/min to which 7′ of stack losses were added:

Loss in hood entry	0.44
Loss in duct entry	0.08
Loss in 20 ft of 11″*D* straight duct	0.14
Loss in elbow	0.08
Loss in tapered contraction	0.07
Loss in 20 ft of 8″*D* straight duct	0.69
Loss for 7 ft of 8″*D* exhaust stack	0.24
Sum of losses	1.75 ″wg

1. What are the values of the two parameters needed to select a fan to operate this system?

   a. _____ cfm b. _____ ″wg
2. At what RPM should a Chicago Blower Corp. SQA $12\frac{1}{4}$ airfoil fan be run to operate this system? One Q value in Figure 16.27 is close enough to 1500 cfm to forego interpolation but sticklers can add 3 rpm.

   _____ rpm
3. Use the fan's outlet area of 127.9 sq.in. = 0.88 sq.ft. in Figure 16.27 for the following calculations.

   a. What are the fan's outlet velocity and VP_{outlet}? (The "OV" in the fan tables is the fan's outlet velocity).

   $V_{fan\ outlet}$ = _____ fpm

   VP_{outlet} = _____ ″wg

   b. What is the FanTP for this fan?

   FanTP = _____ ″wg
4. What size fan motor does the fan table indicate is needed to operate the fan? [BHP means "brake horsepower" to indicate a way of testing.]

   _____ bhp
5. What is the apparent mechanical efficiency of that fan using (looking ahead to) Equation 17.4a?

   $\%_{mechanical\ efficiency}$ = _____
6. Compare the above efficiency to the efficiency of the Chicago Blower SQI 11LS radial fan (Figure 16.25) that to move 1500 cfm against a 1.75 ″wg FanSP will generate a FanTP of 2.07 ″wg (because of its smaller outlet area) at 1056 rpm and will consume 0.79 bhp. This operating point is just outside the low end of its optimum performance envelope.

   $\%_{mechanical\ efficiency}$ = _____
7. At what RPM should the more efficient SQA $12\frac{1}{4}$ fan be operated in order to obtain 1500 cfm and 1.75 in. FanSP at Logan Utah's altitude of 4700 ft? At what horsepower will this fan be running at this new RPM after adjusting for the low air density at this altitude?

   FanSP used in fan table = _____ ″wg

   "density correction factor" = _____

   RPM at 4700 feet = _____

   horsepower in the table at the above RPM = _____ hp

   operating horsepower adjusted for air density = _____ hp

REFERENCES

1. Committee on Industrial Ventilation, *Industrial Ventilation: A Manual of Recommended Practice*, 24th ed., American Conference of Governmental Industrial Hygienists, Cincinnati, OH, 2001.
2. *Fans and Systems*, AMCA Publication 201-02, Air Movement and Control Association International, Inc., Arlington Heights, IL, USA.
3. *Certified Ratings Programme — Product Rating Manual for Fan Air Performance*, AMCA Publication 211-05, Air Movement and Control Association International, Inc., Arlington Heights, IL, USA.
4. AIHA/ANSI Z9.1, *Practices for Ventilation and Operation of Open-Surface Tanks*, American Industrial Hygiene Association, Fairfax, VA, 1991.
5. Vatavuk, W. M., *Estimating Costs of Air Pollution Control*, Lewis Publisher, Chelsea, MI, 1990.
6. Vanderburgh, C. R. and Paulauskis, A., The causes of unwanted results of aerodynamic system effect, *ASHRAE J.*, 36(2), 24–29, 1994.

17 Ventilation Operating Costs

The learning goals of this chapter:

- ☐ Know how to use the nominal fan electricity annual costs of \$100 per 1000 ft^3/min per "wg to estimate the fan operating costs for LEV systems.
- ✍ Understand the concept of heating degree days and cooling degree days and how they are used to estimate the make-up air conditioning costs.
- ✍ Be familiar with the differences in heating costs among heating sources and between heating and fan electrical costs.
- ☐ Be able to use either Equation 17.9a and Equation 17.9b or Equation 17.11 to estimate the make-up air heating costs at a given flow rate Q.
- ☐ Be able to use the thermal coefficient table to estimate the cooling costs at a given flow rate Q.

The money to design a local exhaust ventilation system, to buy its materials (capital costs), and to build and install everything (sometimes called labor costs) is all spent before someone flips the switch to turn it on. The money to operate the system is paid year after year. The following activities each contribute to exhaust ventilation operating costs. Only the energy-related costs are covered within this book.

- Air pollution permits, fees, compliance monitoring, and employee education.
- Energy to operate the fan and to heat, cool, and/or otherwise condition the make-up air.
- Maintenance of hoods, inlets, motors, air cleaners, flexible collars, hoods, etc.
- Air cleaner waste handling and recycling or disposal.

I. FAN EFFICIENCY AND POWER CONSUMPTION

A fan's efficiency affects how much electrical power it will consume. An inefficient fan will use more electricity and cost more to run than an efficient fan. As suggested in the previous chapter, a fan's efficiency is one variable that can and should be used to select an exhaust fan.

1. The energy consumed by any machine is always more than the energy actually put into whatever the machine is supposed to do. Efficiency is the fraction (usually expressed as a percent) of the energy or power consumed by the machine that actually does whatever it is supposed to do. (Recall from physics that power is the rate that energy is used.) Efficiency can be defined for a mechanical system (such as a fan), for an electrical system (such as a motor or heater), or for a thermal system (such as a furnace or refrigerator). The overall mechanical efficiency of a fan defined by Equation 17.1 includes both its electrical and its mechanical components. Fans are not very efficient (50–80%) in comparison to electric motors (>95%) and belt drive systems (90–95%). Most of the inefficiency of fans is due to the way the fan blades interact with the air

as discussed in Chapter 16.

$$\%_{\text{mechanical efficiency}} = \frac{\text{power imparted to the air}}{\text{power consumed by the fan}} \times 100 \tag{17.1}$$

2. The power imparted to the air relates back to Principle 10 in Chapter 10 (that the fan must replace the lost energy and create the VP of the air that leaves the ventilation system), to the definition in Chapter 11 of Q as the volume per minute (ft^3/min or m^3/min) of air being moved, and to the definitions of static pressure, velocity pressure, and total pressure as expressions of energy per unit of volume. Recall from Chapter 16 that FanTP is the difference in the total pressure (TP_{outlet} minus TP_{inlet}) imparted by the fan to each unit volume of the air. Thus, the product of $Q \times$ FanTP is the rate per minute at which energy is transferred into the moving air by the fan, driven by a fan motor, and usually powered by electricity (or occasionally by steam).

$$\text{Power imparted to air} = Q \times \text{FanTP} = \frac{\text{Energy}}{\text{Volume}} \times \frac{\text{Volume}}{\text{minute}} = \frac{\text{Energy}}{\text{minute}} \tag{17.2a}$$

While the power imparted to the air could be expressed in various units, motor or engine power in the USA, is commonly defined in units of horsepower (hp) defined as 33,000 ft-lb/min.[a] Horsepower is sometimes also called brake horsepower and hence is also abbreviated as either bhp or BHP because of the way motors were historically tested in the lab. The following equivalence between horsepower and the product of $Q \times$ FanTP assumes ideal or 100% fan efficiency.[b]

$$hp_{\text{imparted to the air}} = \frac{\text{FanTP}\ ['' \text{wg}] \times Q\ [\text{ft}^3/\text{min}]}{6356} \quad \text{at 70°F air and one atmosphere} \tag{17.2b}$$

Example 17.1

What is the power imparted to the air in Example 14.2 where $Q = 1350$ ft^3/min and the FanSP = 1.63 "wg (at sea level).

a. If a fan has not yet been selected, one can assume that the fan will have about the same inlet and outlet areas as chosen duct. In the 8 in. duct of Example 14.2, $V_{outlet} = 3867$ fpm $\times$ (0.349/0.660) = 2045 fpm and $VP_{outlet} = 0.261$ "wg. Therefore,

$$\text{FanTP} = \text{FanSP} + VP_{outlet} = 2.00 + 0.261 = 2.26\ ''\text{wg} \quad \text{(using Equation 16.4b)}$$

$$hp_{\text{imparted to the air}} = \frac{Q \times \text{FanTP}}{6356} = \frac{1350\ \text{cfm}(1.63 + 0.932\ ''\text{wg})}{6356} = 0.544\ \text{hp} \tag{17.2b}$$

[a] Because Q is expressed in ft^3/min and FanTP is expressed in inches of water gauge, the units of power imparted to the air could (hypothetically) be expressed as "wg $\times$ ft^3/min. If ventilation pressure were expressed as pounds per square foot (which it is not but could be), the power imparted to the air would be in units of ft-lb/min. A foot $\times$ pound per minute is a nonmetric unit for power often used in U.S. mechanical engineering design.

[b] The *Manual of Industrial Ventilation*'s use of a conversion value of 6362 in its Equation 6.4[1] corresponds to a temperature of 25°C (77°F) which is neither its usual normal temperature nor the value of 6356 used in Equation 17.2a, 17.2b, Equation 17.4a, Equation 17.4b and Equation 17.4c and most other texts. A table in the Appendix to Chapter 11 explains the derivation of these conversion values.

b. If the fan has been selected, its outlet area is known. For instance, the outlet area of the radial blade fan in Figure 16.25 is 0.66 ft^2 (near the top of the page; these fan tables also provide the outlet velocity in the second column of data).

$$\text{hp}_{\text{imparted to the air}} = \frac{Q \times \text{FanTP}}{6356} = \frac{1350\ \text{cfm} \times (1.63 + 0.261\ ''\text{wg})}{6356} = 0.401\ \text{hp} \tag{17.2b}$$

3. The power consumed by the fan is almost always supplied by an electric motor. Thus, the fan motor has to be rated for (capable of producing) the amount of power to be consumed by the fan. Electrical power is quantified in units of watts (W) and kilowatts (kW). A watt is the product of volts × amps (and phase factor in alternating current), and thus occasionally power is also written as volt-amperes.

$$1\ \text{kW}_{\text{consumed}} = \text{amps} \times \text{volts} \times [\text{"phase factor" in AC}]/10^3 \tag{17.3a}$$

For instance, a light bulb that draws 0.83 A (with a phase factor of one) on a North American 120 V AC circuit consumes 0.83 × 120 = 100 W of power. Of course, there are also equivalences between watts, kilowatts, and horsepower. That 100 W × $\text{kW}/10^3\text{W}$ = 0.1 kW = 0.134 hp.

$$1\ \text{kW}_{\text{consumed}} = 44{,}253\ \text{ft} \times \text{lb/min} = 1.341\ \text{hp}_{\text{consumed}} \tag{17.3b}$$

A phase factor accounts for a temporal or phase difference that can exist in alternating current (AC) between the sinusoidal cycles of volts and amps. The impedance (very much like resistance) in an induction load is high when the volts are changing rapidly (as they cycle through zero) and low when the volts pass through a peak or valley. Thus, the amps that pass through an induction load (such as a large electric motor, a transformer, or an induction heater) lag the volts. The impedance in a capacitive load is high when the volts do not change much as they pass through their peak and valleys, but is low when they change rapidly (again when they pass through zero). Thus, the amps that pass through a capacitive load lead the volts. A large induction electric motor might have a phase factor of about 0.80, while a small capacitor motor can have a phase factor near one. Transformers or induction heaters (which have nothing to do with ventilation fans) can have phase factors of less than 0.5. A phase factor will be used in Example 17.2, but fortunately for most of us, phase factors are not an IH responsibility.

4. The above relationships result in the three potentially useful equations listed below for a fan's mechanical efficiency [$\%_{\text{mechanical efficiency}}$] and power based on ventilation fan total pressure (FanTP) in units of "wg and flow rate (Q) in ft^3/min. The difference between FanTP and FanSP is the VP at the fan outlet (Equation 16.4). For a V_{outlet} of around 4000 fpm, this difference would be around 1 "wg.[b]

$$\%_{\text{mechanical efficiency}} = \frac{Q \times \text{FanTP} \times 100}{6356 \times \text{hp}} = \frac{Q \times (\text{FanSP} + \text{VP}_{\text{outlet}}) \times 100}{6356 \times \text{kW} \times 1.34} \tag{17.4a}$$

$$\text{hp}_{\text{consumed}} = \frac{Q \times \text{FanTP} \times 100}{6356 \times \%_{\text{mechanical efficiency}}} = \frac{Q \times (\text{FanSP} + \text{VP}_{\text{outlet}}) \times 100}{6356 \times \%_{\text{mechanical efficiency}}} \tag{17.4b}$$

$$kW_{consumed} = \frac{Q \times FanTP \times 100}{8523 \times \%_{mechanical\ efficiency}} = \frac{Q \times (FanSP + VP_{outlet}) \times 100}{8523 \times \%_{mechanical\ efficiency}} \tag{17.4c}$$

II. ELECTRICITY COSTS FOR FANS

1. Electricity consumers are billed per kilowatt hour [kWh]. A kWh is the energy consumed by one kilowatt of power in 1 h. The 0.1 kW light bulb running for 24 h would consume 2.4 kWh of electrical energy. Electricity in the U.S. is sold for between $0.05 to 0.10 per kWh depending upon the supplier's location and the user's consumption rate. Some utilities also charge for a consumer's phase factor. Large industrial users can purchase electricity (with some restrictions) for 50 to 90% of the residential rate (see, for example, www.eia.doe.gov/cneaf/electricity/epav1/fig12.html for rates by state and www.eia.doe.gov/cneaf/electricity/epm/table5_6_a.html for rates by sector). At a rate of 7¢ per kWh, that 2.4 kWh would cost a consumer 17¢ per day, $5 per month, or approximately $60 per year.
2. Since electricity consumption is charged on the basis of kilowatt-hours, the cost to run a ventilation fan (excluding heating and cooling) depends on the fan's power consumption (kW), how long the system is operated (hours), and the cost of electrical energy ($/kWh), as shown in Equation 17.5:

$$\begin{array}{lcccccc} \text{cost} & = & (\text{power consumption}) & \times & (\text{usage}) & \times & (\text{energy cost}) \\ \$_{fan} & = & kW & \times & \text{hours} & \times & \$/kWh \end{array} \tag{17.5}$$

Example 17.2

Estimate the annual operating costs if a radial fan were used to operate the bag emptying LEV system designed in Example 14.2 where $Q = 1350$ ft³/min and the FanSP = 1.3 "wg (at sea level).

In order to calculate the FanTP from a FanSP using Equation 17.4b, one needs to know the VP_{outlet}. After a fan is selected, both its horsepower requirement and its inlet and outlet sizes would be known from the manufacturer's literature (as in Figure 16.25, etc.), and if we wanted to know a given fan's efficiency, one could calculate it from Equation 17.4a using the manufacturer's data. But we will not let that interfere with a good learning example!

a. Being asked to estimate means that a fan has not yet been selected. In that case, one can make the same assumptions about the fan inlet and outlet areas as used in the first solution to Example 17.1. One good way to estimate is to assume a middle value. An efficiency of 70% is midway between the radial and backward inclined fans described in subsection VI.2 of Chapter 16. The size of the fan motor can then be estimated using Equation 17.4b:

$$hp = \frac{Q \times FanTP \times 100}{6356 \times \%_{mechanical\ efficiency}} = \frac{1350\ cfm \times (1.30 + 0.93)\ "wg \times 100}{6356 \times 70\%} = 0.68\ hp \tag{17.4b}$$

The electrical power consumed can be predicted using Equation 17.4c or derived by converting hp:

$$kW_{consumed} = 0.68\ hp \times \frac{1\ kW}{1.34\ hp} = 0.50\ kW \qquad \text{(using Equation 17.3b)}$$

b. The amperage with a phase factor of 0.95 can be determined using a variation of Equation 17.3a:

$$\text{Amps} = 500\ \text{watts}/(120 \times 0.95) = 4.4\ \text{amps for 120-volt wiring or}$$
$$\approx 500/(220 \times 0.95) = 2.4\ \text{amps for 220-volt AC power}$$

(from Equation 17.3a)

c. The annual costs calculated using Equation 17.5 assuming continuous operation (365 days/year × 24 h = 8760 h/year):

$$\$_{\text{fan}} = (0.50\ \text{kW}) \times (8760\ \text{h/year}) \times (0.07\ \$/\text{kWh}) = \$307/\text{year}$$

(using Equation 17.5)

If on the other hand, the fan were only run during normal business hours (5 days/week × 50 week/year = 250 days/year × 9 h/day = 2250 h/year or approximately 25% of 8760 h/year of continuous use):

$$\$_{\text{fan}} = (0.50\ \text{kW}) \times (2250\ \text{h/year}) \times (0.07\ \$/\text{kWh}) = \$70/\text{year}$$

(using Equation 17.5)

3. The cost to move 1000 ft^3/min with a FanTP of 1 "wg and a 70% efficient fan (midway between radial and backward inclined fans) full-time (8760 h/year) at 0.07 \$/kWh is a useful point of reference. This cost approximates the power to operate a 150 W light bulb (or two 75 W light bulbs, etc.)

$$\text{kW} = \frac{Q \times \text{FanTP} \times 100}{8523 \times \%_{\text{mechanical efficiency}}} \quad \frac{1000 \times 1 \times 100}{8523 \times 70\%} \approx 0.17\ \text{kW}$$

(again using Equation 17.4c)

$$\$_{\text{fan}} = 0.17\ \text{kW} \times 8760\ \text{h/year} \times 0.07\ \$/\text{kWh} = \$103\ \text{per year}$$

(again using Equation 17.5)

✓ The electricity to power a fan moving 1000 ft^3/min at 1 in. Fan TP is approximately \$100 per year.

This \$100 per year can be scaled up to any larger LEV system in proportion to its Q (in multiples of thousands) and/or its FanTP. For instance, a large plant exhausting 1,000,000 ft^3/min at an average FanTP of 4 "wg would be paying \$400,000 per year just to run the fans. However, the next sections will show that the cost to heat and cool the make-up air commonly exceeds the cost just to move the air.

III. MAKE-UP AIR

Fresh air entering a building is called make-up air. This section describes reasons why the flow of make-up air should be controlled rather than relying on infiltration. The flow of make-up air can be controlled by providing a separate fan (and often ducting) to force air into the building. Infiltration is the natural flow of air into a building through random leaks in doors, windows, vents, etc.

1. Resistance to the flow of fresh air into a building via uncontrolled infiltration will generally be higher than if make-up air is provided by a fan. Resistance to infiltration is just another loss that will increase the FanSP (and FanTP) required of the exhaust fan. A higher than expected loss will lower the flow below that for which the system was designed and perhaps below that needed for employee protection.
2. Trying to exhaust air against the resistance of infiltration will create a negative difference between the pressure inside the room or building being exhausted and the atmospheric pressure outside the building. Adverse effects start to happen when the negative pressure exceeds the following recommended limits[1]:
   a. 0.01–0.05 "wg can impair the normal exhaust flow up short natural-draft chimneys.
   b. 0.03–0.1 "wg can impair the normal exhaust flow of propeller fans (perhaps severely).
   c. 0.1–0.25 "wg can impair the normal flow of centrifugal fans (noticeably).
3. Negative building pressure can cause back-flow to occur in other exhaust systems. The weakest exhaust systems are most commonly a building's space and water heater that only rely on thermal buoyancy (the "natural draft chimney" described in subsection III.3 of Chapter 15). The failure of such chimneys to exhaust combustion products will result in carbon monoxide being dispersed into the building (high humidity from water vapor also produced by combustion can also create condensation problems).
4. A negative building pressure can also result in back-flow of disagreeable sewage odors through water traps. Building codes specify that the height of a water trap (such as shown in Figure 17.1) should be $1\frac{1}{2}$ to 2 in. depending upon the fixture type. However, the minimum pressure difference (ΔP) necessary to allow sewer gas to flow into a building can be much less than the trap's height, due either to the hydrodynamics that can occur within the trap because of pressure fluctuations or to evaporation from traps to which not much water is added routinely (such as in floor drains).
5. The forces created by larger negative building pressures can become more than a nuisance. Recall from Equation 2.3b that force $= \Delta P_{building} \times$ Area. For example, if a 1 "wg $\Delta P_{building} = 5.211$ psf were applied to a 20 ft^2 door (3×6.5 in.), the force would be more than 100 lb! A more common $\Delta P_{building}$ of 0.05 to 0.10 "wg causes 5 to 10 lb of force, which is noticeable, undesirable, capable of causing minor injuries, but often tolerated for years.

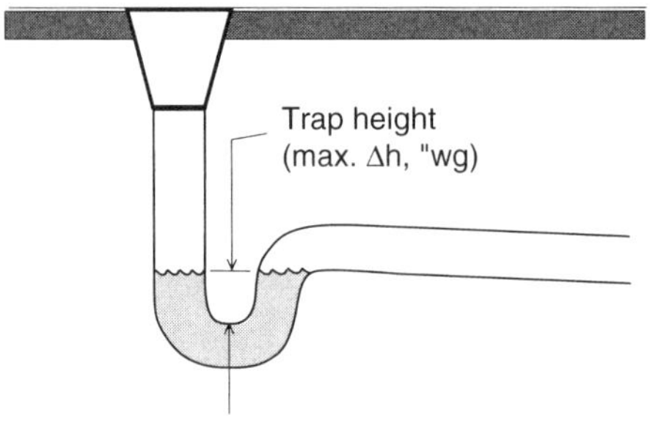

FIGURE 17.1 The height of a water trap.

6. Uncontrolled fresh air (infiltration) is not conditioned in terms of temperature, humidity, or dust level. Untempered air may be too cold causing drafts, too hot for employee comfort, or adversely affect product quality. Dirt and dust in unfiltered air may also cause productivity problems from dust deposition, increased housekeeping costs for cleaning, or the growth and dissemination of microbial aerosols (one source of the "sick building" syndrome).
7. Finally, some OSHA ventilation standards such as exhaust hoods for dipping and coating operations in 1910.124(d)(1) and 1926.57(i)(8) specify that between 90 and 110% of the volume of exhaust air must be provided as make-up air (implying mechanically controlled air flow); however, specifying make-up air volume is not a common OSHA requirement.

Whether the make-up air is controlled or not, the mass and volume of air being exhausted out by fans will always equal the mass and volume of air flowing in. The inlet for the make-up air supply should not be located where it can draw air contaminants into the building along with the fresh air. Such contaminants can come from the building's own exhaust stacks (discussed in Section VI of Chapter 15), from an emission from a neighbor, from an intermittent source such as a truck loading dock, or from an unexpected source such as an employee smoking break area. Only recently has the vulnerability of air inlets to someone purposefully adding a hazardous chemical or biological agent to a building's fresh air supply been recognized. NIOSH recommends several simple actions that could be taken to thwart someone attempting to sabotage a building's ventilation system.[2]

IV. HEATING COSTS FOR MAKE-UP AIR

1. The thermal efficiency of heaters as defined in Equation 17.6 is conceptually very similar to the mechanical efficiency of a fan (as defined in Equation 17.1). In other words, heaters typically consume more thermal energy than is actually transferred to the make-up air or delivered into the room as heat.

$$\%_{\text{thermal efficiency}} = \frac{\text{Heat transferred to the air}}{\text{Heat consumed by the furnace}} \times 100 \tag{17.6}$$

Thermal efficiencies vary among heating systems, particularly between the two broad categories of direct fired vs. indirect fired furnaces. Heat pumps comprise a totally separate category that can have an efficiency equivalent to 200 to 300% but are expensive to purchase (see Section IV.A.3 in Chapter 20).

a. In a direct fired furnace, all products of combustion are released directly into the hot air stream going into the room. Direct fired furnaces may be either permanent (e.g., large heating units used in a heavy industry factory) or mobile (e.g., small heating units commonly used in the construction industry, often called a salamander heater). The advantages of direct fired heaters include the fact that they are nearly 100% efficient and a chimney is not required. The major disadvantage of direct fired heaters is the fact that combustion products (primarily water [H_2O] and carbon dioxide [CO_2] and a small amount of carbon monoxide [CO]) accumulate within the room air. High efficiency heaters normally produce very low concentrations of CO. Thus, the condensation of water vapor on cool surfaces can actually be a bigger problem than the build-up of CO or even CO_2. Combustion products of any chlorinated hydrocarbon vapors present in the room air include corrosive hydrochloric acids. In addition, direct fired heaters are largely limited to using gaseous fuels such as natural gas or propane, and they may be regulated by local ordinances.

b. In an indirect fired furnace, the products of combustion are kept separate from the heated air entering a space. Heat passes from the combustion products to the room air via a heat exchanger (typically a series of metal tubes often with fins). Most commercial and nearly all institutional and residential heaters use indirect fired furnaces. Their advantages include avoiding the release of combustion products into occupied heated zones and allowing the use of oil or even coal as a fuel. The main disadvantages of indirect fired heaters include the fact that large size units are generally less efficient than direct fired heaters and all indirect furnaces require a chimney.

2. The amount of heat transferred either to or from any gas can be calculated either by its change in enthalpy or by using its specific heat and change in temperature. Enthalpy is a measure of the total energy content of a parcel of a gas such as air. Specific heat defines the energy needed to change the temperature of one unit mass of any substance by one degree. The specific heat of air varies strongly with its water content (its absolute humidity) and slightly with temperature, but because cold air cannot hold much water, its specific heat approximates a constant value.[c] Thus, the heat transferred to make-up air could be calculated as the product of the exhaust air flow rate, the air's specific heat, and the temperature difference between outside and inside conditions at any given time. Unfortunately, outside air temperature changes throughout a day and from day to day with the seasons.

   The concept of annual heating degree days is a clever way that allows one to calculate the total amount of energy needed to heat make-up air over an entire heating season. Annual Heating Degree Days (typically denoted as AHDD but denoted as "dg" in the *Manual of Industrial Ventilation*[1]) is the total number of heating degree days in a given or average year at a given geographic location. A traditional heating degree day is the number of degrees that the time-weighted average outside air temperature for any given day is below 65°F. A temperature of 65°F (18°C) is a historic rule-of-thumb for the temperature where a building's heating requirement just balances its cooling load from internal heat sources such as lights, machinery, and people. Any day whose average temperature is above 65°F has zero heating degree days and contributes nothing to the total AHDD. Neither wind speed nor the wind chill factor is used to calculate a heating degree day. Appendix C to this Book lists 30-year average AHDD values tabulated by the U.S. National Climatic Center for selected locations in units of x days/year. The NOAA climatography table also lists AHDD values at both higher and lower temperature bases.[d,3]

3. Because a heating degree day is a 24-h average, AHDDs are best suited to calculate the annual energy needed to heat make-up air for exhaust ventilation that is run continuously (24/7). Equation 17.7b, Table 17.2 herein, Table 7.7 in the *Manual of Industrial Ventilation*, and many other "book" solutions all assume that the amount of heat needed for part-time operations is proportional to the fraction of time that the heat is needed, independent of the portion of the day that it is needed. For make-up air in a commercial building, the "fraction of time running" in Equation 17.7b should be estimated as the

[c] Technically, gases (including air) have two specific heats. C_p is for heating at constant pressure (such as an open atmosphere or building) and C_v is for heating at constant volume (such as within a gas cylinder). Their values differ by approximately 40%. Both coefficients vary slightly with temperature, but they can be approximated by a constant value within the temperature range of human work and comfort as used here. C_p for dry air is approximately 0.24 Btu/lb/°F and $\rho \times C_p = 0.018$ Btu/ft^3/°F.

[d] Some modern commercial buildings use a lower basis for both heating and cooling because they have better insulation and more internal sources of heat than when the value of 65°F was first estimated. Such buildings must continue to be cooled mechanically even when the outside air temperature dips below 65°F. NOAA, and even some local heating companies, also maintain current seasonal running totals, see www.cpc.ncep.noaa.gov/products/analysis_monitoring/cdus/degree_days/.

ratio of the number of hours the exhaust system is used per week over 168 h in a full 7-day week, as shown in Equation 17.9b. The typical 5 days per week 8-to-5 shift is 45 h out of 168 h per week or roughly 25%.[e]
For heating:

$$\text{Btu/year} = [\text{air flow rate}] \times [\text{air density}] \times [\text{specific heat}] \times [^\circ\text{F}\times\text{days/year}] \times [\text{time}]$$

$$= Q[\text{cfm}] \times \rho[\text{lb/ft}^3] \times C_p[\text{Btu/lb/}^\circ\text{F}] \times \text{AHDD} \times (60\times 24[\text{min/day}])$$

$$\text{Btu/year} = Q[\text{ft}^3/\text{min}] \times 0.018[\text{Btu/ft}^3/^\circ\text{F}] \times 1440[\text{min/day}] \times \text{AHDD}[^\circ\text{F day/year}] \quad (17.7\text{a})$$

$$\text{Btu/year transferred to the air} = Q \times 25.9 \times \text{AHDD} \times \text{fraction of time running} \quad (17.7\text{b})$$

EXAMPLE 17.3
What is the annual energy needed to heat 1000 ft^3/min of continuous make up air in the northern Wasatch region? The annual heating degree days listed in Appendix C for cities in the Wasatch front is approximately 6500°F days.

$$\text{Btu heat/year} = Q[\text{cfm}]\times 25.9[\text{Btu/ft}^3/\text{day/}^\circ\text{F}]\times\text{AHDD} \qquad (\text{using Equation 17.7b})$$

$$\text{Btu heat/year} = 1000 \times 25.9 \times 6500 = 168 \times 10^6 \text{ Btu/year, delivered to the air.}$$

4. The heat consumed (and usually purchased) can be calculated from the heat delivered as found above by rearranging the thermal efficiency equation for furnaces (Equation 17.6) into Equation 17.8.

$$\text{Heat } \textit{consumed} \text{ by the furnace} = \frac{\text{Heat transferred to the make-up air}}{\text{Furnace}\%_{\text{thermal efficiency}}/100} \quad (17.8)$$

5. However, unlike fan energy that is almost always powered by electricity, there are several common sources of heat (broadly called fuels herein) that differ considerably in the units in which they are bought and delivered, in their cost/Btu, and in the efficiency of the furnace or how well the energy is transferred into the air. One of the versions of Equation 17.9 with appropriate values should be usable in any climatic region, using any heating technology, and in changing market prices.

$$\text{Annual } \$_{\text{heating}} = \frac{\text{Heat put into air Btu/year}}{\%_{\text{thermal efficiency}}/100} \times \frac{\$/\text{unit of fuel}}{\text{Btu/unit of fuel}} \times \begin{array}{c}\text{fraction of}\\ \text{time in use}\end{array} \quad (17.9\text{a})$$

$$\text{Annual } \$_{\text{heating}} = \frac{Q \times 25.9 \times \text{AHDD}}{\%_{\text{thermal efficiency}}/100} \times \frac{\$/\text{unit of fuel}}{\text{Btu/unit of fuel}} \times \frac{\text{hours used/week}}{168 \text{ h/week}} \quad (17.9\text{b})$$

[e] The use of a fraction of the AHDDs (as in Equation 17.7b) will overestimate the actual make-up air heat required for an exhaust ventilation system that is only run during the day shift because those hours usually comprise the warmest part of a day. Less heat will also be needed for a continuously operated system if the building thermometer automatically sets back to a lower temperature at night (and resets to a higher setting before work starts the next morning).

To calculate the annual cost to heat make-up air using Equation 17.9a and Equation 17.9b, one would need to know the following information (cf., an alternative (somewhat generic) method using Equation 17.11 to follow):

a. The flow rate of make-up air (Q) to be heated. The make-up air flow rate will equal the exhaust air flow rate either calculated during design or measured on one or more existing systems. The less the Q, the cheaper the cost; but too little Q might mean inadequate employee protection.
b. The annual heating degree days of the location. The AHDD reflects the location's climatic environment in °F × days/year. A value for AHDD can be found in Appendix C.
c. The furnace's combustion or thermal efficiency. Combustion efficiency varies with the fuel and the furnace or the central heating plant's manufacturer and perhaps their model. A generic value from Table 17.1 can be assumed until a more specific value is provided by a manufacturer.
d. The cost of the fuel in $ per unit in which the fuel is bought. Both the price and units vary among the fuel sources, and the price is subject to both the local fuel supplier and market fluctuations.
e. The fuel's energy content, Btu per unit of fuel. Energy content varies broadly among fuels, but a generic value from Table 17.1 is probably as good as you need or can get from a supplier.
f. The fraction of time (e.g., hours per week) the exhaust fan is to be run. The running time is determined by the user and/or use conditions.

Equation 17.11 represents an alternative method to calculate the cost per year if you already know the Btu/year, for instance from Equation 17.7a and 17.7b. Table 17.1 shows recent costs calculated using Equation 17.10 for four common heating sources. This table consolidates the cost of the fuel, its energy content, and furnace efficiencies for each fuel to yield the "cost per 10^6 Btu". Equation 17.11 can be used with the cost per million Btu from either the last line from Table 17.1 using recent USA values or from Equation 17.10 using other market and/or technology conditions.

$$\text{Cost}/10^6\ \text{Btu delivered} = \frac{100 \times \text{Cost of fuel [\$/unit]}}{\text{Energy content [Btu/unit]} \times \%_{\text{thermal efficiency}}} \quad (17.10)$$

$$\text{Annual } \$_{\text{heating}} = \frac{\text{Btu/year from Equation 17.7}}{10^6\ \text{Btu}} \times \text{Cost per } 10^6\ \text{Btu} \quad (17.11)$$

TABLE 17.1
Examples of Heating Costs per Million Btu Delivered to the Air

	Coal[a]	Natural Gas[b]	Fuel Oil[c]	Electricity
Cost of fuel ($/unit)	45 $/ton	0.005 $/ft^3	1.00 $/gal	0.07 $/kWh
Energy content (Btu/unit)	13,000 B/lb	1000 B/ft^3	140,000 B/gal	3413 B/kWh
Furnace efficiency	~45%	~90%	~75%	~100%
Cost per 10^6 Btu delivered	3.85 $	5.56 $	9.52 $	20.51 $

[a] Coal is priced per ton (2000 lb), but its heat content is rated per pound (ranges 12,000 to 15,000 Btu/lb).
[b] Range of 1000 to 1100 Btu/ft^3. Natural gas may be priced per "Therm" which equates to approximately 100 ft^3.
[c] The viscosity and heat content of fuel oils vary by grade from approximately 137,000 to 152,000 Btu/gal.

TABLE 17.2
The Annual Costs to Heat 1000 ft^3/min of Make-up Air in a 6500 AHDD Region Using Various Fuels and for Electricity to Run the Fan against a 2 ″wg FanTP

	Continuous Operation	25% Operation (for 25% Use)
Electrical heating costs	\$3445	\$861
Fuel oil heating costs	\$1600	\$400
Natural gas heating costs	\$935	\$234
Coal heating costs	\$650	\$163
Fan (2 ″wg) motor electricity costs	+\$200	+\$50

EXAMPLE 17.4

Calculate the annual cost to heat the 1000 ft^3/min in Example 17.3 if heating with natural gas in a 6500 annual heating degree day environment typical of northern Utah.

$$\text{Annual }\$_{\text{heating}} = \frac{1000 \times 25.9 \times 6500}{90/100} \times \frac{0.005/\text{ft}^3}{1000\ \text{B}/\text{ft}^3} \times \frac{168}{168} = \$935$$

(using Equation 17.9b)

Alternatively (using 168×10^6 Btu/year to heat air 1000 ft^3/min in this region from Example 17.3):

$$\text{Annual }\$_{\text{heating}} = 168 \times 10^6 \text{Btu/year} \times 5.56\ \$/10^6\text{Btu} = \$935$$

(using Equation 17.11)

6. Table 17.2 compares the nominal costs per 1000 ft^3/min of heating make-up air in a 6500 AHDD environment by the various fuels. The cost of electricity for the fan motor in this comparison assumes a more typical FanTP = 2 in. = FanSP + VP_{outlet} than the 1 in. FanTP assumed in Section II. The cost to move the air through a ventilation system is typically much smaller than the cost to heat the make-up air in most climatic regions. These heating costs can be scaled to another system and climate in direct proportion to the two system's flow rates and AHDD climates, while these fan operating costs need to be scaled only in proportion to the two systems' FanTPs.

   Electricity is always 100% thermally efficient but is very expensive per Btu or kcal. Heating oil is more expensive per Btu than natural gas and its furnaces are not as efficient (75%), but it is more convenient than coal. Natural gas burns with high (90%) efficiency, is convenient to handle, but not available everywhere. Coal is largely limited to a centralized steam heating plant where it might be 50% efficient; even then it tends to be the least expensive source of heat. To put these costs in perspective, the above 1000 ft^3/min is about equivalent to a 5 mph wind blowing through a 2 ft^2 open window (did someone say close the window?). At the other end of the spectrum, a modest size manufacturing plant exhausting 100,000 to 500,000 ft^3/min would have a heating bill of \$100,000 to \$1,000,000 per year. A research university campus with numerous laboratories is in this latter category.

V. COOLING COSTS FOR MAKE-UP AIR

Calculating cooling costs is both somewhat more difficult and somewhat easier than calculating heating costs. It is more difficult because both air's enthalpy and its specific heat vary with its moisture content (humidity) and the moisture content of warm air can vary quite broadly.

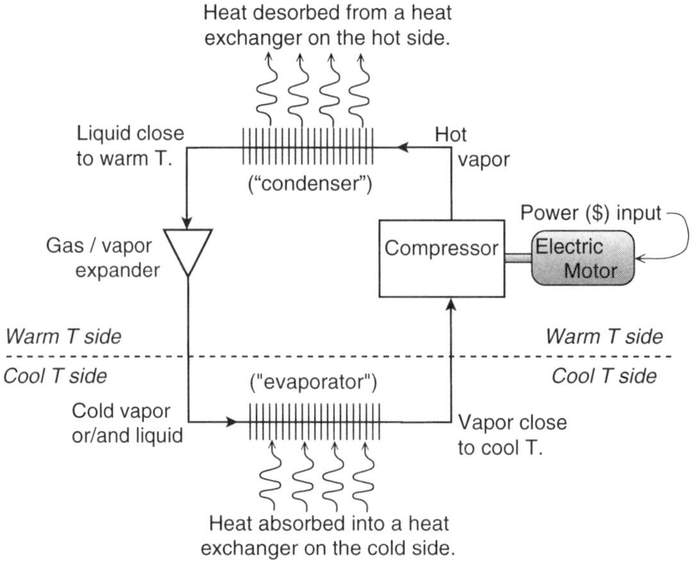

FIGURE 17.2 Schematic diagram of components of a refrigeration cycle.

It is easier because virtually all cooling systems are run by electricity (cf., the variety of fuels used to heat air). A method is presented herein that allows cooling cost calculations to be made similar to the way heating costs were estimated in the previous section of this chapter. However, this discussion will begin by explaining refrigeration systems, their components, and energy cycles.

1. The purpose of refrigeration is to take heat away from where the temperature is cold (but not as cold as desired) and transfer it to where the temperature is already warm if not hot (usually into the outdoor environment). When refrigeration is used to cool make-up air, it cools the warm outdoor air as it enters the building (the "cool side" in Figure 17.2) and dumps heat outdoors where the air is already too warm. This is opposite to the direction that heat wants to move by itself, viz., from a high to a low temperature.

   While various refrigeration cycles exist, the most common is a two-phase vapor compression cycle depicted in Figure 17.2.[4–6] A compression process (on the right side of Figure 17.2) raises the temperature of any gas. The refrigerant gas and its high pressure are carefully selected by a manufacturer to condense and give off its heat of evaporation into the warm air via a finned heat exchanger (depicted at the top of the figure). The liquid refrigerant that leaves the condenser is cooler than the gas when it arrived but it can never be cooler than the ambient warm air.

   The pressure reduction in the rapid expansion that occurs on the left side of Figure 17.2 will cool any gas or vapor. However, a good refrigerant will absorb heat on the cool side of the system due to its ability to evaporate at a lower temperature at this lower pressure, and thus absorb its heat of evaporation from the already cool air outside the evaporator. The heat absorbed by the refrigerant makes the cool side colder. The vapor phase refrigerant will leave the evaporator and returns to the compressor to repeat the cycle at or above the temperature outside the evaporator.
2. Instead of thermal efficiency (as in Equation 17.6 for heaters), efficiency in a refrigeration system (or for heat pumps) is described by either a coefficient of performance (COP) or

an energy efficiency ratio (EER). Equation 17.12 is stated in a way that defines both terms.[f]

$$\mathrm{COP} = \frac{\text{Heat transferred from the cool air}}{\text{Energy consumed to drive the transfer}} = \frac{\mathrm{EER}}{3.414} \quad (17.12)$$

A cooling system's COP has a theoretical upper limit defined by its chemical refrigerant, its mechanical components, and the ambient temperature. A good room air conditioner can have a COP of approximately 3, while a good central air conditioner might have a COP of 4 to 6. Systems built prior to 1990 might have a COP as low as 2. One can estimate costs for a new system based on an optimistic COP of 4. A COP of 4 is equivalent to a thermal efficiency of 400%. The equivalent efficiency of all of these refrigeration systems (and heat pumps) is much higher than furnace efficiencies whose combustion efficiency can be no higher than 100%.

3. Calculating a building's total cooling requirement and costs is complicated because of solar heat absorption and heat conduction through exterior walls (determined by a building's design), by the many sources of heat internal to most buildings (determined by its occupancy), and the temperature and the latent heat of moisture in the outside air (determined by the climate in which it is located). The following calculations will only consider cooling the make-up air.

   The use of annual cooling degree days (ACDD) allows the same method to be used to calculate the costs to cool make-up air as was used above to heat make-up air. A cooling degree day is the number of degrees that the time-weighted average outside air temperature for that day is above the same 65°F used in heating degree days, and in an analogous way temperatures below 65°F do not contribute to a cooling degree day. This 65°F (18°C) temperature reflects the same historic rule-of-thumb estimate for where a typical building's heating requirement just balanced its cooling load. The same rationale to use one of the lower temperature bases listed in the NOAA data source can be justified if a building has better insulation and/or more internal sources of heat than buildings in the past.

   However, because air that needs cooling typically has more water vapor (humidity) than air that needs heating, more climatological information than just the local ACDD is needed to calculate the annual energy needed to cool make-up air for exhaust ventilation. The fact that the specific heat of water vapor is approximately 15% higher than that of dry air is a small difference in comparison to the dramatic increase in energy required if some of the outside air's humidity is removed via condensation. When an air-conditioning system removes moisture, it must remove not only the "sensible heat" associated with the air's temperature but also some of the latent heat associated with its humidity.

   Figure 17.3 depicts a psychrometric chart to which an enthalpy scale was added.[g] Lines of constant enthalpy nearly parallel the wet bulb temperature. The water content of even 100% saturated cold winter air (in the lower left corner of Figure 17.3) is very low, which is why a single term for the specific heat in Equation 17.7a,b can adequately predict the energy needed to heat dry air to which no additional humidity is added. However, the heat content of hot summer air (comprising the right side of Figure 17.3)

[f] The term EER is relatively new, being formulated by the U.S. Department of Commerce (with the concurrence of the Department of Energy) in response to 1992 Congressional legislation. The unit conversion of 3.414 Btu/W was built into this definition. Because a system's EER varies seasonally and with the frequency of its start-up cycles, commercial systems are rated by their seasonal energy efficiency rating or SEER.

[g] If you are not already familiar with a psychrometric chart from the topic of heat stress or elsewhere, you might want to review its structure and use in general ventilation in Section VI.7 in Chapter 20 of this book.

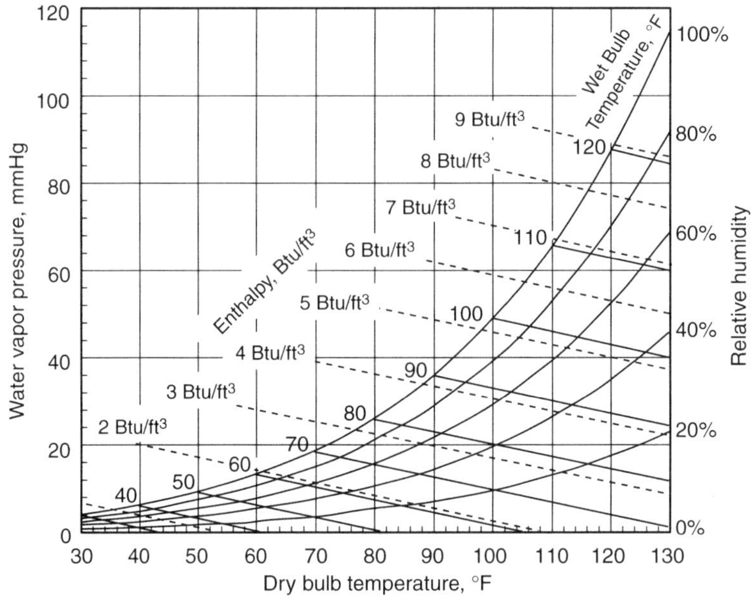

FIGURE 17.3 A psychrometric chart with an enthalpy scale.

and its specific heat depend upon both its temperature and its moisture content or humidity. Equation 17.13a and Equation 17.13b allow the enthalpy of moist air at normal pressure to be estimated quite closely by its temperature and the partial pressure of water in that air (adapted from[7]).

$$\text{Enthalpy [Btu/ft}^3] = 0.01775 \times {}^\circ\text{F}\left(\frac{P_{\text{water}}}{P_{\text{air}} - P_{\text{water}}} \times (48.81 + 0.0204 \times T_{\text{air in }^\circ\text{F}})\right) \tag{17.13a}$$

$$\text{Enthalpy [kJ/m}^3] = 1.185 \times {}^\circ\text{C} + \left(\frac{P_{\text{water}}}{P_{\text{air}} - P_{\text{water}}} \times (1843 + 1.371 \times T_{\text{air in }^\circ\text{C}})\right) \tag{17.13b}$$

where P_{air} and P_{water} are the air pressure (the barometric pressure, mmHg) and the partial pressure of water vapor (its absolute humidity, mmHg), respectively.

4. The three progressive forms of Equation 17.14 are shown below to try to explain how the values in Table 17.3 were derived. The ΔEnthalpy/ΔT values in Table 17.3 (the change in enthalpy per cubic foot per degree change in air temperature) approximate the air's density times its specific heat in Equation 17.7a. The thermal coefficient for ACDD equates to the 25.9 used with AHDD in Equation 17.7b. Values for both terms in Table 17.3 were derived for an array of four starting temperature and three humidity conditions each ending with the air cooled to 65°F.

   For cooling:

$$\text{Btu/year} = [\text{flow rate}] \times [\text{density}] \times [\text{specific heat}] \times [\text{time}] \times [{}^\circ\text{F} \times \text{days/year}] \tag{17.7}$$

$$\text{Btu/year} = Q[\text{cfm}] \times \frac{\text{Enthalpy}}{\Delta T_{\text{cooling}}}[\text{Btu/ft}^3/{}^\circ\text{F}] \times (60 \times 24[\text{min/day}]) \times \text{ACDD} \tag{17.14a}$$

TABLE 17.3
Examples of Values for ΔEnthalpy/Δ*T* (Shortened to Δ*E*/Δ*T*) to Be Used in Equation 17.14a and the Thermal Coefficient to Be Used in Equation 17.14c at Three Humidity Ranges

	Dry (RH < 30%)		Moderate (30–60%)		Humid (RH > 60%)	
Outside *T*	Δ*E*/Δ*T*	Thermal Coefficient	Δ*E*/Δ*T*	Thermal Coefficient	Δ*E*/Δ*T*	Thermal Coefficient
100°F 38°C	0.0179	25.8	0.0308	44.4	0.0665	95.8
90°F 32°C	0.0179	25.8	0.0194	27.9	0.0552	79.5
80°F 27°C	0.0179	25.7	0.0181	26.1	0.0422	60.8
70°F 21°C	0.0178	25.7	0.0180	25.9	0.0182	26.2

$$\text{Btu/year} = Q[\text{cfm}] \times 1440[\text{min/day}] \frac{\text{Enthalpy}}{\Delta T_{\text{cooling}}} [\text{Btu/ft}^3/\text{day/°F}] \times \text{ACDD} \tag{17.14b}$$

$$\text{Btu/year} = Q \times \text{"thermal coefficient"} \times \text{ACDD} \times \text{fraction of time running} \tag{17.14c}$$

Notice that the $\Delta E/\Delta T$ and thermal coefficient to cool dry air (the left columns of Table 17.3) closely approximates the corresponding values for heating dry air (viz. the 0.018 Btu/ft^3/°F within Equation 17.7a and 25.9 Btu/ft^3/°F/day within Equation 17.7b, respectively) and that both of these values increase as the air temperature of humid air increases (in the right columns of Table 17.3). This pattern reflects the dramatically higher energy required to cool humid air rather than trying to cool dry air.

To make a cooling calculation with ACDD, one must have both a thermal coefficient from Table 17.3 and a number of annual cooling degree days from Appendix C. The former is not the peak daily temperature and humidity but should be representative of typical conditions during the cooling season and times of the day when the system would be running based on either familiarity or climatological data (e.g., www.cpc.ncep.noaa.gov). The thermal coefficient does not vary greatly with representative temperatures in dry or even most moderately humid climates (except for moderate humidity at 100°F), but the values in high humidity climates vary almost fourfold across the temperatures in Table 17.3, reflecting the increasing amount of water vapor in the air that will condense during cooling to 65°F (18°C). The examples below will demonstrate the implications of these differences.

5. Despite the above complication, cooling does permit one simplification because basically all refrigeration systems use electricity for an energy source (although like a fan, they can be run by steam or an engine). Thus, there is only one equation for cooling costs equivalent to the four versions of Equation 17.11 for heating costs. Equation 17.15b is simply Equation 17.15a into which a value from one of the versions of Equation 17.14 can be inserted. The hours of cooling per week can be estimated in a manner similar to heating except that the hottest outside temperatures and highest cooling loads occur during the normal day-time business hours (cf. the coldest temperatures for heating that occur at night); thus, it is possible for a refrigeration system to have to cope with approximately 70% of the ACDD during the 25% of the time it is running from 8 a.m. to 5 p.m. 5 days per week.

$$\text{Annual }\$_{\text{cooling}} = \frac{Q \times \text{"thermal coefficient"} \times \text{ACDD}}{\text{Coefficient of Performance}} \times \frac{\$/\text{kWh}}{3414\,\text{Btu/kWh}} \times \frac{\text{hours used/week}}{168\,\text{h/week}} \tag{17.15a}$$

$$\text{Annual } \$_{\text{cooling}} = \frac{\text{Btu/year}_{\text{cooling}}}{\text{COP}} \times \frac{\$/\text{kWh}}{3414\,\text{Btu/kWh}} \times \frac{\text{hours used/week}}{168\,\text{h/week}} \quad (17.15\text{b})$$

Example 17.5

What is the annual energy needed and cost to cool 1000 ft^3/min of make-up air in a hot (80–90°F), arid climate like a summer day in the Rocky Mountains?

An appropriate thermal coefficient value from Table 17.3 for this problem would be 25.8 Btu/ft^3/day/°F corresponding to a dry climate and 80–90°F typical of midday summer temperatures in the region; again the choices in the dry column do not vary very much. This value corresponds to the cooling effect of removing enough enthalpy to move air from position A to position B on the psychrometric chart in Figure 17.4. In this example, there is no condensation.

The ACDD value could be from a specific location listed in Appendix C to this book, but in this case an eyeball average from among various locations in Utah and Colorado would be adequate. A value of 1000°F × days is representative of many urban locations in this region.

$$\text{Btu year}_{\text{cooling}} = Q[\text{cfm}] \times 25.8[\text{Btu/ft}^3/\text{day/°F}] \times \text{ACDD} \quad \text{(using Equation 17.14c)}$$

$$\text{Btu/year}_{\text{cooling}} = 1000 \times 25.8 \times 1000 = 26 \times 10^6 \text{ Btu/year removed from the air}$$

$$\text{Annual } \$_{\text{cooling}} = \frac{26 \times 10^6 \text{ Btu/year}}{4[\text{assumed COP}]} \times \frac{\$0.07/\text{kWh}}{3414 \text{ Btu/kWh}} = \$132/\text{year}$$

(using Equation 17.15b)

Example 17.6

What is the annual energy needed and cost to cool 1000 ft^3/min in a humid climate like the southeastern USA? Compare these values to those in the previous example.

Cooling from 85 to 65°F will remove enough enthalpy to move the air from position C to position D in Figure 17.4. Condensation will occur at the air's dew point or beginning just below

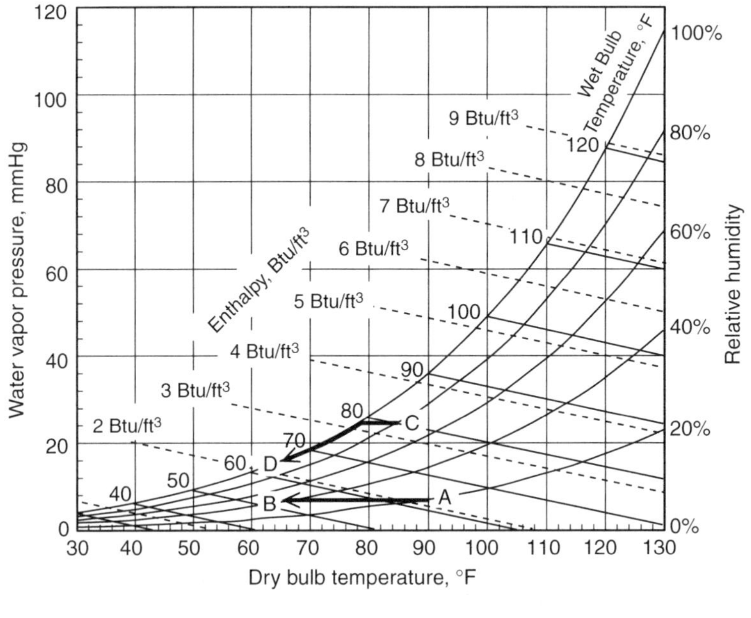

FIGURE 17.4 The position on a psychrometric chart of the air in Example 17.5 and Example 17.6.

tion Service, National Climatic Data Center, Asheville, NC, 2002 (available at
gov/climatenormals/clim81_supp/CLIM81_Sup_02.pdf).
C *Principles and Applications Manual*, McGraw-Hill, New York, 1997.
ndbook of Air Conditioning and Refrigeration, 2nd ed., McGraw-Hill, New York,
d Wilson, C. L., *HVAC Systems Design Handbook*, 4th ed., McGraw-Hill, New York,
cQuiston, F. C. and Parker, J. D., *Heating, Ventilating, and Air Conditioning Analysis*
ley, New York, pp. 54–55, 1988.
dard 55-2004 — Thermal Environmental Conditions for Human Occupancy, American
ing, Cooling, Refrigerating and Air Conditioning Engineers, Inc., Atlanta, GA, 2004.

80°F in this case. The appropriate thermal coefficient for 85 and 90°F and high humidity air from Table 17.3 would be 70.1 Btu/ft^3/day/°F. Using sites in Alabama listed in Appendix C as representative of the region, an ACDD for this region might be about 2000°F × days.

$$\text{Btu/year}_{\text{cooling}} = Q[\text{cfm}] \times 70.1[\text{Btu/ft}^3/\text{day}/°\text{F}] \times \text{ACDD} \qquad \text{(using Equation 17.14b)}$$

$$\text{Btu/year}_{\text{cooling}} = 1000 \times 70.1 \times 2000 = 140 \times 10^6 \text{ Btu/year removed from the air}$$

$$\text{Annual } \$_{\text{cooling}} = \frac{140 \times 10^6 \text{ Btu/year}}{4[\text{assumed COP}]} \times \frac{\$0.07/\text{kWh}}{3414 \text{ Btu/kWh}} = \$720/\text{year}$$

(using Equation 17.15b)

The make-up air cooling cost estimated for this humid environment is approximately five times the cost in the arid environment (despite a slightly lower representative outside air temperature). Thus, the annual cost to cool make-up summer air in a humid environment can easily exceed the cost to heat winter air in the same region. Despite the AHDD in the southeast region of the USA being around one third the AHDD values in the Rocky Mountain region, half the ACDD and much lower cooling costs in a dry summer climate tend to balance out the apparent difference.

6. A system's cooling capacity is sometimes expressed in terms of "tons of refrigeration." The capacity of a refrigeration unit is the rate at which it can remove heat. A "ton" has an interesting history. One ton of refrigeration is the cooling rate produced by melting one ton of ice at 32°F (0°C) in 24 h. This historical way of cooling is equivalent to 12,000 Btu/h or 200 Btu/min. People actually cooled buildings (or at least some rooms) by storing ice from the winter and melting it in the summer. While that era has passed, the unit is still used within the air conditioning industry.

VI. ENERGY CONSERVATION

The pollution prevention paradigm places energy conservation activities in the following priorities: prevention, minimization, recycle, and disposal. While virtually all of these options go beyond the authority of industrial hygienists, our interests in local exhaust ventilation impacts these energy costs, and those who are interested in energy are potential allies in controlling health hazards in the workplace.[h]

1. The first option is to prevent the consumption of nonrenewable energy for heat. Controlling the need for energy is well beyond the pale of most hygienists. Many of the best options go back to designing the building with summer solar shading and winter solar collection; however, some options are amenable to renovation projects. Modern solar collectors can avoid the need for or at least reduce heating and (perhaps surprisingly) even some cooling costs.
2. Hygienists can do some things to minimize energy demand within our typical realm of responsibility. One option is to look for ways to reduce the volume of heated or cooled exhaust air. The more air that is exhausted, the more make-up air must be conditioned. An IH can prepare an inventory of exhaust hoods and suggest changes to delete unnecessary face areas, optimize the areas that are necessary, and obtain the appropriate face velocities (i.e., make the hood more efficient at collecting contaminants). The use of low-volume/high-velocity hoods (see Section V.6 in Chapter 13) will reduce exhaust air volume, as will the use of variable volume fans that respond to demand via a sensor

[h] See also the discussion in the *Manual of Industrial Ventilation* Sections 7.11–7.13.

(see for instance, the Phoenix hood control feedback system at www.phoenixcontrols.com/index.html).

Energy savings are almost always possible by reducing the degrees of heating and cooling required, and are especially easy to implement where occupancy frequency or duration is very low. Setting the thermostat to a lower heating temperature can easily be compensated for by wearing (or providing) warmer clothing. Slightly higher cooling temperatures cannot be compensated for so easily, but thermostat settings during the summer are often cooler than necessary. The ASHRAE Standard 55–2004 *Thermal Environmental Conditions for Human Occupancy* now describes an adaptive comfort option to the old fixed thermal comfort zone.[8] Another step is to reduce the loss of heat through walls and ceilings. Better building insulation is an up-front capital cost whose NPV or payback time can easily be determined by contractors.

3. Heat exchangers are one way to recycle excess heat. Heat exchangers can be either passive storage or active devises that transfer heat between fluids (e.g., transferring warm indoor exhaust air into cold outdoor make-up air in the winter or transferring cool summer exhaust air into hot outdoor make-up air in the summer). An example of an active device applicable to exhaust ventilation is a mechanically rotated media (what the *Manual of Industrial Ventilation* in Section 7.13 of Chapter 7 calls "heat wheels").[1] Fixed plate heat exchangers rely on the passive transfer of heat through a thermally conductive barrier between two fluid streams flowing in either countercurrent (opposite directions), concurrent (the same direction), or crosscurrent flow patterns.

Some buildings are better candidates for heat exchangers than others. Modern well-insulated commercial and light industrial buildings often have enough internal electrical heat sources to actually generate excess heat even in the winter. The problem in such buildings is simply to provide good mixing of the return air. Some manufacturing processes produce huge quantities of excess heat in localized portions of a building. The problem in these buildings is to distribute that local heat to other portions of the plant either directly as hot air, via a return air mixing system, or via a liquid intermediary. Thermal recirculation is additionally challenged by intermittent operations and the need to bypass a heat recirculation system during the summer or winter season as appropriate.

It is also possible to recycle heat by recirculating cleaned exhaust air directly back into the workroom. However, the potential to also recirculate some toxins back into the workplace puts serious constraints on when and where exhaust air recirculation is permissible or even advisable. Criteria for the acceptable recirculation of exhaust air include the following:

a. A safe exposure level is known from toxicology or epidemiology. Such threshold knowledge is needed for both the major contaminants (e.g., metal fumes or solvent vapors) and any minor contaminants (smoke, odors).
b. A suitable air cleaner is available to reduce the air contaminant to well below the applicable exposure limit(s).
c. A monitoring system is available to detect and warn of air cleaner breakthrough or other such failures. Employee education and posted signs are recommended that indicate the purpose of such monitoring and the action to be taken in the event of such a warning.
d. Some assurance that the failure of both the cleaning and warning systems will not cause serious or permanent health damage (e.g., a "fail safe" system or only irritation effects).
e. Local or other government regulations do not prohibit such recirculation.

Given that the high costs of en energy conservation offer plenty of almost immediate economic benefits reviews.

PRACTICE PROBLEMS

1. The previous problems found tha moving 1500 ft^3/min against a loss cost (disregarding any heating costs electricity were to cost \$0.07 per kW
2. How much would it cost per year to he coal at \$50 per ton in a 50% therm AHDD = 7109?
3. How much cheaper would it be to heat t Waco Texas where AHDD = 2164?
4. How much would it cost per year to cool 15(system with a COP of 4 in Waco, Texas wh (cooling) weather is approximately 80°F and
5. How much cheaper would it be to cool the sar Pocatello Idaho where ACDD = 387 and th approximately 80°F and 30% humidity?
6. Suppose there are 400 laboratory hoods on campus 30 in. tall by 4 ft wide with an 80 fpm face veloci campus exhausting each minute?
7. Approximately what does it cost the campus to heat air per year if each of these hoods were r up) approximately 1 "wg of pressure losses, and the camp the heating costs per 10^3 ft^3/min in Table 17.2?

REFERENCES

1. *Industrial Ventilation: A Manual of Recommended Practice*, 2 Governmental Industrial Hygienists, Cincinnati, OH, 2001.
2. Guidance for Protecting Building Environments from Airborne Che Attacks, DHHS (NIOSH) Publication No. 2002-139. May, 2002, bldvent/2002-139.html.
3. *Climatography of the United States No. 81, Supplement No. 2, Annual 1971–2000*, National Oceanic and Atmospheric Administration, Nat

Data, and Informa www5.ncdc.noaa.
4. Mull, T. E., HVA
5. Wang, S. K., Ha 2000.
6. Haines, R. W. an 2003.
7. Adapted from M and Design, W
8. ASHRAE, Stan Society of Hea

18 LEV System Management

The learning goals of this chapter:

✍ Know what is the most definitive method to evaluate the effectiveness of an LEV system.
✍ Know how Q, V, and SP can be monitored, and understand why monitoring them periodically is beneficial.
✍ Be able to identify (troubleshoot) changes in the FanSP and Q as being caused by changes in either the fan or the hood–duct system.
✍ Know how to use the fan laws (especially the effect of RPM on fan Q) to make postinstallation adjustments.
☐ Know the effects that changing the flow rate (Q) will have on a system's V, VP, Σ losses, fan operating cost, and make-up air heating and cooling costs.

Thus far, the major emphasis on ventilation within this book has been on system design, but the IH's role in ventilation does not stop with design, not even with system installation. As previously pointed out, a control velocity cannot be used to predict a quantitative level of exposure before a system is built. Thus, an exposure assessment must be made after an LEV system has been built, and like any other control, even an adequately performing ventilation system needs to be managed. "To manage" broadly means to monitor a system's performance with the goal of at least maintaining if not improving its performance. In order to detect a change in system performance, one needs an ongoing assessment procedure. This chapter will discuss the timing, frequency, location, and methods by which an IH can conduct LEV system performance assessments, followed by procedures to diagnose causes for deterioration and tools to make improvements.

I. WHY MONITOR SYSTEM PERFORMANCE?

Two of the advantages of ventilation control over administrative or personal protective equipment controls are its reliability and its independence from the active participation of the user. If the results of air sampling determined that a ventilation system was able to protect employees at one point of time, that system should be able to continue to protect employees until either the source, the system, or the control criteria change.

1. Chapters 2, 3, 5, 6, 8, and 9 discussed the nature of gas, vapor, and aerosol sources. A change in the amount of chemical used is just one of many causes for exposures to change. Equation 5.13 identified several variables that can cause a source either to generate more vapor or to generate it faster and which should be recorded any time exposures are evaluated. Recall too from Chapter 2 how P_{vapor} changes with a source's temperature.

$$\text{Evaporation rate } G_{moles} = (\text{geometry coef.}) \times (\text{area}) \times (V^{0.5}) \times P_{vapor} \quad \text{(Equation 5.13)}$$

Changes in use conditions should initiate a reevaluation of exposures and ventilation system performance. Similarly, use conditions should be investigated as a cause of measured changes in employee exposures. A good IH management program would be able to store, retrieve, and link use condition data both to personal exposure assessments and to exposure control system audits.

2. A combination of source conditions, work practices, and ventilation are responsible for the amount of dilution that occurs between a source and someone's breathing zone in the exposure scenarios presented in Chapter 5. Equation 5.7 defined this dilution as the breathing zone dilution ratio.

$$\text{Breathing zone dilution ratio} = \frac{P_{\text{vapor}}}{P_{\text{BZ}}} = \frac{\text{Vapor pressure concentration}}{\text{Breathing zone concentration}}$$

(Equation 5.7)

For the same source and work practices, a change in the amount of dilution means that the ventilation flow rate (Q) and/or the airflow pathway that create the Q_{apparent} in Equation 5.14 must have changed. A change in airflow is due either to a change in the hood and duct system's total losses or to a change in the fan's performance. These potential changes suggest the kind of system measurements that should be collected.

3. Health control criteria can also change. TLVs® and PELs are not objective physical conditions like an LFL or P_{vapor}. Exposure limits are set by committees, hearings, evolving data, and new policies. Thus, if "recognition" of new toxicity data were to cause a chemical's exposure limit to be reduced, the same LEV system and work practices that create a given breathing zone dilution ratio (EDR_{BZ}) may no longer achieve an acceptable exposure concentration. The concept that compliance with an exposure limit is the appropriate balance between the vapor hazard ratio (VHR) of a given chemical and the EDR_{BZ} that a given environment can create was introduced via Equation 5.20.

✍ $$\text{Compliance ratio} = \frac{C_{\text{measured in the BZ}}/P_{\text{vapor}}}{\text{TLV or PEL}/P_{\text{vapor}}} = \frac{\text{VHR}}{\text{EDR}_{\text{BZ}}} \leq 1 \text{ to be acceptable}$$

(Equation 5.20)

Equation 8.6 expressed this same concept by stating that the minimum amount of dilution needed for someone to safely use a given chemical is determined by that chemical's vapor hazard ratio.

$$\text{Breathing zone dilution ratio} = \frac{P_{\text{vapor}}}{P_{\text{partial,BZ}}} \text{ should be} \geq \frac{P_{\text{vapor}}}{\text{TLV}} = \text{the chemical's VHR}$$

(Equation 8.6)

A decrease in a chemical's TLV will increase its VHR. If the chemical's new VHR exceeds the environment's ability to dilute the chemical that reaches someone's breathing zone sufficiently, then their exposure concentration will be out of compliance with the control criterion. A good IH management program will systematically track changes in exposure limits. While such changes may not necessitate further system monitoring, they may trigger the need for system improvements.

II. WHEN TO MONITOR SYSTEM PERFORMANCE

1. For management to evaluate the consistency or to detect a change in system performance, a baseline of system performance must exist. As described above, measuring system performance should be an integral part of an IH exposure assessment program. A baseline should be established at system start-up, but it could be established anytime.

It is not nearly as important when a baseline is established as it is that a baseline be established. Continued system monitoring can provide a documented history.

If exposures were acceptable at the baseline and/or within that history, periodically remonitoring the system is a comparatively inexpensive way both to provide some level of assurance that control is being maintained and to detect a change that can forewarn of over-exposures. Monitoring ventilation system performance periodically yields quicker results and is cheaper than running as many exposure evaluations. Detecting a change in performance could trigger more exposure monitoring, and baseline measurements within a system can also aid in diagnosing the cause of a change in system performance.

2. The frequency of measuring system performance should reflect the probability of change (based on a history of system reliability and/or production stability) and the severity of the adverse health outcome if the ventilation system were to deteriorate (based on the chemical's toxicity). Exhaust hoods should be assessed at least annually. Hoods with highly toxic chemicals could be assessed at quarterly or monthly intervals.[a] A risk-benefit analysis could be conducted to compare the costs of more frequent monitoring with the costs of an adverse health effect from an acute over-exposure as the result of an undetected system malfunction between assessments. Assessments more frequently than once per month will likely exceed the cost of purchasing and installing a continuous performance monitor. While continuous monitoring of either the control velocity outside of a collection hood or the face velocity of a containment hood is usually not feasible because it would interfere with the production work, an electronic anemometer can be placed in a calibrated orifice inserted into the wall of either form of hood (e.g., TSI's Model 8612 EverWatch® Fume Hood Monitor).
3. No matter how frequently performance will be monitored, a management procedure needs to be established and promulgated to define:
   a. Who is going to check the monitor reading? Operators? Supervisors? IH field techs?
   b. When or how often do they check the reading? Hourly? Daily? Weekly? Monthly?
   c. How can they recognize an "out of limit" indication or reading? Upper and/or lower bounds?
   d. What do they do if a reading is out of limits? Call their supervisor? Maintenance? IH?

   The users of a hood with a locally read continuous monitor need to have a written protocol and training for them to know how to deal with the above questions. Flow sensors that are monitored in a central control room, maintenance dispatch office, or security center offer some assurance that problems will be detected and a corrective response will be mounted in a timely manner; however, such remotely monitored systems are not cheap.

III. WHAT, WHERE, AND HOW TO MONITOR SYSTEM PERFORMANCE

The most fundamental requirement of a local exhaust ventilation system is to protect the employees. Assessing employee exposure requires exposure measurements.[b] Exposure assessment is a vital phase of IH practice, but it is also both time consuming and expensive. In order to obtain the most value from the cost of evaluation, one should collect and record as much concurrent information

---

[a] OSHA 1910.94(j)(1) states that hoods and ductwork for dipping and coating operations must be inspected for corrosion or damage at least quarterly (three-month intervals) during operation. Z9.1-1991 avoided recommending any inspection frequencies, but the next version to follow is likely to include a hierarchy of inspections ranging from monthly to annually.

[b] Recall following the formal definition of industrial hygiene given in Chapter 1, that the practice of industrial hygiene is not just recognition, evaluation, or control. LEV system management is a good example of how each of these functions involves the other to one extent or another.

about the working environment as one can. Table 18.A1 in the appendix to this chapter lists common and virtually required data elements to be recorded for each air sample. Table 18.A2 lists optional data that includes ventilation conditions, the employee and/or sample location, employee movements or mobility, and external circumstances about the collected sample, and Table 18.A3 lists optional entries oriented around dermal exposures. Such data can not only document the effectiveness of environmental exposure controls, but they can also be used to help set occupational health policies and priorities in conjunction with other components of priority setting discussed in Section VIII of Chapter 8 of this book. No IH ever says they have too many field notes, only too few.

LEV Principles 1 and 5 in Chapter 10 state that exposure control is based on flow rates that depend upon small pressure differences. However, in Chapter 11, we found that velocity is much easier to measure than either a flow rate or a pressure difference. These relationships suggest four approaches that could be used to assess ventilation system performance:

- Control velocity is most directly relatable to exposure, but where to monitor the control velocity outside of a collection hood is not defined universally. See subsection III.1 below.
- Hood face velocity is appropriate for containment hood performance, but it is not a complete indicator of a collection hood's performance. See subsection III.2 below.
- Duct velocity is convertible to total flow via $Q = V \times A$ as long as the system has no large leak or secondary flow of air bypassing the hood. See subsection III.3 below.
- Static pressure is affected by VP and by losses which are also affected by VP, making the relationship between SP and the effectiveness of control somewhat circumspect. See subsection III.4 below.

1. The second best metric of system performance (after exposure monitoring) is to evaluate the control velocity. Because V_{control} for a collection hood varies with distance and angle in front of the hood, the best place to measure it in any given setting is not well defined. One place might be the edge of the source farthest from the hood's face. Another place might be at Hemeon's conceptual "null point" where if the plume were any closer it would be collected into a hood or if further away it would escape into the room. The virtual boundary in the Hemeon method described in Section VI.3 of Chapter 13 is an array of such points across which the control velocity should be maintained. Measuring only the face velocity of a collection hood would not account for an external interference with that control velocity or for a shift in flow direction that could be caused by a change in the cross-draft conditions caused perhaps by an open door or window.

   The range of options of where to measure velocity makes it doubly important to record the position(s) of your control velocity measurements relative either to the collection hood, to the source, and/or to some other permanent and uniquely identifiable feature. As implied earlier, field notes can never be too extensive. Do not trust your memory to something that you may not see or revisit for another year.

   Eventually, the monitoring program must include an action criterion at which point either further investigations or a corrective action might be initiated. For example, a decrease in V_{control} by as little as 15% from either the designed or previously measured V_{control} might trigger a follow-up investigation, perhaps an exposure assessment. A decrease in V_{control} by as much as 25 or 30% should trigger an exposure assessment and/or a troubleshooting investigation described later in this chapter.
2. The face velocity is the control velocity of a containment hood and an indicator of the potential control velocity outside the face of a collection hood (via the DallaValle equation). However, obstacles near the face, incorrectly placed baffles internal to the

hood, and even the *vena contracta* itself can cause variations in the velocity across the face of a hood. Thus, to evaluate face velocity, one should measure the velocity within at least every square foot across an open hood's face (see subsection 12.VI.2.c). The corrective action criteria suggested below for both the average and the minimum velocities of laboratory hoods could be applied directly or adapted to other control velocities:

a. An unacceptable average velocity might be $<$approximately 80 fpm if there are no sources of turbulence are close to the hood, or $<$100 fpm if nearby turbulent sources could disrupt the air's flow through the hood's face.
b. An unacceptable minimum velocity might be $<$70 fpm. That minimum is the same as any V less than 15% below the lower average (viz. 85% of 80 fpm) or 30% below the higher average (viz. 70% of 100 fpm).

No general guidance of the maximum velocity for a containment hood is available, but more is not necessarily better. An increasing exhaust velocity not only increases the cost of make-up air but also induces more turbulence that can cause higher levels of exposure than the optimum velocities.

3. Monitoring duct velocity is an excellent indicator of Q as long as the ducting is intact between the hood and that monitoring point (and between the fan and monitoring point if monitoring on the outlet side of the fan). The monitoring location should be at least eight diameters downstream of a major change in flow conditions. The duct centerline velocity is easy to measure and is the most stable velocity with respect to radial position at a given location. If one had previously conducted a duct traverse and measured $V_{duct℄}$ as a supplement to (but not used as an actual part of) that traverse, one could validate that $V_{duct\ average} \approx 0.93V_{duct℄}$ (Equation 12.4) and thereafter use $V_{duct℄}$ at that monitoring location as a good approximation of airflow.
4. In principle, SP can be used to monitor airflow. SP can be measured within most straight ducts either by attaching a manometer onto a static port (a smooth (burr-free) hole drilled into the side of a duct) or by installing a Pitot tube inside the duct (as depicted in Figure 11.5).[c] A change in SP indicates that the velocity has changed somewhere, but this author does not recommend relying upon SP as the primary means to monitor system performance, cf., monitoring velocity. Remember, that SP and VP are related, most notably by Equation 11.15 and Equation 11.16b for SP before and after the fan, respectively, and by Equation 14.3 that relates losses to VP. Thus, the SP that one might monitor is proportional to both VP and to the loss factors (the standard symbol " $\propto$ " in Equation 18.1 means "proportional to").

$$-SP_{before\ the\ fan} = VP_{before\ the\ fan} + \Sigma\ Losses_{since\ entry} \quad \text{(Equation 11.15)}$$

$$SP_{after\ the\ fan} \approx \Sigma\ Losses_{until\ exit} \quad \text{(Equation 11.16b)}$$

$$Loss_i = Loss\ factor_i \times (VP_i\ or\ \Delta VP_i) \quad \text{(Equation 14.3)}$$

$$SP \propto VP \times \Sigma\ (Loss\ factors_i) \quad (18.1)$$

These relationships tell us that monitoring SP can detect a change in either VP or Σloss factors, but it cannot consistently tell us if the employee is being protected. The following scenarios can all result in decreased employee protection but present strikingly different

[c] A 1/16 to 1/4 in. hole **with no burrs** drilled through the duct wall would suffice; a special tool can remove burrs created while drilling from outside, or a prefabricated, smooth fitting can be purchased (see *Manual of Industrial Ventilation* Figure 9-7).[1]

patterns of changes in SP and Q to the field investigator (see also Figure 18.1 to be discussed later):

- Blockage or leakage downstream of the SP monitoring site (e.g., from a closed blast gate, the accumulation of debris, or a clogged filter or other air cleaner) will decrease $V_{control}$, the losses since entry (from the hood to the monitoring site), and the magnitude of the SP being monitored.
- Blockage upstream of the SP monitoring site (e.g., from particle deposition on the lips of inlet slots, shifted baffles, or unseen obstacles) will decrease $V_{control}$, the airflow Q, and VP at the site as above but will increase the losses since entry. The net change in SP being monitored ($\text{VP} + \Sigma\text{losses}_i$) will depend upon both the ratio of the upstream to downstream losses and the slope of the fan curve at the initial operating point, but SP will generally increase.
- Leakage into an unplanned opening upstream of the SP monitoring site (e.g., caused by an access panel left off of a hood) will decrease the losses and $V_{control}$ but increase the Q and V being monitored. Again, the net change in the SP being monitored would depend upon both the ratio of the upstream to downstream losses and the slope of the fan curve.

5. A qualitative assessment of airflow can be a useful adjunct to all the previous quantitative monitoring (see Section I of Chapter 12). For example, smoke tubes may be used qualitatively to detect a crosswind deflecting a plume, to get a sense for the magnitude of turbulence in the plume and/or exhaust air, and perhaps even to detect the wrong direction of flow (viz., outward instead of inward) if using a nondirectional anemometer or if the control velocity is too low to be sensed by hand. Quantifying turbulence via the systematic statistical analysis of the fluctuations in air velocity from a hot wire thermal anemometer is a technique that is still too much in the research realm to be of practical IH field use.

IV. LEV TROUBLESHOOTING

Imagine getting a call that a continuous airflow monitor at your facility has decreased to an action criterion. Or suppose you are evaluating an LEV because of an unacceptable increase in measured employee exposure level. Or suppose you are coming back on a routine schedule to evaluate an LEV that was working fine the last time you tested it (or it was installed) but is not now. If either continuous or periodic velocity data has revealed a decrease in LEV performance, then one is faced with several questions. Was the change caused by the fan or by the hood–duct system? If the cause is within the system, where is the change located? And if the cause is with the fan, can it be fixed or must the fan be replaced? Guffey has published some good engineering approaches to troubleshooting LEV systems[2–4] and AMCA has their troubleshooting publication,[5] but a logical approach is simply to start with the easiest option first.

1. The easiest option is to look for causes that can be detected visually. As Yogi Berra said, "You can observe a lot just by watching."[6] The following discussion separates visually detectable causes into obstructions and openings because they cause different symptoms.
   a. Obstructions to flow can occur either at the hood, along the ducting leading to the fan, or beyond the fan in the exhaust stack. Obstructions will cause an increase in resistance and pressure losses, and a decrease in airflow and employee protection.
      - Corrosion or the build-up of particles on the edges of slotted hoods can decrease its effective face opening, increase the hood entry losses, and cause a decrease in the exhaust airflow rate (Q). A containment hood's face can be partly closed due to a collision by a vehicle or other moving piece of equipment in a factory. A vehicle

colliding with a duct can also create a local constriction that (just like a bent soda straw) will increase losses and decrease Q. The loss of a previously installed flange (that may be visually obvious only by knowing that it used to be there) will result in a decrease in control velocity for a roughly similar airflow rate (the loss of a flange will also increase a hood's entry loss although it is likely to be a minor (perhaps undetectable) contributor to the sum of all the system's losses).

- Visually inspect the system for the presence of dampers (blast gates) and note their position. The presence of all dampers may not be readily apparent, and their appropriate position may be unknown without a map of the complete ventilation system. Dampers might have been placed where they are not easily accessed to help prevent unauthorized adjustments, complicating a visual inspection. A closed damper in the duct of concern or an open damper in a parallel duct of a branched system can both decrease the desired exhaust airflow rate.

b. The unintended openings on the suction side of the LEV system (either at the hood or along the ducting leading to the fan) will increase the quantity of air that reaches the fan but will decrease either the desired control velocity or that which was present the last time the system was inspected.

- A missing hood maintenance access panel that lets air into the hood through an opening other than its face will decrease the control velocity, although the exhaust airflow rate reaching the fan may either be the same or increase slightly. Some hoods are made to be at least partially taken apart for periodic maintenance or to change-out tools or patterns. If a removed panel cannot be reinstalled easily, sooner or later it is likely to be left off. On the other hand, if the panel can be removed too easily, it is subject to unauthorized tampering or it may just plain fall off. Selecting the right level of system flexibility for a given setting is an art in itself.
- Gaps in duct sections further downstream are not hard to find if they exist. Sections of ducting can separate due to broken or stretched duct hangers (a problem often induced by corrosion and/or the weight of dust that can accumulate in a long horizontal run). The ducting is often connected to its fan by a flexible sleeve either to reduce secondary noise from propagating through the duct structure and/or to bridge a slight misalignment between the duct and the fan inlet. These sleeves can deteriorate due to weather or chemicals in the exhaust, creating an air gap (leak) into the duct right before the fan. Getting to the fan may mean the investigator must climb a ladder or access a roof.
- An open duct inspection plate, cleaning port, or bleed-in are not necessarily easily visible. Thus, the visual inspection needs to be conducted completely and patiently. Take a flashlight if the ducts are above the ceiling light fixtures or under perforated floors, and in the latter case anticipate high concentrations if contaminants could have accumulated due to leakage or spills.

c. If the system is branched and comprises multiple hoods, continue to check visually for changes in other branches that might cause an imbalance among the branches. For example, a new opening may have been added in another branch or a new hood and/or branch may have been added to the system without authorization.

2. A useful step at this point (if not before) in troubleshooting an inadequate LEV system is to gather whatever prior information is available about the system. The information available will vary with the age of the ventilation system, the history of prior IH services, and the quality of the organization's recordkeeping.

a. Troubleshooting is easier if construction drawings or the designer's notes are available. If construction documents or "as built" plans are not available, make a diagram (sketch) of the duct system while carrying out Step number 1. Determine the duct diameters, lengths, junctions, etc. Measuring the diameter of a round duct is simplified by remembering that the diameter of a circle equals its perimeter $\div \pi$ (from which two wall thickness can be either subtracted or disregarded).

If Step number 1 is unsuccessful, these construction drawings or field notes will be used later (in Step number 4) to identify sections vulnerable to leakage, blockage, or deterioration (such as elbows, constrictions, or long horizontal sections of ducting, respectively) and to calculate expected losses in these sections.

b. Maintenance records are another source of potentially useful information. Check with maintenance about both scheduled and unscheduled service work carried out on the exhaust ventilation system since the last LEV system evaluation. A dirty (clogged) air cleaning filter is not easy to see, but its potential impact suggests a good question to ask maintenance. Also ask if any other maintenance work was carried out on the fan or its motor (such as rewiring). Do they maintain service logs that could be used for determination or verification?

3. If the cause of the trouble was not detected visually, then the cause is likely to be internal either with the hood–duct system or with the fan and thus not visible. The key to being able to differentiate between a change in the exhaust system and a change in the fan is to examine the pattern of change. The following assumes that you either have measured or can measure Q and the FanSP. Their measurements are reviewed in footnotes.[d,e]

   Recall from Figure 16.6 that any ventilation system operates at the intersection of a fan performance curve and a system loss curve. The slope of a fan's performance curve will always be negative within its recommended operating range, meaning that Q will decrease as the FanSP increases. The system loss curve always has a positive slope because the sum of its losses will increase as Q^2 increases (per Equation 14.4). The principle underlying the patterns depicted in Figure 18.1 and the examples in Table 18.1 is that *any change in one LEV component will cause the system's operating point to move along the curve of the other LEV component.*

   Thus, the investigator must ask, has the fan FanSP and Q changed in the same direction or in opposite directions? A change in the duct system (for an unchanged fan) will cause the operating point to slide up or down along the negative slope of the fan curve where FanSP would vary inversely with a change in Q. These symptoms are diagnosed in Step 4. Alternatively a change in the condition of the fan (for an unchanged hood–duct system) will cause the operating point to slide up or down along the positive slope of the system requirements curve where the FanSP varies directly with a change in the square of Q. These symptoms are diagnosed in Step 5. It is also possible for both the duct system and the fan to have changed; however, that dual-cause diagnosis should be deferred until both of the single causes for a change in system performance have been investigated and resolved.

4. If the Q and losses appear to have moved in opposite directions (one is increasing and the other decreasing), then the likely cause is something in the hood or duct system and not in the fan. The flow rate and losses change in opposite directions as the system's operating point moves along a normally behaving fan curve (whether the exact shape of that fan curve is known or not). In this case, continue to troubleshoot by measuring the pressure loss across various components of the ventilation system, and comparing the

[d] Q can be measured by a velocity traverse across a known area where the flow is undisturbed, as recommended in Chapter 12. Measuring Q at the hood face is only acceptable if the flow is slow enough to not create a *vena contracta*. If there are no leaks, Q will be constant (as long as Q does not have to be adjusted for changes in air density if the SP is over $\pm 20''$ wg using Equation 11.19). Thus, one measurement of Q and a visual inspection for major leaks can suffice, or a second downstream measurement of Q can confirm that there are no leaks.

[e] FanSP can be measured using Equation 16.4 $= SP_{outlet} - SP_{inlet} - VP_{inlet}$. The first two terms can be measured just after and just before the fan, respectively. VP_{inlet} can be determined by a duct traverse near the fan inlet, but it is more commonly calculated using a Q value measured at some more convenient point (when there are no leaks in the system).

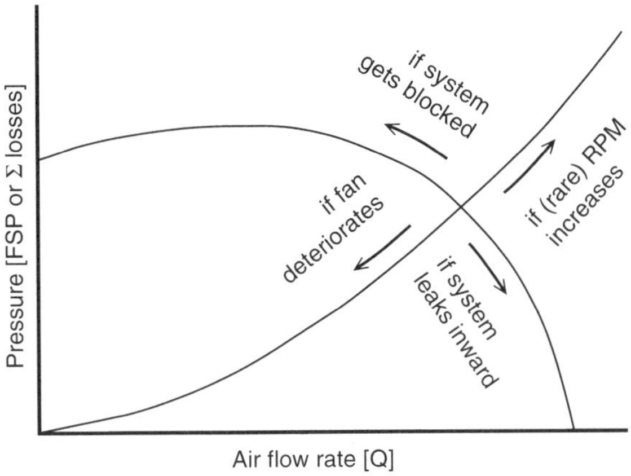

FIGURE 18.1 Troubleshooting trend analysis.

measured loss to that expected. (This part of the investigation may be tedious but necessary.) The actual pressure loss in each component can be determined by measuring either SP or TP as described below:

- Measure SP before and after a system component. If the duct diameter before the fan changes across the component, calculate the actual loss after adjusting for the change in VP using Equation 18.2a

$$\text{Loss}_{\text{from 1 to 2}} = (SP_1 + VP_1) - (SP_2 + VP_2) \quad (18.2a)$$

  Or if the duct is after the fan or its size constant, calculate the actual loss using Equation 18.2b

$$\text{Loss}_{\text{from 1 to 2}} = (SP_1 - SP_2) = \Delta SP \text{ if V is constant} \quad (18.2b)$$

- Alternatively, measure TP before and after a component. (Recall that TP can be measured by comparing the tip of a Pitot tube to the local atmospheric pressure as in Figure 11.5.) Calculate the actual loss using Equation 18.2c. This alternative calculation does not require the separate use of VP whether before or after the fan.

$$\text{Loss}_{\text{from 1 to 2}} = TP_1 - TP_2 = \Delta TP \quad (18.2c)$$

a. One can start checking for aberrant internal losses anywhere. The list of segments or components vulnerable to blockage developed in Step number 2 suggests some

TABLE 18.1
Patterns and Causes of Changes in *Q* and *V* versus Pressure Losses and FanSP

	Σ Losses and FanSP will Increase ...	**Σ Losses and FanSP will Decrease ...**
Q and *V* will both increase ...	Only if the fan RPM is increased. This change follows the system curve. Go to Step 5.	If a blast gate, damper, access panel, or bleed-in is left open. This change follows the fan curve. Go to Step 4.
Q and *V* will both decrease ...	If a blast gate is closed or a blockage occurs in the hood or duct. This change follows the fan curve. Go to Step 4.	If the fan RPM is decreased or the fan physically deteriorates. This change follows the system curve. Go to Step 5.

good places to start checking. The following subsections offer explanations of why the usual suspects clog.

- Elbows, dampers or blast gates, and contractions can all clog with debris. Small debris can accumulate slowly while large debris may occasionally become entrained within the exhaust air and be carried at high-speed partway down the duct until it impacts on an internal obstruction. Elbows that leak due to erosion on their outer wall and unauthorized adjustments to blast gates should have been detected during the visual inspection at Step number 1.
- Air cleaners are especially susceptible to being clogged or blocked whether they were designed for particulate matter or not. Most commercial vendors can provide assistance to owners in detecting and diagnosing problems in their air cleaners.
- Resistance can increase slowly from sedimentation building up in long horizontal runs of ducting. Verifying such a build-up usually requires opening inspection or cleanout ports (see Section I of Chapter 10). The weight of such a build-up of dust (perhaps coupled with concurrent corrosion) can also cause a section of the duct to sag or separate, but again these latter problems should have been detected visually.

b. If the source of the problem is not encountered at a likely or vulnerable segment, one is pretty much forced to continue this diagnostic exploration sequentially from one end of the system to the other. The hood's accessibility makes it a convenient location to start checking losses, but at this point the sequence is arbitrary.

At each segment or component, compare the measured loss with that predicted by one of the methods covered in Chapter 14. Because we determined at the beginning of this step that the change in operating condition was following the fan curve, the relative difference in the flow rate due to the change in any particular system loss will be determined by the steepness of the particular fan's curve. As a gross generalization, the flow rate through a radial blade fan will change in approximate proportion to the change in system losses, while the flow rate through a backward inclined and tube axial fan will only be half as large as the relative change in system losses. Thus, an arbitrary 15% decrease in the flow rate that might have triggered a system investigation would correspond to anywhere from a 15 to 30% increase in the total losses. For one segment of a system to cause such an increase in the sum of all the losses, a change in the loss at that single component is likely to be on the order of 2× or more from what was expected, a difference that is easily detectable without any tests of statistical significance.

Once the suspect internal blockage is detected, clearing it might require either opening cleanouts (if they are installed), perhaps disassembling some ducting, and possibly replacing some system duct components. Keep in mind that making a change in one branch of a multi-branch system will have the opposite effect on another branch (and possibly start another cycle of investigation).

5. If the measured Q and pressures appear to have moved in the same direction, then the likely cause of change is in the fan (and not in the hood–duct system). The flow rate and losses change in the same direction (both increase or both decrease) as the system moves along the system requirement curve, and the hood and duct system appears to be performing normally. Changes in fan performance can be caused by a change in its RPM, a reversal in its direction of rotation, or its physical deterioration. The following explorations and remediation steps assume that troubleshooting Step number 3 concluded that the source of low Q is fan damage or a fan malfunction.

a. If electrical maintenance has been (or could have been) carried out on the fan's motor, check for the proper rotational direction of the motor and fan blades. Only in the case of an axial fan running in the wrong direction will the air actually be

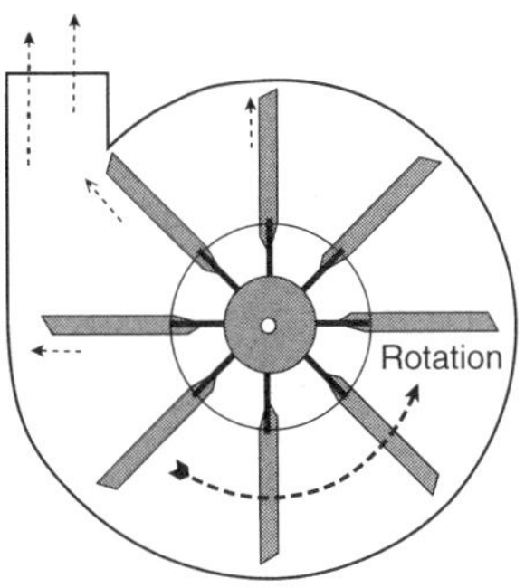

FIGURE 18.2 Radial fan running in reverse.

flowing backward! Improper rotation of a centrifugal fan impeller (counter-clockwise in Figure 18.2) will cause the airflow to decrease but not reverse. The blades still fling the air out radially, just not in the direction of the exhaust outlet. Backwards rotation of a radial fan may reduce Q by only 30–50%, while backwards rotation of an inclined fan may reduce Q by $>50\%$. Interchanging any two electrical wires on a three-phase motor will cause it to run backward. Thus, a simple error is a likely cause for a decrease in flow subsequent to electrical maintenance or for only one out of a series of new and nominally identical LEV systems to not perform as expected upon system start-up. Fixing the error is equally simple.

b. While (or after) checking for the proper rotation direction, have someone check the fan's RPM. An electronic speed control can be installed on a fan motor that is either directly connected to the fan or connected via a fan belt and pulleys on the fan motor and fan shafts. An electronic speed control can fail without shutting a system down or giving a warning or other indication. The ratio of the diameter of the pulley on the fan motor to the diameter of the pulley on the fan shaft determines the ratio of the fan motor RPM to the fan RPM. If an adjustable pulley is not properly tightened, its diameter could possibly shift while the fan is running. However, purposefully changing the ratio or changing the pulleys requires that the fan be shut down. Either of these changes should be in a maintenance log. Changes to the fan's RPM will be discussed further in subsection 18.V.1.
c. If the fan has been running for quite some time without maintenance, arrange for someone to check the status of the fan blades for clogging, corrosion, and/or erosion. Since this will require both shutting the fan down and some disassembly, employee protection will be interrupted. If that maintenance cannot be conducted during nonproduction hours, this interruption is a valid reason for the temporary use of personal protection by the production workers. The maintenance workers may also need some degree of personal protective equipment against the chemicals inside of a contaminated fan. Selecting a replacement for a worn fan is similar to choosing a new fan.

V. POST-INSTALLATION LEV ADJUSTMENTS

Adjustments to an existing LEV system may need to be made for several reasons. One reason is if the system is generating more pressure losses than expected and these losses cannot be fixed.

Another reason is to increase the control velocity above that originally either expected or needed, e.g., if a given chemical's PEL or TLV is decreased or the original chemical is changed to one with a higher VHR. The easiest adjustment is often to increase the fan speed.[f] More control can also be achieved with somewhat more effort by increasing the control efficiency of the hood or by decreasing the resistance of the ducting.

1. If the fan appears to be performing as expected, a higher control velocity and potentially improved employee protection may be obtained by increasing the fan speed RPM. The fan laws allow quite reliable predictions of the multiple effects of changing the fan RPM to be made.
   a. The first of the fan laws was incorporated into Equation 18.3 that may be used to predict a new RPM_{new} sufficient to achieve a Q_{new} based on a previously measured Q_{old} at an RPM_{old}.

$$\boxed{RPM_{new} = RPM_{old} \times Q_{new}/Q_{old}} \tag{18.3}$$

   b. The second of the fan laws incorporated into Equation 18.4 predicts that the FanSP will change in proportion to the square of the RPM_{new}/RPM_{old} ratio determined above. However, rather than the fan law portion of Equation 18.4, the change of FanSP in proportion to $(Q_{new}/Q_{old})^2$ is boxed to be remembered because at this point the RPM ratio was already determined by Equation 18.3.

$$\boxed{FanSP_{new} = FanSP_{old} \times (Q_{new}/Q_{old})^2} = FanSP_{old} \times (RPM_{new}/RPM_{old})^2 \tag{18.4}$$

   c. Because fan power equals $Q \times$ FanSP, the cubic relationship in the Equation 18.5 version of the third fan law is the result of multiplying Equation 18.3 times Equation 18.4. Both the new fan motor power requirement and its electrical costs will increase with the cube of the Q ratio. While only one version of Equation 18.5 is boxed, the same concept applies to all three versions of Equation 18.5.

$$\boxed{\text{fan kW}_{new} = kW_{old} \times (Q_{new}/Q_{old})^3} = \text{fan kW}_{old}(RPM_{new}/RPM_{old})^3 \tag{18.5a}$$

$$\text{or } hp_{new} = hp_{old} \times (Q_{new}/Q_{old})^3 = hp_{old}(RPM_{new}/RPM_{old})^3 \tag{18.5b}$$

$$\text{or } \$_{fan\ new} = \$_{fan\ old} \times (Q_{new}/Q_{old})^3 = \$_{fan\ old}(RPM_{new}/RPM_{old})^3 \tag{18.5c}$$

   And if the horsepower at the desired RPM_{new} exceeds the power that the existing motor is capable of producing, it must be replaced to prevent the motor from burning out.
   d. In contrast, recall from Chapter 17 that the annual operating costs are typically dominated by heat and cooling costs which are just proportional to Q (assuming

[f] The fan laws were defined in Table 16.1, and an explanation to their logic was provided in Section 16.VII. The Fan Laws may also be used to decrease a fan's performance perhaps to reduce the adverse effects on the product or to decrease the cost of make-up air (this latter change is only viable if the decrease can still maintain adequate employee protection).

constant AHDD and ACDD values).

$$\text{Annual } \$_{\text{heat new}} = \text{Annual } \$_{\text{heat old}} \times (Q_{\text{new}}/Q_{\text{old}}) \quad \text{(Equation 17.9)}$$

Two examples should help clarify these changes in fan performance. The first example is a generic solution (without numbers) using the fan laws. This same example is repeated for a specific fan using the numbers in its fan tables.

Example 18.1

Suppose that an LEV fan has successfully served one hood, just as it was designed to do some time ago. Now due to production increases, the existing fan needs to serve a hood twice as large. By how much would the fan speed have to be increased to move twice as much air?

Clearly, $Q_{\text{new}} = 2 \times Q_{\text{old}}$. And using the first of the fan laws shown in Table 16.1 or Equation 18.3 above, we can predict that the fan should move twice the airflow if its fan speed were doubled.

$$\text{RPM}_{\text{new}} = \text{RPM}_{\text{old}} Q_{\text{new}}/Q_{\text{old}} = \text{RPM}_{\text{old}} \times 2Q_{\text{old}}/Q_{\text{old}} = 2 \times \text{RPM}_{\text{old}} \quad \text{(using Equation 18.3)}$$

This adjustment is only a pathway to solving the basic problem. What happens when we attempt to force twice as much air through the existing duct system? The second of the fan laws (Equation 18.4 as used below) indicates that the Σlosses will increase in proportion to the square of Q as depicted in Figure 18.3. We do not necessarily have to calculate this $4\times$ change in each loss; it will just happen if the fan and motor are sufficiently capable of moving twice as much air through

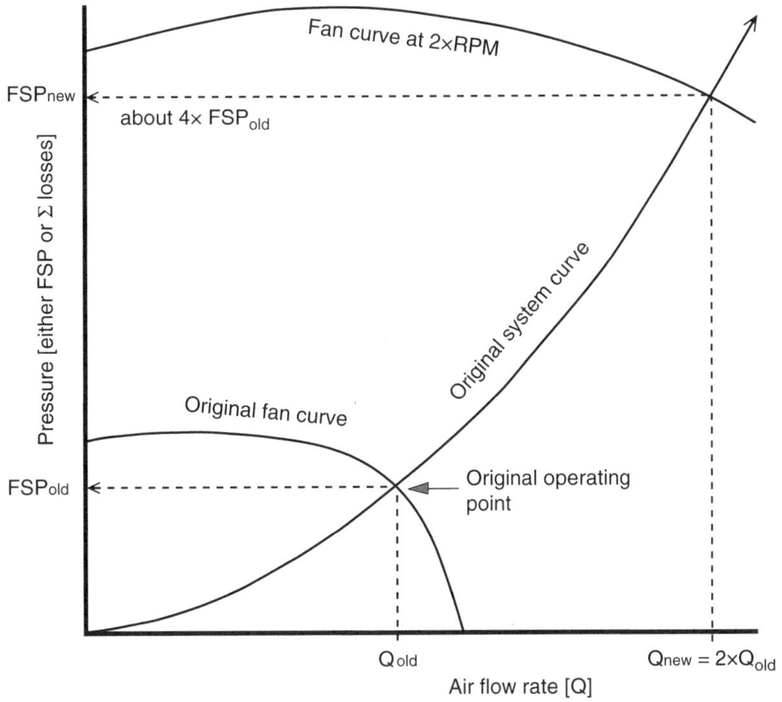

FIGURE 18.3 Depiction of making $Q_{\text{new}} = 2 \times Q_{\text{old}}$.

the original system.

$$\text{FanSP}_{\text{new}} = \text{FanSP}_{\text{old}}\left(\frac{\text{RPM}_{\text{new}}}{\text{RPM}_{\text{old}}}\right)^2 = \text{FanSP}_{\text{old}}\left(\frac{2\ \text{RPM}_{\text{old}}}{\text{RPM}_{\text{old}}}\right)^2 = 4 \times \text{FanSP}_{\text{old}} \quad \text{(using Equation18.4)}$$

The third of the fan laws says that the power consumption will increase by the product of the higher Q times the higher FanTP. The eight-fold increase in horsepower that is predicted by Equation 18.5 for this example is a large change that might exceed the power of the existing motor.

$$\text{kW or hp}_{\text{new}} = \text{hp}_{\text{old}}\left(\frac{\text{RPM}_{\text{new}}}{\text{RPM}_{\text{old}}}\right)^3 = \text{hp}_{\text{old}}\left(\frac{2\ \text{RPM}_{\text{old}}}{\text{RPM}_{\text{old}}}\right)^3 = 8 \times \text{hp}_{\text{old}} \qquad \text{(using Equation18.5)}$$

EXAMPLE 18.2

Now suppose that the above generic fan were the specific radial fan whose performance is shown in Figure 16.25 and that we found in Example 16.3 could move 1350 ft^3/min against a FanSP of 2$''$wg at 1062 rpm = by how much would that specific fan's initial speed have to be increased to move twice as much air through the same system?

Equation 18.3 concludes the needed $Q_{\text{new}} = 2\ Q_{\text{old}} = 2 \times 1350 = 2700$ ft^3/min. While one could calculate the sum of the losses at this new flow rate, one could use Equation 18.4 to predict that the new losses and FanSP$_{\text{new}}$ should be approximately $2^2 \times 2''\text{wg} = 8''\text{wg}$. Looking in the 8$''$SP column of Figure 16.25, the manufacturer determined that at 2124 rpm (exactly twice the 1062 rpm) the fan will move 2700 ft^3/min (exactly the $2 \times Q_{\text{new}}$ predicted by the fan law).

$$\text{Interpolated RPM} = 2112 + \frac{(2700 - 2640)}{(2772 - 2640)} \times (2112 - 2139) = 2112 + 12 = 2124\ \text{rpm}$$

$$\text{Interpolated hp} = 5.80 + (0.462 \times (6.22 - 5.80)) = 5.80 + 0.19 = 5.99\ \text{hp}$$

The horsepower data also in the manufacturer's fan performance table indicates that the actual power to drive the fan will increase from 0.75 hp at 1062 rpm at NTP to 6 hp at 2124 rpm (an increase again of the 8 × predicted by the third fan law), probably necessitating a new motor. Converting 6 hp into 8040 W (using Equation 17.3b) means that a 120 V AC electric line would have to carry 67 A, an unacceptably high current. The building maintenance department, engineering department, or contractor would probably opt for installing a new 220 or 440 V AC line for this level of power.

Thus, the fan laws gave us a good estimate of this fan's actual performance. The fan performance table will generally also indicate to the user whether the fan will still be within its recommended operating range at the new RPM$_{\text{new}}$. We were fortunate to find that this particular fan would still be operating within its recommended operating range (albeit much noisier).

2. If the LEV ducting and air cleaner system are functioning normally and if (for whatever reason) no changes can be made to the fan or its operating conditions (as were just described above), the only remaining way to improve the system's performance is to change one or more pieces of the system's hardware. The following options are referenced to prior sections of this book.
   a. Consider moving a collection hood closer to the source. Reducing the existing distance "X" in the DallaValle equation (Section III of Chapter 13) may or may not be acceptable to the production department and its workers, but any reduction in X will have a large increase in the control velocity.

b. If the hood is free-standing and without a flange that would create a reduced "K_{hood}" in the DallaValle equation (subsection 13.III.4), consider adding a flange to block flow from unimportant areas around a collection hood or to enclose the source more completely.
c. Consider reducing the system's flow resistance by replacing highly turbulent or restrictive duct components. Examples from Chapter 14 include using fewer bends, less tight bends (with a larger *R*/*D*), or perhaps larger ducting.
d. If the fan has any obstructions or a duct elbow near its inlet or outlet, or has no duct at all attached to its outlet, its performance could be increased by reducing the "system effect loss factors" (see Section 16.VIII.2).

3. Changing the LEV system hardware is usually more expensive than fixing a leak or duct obstruction. But changing a portion of an existing system may be less expensive than changing a fan motor. Unfortunately, the fan laws *cannot predict* the appropriate RPM if the system either changes on its own (e.g., from some damage that is not fixed) or is changed on purpose (e.g., to increase the flow by decreasing the resistance as in the following example). A changed system creates a totally new system requirements curve, thus violating an underlying assumption of the fan laws that the Σ losses are proportional to Q^2. Attempting to predict the RPM needed for a changed system is a good example of "not being able to get there from here." The only way to predict the operating RPM for a new system is to calculate new losses from scratch and use the manufacturer's fan performance data.

Example 18.3

Suppose the production department that wants to double the airflow in the above examples is willing to install two parallel ducts to a branch junction close to the existing fan. By doubling the flow area throughout most of the system, most of the VP values will stay the same, which in turn will keep the losses the same (disregarding a little more loss for the branch junction). This change causes the system's desired operating point depicted as point A in Figure 18.4 to move horizontally from Q_{old} to Q_{new}. With no other change to the fan, doubling the flow area would first cause the initial operating point to move along the fan curve to point B, well less than double the Q_{old} as desired at point C.

Because the system has changed, the fan laws no longer apply. The only way to predict the RPM to achieve the desired $Q_{new} = 2 \times Q_{old}$ in this modified system would be to use the manufacturer's fan performance data. Looking down the same 2" SP column in Figure 16.25 (for virtually the same sum of

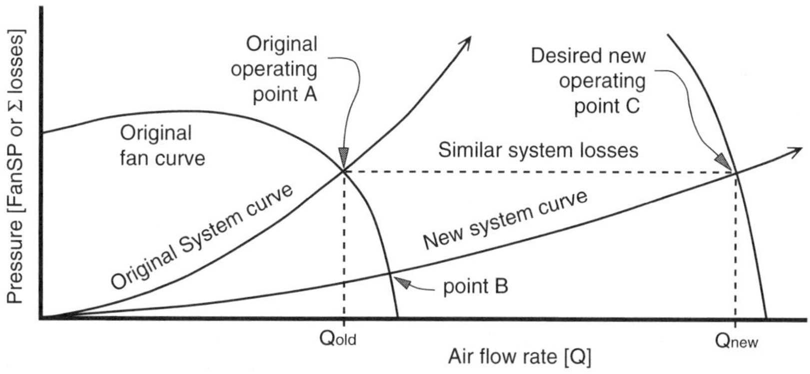

FIGURE 18.4 A depiction of changing the LEV system.

losses) and interpolating, one can find that 1415 rpm would be necessary to double the Q to 2700 ft^3/min at the same 2" FanSP.

$$\text{Interpolated RPM} = 1397 + \frac{(2700 - 2640)}{(2772 - 2640)} \times (1437 - 1397) = 1397 + 18 = 1415 \text{ rpm}$$

4. If neither the existing fan can be run fast enough nor losses in the system can be reduced enough, one can consider buying a new larger fan and motor. That means another trip to the manufacturer's performance data catalogue.

If that does not work or is not an option, then perhaps local exhaust ventilation is not a viable control for the hazard. Remember that source control (Chapters 8 and 9) is usually a better option than ventilation. Respirators are still a possible control, but the respirator selection, training, and supervisory requirements and limitations to be described in Chapter 22 may well convince the reader to consider them a last resort.

By this time, a thoughtful reader should have all the tools needed to design, evaluate, or troubleshoot a local exhaust ventilation system. While not all hygienists actually have to make all of these detailed design calculations, it can help to know what the responsible person is (or should be) doing. A new hygienist may not know all the nuts and bolts (or rivets and screws, in this case) of ventilation hardware, but a hygienist should be involved in the conceptual design of an LEV system, should understand the principles that will be employed and the terminology used in such a system, should be able to anticipate weaknesses or inadequacies in the typical approach taken by someone without health protection training or experience, and should be willing and capable of suggesting better (nonventilation) solutions. It can feel really good at the end of a project to have been able to solve a problem, to make someone else's work a bit safer, and hopefully to improve everyone's profitability in the long run.

APPENDIX TO CHAPTER 18

Data to record in conjunction with air and dermal exposure assessments.

TABLE 18A.1
Common and/or Basic Characteristics of the Exposure Setting to Record as Part of an Environmental Sample Collection Database (This Group Is Similar to Patty, 1991, Table 25.1)[7]

(1) SAMPLE COLLECTION DATE in a specified database-compatible format.
(2) CHEMICAL NAME of the agent(s) that is(are) sampled (a consistent list of names should be specified).
(3) A systematic and unique SAMPLE NUMBER identifier (the identifier might incorporate items 1 and 4).
(4) The PLANT NUMBER and/or PLANT NAME for the respective location or the SITE OR LOCATION NAME in which the employee being sampled is during the sampling period.
(5) SAMPLE TYPE such as Breathing Zone, Area, Bulk, Source, etc.
(6) EMPLOYEE NAME and/or Employee identification NUMBER if a employee is being sampled.
(7) JOB CODE, e.g., a Department of Labor job code or a Plant Job Code of the employee being sampled.
(8) TASK CODE that the employee is conducting during the sampling period.
(9) OPERATIONAL STATUS such as normal or during start-up, shutdown, spill or upset, and maintenance,
(10) SAMPLE INITIATOR CLASSIFIER such as routine, programmed random, targeted heavy exposure, or conducted in conjunction with an external agency, at an employee's request, duplicate, etc.
(11) The SHIFT or SAMPLING TIME during which the sample was collected (if not a part of 14).
(12) SAMPLE METHOD for the sample being collected and its analytical method (from a consistent list of methods). A Limit of Detection may also be added or recorded elsewhere for this method.
(13) Record either the SAMPLE VOLUME (l) or the pump's calibrated FLOW RATE (lpm) and the SAMPLE TIME (min).
(14) Allow a free field for COMMENTS not otherwise categorized.

TABLE 18A.2
Examples of Optional Environmental or Sample Characteristics That Can Affect Airborne Exposures. Many of These Characteristics Can Be Written as Comments; However, They Cannot Be Integrated into Control Decisions unless They Are Entered into a Systematic Database

(15) General Ventilation in area:
- 0 = no special ventilation provisions.
- 1 = fan blowing room air at operator (movable and/or variable).
- 2 = recirculated "cleaned" room air blowing at operator (ducted).
- 3 = fresh (outside) air blowing at operator (either ducted or unducted).

(16) Local Exhaust Ventilation at the work station:
- 0 = no local exhaust system in place.
- 1 = some local exhaust but not well directed or enclosed.
- 2 = local exhaust with well placed hood or enclosure but a low control velocity.
- 3 = local exhaust with well placed hood or enclosure and an adequate control velocity.

(17) Weather conditions (especially OUTDOOR TEMPERATURE) which could affect exposure concentrations, e.g., the amount of natural ventilation from open doors and windows.

(18) An identifier that describes the RESPIRATOR WORN by the employee being sampled, e.g., APR, PAPR, SAR, SCBA (see Chapter 22 or the list of abbreviations in this book).

(19) PLANT FLOOR LEVEL to locate/identify where vertically the person being sampled is working. For example, +1 for an employee working one (1) level above the plant's main (ground) level (which could be either 0 or G), and NA or X = for mobile, multi-floor tasks, jobs, or job codes.

(20) PLANT GRID-LOCATION IDENTIFIER, for example a bay and row number or similar *X* and *Y* coordinate assigned using a prepared plant map or transparency. Coordinates with numbers one way and letters the other are a good way to avoid interchanging and/or to identify interchanged coordinates. The same code used above for plant floor level could be used here for mobile tasks, jobs, or job codes.

(21) A FREQUENCY-MOVEMENT CLASSIFIER indicating how often the person typically takes three or more steps (moves more than approximately 10 ft) away from his/her work station based on a random observation period while being sampled. Entering 1 to 9 movements per 10 min interval has worked well.

(22) A DISTANCE-MOVED CLASSIFIER to identify the average distance the employee moves around or away from his/her work station while being sampled. Entering 10 ft or increments seemed to work well.

(23) TIME ON STATION estimates the total daily time that an employee spends at a work station completing the tasks that are being sampled (an alternative to the Frequency and Distance classifiers above).

(24) SAMPLED BY, to identify the person conducting the sampling.

(25) The assigned PUMP NUMBER of the personal sampling pump used to collect the sample. A code should be entered even if the sample is collected passively (no pump) to avoid errors of omission.

(26) SAMPLE VALIDITY CLASSIFIER such as pump or dosimeter malfunction, postcalibration differed from precalibration by $>10\%$, pump did not run for entire sampling period, etc.

TABLE 18A.3
Examples of Optional Task Characteristics Affecting Dermal Exposure Such as from Work Involving Metal Working Fluids*

(27) Machine Shielding for Liquid Spray or Mist:
- 0 = unshielded.
- 1 = partial shield (e.g., flexible rubber shield, etc) between machine and operator.
- 2 = enclosed with doors.

(28) Degree of Liquid Contact:
- 0 = a dry job.
- 1 = indirect contact (e.g., handling wet parts after machining).
- 2 = intermittent direct contact (wetting skin 1–10 times/day).
- 3 = frequent direct contact (wetting skin 11–30 times/day).
- 4 = continuous direct contact (wetting skin >30 times/day).

(29) Dermal Personal Protection from Chemicals:
- G = chemically resistant gloves in use.
- P = chemically protective clothing for a portion of the body (e.g., an apron) in use.
- B = both chemically protective gloves and partial clothing in use.
- W = whole body chemically protective ensemble in use.

* These characteristics and the Local Exhaust Ventilation Characteristics (such as [16] in Table 18A.2) can also have a strong influence on exposure to mists.

PRACTICE PROBLEMS

1. Suppose that a LEV system had been moving 20,000 cfm against a pressure loss of 4 "wg with a Chicago SQA36$\frac{1}{2}$ airfoil fan running at 961 rpm and consuming 17 hp. However, measurements taken during an annual inspection found that Q was 12,000 cfm and the FanSP was 1.5 "wg.
   a. Is the cause likely to be due to the fan or the system? Explain your rationale in terms of Figure 18.1. And if the cause is the fan, what could be causing this poor fan performance? What if the fan were new?
   b. Increasing the fan's RPM should increase the air flow rate through this system, but why in this case can't the fan's new rpm necessary to reset to achieve the desired Q be predicted using either the fan laws or the fan's performance table?
2. Suppose that the Chicago SQA12$\frac{1}{4}$ airfoil fan running at 2045 rpm (as chosen in the Practice Problems for Chapter 16) was previously tested and shown to be moving 1500 cfm against a total pressure loss of 1.75 "wg through the LEV system designed in Practice Problems for Chapter 14 including the exhaust stack. However, upon a later annual inspection, measurements found that Q was 1375 cfm and the FanSP was 2.5 "wg.
   a. Is the cause likely to be due to the fan or the system? Explain your rationale in terms of Figure 18.1.
   b. Although reducing the resistance is an option and assuming that the fan's RPM is as it was designed, to what new RPM should the fan be set to increase the actual $Q = 1150$ cfm back up to the desired $Q = 1500$ cfm on the Chicago SQA12$\frac{1}{4}$ airfoil fan?

RPM predicted by the Fan Laws = _____ rpm
predicted FanSP = sum of new losses at 1500 cfm = _____ "wg
RPM on the fan table for 1500 cfm and a FanSP of 3 "wg = _____ rpm
predicted horsepower = _____ hp
fan BHP from the fan table = _____ hp

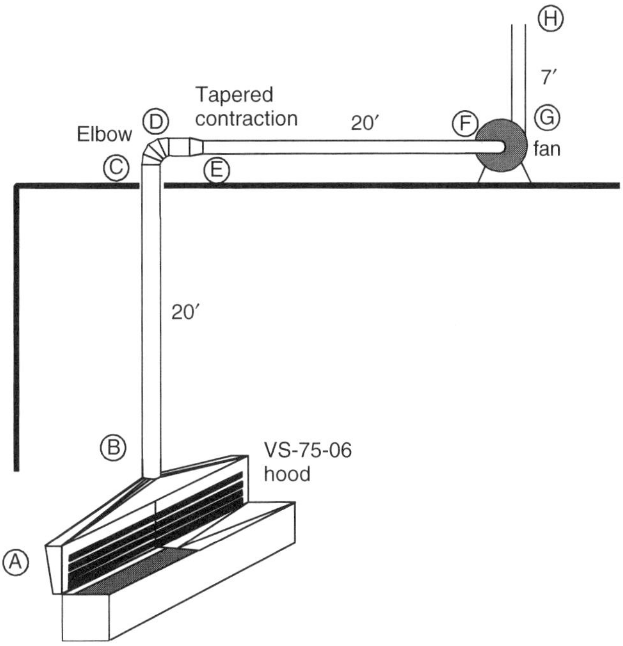

FIGURE 18.5 Diagram of the system designed in Chapter 14, for which the fan was selected in Chapter 16, and which now is not working as well as it was.

3. Suppose that before changing the RPM, you work with maintenance to measure the SP at the locations shown in Figure 18.5 in attempt to identify the segment with the unexpectedly high loss (perhaps caused by an internal blockage of some sort). The table below shows a summary of the pressures calculated during design (on the left) and those measured in the field (on the right). The VP values were estimated from the measured Q. In order to compare the actual loss in each segment to that predicted, you must first calculate the cumulative loss up to each point (perhaps using Equation 11.15 or Equation 11.16) and the loss in each segment.
   a. Complete the data in the table provided.
   b. Identify the suspected segment with an unexpectedly high loss.

   With the guilty segment identified, repairs can now be conducted to remove the blockage.

		Normal (Expected during Design) $Q_{expected}$ = 1500 cfm				Measured in the Field $Q_{measured}$ = 1150 cfm			
ID	Segment Description	Loss in Segment	Cumul. Loss	VP	SP	SP	VP for Q	Cumul. Loss	Loss in Segment
A	Slot loss	0.444	0.444	0	−0.444	−0.26	0		
B	Duct entry loss	0.08	0.524	0.322	−0.846	−0.50	0.19		
C	20′ of 11″ duct	0.138	0.662	0.322	−0.984	−0.58	0.19		
D	Elbow loss	0.077	0.739	0.322	−1.061	−2.10	0.19		
E	Contraction loss	0.07	0.809	1.151	−1.960	−2.63	0.68		
F	20′ of 8″ duct	0.695	1.504	1.151	−2.655	−3.04	0.68		
G	7′ of 8″ duct	0.278	1.782	1.151	0.278	0.14	0.68		

This room has no emergency backup electricity generator. If the power were to fail during a day shift when the lid was off (and the room was at steady state as in question 1), about how long would it take for the room average concentration to reach toluene's LFL? (Without a fan or activity to mix the room air, vapor uniformity is questionable, but what is likely to happen when the electric power comes back on? Lights anyone? The fan motor could spark too, but recall that it was placed outside the building.)

C_{LFL} = _____ mg/m^3

$t_{\text{to reach LFL}}$ = _____ min

or _____ days and _____ h

4. What would happen if the room ventilation or the electric power were to fail at night? Assume that the lid is on but some evaporation will still occur and in the worst case (maybe after a 3-day weekend) could reach vapor equilibrium. Use the appropriate exposure scenario in Chapter 5 to estimate the equilibrium air concentration in ppm at NTP.

_____ ppm

5. Now assume that the room concentration had reached that equilibrium by the next time the employees arrive. If G with the lid on is low enough to approximate zero with the ventilation system running, how long would it take for the concentration to drop below the LFL after the ventilation system was re-started?

_____ min

How long would it take for the concentration to drop below the 50 ppm TLV?

_____ min

or _____ h

REFERENCES

1. Committee on Industrial Ventilation, *Industrial Ventilation: A Manual of Recommended Practice*, 24th ed., American Conference of Governmental Industrial Hygienists, Cincinnati, OH, 2001.
2. Guffey, S. E., Quantitative troubleshooting of industrial exhaust ventilation systems, *Appl. Occup. Environ. Hyg.*, 9(4), 267–280, 1994.
3. Guffey, S. E. and Booth, D. W. Sr., Evaluation of industrial ventilation troubleshooting methods in experimental systems, *Am. Ind. Hyg. Assoc. J.*, 62(6), 669–679, 2001.
4. Booth, D. W. Sr. and Guffey, S. E., Field evaluation of methods for determining the obstructed section of branches of industrial ventilation Systems, *J. Occup. Environ. Hyg.*, 1(4), 248–255, 2004.
5. AMCA 202 Committee, *Troubleshooting*, *AMCA Publication 202-98*, Air Movement and Control Association International, Inc, Arlington Heights, IL, 1998.
6. Berra, Y., *The Yogi Book*, Workman Publ. Co., New York, 1998.
7. Phillips, C. F. and Halley, P. D., Industrial Hygiene Records and Reports, *Patty's Industrial Hygiene and Toxicology*, 4th ed., Vol. I, Part B, Clayton, G. D. and Clayton, F. E., Eds., John Wiley & Sons, New York, pp. 7–33, chap. 25, 1991.

19 General Ventilation and Transient Conditions

The learning goals of this chapter:

- ☐ Understand the concept behind the two perspectives of the dilution ventilation mixing factor: $K_{\text{room mixing}} = Q_{\text{actual}}/Q_{\text{effective}} = C_{\text{in work zone}}/C_{\text{in exiting air}}$.
- ✍ Be familiar with the general range of $K_{\text{room mixing}}$ values and the factors that affect its value.
- ✍ Understand the significance and use of the terms $C_\infty = G/(Q/K_{\text{room mixing}})$ and $\tau_{\text{effective}} = \text{Vol}/(Q/K_{\text{room mixing}})$.
- ✍ Be able to use a solution to the dilution ventilation model either to find C at a given time (as shown here) or to find the time needed to reach a given C:
- ☐ dilution ventilation increase $C = C_\infty(1 - \exp(-t/\tau_{\text{effective}}))$.
- ☐ dilution ventilation decrease $C = C_0 \exp(-t/\tau_{\text{effective}})$.
- ☐ increase with no ventilation $C = (G/\text{Vol}) \times t$.
- ✍ Be able to differentiate a transient problem from a steady state problem.

I. GENERAL VENTILATION AND CHEMICAL CONTROL

1. In contrast to the specialized nature of local exhaust ventilation, general ventilation is designed into virtually all commercial and institutional buildings (and into any enclosed industrial manufacturing building) to provide fresh air and thermal comfort to its occupants. Ventilation in such buildings relies on a mechanical system to exchange air between indoors and outdoors.[a] The minimum required rate at which these systems must provide outdoor air has been established for a range of defined occupancy classifications. The ability of general ventilation to "dilute" an airborne contaminant that is generated indoors causes industrial hygienists to sometimes call general ventilation "dilution ventilation." To industrial hygienists, the terms are interchangeable. However, problems arise when the general ventilation system of a given building is not able to dilute the airborne chemical or microbiologic contaminants that may be released into its indoor environment to acceptably low concentrations.
   a. Recall from Chapters 2 and 5 that a high concentration plume will always occur near a source of a contaminant, whether it is emitted as a gas, a vapor, or an aerosol. General ventilation does virtually nothing to change the near-field plume.
   b. A mechanical fan or other source of turbulence can be used to increase the plume's rate of dispersion; however, most general ventilation systems are designed to avoid high velocity air to prevent uncomfortable drafts. Also, even if such a fan disperses a plume, the contaminant stays within the room; it will accumulate and expose everyone within a room to some concentration.

[a] Only in residential buildings, in buildings with low occupancy density, or in a building with a high wall-to-floor area ratio is the provision of fresh air and the dilution of internally generated contaminants left to random air leakage (called "infiltration").

For both of the above reasons, dilution ventilation is a poor substitute for local exhaust ventilation to control toxic contaminants released repeatedly or continuously, especially from a point source, and is an unacceptable substitute when the chemical is highly hazardous.

On the other hand, general ventilation can control multiple sources of low hazard and low risk contaminants such as custodial cleaners, emissions from interior finishes, or ozone produced by electrical equipment. Many principles of LEV apply to dilution ventilation, but the emphasis of dilution ventilation is on supplying fresh air rather than exhausting contaminated air.

This chapter will discuss how a contaminant accumulates within a ventilated room until it reaches some steady state concentration (when concentration does not change with time). The dilution model will show that the room volume (or more technically the volume of air initially within the room) is only important during what will be called a "transient" condition (where the concentration does change with time). Room volume plays no role in the average or "steady state" room concentration. At steady state, the contaminant leaves the room at the same rate that it enters and/or is generated within that room.

2. At least three criteria have to be met for dilution ventilation to be an acceptable control for workplace exposures to toxic chemicals.
   a. The plume dispersion pattern must be such that nobody is overexposed to the most hazardous plume concentration near the source. One scenario that would violate this criterion is if someone were working in the plume downwind of the source, such as person B in Figure 5.5. A common adage to avoid that scenario is that fresh air should always flow first to and through the breathing zone before it passes through the plume near a source. However, even the dilution created in a breathing zone upwind of a source (such as person A in Figure 5.5) can be insufficient to avoid overexposing people to some chemicals. Meeting this criterion is especially challenging if the chemical has a ceiling exposure limit above which no exposure is acceptable.
   b. For dilution to be acceptable, a non-zero exposure threshold must exist for the particular chemical. Rightly or wrongly, carcinogenic chemicals are often assumed to have no threshold. The threshold for a sensitizer is usually unknown and may not exist. In both of these special cases, if the total amount of chemical released remains the same, then the liability (calculated as the product of the larger number of exposed people times the collective probability of someone responding to their individually lower dose) can remain the same. This limitation will also apply to some administrative controls in Chapter 21.
   c. The amount of toxin dispersed into the air must be sufficiently small to permit dilution to below its acceptable exposure threshold with a sufficiently small volume of fresh (outside) air to be economically feasible. As shown in Chapter 17, ventilation costs are proportional to the air volume being moved and the amount of heating and/or cooling needed in a given climatic setting. Moreover, the cost of cleaning a contaminant from the exhaust air increases inversely with its initial concentration (in other words, it costs more to remove a given amount of dilute contaminant than a more concentrated contaminant). Thus, at some point, dilution ventilation will become cost prohibitive or at least less cost effective than other controls.

II. COMPONENTS OF THE DILUTION VENTILATION MODEL

The following single compartment model is not the most quantitatively accurate predictor of concentrations and exposures within a room supplied with general ventilation,[1,2] but it is sufficiently accurate to allow the reader to envision the dynamics of accumulation and dissipation within the transient conditions presented in this chapter and to yield some quantitative tools to help the reader analyze steady state conditions presented in the next chapter.

1. The dilution ventilation model is based on the principle of a mass balance as depicted schematically in Figure 19.1.[b] In this mass balance, none of the airborne contaminant is destroyed (at least within the confines of the space being modeled). Therefore, all of the mass of the contaminant generated within this zone, plus any entering into the zone with the makeup air, must equal or balance the mass that either exits from that zone or is stored within it. The pseudo-mathematical form of Equation 19.1a expresses each of these masses in terms of their rate or mass per unit of time.

$$\boxed{\begin{array}{c}\text{Rate of}\\ \text{Generation}\end{array}} + \boxed{\begin{array}{c}\text{Rate of}\\ \text{Entry}\end{array}} = \boxed{\begin{array}{c}\text{Rate of}\\ \text{Removal}\end{array}} + \boxed{\begin{array}{c}\text{Rate of}\\ \text{Accumulation}\end{array}} \quad (19.1a)$$

The boundary for this mass balance will be the walls of a room (creating a single compartment). Applying this principle to the boundaries of a room, such as that depicted in Figure 19.2, means that the sum of the rate at which a contaminant is released or generated within a room plus the rate at which the contaminant may come into the room with the fresh air must equal (balance) the sum of the rate at which the contaminant is leaving the room with the contaminated room air plus the rate at which it is (or may be) accumulating within it.

2. The five components of the standard dilution ventilation model that characterize either the transient and/or the steady-state concentration of a contaminant are listed in Table 19.1. The first four have been used many times before in this book; the fifth is new and deserves some explaining.

The empirical coefficients "$K_{\text{room mixing}}$" or "m" are merely reciprocals of each other ($m = 1/K_{\text{room mixing}}$).[c] Both coefficients characterize how poorly the fresh air entering a room mixes with contaminants being generated within the room before they both leave the room. Ideal mixing would be complete and instantaneous ($K_{\text{room mixing}} = m = 1$). Both coefficients can be viewed from two different perspectives. From the perspective of designing a dilution ventilation system, $K_{\text{room mixing}}$ is the design safety factor expressed

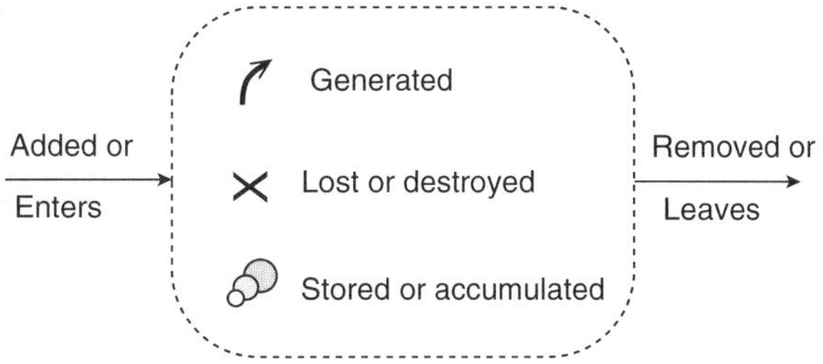

FIGURE 19.1 Conceptual mass balance.

---

[b] To put the concept of "balances" in perspective, a related "force balance" was used in Chapter 3 leading to the formula for a particle's falling velocity (Equation 3.11). The same "mass balance" principle was used in Chapter 11 leading to $Q = A_1V_1 = A_2V_2$ (Equation 11.7). And an "energy balance" was used in Chapter 11 to explain the components of Total Pressure. This "balancing" principle was also implied but not used to explain mechanical and thermal "efficiency" in Chapters 16 and 17, respectively. And it is used again here.

[c] Most other texts just use a plain K for not only this factor, but for practically all other empirical adjustment factors such as the K_{bulkflow} in Chapter 12 and the K_{hood} in the DallaValle equation in Chapter 13. The *Manual of Industrial Ventilation* also uses plain K as a turbulent duct loss factor.[3]

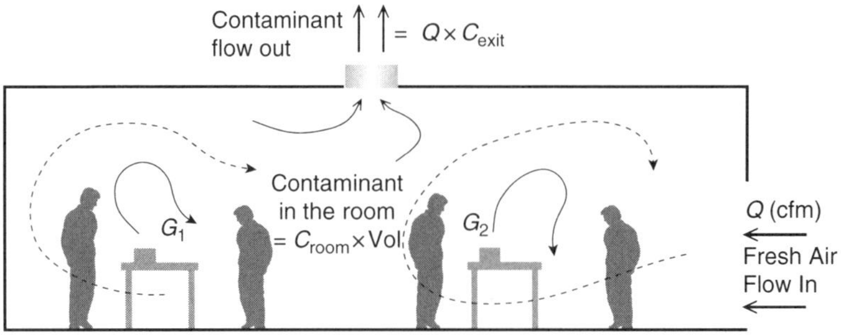

FIGURE 19.2 Depiction of contaminants generated, circulating within, and leaving from a generic workroom.

as the ratio in Equation 19.2a. From this perspective, a system designer should always use a $K_{\text{room mixing}}$ greater than 1 to predict or allow for more outdoor air (the Q that actually enters and leaves a room in the numerator) to achieve a desired dilution of a contaminant.

$$\boxed{K_{\text{room mixing}} = \frac{\text{actual } Q \text{ moving into and out of the room}}{\text{the effective } Q \text{ based on the room air's behavior}}} \tag{19.2a}$$

From the perspective of protecting employees, the concentration of the contaminant in the occupied portion of the room outside of the plume (C_{room}) will be greater than the concentration in the air that exits the room (C_{outlet}), as expressed quantitatively by Equation 19.2b. From this perspective, a room with a good dilution ventilation system will have its $K_{\text{room mixing}}$ and m as close to 1 as possible, and an "m" less than 1 is the fraction of the actual room ventilation air that reaches and dilutes the contaminant.

$$\boxed{K_{\text{room mixing}} = \frac{C \text{ in the general work zone}}{C \text{ in the air leaving room}}} = \frac{C_{room}}{C_{outlet}} \tag{19.2b}$$

a. In both of the above perspectives, $K_{\text{room mixing}}$ is simply an empirical adjustment to account for the less than ideal mixing of the fresh (outdoor) air with the contaminant generated within a room or work zone. Incomplete mixing is sometimes called "shunting" or "short circuiting," when much of the incoming fresh air moves toward

TABLE 19.1
The Name, Symbol, and Units of Components within This Single-Compartment Model

Generation rate of the contaminant	G	mg/min	The rate of generation can be an evaporation rate, the rate of gas leakage, the fume generation rate, etc.
Volume of the room	Vol	m^3	The volume of air in most rooms can be approximated by its internal dimensions; the volume of objects within the room is usually neglected
Concentration within the general room air	C	mg/m^3	The airborne concentration of chemical in the occupied portions of the work zone exclusive of the plume
Rate of general room ventilation	Q	m^3/min	The rate at which nominally fresh air is flowing into the room; units in the U.S. are ft^3/min although m^3/min are more compatible with C within this model
Design safety factor or 1/mixing factor	$K_{\text{room mixing}}$	—	See text explanation. $K_{\text{room mixing}}$ is just K in some other texts, and "mixing factor" is typically abbreviated m

and out of the return air or exhaust vents by some path that bypasses the occupied work zone. Shunting can occur in either a horizontal or a vertical plane.

Figure 19.3 depicts shunting in a horizontal plane as if looking down into the room. In this case, the general ventilation inlet diffusers and return vents are placed too close to each other, allowing the fresh air to take the shortest path from the inlet to the outlet and not mix well with the contaminants generated within the room. Partitions, or large equipment, can also cause horizontal shunting by blocking fresh air from mixing well with the contaminated air in much of the room.

Vertical shunting is typically caused by a temperature inversion when warm air is trapped near the ceiling. Vertical stratification is common in winter when heating the fresh air supply and is exacerbated when the warm air enters above the occupied portions of the room and/or when ceiling exhaust vents are inadequate. This problem is common in factories but can also occur in a commercial building or even a home. An indoor temperature inversion acts much like an outdoor temperature inversion to prevent a warm plume from rising above the occupants or from reaching ceiling exhausts, as depicted in Figure 19.4. An inversion not only contributes to poor air mixing and a larger than desired $K_{\text{room mixing}}$ value, but it also traps contaminants closer to the breathing zone.

b. The following subsections describe three general ventilation conditions or situations for which $K_{\text{room mixing}}$ does not and cannot account.

The first is a common misconception that $K_{\text{room mixing}}$ has something to do with the plume. A plume will always exist near a source (as in Figure 5.6 and Figure 5.8). Unless it is locally exhausted, the concentration in the plume near the source will (by definition) be higher than the concentration to which most people in the room are exposed. However, the value of $K_{\text{room mixing}}$ is unrelated to the elevated concentration within the plume, as implied by Equation 19.3.

$$K_{\text{room mixing}} \neq \frac{C \text{ in vicinity of source}}{C \text{ in general work zone}} \tag{19.3}$$

Second, any compartmentalized model will only work if some mixing occurs. If the incoming air moves in a laminar fashion across or through the room with very little to no mixing, the appropriate mixing factor ($K_{\text{room mixing}}$) would be infinite. Similarly, accumulation cannot occur without mixing.

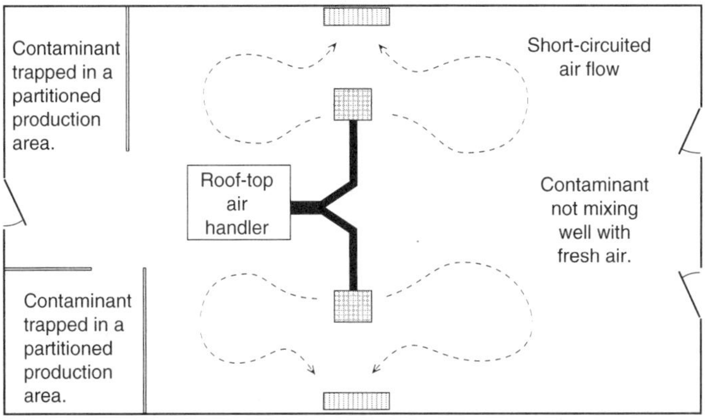

FIGURE 19.3 Plan view (looking down into a room) with horizontally short-circuited ventilation flow.

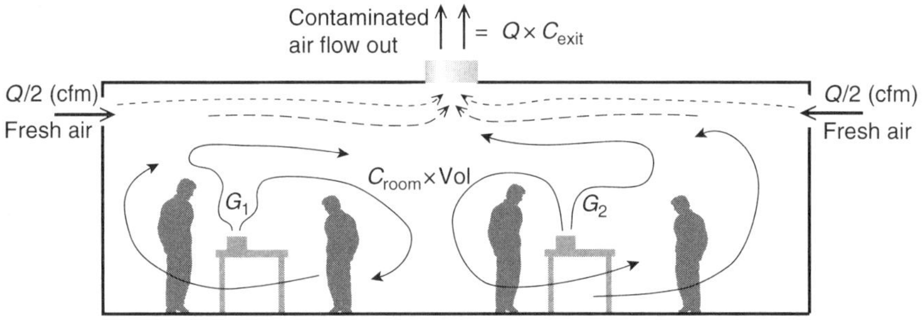

FIGURE 19.4 Vertical stratification creating short-circuiting.

A room with no mixing obviates the need for such an accumulation model. A room with laminar air flow has some advantages and disadvantages. Laminar flow can be created purposefully by bringing the makeup air into the room at a uniform, slow velocity across one wall and exhausting it uniformly on the opposite wall (or moving it from ceiling to floor or *vice versa*). An example of downward laminar flow is depicted in Figure 19.5. Such laminar flow would sweep across a room and completely replace the room's air (and remove its contaminants) in one pass without any turbulent mixing. Downward laminar flow would keep most of the plume out of an employee's breathing zone. Upward laminar flow of about 100 fpm can keep particles up to 100 μm from settling onto horizontal surfaces. Horizontal laminar flow is also possible.

However, creating laminar flow within a room requires considerable effort (meaning it is costly) and can usually only be justified in highly valuable processes such as to isolate people or laboratory animals from infectious or communicable disease, or where the product being made must be protected from human contamination such as in computer chip manufacturing or satellite assembly (see Section 10.10 and VS-10-01-VS-10-03 in the *Manual of Industrial Ventilation*).[3]

The third condition not included within this model is the existence within the room of a separate reservoir into which contaminants could be adsorbed (what might also be called a sink). Some contaminants (especially aerosols and hydrophilic gases or vapors) can be adsorbed onto surfaces within the room. Reservoirs for adsorption include walls, carpeting, fabric surfaces, and water or wet substrates.

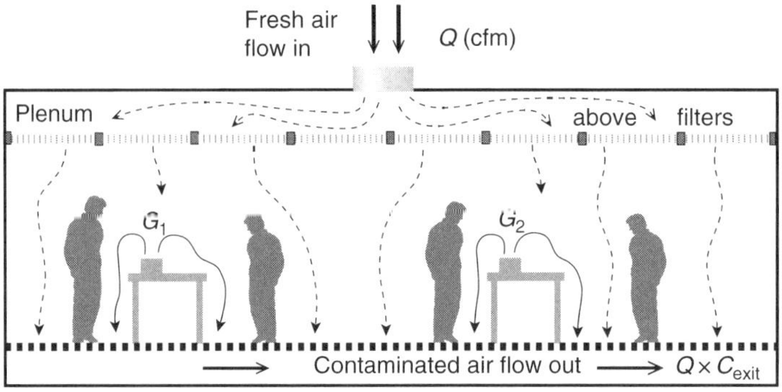

FIGURE 19.5 Vertical (downward) laminar flow ventilation.

TABLE 19.2
Descriptive Ventilation Conditions

	$K_{room\ mixing}$	m	Typical of
An ideally well-mixed room	1	1	
An open workspace with well distributed diffusers and return vents and no point sources (Figure 19.6)	1.05–1.1	0.95–0.9	Office settings
Typically distributed diffusers and return vents, few partitions, and no point sources	1.1–1.3	0.9–0.8	Office settings
Typically distributed vents with some partitions or small point sources (Figure 19.7)	1.3–1.5	0.8–0.7	Offices or light manufacturing settings
Heavily partitioned office; poorly distributed return vents or point sources of contaminants (Figure 19.8)	1.5–3	0.7–0.3	Offices; light to moderate manufacturing settings
A trunk system with diffusers (Figure 19.9)	4	0.25	Manufacturing
Natural ventilation with ceiling exhaust fans	6	0.17	Large manufacturing
Large room with infiltration and natural draft exhaust	10	0.1	Heavy industry

The existence of another storage compartment within the room (in addition to the room air) would make the contaminant appear leave the room at a faster rate than expected from dilution ventilation alone. Such reservoirs are sometimes hard to anticipate, but their real risk is from later changes to these reservoirs (such as a clean-up or renovation) that can cause a rapid release of such gases or dusts back into the breathing zone of unsuspecting workers.

c. Values for $K_{room\ mixing}$ were originally proposed by Brief (1960) to predict contaminant behavior in large manufacturing buildings ("rooms" in a very broad sense of the word).[4] Since then, $K_{room\ mixing}$ factors have been used by others for indoor air quality (IAQ) studies in commercial (office) buildings.[2,5–8] The values of $K_{room\ mixing}$ factors in commercial buildings with more extensive ventilation systems are much lower than the $K_{room\ mixing}$ that typically occurs in large factories, as can be seen in Table 19.2. Figure 19.6 to Figure 19.9 were created to acquaint the reader with some of the ventilation air distribution systems represented by the mixing factors listed in Table 19.2. In these diagrams, the gray lines represent supply air ducts terminating at diffusors; the dashed lines represent return air ducts going to a roof-mounted air handler and conditioner.

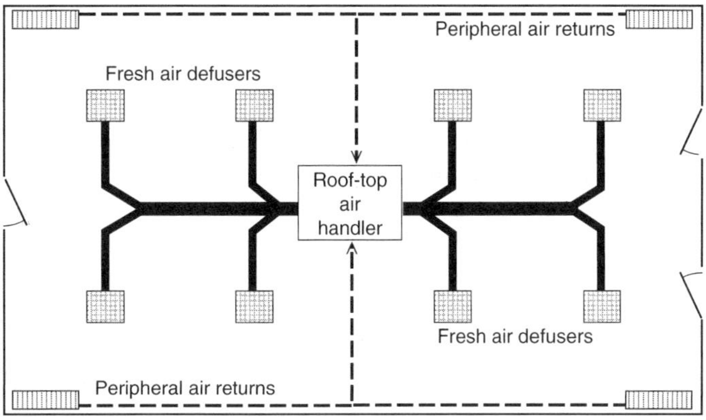

FIGURE 19.6 A well ventilated building with $K_{room\ mixing} \approx 1.1$ to 1.3.

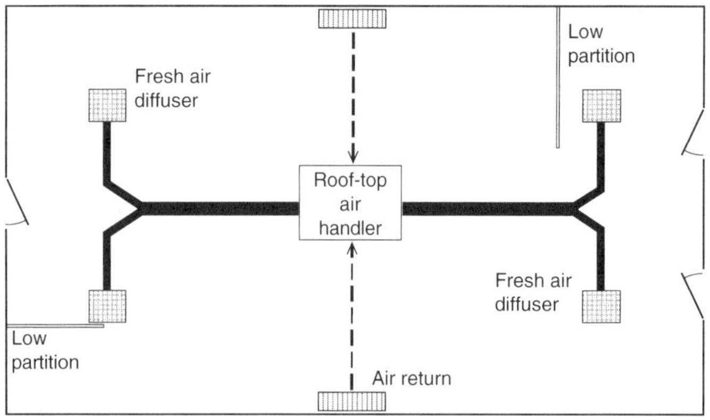

FIGURE 19.7 A typically ventilated building with $K_{\text{room mixing}} \approx 1.3$ to 1.8.

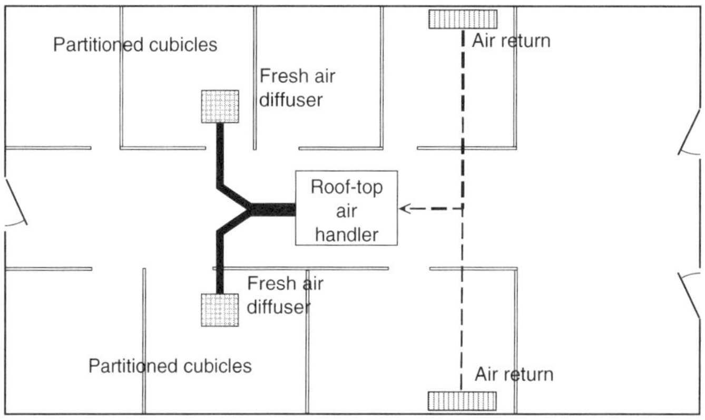

FIGURE 19.8 A poorly ventilated and partitioned building, $K_{\text{room mixing}} \approx 2$ to 3.

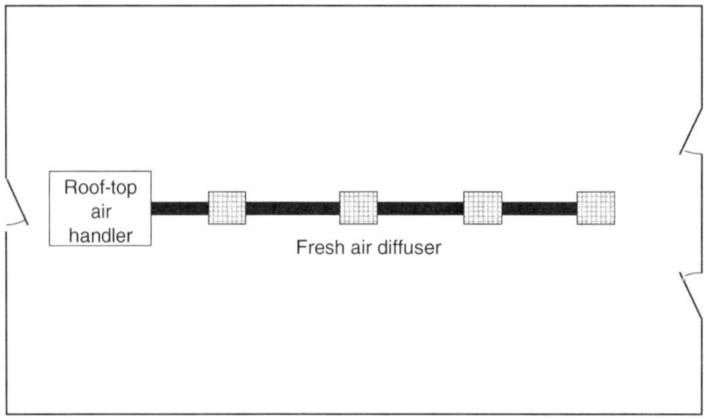

FIGURE 19.9 A trunk ventilation system with one return air path directly into the air handler, $K_{\text{room mixing}} \approx 3$ to 5.

3. The following assumptions and details of the model apply to the contaminant's generation rate, its accumulation within the room, its entry into the room from outside, its mixing within the room, and its removal from the room.
   a. More complicated forms of this model can accommodate a contaminant being generated in any temporal pattern, but the simplest case is to assume that the rate at which a contaminant is generated (G) is reasonably constant. The actual value of G may be known; it may be unknown; it may start from zero or stop and return to zero; but the model used herein assumes that when the source is emitting, the rate of generation is a constant value defined by G.
   b. The initial concentration of the contaminant within the room (C_0) caused by whatever reason or source, will be assumed to be uniform within the occupied portions of the room. Furthermore, the concentration of the contaminant generated within this room after startup will also be assumed to be uniform. The expectation that everyone within a room will really be exposed to the same contaminant concentration is a somewhat unrealistic but necessary assumption to make this model simple (although with less accuracy than may be desired for truly quantitative predictions).[d]
   c. The amount (mass) of contaminant within the room at any time is the air Vol $\times$ C. This is the third (but not the last) time the reader has encountered the relationship originally stated in Equation 5.5b.

$$\boxed{\text{mass} = C \times [\text{air Vol}]} \qquad \text{(Equation 5.5b)}$$

   The rate at which the contaminant either accumulates within or dissipates from the room is the change in Vol $\times$ C within a given time interval. In calculus, a rate of change with respect to time is called "a differential with respect to time." Unless the room is a balloon or an inflatable building, the room volume is constant. Although C is assumed to be uniform within the occupied portions of the room, it can change over time during a transient increase or decrease. Thus, the rate of change of the mass of contaminant within the room concentration will be Vol $\times$ $\delta C/\delta t$.
   d. The rate that an additional mass of contaminant might enter into a room from outside is the product of the rate that makeup air flows into the room (Q) times the concentration of the contaminant within that makeup air ($C_{\text{make up}}$). $C_{\text{make up}}$ entering most rooms is zero or nearly so, in which case it can be disregarded. Carbon dioxide is one exception because its background concentration is about 370 ppm in fresh air (recall Table 2.2 and Figure 2.3). Another source of contaminant in the makeup air is re-entrainment (see Section VI in Chapter 15). Re-entrainment rarely involves all of the exhaust air. The concentration of the re-entrained contaminant in the makeup air might be anywhere from $10\times$ to $10^4\times$ below its concentration in the exhaust air.

   The lack of good mixing between the contaminant and fresh air also implies that the fresh air flowing into the room does not mix well with the air in the occupied portions of the room before it exits. Thus, the mixing term ($K_{\text{room mixing}}$) should also be applied to the mass of contaminant entering into the occupied portion of the room with the makeup air, reducing the effective rate at which any of this mass enters into, and is dispersed within, the occupied portions of the room to which the concentration "C" applies. Thus, the rate that an outside mass of contaminated enters into the occupied portion of the room equals $Q \times C_{\text{makeup}}/K_{\text{room mixing}}$.

[d] Where this assumption is clearly not valid, the reader may want to use a two-compartment or multi-compartment model.[2]

e. In a similar manner, the rate at which the contaminant is removed from the room is Q times the concentration of the contaminant within that outgoing air (C_{outlet}), viz. $Q \times C_{outlet}$. However, C_{outlet} cannot always be measured, nor is it the concentration of interest to employee health. For these reasons, the model relies on the second definition of $K_{room\ mixing} = C/C_{outlet}$ (Equation 19.2b). Thus, the rate of removal of the contaminant from the occupied workroom equals $Q \times C/K_{room\ mixing}$. Because the room's volume and pressure remain constant, the rate that air flows out of the room is the same as the rate that it flows into the room.

f. The model assumes that $K_{room\ mixing}$ is a constant within a given room. In reality, $K_{room\ mixing}$ is probably a weak inverse function of Q. That is, increasing Q will probably increase the amount of mixing of the fresh air with the contaminated air which means $K_{room\ mixing}$ will decrease to some degree. Assuming that $K_{room\ mixing}$ is constant and independent of Q is a conservative assumption in that any decrease in $K_{room\ mixing}$ as Q is increased to protect the occupants will beneficially lower C even further than predicted by assuming a constant $K_{room\ mixing}$.

III. THE GENERAL SOLUTION TO THE DILUTION VENTILATION MODEL

1. The mass balance equation will first be stated in Equation 19.4a as it was defined conceptually in Equation 19.1a, then rearranged incrementally into Equation 19.5 to facilitate the calculus that will eventually result in the solutions shown in Equation 19.11 and Equation 19.12a,b. The intervening steps between Equation 19.5 and Equation 19.12a,b are provided primarily as background for the reader and as an example of applied calculus. Such equations are not solved in routine industrial hygiene (IH) practice.

$$\text{Rate of Generation} + \text{Rate of Entry} = \text{Rate of Removal} + \text{Rate of Accumulation} \tag{19.1a}$$

$$G + \frac{Q \times C_{makeup}}{K_{room\ mixing}} = \frac{Q \times C}{K_{room\ mixing}} + \text{Vol}\ \delta C/\delta t \tag{19.4a}$$

$$\text{Rate of Generation} + \text{Rate of Entry} - \text{Rate of Removal} = \text{Rate of Accumulation} \tag{19.1b}$$

$$G + Q \times C_{makeup}/K_{room\ mixing} - Q \times C/K_{room\ mixing} = \text{Vol}\ \delta C/\delta t \tag{19.4b}$$

$$[G \times K_{room\ mixing} - Q \times (C - C_{makeup})]\delta t = K_{room\ mixing} \times \text{Vol} \times \delta C \tag{19.5}$$

2. Equation 19.5 is called a first order differential equation because the change in C (δC on the right-hand side of Equation 19.5) is proportional to itself on the left-hand side of that equation to the first power. We will use a standard calculus substitution method to integrate this differential equation. Let $X = G \times K_{room\ mixing} - Q(C - C_{makeup})$, the polynomial on the left. Then $\delta X = -Q\,\delta C$ (because everything else in that polynomial is made up of constants) and $\delta C = -\delta X/Q$. The following two equations result from substituting these values into Equation 19.5 and regrouping the variables:

$$X\,\delta t = [-K_{room\ mixing} \times \text{Vol}/Q]\delta X \tag{19.6}$$

$$\frac{\delta X}{X} = \left[\frac{-Q}{(K_{room\ mixing} \times \text{Vol})}\right]\delta t \tag{19.7}$$

In calculus, the integral of $\delta X/X = \ln[X]$, the natural logarithm of the variable being integrated. Thus, Equation 19.8 is the integral of Equation 19.7 between an initial concentration $C = C_0$ (and $X = X_0$) when time $t = 0$ and some concentration C at any later time t. We will also abbreviate $K_{\text{room mixing}}$ to K_{mix}.

$$\ln[X] - \ln[X_0] = \ln\left[\frac{X}{X_0}\right] = \frac{-Q}{K_{\text{room mixing}} \times \text{Vol}} \times t \tag{19.8}$$

$$\frac{X}{X_0} = \frac{G \times K_{\text{mix}} - Q(C - C_{\text{makeup}})}{G \times K_{\text{mix}} - Q(C_0 - C_{\text{makeup}})} = \exp\left[\frac{-Qt}{K_{\text{mix}} \times \text{Vol}}\right] \tag{19.9}$$

$$C\left[\frac{Q}{GK_{\text{mix}} - Q(C_0 - C_{\text{makeup}})}\right] = \frac{GK_{\text{mix}} + QC_{\text{makeup}}}{GK_{\text{mix}} - Q(C_0 - C_{\text{makeup}})} - \exp\left[\frac{-Qt}{K_{\text{mix}}\text{Vol}}\right] \tag{19.10}$$

$$C = \frac{GK_{\text{mix}} + QC_{\text{makeup}}}{Q} - \frac{GK_{\text{mix}} - Q(C_0 - C_{\text{makeup}})}{Q}\exp\left[\frac{-Qt}{K_{\text{mix}}\text{Vol}}\right] \tag{19.11}$$

3. Equation 19.12 is the general solution for the ventilated room concentration C for any time t. The two starting times of interest to industrial hygienists will be either the time that a source begins to generate a contaminant (e.g., a process is turned on or a container is opened) or the time that the contaminant stops being generated (e.g., a process is turned off or the container is closed again). This general solution can be viewed in three parts as marked by the brackets inserted below Equation 19.12a. Each part will be discussed briefly in the paragraphs below, and in more detail in separate sections later in this chapter.

$$C = \underbrace{\frac{GK_{\text{mix}} + QC_{\text{makeup}}}{Q}}_{\text{see text 3.c below}} \times \underbrace{\left(1 - \exp\left[\underbrace{\frac{-Qt}{K_{\text{mix}}\text{Vol}}}_{\text{see text 3.a below}}\right]\right)}_{\text{see text 3.b below}} + \underbrace{C_0 \exp\left[\underbrace{\frac{-Qt}{K_{\text{mix}}\text{Vol}}}_{\text{see text 3.a below}}\right]}_{\text{see text 3.b below}} \tag{19.12a}$$

$$C = C_\infty \times \left(1 - \exp\left[\frac{-t}{\tau_{\text{effective}}}\right]\right) + C_0 \exp\left[\frac{-t}{\tau_{\text{effective}}}\right] \tag{19.12b}$$

a. Time (t) only appears in the exponential terms. Each exponential term involves the cluster $K_{\text{room mixing}} \times \text{Vol}/Q$ that defines the time required for one effective room air change. The time for each effective room air change (the "time constant" $\tau_{\text{effective}}$ as defined in Equation 19.13) is the time in the units of Q for the fresh air entering the room to have the effect on C of equaling the room's volume.

$$\tau_{\text{effective}} = \frac{K_{\text{room mixing}} \times \text{Vol}}{Q} = \frac{\text{Vol}}{Q/K_{\text{room mixing}}} \tag{19.13}$$

One effective air change occurs in every $\tau_{\text{effective}}$ of elapsed time. The value $t/\tau_{\text{effective}}$ is the number of effective air changes, sometimes called N_{changes}.[e]

[e] Some other text uses N_{changes} to mean "air changes per hour", a term abbreviated by ACH within this and ASTM texts.

b. As just mentioned above, Equation 19.12a,b has two groups of exponential terms. The middle group of terms inside the parentheses describes how an initial C will increase exponentially with time toward a steady state concentration as the airborne contaminant accumulates. This portion of the model is rarely used in actual field practice to predict C at any point in time during a transient accumulation. However, Section IV of this chapter will describe the concept of exponential accumulation and how the value of $\tau_{\text{effective}}$ can be used to estimate how short a time that a room will generally be in a transient accumulation condition before reaching a steady condition.

The group of terms on the right side of the Equation 19.12a,b describes how dilution will cause an initially high room concentration (C_0) to decrease exponentially with time toward zero (or C_{makeup} if it is above zero). Section V of this chapter will use this portion of the model to predict the concentration that will result from using dilution ventilation to purge an initially highly concentrated contaminant from a workspace or a confined space.

c. The first term on the left is the only part of the model that does not involve time (t). Notice that it also does not involve the room volume. This first term will be the only term left after the value of the two negative exponential terms approach zero. Recall that "steady state" means a condition that does not change with time. Thus, the first term is the steady state concentration (C_∞).[f] Equation 19.14 implies that the steady state concentration is equivalent to diluting the contaminant generated within a given time period (G) into the effective fresh air entering the room within that time period.[g]

$$C_\infty = \frac{G}{Q/K_{\text{room mixing}}} + C_{\text{makeup}} \tag{19.14}$$

As will be discussed in Section IV and used in the next chapter, the room concentration (C) is practically already a constant steady state concentration by the time t is only $2 \times \tau_{\text{effective}}$. In most workplaces, $\tau_{\text{effective}}$ is only 10 to 15 min. If the process that generates the source continues for most of the workday, this steady state concentration (C_∞) approximates what hygienists call the time weighted average concentration (C_{TWA}). The fresh air entering a room is usually uncontaminated unless the incoming air is affected either by re-entrainment, by a second source, or by a non-zero background level in the ambient air, such as from carbon dioxide. Thus, in the typical case where $C_{\text{makeup}} = 0$, this $C_{\text{TWA}} \approx G \times K_{\text{room mixing}}/Q$, a result that comprises a very simple yet very powerful description of steady state dilution ventilation conditions that will be the focus of Chapter 20.[h]

As just implied, this general solution to the dilution ventilation model is applicable to both the steady state or TWA concentration, and to two of the three transient conditions of practical interest to industrial hygienists. The next three sections of this chapter will discuss three transient

[f] The subscript ∞ means after a long enough time to reach steady state conditions when things no longer change with time. As time increases, $\exp(-t/\tau_{\text{effective}})$ decreases. Just as knowing that $e^0 = 1$, it is useful to remember that $e^{-\infty} = 1/e^\infty = 0$. The latter is used in Equation 19.14 and Equation 20.1 to find the C when time gets very large, such as shown in Table 19.3.

[g] This dotted boxed equation will become solid boxed Equation 20.1 for the steady state conditions to be discussed in the next chapter.

[h] The notation $Q_{\text{effective}} = Q/K_{\text{room mixing}}$ will be used in the next chapter to mean the effective rate of general ventilation (in comparison to the shorter but less descriptive symbol Q' used in the ACGIH *Manual* Figure 2.1 and some other texts.

conditions. The first two transient conditions directly utilize the above general solution, while the derivation of a solution to the third transient condition has to start back at the mass balance for reasons to be explained. Steady state conditions will be discussed in the next chapter. In each case, the specific solution becomes much simpler than the general solution.

The reader is again cautioned not to view the results of this model (especially during the transient conditions) as quantitatively precise. The limitations and assumptions presented above are almost never matched in a given real-world case. Thus, view the following results as broad patterns that you might expect to encounter if you were to investigate or anticipate conditions in any real-world setting.

IV. CONCENTRATIONS DURING A TRANSIENT INCREASE

In a transient accumulation scenario, a production process (source) that is initially "off" begins to emit a contaminant at a constant rate (G). The initial concentration (C_0) will be assumed to be zero such as at the start of a workday. The airborne concentration will begin to accumulate until a steady state occurs (a condition where the concentration does not change over time). The accumulation problem could also be applied to a substantial increase in G, a decrease in Q, or where C_0 is not zero (perhaps from an intermittent process), but such problems go beyond the more limited learning goals of this chapter.

1. Inserting the initial condition that $C_0 = 0$ at $t = 0$ into the general solution (Equation 19.12a,b) removes the right-most term and yields the following equation that predicts how fast and how high the concentration could get at any time after start-up:

$$C = \frac{GK_{\text{room mixing}} + QC_{\text{makeup}}}{Q}\left(1 - \exp\left[\frac{-Qt}{K_{\text{room mixing}}\text{Vol}}\right]\right) \tag{19.15a}$$

Equation 19.15a can be made to appear simpler if the first group of terms is replaced by the symbol C_∞ (as defined in Equation 19.14) and the group in the exponential term ($K_{\text{room mixing}}$ Vol/Q) is replaced by the symbol $\tau_{\text{effective}}$ (as defined by Equation 19.13). Equation 19.15b is identical to Equation 19.15a; it just looks simpler.

$$C = C_\infty\left(1 - \exp\left[\frac{-t}{\tau_{\text{effective}}}\right]\right) \quad \text{when } C_0 = 0 \tag{19.15b}$$

The above equation can also be solved for time, yielding Equation 19.16. This solution can be used to predict how long it might take before the concentration exceeds some exposure limit (such as a STEL).

$$t = -\tau_{\text{effective}} \times \ln\left[1 - \frac{C}{C_\infty}\right] \quad \text{when } C_0 = 0 \tag{19.16}$$

2. The most important use of these solutions is to realize how quickly a transient accumulation can approximate the steady state concentration. The duration of a transient accumulation is often between 10 to 15 min, almost always less than 30 min, and very much shorter than an 8-h workday.

   The calculated results of all equations where the dependent variable of interest equals 1 minus a negative exponent of the independent variable (such as time in Equation 19.15) follow a very predictable pattern. The points of interest for this pattern occur at integer values of X in e^{-X}, viz., e^{-1}, e^{-2}, e^{-3}, etc. At each integer value, the result repeatedly advances 63% of its way toward steady state. In other words, over each interval of $\tau_{effective}$ minutes, C gets 63% of the way from the concentration at which it starts that interval towards its eventual steady-state C_∞. The fractional change in the room concentration from 0 to 1, as given by Equation 19.15, can be visualized in Figure 19.10 and Table 19.3.

a. The first point of reference is $1 - e^{-1}$ which equals $1 - 1/e = 1 - 1/2.718 = 0.632$ or approximately 63% of the way toward steady state (or see this example as an equation below). Thus, after a source begins to generate a contaminant at a constant rate G and is diluted into a constant fresh air flow rate of Q, C will be within 63% of its steady-state concentration within one equivalent air change.

$$\text{When } t = \tau_{effective}, \quad C = C_\infty[1 - \exp(-1)] = 0.632C_\infty \qquad \text{(using Equation 19.15b)}$$

b. After $t = 2 \times \tau_{effective}$, C advances another 63% of the remaining way (or a total of 86%) toward its eventual C_∞. The remaining 14% to reach steady state is difficult to detect by chemical measurements. The remaining 5% difference after $t = 3 \times \tau_{effective}$ is almost impossible to detect.

✓ Practically speaking, if C_∞ is an airborne hazard, then 86 or 95% of C_∞ is already also an airborne hazard. Even 68% of C_∞ is likely to exceed a concentration at which the duration of exposure should be limited. And $\tau_{effective}$ values are typically as short as 5 to 15 minutes.

Thus, if a workplace is subject to intermittent excursions into acutely toxic concentrations, employees must be able to leave (or don respirators) within less than one effective air change. The model is not precise enough to estimate how much shorter that one air change they should leave. Nor is the model precise enough to anticipate the exact concentration even at steady state. However, this model should yield good approximation.

3. A very secondary use of the transient accumulation solution is to envision the effect of transient excursions in source emission on an accumulating transient hazard and/or on the time-weighted average.

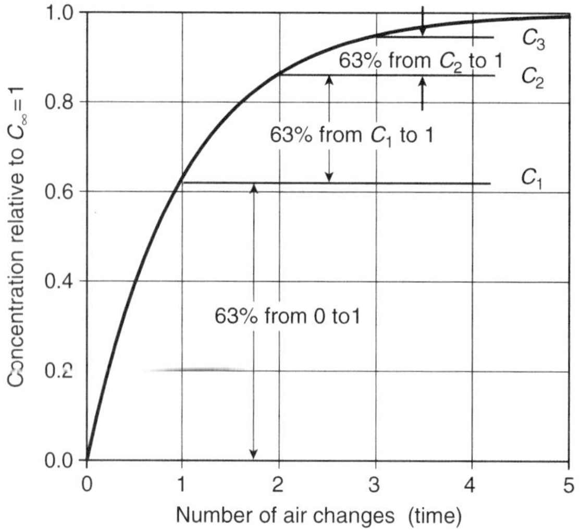

FIGURE 19.10 The pattern of one minus a negative exponential accumulation ($Y = 1 - e^{-X}$).

TABLE 19.3
The Characteristic Progression of All Exponential Changes

Number of Air Changes $N_{changes} = t/\tau_{effective}$	Fraction Left $= 1 - C/C_\infty$ $= e^{-t/\tau_{effective}}$	Fraction Changed $= C/C_\infty$ $= 1 - e^{-t/\tau_{effective}}$	Fractional Change $= \left(\frac{C_{t=N\tau} - C_{t=(N-1)\tau}}{C_{t=\infty} - C_{t=(N-1)\tau}}\right)$
1	0.3679	0.632	$\left(\frac{0.632 - 0}{1 - 0}\right) = 0.632$
2	0.1353	0.865	$\left(\frac{0.865 - 0.632}{1 - 0.632}\right) = 0.632$
3	0.0498	0.950	$\left(\frac{0.950 - 0.865}{1 - 0.865}\right) = 0.632$
4	0.0183	0.982	$\left(\frac{0.982 - 0.950}{1 - 0.950}\right) = 0.632$
5	0.0067	0.993	$\left(\frac{0.993 - 0.982}{1 - 0.982}\right) = 0.632$

a. The value of $\tau_{effective}$ as defined in Equation 19.13 predicts the duration of a transient condition. It is only during this "transient" stage that room volume can play a part in controlling exposure.

 The larger a room and/or the $K_{room\ mixing}$ value, the longer it will take for the room concentration to reach steady state. In other words, a contaminant can be dispersed into a large room volume for a longer time before reaching or approaching a steady state hazard than it can in a small room.

 A fan can potentially redirect a plume away from someone's breathing zone, but it will also improve room mixing, decrease $K_{room\ mixing}$ (as in Equation 19.2a,b), and decrease C_∞ (as in Equation 19.14). The first effect can benefit someone in the near-field of the plume, while the latter effects may or may not be beneficial to everyone else in the room. A lower $K_{room\ mixing}$ will shorten $\tau_{effective}$ and the time for the room to reach steady state (as in Equation 19.16). Whether or not the lower C_∞ is an acceptable concentration depends upon the circumstances. Without a more detailed analysis of specific circumstances, using a fan to disperse a plume should be limited to protecting someone in the near-field of an unacceptable plume for a short time.

b. Transient exposures can acceptably exceed the allowed 8-TWA limit by several fold as long as the highest concentration does not exceed a ceiling limit, a STEL, a maximum value from OSHA Table Z-2, an IDLH, or any other limit imposed by an acute toxic effect. The duration of an acceptable transient accumulation condition should be limited by the time that the excursion to an elevated concentration can be balanced by a generally longer interval at a lower concentration, such that the time weighted average concentration over the workday does not exceed the 8TWA exposure limit. The transient accumulation model could be manipulated to set explicit limits on either the transient peak concentration or its duration, but specifying enough information and constraints again goes beyond the learning goals of this section. Equation 19.17 only outlines a very general relationship between the magnitude of the excursion ($C_{excursion} \times t_{excursion}$) and the lower concentration that can be allowed to exist for the rest of the workday when the contaminant is not being generated.

$$C_{excursion} \times t_{excursion} \leq [480 \times C_{TWA\ exposure\ limit}] - [C_{\text{"lower"}} \times (480 - t_{excursion})] \quad (19.17)$$

4. The following example of an accumulating transient hazard is intended to demonstrate how the accumulation model can be used either to predict a hazard or to analyze a past episode. Notice that in order to solve a transient concentration problem using either Equation 19.15 or Equation 19.16, one must first predict the steady state concentration (the C_∞ defined in Equation 19.14).

EXAMPLE 19.1

Small gasoline engines are often used to power equipment such as an electric generator, compressor, power trowel, or power washer where conventional AC electricity is not available. Estimates of carbon monoxide (CO) emissions from such engines have ranged from approximately 0.18 ft^3/min CO/bhp to 0.34 ft^3/min CO/bhp (the latter is 0.67 kg/bhp/h).[9] The TLV® for CO is 25 ppm (29 mg/m^3). There is no STEL for CO, but NIOSH recommended a 200 ppm ceiling; and CO has a 1200 ppm IDLH. Suppose that a farmer were to use a 2 bhp engine that produced approximately 0.34 ft^3/min CO/bhp to power wash and disinfect a 20 × 40 × 8 ft high animal production room (such as a hog farrowing room) that has one 1200 ft^3/min wall fan providing ventilation. In this setting, $K_{room\ mixing}$ might equal 3. It will help to recall that 1 ft^3 = 0.0283 m^3 or 1 m^3 = 35.5 ft^3.

a. How long will it take to be within 14% of steady state? To be 14% from steady state means the concentration will be 100 − 14 = 86% of the way toward steady state. As shown in Table 19.3 above, $C/C_\infty = 0.865$ when $t/\tau_{effective} = 2$. Thus, the question can be restated as, how long is two effective room air changes?

   To solve virtually any transient problem, the value for the time constant $\tau_{effective}$ must be found using the definition of $\tau_{effective} = K_{room\ mixing} \times Vol/Q$ given in Equation 19.13. Of course to do that, one needs values for each variable within that equation.

$$Vol = 20 \times 40 \times 8 = 6400\ ft^3 \times 0.0283\ m^3/ft^3 = 181\ m^3$$

$$\tau = Vol/Q = 6400\ ft^3/1200\ ft^3/\min = 5.3\ \min$$

$$\tau_{effective} = K_{room\ mixing} \times \tau = 3 \times 5.3 = 16\ \min$$

   In this case, the desired time = $2 \times \tau_{effective} = 2 \times 16 = 32$ min.

b. What would the concentration of CO be after $t = 2 \times \tau_{effective} = 32$ min ? To find C in transient accumulation problems generally requires solving for the steady state concentration C_∞ as an intermediate step. Two ways to solve such a steady state problem will be presented.

   The metric solution for C_∞ is used to calculate G as mass per minute.

$$G = 0.67\ kg/bhp/h \times 2\ bhp/60\ \min/h = 0.022\ kg/\min = 22 \times 10^3\ mg/\min$$

   For convenience, the Q will be converted from ft^3/min to m^3/min:

$$Q = 1200 \times 0.0283\ m^3/ft^3 = 34\ m^3/\min$$

And these values can then be used to solve for C_∞ using Equation 19.14:

$$C_\infty = \frac{G \times K_{mix}}{Q} = \frac{22 \times 10^3\,\text{mg/min} \times 3}{34\,\text{m}^3/\text{min}} = 1941\,\text{mg/m}^3 \times \frac{24.45}{28} = 1695\,\text{ppm} \qquad (19.14)$$

An alternate solution for the steady state concentration is possible using Equation 5.8 using the volume of the contaminant being generated (remember that ppm is also a volume to volume concentration). The volume of CO gas being emitted equals 2 bhp × 0.34 ft^3/min/bhp = 0.68 ft^3/min of CO. Note in the solution below that the numerator is multiplied by one million to yield ppm:

$$C_\infty = \frac{G \times K_{\text{room mixing}}}{Q} = \frac{0.68\,\text{ft}^3/\text{min} \times 3 \times 10^6}{1200\,\text{ft}^3/\text{min}} = 1700\ \text{ppm steady state}$$

(using Equation 5.8)

In either case, C_∞ is then used in Equation 19.15b, to find C when $t = 2 \times \tau_{\text{effective}} = 2 \times 16 = 32$ min.

$$C = C_\infty(1 - \exp(-t/\tau_{\text{effective}})) = 1695(1 - \exp(-2)) = 1695 \times 0.865 = 1466\,\text{ppm} \qquad (19.15)$$

c. Since the above concentration would be unacceptable, just how long can the work be done before reaching the NIOSH recommended 200 ppm ceiling? The time can be calculated using Equation 19.16 where C_∞ was found above to be 1700 ppm.

$$t = -\tau_{\text{effective}} \ln\left[1 - \frac{C}{C_\infty}\right] = -16 \ln\left[1 - \frac{200}{1700}\right] = 16 \times 0.125 = 2\,\text{min} \qquad (19.16)$$

What are the options to avoid a hazard? Source, pathway, or receiver controls? What would you recommend? What might the farmer most accept?

V. CONCENTRATIONS DURING A TRANSIENT DECREASE

Examples of a transient decrease can be found in several scenarios:

- After a process has been running and emitting a contaminant for some time, it is stopped, turned off, or closed up. Thereafter, airborne concentrations begin to dissipate at some rate back toward clean air.
- One wants to enter a confined space in which a toxic gas or vapor has accumulated slowly over time to unacceptable levels. In order to allow safe entry, the contaminant must be purged by ventilation to some lower acceptable level.
- To test the effective ventilation, a tracer gas is purposefully released and mixed into room air and the concentration is monitored while it dissipates toward zero.

Typical IH questions related to a transient decrease include "What will C be after a given time?" and "How long does it take to get from an initial unacceptably high C_0 to an acceptable C (such as a PEL or TLV)?"

1. The elevated concentration at which the contaminant starts out (C_0) may be either predicted or measured. For the scenarios to be of IH interest, let C_0 be well above an acceptable concentration. For this model to work well, C_0 should be reasonably well mixed into the room air (although it need not have reached a steady state). In the solution

that follows, G is assumed to be zero. When the very first term in the right side of the general solution of Equation 19.12a is zero, Q in the numerator and denominator cancel, as shown below (again letting $K_{\text{room mixing}} = K_{\text{mix}}$ for space).[i]

$$C = \frac{0 + \cancel{Q} C_{\text{makeup}}}{\cancel{Q}} \times \left(1 - \exp\left[\frac{-Q \times t}{K_{\text{mix}} \times \text{Vol}}\right]\right) + C_0 \exp\left[\frac{-Q \times t}{K_{\text{mix}} \times \text{Vol}}\right]$$

(using Equation 19.12a)

In order to simplify what is left to just one exponential term, the two coefficients of that exponent (C_{makeup} in the first term and C_0 before the last term) will be combined, yielding Equation 19.18a.

$$C = C_{\text{makeup}} + [C_0 - C_{\text{makeup}}] \exp\left(\frac{-Q \times t}{K_{\text{room mixing}} \times \text{Vol}}\right) \quad (19.18\text{a})$$

This equation can also be further simplified (as in the previous case of an increasing concentration) when the fresh air entering the room is uncontaminated (i.e., when $C_{\text{makeup}} = 0$) and by again letting $\tau_{\text{effective}} = K_{\text{room mixing}} \times \text{Vol}/Q$. The C resulting from Equation 19.18b (analogous to Equation 19.15b for accumulations) is depicted in Figures 19.11a and b over five and 10 air changes, respectively.

$$C = C_0 \exp\left(\frac{-t}{\tau_{\text{effective}}}\right) \quad \text{when } G \text{ and } C_{\text{makeup}} = 0 \quad (19.18\text{b})$$

Equation 19.19 (analogous to Equation 19.16 for accumulations) can be used to find the time for the initial high C_0 concentration to be diluted to a lower C. Notice in the second version of Equation 19.19 that the negative sign has been deleted because the ratio is inverted.

$$t = -\tau_{\text{effective}} \ln\left[\frac{C}{C_0}\right] = \tau_{\text{effective}} \ln\left[\frac{C_0}{C}\right] \quad \text{when } G \text{ and } C_{\text{makeup}} = 0 \quad (19.19)$$

2. The temporal pattern of decreasing C given by Equation 19.18b and shown in Figure 19.11 follows the same 63.2% per $\tau_{\text{effective}}$ characteristic of accumulations and any other negative exponential relationship. However, the practical duration of a transient decrease is not limited to any fixed number of air changes. The number of air changes to dilute an initial hazard depends on how much the initially hazardous C_0 must be diluted to reach an acceptable concentration. The solution (Equation 19.19) indicates that the duration of a transient decrease depends upon the logarithm of the ratio of the initial acute hazard defined by the high C_0 to the lower acceptable exposure level. The appropriate lower acceptable exposure level depends upon the activity in which one wants to engage. Thus, an acceptable exposure level might be an LFL, an IDLH, an STEL, an 8-TWA PEL, or TLV, etc. In an extreme case (such as to reach a toxicologically derived airborne "no observable effect level" [NOEL]), the duration of a transient decrease might even need to be extrapolated below the contaminant's limit of detection.

[i] In the case of a confined space, the contaminant may still continue to be emitted slowly, but G can be approximated as being zero if it took many days to reach the initially hazardous C_0.

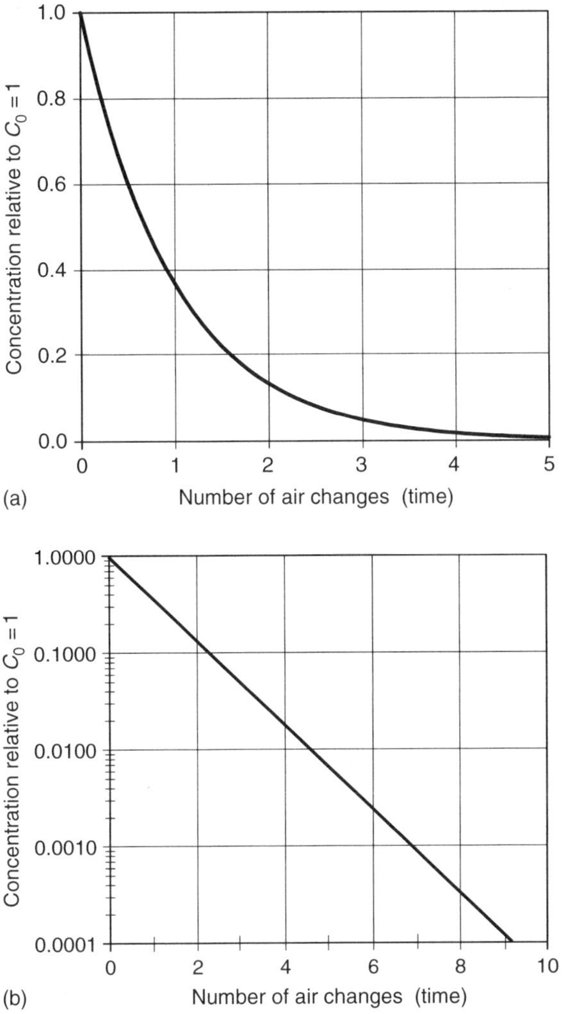

FIGURE 19.11 (a) Exponential decrease on a linear scale. (b) Exponential decrease on a logarithmic scale.

3. The following example of a decreasing hazard continues from where the preceding example left off:

Example 19.2

Suppose another farmer was working at the same task but left the gasoline engine running for more than 30 min (long enough to have reached the C_∞ previously found to be approximately 1700 ppm). At some point he felt too ill to continue working, so he shut the engine down and left the building. He soon felt better and realized that his symptoms were probably from the engine's exhaust (which, heaven forbid, he would probably call exhaust fumes).

a. What would the carbon monoxide concentration be if he decided to go back into the building after only a half-hour?

$$C = C_0 \exp\left(\frac{-t}{\tau_{\text{effective}}}\right) = 1700 \exp\left(\frac{-30}{16}\right) = 1700 \times 0.153 = 261 \text{ ppm}$$

(using Equation 19.18b)

b. How long would he have to wait for the CO to drop to the TWA-TLV of 25 ppm?

$$t = \tau_{\text{effective}} \ln\left[\frac{C_0}{C}\right] = 16 \ln\left[\frac{1700}{25}\right] = 16 \times 4.22 = 68 \text{ min}$$

(using Equation 19.19)

4. This transient decreasing dilution model is also used with a tracer gas as an investigative tool to measure the effective ventilation rate ($Q/K_{\text{room mixing}}$) and, if Q can be measured independently, to also measure $K_{\text{room mixing}}$ itself. "Tracer decay" is one of several techniques described in the ASTM standard test method to determine the air change rate in a single zone.[10] See Sherman (1990) for a discussion of the constant injection, constant concentration, and other techniques.[11]
   a. The first step in this test method is to release a sufficient amount of a detectable gas to create an initial steady state condition C_0 at $t = 0$. The tracer might be SF_6 or CO_2. Both tracer gases are detectable at a concentration that is 10^5 below their TLV exposure limits of 1000 ppm and 5000 ppm, respectively. The minimum mass necessary to achieve any desired initial C_0 can be found using the Complete Evaporation Case 3 discussed in Section IV of Chapter 5 in this book. The tracer gas can be well mixed into a small room by any convenient form of mechanical mixing.[j]

$$\text{mass released (mg)} = C_0(\text{mg/m}^3) \times \text{room volume (m}^3) \quad \text{(using Equation 5.5b)}$$

$$\text{mass released (g)} = C_0(\text{mg/m}^3) \times \text{room volume (m}^3)/10^3 \quad \text{(using Equation 5.5b)}$$

   Alternatively, the released volume of a gas can be found using the same principle but expressing C_0 as a ppm concentration (viz., ppm_0):

$$\text{released volume (ft}^3) = \text{ppm}_0 \times \text{room volume (ft}^3)/10^6 \quad \text{(using Equation 5.7)}$$

$$\text{released volume (m}^3) = \text{ppm}_0 \times \text{room volume (m}^3)/10^6 \quad \text{(using Equation 5.7)}$$

$$\text{released volume (L)} = \text{ppm}_0 \times \text{room volume (m}^3)/10^3 \quad \text{(using Equation 5.7)}$$

   b. Release of the tracer gas is stopped after reaching the desired C_0. The subsequent dilution of the tracer gas can be tracked by recording the room concentration either periodically or continuously until either the room concentration is reduced to an unmeasurable concentration or enough data are collected for statistical confidence. The resulting time series data is then analyzed via log-linear regression as defined in Equation 19.20 that was derived by taking the natural logarithm of both sides of Equation 19.18d (which in turn was derived by subtracting C_{makeup} from both sides of Equation 19.18a).

$$C - C_{\text{makeup}} = [C_0 - C_{\text{makeup}}]\exp\left(\frac{-Q \times t}{K_{\text{room mixing}}\text{Vol}}\right) \qquad (19.18\text{d})$$

[j] Pavelchak et al. (2002) showed that artificially generated smoke is an acceptable alternative tracer whose concentration can be monitored with a direct reading, handheld, light-scattering airborne particle counter.[12]

$$\ln[C - C_{\text{makeup}}] = \ln[C_0 - C_{\text{makeup}}] + \left[\frac{Q}{K_{\text{room mixing}} \times \text{Vol}}\right] \times t \qquad (19.20)$$

$$\text{Y} \quad = \quad \text{A} \quad + \quad \text{B} \quad \text{X}$$

c. Calculate the linear regression coefficients "A" and "B" as shown below[k]:

$$\text{A} = \ln[C_0 - C_{\text{makeup}}] \quad \text{(a result that is usually not too interesting)} \qquad (19.21a)$$

$$\text{B} = (-Q/K_{\text{room mixing}} \times \text{Vol}) = (-1/\tau_{\text{effective}}) \qquad (19.21b)$$

d. Use the slope "B" and the measured value for the room volume (Vol) to find:

1. the time in minutes for one *effective* room air change $= \tau_{\text{effective}} = -1/\text{B}$ (19.22a)
2. the number of *effective* room air changes per minute $= (1/\tau_{\text{effective}}) = -\text{B}$ (19.22b)
3. the number of effective room air changes per hour $[\text{ACH}] = (60/\tau_{\text{effective}}) = -60\text{B}$ (19.22c)
4. the effective ventilation rate $[Q_{\text{effective}}] = Q/K_{\text{room mixing}} = \text{Vol}/\tau = -\text{Vol} \times \text{B}$ (19.22d)

e. If the actual ventilation rate (Q) can be measured, use the above results to find the mixing coefficient

$$K_{\text{room mixing}} = Q/(\text{Vol}/\tau_{\text{effective}}) = -Q/(\text{B} \times \text{Vol}) \qquad (19.22e)$$

VI. ACCUMULATION WITH NO VENTILATION

The third transient condition to be presented here is where an emitting source (G) is present (as in the accumulation case) but there is no room ventilation [i.e., when $Q = 0$]. This scenario is almost guaranteed to create a dangerous condition if given enough time. Examples include an initially clean confined space with a continuously emitting source, a small well-sealed cab or control booth whose ventilation has just failed, or a portion of a mine or tunnel whose ventilation was just sealed-off by a collapse.

1. To model this condition and avoid dividing by zero (as would happen in Equation 19.11 if Q were set equal to 0), one must go back and take the fresh air Q out of Equation 19.5 (as shown below).[l] Integration of Equation 19.23a yields the simple linear equation shown in Equation 19.23b, and the IH solution shown in Equation 19.24 in which C

[k] Such data should fit a log-linear regression quite well (with an $r^2 > 0.9$), but visually inspecting a plot of $Y = \ln[C - C_{\text{makeup}}]$ versus $X = t$ is recommended. Check that the plot line is straight, particularly near its bottom where delayed re-entrainment can cause C_{makeup} to increase and C to flatten out at later times. An interesting alternative to regression is to measure C at uniform intervals Δt; for each interval, calculate the $\ln[C_{i+1}/C_i] = \ln[C_{i+1}] - \ln[C_i] = \Delta\ln[C]$; then find the average of $\Delta\ln[C]/\Delta t$ which also equals the same slope $\text{B} = (-Q/K_{\text{room mixing}}\text{Vol}) = (-1/\tau)$. Changes in $\Delta\ln[C]/\Delta t$ values late in the study can also be used as a check for re-entrainment.

[l] Two technicalities are skipped. One is if G were vapors formed by evaporation within the confined airspace, G would not be constant but would decrease as C approached P_{vapor} according to Equation 5.3. Second, this model requires that $K_{\text{room mixing}}$ is finite. No particular value of $K_{\text{room mixing}}$ needs to be assumed but the combination of air space and mixing should be small enough to be reasonable homogeneous.

builds up uniformly over time after starting from $C = C_0$ at $t = 0$:

$$[GK_{\text{room mixing}}]\delta t = K_{\text{room mixing}} \times \text{Vol}\ \delta C \qquad \text{(modified Equation 19.5)}$$

$$(G)\delta t = \text{Vol}\ \delta C \qquad (19.23a)$$

$$G \times t = \text{Vol} \times (C - C_0) \qquad (19.23b)$$

$$\text{for no ventilation, } C = C_0 + \left[\frac{G}{\text{Vol}}\right]t \qquad (19.24)$$

Equation 19.25 is the equivalent equation to find the time to reach any chosen concentration, C.

$$\text{for no ventilation, } t = (C - C_0)\left[\frac{\text{Vol}}{G}\right] \qquad (19.25)$$

Example 19.2

Assume that a well sealed, air conditioned control cab was built to protect a foundry shakeout manipulator arm operator from dust and noise. The volume of air inside the cab is approximately 3 m^3 (85 ft^3). If the control cab's ventilation system were to fail or be turned off without the operator's knowledge, how long would it take for carbon dioxide from respiration to accumulate to a dangerous level? To answer this question implies answering several intermediate questions.

a. What is the G value for carbon dioxide emitted in human exhaled air?
   The G value for human CO_2 can be calculated using data from the work physiology literature. The rate of oxygen consumption can be equated to the body's metabolic work rate at a ratio of 0.205 L O_2/kcal ($\pm$ 5%). The respiratory quotient (volume of CO_2 produced to O_2 consumed) varies with ones diet, the level of exercise, etc., but a value of 0.85 is typical. Using a metabolic rate for the operator of 2.5 kcal/min, $G = 0.85 \times 0.205 \times 2.5 = 0.435$ L of CO_2/min. This gaseous volume can be converted to mass using Equation 2.20b, 44 g/mole, and 24.45 L/mole, to yield 0.784 g/min.
   As an alternative, Bearg (1993) cites ASHRAE's rate of carbon dioxide emission for office work as 0.3 L/min (0.0106 ft^3/min).[13] Table 19.4 is based on more data from Bearg. For this example, we will assume "light work" and use a $G = 0.38$ L/min $= 0.684$ g/min.
b. What is a dangerous level of CO_2?
   This is a judgement call. Among the options is the STEL = 30,000 ppm; the recently reduced IDLH = 40,000 ppm; and the $LD_{50} \approx$ 100,000 ppm (from RTECS). The lowest value is the STEL that is equivalent to $30{,}000 \times 44/24.45 = 54{,}000$ mg/m^3.

TABLE 19.4
Rates at Which Humans Generate Carbon Dioxide (from Bearg, 1993)[13]

Activity	L/min	ft^3/min	g/min	m
Resting	0.20	0.0071	0.36	0.8
Sitting	0.25	0.0088	0.45	1.0
Light work	0.38	0.0135	0.68	1.2
Manual work	0.50	0.0177	0.90	1.6

c. What is the initial (C_0) concentration of CO_2?
The standard atmosphere listed in Table 2.2 contains 314 ppm CO_2. Because 314 ppm is only 1% of the STEL, it could probably be neglected. Its currently measured background level of approximately 370 ppm (from Figure 2.6) is just as negligible; however, for other problems keep in mind that the background concentration inside of any occupied room or building is likely to be higher than the outdoor background concentration. To use this background level, 370 ppm × (44/24.45) is equivalent to 666 mg/m^3.

d. How long would it take for carbon dioxide to increase from 666 mg/m^3 (or zero) to the 54,000 mg/m^3 STEL if the person were emitting CO_2 at a rate of 684 mg per minute? The solution for the time "t" can be found using Equation 19.25 and the above values:

$$t = (C - C_0)\frac{\text{Vol}}{\text{G}} = (54,000 - 666\ \text{mg/m}^3)\frac{3\ \text{m}^3}{684\ \text{mg/min}} = 234\ \text{min} \approx 4\ \text{h}$$

Another calculation without the 666 mg/m^3 would show that disregarding the initial background of CO_2 would increase the estimate by less than 3 min. Just over 13 h would be required for the enclosure to reach the LD_{50} that might cause the person to suffocate. How much quicker would a person suffocate if they were to place their head inside a plastic bag with maybe 3 L of air? A computational solution is much preferred over an experimental test. In other words, do not try this at home.

PRACTICE PROBLEMS

Remember in the Practice Problems for Chapter 5 where 1/2 L of toluene was spilled in a closed solvent storage shed (20 ft × 20 ft × 10 ft) with no ventilation (a complete evaporation scenario)? Some time after that spill, someone decided to renovate the room and install an old 200 gallon capacity toluene parts washer and a 3000 ft^3/min general room ventilation system whose fan motor is outside the building. Because of the ventilation diffuser's placement and the small size of the room, it would probably have a room mixing factor (K_{mix}) of approximately 2. If you assume that evaporation from the parts washer would be approximately the same as it was before it was moved (which you were told was about 1/2 L per hour under normal conditions), you can use some of your vapor answers from Practice Problem 5. The following data pertain to toluene:

TLV	50 ppm (A4, skin)	MW	92.1	P_{vapor}	28.2 mmHg (at 77°F)
OSHA TWA PEL	200 ppm	Boiling Point	232°F	P_{vapor}	50.4 mmHg (at 98°F)
Ceiling	300 ppm	Solub. (74°F)	0.07%	FRZ	− 139°F
For	10 min	Flash Point	40°F	UFL	7.1%
Max. Peak	500 ppm	Specific Gravity	0.866	LFL	1.1%

1. What would be the steady state concentration (C_∞) while the ventilation was running and the solvent is evaporating normally? How does this compare to its TLV?
G[mg/min] = Q[m^3/min] = C_∞ = _____ mg/m^3 = _____ ppm
2. If the lid to the parts washer were closed overnight to reduce evaporation, how long would it take after the lid was taken off in the morning for the vapor concentration in the room to get from nearly clean air ($C_0 = 0$) to within 14% of its steady state concentration?
a. A good starting point is to get the room volume
_______ m^3

b. Next, how long is each effective room air change?

______ min

c. How many effective room air changes would it take to reach 86% of steady state?

$N_{changes}$ = _______

d. Therefore, the time to reach to within 11% of its steady state

_______ min

3. As part of your risk analysis, you speculate what would happen if the room ventilation were to break down or if the electric power were to fail. This room has no emergency backup electricity generator. If the power were to fail during a day shift when the lid was off (and the room was at steady state as in question 1), about how long would it take for the room average concentration to reach toluene's LFL? (Without a fan or activity to mix the room air, vapor uniformity is questionable, but what is likely to happen when the electric power comes back on? Lights anyone? The fan motor could spark too, but recall that it was placed outside the building.)

C_{LFL} = ______ mg/m^3

$t_{to\ reach\ LFL}$ = _______minutes

or ____ days and ____hours

4. What would happen if the room ventilation or the electric power were to fail at night? Assume that the lid is on but some evaporation will still occur and in the worst case (maybe after a 3-day weekend) could reach vapor equilibrium. Use the appropriate exposure scenario in Chapter 5 to estimate the equilibrium air concentration in ppm at NTP.

______ ppm

5. Now assume that the room concentration had reached that equilibrium by the next time the employees arrive. If *G* with the lid on is low enough to approximate zero with the ventilation system running, how long would it take for the concentration to drop below the LFL after the ventilation system was re-started?

______ minutes

How long would it take for the concentration to drop below the 50 ppm TLV?

_____ minutes

or ____ hours

6. What "class" and "division" of electrical hazard would the above condition constitute for purposes of electrical safety? (Remember this topic from Chapter 4?)

REFERENCES

1. Reinke, P., *Box Models. Mathematical Models for Estimating Occupational Exposure to Chemicals*, AIHA, Fairfax, VA, pp. 25–40, chap. 5, 2000.
2. Nicas, M., *Two Zone Model. Mathematical Models for Estimating Occupational Exposure to Chemicals*, AIHA, Fairfax, VA, pp. 51–56, 2000.
3. *Industrial Ventilation: A Manual of Recommended Practice*, 24th ed., American Conference of Governmental Industrial Hygienists, Cincinnati, OH, 2001.
4. Brief, R. S., Simple way to determine air contaminants, *Air Eng.*, 2, 39–41, 1960.
5. Repace, J. L. and Lowrey, A. H., Indoor air pollution, tobacco smoke, and public health, *Science*, 208, 464–472, 1980.
6. Burton, D. J., ASHRAE's indoor air quality procedure determines the outdoor air requirement, *Occup. Health & Safety*, 62(2), 18, 1993.
7. Burton, D. J., ASHRAE's indoor air quality procedure generates the dilution ventilation rate, *Occup. Health & Safety*, 62(4), 18, 1993.

8. Burton, D. J., General methods for the control of airborne hazards, In: *The Industrial Environment, Its Evaluation and Control.* DiNardi, S. Ed. American Industrial Hygiene Association, Fairfax, VA, chap. 31, 1997.
9. Earnest, G. S. and Mickelson, R. L., Carbon monoxide poisonings from small gasoline-powered, internal combustion engines: just what is a "well-ventilated area"? *Am. Ind. Hyg. Assoc. J.*, 58, 787–791, 1997.
10. ASTM, *E741-00: Standard Test Method for Determining Air Change in a Single Zone by Means of a Tracer Gas Dilution*, ASTM International, West Conshohocken, PA, 2000.
11. Sherman, M. H., *Air Change Rate and Airtightness in Buildings*, ASTM, Philadelphia, PA, 1990.
12. Pavelchak, N., Palmer, W., DePersis, R. P., and London, M. A., Simple and inexpensive method for determining the effective ventilation rate in a negatively pressurized room using airborne particles as a tracer, *Appl. Occup. Environ. Hyg.*, 17(10), 704–710, 2002.
13. Bearg, D. W., *Indoor Air Quality and HVAC Systems*, Lewis Publisher, Boca Raton, FL 1993.

20 General Ventilation in Steady State Conditions

The learning goals of this chapter:

- ☐ Know the concept leading to $C_\infty = G/Q/K_{\text{room mixing}}$, and how and why C_∞ relates to the familiar C_{TWA}.
- ☐ Know how to predict the steady state C_{TWA} if $Q_{\text{effective}}$ (or $Q/K_{\text{room mixing}}$) and G are known.
- ✍ Understand why neither $\tau_{\text{effective}}$ nor ACH is a good predictor of C_{TWA}.
- ✍ Be familiar with methods to predict G via reported emission rates, model equations, and consumption.
- ✍ Know how to predict the necessary change in Q or $K_{\text{room mixing}}$ to achieve an acceptable reduction in C_{TWA}.
- ✍ Be familiar with the range of causes of HVAC problems.
- ✍ Be able to use a psychrometric chart to find the relative humidity if outside air is heated or cooled at a constant absolute humidity (water content).

I. THE CONCENTRATION IN STEADY STATE CONDITIONS

1. An unstated learning goal of Chapter 19 was to realize how quickly an airborne contaminant can accumulate to a steady state concentration. The general solution to the dilution ventilation model is repeated here to begin a review of that point.

$$C = C_\infty \times \left[1 - \exp\left(\frac{-t}{\tau_{\text{effective}}}\right)\right] + C_0 \times \exp\left[\frac{-t}{\tau_{\text{effective}}}\right] \qquad \text{(Equation 19.12b)}$$

As t gets larger, $e^{-t/\tau_{\text{effective}}}$ gets smaller. As $e^{-t/\tau_{\text{effective}}}$ approaches zero, the influence of C_0 decreases and C approaches C_∞, a constant *steady state* concentration within the room. According to Equation 19.12b, C is more than halfway to C_∞ by the time $t = \tau_{\text{effective}}$.

$$C \text{ at time} = \tau_{\text{effective}} = C_\infty \times [1 - \exp(-1)] = C_\infty \times [1 - 0.368] = 0.63\ C_\infty$$

(Using Equation 19.12b)

From a practical perspective, if C_∞ is going to be hazardous, then the room air is probably already hazardous after only one or at most two $\tau_{\text{effective}}$ time intervals. Because $\tau_{\text{effective}}$ can be as short as 5 to 10 min in a reasonably well ventilated room (although it can also be as long as 30 min in a large, weakly ventilated room), a room with an emission source can be nearly in steady state in as little as 10 to 20 min (and almost always within an hour).

2. If G and Q are roughly constant values throughout a workday, then the 8-h time-weighted average concentration [C_{TWA}] will approximate the steady state concentration [C_∞]. Equation 20.1a was originally stated without derivation as Equation 5.14 and was later

derived as Equation 19.14.

$$C_\infty = \frac{G \times K_{\text{room mixing}}}{Q} + C_{\text{makeup}} \approx C_{\text{TWA}} \tag{20.1a}$$

If the fresh air entering the room is clean (namely, if $C_{\text{makeup}} \approx 0$), then Equation 20.1b states this concept in its simplest and most memorable form. The term $Q_{\text{effective}} = Q/K_{\text{room mixing}}$ is used within Equation 20.1b and will be used many more times throughout this chapter to explain the effect of general ventilation on a room's average concentration.

$$\boxed{C_\infty = \frac{G}{Q/K_{\text{room mixing}}} = \frac{G}{Q_{\text{effective}}} \approx C_{\text{TWA}}} \quad [\text{if } C_{\text{makeup}} \approx 0] \tag{20.1b}$$

Equation 20.1 can be accurately viewed as if the room's C_{TWA} is the result of diluting the mass of contaminant generated per minute [G] into the effective volume of *fresh air* entering the room every minute [$Q_{\text{effective}}$]. From this perspective, Equation 20.1 is analogous to Equation 5.6a for complete evaporation with no ventilation (C = mass/air Vol) except that both the contaminant mass and air volume are rates. Equation 20.1 can be used to estimate C_∞ for past, present, or future conditions. For instance, some research specialists and expert witnesses are paid well to reconstruct historic exposures using this and similar models. Time and money can be saved if calculations can show that current exposures are minimal before bothering to measure them. It can also be used to anticipate hazards before a new production facility is built and a change can still be made on paper (paper changes are always cheaper than renovations and additions made after a facility is built). Thus, Equation 20.1 can tell if an airborne chemical was, might now be, or will become hazardous.

3. Dilution ventilation is a pathway control. The essence of dilution ventilation is to dilute the mass of contaminant generated per unit of time [G] into the effective amount of general ventilation air that is circulated into and out of the occupied portion of a room in the same unit of time [$Q_{\text{effective}}$]. Equation 20.2 is Equation 20.1 solved for the ventilation rate in a way analogous to Equation 5.6c (air Vol = mass/C) for complete evaporation with no ventilation. However, in order to use Equation 20.2 to predict an adequate rate of outdoor air ventilation, both G and $K_{\text{room mixing}}$ must be known.

$$\text{theoretically acceptable } Q \geq \frac{G \times K_{\text{room mixing}}}{\text{TLV}^{\circledR} \text{ or other limit}} \tag{20.2}$$

Where $K_{\text{room mixing}}$ is unknown and/or G is unknown but is left unchanged, one can still compare conditions before (pre) and after (post) the amount of effective dilution ventilation is changed, as shown in Equation 20.3. This equation leads to the more useful predictions shown in Equation 20.4.

$$G = \text{a constant} = \frac{C_{\text{TWA, post}} Q_{\text{post}}}{K_{\text{room mixing,post}}} = \frac{C_{\text{TWA, pre}} Q_{\text{pre}}}{K_{\text{room mixing,pre}}} \tag{20.3}$$

Of particular interest is the amount of dilution ventilation [Q_{post}] needed to improve the air concentration from some unacceptable precontrol concentration [C_{pre}] down to some desired or target exposure limit (C_{target}) that could be a TLV, a fraction of a TLV, or any other limit.

$$\text{acceptable } Q_{\text{post}} \geq Q_{\text{pre}} \frac{K_{\text{room mixing,post}} \times C_{\text{TWA,pre}}}{K_{\text{room mixing,pre}} \times C_{\text{target}}} \tag{20.4a}$$

Equation 20.4b is a good first approximation if one can assume that $K_{\text{room mixing}}$ is constant.[a]

$$\text{acceptable } Q_{\text{post}} \geq Q_{\text{pre}} \frac{C_{\text{TWA, pre}}}{C_{\text{target}}} \quad \text{if } K_{\text{room}} \text{ mixing is constant} \qquad (20.4b)$$

The following sections will show why the ventilation rate necessary to dilute hazardous chemicals is often much greater than that needed to meet normal general ventilation requirements, making dilution ventilation an unacceptably expensive control. The reverse is also true: the motivation to reduce heating and cooling costs by reducing general ventilation can result in overexposures or at least to complaints of overexposure to normal indoor air constituents.

II. NORMAL BUILDING VENTILATION REQUIREMENTS

1. The general ventilation rates in Table 20.1 use Equation 20.1 or Equation 20.2 in one way or another. The dilution of carbon dioxide [CO_2] generated by human respiration is a straightforward application of Equation 20.2. The replacement of the oxygen [O_2] consumed by human respiration is an indirect application of dilution ventilation. Body odors and tobacco smoke are additional emissions (beyond CO_2) that require more dilution ventilation.
2. Table 20.2 summarizes the history of recommended minimum ventilation rates. These recommendations reached a historic low in 1973 subsequent to major increases in world oil prices in 1972. However, complaints and symptoms from occupants in buildings designed with less than 10 ft^3/min per person led to the realization that too little air can cause poor indoor air quality, a result largely predictable from Table 20.1.
3. Current ventilation rates in the U.S. are specified by two organizations in a way roughly analogous to how ACGIH specifies TLVs and OSHA sets and enforces PELs:
   - The American Society of Heating Refrigeration and Air Conditioning Engineers (ASHRAE) publishes recommendations much as the ACGIH publishes TLVs. While ASHRAE calls such recommendations standards, they are not strictly enforceable unless they are referenced by some other administrative authority. For instance, ASHRAE Standard 62-2001 "Ventilation for acceptable air quality" is a voluntary consensus standard that provides two design options: the "Ventilation rate procedure" and the "Indoor air quality procedure" to be outlined below.[2]
   - A family of *International Building Codes* (known collectively as the IBC) is a comprehensive guideline for building design and construction.[3] The IBC is not legally enforceable in-and-of itself. However, it (or a similar predecessor called the *Uniform Building Code* or UBC) has been adopted by the vast majority of county and city building departments in the U.S. who do have the legal authority to enforce these building codes through a process of building permits and inspections.[b] The *International Mechanical Code* (a member of the IBC family of codes) utilizes

[a] As the room air will probably be better mixed as Q is increased, $K_{\text{room mixing}}$ will probably decrease slightly as Q is increased; but this is a beneficial side-benefit of the simpler model represented by Equation 20.14b when $K_{\text{room mixing}}$ is assumed to be constant.

[b] The IBC has replaced the Uniform Building Codes (known informally as the UBC) and a few other codes used through the 1990s. The IBC is published by the International Code Council, an affiliation of the Building Officials and Code Administrators International, Inc (BOCA), the International Conference of Building Officials (ICBO), and the Southern Building Code Congress International (SBCCI).

TABLE 20.1
Theoretical Criteria for Setting General Ventilation Rates[1]

To Control for	ft^3/min/person
$O_2 > 18\%$ (the TLV in the 1970s)	1
$CO_2 < 0.5\%$ (the TLV then and now)	3.5
Body odors for low physical activity	10

TABLE 20.2
A Short History of ASHRAE Minimum Ventilation Recommendations

1895	30 ft^3/min/person
1936	10 ft^3/min/person
1973	5 ft^3/min/person (subsequent to the 1972 OPEC oil crisis)
1989	15 ft^3/min/person minimum (25 ft^3/min/person in areas where people are active) (60 ft^3/min/person in areas where smoking is permitted)

large parts of ASHRAE and other voluntary consensus standards[4]; however, the IBC (at least through the 2003 edition) only allows ASHRAE's "Ventilation rate procedure."

4. The "Ventilation rate procedure" specifies the minimum flow rate of fresh air to be supplied to each room based on the activity(s) to be conducted within that room (its "occupancy classification"). The ASHRAE and IBC standards cover nearly 100 occupancy classifications. The minimum ventilation rates are stated on the basis of either the room's expected occupancy (ft^3/min/person as in Table 20.3) or the size of the room ($ft^3/min/ft^2$ as in Table 20.4). Expected occupancy tables also include an estimated maximum number of occupants per room. Both procedures assume that the only significant source of airborne contaminants are those normally associated with the

TABLE 20.3
Examples of Minimum Required Outdoor Air Ventilation Rates per Person from the 2003 *International Mechanical Code*[4]

ft^3/min/person	Occupancy Classifications
15	Barber shop, classroom, florist, hardware, supermarket, auditorium, hotel lobby, waiting room, bank vault, pharmacy, photo studio
20	Theater lobby, office or conference room, cafeteria, campus laboratory, gymnasium playing floor
25	Beauty shop, ballroom, commercial laundry, disco, bowling alley, game room, hospital patient room
30	Bar or cocktail lounge, commercial dry cleaner, gambling casino, hospital operating room
60	Smoking lounge[a]

[a] Recirculation of air exhausted from such areas is prohibited.

TABLE 20.4
Examples of the Required Outdoor Air Ventilation Rate per Unit Area from the *International Mechanical Code*[4] as Adopted from ASHRAE Standard 62-2001[2]

$ft^3/min/ft^2$	Occupancy Classifications
0.05	Public corridors, retail or storage warehouse
0.10	Education corridors, retail florists, hardware, fabrics, supermarket
0.15	Library, retail shipping and receiving
0.20	Arcade or mall retail showroom
0.30	Basement or street-level retail showroom, clothier, furniture shop
0.50	Locker room,[a] darkroom, duplicating or printing room
1.00	Commercial dry cleaner, conference room, elevators, pet shop
1.50	Automotive service station, repair garage, enclosed parking garage

[a] Recirculation of air exhausted from such areas is prohibited.

occupancy classification. This procedure constitutes a specification standard as defined in Section I.2 of Chapter 4.

5. The ASHRAE "Indoor air quality procedure" is a newer, performance-based approach to setting minimum outdoor air ventilation rates.[2] The "Indoor air quality procedure" requires four steps.
   a. One evaluates the pertinent emission data to estimate G. One source of such data is the emission rate of vapors and gases from furnishing or construction materials for which dilution ventilation is ideally suited. Section III of this chapter summarizes this and various other methods that could be used to estimate G.
   b. One sets a target air quality concentration level [C_{target}]. For chemicals that have a TLV, a value of 10% of the TLV is sometimes chosen as an acceptable target steady state concentration [C_{target}], despite the fact that the philosophy used when setting a TLV is not necessarily applicable to nonindustrial settings nor are the temporal exposure patterns in many nonindustrial settings similar to those contemplated when setting a TLV.
   c. One chooses an appropriate room mixing factor [$K_{room\ mixing}$]. ASHRAE claims $K_{room\ mixing}$ should be 1.0 to 1.2 in good office settings, while $K_{room\ mixing}$ could easily be 1.5 to 2.0 in bad offices (e.g., crowded, many partitions, and poor supply and return vent locations). The values in Table 19.2 span a wider range of $K_{room\ mixing}$ values appropriate to a more diverse range of settings.
   d. Finally, one uses the steady state air quality model as defined by Equation 20.5 to find the required outdoor air ventilation rate (Equation 20.5 is identical to Equation 20.2 except the denominator is an explicit target concentration instead of the TLV or other limit).

$$\text{ASHRAE performance-based } Q = G \times K_{room\ mixing}/C_{target} \tag{20.5}$$

6. While decreasing G or increasing Q can control C_{TWA}, room volume plays virtually no role in either predicting or controlling steady state concentrations. Recall in Chapter 19, Section III.3 that room volume only appears in the general solution to the dilution model as one component of $\tau_{effective}$, the time for an effective room air

change as defined in Equation 19.13.

$$\tau_{\text{effective}} = \frac{\text{Vol}}{Q/K_{\text{room mixing}}} = \frac{\text{Vol}}{Q_{\text{effective}}} \quad \text{(repeat of Equation 19.13)}$$

Without knowing $K_{\text{room mixing}}$, one is left with the theoretical definition of "one room air change" shown in Equation 20.6. The range of values of $K_{\text{room mixing}}$ in Table 19.2 reflects the potential magnitude of the error of not using $K_{\text{room mixing}}$ (or of assuming it is one when it is not).

$$\tau = \frac{\text{Vol}}{Q} = \text{the time for one room air change} \tag{20.6}$$

A related term room "air changes per hour" (ACH) is simply the number of τ intervals within an hour, as stated in Equation 19.22c and repeated here as Equation 20.7 (see also the *Manual of Industrial Ventilation*, Section 7.8).[5]

$$\text{ACH} = 60\, Q/(\text{Vol}) = 60/\tau \tag{20.7}$$

Any specification of a room's or building's ventilation by its ACH should be viewed with suspicion. The $\text{ft}^3/\text{min}/\text{ft}^2$ values in Table 20.4 if multiplied by 60/ceiling height could yield an equivalent ACH that can vary from about one to ten air changes per hour. The corresponding time for one room air change derived from Equation 20.6 can range broadly from 60 down to 6 min. Although such rules of thumb should not be used for design, they are often referred to verbally and within some of the ventilation literature.

From an IH perspective, neither τ nor effective room air changes per hour provides a good basis either to predict the steady state air quality or to investigate an indoor air quality problem. The number of room air changes per hour is related to how rapidly a room concentration changes (the duration of the transient excursions discussed in Chapter 19), but it is not related to the steady state condition characterized by Equation 20.1. The steady state condition [C_∞] shown in Equation 20.1 depends upon diluting the G generated into the $Q/K_{\text{room mixing}}$ effective fresh air provided and has nothing to do with room volume. Only if the room "Vol" is specified or at least constrained within a narrow range could one apply room air changes per hour to steady state conditions. Thus, one should avoid misusing room air changes per hour data to interpret or compare steady state conditions when either the room volumes or the room mixing conditions (underlying the value of $K_{\text{room mixing}}$) can differ widely.

III. METHODS TO ESTIMATE A CONTAMINANT GENERATION RATE

Although Equation 20.4 does not require a known value for G, Equation 20.1, Equation 20.2, or Equation 20.5 do (the latter is the "Indoor air quality procedure"). The following subsections describe six methods that might be used to estimate the contaminant generation rate [G]. The first method uses tabulated rates of emission. The next two methods use the basic evaporation model from Chapter 5 and the physical displacement of vapors while filling containers from Chapter 9, respectively. The last three methods include measuring the plume, accessing purchase records of chemicals, and using a tracer gas.

1. The interest in indoor air quality in the 1980s included sources caused by occupants as well as passive sources intrinsic to furnishings and construction materials.

Table 19.4 contains data that can be used to predict the rate at which human occupants generate carbon dioxide. Table 20.5 contains emission rates (a process called off-gassing) from selected building materials tabulated by Burton.[6,7] Such emission rate data can be multiplied by the total surface area of the material within a given room or building to determine a value for G. Since the early 1990s, interest has focused mainly on tobacco smoke, carbon monoxide, and mold, with some lingering interest on formaldehyde, pesticides, and other volatile organic compounds (VOC).[8–10]

2. A value for the rate [G] at which a volatile solvent could evaporate from a spill or similar uncontained surface can be predicted using the basic evaporation equation (Equation 5.13) repeated below.

$$G = (\text{geometry coefficient}) \times (\text{Area}) \times V^{0.5} \times P_{\text{vapor}} \qquad \text{(repeat of Equation 5.13)}$$

Values for three of the four variables found within this equation are easy to find; however, a value for the geometry coefficient is not easy to find. A specific example of this equation applicable to the evaporation of volatile organic solvent spills was presented in Section I.4 of Chapter 5.[11]

$$G\ [\text{mg/min}] = 0.0706 \times \text{MW} \times A \times V^{0.625} \times P_{\text{vapor}} \qquad \text{(repeat of Equation 5.4b)}$$

Values of the coefficient for other geometries could be found by determining G experimentally and then back-calculating using Equation 20.8.

$$\text{geometry coefficient} = \frac{\text{experimentally determined } G}{\text{Area} \times V^{0.5} \times P_{\text{vapor}}} \qquad (20.8)$$

Table 20.6 provides some examples of G [mg/min per square foot of liquid source area] calculated using Equation 5.4b for four chemicals that span a wide range of MW and P_{vapor} values and at four air velocities [V]. The chemicals on the right-hand side of Table 20.6 have vapor pressures within $\pm 20\%$ of the chemical listed on the left-hand side of that table, although they do not necessarily have similar MWs (the latter are shown in parentheses). A close comparison will show that the lowest emission rate from a liquid in Table 20.6 exceeds the highest emission rate in Table 20.5 from building materials by a factor of just over 10^4 while the highest extremes differ by a factor of 10^9. While these differences must be viewed in light of the much larger surface areas expected for many of the building materials, the ratio of their surface areas will probably rarely exceed 10^6 or 10^7. Thus, these data illuminate the potential for the vapors emanating from an uncontained volatile liquid to dominate completely the requirement for dilution ventilation and to exceed the normal ASHRAE recommendation levels.

3. Empirical models that predict vapor emissions from filling containers were discussed early in Chapter 9. Examples of vapor emission rates calculated using Equation 9.4 are tabulated in Table 20.7 for the same chemicals as in Table 20.6 but expressed as g/min (vs. mg/min/ft^2 above).[12]

$$G\,[\text{g/min}] = \frac{f_{\text{head space}} \times \text{MW [g/mole]} \times \text{Vol [liquid L]} \times P_{\text{vapor}} \text{ [mmHg]}}{t \text{ [minutes]} \times 760 \text{ [mmHg]} \times 24.45 \text{ [L/mole]}} \text{ at NTP}$$

(using Equation 9.4)

After adjusting for the difference in units, the rates at which displaced vapors are emitted in the conditions described in Table 20.7 are between 10 and 30 times higher than the uncontained liquid evaporation rates for the same chemicals in Table 20.6. Again, the same conclusion can be drawn that active uses of chemicals such as pouring processes

TABLE 20.5
Examples of Reported Vapor Emission Rates at Various Ages of the Construction Products at the Time of the Test. The Original Units of lb/min-ft^2 × 10^{-9} Were Converted Herein[6,7]

Materials	Vapor	Age	lb/min-ft^2 × 10^{-9}	μg/min/ft^2	Age	lb/min-ft^2 × 10^{-9}	μg/min/ft^2
Medium density fiberboard	HCHO	Newer	7.4	3.4	Older	2.3	1.0
Hardwood plywood paneling	HCHO	Newer	4.8	2.2	Older	0.2	0.091
Particle board	HCHO	Newer	6.8	3.1	Older	0.3	0.14
Urea-formaldehyde foam insulation	HCHO	Newer	2.7	1.2	Older	0.2	0.091
Softwood plywood	HCHO	Newer	0.1	0.045	Older	0.03	0.014
Clothing	HCHO	Newer	0.07	0.032	Older	0.05	0.023
Silicone caulk	TVOC	< 10 h	44	20	10–100 h	< 6.8	3.1
Floor adhesive	TVOC	< 10 h	750	341	10–100 h	< 17	7.7
Floor wax	TVOC	< 10 h	270	123	10–100 h	< 17	7.7
Wood stain	TVOC	< 10 h	34	15	10–100 h	< 0.3	0.14
Polyurethane wood finish	TVOC	< 10 h	31	14	10–100 h	< 0.3	0.14
Floor varnish or lacquer	TVOC	10 h	3.4	1.5			
Particle board	TVOC	2 yr	0.7	0.32			
Latex-backed carpet	4-PC	1 week	0.5	0.23	2 weeks	0.3	0.14
Dry-cleaned clothes	PCE	0–1 day	3.4	1.5	1–2 days	1.7	0.8

HCHO, formaldehyde; PCE, perchloroethylene; 4-PC, 4-phenyl cyclohexene; TVOC, total volatile organic chemicals.

TABLE 20.6
Example Values of G (mg/min per ft^2) Predicted Using Equation 5.4b

Chemical	MW	P_{vapor}	Air Velocity (fpm) 50	100	500	2000	Chemicals with a Similar P_{vapor} (MW in Parentheses)
Methylene chloride (dichloromethane)	84.9	435.0	30,065	46,366	126,782	301,540	Ethyl ether (74), carbon disulfide (76)
MEK	72.1	95.0	5576	8599	23,514	55,925	Acrylonitrile (53), benzene (78), ethyl acetate (88)
Ethyl benzene	106.2	9.6	830	1280	3500	8324	Chlorobenzene (112), xylene (106)
p-Dichlorobenzene	147.0	1.0	120	185	505	1,200	2-butoxyethanol (118), benzyl chloride (127), ethylamine (45)

can emit much more vapor than is expected from building and finish materials that the ASHRAE "Indoor air quality procedure" is intended to control.

4. The rate at which contaminants are generated can sometimes be calculated from a chemical analysis of the emitted plume. The emission rate of a thermally warm plume (including fumes and exothermic reaction products) is relatively easy to measure by collecting it into a temporary canopy hood. An isothermal plume is more difficult to collect, but research methods have been used to estimate the emission rate from plume dispersion measurements.[13–17]
   a. For small plumes, all of the emissions for a given time period may be captured either onto a filter or into some gaseous collecting media that can be analyzed quantitatively. The result is then translated into G using Equation 20.9.

$$G\,[\text{mg/min}] = \frac{\text{mass [mg]}}{\text{time [min]}} \tag{20.9}$$

   b. For larger plumes, emissions may be channeled through a duct in which the volume flow rate (Q expressed in m^3/min) and the hazardous chemical concentration (C in mg/m^3) can be measured (using a continuous sampler such as a nephelometer or gas analyzer). The result can then be translated into G using a slight adaptation of Equation 9.1. This option may require achieving a certain level of homogeneity within

TABLE 20.7
Example Values of G [g/min] Predicted Using Equation 9.4 for Filling with $f_{head\ space} = 1$

Chemical	MW	P_{vapor}	Filling Rate (L/min) 50	100	500	2000	Chemicals with Similar P_{vapor} (MW in Parentheses)
Methylene chloride (dichloromethane)	84.9	435.0	99	199	994	3975	Ethyl ether (74), carbon disulfide (76)
MEK	72.1	95.0	18	37	184	737	Acrylonitrile (53), benzene (78), ethyl acetate (88)
Ethyl benzene	106.2	9.6	2.7	5.5	27	110	Chlorobenzene (112), xylene (106)
p-dichlorobenzene	147.0	1.0	0.4	0.8	4.0	16	2-butoxyethanol (118), benzyl chloride (127), ethylamine (45)

the collected gas stream, for example by using a static mixer within the duct through which the gas stream is channeled.

$$G[\text{mg/min}] = C[\text{mg/m}^3] \times Q[\text{m}^3/\text{min}] \qquad \text{(basically Equation 9.1)}$$

5. If the emitted contaminant is a chemical that is purchased, then G may be estimated from purchase records, perhaps adjusted by the fraction of the contaminant that is either incorporated into the end product and/or destroyed (e.g., by combustion).
   a. If the purchased gases and solvents are not chemically bound within the finished product or somehow consumed, then the time-weighted average emission rate G must equal the rate that the chemical is purchased or consumed. Estimating G from purchase or consumption records is the easiest option for multiple small sources in a workroom.
   b. The emission rate based on purchase records may need to be reduced via Equation 20.10 in proportion to the percentage of the purchased chemical that is either incorporated within the end product or destroyed during its manufacture. That fraction can be based on a chemical analysis of the product or, in some cases, only an estimate is adequate.

$$G = [\text{Purchase rate}] \times \frac{100 - [\%_{\text{incorp.}}] - [\%_{\text{destroyed}}]}{100} \qquad (20.10)$$

where

Purchase rate = the amount of the chemical purchased and used over a typical, known period of time converted to units of mass/time (e.g., pounds/month or kg/day).

$\%_{\text{incorporated}}$ = the mass fraction of contaminant chemical incorporated into the product based on some form of analysis.

$\%_{\text{destroyed}}$ = the mass fraction by which the purchased chemical is destroyed within the room or process via any known chemical reaction including combustion.

6. In theory, one could also use a G imputed by using either of the previous four methods for one chemical (subscripted 1), then adjust that rate for a second chemical (subscripted 2) based on differences between the chemicals' vapor pressures using the modification of the basic evaporation equation, Equation 5.13 shown in Equation 20.11.

$$G_2 = G_1 \frac{P_{\text{vapor, 2}}}{P_{\text{vapor, 1}}} \qquad (20.11)$$

7. In the least direct approach, the G for a continuous generation process in a particular setting can be imputed using Equation 20.12 by measuring the C_{TWA} in the working zone and the effective ventilation rate $Q/K_{\text{room mixing}}$. One could estimate $Q/K_{\text{room mixing}}$ by measuring the buildup of either the chemical of interest or a tracer or a surrogate contaminant released at a known rate in a manner similar to the chemical of interest (although this technique has not been reported). Alternatively, Q can be measured directly (see Section V.1 in this chapter) and $K_{\text{room mixing}}$ can either be measured using the tracer gas method described in Section V.4 of Chapter 19 or estimated from Table 19.2. Equation 20.12 is an application of Equation 9.1.

$$G = C_{\text{TWA}} \frac{Q}{K_{\text{room mixing}}} = C_{\text{TWA}} \times Q_{\text{effective}} \qquad (20.12)$$

In addition to providing the reader with an array of tools for estimating the contaminant generation rate, the conclusion to be drawn from this section is the limitation for general ventilation to dilute the occupational use of a volatile chemical. Consider for a moment the impact of the vapor generation rates in Table 20.6 or Table 20.7 on either the design of a new system or the modification of an existing HVAC system that may just meet the ASHRAE recommendations in either Table 20.3 or Table 20.4. If an occupational activity were to emit a chemical that required the general ventilation flow rate to be doubled (i.e., $Q_{new} = 2 \times Q_{old}$), the ventilation heating/cooling costs would double, which might be financially painful but tolerable. But increasing the flow by 2× will quadruple the pressure losses (that are proportional to Q^2), increase the fan power requirements by 8× (or 2^3 according to the fan laws), and increase the noise level of the moving air by 16 to 18 dB. Such large changes could exceed the ability of an existing system. Also, if the chemical use required 10× the flow rate, the outdoor air flow rate would be unrealistically high, the costs would probably be exorbitant, and the hazard could almost certainly be dealt with more cost effectively by local exhaust ventilation. Many of the evaporation rates in Table 20.6 and Table 20.7 for active chemical uses could easily require dilution ventilation rates that exceed those required by building codes for normal occupancies by more than 100×. Such high rates of ventilation are not feasible.

Example 20.1

Envision a print shop 24 × 36 ft with an 8-ft ceiling. Vapors and odors from inks and cleaners without a TLV cause irritation to several of the employees who spend up to 8 h in the work area. The literature is minimal on the chemicals used, but experience suggests that irritation would be reduced at half the concentration and virtually gone at one third.

The first step might be to determine Q_{pre} by one of the methods described above. Suppose the existing Q_{pre} is found to be 900 ft³/min (just at the recommended 1 ft³/min/ft² in Table 20.4). Equation 20.4a predicts that the concentration will be reduced to one third if Q_{post} were to be increased to $3 \times Q_{pre}$ or from 900 to 2700 ft³/min.

The first problem in trying to achieve such an increase in general ventilation would be getting that much air to flow through the HVAC system.

$$\boxed{\text{losses} \propto \text{VP (and to } Q^2)} \text{ means losses}_{post} = \text{losses}_{pre} \times \left(\frac{Q_{post}}{Q_{pre}}\right)^2 = 9 \times Q_{pre} \qquad \text{(Equation 14.2)}$$

and

$$\boxed{\text{hp and kW(\$)} \propto (Q_{new}/Q_{old})^3} \text{ means hp}_{post} = \text{hp}_{pre} \times \left(\frac{Q_{post}}{Q_{pre}}\right)^3 = 27 \times \text{hp}_{pre} \qquad \text{(Equation 18.3)}$$

Twenty-seven times the power is probably too much of an increase for the fan motor; it will need to be replaced. If V_{pre} were 2000 fpm, a V_{post} of 6000 fpm could produce another 30 dB of noise (going from maybe 35 to 40 dBA to about 65 dBA), and the high velocity at the diffusers can produce both more noise and uncomfortable drafts. If that were not enough, the annual costs for heating, cooling, and running the fan are likely to go up by almost $3000 per year, which is a lot of money for a small print shop. This control is unlikely to be acceptable from a comfort perspective let alone probably not be cost-effective compared with local exhaust or even respirators.

IV. GENERAL VENTILATION (HVAC) SYSTEMS

An HVAC system is virtually required in institutional, commercial, and high-tech manufacturing buildings to meet the ASHRAE 62-2001 and building code requirements described in Section II.[c] HVAC is the acronym for heating, ventilating, and air conditioning. Conditioning includes heating, cooling, dehumidifying or humidifying, and cleaning (filtering) the air.

---

[c] Technically, Section 402 of the *International Mechanical Code* of the *IBC* allows a space to be naturally ventilated if its openable windows, doors, louvers, etc are readily controllable by the building's occupants and comprise at least 4% of the floor area being ventilated, the openings between interior rooms of such a building and adjoins exterior rooms comprise at least 8% of the combined floor area, and the horizontal clear space in front of openings below grade is sufficiently wide.[4]

HVAC systems are more complex than the ventilation system in most homes or apartments. HVAC systems supply outside air (cf. most homes that rely on infiltration for their fresh air), usually supply that air to multiple zones (cf. one thermostat at home), and have to include life-safety components such as smoke detectors and fire dampers in their return air ducts.[d]

1. HVAC systems, as a whole, can be broadly categorized as either an air system, a hydronic or steam system, or a unitary system. Installed systems cannot always be categorized so simply.
   a. Air systems provide heating or cooling directly to the air that passes through a central air handling unit. This conditioned air is circulated via a fan in the central air handling unit through supply air ducts to the occupied spaces and return air ducts back to the central air handler. Air systems typically consist of an air handling unit (AHU) and an air distribution network of ducting and regulators that usually only support one building. Most home furnaces are air systems.
   b. Hydronic systems provide heating and cooling indirectly to the air via heated or chilled water (or perhaps low pressure steam in place of heated water). Heated water or steam is usually generated by a boiler. Chilled water may be generated by a chiller or a cooling tower. Boilers and chillers are usually located in a main central plant that can support multiple buildings or a campus. Cooling towers can be central but are commonly associated with each building. The heated or cooled water is piped to and from either each building's air handling unit that supports all air conditioned zones or multiple heat exchangers such as individual fan-coil units, unit heaters, or radiators that support individual air conditioned zones.
      1. Fan-coil units comprise a finished cabinet that encloses a small fan to circulate room air (and perhaps some proportion of fresh outside air) over a heating or cooling coil (see Figure 20.3).
      2. Unit heaters also have a small fan but only circulate room air over a heating coil.
      3. Radiators (usually finned tube radiators in modern systems) rely for heat exchange on natural (passive) air convection caused by the air density differences discussed in Section VI of Chapter 2.
   c. Unitary or package systems comprise a complete heating and cooling unit in a single package or cabinet. Unitary systems have low installation costs but do not provide precise temperature control or handle high humidity cooling loads very well. Since they must be placed along the periphery of a building, unitary systems are more prone to reintrainment as discussed in Section VI of Chapter 15. Examples of unitary systems include window air conditioners, through-the-wall air conditioners, and package rooftop air conditioners.

      Unitary systems can take advantage of a heat pump. A heat pump is a piece of refrigeration equipment that can be used either in its normal mode to cool air or water or in a reverse mode to pull heat out of relatively cool air and transfer it to warmer air or water, thus heating it. One disadvantage of a heat pump is that it initially costs more than a heater. One advantage of a heat pump is that it can heat more efficiently than burning a fossil fuel (this advantage decreases as the temperature of the cool air side decreases and is often lost below about 30°F). An even greater advantage can be gained if a heat pump can be used both to cool warm zones and to heat cool zones simultaneously by connecting the refrigerant evaporation and condensing coil in separate zones. (The latter condition is also compatible with a circulating water circuit with heat exchangers on both ends; see hydronic systems above.)

[d] In HVAC, a zone is an area regulated by one sensor such as a thermostat. Interior zones within a building often just need cooling while perimeter zones alternate their needs for heating and cooling at least seasonally if not daily.

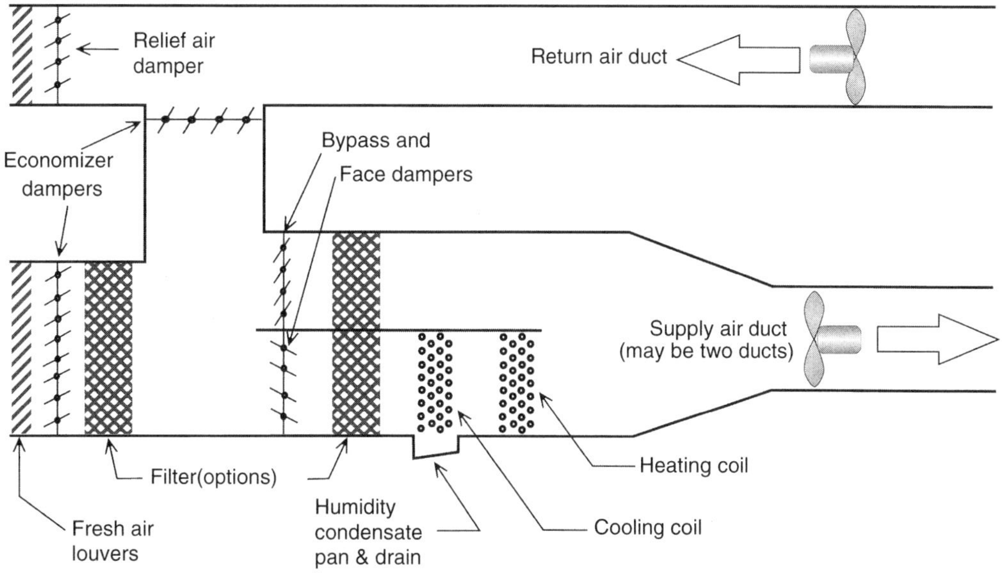

FIGURE 20.1 A schematic depiction of a generic air handling unit with several optional components.

2. A typical air handling unit (AHU) has a fan, filter, a cooling and/or heating coil, and perhaps a humidifying device, an economizer, and a bypass. Most air handling units not only condition the fresh air brought in from outside the building but also recirculate and recondition a proportion of the air from within the building (called return air) back into the building's supply air. All of these components are depicted in Figure 20.1. A commercial system may not have two fans to handle the return and supply air separately. Similarly, not every air handler will have a bypass damper and/or an economizer; and some systems have separate cold and hot air supply ducts.
   a. The air flow rate may be controlled either by a variable speed fan or by a series of automatic, pneumatically powered dampers. The commanded fan speed or position of a damper is derived from temperature and/or humidity sensors and more recently also by a carbon dioxide sensor. (Multiple sensors are typically integrated through a computer interface.) A single damper is depicted in Figure 20.2; several parallel arrays of dampers are depicted in Figure 20.1. Mispositioned or malfunctioning dampers can cause air quality problems (to be discussed in Section VI.2). A characteristic of such

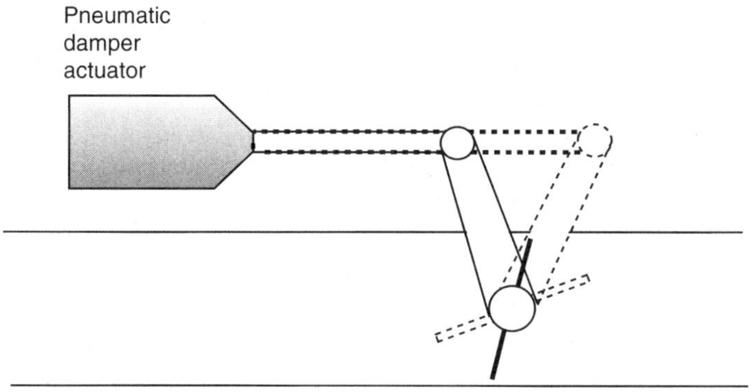

FIGURE 20.2 Depiction of an air damper actuator in both the nearly closed position (solid lines) and nearly open position (dashed lines).

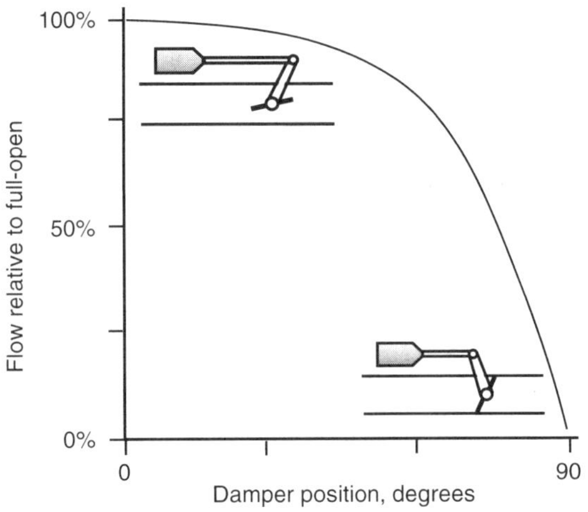

FIGURE 20.3 Depiction of the effect of air damper position on flow rate through the damper.

dampers, which can be potentially important during troubleshooting an HVAC system, is a damper's small effect on the air flow rate until it approaches its closed position and its relatively large effect on flow in positions beyond half closed (see the depiction in Figure 20.3).

b. One or more banks of filters are usually present to remove large dust particles, insects, leaves, etc. Such filters should be inspected frequently to assure they can pass air easily (recall "permeability" from Chapter 15) and cleaned or replaced if they show any sign of microbial growth. Inadequate cleaning and poor filter maintenance can allow microbial growth to go undetected or reduce air flow rates. Either problem can lead to air quality complaints (to be discussed in Section VI.3 below).
c. Air conditioning units will have one or more sets of cooling and/or heating coils. Some systems put the cooling and heating coils in separate ducts that can join either within the AHU or in a terminal regulator within each conditioned zone. In regions with hot, humid summers, the air will often be super-cooled below its eventual delivery temperature to reduce its absolute humidity. If the air were delivered at its dew point, its relative humidity would be 100%. Super-cooling will take out more water vapor, and then partially reheating the air will decrease its relative humidity at its delivered temperature. A drain pan should be present to collect the condensate and remove it from the system. The combination of condensed water and organic matter that either passed through the filter or was carried in the return air from the occupied building to the condensate provides a good media for microbial growth (as will be discussed in Section VI.4 below).
d. Low temperature air cannot hold much water. When winter outdoor temperatures are near or below freezing, some means (such as a water mist, a heated pan, or steam injection) is sometimes used to increase both the absolute and relative humidity in the heated air delivered to the occupants of the building. Problems caused by either uncomfortably low or high humidity will be discussed in Section VI.7.
e. The temperature of the conditioned air can be controlled either by modulating the flow of the heating or coolant media (with a thermostatically controlled valve), by changing the fan speed, or by use of a bypass. A bypass diverts a proportion of the supply air around the heating and/or cooling coils. One advantage of bypass dampers is to prevent really cold air from freezing the water or steam within coils if low heat

demand allows the flow to be virtually shut down. Face and bypass dampers can also be used in conjunction with reheat coils in multizone air handling units.

f. An "economizer" increases the ratio of outside to return air that is supplied to the building when temperature conditions are favorable. An economizer works well to supply cooling air to meet internal heating loads when the outdoor temperature is cooler than the return air temperature. The economizer dampers open and the outside air flow rate is increased as the return air flow rate is decreased (via the relief air dampers). At other times when the outside air temperature is either too cold or too hot, the outside air flow rate should not decrease below the minimum called for by the codes (such as Table 20.3 and Table 20.4).

3. HVAC systems can use one of the three major methods listed below to distribute air throughout a building.
   a. Single-duct constant air volume (CAV) systems supply air at a constant flow rate through one duct or pathway (e.g., see Figure 8.1, Figure 8.2, and Figure 8.7 of the *Manual of Industrial Ventilation*).[5] The air temperature delivered to a single zone by a CAV system can be controlled by changing the flow rate of the thermal media (water or air) passing through the air handling unit. The air temperature in CAV systems serving multiple zones can be varied either by face and bypass dampers or by maintaining the air in the duct either hotter in the winter or cooler in the summer than needed and using a small cooling/heating coil just before the air enters the room or zone to tweak the temperature as demanded by a thermostat. Single-duct CAV systems were common but are one of the lowest energy inefficiency.
   b. Dual-duct constant air volume systems supply both cool and warm air in parallel ducts to the conditioned zones where the volume flow rates of the two air supply ducts are blended as needed (usually at a constant air volume) just prior to the air diffuser (see Figure 20.4 herein or Figure 8.6 of the *Manual of Industrial Ventilation*).[5] Dual-duct systems are energy efficient but more costly to install than single-duct systems.
   c. Single-duct variable air volume (VAV) systems supply air at a constant temperature (see Figure 8.3, Figure 8.4, and Figure 8.5 of the *Manual of Industrial Ventilation*).[5] Dampers just before the diffuser modulate the air volume delivered to each room or zone. The air temperature in VAV systems is usually cooled to a constant low temperature. Heat may be provided by a nonair system such as baseboard radiators or by a local reheat coil within the VAV terminal. The air volume delivered by the fan can be controlled (in order of decreasing energy efficiency) either by a variable frequency motor drive, by inlet vanes, or by a discharge damper. VAV systems are quite common in newer buildings.

   Figure 20.4 depicts a single-zone dual-duct HVAC system that again artificially combines features from several types of systems. No real system will have all of the features shown. The central air handling unit is all of the apparatus to the left of the room in the diagram. The dashed lines through the walls at the top and bottom of the diagramed room represent infiltration and exfiltration into and out from the building, respectively. One or the other will usually occur in any building, but they are not designed as a specific component of a central air system. Infiltration will occur if the room or building's static pressure is slightly less than the outdoor total pressure. Wind can cause infiltration to occur on one wall while exfiltration may be occurring through other walls. A tight building is designed and built to minimize both flows.

4. The return air should be ducted as it moves from the conditioned zones back to the air handler (except for fan coil units in which the entire unit and air stays within the room). However, older buildings (especially from the 1970s) sometimes use the plenum above a suspended acoustical ceiling to convey the return air back to the air handler (or occasionally to convey the fresh air). Even worse is if the return air travels through

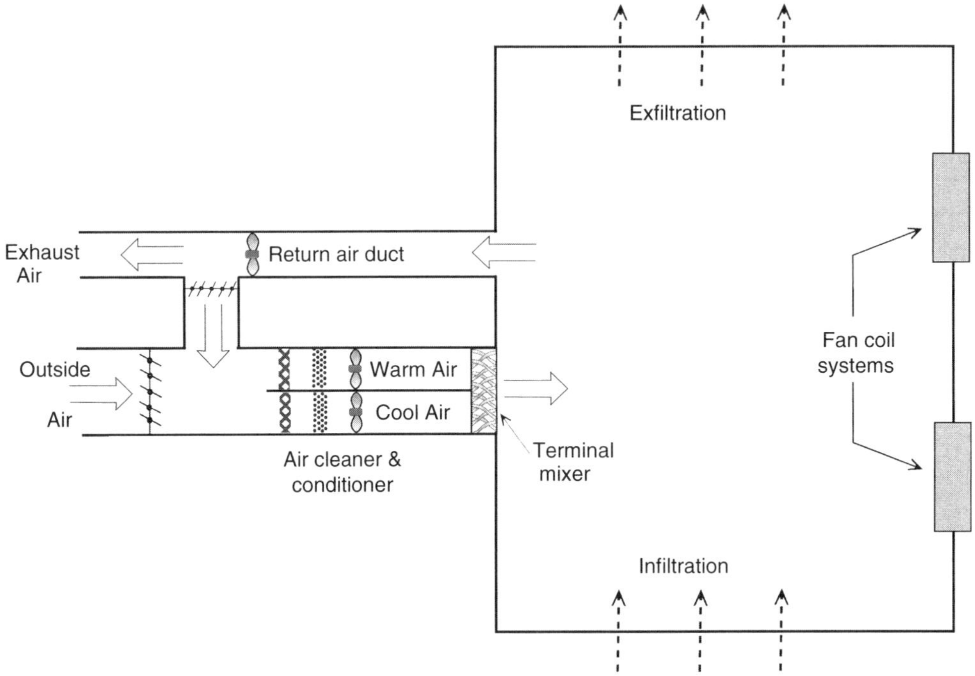

FIGURE 20.4 A depiction of a single-zone dual-duct HVAC system.

an open hallway. The use of ceiling plenums and hallways or corridors to move air provides no way to block the distribution of smoke from a fire in one zone of a building from spreading to other zones; smoke can prevent emergency egress of people through these hallways. The options of how fresh air can be routed and mixed may also be regulated by local building codes.

V. CONTROLLING EXPOSURES VIA DILUTION VENTILATION

To an IH, the goal of dilution ventilation is to maintain the room's concentration of unwanted contaminants at or below some acceptable concentration by manipulating Q or/and $K_{\text{room mixing}}$. The notation C_{pre} in Equation 20.13 and the following discussion refers to some existing but assumed to be unacceptable TWA or steady state concentration in need of control. The notation C_{post} refers to the TWA or steady state contaminant concentration that is equal to or below the target concentration limit after a control is implemented.

$$\text{acceptable } Q_{\text{effective,post}} \geq Q_{\text{effective,pre}} \frac{C_{\text{TWA,pre}}}{C_{\text{target}}} = \frac{Q \times C_{\text{TWA,pre}}}{K_{\text{room mixing}} \times C_{\text{target}}} \tag{20.13}$$

The question is by how much does the initially inadequate $Q_{\text{effective,pre}}$ that caused C_{pre} to exceed C_{target} need to be increased to create and maintain an acceptable postcontrol concentration C_{post} that is equal to or less than C_{target}? According to the dilution ventilation model developed herein and Equation 20.13, this change to an acceptable $Q_{\text{effective,post}}$ can be obtained by either increasing the actual Q or decreasing $K_{\text{room mixing}}$.

1. Several techniques are available to determine the general ventilation air flow rate that is supplied by an HVAC system. The simplest way is to ask the plant engineer. An engineer

should know by using one of the following techniques. If they do not know, a ventilation consultant could be hired. Or you (if you are adventuresome) might use one of the following techniques.

a. The enthalpy method is a traditional engineering technique that calculates Q using an energy balance based on the measured air temperature and humidity in the fresh air inlet and return air outlet.[18] The enthalpy method is made complex by the need to include humidity and the effect of latent heat cf., sensible heat. (Recall that Equation 17.13 calculates the enthalpy of moist air.)
b. The most direct method is to measure the velocity through a measurable flow area. If the room pressure is slightly positive relative to the outdoor air pressure or to neighboring hallways or rooms (meaning air will leak out through any and all openings), a duct traverse should be conducted in each of a room's fresh air inlet ducts. The inlet Q is the product of the average V in each inlet times the cross-sectional area of that duct. (This product would need to be summed for each inlet for a trunk system, if applicable.) Similarly, if the room is negatively pressurized (meaning air can be entering the room through any and all openings), the A and V in each of the room's return air ducts should be measured to derive Q flowing out of a room. In each case, assessing all of the pathways of the opposite flow (e.g., the outward flow in the case of a positively pressurized room) would be difficult. In practice, access to room return air outlets may be obstructed, and the fresh air inlet from an HVAC duct usually has a diffuser to disperse what would otherwise be a draft but which also causes the velocity to vary laterally across the diffusor.

   Commercial instruments (such as the Shortridge Instrument's Flowhood shown in Figure 20.5) are made that convey all of the air exiting from a diffuser via a tapered adaptor of various shapes and/or sizes through the instrument's throat that comprises a fixed known flow area. An anemometer in the throat measures the

FIGURE 20.5 Commercial flow rate meters. Courtesy of Shortridge Instruments, Inc.

velocity and the meter derives and directly reads out the flow rate Q in ft^3/min or L/sec.

c. Human physiology underlies a third method to estimate $Q_{effective}$. One can measure the carbon dioxide concentration from which the effective Q can be calculated using Equation 20.14 where the rate that CO_2 is generated physiologically by the room's occupants is taken from data in Table 19.4.[19,20]

$$Q_{effective} = \frac{Q}{K_{room\ mixing}} = \frac{G_{CO_2}}{C_{CO_2}} \tag{20.14}$$

ASTM D 6245-98 provides a more detailed method to estimate carbon dioxide generation rates, guidelines for acceptable CO_2 concentrations, and a method to evaluate ventilation based on indoor carbon dioxide concentrations.[21]

d. The use of a tracer gas to estimate the effective fresh air flow rate [$Q_{effective}$] was described in Section V.4 of Chapter 19. A tracer gas works best in a room within a building rather than in a whole building.[22]

e. A small building or home without HVAC has no controlled air inlet and relies on infiltration for its fresh air. The infiltration rate can be equated to the flow through a single orifice (opening) in a wall of that building (see orifice in Subsection VII.1.c of Chapter 12 in this book). The equivalent diameter of that opening can be estimated by measuring the flow rate Q needed to obtain a small defined positive pressure difference [ΔP] within the building.[23,24]

Unfortunately, Q is directly proportional to $ (or the local currency of money) for fan electricity and make-up air conditioning (e.g., heating, cooling, filtering, or dehumidifying) as described in Chapter 17, not to mention more costly emission controls that would be needed to treat a larger volume of air with more dilute contaminants described in Chapter 17. As demonstrated in Example 20.1, unless Q is initially very low, increasing it in excess of 10 air changes per hour for most size rooms is too costly to be a financially practical means of controlling a toxic source.[25]

2. Decreasing the value of $K_{room\ mixing}$ is a potential alternative to increasing Q. The actual value for $K_{room\ mixing}$ may not be known, but at least some qualitatively predictable control can be achieved by reducing $K_{room\ mixing}$ for a constant G and Q by providing better mixing of the room's air. The degree of mixing that can be achieved by a given action (such as adding one or more fans) is not accurately predictable (see $K_{room\ mixing}$ values in Table 19.1), but the effect of a given change can easily be predicted by Equation 20.15b:

$$\frac{K_{room\ mixing,post}}{K_{room\ mixing,pre}} = \frac{C_{post}}{C_{pre}} \tag{20.15a}$$

$$\text{acceptable } K_{room\ mixing,post} = K_{room\ mixing,pre}\frac{C_{target}}{C_{pre}} \tag{20.15b}$$

If placed correctly, one or more circulating fans can not only increase mixing (thereby reducing both $K_{room\ mixing}$ and C) but can also help to redirect the source's plume away from the breathing zone. On the other hand, a fan placed incorrectly can blow a plume into an employee's breathing zone. A fan placed incorrectly may also interfere with the control efficiency of a local exhaust hood intended to collect and remove a contaminant at the source. Also, fans almost always have a cooling effect that may or may not be comfortable for all employees or in all seasons. Adversely affected employees often block cold air motion in the winter, which will diminish the goal of increasing the effectiveness of whatever existing dilution ventilation is present. Perhaps, most importantly, decreasing $K_{room\ mixing}$ will have a significant effect only if $K_{room\ mixing}$ is initially much larger than 1. In other words, increasing the air mixing in a setting that is already fairly well mixed will have a small effect on the room concentration.

In summary, while decreasing G is always a good idea, the options that will improve the effective ventilation are limited. Decreasing $K_{room\ mixing}$ is only effective if the preexisting $K_{room\ mixing}$ is large. Increasing the outdoor air Q tends to be an expensive exposure control. These limitations (plus those presented in Section I of Chapter 19) make source control or local exhaust ventilation look more attractive.

VI. SOURCES OF HVAC AND INDOOR AIR QUALITY PROBLEMS

General ventilation should be of some concern to an IH even if it is not used for dilution control of an occupational chemical. Data have shown that many common respiratory irritants are generated within homes and commercial buildings. Indoor sources include microbiologic organisms, furnishings and building construction components, home and small business appliances (e.g., furnaces and cooking stoves), and animals (human, pets, or livestock). While in the past, liberal ventilation or natural leakage kept levels of airborne irritants below noticeable thresholds, when buildings are well sealed and ventilation rates are reduced (usually as a result of energy conservation efforts), air concentrations can rise to create what has been termed a "sick building." A summary of NIOSH findings of causes for sick building complaints is shown below.[26] Other publications on complaint investigations and system management are available.[27–31e]

Inadequate ventilation	53%
Indoor contaminants	15%
Unknown	13%
Outdoor contaminants	10%
Biological contaminants	5%
Building materials	4%

1. Most of the causes found by NIOSH are related to the ventilation system. Outdoor air may be either inadequate, contaminated, or poorly distributed.
   a. Inadequate supply of outdoor air: Inadequate means fresh air flow rates below those currently recommended (e.g., those in Table 20.3 and Table 20.4). Inadequacies can lead to increased complaints at levels of known contaminants below those currently thought to be harmful. CO_2 is not the cause of complaints, but complaints start to increase when the concentration from human occupancy reaches the 800 to 1000 ppm range.
   b. Contaminated outdoor air supply: Complaints can arise if the outdoor air intake for make-up air is drawn from near an emission source. Examples of sources include auto or diesel exhaust from a loading dock or cigarette odors from a smoking area. See also discussion in Section III of Chapter 17.
   c. Poor distribution of outdoor air: Poor mixing is the result of fresh air reaching the exhaust vents without mixing with the air within the occupied areas of the room.

[e] I-BEAM[31] is a computer-driven guidance package that integrates indoor air quality, energy efficiency, and building economics into a building management tool. I-BEAM is available at http://www.epa.gov/iaq/largebldgs/i-beam_html/ibeami.htm and on CD ROM.

Poor mixing is the cause of a large $K_{\text{room mixing}}$ value. Specific causes covered in Section II of Chapter 19 include:

1. Intraroom obstructions such as the partitions of an open office layout.
2. Poor placement of large objects such as cabinets, bookcases, etc. that block supply air diffusers or/and return air grilles.
3. Damage to general ventilation outlets (diffusers) that block or redirect the fresh air.
4. Thermal stratification typically caused by warm fresh air entering high in a room and passing over cooler contaminated air in the winter.

2. Dampers (as shown in Figure 20.2) can fail to operate properly for any of several reasons. Failure, deterioration, or malfunctioning of dampers can result in inadequate fresh air. One common mistake while adjusting or inspecting dampers is to assume that the flow rate through the damper will vary linearly with the position of the damper vane. Because most of the change in the damper's resistance to flow (its loss factor) will occur in the smallest quarter of the damper's opening (see Figure 20.3) and the change in a fan's flow rate only varies with the change in the sum of all existing losses, the damper may have to be nearly closed to have much of an effect on the flow rate.
3. Air cleaners can either block the free flow of fresh air, fail to clean the incoming fresh air, or actually be a source of airborne microbes.
   a. Inadequate cleaning of accumulated leaves or pollen from inlet louvers or inlet filters will increase flow resistance and decrease air flow. Using replacement air filters that are too small will also increase flow resistance and decrease air flow.
   b. Inadequate filtration of either incoming outdoor air or recirculated return air can distribute dusts, pollens, or organic aerosols throughout a building. The cleaning (stripping) of gaseous contaminants such as ozone, vapors, or odor from makeup air is costly and very rare.
   c. Inadequate maintenance of makeup air filters and plenums can lead to a proliferation of microbes that will reduce the effectiveness of heater/cooler coils, eventually block their heat transfer, and be a source of microbial aerosols within the building.
4. Water on the cooling coil or in its condensate tray (where condensed water is collected and normally routed for disposal) can also be a source of microbial growth (much like the air cleaners). Although wet growth is not friable and will tend to stay in place, when the air dries the slim, mold spores can be dispersed throughout the building with the circulating conditioned air.
5. In addition to the potential problems of punctured, leaking, or separated ducting that HVAC ducting has in common with LEV ducting, HVAC ducting is run at a lower velocity and sometimes lined to reduce noise and minimize heat exchange. The ACGIH *Ventilation Manual* recommends against the common commercial practice of lining ducts with fiberglass padding for internal noise suppression.[5] Condensed moisture in such padding is a somewhat sheltered environment for microbial growth (even though microbes can also grow where such shelter is not available).
6. Contaminants can migrate from one room to the next. Migration is often a problem in commercial and institutional settings in which offices and laboratories are intermixed. The movement of indoor contaminants between rooms or between rooms and their adjoining corridors is caused by (and can be controlled by) a small relative pressure difference between these zones. The direction of the pressure difference between zones (e.g., between rooms or between rooms and a hall, etc.) can be either positive or negative, as discussed below:
   a. A *positive* pressure balance (where the pressure in one room is slightly greater than the pressure in a neighboring room or hallway) works well to prevent inflow from other parts of a building into more sensitive areas, such as to protect a product or

patient from outside contaminants. However, odors and/or chemicals will migrate out from that positively pressurized room into the other potentially occupied spaces.

b. A *negative* pressure balance (where the pressure in one room is slightly less than the pressure in a neighboring room or hallway) works well to prevent the outflow of objectionable chemicals or airborne communicable diseases to other parts of the building. However, negatively balanced rooms will draw in outside contaminants of all sorts.

A pressure difference as small as 0.005 "wg can be achieved and maintained either via sensors and active controls or by purposefully setting the HVAC inlet air and outlet air flow rates to differ by perhaps 5% but not less than 50 ft^3/min from each other. The latter method is cheaper but can be disrupted if the air flow either into or out from one of the two zones changes due to transient processes or factors independent of the two zones (this topic is discussed briefly in the Section 7.5 of the *Manual of Industrial Ventilation*).[5]

7. Air moisture can be a source of personal discomfort if it is either too high or too low. The moisture content is usually expressed as humidity in either absolute or relative units. (Do you remember these terms from Chapter 2?) Humidity was also encountered in Chapter 17.
   - Absolute humidity (AH) is the concentration of water vapor present in the air. AH is meaningful to comfort, heat stress, and HVAC costs. Absolute humidity can be expressed as a partial pressure in units such as mmHg but is sometimes expressed on a weight/volume (mg/m^3 or $grain/ft^3$) or volume/volume (ppm) basis. The old unit of $grain/ft^3$ is equivalent to 2288 mg/m^3. The absolute humidity of air on the psychrometric chart in Figure 20.6 are the horizontal lines at the P_{water} (mmHg) in the air.
   - Relative humidity (RH) is the ratio of the actual amount of water vapor present (its AH) to the maximum amount of water vapor that the air can hold at that temperature (equal to water's vapor pressure at the air's temperature as described in Section II of Chapter 2). An Antoine Equation for water between $0° < T < 60°C$ is

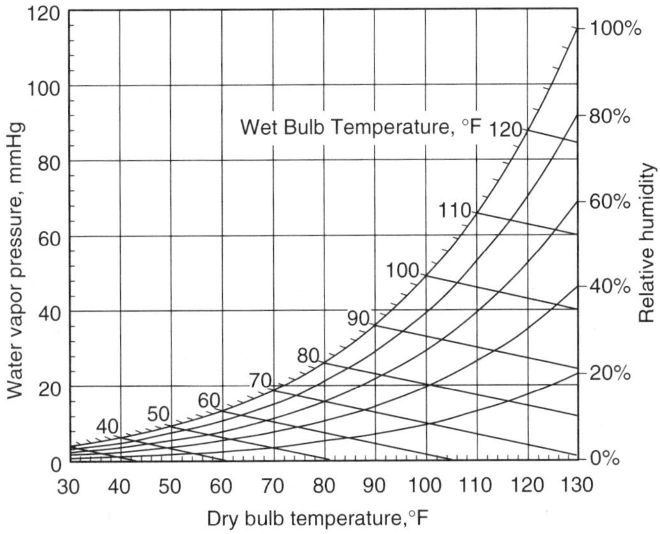

FIGURE 20.6 A simplified psychrometric chart. Air temperatures are on the *X*-axis, absolute humidity is on the *Y*-axis, wet bulb temperatures are the diagonal straight lines, and RH is denoted by the curved lines.

shown below.[32]

$$\log(P_{\text{vapor}} \text{ of water [mmHg]}) = 8.10765 - \frac{1750.286}{T_{\text{water}}[^\circ\text{C}] + 235.0} \quad \text{(using Equation 2.10)}$$

The relative humidity of air corresponds to the curved line on a psychrometric chart that passes through the intersection of a particular air temperature (vertical line) and absolute humidity (horizontal line). RH is usually expressed as a percent, as in Equation 20.16. Air at 100% RH is said to be "saturated" meaning it contains as much water vapor as it can hold at that temperature. RH is common in weather reports because it can predict the formation of fog and clouds (how close the air is to RH = 100%), but it is not too useful in either HVAC calculations or assessing comfort unless the air temperature is also defined.

$$\text{RH\%} = 100 \times \frac{\text{ambient partial } P_{\text{water}}}{P_{\text{vapor}} \text{ of water at ambient } T} \quad (20.16)$$

a. Both high and low relative humidities have been found to be associated with an elevated prevalence of symptoms of allergic rhinitis and asthma.

RH that is too high (above 60%) is associated with elevated airborne concentrations of mites, fungi, bacteria, and viruses. As warm air can contain more absolute humidity than cooled air, the major problem in the summer tends to be high RH, especially when warm moist air is cooled. High humidity is also a contributor to thermal stress (but that is the topic of another course).

RH that is too low (below 40%) is associated with discomfort, respiratory infections, and with higher than normal dust levels and ozone production. Owing to cold air only containing a small amount of water, the major problem in the winter tends to be low RH, especially when cold outside air is heated.

Thus, the RH (and other general ventilation parameters) recommended by ASHRAE 55-1992 for comfort vary with the season from not being too low in the winter heating season to not being too high in the summer cooling season, as summarized in the following Table.[33]

Comfort Guidelines	**Heating Season**	**Cooling Season**
Air temperature	67 –76°F	72–81°F
Relative humidity	20–30%	<60%
Average air velocity	≤30 fpm	≤50 fpm

b. The cooling of makeup air to its dew point can create serious problems. Simple heating or cooling of a parcel of air does not change its AH. Therefore, as air is cooled, its condition within a psychrometric chart (Figure 20.6) will move horizontally to the left as the temperature drops. The temperature at which a parcel of air reaches 100% RH is called its dew point. The dew point of any parcel of air is a fixed temperature determined by its absolute humidity. If that parcel is in contact with a cooler surface (e.g., a cooling coil below its dew point), moisture will begin to condense out of that air and be deposited onto that surface. The moisture from outdoor air can also diffuse into (or through) a building's structural wall that has either an improperly placed or missing vapor barrier; if the indoor temperature is cooled below the dew point, again, condensation can occur inside of the wall. Any condensed moisture can support microbial growth.

EXAMPLE 20.2

How much water is in air at 88°F and 80% RH? At what temperature will it start to condense? The solution can be found on a psychrometric chart such as depicted in Figure 20.6. The air's starting condition can be found by following the dry bulb temperature of 88°F vertically up until it intersects the curved relative humidity [RH] line of 80%. The absolute humidity [AH] is the *Y*-axis value.

$$\text{AH} = 26\ \text{mmHg}$$

The partial pressure of water can (optionally) be translated into concentration using Equation 2.8.

$$C_{\text{water}} = \frac{P_{\text{partial}} \times \text{MW}_i \times 10^6}{760 \times 24.45} = \frac{26 \times 18 \times 10^6}{760 \times 24.45} = 25,200\ \text{mg/m}^3 = 0.025\ \text{g/L}$$

As the air is cooled, its condition on the psychrometric chart will move horizontally to the left until it encounters the saturation line (100% relative humidity at its dew point) at which temperature the water vapor will start to condense.

$$\text{Dew Point} = 80°\text{F}$$

c. The heating of typical winter air can also create problems although ones that tend to be a little easier to control. Cold air can hold very little water vapor; therefore, it has low absolute humidity. For instance, the water vapor at 100% RH saturated air below 40°F is less than 5 mmHg (cf. 42 mmHg AH at body core temperature for air within the lungs). As this air is heated with no additional water added, its absolute humidity will again stay the same, only this time its condition on a psychrometric chart will move horizontally to the right and to a lower RH.[f] As a result, warmed but unhumidified winter air has such a low RH that it can be uncomfortable or irritate respiratory tissue. Some HVAC air handling units humidify the incoming air by adding heated water or steam to the heated air.

EXAMPLE 20.3

What is the maximum amount of water that can be in air at 40°F? (In other words, what is the saturation concentration at that temperature?) What will the RH of this air be if it is heated to 70°F? The first answer can be found by reading the *Y*-axis value at the intersection of the 40°F and the 100% relative humidity line.

$$\text{AH}_{\text{saturation}}\ \text{at}\ 40°\text{F} = 6\ \text{mmHg}$$

As this air is heated, its condition will move horizontally to the right. The relative humidity drops continuously as its temperature is raised at a constant AH. Heating winter air that starts at less than 100% RH and 40°F will typically result in relative humidities of less than 20%.

$$\text{RH at 6 mmHg and } 70°\text{F} = 30\%$$

The HVAC systems described herein are simplifications of many real-world systems. Many more books seem to have been written about HVAC than about LEV. Some recent examples are cited.[34–38] Of these, this author would recommend the book by Thomas Mull as adequately complete, the most readable, and probably most appropriate for readers of this book.[34] While HVAC can take care of routine airborne contaminants (those generated by normal human habitation), they are usually a very expensive control for pathogenic organisms or toxic or irritating chemicals. Unwise attempts to save costs either by reducing fresh air flow rates below recommended values and/or by deferring maintenance of filters, fans, heat exchange coils, drain pans, etc. can easily lead to poor air quality and employee complaints. The guidance in this chapter or in these other references should provide some insights to help troubleshoot the source of complaints and diagnose improperly operating systems.

PRACTICE PROBLEMS

These problems will use a room twice as large as the room in the Practice Problems from Chapter 5 and Chapter 19. A building 40 ft × 20 ft × 10 ft ceiling is large enough to conduct some production activities. Suppose that it had the same 375 cfm general room ventilation system whose fan and motor is outside and it has the same room mixing factor [$K_{\text{room mixing}}$] of about 2. Thus, you can use the same $Q_{\text{effective}} = 5.31\ \text{m}^3/\text{min}$ from the Practice Problem in Chapter 19.

1. If six employees were assigned to work in this building on some days and the activity involved light work" as defined by Table 19.4, what would be the steady state carbon dioxide concentration due to their respiration? Convert that to ppm, then add an assumed clean-air $C_{\text{make-up}}$ of 380 ppm CO_2

$$G\ [\text{mg/min}] = \_\_\_\_\_$$

$$C_{\text{TWA from source}} = \_\_\_\_\_ = \_\_\_\_\_\ \text{mg/m}^3 \times \_\_\_\_\_ = \_\_\_\_\_\ \text{ppm}$$

$$C_{\text{TWA}} = \_\_\_\_\_\ \text{ppm}_{\text{from source}} + \_\_\_\_\_\ \text{ppm}_{\text{in make-up}} = \_\_\_\_\_\ \text{ppm}$$

How would this ventilation rate compare with that recommended in Table 20.3 for a campus laboratory with this occupancy? The IBC Section 1004 states that an "educational shop" should be designed for as many as 16 occupants (at 50 ft^2 per occupant), which is nearly three times as many as can work in this room at once.

$$\frac{Q_{\text{actual}}}{Q_{\text{recommended for 6 people}}} = \_\_\_\_\_\_\_\_$$

2. Suppose that the ceiling of the building were constructed from unfinished particle board that gave off formaldehyde as tabulated in Table 2.5. What would be the concentration of formaldehyde when the ceiling was "newer"? The MW of formaldehyde is 30.03, and its TLV is a 0.3 ceiling.

$$G\ [\text{mg/min}] = \_\_\_\_\_$$

$$C_{\text{TWA}} = \_\_\_\_\_ = \_\_\_\_\_\ \text{mg/m}^3 \times \_\_\_\_\_ = \_\_\_\_\_\ \text{ppm}$$

And what would be the concentration of formaldehyde when the ceiling was "older"?

$$G\ [\text{mg/min}] = \_\_\_\_\_$$

$$C_{\text{TWA}} = \_\_\_\_\_ = \_\_\_\_\_\ \text{mg/m}^3 \times \_\_\_\_\_ = \_\_\_\_\_\ \text{ppm}$$

The previous problems demonstrate what general ventilation can handle. The following problems demonstrate how easy it is for routine uses of a chemical to exceed the capabilities of general ventilation; but keep in mind that these operations could be tolerated within transient excursions if they were short term.

MEK (methyl ethyl ketone) [CAS 78-93-3]

MW = 72.1	Boiling Pt = 175°F	TLV = 200 ppm (590 mg/m^3)
Flash Pt = 16°F	LEL = 1.4%	PEL = 200 ppm
Sp. Gr. = 0.81 g/mL	Vapor Pressure = 95.3 mmHg	IDLH = 3000 ppm

3. Suppose that on the days they worked in this room, the employees were conducting stripping and cleaning operations using MEK. What would be the average room concentration of MEK if the operation results in a total of 1 square foot of liquid MEK available to evaporate per Equation 5.4b and the equivalent air velocity over the liquid of 100 fpm (due to a combination air and wetted cleaning cloth movements)?

$$G\ [\text{mg/min}] = \_\_\_\_\_$$

$$C_{\text{TWA}} = \_\_\_\_\_ = \_\_\_\_\_\ \text{mg/m}^3 \times \_\_\_\_\_ = \_\_\_\_\_\ \text{ppm}$$

Although a reduction in exposure could be achieved by reducing the wetted surface area ("source control"), how much more ventilation $[Q_{post}/Q_{pre}]$ is needed for the average C_{room} to comply with MEK's exposure limit?

$$\frac{Q_{post}}{Q_{pre}} = \_\_\_\_\_\_ = \_\_\_\_\_\_$$

4. On other days the employees might be repackaging MEK from large into smaller containers at an average rate of 5.28 gallons per minute. What would be the average room concentration of MEK according to Equation 9.4 if the transfer process emitted only vapors?

$$G\ [\text{mg/min}] = \_\_\_\_\_$$

$$C_{TWA} = \_\_\_\_\_ = \_\_\_\_\_\ \text{mg/m}^3 \times \_\_\_\_\_ = \_\_\_\_\_\ \text{ppm}$$

Although a reduction in exposure could be achieved by reducing the pouring rate ("source control"), how much more ventilation $[Q_{post}/Q_{pre}]$ is needed for the average C_{room} to comply with MEK's exposure limit?

$$\frac{Q_{post}}{Q_{pre}} = \_\_\_\_\_\_ = \_\_\_\_\_\_$$

5. Using a concept linked to Equation 5.6b, what is the maximum average rate at which liquid MEK could be consumed via evaporation within this room for the average room concentration to comply with its 200 ppm TLV and PEL?

$$\text{max. acceptable } C\ [\text{mg/m}^3] = \_\_\_\_\_\ \text{mg/m}^3$$

$$Q_{effective}\ [\text{m}^3/\text{min}] = \_\_\_\_\_\ \text{mg/m}^3 \text{ (previously given)}$$

$$\text{max. acceptable } G\ [\text{mg/min}] = \_\_\_\_\_ = \_\_\_\_\_\ \text{mg/min}$$

$$\text{max. acceptable liquid } G \text{ consumed} = \_\_\_\_\_\ \text{mL/min} = \_\_\_\_\_\ \text{L/8-hour day}$$

REFERENCES

1. *Industrial Ventilation: A Manual of Recommended Practice*, 17th Ed., American Conference of Governmental Industrial Hygienists, Cincinnati, OH, 1982, Figure 3.1.
2. ASHRAE, *Standard 62-2001 — Ventilation for Acceptable Indoor Air Quality*, American Society of Heating, Cooling, Refrigerating and Air Conditioning Engineers, Inc., Atlanta, GA, 2001.
3. *International Building Code, 2003*, International Code Council, Inc., Falls Church, VA, 2003.
4. *International Mechanical Code, 2003*, International Code Council, Inc., Falls Church, VA, 2003.
5. *Industrial Ventilation: A Manual of Recommended Practice*, 24th ed., American Conference of Governmental Industrial Hygienists, Cincinnati, OH, 2001.
6. Burton, D. J., ASHRAE's indoor air quality procedure determines the outdoor air requirement, *Occup. Health & Safety*, 62(2), 18, 1993.
7. Burton, D. J., ASHRAE's indoor air quality procedure generates the dilution ventilation rate, *Occup. Health & Safety*, 62(4), 18, 1993.
8. Tichenor, B. A., Guo, Z., and Sparks, L. E., Fundamental mass transfer model for indoor air emissions from surface coatings, *Indoor Air,* 3(4), 263, 1993.
9. Hirvonen, A., Pasanen, P., Tarhanen, J., and Ruuskanen, J., Thermal desorption of organic compounds associated with settled household dust, *Indoor Air*, 4(4), 255, 1994.
10. Won, D., Corsi, R. L., and Rynes, M., Sorptive interactions between vocs and indoor materials, *Indoor Air*, 11(4), 246, 2001.
11. Caplan, K., *Evaporation Rate of Volatile Liquids*, Final Report, EPA Contract Nos. 68-2-4248, 1988 and 68-D8-0112, 1989.
12. U.S. Environmental Protection Agency, *Compilation of Air Pollutant Emission Factors*, Vol. I: Stationary Point and Area Sources, Research Triangle Park, NC (AP-42) USEPA, chap. 5, Petroleum Industry, Available at www.epa.gov/ttn/chief/ap42/index.html, 1995.

13. Keil, C. B., Wadden, R. A., Scheff, P. A., Franke, J. E., and Conroy, L. M., Determination of multiple source volatile organic compound emission factors in offset printing shops, *Appl. Occup. Environ. Hyg.*, 12(2), 111–121, 1997.
14. Scheff, P. A., Friedman, R. L., Franke, J. E., Conroy, L. M., and Wadden, R. A., Source activity modeling of freon emissions from open-top vapor degreasers, *Appl. Occup. Environ. Hyg.*, 7(2), 127–134, 1992.
15. Wadden, R. A., Scheff, P. A., and Franke, J. E., Emission factors for trichloroethylene vapor degreasers, *Am. Ind. Hyg. Assoc. J.*, 50(9), 496–500, 1989.
16. Wadden, R. A., Hawkins, J. L., Scheff, P. A., and Franke, J. E., Characterization of emission factors related to source activity for trichloroethylene degreasing and chrome plating processes, *Am. Ind. Hyg. Assoc. J.*, 52(9), 349–356, 1991.
17. Wadden, R. A., Baird, D. I., Franke, J. E., Scheff, P. A., and Conroy, L. M., Ethanol emission factors for glazing during candy production, *Am. Ind. Hyg. Assoc. J.*, 55(4), 343–351, 1994.
18. Energy Estimating Methods, *1989 ASHRAE Handbook Fundamentals*, American Society of Heating, Refrigerating and Air Conditioning Engineers, Atlanta, GA, chap. 28, 1989.
19. Jankovic, J. T., Ihle, R., and Vick, D. O., Occupant generated carbon dioxide as a measure of dilution ventilation efficiency, *Am. Ind. Hyg. Assoc. J.*, 57(8), 756–759, 1996.
20. Auger, M. R. and Farant, J. P., Measurement of dilution ventilation efficiency using CO_2 as a tracer gas, *Am. Ind. Hyg. Assoc. J.*, 57(11), 1054–1058, 1996, *Am. Ind. Hyg. Assoc. J.*, 58(11), 787–791, 1997.
21. ASTM 6245-98(2002), *Standard Guide for Using Indoor Carbon Dioxide Concentrations to Evaluate Indoor Air Quality and Ventilation*, ASTM International, West Conshohocken, PA, 2002.
22. Leaderer, B. P., Schaap, L., and Dietz, R. N., Evaluation of the Perfluorocarbon tracer technique for determining infiltration rates in residences. *Environ. Sci. Technol.*, 19, 1225–1231, 1985.
23. Sherman, M. H., *Air Change Rate and Airtightness in Buildings*, ASTM, Philadelphia, PA, 1990.
24. ASTM E1827-96(2002), *Standard Test Methods for Determining Airtightness of Buildings Using an Orifice Blower Door*, ASTM International, West Conshohocken, PA, 2002.
25. Goldfield, J., Contamination Concentration Reduction: General Ventilation versus Local Exhaust Ventilation, *Am. Ind. Hyg. Assoc. J.*, 41: 812–818, 1980.
26. Crandall, M. S. and Sieber, W. K., The National Institute for Occupational Safety and Health Indoor Environmental Evaluation Experience Part One: Building Environmental Evaluations, *Appl. Occup. Environ. Hyg.*, 11(6), 533–539, 1996.
27. *Building Air Quality: A Guide for Building Owners and Facility Managers*, EPA Publication No. 400/1-91/003 and DHHS (NIOSH) Publication No. 91-114.
28. Roberston, L. D., The identification of baseline indoor air quality parameters for a renovated building prior to occupancy, *Am. Ind. Hyg. Assoc. J.*, 57(11), 1058–1061, 1996.
29. *Building Air Quality Action Plan*, EPA Publication No. 402-K-98-001 and DHHS (NIOSH) Publication No. 98-123, See www.ded.gov/niosh/98-123a.html, June, 1998.
30. Benda, G. and Melson, G., *Indoor Air Quality Case Studies Reference Guide*, The Fairmont Press, Lilburn, GA, 1999.
31. Indoor Air Quality Education and Assessment Guidance (I-BEAM), EPA Publication (CD ROM) No. 402-C-01-001.
32. Dreisbach, R., Physical properties of chemical compounds — III, *J. Am. Chem. Soc.*, 1961.
33. ASHRAE, *Standard 55-2004 — Thermal Environmental Conditions for Human Occupancy*, American Society of Heating, Cooling, Refrigerating and Air Conditioning Engineers, Inc., Atlanta, GA, 2004.
34. Mull, T. E., *HVAC Principles and Applications Manual*, McGraw-Hill, New York, 1997.
35. Burton, J. L., *Fundamentals of H.V.A.C.*, Prentice Hall, Upper Saddle River, NJ, 1999.
36. Bell, A. A. Jr., *HVAC: Equations, Data, and Rules of Thumb*, McGraw-Hill, New York, 2000.
37. Wang, S. K., *Handbook of Air Conditioning and Refrigeration*, 2nd ed., McGraw-Hill, New York, 2000.
38. Haines, R. W. and Wilson, C. L., *HVAC Systems Design Handbook*, 4th ed., McGraw-Hill, New York, 2003.

21 Administrative Controls and Chemical Personal Protective Equipment

The learning goals of this chapter:

- Be able to define or identify an administrative control.
- Be aware of problems inherent in administrative controls.
- Know the generic OSHA criteria for using PPE.
- Be familiar with the six program elements required by OSHA for any PPE.
- Understand the definition and use of breakthrough and permeation rate.
- Know what "Skin" in TLV® (or PEL) means to hazard evaluation and control.
- Know how to use at least one glove selection guide (possibly one for an exam).
- Be familiar with the important differences among levels of CPC ensembles.

I. ADMINISTRATIVE CONTROLS

1. The term "administrative control" can be defined either very broadly or rather narrowly. For instance, the White Book defines administrative controls in its broader sense to include management involvement in the training of employees, rotation of employees, air or biological sampling, medical surveillance, and housekeeping. In this sense, almost all forms of source control (except substitution) involve one or more components of administrative control as would all receiver controls and even some evaluations.

   This Book will use a more narrow definition of administrative control, meaning to reduce the dose of a hazardous agent to one or more employees by limiting the duration that each individual is exposed to that agent. Implicit in this definition of administrative control is an increase in the portion of the workday that each individual spends in low or no exposures. Limiting the duration of exposure can be used to reduce an employee's time-weighted average exposure to either airborne chemicals, dermally deposited chemicals, or physical agents such as noise or radiation. Administrative controls rely on two concepts or principles: (1) that dose is a product of the concentration or intensity of the hazardous agent in the environment times the duration of exposure (expressed later as Equation 21.1), and (2) that "dose makes the poison" (expressed later as Equation 21.2).

   a. If the high-hazard or high-concentration condition exists continuously in a workspace that requires continuous or frequent employee attention, then administrative control implies employee rotation. In a classic employee rotation control, different people are cycled into the high-hazard area but each employee is allowed to spend only a defined short length of time exposed to the high-hazard (usually on a per day basis for chemicals, although ionizing radiation is limited by Federal regulation on the basis of a dose per quarter, per year, or even per working life).

b. If the high-hazard or high-concentration condition exists only when certain definable and/or predictable tasks or phases of production occur (intermittent hazards), then an administrative control can be to reschedule either the work or the employees to eliminate, or at least to minimize, the number of people present during the task or phase that generates the high-exposure or high-concentration. Rescheduling the work means to conduct the high-hazard task or phase of production only at a time when no, or a least only a few, employees are present. Rescheduling the employees may be limited to avoiding the unnecessary exposure to employees who might otherwise be present but are unrelated to the accomplishing the high-hazard task or generating the high-concentration. Rescheduling is only feasible for intermittent operations.

2. Employee rotation and work or employee rescheduling appear at first glance to be simple (seen by management as cheap), but such administrative controls may not always be either possible or effective. Administrative controls can, and often do, fail either because they are intrinsically flawed due to toxicity considerations and/or because they are inadequately supervised (probably the most common flaw). The following discussion elaborates on four factors that can undermine the effectiveness of an administrative control.

   a. A threshold dose for an adverse toxic response must exist for an administratively maintained lower dose to be an effective control. If a toxicity threshold does not exist, then the collective risk, both to the employer and to the group of exposed employees, remains the same but is merely spread over more people. This conclusion is based on one assumption and three principles. The assumption is that the total time to complete a task will be the same whether it is carried out by one person (or one group of people) or by several people (or different groups of people). The principles underlying administrative control include the concepts that the total dose equals the exposure concentration or intensity multiplied by the duration of the exposure (implied by Equation 21.1), that the probability of a toxic response is proportional to the dose (Equation 21.2),[a] and that the number of cases or adverse responses equals the probability or predicted incidence rate times the number of people exposed or at risk (Equation 21.3).

$$\text{Dose} = \text{Concentration or Intensity} \times \text{time} \tag{21.1}$$

$$\text{Probability or Incidence} = \kappa \times \text{Dose} \tag{21.2}$$

$$\text{Number of cases} = \text{Probability or Incidence} \times \text{Number at risk} \tag{21.3}$$

   The first principle implies that rotating multiple employees (or groups of employees) into a given task will reduce the time of exposure to each individual employee and their individual dose in proportion to the number of individual employees (or groups) being rotated ($N_{\text{individuals}}$ in Equation 21.4). The second principle implies that each individual's probability of developing an adverse toxic response is reduced in proportional to their individual dose and hence inversely to the number of individual

[a] The second of these two principles is a quantitative expression of Paracelsus' principle that "the dose makes the poison." A linear $P = \kappa \times D$ is the simplest quantitative toxicology model but is probably inaccurate. While the most accurate model would depend upon the chemical, Roach (1994) suggested that $P = \kappa \times D^2$ is a more accurate model than the linear model for chemicals in general.[1] The incidence and probable number of adverse responses at a dose lowered by control predicted by the linear model will always be higher than that predicted by the more accurate dose-squared model.

employees (or groups of employees) being rotated (Equation 21.5).

$$\text{Dose}_{\text{individual}} = \text{Dose}_{\text{total}}/N_{\text{individuals}} \tag{21.4}$$

$$\text{Probability}_{\text{individual}} = \kappa \times \text{Dose}_{\text{individual}} = \kappa \times \text{Dose}/N_{\text{individuals}} \tag{21.5}$$

Such a condition is typified by governmental policies for controlling hazards from ionizing radiation and might apply to the probability of developing mesothelioma from asbestos. However, because the total probability of an adverse response within all of the individuals being rotated in and out of the high exposure setting is the sum of all probabilities (the $\Sigma(\text{Probability}_{\text{individual}})$ as shown in Equation 21.6), the total probability of having an adverse outcome with rotation (Equation 21.6) is the same as the probability without rotation (as previously shown in Equation 21.2).

$$\begin{aligned}\Sigma(\text{Probability}_{\text{individual}}) &= N_{\text{individuals}} \times \kappa \times \text{Dose}_{\text{individual}} \\ &= N_{\text{individuals}} \times \kappa \times \text{Dose}/N_{\text{individuals}} = \kappa \times \text{Dose}\end{aligned} \tag{21.6}$$

Similarly in an epidemiologic sense, the total number of adverse cases or number of people responding adversely (N_{cases}) is also the same with rotation (as shown in Equation 21.7) as the number of cases responding adversely without rotation (as previously shown in Equation 21.3).

$$\begin{aligned}N_{\text{cases}} &= \text{Incidence} \times N_{\text{individuals}} = \kappa \times \text{Dose}_{\text{individual}} \times N_{\text{individuals}} \\ &= [\kappa \times \text{Dose}/N_{\text{individuals}}] \times N_{\text{individuals}} = \kappa \times \text{Dose}\end{aligned} \tag{21.7}$$

Thus, while administrative control via employee rotation for a chemical or other agent whose toxic effect has no threshold does reduce each individual's probability of responding adversely, it does not change the total impact on either the whole group of employees or the employer's total liability. The best way to avoid this toxicological limitation is to not use an administrative control for a chemical without a known threshold below which an adverse response should not occur. An alternative but less certain way to avoid this limitation is to restrict administrative control to those chemicals where the probability of response decreases more rapidly than the dose (as described in the previous footnote).

b. Employee rotation may keep each individual's time-weighted average exposure within compliance limits, but poorly controlled hazards may still allow conditions to exceed a chemical's short-term exposure limit within a rotation cycle. The simplest example is a chemical with a ceiling (*C*) exposure limit: because no exposure is allowed above a ceiling limit, limiting the time of exposure above a *C* value is never an acceptable option for a chemical with a ceiling exposure limit. As another example, a TLV-STEL® sets an explicit upper limit to exposure conditions for 15 min which are in principle always less than (and in practice are well less than) 32 times the TWA-TLV (8 h × 60 min/h divided by 15 min). Of course, exposure should never exceed any unregulated acute effect (especially a lethal effect such as an LC_{low}).

In contrast to the intrinsic flaw due to the lack of a chronic toxicity threshold, the risk from an acutely hazardous condition can be minimized by using continuous monitoring with direct reading meters and implementing a policy that certain operations be terminated (or at least postponed) should an acutely toxic threshold be detected. Direct reading monitors are an added expense and are not available or feasible for all chemical hazards. Data logging can document compliance (or non-compliance), but a real-time warning is necessary for an employee to exit the area in time to avoid an acute adverse reaction. Training and continuous monitoring can decrease the probability of acute hazards but perhaps not eliminate them. Additional source controls are usually necessary to completely avoid excursions in contaminant concentration from exceeding these high limits.

c. Conflicts between an employer's administrative control policy and the priorities of either the exposed employees and/or their supervisors can cause them to disregard or violate exposure time limits. An example of such conflicts is when the time required to conduct hazardous work is underestimated. Can the original completion date be allowed to slip? Who is going to pay for more employees? The desire to complete a task on time and within budget can easily outweigh a supervisor's desire to comply with a time-limiting policy. Employee incentives through pay, perceptions of promotion or job retention, or peer pressure are other sources of conflict that can affect both parties.

d. Even if there is a desire to comply, resources can be insufficient to sustain a rotation policy. For instance, employee rotation is unlikely to succeed if there are not enough adequately skilled or experienced employees available to rotate into a hazardous area to complete a specific task. Similarly, employee rotation is likely to be too expensive if there are too few alternative nonexposed tasks available into which exposed employees can be rotated for the remainder of their work time.[b] Pay scales and seniority or job classification clauses within a labor agreement can restrict the flexibility of management to rotate job assignments, and thus create an artificial resource conflict.

The failure to properly evaluate toxicological factors can be avoided by a competent IH review. Most failures of administrative controls can be traced to inadequate supervision. A proposal to control a hazard administratively should anticipate potential conflicts, evaluate their influence, and perhaps change some of the conditions that create or drive such conflicts. The influence of some conflicts affecting production employees can be reduced by ensuring that supervisors have adequate authority to enforce compliance with time-of-exposure limits.[c] Setting and publicizing a clear and strong enforcement policy for violations increase their authority. The influence of conflicts within supervisors can be reduced by creating a positive incentive for them to use that authority, for instance, by including measures of safety and health performance in their periodic performance evaluation. All such policies need to be developed with the support of upper managers, with the involvement of production managers, and in coordination with the employer's personnel or human resources department. The full package is necessary to ensure that both the health of the employees and the employer's ultimate responsibility are protected.

3. In order for an administrative control to succeed, clear management policies and strong supervision needs to be in place. OSHA does not specify the program elements they require for an acceptable administrative control, but the following recommended elements of an acceptable administrative control program can be deduced from their noise standard (1910.95 where reduced time of exposure over the 90 dBA 8-h limit is routine) and various PPE standards (1910.132 through 1910.139).[2]

---

[b] One solution to this latter limitation is to contract intermittent high hazard work to an outside employer. The risks of this solution include the potential either for contract employees to be inadequately trained about the hazards and the importance of limiting exposure time or for the contractee (the company with the hazard) to lack adequate authority to ensure that the contractor provides adequate supervision to their own employees.

[c] Supervisors should have a written enforcement policy on which they can rely. Enforcement policies can be as strict as termination upon one violation or progressive in nature that might start with formal re-instruction and a written reprimand, docking (withholding) a portion of a violator's pay on the next violation, and terminating employment on the third violation.

a. *Train supervisors and employees.* Training on the recognition of the hazardous conditions to be administratively controlled is really an extension of, or emphasis within, a normal hazard communication program. Document the training (what was covered and who got it).

Within U.S. legislation and case law, employers are virtually always liable for the actions and welfare of their employees while at work. Operationally, administrative controls put full responsibility for control on the actions of supervisors.

In order for management to assure compliance with the administrative control policy, the personnel evaluations of supervisors should be structured to include explicit measures of their performance in carrying out health and safety responsibilities on behalf of their employer (acting in what is called a "fiduciary" capacity). Similar measures can be used to evaluate and reward the health and safety performance of individual employees.

b. *Document the exposure.* "Document" means to create and keep retrievable records of an activity. If continuous monitoring of exposure conditions (concentration or intensity) is not feasible, frequently measure the concentration (or intensity of exposure) within the area under administrative control. The more people that are exposed, the more sampling or monitoring is justified. Document employee exposure times individually.

Monitor and keep records of the airborne chemical exposure level and/or the intensity of certain physical agents via either periodic evaluations with samplers or detectors or some form of continuous monitoring with detectors or dosimeters.

Keep good records of the time of exposure received by each individual. For example, establish a sign in/out log for hazardous restricted entry areas, maintain logs of active production periods (by hours or days), and/or keep a record of who is assigned to which jobs.

Consider biological monitoring as an adjunct to the above (e.g., BEIs). This easily documented monitoring is an added assurance that the administrative control program is working (and possibly an added incentive for individual compliance if the results are considered a part of the person's performance evaluation). Several reviews of biological monitoring methods are suggested.[3–6]

c. *Review the program periodically.* Such a review might start off monthly or quarterly but is probably most often scheduled annually. Each cycle of the review should itself document that the above monitoring data has been reviewed, who the reviewer was and perhaps the review method or process, and describe the appropriate corrective action that was taken and/or is being taken. A failure by management to carry out their own employee protective procedure or process can have its own legal consequences; or to paraphrase Shakespeare's words, failure to follow an employer's own standards is to risk being "hoisted by one's own petard." [Hamlet, III]

II. PERSONAL PROTECTIVE EQUIPMENT

Personal protective equipment (PPE) can be categorized into at least eight types of equipment including gloves, protective clothing (thermal, chemical, etc.), respirators, eye and face guards, ear protection, head protection, foot protection, and various other body restraints, supports, and special equipment. Several of these types of equipment can be and often are worn together (simultaneously) to form an ensemble. This section will merely present the reasons for one to use each of these individual types of protective equipment.

1. Gloves can be selected (in accordance with 29 CFR 1910.138 and Subpart I Appendix B §11) to prevent or reduce a wide range of adverse effects such as:

a. dermatitis from dusts, liquids, or mechanical abrasion,
b. systemic toxicity from dermally absorbed chemicals,
c. traumatic injury (e.g., cuts or major abrasions),
d. thermal discomfort or damage (frost bite or burns),
e. electrical shock and burns (§1910.137),
f. Vibratory White Finger from repeated exposure to segmental vibration.

2. Protective clothing can be selected to prevent or reduce the adverse effects from the hazards listed below. OSHA only specifies clothing as part of ensembles within 1910.120. Notice how similar this list is to the above list for gloves and where it differs:
   a. dermatitis from dusts or liquids,
   b. systemic toxicity from dermally absorbed chemicals,
   c. thermal imbalances from environmental heat or cold,
   d. product contamination (e.g., boots, hats, masks, and garments worn in clean rooms to protect the product or in hospitals to protect the patient), or
   e. the accumulation and transportation of stable toxins off-site (e.g., radioactive compounds, asbestos, lead, or pesticides).
3. Respirators can be selected (in accordance with 29 CFR 1910.134) either:
   a. to remove the contaminant from the air in the breathing zone (what in the next chapter will be called an "air purifying respirator") or
   b. to provide clean air with oxygen into a highly contaminated and/or an oxygen deficient atmosphere (what in the next chapter will be called an "atmosphere supplying respirator").

   The distinction between cleanable air versus an oxygen deficiency will play a critical role in the respirator selection protocol to be covered in detail within the next chapter.
4. Eye and face guards can be selected (in accordance with 29 CFR 1910.133 and §8 to Appendix B of 1910 Subpart I) to protect the wearer from any combination of the following:
   a. chemicals: Face shields or goggles for handlers are typically made of either polycarbonates or poly-methyl methacrylate (also known as PMMA or acrylic). However, chemical reactions from some vapors can cause normally transparent face shields (including those on full facepiece respirators) to become opaque rapidly.
   b. physical agents: Examples include special filters for welders, glasses and goggles for lasers, and creams for UV protection.
   c. traumatic injury: Protection may be provided by the same face shields and goggles that provide chemical protection. Safety glasses (preferably with side shields) provide very little chemical protection. Safety glasses are identifiable by a small but visible manufacturer's mark on the lens.
5. Ear protection such as ear plugs and muffs selected (in accordance with 29 CFR 1910.95) to prevent hearing loss. Although hearing loss may not have a physiological threshold, it does have a practical enough threshold to allow administrative controls to be an explicit part of the noise control standard.
6. Head protection selected (in accordance with 29 CFR 1910.135) to prevent traumatic injury. Protective helmets (commonly called "hard hats") are categorized in Appendix B to 1910 Subpart I into the following three classes:
   a. Class A: provides general impact and penetration resistance and electrical protection to 2200 volts (low voltage).
   b. Class B: provides the above plus high-voltage protection to 20,000 volts (designed for electrical utility workers).
   c. Class C: special service helmets provide general impact and penetration resistance but no voltage protection. They are usually made of aluminum and are allowable only where no danger from electrical hazards or corrosion exists (e.g., for lumberjacks).

7. Foot protection in the form of steel or hard-toed safety shoes with or without added metatarsal protection are selected (in accordance with 29 CFR 1910.136) most often to prevent traumatic injury from impact, compression, and puncture but can also be chemically resistant and sometimes protect against electrical hazards.
8. The broad definition of PPE also includes various body restraints, supports, and other specialized pieces of equipment that have nothing to do with protection against chemicals. Examples include the following:
   a. Fall and rescue protection using a life line or lanyard and a body harness (safety nets are a nonpersonal alternative or a supplement to a life line).
   b. Segmental protection in the form of kidney support (for vibration), lower back support (for repetitive or heavy lifting), or wrist support (for carpal tunnel protection).
   c. High visibility clothing (e.g., light reflecting or fluorescent vests or appliqués for highway construction or airport flight line workers).
   d. Life jackets or buoyant work vests (note that a life jacket could be appropriate for construction next to or over water without ever involving a boat).
   e. Firefighters' turnout or bunker coat and pants to protect against heat, hot water, and some particles. A proximity garment uses aluminized Nomex® to protect further against thermal radiation.
   f. Body coolants (e.g., a passive solid or a circulating liquid) may be used either with or without the previous thermal shield protection to absorb heat from inside clothing.
   g. Blast and fragment suits absorb and deflect the impact from the blast of a small amount of explosives or a small bomb (in this day-and-age "small" should probably be emphasized).

III. BASIC PPE PROGRAM MANAGEMENT

1. While management is much more than just meeting an OSHA requirement[2] or consensus standard,[7] such requirements can comprise a starting point for PPE program management. OSHA 29 CFR 1910 Subpart I contains the following eight separate standards and appendices that cover most of the above categories of personal protective equipment. These standards vary greatly in their complexity. Some are mandatory (meaning a lack of compliance is a citeable violation); others are nonmandatory (meaning they are only advisory). This chapter will only summarize some of these standards; however, the next chapter on respiratory protection will cover OSHA 1910.134 in considerable detail.

1910.132	General Requirements.	(see Appendix to this chapter)
1910.133	Eye and Face Protection.	
1910.134	Respiratory Protection (and its Appendices A - D).	(see Chapter 22)
1910.135	Head Protection.	
1910.136	Occupational Foot Protection.	
1910.137	Electrical Protective Devices.	
1910.138	Hand Protection.	(see Appendix to this chapter)
1910 Subpart I App A	References for Further Information (Non-mandatory)	
1910 Subpart I App B	Nonmandatory Compliance Guidelines for Hazard Assessment and Personal Protective Equipment Selection.	(see Appendix to this chapter)

2. The material contained in Appendix B to 1910 Subpart I contains very useful but nonmandatory "Compliance Guidelines for Hazard Assessment and Personal Protective

Equipment Selection." These guidelines largely affirm OSHA's agreement with the classic IH control paradigm, while clarifying the roll of cost in feasibility. To paraphrase this appendix, one of the following three situations would have to exist for the use of PPE to be acceptable:

a. In emergency situations (examples include rescue, fire, chemical spills, or electrical failure);
b. As a temporary control (classic examples include while awaiting repairs or modification, when investigating an unknown hazard (that may in fact be found to not require PPE), and spill mitigation (as needed to contain or decontaminate); or
c. When PPE is the only feasible means of control based on the costs of alternatives. Feasible implies an ill-defined economic criterion. Feasible conditions often overlap with emergency and temporary conditions but might include:
   1. work at a remote or temporary location,
   2. a task that requires personal mobility precluding fixed controls,
   3. a task of short duration (even if repetitive but at widely separated intervals),
   4. decontaminating other employees who are in PPE or their contaminated equipment.

These three conditions for the use of PPE to be acceptable have almost become another IH paradigm. These acceptable use criteria should be part of any PPE management program and are largely mandatory for respirators under OSHA within 1910.134(a), as covered in the next chapter.

3. Much like administrative controls, clear policies and strong supervision needs to be in place in order for a personal protection control to succeed. Most professional hygienists would probably consider the following six generic program elements to be also essential to their own programs and are contained within the ASTM Standard Practice for a Chemical Protective Clothing Program.[7] However, only half of these elements are explicitly covered within OSHA 1910.132. Requirements conveyed by the legally important word "must" are mandatory, while those conveyed by "should" are nonmandatory. OSHA's Respiratory Protection standard not only requires the other three program elements (makes them mandatory), but also adds several specialized requirements lumped here into a seventh catch-all program element. The order in which the program elements are listed below are the order in which they (and PPE) would be used (which differs from the sequence within 1910.132).
   a. The employer must identify an individual employee responsible to manage the particular PPE program. The employer should have a written program wherein this person is identified. Although a written program is so far only explicitly required by §1910.134 for respirators, a written program is by far the simplest way to document all the other procedures described in 1910.132 and implied by other specific PPE standards. See for example the CDC Office of Health and Safety Personal Protective Equipment Program at www.cdc.gov/od/ohs/manual/pprotect.htm.
   b. The employer must have an established protocol for selecting equipment to assure that it is appropriate to the hazard based on measured or predicted levels of exposure, that it fits the user, and that it is safe and appropriate for the hazard (specified in §1910.132(c) and (d), respectively). Subpart I, Appendix B §3 and §4, and some other specific sections in 1910.134 elaborate on how a hazard assessment might be accomplished. They need not be elaborate, but they must be documented. In contrast, the OSHA protocol for fitting a typical respirator to each user are quite detailed.
   c. The employer must have a defined training protocol for users (§1910.132(f) and .134(k)) including:
      1. why, when, and what PPE is necessary,
      2. the proper equipment donning and doffing procedures,

3. the limitations of their PPE, how to recognize a failure of the PPE, and actions to be taken if a failure is found,
4. proper use, care, inspection, maintenance, and possible decontamination of their PPE, and
5. hands-on familiarity training in a nonhazardous environment.

Such training needs to be applicable to the user's setting. The use of training aids (provided by most PPE equipment vendors) also helps to make the training applicable to the equipment in use. The employer must be able to document that each affected employee both received and understood that training. Such documentation cannot only satisfy OSHA, but it also helps to adequately protect the user and can protect an employer from a unwarranted legal claim of "failure to warn." Validating employee learning by their behavior and use of PPE in the workplace is just one more function of good supervision.

d. The employer should provide adequate supervision of users and of use conditions.
   1. Users should be encouraged to report any perceived problems or difficulties to their supervisor and these reports should also be passed on to the program manager.
   2. Supervisors should also inform program managers of changes in use conditions that might warrant a reassessment of the program requirements for protection. It is far better to have many trained supervisors reporting changes than expect the small number of people on most health and safety staffs to detect changes in the many sites they might serve.

   The supervision of supervisors of respirator users is specified in §1910.134(g) but is not specified in §1910.132. The employer's liability for supervising PPE is fostering a growing interest in biological monitoring as a means of assuring compliance or at least detecting noncompliance.
e. The employer should define procedures to maintain and care for PPE equipment (e.g., decontaminate, disinfect, inspect, and repair) that specify when, how, and by whom these procedures will be carried out. For example, the frequency of inspections can vary from before and/or after each use, to monthly, annually, etc.). The kind of maintenance that may be carried out by an individual typically depends upon their level of training, e.g., some maintenance can be done by any trained user or wearer, some only by an on-site trained and equipped maintenance person, and some only by the factory or its authorized repair specialist.
f. The employer should conduct some form of periodic program evaluation, review, or audit (e.g., annually or any time use or work conditions change). Documentation does no good unless it is periodically used to identify weaknesses and action that needs to be (or has been) taken to improve that program. §1910.134(m) refers to the employer auditing the adequacy of (their respirator) program, but neither requires nor specifies a frequency.
g. A seventh program element is listed here as a surrogate for any and all additional specialized program requirements that are imposed on some forms of PPE by OSHA standards or good IH practices. The following are examples of specialized requirements that pertain to respirators:
   1. Where §1910.132(c) only states that such equipment will be safe and appropriate for the hazard, §1910.134(d) specifies the use of NIOSH certified respirators.
   2. §1910.134(e) requires that employers provide a medical evaluation to all potential respirator users before they are fit tested or required to wear a respirator in the workplace.
   3. §1910.134(i) requires the employer to assure that the supplied breathing air, if applicable, meets certain ANSI and other quality criteria.

IV. TERMS AND CONCEPTS REGARDING CHEMICAL PPE

1. Breakthrough time is that time following an initial and generally continuous chemical contact before a given chemical can be detected or measured on the back or the wearer's side of a protective membrane or other barrier, as depicted in Figure 21.1. The delay is caused by the time required for the chemical to pass through the protective material. Up until the breakthrough time, the material is virtually completely (100%) protective. The concept of breakthrough applies to membranes used for gloves and protective clothing (even to skin) and to chemical adsorbents used in air purifying respirators, air samplers, and air cleaning devices.[8]

   In laboratory practice, the breakthrough time would depend upon the sensitivity of the instrument used to detect or measure the chemical on the back side of the membrane (the "normalize breakthrough time" defined as term 4 below gets around this problem). In field use, breakthrough time is virtually synonymous with the service life of a particular piece of PPE and is therefore related to its replacement schedule (see term 5).
2. Permeation describes the net movement of the molecules of one chemical (either gaseous or liquid) through another solid chemical. In the context of dermal chemical protection, the solid chemical is an intact protective membrane (one without holes or other openings). The specific parameter of interest is the steady-state permeation rate ($\mu g/cm^2/min$), the rate at which the toxic chemical will pass through each cm^2 of the membrane after breakthrough. A chemical's permeation rate is sometimes called its flux rate (flux sounds more quantitative but really means the same thing). Figure 21.2 depicts an idealized pattern of breakthrough followed by a transition to steady-state permeation. The permeation rate in Figure 21.2 turns out to also be the slope of the data in Figure 21.1; thus, the data in Figure 21.1 is really the net effect of the permeation rate in Figure 21.2 over time (in other words, calculus integration in action!).

   Permeation through a membrane is due mostly to diffusion. (Because the membrane is a solid, albeit flexible media, the process is called solid state diffusion.) The chemical process of diffusion of any chemical through a given protective material is affected by the following four factors:
   1. the thickness of the protective material (a thicker material will have a lower permeation rate),
   2. the liquid temperature (a higher temperature will increase the permeation rate),

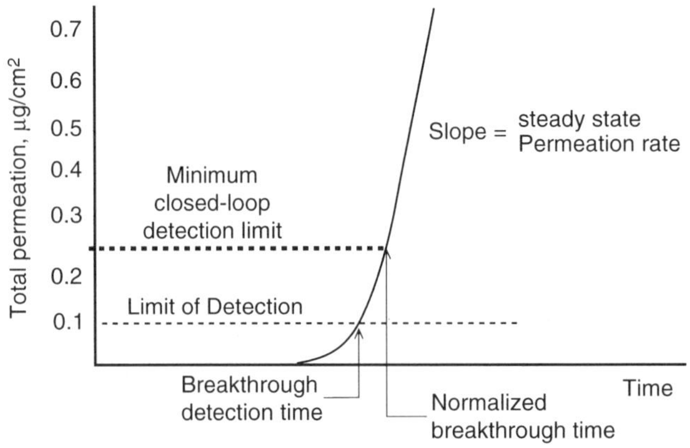

FIGURE 21.1 A depiction of the total amount of chemical permeating through a membrane that could be measured by a closed-loop permeation system.

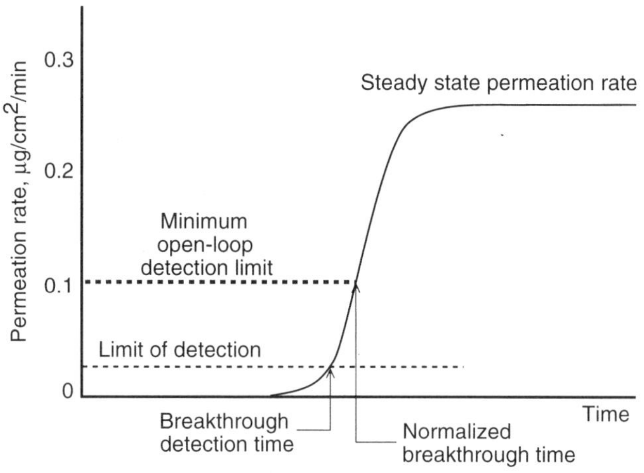

FIGURE 21.2 The permeation rate of a chemical that could be measured by an open-loop test system.

3. the presence and nature of other chemical components within a permeating mixture (co-permeation is a problem when one rapidly permeating chemical component increases the permeation rate of another otherwise slowly permeating chemical), and
4. the quality of the material (e.g., the membrane material density or the presence of and size of micropores within a membrane that can vary within, and does vary between, manufacturers).

3. In order to provide consistently definable values for breakthrough time and permeation rate, glove manufacturers formed a joint committed to define a uniform laboratory testing method described in ASTM F739-99a.[9] This method uses a Permeation Test Cell, commercially available from vendors such as from Pesce Lab Sales (see www.pescelabsales.com) and depicted functionally in Figure 21.3.[d] A specimen of the membrane to be tested (approximately 3 in. (75 mm) in diameter) is clamped between the two halves of the permeation test cell with its front side on the challenge chamber side of the cell and its back side on the collecting side of the cell. The chemical to be tested (either liquid or gas) is delivered into the challenge chamber (shown on the right side of Figure 21.3). A collection medium (usually either air or preferably distilled water) is delivered to the other side of the cell. The concentration of the test chemical in the collection medium can be determined either by circulating the medium through a continuous, real-time analytical detector and back into the collection chamber (called a closed-loop system as depicted in Figure 21.4a) or by withdrawing, analyzing, and discarding a small portion of the medium from the collection side (called an open-loop system as depicted in Figure 21.4b). The data generated by the closed-loop system depicted in Figure 21.4a would typically be plotted as shown in Figure 21.1. The concentration measured in an open-loop system withdrawn at a fixed flow rate would typically be plotted as shown in Figure 21.2.
4. ASTM F739-99a defines a normalized breakthrough time as the time at which either the permeation rate reaches 0.1 $\mu g/cm^2/min$ or the total mass of chemical that has permeated reaches 0.25 mg/cm^2.[9] Because detection limits vary widely with the analytical

[d] The procedure and apparatus to determine the breakthrough time of a gas or vapor through an absorbent respirator cartridge are similar to but simpler than that for testing chemically protective membranes. The rate that a gaseous contaminant will permeate through a porous respirator adsorbent after breakthrough is typically so high as to be unacceptable. Therefore, a respirator air purifier must be replaced after breakthrough.

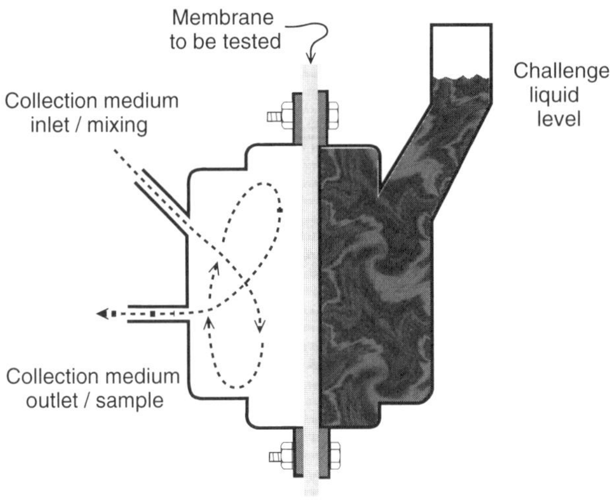

FIGURE 21.3 Functional depiction of an ASTM permeation test cell.

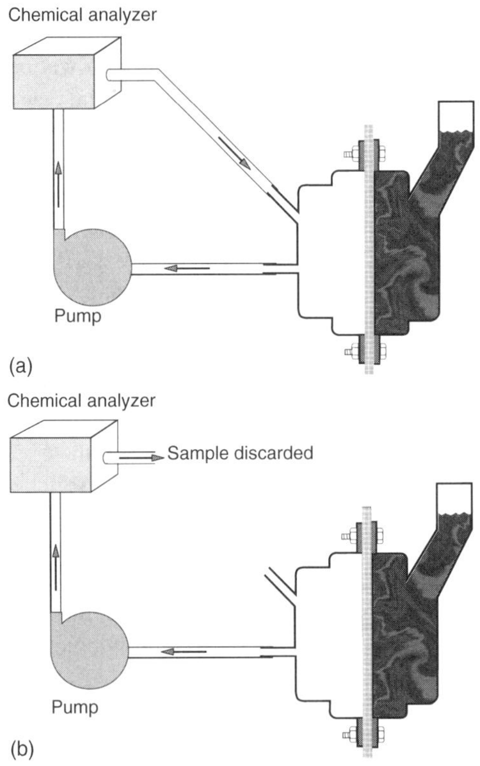

FIGURE 21.4 (a) A schematic depiction of a closed-loop permeation test system. (b) A schematic depiction of an open-loop permeation test system.

technology, with the challenge chemical being tested, and with the collecting medium (viz., air versus water), the breakthrough time using one definition does not always agree with the time found using the other definition. However, the criteria for the normalized values can be applied to most chemicals, and the value of consistency outweighs the alternative of users having to contend with uncomparable performance data (the use of performance data will be discussed in more detail in Section VI.4).

5. Service life means the longest period of time that a glove, filter or sorbent, or any piece of personal protective equipment is expected to provide adequate protection to the wearer. Breakthrough often determines the service life of chemical PPE, but other factors such as degradation (due to a chemical, heat, or sunlight), abrasion, or puncture (cut or tear) frequency can shorten a device's service life.
6. A replacement period or change schedule is the nominal time that is less than or equal to the service life at which the protective element within the PPE should be replaced. The change schedule for each piece of PPE should be set as a part of the PPE program's management policy. The replacement period for a glove (the time from the first chemical contact until the user should exchange the gloves for a fresh pair) can be based either solely on the breakthrough time or it can be longer if the rate of chemical permeation through the glove is acceptable. The service life of a respirator's air purifying element will be discussed at considerable length in Chapter 22.
7. Directions for the use of PPE (such as in an MSDS) often suggest wearing impermeable gloves. No chemically protective material is impermeable. Most chemicals will eventually break through any material. The protective capability of any PPE will decrease rapidly after the breakthrough time. Exposure to the wearer from permeation after breakthrough and/or chemical degradation may be acceptable, but one should assume that protection with permeation is inadequate until the rate of exposure is evaluated. Users must be made aware of the temporal limits for any PPE.
8. Chemical degradation, when applied to clothing or gloves, means some sort of interaction between the chemical and the PPE membrane that either increases its permeation or weakens the material enough to cause penetration. Chemical degradation can also cause the functional loss of another component such as causing a visor to become opaque or a respirator's harness to lose its elasticity. The early signs of chemical degradation can often be seen as material discoloration or swelling. If dealing with a mixture of chemicals, chemical degradation can be caused by a component other than the primary toxic chemical of interest. The classic example is water degrading polyvinyl alcohol (PVA).
9. Penetration is the passage of vapor, liquid, or solids through openings in the chemical protective equipment (cf. permeation being the passage through the protective material itself). Penetration may be the result of any of the following four causes:
   - openings designed into the equipment such as at the neck, sleeve, cuff, or stitched seams. The seams in most clothing are stitched; most chemically protective gloves are made without seams.
   - holes caused by punctures, abrasion, or wear duration. Such damage varies with the protective material, use conditions, and employee practices. Chemical degradation generally decreases a membrane's ability to tolerate punctures, abrasion, or wear. Damage sufficient to cause penetration can usually (but not always) be detected upon inspection.
   - poor quality of the PPE membrane that allows virtually invisible pin-holes or other imperfections.
   - cross contamination from the outside of contaminated clothing onto the users skin. Cross contamination can occur while doffing a recently contaminated piece of PPE or from contact with an old contaminant while donning previously used protective material.

   Most penetration is caused by either poor equipment design, a poor match between the protection selected and usage conditions, or poor user practices. These causes are less quantifiable than early breakthrough and/or high permeation, but almost as predictable where program management is poor.

10. Decontamination means the removal of a chemical contaminant from a surface. In principle, helmets, gloves, clothing, or respirators could be decontaminated, but the lack of a defined acceptable level of a residual contaminant in any surface often limits the viable use of decontamination procedures for PPE. Does an acceptable residue mean nondetectable? Does decontamination decrease a subsequent chemical's breakthrough time? Useful descriptions or reviews of decontamination procedures for clothing are limited.[10–12]
    a. *Detergents*: Generally limited to removing only surface contaminants or those that are water soluble or emulsifiable.
    b. *Solvents*: Solvents can remove more hydrophobic contaminants, but they generate large volumes of contaminated solvent and may degrade the membrane material's protective qualities.
    c. *Low heat*: For example, approximately 50°C for 24 h worked well on butyl rubber CPC; however, excessive heat can damage polymers.
    d. *Natural Aeration*: The effectiveness of leaving the contaminant to evaporate at room temperature is a function of aeration time, contaminant volatility, and its solubility in the clothing (partitioning), although Perkins concluded that aeration was slow at best and rarely feasible.[11]
11. The distinction between disposable vs. reusable PPE is often vague. The distinction can best be based upon cost. Specifically, does it cost more to decontaminate and reuse the particular PPE than to dispose of the contaminated PPE and purchase another new one?
    a. Disposable (or limited use) PPE is generally lightweight, inexpensive, has a comparatively short time to breakthrough, and costs more to decontaminate and prepare for reuse than to replace.
    b. Reusable PPE is generally more rugged, more expensive, and capable of being laundered or otherwise decontaminated for less than it costs to replace it.

    The cost of decontamination and the lack a criteria for "acceptably clean" may render any garment disposable. A further discussion of this trade-off will be presented in Section VI on Gloves.
12. While the previous section listed at least eight categories of PPE, one should think of how the various components function together as an ensemble. The components of a chemical PPE ensemble could include all the components previously discussed or only selected components. Personal protection depends upon protecting all exposed or potentially exposed parts of the body. Thus, the appropriate components of a chemical protection ensemble depend on the specific setting in which they will be used. The total effect can only be determined when all parts of the ensemble are considered as a whole.
13. An ensemble is either fully encapsulating or it is nonencapsulating. There is no in-between:
    a. A fully encapsulating suit (a vapor protective suit or a "moon suit") protects against splashes, dust, gases, and vapors. To protect against gaseous contaminants, a fully encapsulating suit requires gas-tight integrity. A fully encapsulating suit completely covers both the wearer and their SCBA breathing apparatus.
    b. A nonencapsulating suit protects against splashes and dust but not against gases and vapors. Gaps in such an ensemble may include the collar, wrists, ankles, under the bottom of the jacket, and the breathing apparatus (thus, the SCBA might be worn outside of nonencapsulating clothing).
14. Some protective equipment can create its own intrinsic hazard. Examples are listed below. The potential for these or similar intrinsic hazards should become part of the PPE

selection criteria and/or should cause limitations or restrictions to be placed on the use of PPE for some tasks or in some settings:

a. Chemical protective clothing can cause heat stress to the wearer due to a decrease in the evaporative cooling of perspiration and possibly an increase in metabolic heat generated by a wearer to overcome the added weight and inflexibility of the protective clothing.
b. Chemical protective clothing can impair a wearer's vision, mobility, and/or flexibility. One manifestation of these limitations could be the wearer's failure to be able to mechanically actuate some tool or device. Another manifestation could be ergonomic stress due to decreased flexibility and to the decreased tactile sense that may cause the wearer to hold onto objects tighter.
c. Communication is often impaired (especially due to a respirator or hearing protection). Impaired communication can both cause psychologic stress and contribute to safety problems.
d. The unexpected reactivity between the chemical and PPE components can rapidly degrade their function. Chemical reactivity is especially a problem for respirator exhaust values and face shields on people responding to emergency or unknown chemicals.
e. Many wearers experience psychologic stress due to decreased vision, hearing, and/or tactile sensations especially if they are claustrophobic. Psychologic stress can often be alleviated (but not eliminated) by a good training program.

The required program management elements described in Section III, the physical limitations (such as permeation, penetration, and a limited service life) described in Section IV, and the intrinsic hazards described above all contribute to the indirect costs of PPE control. Unrecognized costs can make receiver controls appear to be much cheaper than they really are. A control program with unexpected and unmet limitations will protect neither the employees nor the employer.

V. RECOGNIZING DERMAL HAZARDS

1. The generic term "chemical dermal hazard" refers to chemicals that can either:
   - cause dermatitis or otherwise damage the skin, or
   - enter the body through intact skin where it can then cause internally toxic effects.

   While both of the above effects are of concern in the workplace, the *Skin* notation in the TLV® and PEL lists is only based on the second of these effects. The Introduction to the Chemical Substances section of the annual ACGIH TLV booklet states that *Skin* notation is given only to toxic chemicals that can enter the body through intact skin (including mucous membranes and the eyes). Properties such as causing irritation, dermatitis, and sensitization in workers are literally not considered relevant when assigning the *Skin* notation. "Use of the *Skin* notation is intended to alert the reader that air sampling alone is insufficient to accurately quantify exposure and that measures to prevent significant cutaneous absorption may be required."[13]

   Approximately 95 OSHA PELs (29 CFR 1910.1000) and 160 TLVs (approximately 25%) have the *Skin* notation.[2] The difference, of course, is due to the evolution of the TLVs since the 1968 list was adopted into OSHA 1910.1000 Tables Z-1 and Z-3 (cf., Table Z-2 that came from the generically quite different ANSI standards).
2. Dermal absorption of chemicals is typically modeled as diffusion through the epidermis and absorption into blood within the skin's capillaries.[14–18] However, the epidermis is neither a homogeneous nor a static membrane. The epidermis is a good water barrier and

can serve as a lipophilic reservoir. The skin's enzymes can even metabolize some chemicals. Its permeability changes in response to both chemical and physiologic stimuli, and blood absorption is strongly influenced by blood flow and a chemical's water solubility. The chemical characteristics that aid diffusion through the lipids of the epidermis are quite different from those that aid its solubility in aqueous blood. The TLV Committee uses Equation 21.8 to model dermal absorption.[19]

$$\text{Flux Rate } [\text{mg/cm}^2/\text{h}] = \frac{C_{\text{sat.H}_2\text{O}}}{15} \times (0.038 + 0.153\, K_{\text{OW}}) \times e^{-\text{MW}/62.5} \qquad (21.8)$$

where

flux rate = the dermal absorption rate in $\text{mg/cm}^2\text{/h}$;

$C_{\text{sat.H}_2\text{O}}$ = the chemical's water solubility expressed as its saturated aqueous concentration [mg/mL];

K_{OW} = the partition coefficient or ratio between the chemical's equilibrium concentration in octanol and in water.

The chemicals that are absorbed most rapidly through the skin are neither too lipophilic, too lipophobic, nor too big. The TLV Committee will designate a chemical as "skin" when its flux rate (using Equation 21.8) through 360 cm^2 of skin (equivalent to approximately one hand or approximately 2% of the body's surface area) exceeds 30% of the dose from working in a TLV concentration while inhaling at 0.9 $\text{m}^3\text{/h}$. This critical flux rate ($\text{mg/cm}^2\text{/h}$) works out to equal the TLV (in mg/m^3)/1333.

While dermal flux in units such as those above is appropriate for the steady-state permeation of chemicals through a membrane such as skin, exposures in industry are often not the steady state that might result from immersion or repeated contact with wet parts. Dermal exposure is often localized (as from splashes or a spray) and transient (intermittent and either wiped or evaporated off). The percent absorption (%) of a transient or dispersed dermal dose is a more common and more practical unit, albeit less scientifically defensible. Based on past field research with agricultural pesticides, dermal exposure is likely to be of concern if a compound's dermal absorption is 10% or more and unlike to be of concern if a compound's dermal absorption is less than 1%. If a compound's skin absorption falls between 1 to 10%, the balance will depend upon the compound's volatility and the ambient ventilation (i.e., how the vapor exposures are controlled). While not covered herein, the interested reader is directed to reviews of methods to evaluate dermal exposure.[20–23]

VI. CHEMICAL PROTECTIVE GLOVES

1. The goal of chemically protective gloves is to protect the the wearer's hands from the chemical for at least a portion of the work task. A glove that has reached breakthrough can be replaced as frequently as needed to protect those hands for the entire work task. The glove must allow the wearer to accomplish the task and should resist damage by either chemical or nonchemical factors associated with the task or the use environment (such as abrasive or sharp objects). Also, the glove must be comfortable enough to at least allow, if not encourage, users to wear it. The most appropriate glove will accomplish the above basic goals for the lowest cost. Chemical protective gloves are commonly made from the following materials, some examples of which are

FIGURE 21.5 Examples of chemically protective gloves (from left to right): nitrile, Neoprene®, PVA, and a laminate.

shown in Figure 21.5. Glove color is not standardized among manufacturers:

a. Natural rubber — (initially poly-isoprene obtained from rubber plants that is refined and dispersed in water to form a latex that adheres to a mold and coagulates into a semi-elastic membrane) provides good protection from bases, alcohols, and aqueous mixtures but is poor for oils and organic solvents.
b. Butyl rubber — (a relatively inelastic isobutylene-isopropylene copolymer) good for glycol ethers, ketones, and esters but poor for hydrocarbon and chlorinated solvents.
c. PVC — (a somewhat elastic polyvinyl chloride) good for strong acids, bases and salts (most aqueous mixtures) and alcohols but not good for many organic solvents.
d. Nitrile — (an elastic acrylonitrile and butadiene copolymer, also known as Buna-N or NBR rubber, shown in Figure 21.5) good for oils, xylene, and some chlorinated solvents.
e. Neoprene® — (a semi-elastic polychlorinated iso-propylene shown in Figure 21.5) good for oxidizing acids, aniline, phenol, and glycol ethers.
f. PVA — (polyvinyl alcohol; inelastic; and expensive; shown in Figure 21.5) good for aliphatics, aromatics, chlorinated solvents, ketones (except acetone), esters, and ethers, but cannot tolerate aqueous solutions.
g. VITON® — (a family of expensive, proprietary fluorinated copolymers such as hexafluoropropylene and vinylidene fluoride or vinyl fluoride; produced as a glove by North Safety Products) good for aromatics, chlorinated solvents, aliphatics, and alcohols. See www.pspglobal.com/nfvitongrades.html.
h. laminates — (bonded layers of different polymers such as polyethylene and ethylene vinyl alcohol (PE/EVAL); trade names include Silver Shield®, 4H®, and LPC®; also called barrier or flat film gloves; shown in Figure 21.5) excellent to good for a wide variety of solvents, but they are inelastic, are not formed to fit hands, and are relatively easily punctured. These limitations can be largely overcome by double gloving (wearing one of the elastic polymer gloves over the laminate).

Leather, cotton, and canvas gloves are not chemically protective. They can prevent mechanical stress and minor physical injury but have virtually no chemical protective qualities. In fact, they can retain chemicals and continue to dose the wearer long after the chemical use has ended.

The PPE program manager is responsible to select a glove that will meet the basic goal of protecting the wearer's hands from the chemical for at least a portion of the work task and then to ensure that gloves are replaced as frequently as needed to protect those hands for the entire work task. As explained above, glove selection can be based on chemical criteria, use and user criteria, and cost.[24] Protection must be the first priority. Acceptability is a second criterion but is required for their consistent use. Costs can be compared with alternatives using the cost-benefit methods described in Sections III in Chapter 7. The weighting given to each of these criteria will vary from workplace to workplace. Each of these criteria will be discussed below.

2. Chemical criteria for glove selection:
   a. Full protection is provided for the length of time that it takes for the chemical to move through the protective membrane via diffusion until breakthrough. After that lag time, the chemical moves through at a nearly constant permeation rate. Thus, the ideal goal of zero dose is only possible if the use period of each glove is less than its breakthrough time for the chemical. If the breakthrough time is longer than the work or task time, then one pair of gloves can be worn for that entire time, up to 8 h or more (commercial glove test data are generally limited to 8 h as a simple expedient). If the glove use period would exceed the breakthrough time, one is faced with the dilemma of whether or not to replace the gloves at that time or to accept the risks from exposures at the permeation rate (see C below).
   b. Chemical degradation will shorten the breakthrough time, and may by itself preclude some gloves. Chemical degradation ratings can also be used to select between two gloves of equal or nearly equal breakthrough times.
   c. If replacing the gloves at or shortly before their breakthrough time is not feasible, then low permeation could be used as a criterion; however, defining an acceptable level of permeation is not a trivial task.[25] While a full discussion of acceptable permeation is beyond the scope of this book, let us take a simple scenario that assumes the concern is for systemic toxicity rather than dermatitis. Ansell Edmont's category of "excellent permeation rate" is less than 0.9 μg/cm^2/min.[e] If this permeation rate were to pass onto the roughly 1000 cm^2 of the hands, approximately 1 mg would reach the hands per minute of continued exposure. If 10% of this exposure were to be absorbed, the absorbed dose would be approximately 6 mg/hour. This absorbed dermal dose would be equivalent to inhaling an airborne concentration of 6 mg/m^3 at a nominal respiration rate of 1 m^3/h assuming that all of an inhaled vapor were absorbed (which it often is not). A 6 mg/m^3 concentration of a vapor whose MW is 147, is equivalent to an airborne dose of 1 ppm or a modestly low exposure limit. Modifications to the assumptions made above between airborne dose and continued use of gloves past their breakthrough time could change the comparison by 100× or more in either direction.

What is too short a breakthrough time? What is too high a permeation rate? The combination of a high permeation rate and short breakthrough time means that the gloves will have to be changed at the end of each breakthrough time and changed often. This requirement will affect costs. A frequent replacement schedule may make natural or accelerated decontamination cost-effective. The viability of gloves as a control option will eventually become an economic decision.

[e] This is almost 10 times a measured permeation rate of 0.1 μg/cm^2/min that can define a "normalized breakthrough time".

3. Use and user criteria for glove selection:
   a. Select a glove that is available in a range of sizes to match multiple users' hands. While the most preclusive limitation would be not finding gloves large enough for people with large hands (a large hand just cannot get into a small glove), buying only big gloves that do not fit small hands well is a less recognized but at least equally common problem. The solution is to buy and stock a good range of glove sizes. Laminated film gloves have particularly good chemical resistance but seem to fit no one well; a common solution to this problem is to wear a somewhat elastic glove (e.g., PVC) over the laminate to hold it tight against the wearer's hands.
   b. Select a glove style that will minimize penetration (cf. permeation) into the top of the glove. Styles vary among manufacturers and glove materials. In addition to hand sizes, gloves may be manufactured in lengths from roughly 10 to 18 in. A folded cuff on long gloves can keep liquids that run up the glove (when the hands are pointed up) from either reaching the skin above its top or running down inside the glove (when the hands are returned to a down position). Duct tape is heavily used in HazMat operations to minimize penetration of both liquids and vapors. Be aware that chemicals can also reach the skin and/or accumulate within gloves due to carry-over of contamination on the hands while doffing or donning. Techniques to minimize carry-over particularly while doffing should be part of the user training.[26]
   c. The glove must have adequate cut, tear, and abrasion resistance for its use. The inability to withstand mechanical wear can shorten the service life of a glove from that predicted by chemical breakthrough or increase penetration above the permeation rate. Chemical degradation can decrease the ability of a glove to withstand mechanical wear. The actual service life of a candidate glove can usually only be determined by actual field testing on the job.[f] If faced with the dilemma of the only chemically protective glove with an adequate service life being unacceptably resistant to cuts, tears, or abrasion, consider using two layers of gloves with the chemically resistant glove on the inside protected by a more rugged material on the outside.
   d. Select a glove that allows adequate dexterity and tactile sense. The loss of flexibility, touch, or the feeling of texture can all reduce productivity, and thus add to the cost of gloves. These characteristics vary with the pliability of the material, its thickness or with the number of layers of gloves, and with fit. Membrane thickness is often specified in "mils" (1 mil = 0.001 in.).
   e. The importance of selecting a glove that is comfortable cannot be over-rated, especially if that glove is to be worn for long work periods. Chemical users are less likely to wear an uncomfortable glove, and noncompliance is no protection at all. Enclosed hands sweat; accumulated sweat can make the hand very slippery inside of the glove. Some chemical protective gloves are supported, meaning that their inner surface is lined with a sweat absorbent material. To manufacture a supported glove, the membrane material is applied onto a fabric liner placed onto the hand-mold. Some gloves (especially latex gloves) can induce an allergic dermatitis that will add to discomfort. Small changes in the formulation made by a manufacturer or small differences between manufacturers of the same glove material can have big differences in a glove's allergenic properties.

[f] One quick check for the mechanical integrity of a glove is to quickly crimp the glove's cuff closed and compress the rest of the glove to inflate it. A leak can usually be heard and will certainly result in rapid deflation. Blowing into a chemically contaminated glove is not recommended.

4. The cost criterion for glove selection (including the determination of the glove replacement period):
   a. An economic goal is to provide adequate protection with the least cost. While the employer will eventually want to know the cost per year, during glove selection the industrial hygienist might first calculate the cost per person per day using Equation 21.9 and Equation 21.10.

$$N \text{ pairs/day} = \frac{\text{duration that protection is needed}}{\text{Replacement Period}} \tag{21.9}$$

$$\text{Cost/day} = [\text{cost per pair or use cycle}] \times [\text{number of pair/day}] \tag{21.10}$$

   b. Glove costs vary with the material, its thickness, the glove's style (length, ribs, etc.), and the manufacturer (brand). Program costs can be roughly approximated from Table 21.1. The relative costs are less likely to change with time, brand, and supplier than are the prices per pair.
   c. If the gloves are to be decontaminated and reused one or more times, the cost per cycle should be used in Equation 21.10 instead of the cost per pair. The cost per cycle is based on the number of times the gloves are to be re-used and the cost to decontaminate them after each use cycle:

$$\text{Cost per cycle if re-used} = \frac{[\text{purchase price}] + ([\text{cost to decontam.}] \times [\#\text{ of reuse cycles}])}{1 + [\#\text{ of reuse cycles}]} \tag{21.11}$$

An example of the cost per cycle is calculated below for gloves (at a purchase cost of \$20 per pair) that are decontaminated (at a cost per pair of \$10 per cycle) and reused another four times before wear-and-tear justifies their replacement:

$$\text{Cost per cycle if re-used} = \frac{[\$20\text{ price}] + ([\$10\text{ to decontam.}] \times [4\text{ reuse cycles}])}{1 + [4\text{ reuse cycles}]}$$

$$\text{Cost per cycle if re-used} = \frac{\$20 + (\$10 \times 4)}{1+4} = \frac{\$60}{5} = \$12 \quad \text{(using Equation 21.11)}$$

To decontaminate a glove that costs \$5 per pair using the same method at \$10 per cycle for four cycles would result in a per cycle cost of \$9, which is more than these cheaper gloves would have cost to be replaced new. Obviously, for decontamination

TABLE 21.1
Approximate Costs upon Which to Compare Gloves

Glove Material (Nominal Thickness)	Price per Pair (\$)	Price Relative to Natural Rubber
Natural rubber (19 mil)	0.60	1×
Nitrile (11 mil)	1.50	2.5×
PVC	1.80	3×
Neoprene (15 mil)	3.00	5×
Laminated film (2.5 mil)	5.40	9×
PVA	20.00	33×

to be cost effective, the decontamination process must cost less than the original replacement gloves.[27]

d. The replacement period depends upon the temporal exposure pattern, the breakthrough time, and possibly the glove's mechanical service life (but not the permeation rate). Several methods to estimate the replacement period are given below for common exposure scenarios:

   1. Protection from frequent or continuous chemical contact can follow one of two philosophies depending upon the acceptability of a known permeation rate.

      The simplest policy is not to accept any exposure past the breakthrough time, and simply replace the gloves at some nominal time equal to or less than the breakthrough time. This same policy could be used if the permeation rate is too high.

      Continued use of gloves past their breakthrough time may be acceptable if the permeation rate is sufficiently low to reduce actual dermal exposure to below the chemical's toxicity threshold. However, such a policy requires the quantitative interpretation of a calculated dose to the hands equal to that permeation rate [$\mu g/cm^2/min$] times the approximately 1000 cm^2 surface area of both hands times the duration in minutes that the gloves will be worn beyond the breakthrough time.

   2. Protection from short, infrequent chemical contact may follow one of the same policies above. However such a pattern potentially offers the following third cost-effective control possibility: if each of those infrequent exposures is obvious to the wearer, the use of cheap gloves with a short breakthrough time would be allowed if a policy were in place and enforced that the user must change that contaminated glove within its predicted breakthrough time after contact. A cost comparison can be made between using expensive gloves that can last all day and changing cheap gloves several times by using Equation 21.11 where the average time between infrequent exposures is used instead of the replacement period.[g]

   3. Protection from chemical contact that is either short but somewhat frequent, or infrequent but long, is difficult to assess quantitatively. The simplest advice is to chose one of the previous policies corresponding to the scenario this setting most closely matches.

   The current limited ability to predict or to interpret permeation data encourages the conservative approach of using the breakthrough time as the replacement period (or even shorter if the glove's mechanical integrity is compromised). Unfortunately (and much like administrative controls), both the employer's costs and the wearer's inconvenience of having to replace gloves frequently encourages their use beyond breakthrough, often well beyond breakthrough. To overcome PPE's many disadvantages, a replacement policy should use some judgement. For instance, while a replacement period is normally rounded down from a breakthrough time, a rational case for increased compliance can be made for slight extensions to fit the user's natural work cycle, break times, lunch, etc. (as long as the consequences of some permeation are acceptable to all parties). Management might even consider financial or alternative incentives for the user to replace the gloves on whatever time is set.

e. Once the glove selection is made and costs per day are available, the annual costs per year may be calculated for budget or comparison purposes. The two approaches to this calculation shown below both use the cost/day per person from Equation 21.9 based

[g] One could possibly add the breakthrough time to the average interval, but this third option is typically most advantageous when the breakthrough time is much smaller than the interval between infrequent exposures and can be neglected.

on either the price per pair of new gloves or the cost per cycle for decontaminated gloves.

1. The annual cost for continuous production operations can be calculated as either 365 or 250 days per year (actual production days) × the number of shifts per day × number of people per shift × the cost/day per person from Equation 21.9.
2. The annual cost for intermittent production operations can be calculated as the number of persons employed who need protection × number of days per year they need protection × the cost/day per person from Equation 21.9.

Remember that the acceptability of health protection costs should always be viewed relative to the benefits of the commercial value of the productive task, the liabilities of doing nothing, and the cost of alternative options of source or pathway controls. Any control has its price.

5. Glove selection resources include manufacturer's published literature (e.g., the Ansell Edmont Chemical Resistance Guide), their Internet sites, and NIOSH. The following examples provide guidelines and data for selecting the most protective glove for working with hazardous chemicals.
   a. CHEMREST is used by Best Gloves at www.chemrest.com.
   b. SpecWare™ is used by Ansell-Edmont at www.ansell-edmont.com/specware/.

   Both of these sites provide data for either heavy/immersion and intermittent/splash contact with the chemicals. Manufacturers may also run tests on other chemicals upon request in support of future sales.

   c. NIOSH Recommendations for Chemical Protective Clothing — A Companion to the NIOSH Pocket Guide to Chemical Hazards at www.cdc.gov/niosh/ncpc/ncpc2.html.

   At the latter site, NIOSH made the following telling comments regarding glove materials. "For the approximately 450 organic substances where it is recommended in the Pocket Guide to protect the skin, a recommendation for specific glove material could be provided for only 39%. For those substances where a glove type was recommended, 47% of the recommendations for glove material were for PE/EVAL co-laminate, Teflon or Viton polymers. Unfortunately, these latter materials are either uncomfortable to wear, lack good tactility, are fragile and expensive, or (as in the case of Teflon) are presently difficult to purchase. Polyvinyl alcohol (PVA) polymer offers excellent resistance against many organic substances but is highly sensitive to degradation by water and may be ineffective for extended use where perspiration occurs. Thus, a glove material such as natural rubber, polyvinyl chloride, butyl rubber, nitrile, or neoprene is suggested for less than 21% of the organic substances listed in the *Pocket Guide* where preventing skin contact is recommended."

6. And finally, commercially available chemically protective creams and lotions (skin barrier creams) deserve some attention as an alternative to gloves for chemicals that might be irritants or have low dermal toxicity. Most cream vendors have different versions of their products categorized for protection against either water or water soluble agents (lipophilic creams) or against oil or oil soluble agents (hydrophilic creams). Both kinds of creams often require regular (periodic) re-application. Creams were long thought by many to owe their effectiveness to the tendency for the user to wash more frequently (e.g., at breaks, meals, and/or end-of-shift or task) to remove the barrier cream. The efficacy of skin barrier creams is still an open issue.[28–30] If a barrier cream is used, ensure that the particular product is selected to provide protection for the type of chemical being used, and anticipate that even under the best of circumstances it will not provide as much protection as gloves.

VII. CHEMICAL PROTECTIVE CLOTHING

1. The goal of chemically protective clothing (CPC) is (similar to gloves) to keep someone's dermal dose acceptably low over the period that he or she must work with a particular chemical or mixture of chemicals. A dry solid has no permeation potential. It can reach the skin only by penetrating through holes, tears, loose weaves, etc. of clothing made from fibers (whether woven or spun). Therefore, any barrier that will prevent penetration may be used for a dry chemical. A few woven or spun fabric can provide protection from dry chemicals, but only clothing that is made either from a membrane similar to those used in gloves or a fabric upon which a membrane has been laminated can provide protection from liquid or gaseous chemicals. The following materials are used to make chemical protective suits:
   a. Tyvek® — Spun bonded olefin coated with polyethylene. Because Tyvek is not a membrane, it is breathable. It provides no protection against vapors, is only weakly resistant to liquids, but is protective against particulates (powders and fibers). personalprotection.dupont.com/protectiveapparel/index.html.
   b. Nomex® — A woven aromatic polyamide that is resistant to many chemicals (but provides no particular chemical protection by itself) and is a good choice for flame resistance (including electric arcs and flash fires), thermal protection, comfort, and durability. See www.dupont.com/nomex/home/index.html.
   c. Saranex® — A laminated polyvinylidene chloride membrane (basically Saran®, a product of Dow Chemical®) bonded between polyolefins like polyethylene and/or ethylene-vinyl acetate. It can be used to coat a fabric substrate to form a protective cloth.
   d. FEP — A fluorinated ethylene–propylene copolymer (related to Teflon®). FEP can also be applied onto a woven substrate (such as Tyvek) to form either a liquid-resistant film or a membrane.
   e. Gore-Tex® — Woven polytetrafluoroethylene (PTFE, TFE, or Teflon®) resists liquid water permeation (and therefore can protect against aqueous mixtures) but is air and water vapor permeable which is good for cooling but not for protection against organic vapors. Gore-Tex® is also commonly used in sports and recreational garments. www.gore.com/fabrics/index.html.
   f. ProShield® — A proprietary, multi-layer fabric from DuPont® combines a microporous film on a spunbonded polypropylene fabric to create a liquid waterproof barrier that allows water vapor to pass. It is available in a variety of thicknesses for different levels of strength and durability.

   Manufacturers also coat some of the above spun and woven materials with one of the chemically resistant membranes used in gloves such as butyl rubber, chlorinated butyl rubber, neoprene, various polyesters. For instance, Tyvek® QC is coated with 1.25 mils of polyethylene. Tychem™ is a family of Tyvek fabrics with different coatings. Barricade™ (also made by DuPont) is a nonwoven fabric with a multilayer laminate that provides excellent chemical resistance, and Trellchem® is a line of chemical protective suits made of nomex by Trelleborg Protective Products AB coated with various combinations of materials inside and out for protection against a wide range of chemicals.

The most appropriate chemical protective clothing material and design must meet the same basic three criteria as for gloves: providing chemical protection, matching the protection to the use and user, and minimizing cost.[27,31,32] One subtle but important difference between gloves and clothing is that clothing must have seams, making the construction of clothing's seams crucial to the wearer's overall protection. Another difference is that heat stress becomes an important clothing selection criterion because so much more of the body is covered by clothing than by gloves. Also, the per item cost of clothing is so much more than for gloves that decontamination and reuse are more feasible, as described by Schwope and Renard.[27]

2. Chemical criteria for selecting CPC are similar to those for selecting gloves.
   a. The breakthrough time should be longer than the time needed for the garment user to complete a task, and definitely must be longer than the time needed to leave the at-risk area and doff the garment. Keep in mind that the time needed to doff contaminated clothing is much greater than the time to doff gloves. Short breakthrough times are particularly applicable to settings where protection is needed only for the unpredictable rare event where the wearer is able to respond by changing clothing. Short breakthrough times are incompatible with long tasks where changing one's clothing is not an option.
   b. While permeation should be low, quantifying the amount of chemical that might reach the whole body is fraught with the same limitations as with gloves only magnified by variability in exposure pattern expected around the body and by the fact that clothing is not in the same intimate contact with the skin as gloves are to one's hands. And co-permeation is still a potential factor when exposed to multiple chemicals. Thus, any permeation is probably less acceptable.
   c. Good resistance to chemical degradation is just as important for clothing as it is for gloves to avoid increased chemical permeation or actual penetration due to physical damage.
   d. Selecting a material that is resistant to multiple chemicals is much more advantageous for clothing than for gloves because it is much more costly to stock multiple types of clothing than it is to stock multiple types of gloves.

   Three good sources of information on protective clothing for specific chemicals are the NIOSH *Pocket Guide to Chemical Hazards* at www.cdc.gov/niosh/ncpc/ncpc1.html,[33] the NIOSH *Recommendations for Chemical Protective Clothing — A Companion to the NIOSH Pocket Guide to Chemical Hazards* at www.cdc.gov/niosh/ncpc/ncpc2.html, and Forsberg and Mansdorf (1999).[34]
3. Use and user criteria for clothing selection are more diverse than those for selecting gloves.
   a. Clothing styles and construction vary widely in the presence and size of fabricated openings that could allow the penetration of hazardous chemicals. Macro-scale openings (measured in centimeters) may or may not be present at the neck, wrist, waist-line, or cuffs. CPC can also have mini-scale passages (measured in fractions of a millimeter) such as through seams or zippers.
      1. All openings must be sealed for an encapsulating ensemble to protect the wearer's air and skin, especially from toxic vapors not to mention from dusts, liquid splashes, and short term immersion. (See Level A protection in Section VIII.1 below.)
      2. Openings may be present in a nonencapsulating ensemble. Levels B and C protection (to be described in Section VIII.2 and Section VIII.3, respectively) protects the wearer's breathing air with a respirator, but only protects their skin against direct contact and splashes and not against vapors, all dusts, or immersion.

      Seams can be sealed or made without external perforations in a variety of ways as depicted in Figure 21.6. The more completely that a seam is sealed, the more protective and costly it is. A serged seam is a special sewing process that interlocks three threads around the raw edge of the material.

      Some CPC is made to protect only segments of the body such as the arms (sleeve protectors), front torso (aprons), or legs (leggings). Styles can be designed and manufactured to ease donning and doffing of the clothing or with other features suited to various tasks. Manufacturers compete (and purchasers can select) among styles on the basis of cost and the number of such features.

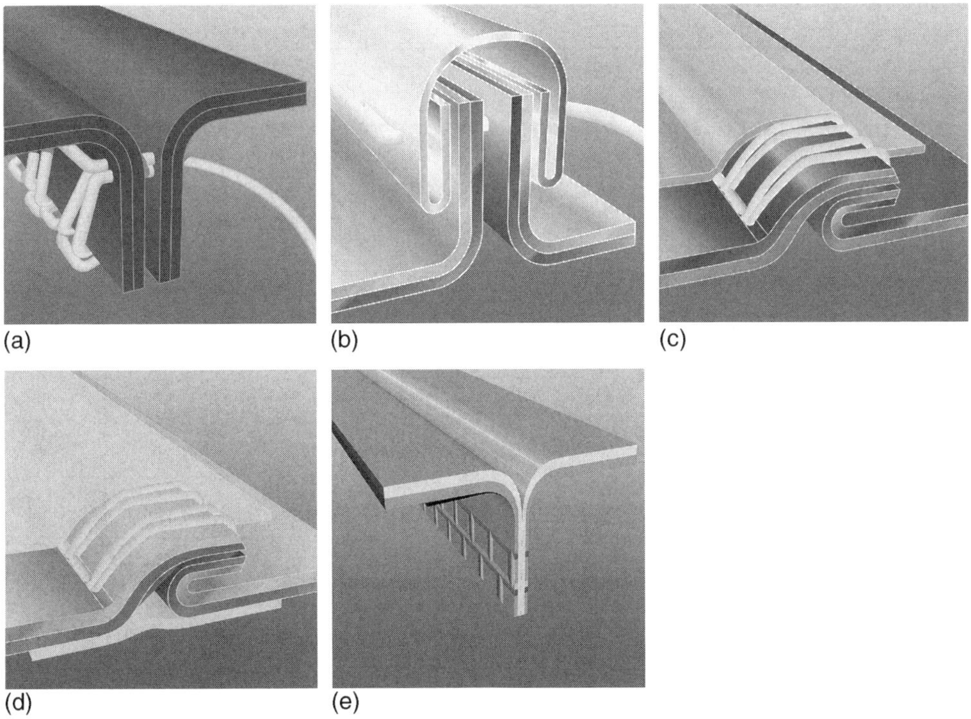

FIGURE 21.6 Various ways of creating a seam (a) a serged seam (strong but not sealed), (b) a bound serged seam (better but not sealed), (c) a folded and taped serged seam, (d) a folded and double-taped serged seam, and (e) a heat-sealed seam (c, d, e are sealed). Diagrams are courtesy of DuPont Corp.

b. The fit of clothing is a function of both the style and the sizes that are available. Size charts for selecting CPC similar to Figure 21.7 are provided by manufacturers. While such charts are based on the wearer's height and weight, the wearer's torso length and waist circumference can also be critical dimensions. Clothing with insufficient torso length or circumference can result in the clothing's back or shoulder seams ripping while bending, thus compromising protection. Although tactile sense is not important for clothing, its size must allow enough dexterity for the requirements of the task. Additional difficulties with size are encountered when the CPC must be worn over (or under) hard hats, winter clothing, respirators, etc. The chosen protective clothing needs to be able to accommodate other elements in the ensemble (such as a respirator or hard hat), and user acceptance is enhanced if it minimizes additional time required for donning/doffing.
c. Comfort is closely related to heat stress from having to wear additional clothing. Clothing adds weight to the work level, increasing the metabolic heat that the body has to dissipate. Metabolic heat build-up within the clothing becomes more important as the level of encapsulation increases. Woven materials generally allow better metabolic heat dissipation (through evaporation of sweat and transpiration through the suit) than nonwoven membranes, but are not recommended for protection from a toxic vapor or serious splashes.

Unacceptable heat stress can sometimes be avoided by working only in the coolest parts of the day or season, by rotating workers through appropriate ratios of short work periods and cool rest periods, or utilizing either active or passive cooling vests.

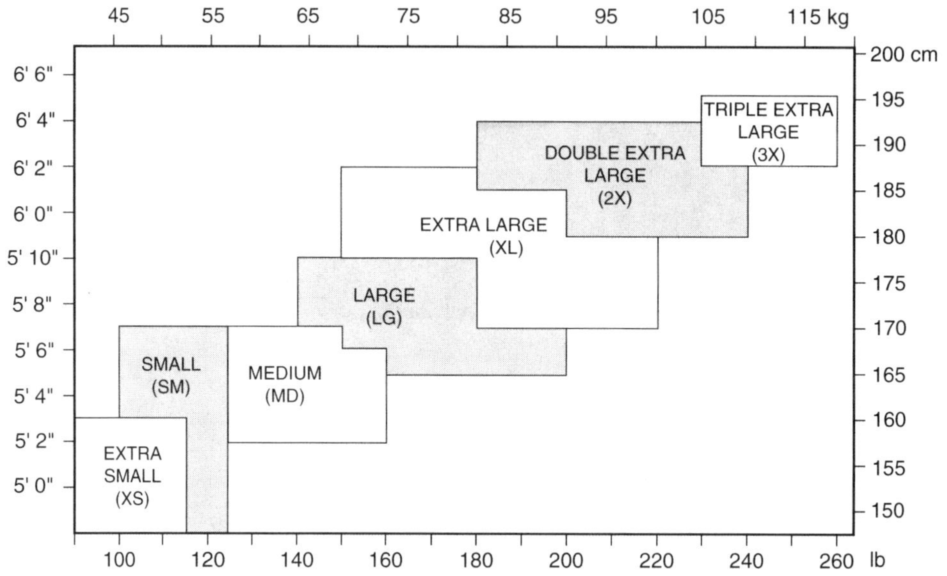

FIGURE 21.7 A typical sizing chart for limited-use or nonencapsulating garments. Sizing charts for fully encapsulating suits differ from the above and between manufacturers.[h]

Several sources of useful information on the evaluation and control of heat stress while wearing CPC are available.[35–38]

d. The importance of clothing material's resistance to cuts and tears varies with the task and setting much like it does for gloves. As with most of the previous criteria, better cut and tear resistance tends to add costs and affects the decision of whether to use short term (single-use or disposable) PPE vs. long term (multi-day or reusable) PPE.

4. Cost is a criterion for CPC selection. Basically all of the cost considerations and equations for gloves are applicable to clothing. The only significant difference is that the initial purchase price of full body CPC is much larger than that for gloves (an encapsulated ensemble can cost several hundred to a thousand dollars per set). Therefore, the cost-effectiveness of decontamination tends to be much greater for clothing than for gloves. Also, although an acceptable residual level of a contaminant is no more defined for clothing than it is for gloves, the minimal direct contact of chemical protective clothing with the wearer's skin greatly increases the accepted practice of decontaminating and reusing chemical protective clothing.

VIII. LEVELS OF PPE ENSEMBLES

Much of the development and terminology pertaining to chemical protective clothing came from the hazardous waste and emergency response industry and from 29 CFR 1910.120, especially its Appendix B. These operations are encompassed by the acronyms "HAZWOPER" for Hazardous Waste Operations and Emergency Response and "HAZMAT" for Hazardous Materials

[h] This chart is a composite of sizing charts available at www.lakeland.com/chemicalsizes.html and www.personalprotection.dupont.com/en/productServices/techinfo/index.shtml.

Response. The concept behind triage is very much in evidence within these operations. For instance, a management plan for an incident response will define at least two, and usually three, areas of control:

Hot zone — the contaminated area of the site where workers must wear the appropriate level of PPE.
Cold zone — the clean area of the site where no PPE is required.
Warm zone — the area where equipment and personnel are decontaminated in transit from the hot to the cold zones.

Controlled entry points are defined between each of the above zones. Training is required regarding the proper doffing sequence and the handling of contaminated CPC within the warm zone in order to prevent the cross-contamination of chemicals to the wearer's body or to others helping in the warm zone. Guidance for good donning and doffing procedures are available.[26] The following four distinct levels of CPC ensembles have become commonly accepted terms and concepts in IH generally.

1. Level A (encapsulation) is the highest level of protection. Level A is required for work with gases, vapors or aerosols that present a skin hazard, where large dermal doses are predictable (as from splashes or immersion), or for work with an unknown chemical hazard (emergency response). If the hazard is unknown, it is prudent to assume that the airborne hazard might be toxic to the skin; thus, Level A protection should be used until conditions permitting a lower level of protection can be assured. Required components of Level A include:
   a. A fully encapsulated, vapor protective suit. NFPA 1991 procedures require that this suit must have passed both a pressure test (exhibiting $<20\%$ drop in 3 min to be fully encapsulated) and a vapor intrusion coefficient test (<5 ppm inside after exercise in a 1000 ppm ammonia test atmosphere, although this does not necessarily protect against other gases and vapors).[39]
   b. Either a positive pressure full facepiece self-contained breathing apparatus (SCBA) or a positive pressure supplied air respirator (SAR) with an escape SCBA (generally the first option is used outside of one's normal worksite) worn inside the encapsulating suit. Respirators will be discussed at great length in the next chapter.
   c. Chemically resistant gloves (usually outer and inner layers for cut/tear protection).
   d. Chemically resistant boots with a steel toe and shank.

   Some of these components are visible in Figure 21.8. Optional Level A equipment includes coveralls or a hard hat inside the suit, disposable clothing, gloves or other gear worn outside the suit to simplify decontamination, and intrinsically safe two-way radio communication. The use of the "buddy-system" is also a recommended work practice in Level A settings.
2. Level B includes respiratory protection similar to Level A but skin protection is reduced to that used in Level C. Level B is allowed for airborne conditions that are hazardous to the respiratory system but are not hazardous to the skin (although Level B will still protect the skin from liquid contact and splashes). Level B protection is allowed in atmospheres with IDLH airborne concentrations including less than 19.5% oxygen. Required components of Level B include:
   a. A chemically resistant, hooded, but not encapsulated suit. Examples include a two-piece liquid splash protective suit, a one-piece disposable hooded coverall (as shown in Figure 21.9), or even overalls and a long-sleeved hooded jacket.
   b. Respiratory protection, gloves, and boots identical to Level A. Notice that the respirator harness in Figure 21.9 is outside the protective clothing.

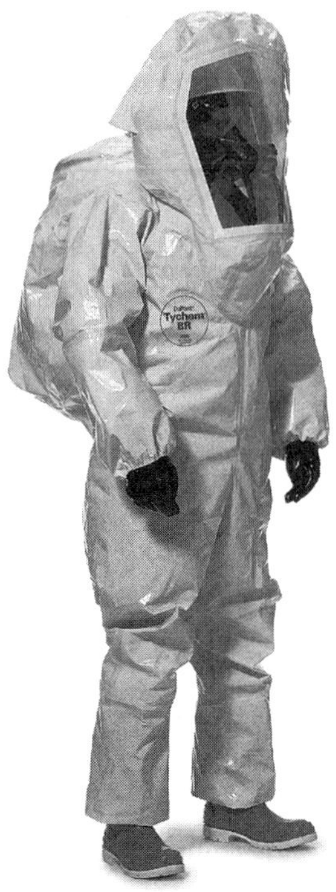

FIGURE 21.8 Level A protection: fully encapsulated with an SCBA respirator, courtesy of DuPont Corp.

FIGURE 21.9 An example of Level B protection, courtesy of DuPont Corp.

Most of the optional gear for Level B protection are also identical to Level A except that outer disposable protective gear are deleted and an optional face shield is added.

Heat stress is a major limitation for both Level A and Level B ensembles. OSHA guidelines suggest limiting body water loss due to perspiration to $\leq$ 1.5% of the individual's body weight (approximately 1 L).[38]

3. Level C includes modest dermal protection similar to Level B but respiratory protection is reduced to an air purifying respirator. The example in Figure 21.10 has a half mask respirator with goggles. Level C is only allowed for hazards that have been characterized sufficiently to permit air-purifying respirators and where liquid splashes of chemicals will not adversely affect or be absorbed through exposed skin. Conditions that allow the use of an air purifying respirator will be discussed in Section VI of Chapter 22. The required components of Level C include:
   a. Hooded chemically resistant clothing similar to Level B,
   b. An air-purifying respirator (either a full-face mask or half-mask may be acceptable),
   c. Two layers of chemically resistant gloves.

   Chemically resistant boots and all other protective gear are optional.
4. Level D allows greatly reduced dermal protection and for respiratory protection to be only optional (but no more stringent than for Level C). Level D is allowed where no significant dermal or respiratory chemical hazard is present, but where more than a basic work uniform is desirable. The only required component of Level D is coveralls. Chemically resistant gloves, boots/shoes, and respirators are all optional, as are safety glasses, chemical splash goggles, etc.

FIGURE 21.10 An example of Level C protection, courtesy of DuPont Corp.

TABLE 21.2
A Summary of the Four Levels of PPE Protection

Level of Protection	Respirator Descriptors: Air Source	Respirator Descriptors: Regulator	Max. Allowed Air Conc.[a]	Clothing Descriptors	Liquid Contact Descriptors[b]
Level A	Air is supplied	Positive pressure	$<2000 \times$ EL or $<10000 \times$ EL	Encapsulated, chemically resistant	Vapors, splash, or immersion
Level B	Air is supplied	Positive pressure	$<2000 \times$ EL or $<10000 \times$ EL	Chemically resistant	Splash only
Level C	Air purifying	None required	$<50 \times$ EL or $<10 \times$ EL	Chemically resistant	Splash only
Level D	No respirator is required			Coveralls	No toxic contact

[a] Air concentration as a multiple of the allowable exposure concentration for that chemical. Concentration must preferably be based on previous and/or real-time monitoring for a known chemical. If the concentration is unknown, it could be based on a worst case scenario such as equilibrium vapor pressure or solubility equilibrium (e.g. contaminated ground water as in Chapter 6).

[b] Degree or frequency of liquid (or vapor) contact with skin. The following scalar ranking of degree of contact has been proposed: (1) splash is a possibility (but rare and there is plenty of time available for evaporation between contacts); (2) intermittent liquid contact (infrequent and some time is available for evaporation between contacts); (3) frequent or continuous liquid contact (too little time is available for evaporation between contacts); (4) high vapor concentration of a dermally absorbable chemical (at circa $100 \times$ EL, skin absorption might approach the absorption of an unprotected lung at the airborne EL).

Table 21.2 provides a summary of the above requirements that can be used to compare the types of protection offered by the various levels of PPE protection.

5. Operations using full body PPE must be well organized. For instance, not only must any contaminated CPC be kept separate from unused or cleaned CPC, provisions must be made to keep the employee's street clothing separate from the CPC. The following other lessons can be gleaned from HAZWOPER operations and decontamination experiences that could be applied to any chemical PPE program:
   a. Minimize the need for decontamination by prior planning.
      Minimize chemical contact within the hot zone. Wear disposable outer garments. Use remote handling techniques where possible (e.g., drum grapples, tools, etc.). Avoid open pockets or folds that can serve to collect contaminants. Teach procedures to minimize wearer contact both with contaminants within the hot zone and with contaminated clothing in the warm zone to prevent the spread of chemicals. Try to protect equipment that is taken into the hot zone. For example, place air monitors (except for ports and sensors) in a protective bag or cover tools with a strippable coating.
   b. Develop a decontamination plan before exposure begins.
      Define the decontamination methods to be used and equipment needed (see the options in Section IV of this chapter.) Lay out and number the decontamination stations in stages of decreasing concentration, preferably in a straight line that branches where the workers and their CPC are separated. Upon exiting the hot zone, the primary focus at station 1 is on gross decontamination of surface contaminants to allow the user to safely doff the CPC at station 2 or 3. Conduct more extensive decontamination processes when time is less pressing. For instance, contaminants that have permeated the protective clothing may need to be decontaminated at a more

developed facility off-site. Designate a place to store unuseable, contaminated material and a method for its disposal.

c. Anticipate the effects of the following factors on the contamination of used personal protective and other equipment taken into the hot zone and/or the necessary efficiency of decontamination.
   - the concentration of the contaminant(s) in the hot zone,
   - the contact time with the contaminant(s) while in the hot zone,
   - the physical state of the contaminant (gaseous or liquid and its viscosity), and
   - the temperature (which can affect the chemical's diffusion into the protective material).
d. While it is difficult to quantify an acceptable level of decontamination, the following methods to test used clothing are available:
   - visual inspection for gross residues or chemical degradation,
   - wipe samples (can also be used to assess contamination on other surfaces[40,41]), and
   - destructive testing (extraction and chemical analysis) of representative samples.

   Biological monitoring may be used to supplement the adequacy of decontamination and the above inspection procedures.
e. Thoroughly document the PPE inspection procedures after decontamination before reuse. Examples of individual actions to be documented include:
   - the ID code of the clothing or equipment (even if assigned to an individual user),
   - the date of the inspection and the person making the inspection, and
   - the results of the inspection including any unusual conditions noted (or lack thereof).
f. Anticipate and plan appropriate emergency responses. For example, what is to be done for an injury or heat casualty to a worker in a contaminated ensemble? What first aid procedures can be carried out on an injured worker in contaminated clothing? How can contaminated clothing be safely removed from an injured worker? What protection should medical responders have?

APPENDIX TO CHAPTER 21

1910.132 — General Requirements

a. Application. Protective equipment, including personal protective equipment for eyes, face, head, and extremities, protective clothing, respiratory devices, and protective shields and barriers, shall be provided, used, and maintained in a sanitary and reliable condition wherever it is necessary by reason of hazards of processes or environment, chemical hazards, radiological hazards, or mechanical irritants encountered in a manner capable of causing injury or impairment in the function of any part of the body through absorption, inhalation or physical contact.
b. Employee-owned equipment. Where employees provide their own protective equipment, the employer shall be responsible to assure its adequacy, including proper maintenance, and sanitation of such equipment.
c. Design. All personal protective equipment shall be of safe design and construction for the work to be performed.
d. Hazard assessment and equipment selection.

   (d)(1) The employer shall assess the workplace to determine if hazards are present, or are likely to be present, which necessitate the use of personal protective equipment (PPE). If such hazards are present, or likely to be present, the employer shall:

(d)(1)(i) Select, and have each affected employee use, the types of PPE that will protect the affected employee from the hazards identified in the hazard assessment;

(d)(1)(ii) Communicate selection decisions to each affected employee; and,

(d)(1)(iii) Select PPE that properly fits each affected employee. Note: Non-mandatory Appendix B contains an example of procedures that would comply with the requirement for a hazard assessment.

(d)(2) The employer shall verify that the required workplace hazard assessment has been performed through a written certification that identifies the workplace evaluated; the person certifying that the evaluation has been performed; the date(s) of the hazard assessment; and, which identifies the document as a certification of hazard assessment.

e. Defective and damaged equipment. Defective or damaged personal protective equipment shall not be used.

f. Training.

(f)(1) The employer shall provide training to each employee who is required by this section to use PPE. Each such employee shall be trained to know at least the following:

(f)(1)(i) When PPE is necessary;

(f)(1)(ii) What PPE is necessary;

(f)(1)(iii) How to properly don, doff, adjust, and wear PPE;

(f)(1)(iv) The limitations of the PPE; and,

(f)(1)(v) The proper care, maintenance, useful life and disposal of the PPE.

(f)(2) Each affected employee shall demonstrate an understanding of the training specified in paragraph (f)(1) of this section, and the ability to use PPE properly, before being allowed to perform work requiring the use of PPE.

(f)(3) When the employer has reason to believe that any affected employee who has already been trained does not have the understanding and skill required by paragraph (f)(2) of this section, the employer shall retrain each such employee. Circumstances where retraining is required include, but are not limited to, situations where:

(f)(3)(i) Changes in the workplace render previous training obsolete; or

(f)(3)(ii) Changes in the types of PPE to be used render previous training obsolete; or

(f)(3)(iii) Inadequacies in an affected employee's knowledge or use of assigned PPE indicate that the employee has not retained the requisite understanding or skill.

(f)(4) The employer shall verify that each affected employee has received and understood the required training through a written certification that contains the name of each employee trained, the date(s) of training, and that identifies the subject of the certification.

g. Paragraphs (d) and (f) of this section apply only to 1910.133, 1910.135, 1910.136, and z1910.138. Paragraphs (d) and (f) of this section do not apply to 1910.134 and 1910.137.

[59 FR 16360, April 6, 1994; 59 FR 33910, July 1, 1994; 59 FR 34580, July 6, 1994]

1910.138 — Hand Protection

a. General requirements. Employers shall select and require employees to use appropriate hand protection when employees' hands are exposed to hazards such as those from skin absorption of harmful substances; severe cuts or lacerations; severe abrasions; punctures; chemical burns; thermal burns; and harmful temperature extremes.

b. Selection. Employers shall base the selection of the appropriate hand protection on an evaluation of the performance characteristics of the hand protection relative to the task(s) to be performed, conditions present, duration of use, and the hazards and potential hazards identified.

[59 FR 16362, April 6, 1994]

1910 Subpart I Appendix B — Non-Mandatory Compliance Guidelines for Hazard Assessment and Personal Protective Equipment Selection

This Appendix is intended to provide compliance assistance for employers and employees in implementing requirements for a hazard assessment and the selection of personal protective equipment.

1. Controlling hazards. PPE devices alone should not be relied on to provide protection against hazards, but should be used in conjunction with guards, engineering controls, and sound manufacturing practices.
2. Assessment and selection. It is necessary to consider certain general guidelines for assessing the foot, head, eye and face, and hand hazard situations that exist in an occupational or educational operation or process, and to match the protective devices to the particular hazard. It should be the responsibility of the safety officer to exercise common sense and appropriate expertise to accomplish these tasks.
3. Assessment guidelines. In order to assess the need for PPE the following steps should be taken:
   a. Survey. Conduct a walk-through survey of the areas in question. The purpose of the survey is to identify sources of hazards to workers and co-workers. Consideration should be given to the basic hazard categories:
      a. Impact
      b. Penetration
      c. Compression (roll-over)
      d. Chemical
      e. Heat
      f. Harmful dust
      g. Light (optical) radiation
   b. Sources. During the walk-through survey the safety officer should observe:
      a. sources of motion; i.e., machinery or processes where any movement of tools, machine elements or particles could exist, or movement of personnel that could result in collision with stationary objects;
      b. sources of high temperatures that could result in burns, eye injury or ignition of protective equipment, etc.;
      c. types of chemical exposures;
      d. sources of harmful dust;
      e. sources of light radiation, i.e., welding, brazing, cutting, furnaces, heat treating, high intensity lights, etc.;
      f. sources of falling objects or potential for dropping objects;
      g. sources of sharp objects which might pierce the feet or cut the hands;
      h. sources of rolling or pinching objects which could crush the feet;
      i. layout of workplace and location of co-workers; and
      j. any electrical hazards. In addition, injury/accident data should be reviewed to help identify problem areas.

c. Organize data. Following the walk-through survey, it is necessary to organize the data and information for use in the assessment of hazards. The objective is to prepare for an analysis of the hazards in the environment to enable proper selection of protective equipment.

d. Analyze data. Having gathered and organized data on a workplace, an estimate of the potential for injuries should be made. Each of the basic hazards (paragraph 3.(a).) should be reviewed and a determination made as to the type, level of risk, and seriousness of potential injury from each of the hazards found in the area. The possibility of exposure to several hazards simultaneously should be considered.

4. Selection guidelines. After completion of the procedures in paragraph 3, the general procedure for selection of protective equipment is to:
   a. Become familiar with the potential hazards and the type of protective equipment that is available, and what it can do; i.e., splash protection, impact protection, etc.;
   b. compare the hazards associated with the environment; i.e., impact velocities, masses, projectile shape, radiation intensities, with the capabilities of the available protective equipment;
   c. select the protective equipment which ensures a level of protection greater than the minimum required to protect employees from the hazards; and
   d. fit the user with the protective device and give instructions on care and use of the PPE. It is very important that end users be made aware of all warning labels for and limitations of their PPE.
5. Fitting the device. Careful consideration must be given to comfort and fit. PPE that fits poorly will not afford the necessary protection. Continued wearing of the device is more likely if it fits the wearer comfortably. Protective devices are generally available in a variety of sizes. Care should be taken to ensure that the right size is selected.
6. Devices with adjustable features. Adjustments should be made on an individual basis for a comfortable fit that will maintain the protective device in the proper position. Particular care should be taken in fitting devices for eye protection against dust and chemical splash to ensure that the devices are sealed to the face. In addition, proper fitting of helmets is important to ensure that it will not fall off during work operations. In some cases a chin strap may be necessary to keep the helmet on an employee's head. (Chin straps should break at a reasonably low force, however, so as to prevent a strangulation hazard). Where manufacturer's instructions are available, they should be followed carefully.
7. Reassessment of hazards. It is the responsibility of the safety officer to reassess the workplace hazard situation as necessary, by identifying and evaluating new equipment and processes, reviewing accident records, and reevaluating the suitability of previously selected PPE.
8. Selection chart guidelines for eye and face protection. Some occupations (not a complete list) for which eye protection should be routinely considered are: carpenters, electricians, machinists, mechanics and repairers, millwrights, plumbers and pipe fitters, sheet metal workers and tinsmiths, assemblers, sanders, grinding machine operators, lathe and milling machine operators, sawyers, welders, laborers, chemical process operators and handlers, and timber cutting and logging workers. The following chart provides general guidance for the proper selection of eye and face protection to protect against hazards associated with the listed hazard source operations.

Eye and Face Protection Selection Chart

Source	Assessment of Hazard	Protection
IMPACT — Chipping, grinding machining, masonry work, woodworking, sawing, drilling, chiseling, powered fastening, riveting, and sanding	Flying fragments, objects, large chips, particles, sand, dirt, etc.	Spectacles with side protection, goggles, face shields[a,c,e,f,j] For severe exposure, use faceshields
HEAT — Furnace operations, pouring, casting, hot dipping, and welding	Hot sparks	Faceshields, goggles, spectacles with side protection. For severe exposure use faceshield[a,b,c]
	Splash from molten metals	Faceshields worn over goggles[a,b,c]
	High temperature exposure	Screen face shields, reflective face shields[a,b,c]
CHEMICALS — Acid and chemicals handling, degreasing plating	Splash	Goggles, eyecup and cover types. For severe exposure, use face shield[c,k]
	Irritating mists	Special purpose goggles
DUST — Woodworking, buffing, general dusty conditions	Nuisance dust	Goggles, eyecup and cover types[h]
LIGHT and/or RADIATION — Welding: Electric arc	Optical radiation	Welding helmets or welding shields. Typical shades: 10–14[i,l]
Welding: Gas	Optical radiation	Welding goggles or welding face shield. Typical shades: gas welding 4–8, cutting 3–6, brazing 3–4[i]
Cutting, Torch brazing, Torch soldering	Optical radiation	Spectacles or welding face-shield. Typical shades: 1.5–3[c,i]
Glare	Poor vision	Spectacles with shaded or special-purpose lenses, as suitable[i,j]

[a] Care should be taken to recognize the possibility of multiple and simultaneous exposure to a variety of hazards. Adequate protection against the highest level of each of the hazards should be provided. Protective devices do not provide unlimited protection.

[b] Operations involving heat may also involve light radiation. As required by the standard, protection from both hazards must be provided.

[c] Faceshields should only be worn over primary eye protection (spectacles or goggles).

[d] As required by the standard, filter lenses must meet the requirements for shade designations in 1910.133(a)(5). Tinted and shaded lenses are not filter lenses unless they are marked or identified as such.

[e] As required by the standard, persons whose vision requires the use of prescription (Rx) lenses must wear either protective devices fitted with prescription (Rx) lenses or protective devices designed to be worn over regular prescription (Rx) eyewear.

[f] Wearers of contact lenses must also wear appropriate eye and face protection devices in a hazardous environment. It should be recognized that dusty and/or chemical environments may represent an additional hazard to contact lens wearers.

[g] Caution should be exercised in the use of metal frame protective devices in electrical hazard areas.

[h] Atmospheric conditions and the restricted ventilation of the protector can cause lenses to fog. Frequent cleansing may be necessary.

[i] Welding helmets or faceshields should be used only over primary eye protection (spectacles or goggles).

[j] Nonsideshield spectacles are available for frontal protection only, but are not acceptable eye protection for the sources and operations listed for impact.

[k] Ventilation should be adequate, but well protected from splash entry. Eye and face protection should be designed and used so that it provides both adequate ventilation and protects the wearer from splash entry.

[l] Protection from light radiation is directly related to filter lens density. See note (d). Select the darkest shade that allows task performance.

9. Selection guidelines for head protection. All head protection (helmets) is designed to provide protection from impact and penetration hazards caused by falling objects. Head protection is also available which provides protection from electric shock and burn. When selecting head protection, knowledge of potential electrical hazards is important.

Class A helmets, in addition to impact and penetration resistance, provide electrical protection from low-voltage conductors (they are proof tested to 2200 V). Class B helmets, in addition to impact and penetration resistance, provide electrical protection from high-voltage conductors (they are proof tested to 20,000 V). Class C helmets provide impact and penetration resistance (they are usually made of aluminum which conducts electricity), and should not be used around electrical hazards.Where falling object hazards are present, helmets must be worn. Some examples include: working below other workers who are using tools and materials which could fall; working around or under conveyor belts which are carrying parts or materials; working below machinery or processes which might cause material or objects to fall; and working on exposed energized conductors. Some examples of occupations for which head protection should be routinely considered are: carpenters, electricians, linemen, mechanics and repairers, plumbers and pipe fitters, assemblers, packers, wrappers, sawyers, welders, laborers, freight handlers, timber cutting and logging, stock handlers, and warehouse laborers.

10. Selection guidelines for foot protection. Safety shoes and boots which meet the ANSI Z41-1991 Standard provide both impact and compression protection. Where necessary, safety shoes can be obtained which provide puncture protection. In some work situations, metatarsal protection should be provided, and in other special situations electrical conductive or insulating safety shoes would be appropriate. Safety shoes or boots with impact protection would be required for carrying or handling materials such as packages, objects, parts or heavy tools, which could be dropped; and, for other activities where objects might fall onto the feet. Safety shoes or boots with compression protection would be required for work activities involving skid trucks (manual material handling carts) around bulk rolls (such as paper rolls) and around heavy pipes, all of which could potentially roll over an employee's feet. Safety shoes or boots with puncture protection would be required where sharp objects such as nails, wire, tacks, screws, large staples, scrap metal etc., could be stepped on by employees causing a foot injury.

    Some occupations (not a complete list) for which foot protection should be routinely considered are: shipping and receiving clerks, stock clerks, carpenters, electricians, machinists, mechanics and repairers, plumbers and pipe fitters, structural metal workers, assemblers, drywall installers and lathers, packers, wrappers, craters, punch and stamping press operators, sawyers, welders, laborers, freight handlers, gardeners and grounds-keepers, timber cutting and logging workers, stock handlers and warehouse laborers.

11. Selection guidelines for hand protection. Gloves are often relied upon to prevent cuts, abrasions, burns, and skin contact with chemicals that are capable of causing local or systemic effects following dermal exposure. OSHA is unaware of any gloves that provide protection against all potential hand hazards, and commonly available glove materials provide only limited protection against many chemicals. Therefore, it is important to select the most appropriate glove for a particular application and to determine how long it can be worn, and whether it can be reused.

    It is also important to know the performance characteristics of gloves relative to the specific hazard anticipated; e.g., chemical hazards, cut hazards, flame hazards, etc. These performance characteristics should be assessed by using standard test procedures. Before purchasing gloves, the employer should request documentation from the manufacturer that the gloves meet the appropriate test standard(s) for the hazard(s) anticipated. Other factors to be considered for glove selection in general include:

    (A) As long as the performance characteristics are acceptable, in certain circumstances, it may be more cost effective to regularly change cheaper gloves than to reuse more expensive types; and,

(B) The work activities of the employee should be studied to determine the degree of dexterity required, the duration, frequency, and degree of exposure of the hazard, and the physical stresses that will be applied.

With respect to selection of gloves for protection against chemical hazards:

(A) The toxic properties of the chemical(s) must be determined; in particular, the ability of the chemical to cause local effects on the skin and/or to pass through the skin and cause systemic effects;
(B) Generally, any chemical resistant glove can be used for dry powders;
(C) For mixtures and formulated products (unless specific test data are available), a glove should be selected on the basis of the chemical component with the shortest breakthrough time, since it is possible for solvents to carry active ingredients through polymeric materials; and,
(D) Employees must be able to remove the gloves in such a manner as to prevent skin contamination.

Cleaning and maintenance. It is important that all PPE be kept clean and properly maintained. Cleaning is particularly important for eye and face protection where dirty or fogged lenses could impair vision. For the purposes of compliance with 1910.132 (a) and (b), PPE should be inspected, cleaned, and maintained at regular intervals so that the PPE provides the requisite protection. It is also important to ensure that contaminated PPE which cannot be decontaminated is disposed of in a manner that protects employees from exposure to hazards.

[59 FR 16362, April 6, 1994]

REFERENCES

1. Roach, S. A., On assessment of hazards to health at work, *Am. Ind. Hyg. Assoc. J.*, 55, 1125–1130, 1994.
2. Chap. 29 of the U.S. Code of Federal Regulations, Part Number 1910, Occupational Safety and Health Standards.
3. Bernard, A. and Lauwerys, R., Biological monitoring of exposure to industrial chemicals, In *Occupational Health Practice*, 3rd ed., Waldron, H. A., Ed., Butterworths, London, pp. 203–216, 1989.
4. Aprea, C., Colosio, C., Mammone, T., Minoia, C., and Maroni, M., Biological monitoring of pesticide exposure: a review of analytical methods, *J. Chromatogr. B.*, 769(2), 191–219, 2002.
5. Teass, A. W., Biagini, R. E., De Bord, G., and Hull, R. D., *Application of Biological Monitoring Methods, NIOSH Manual of Analytical Methods*, 4th ed., National Institute for Occupational Safety and Health, Cincinnati, OH, 2003, Third supplement, pp. 52–62.
6. Harper, M., Assessing workplace chemical exposures: the role of exposure monitoring, *J. Environ. Monit.*, 6(5), 404–412, 2004.
7. ASTM F1461-93(2005), *Standard Practice for Chemical Protective Clothing Program*, ASTM International, West Conshohocken, PA, 2005.
8. Grubner, O. and Burgess, W. A., Simplified description of adsorption breakthrough curves in air cleaning and sampling devices, *Am. Ind. Hyg. Assoc. J.*, 40, 169–179, 1979.
9. ASTM F739-99, *Standard Test Method for Resistance or Protective Clothing Materials to Permeation of Liquids and Gases Under Conditions of Continuous Contact*, ASTM International, West Conshohocken, PA, 1999.
10. Laughlin, J. and Gold, R. E., Evaporative dissipation of methyl parathion from laundered protective apparel fabrics, *Bull. Environ. Contam. Toxicol.*, 42(4), 566–573, 1989.
11. Perkins, J. L., Decontamination of protective clothing, *Appl. Occup. Environ. Hyg.*, 6(1), 29–35, 1991.
12. Raheel, M., Protective clothing: an overview, In *Protective Clothing Systems and Materials*, Raheel, M., Ed., Marcel Dekker, Inc., New York, pp. 1–23, 1994.

13. *Threshold Limit Values for Chemical Substances and Physical Agents*, ACGIH Worldwide, Cincinnati, OH, 2005.
14. Bronaugh, R. and Maibach, H. I., *Percutaneous Absorption*, Marcel Dekker, New York, 1989.
15. Fiserova-Bergerova, V., Relevance of occupational skin exposure, *Ann. Occup. Hyg.*, 37(6), 673–685, 1993.
16. Guy, R. H. and Potts, R. O., Penetration of industrial chemicals across the skin: a predictive model, *Am. J. Ind. Med.*, 23(5), 711–719, 1993.
17. Auton, T. R., Westhead, D. R., Woollen, B. H., Scott, R. C., and Wilks, M. F., A physiologically based mathematical model of dermal absorption in man, *Human Exp. Toxicol.*, 13(1), 51–60, 1994.
18. Wilschut, A., ten Berge, W. F., Robinson, P. J., and McKone, T. E., Estimating skin permeation. The validation of five mathematical skin permeation models, *Chemosphere*, 30(7), 1275–1296, 1995.
19. *Documentation of the Threshold Limit Values for Chemical Substances*, 7th ed., ACGIH Worldwide, Cincinnati, OH, 2001.
20. McArthur, B., Dermal measurement and wipe sampling methods: a review, *Appl. Occup. Environ. Hyg.*, 7(9), 599–606, 1992.
21. Fenske, R. A., Dermal Exposure Assessment Techniques, *Ann. Occup. Hyg.*, 37(6), 687–706, 1993.
22. Ness, S. A., *Surface and Dermal Monitoring for Toxic Exposures*, Van Nostrand Reinhold, New York, 1994.
23. El-Ayouby, N. S., Berardinelli, S. P., and Hall, R. C., Evaluation of the permea-tec pads as new technology for the detection of chemical breakthrough in PPC, *Am. J. Ind. Med. Suppl.*, 1, 128–129, 1999.
24. Mellström, G. A., Wahlberg, J. E., and Maibach, H. I., *Protective Gloves for Occupational Use*, CRC Press, Boca Raton, FL, 1994.
25. Schwope, A. D., Goydan, R., Reid, R. C., and Krishnamurthy, S., State-of-the-art review of permeation testing and the interpretation of its results, *Am. Ind. Hyg. Assoc. J.*, 49(11), 557–565, 1988.
26. *OSHA Technical Information Manual*, Section VIII, chap. 1, Chemical Protective Clothing, at www.osha-slc.gov/dts/osta/otm/otm_viii/otm_viii_1.html#3.].
27. Schwope, A. D. and Renard, E. P., Estimation of the cost of using chemical protective clothing, *Performance of Protective Clothing*, Vol. 4, McBriarty, J. P. and Henry, N. W., Eds., American Society for Testing and Materials, Philadelphia, PA, ASTM STP, 1133, pp. 972–981, 1992.
28. Kütting, B. and Drexler, H., Effectiveness of skin protection creams as a preventive measure in occupational dermatitis: a critical update according to criteria of evidence-based medicine, *Int. Arch. Occup. Environ. Health*, 76(4), 253–259, 2003.
29. Kresken, J. and Klotz, A., Occupational skin-protection products: a review, *Int. Arch. Occup. Environ. Health*, 76(5), 355–358, 2003.
30. Korinth, G., Geh, S., Schaller, K. H., and Drexler, H., In vitro evaluation of the efficacy of skin barrier creams and protective gloves on percutaneous absorption of industrial solvents, *Int. Arch. Occup. Environ. Health*, 76(5), 382–386, 2003.
31. Perkins, J. L., Chemical protective clothing: I. Selection and use, *Appl. Ind. Hyg.*, 2(6), 222–230, 1987.
32. Perkins, J. L., Chemical protective clothing: II. Program considerations, *Appl. Ind. Hyg.*, 3(1), 1–4, 1988.
33. NIOSH *Pocket Guide to Chemical Hazards*, DHHS (NIOSH) 97–140, 1997. (Also on-line at www.cdc.gov/niosh/npg/npg.html).
34. Forsberg, K. and Mansdorf, S. Z., *Quick Selection Guide to Chemical Protective Clothing*, 3rd ed., Van Nostrand Reinhold, New York, 1993, was cited by NIOSH but see also the 4th ed. John Wiley & Sons, New York, 2003.
35. Beaird, J. S., Bauman, T. R., and Leeper, J. D., Oral and tympanic temperatures as heat strain indicators for workers wearing protective clothing, *Am. Ind. Hyg. Assoc. J.*, 57(4), 344–347, 1996.
36. Reneau, P. D. and Bishop, P. A., A review of the suggested wet bulb globe temperature adjustment for encapsulated protective clothing, *Am. Ind. Hyg. Assoc. J.*, 57(1), 58–61, 1996.
37. Reneau, P. D. and Bishop, P. A., Validation of a personal heat stress monitor, *Am. Ind. Hyg. Assoc. J.*, 57(7), 650–657, 1996.

38. *OSHA Technical Information Manual*, Section III, chap. 4, Heat Stress, at www.osha-slc.gov/dts/osta/otm/otm_iii/otm_iii_4.html.
39. NFPA, 1991, *Standard on Vapor-Protective Ensembles for Hazardous Materials Emergencies*, National Fire Protection Association, Quincy, MA, 2005.
40. Caplan, K. J., The significance of wipe samples, *Am. Ind. Hyg. Assoc. J.*, 54(2), 70–75, 1993.
41. Klingner, T. D. and McCorkle, T., The application and significance of wipe samples, *Am. Ind. Hyg. Assoc. J.*, 55(3), 251–254, 1994.

22 Respirator Controls

The learning goals of this chapter:

- ☐ Know the definition of the generic protection factor.
- ✍ Be able to differentiate an atmosphere supplying respirator (ASR) from an air purifying respirator (APR).
- ✍ Be familiar with the various respiratory inlet covering designs and how they differ.
- ☐ Be able to calculate the effect on WPF of not wearing a respirator all the time (ca., Equation 22.7).
- ✍ Know the basic limitations of air purifying respirators that make an ASR the default selection.
- ✍ Know how APF differs from QNFF, and how APF is used to select a respirator.
- ✍ Be familiar with the respirator characteristics that have the greatest influence on protection factors.
- ✍ Know how qualitative fit testing differs from quantitative fit testing.
- ✍ Understand the respirator decision logic. Be able to define and know how the terms oxygen deficiency, IDLH, and "objective information or data" (regarding service life) are used in respirator selection.

The first section of this chapter introduces the terms and concepts that underlie the design, selection, and use of respirators. The next three sections (Section II to Section IV) elaborate on specific ways that the protection afforded by respirators can be measured, affected (made worse or improved), and tested. Section V summarizes the requirements of a good respiratory management program, and the last section takes the reader through a formal process (protocol) by which an appropriate respirator can be selected.

I. RESPIRATOR TERMS AND CONCEPTS

Respiratory protection involves a whole raft of terms and enough abbreviations to earn its own page of abbreviations in the front of this book. The first subsection introduces the concept of protection factors. The next several subsections try to explain a variety of terms that can be used to differentiate or categorize respirators. The last three subsections describe additional features of some respirators. The following respirator terms can also be found in the OSHA Respiratory Protection standard 1910.134,[1] in NIOSH's "Definition of Terms" for their certified equipment (at http://www2.cdc.gov/drds/cel/def.htm), or elsewhere.[2,3]

1. Respirators are worn either to protect the wearer from a toxic chemical and/or to provide oxygen. Of the two reasons, the most common reason is to reduce the wearer's exposure to a toxic airborne chemical in the ambient air by reducing the airborne concentration in

the wearer's breathing zone.[a] This generic reduction in concentration is called the respirator's protection factor and is defined quantitatively by Equation 22.1. A simple concept to learn, and an important one to remember.

$$\text{protection factor} = \frac{\text{contaminant concentration } \textit{outside} \text{ the respirator}}{\text{contaminant concentration } \textit{inside} \text{ the respirator}} \tag{22.1}$$

Each of the following four specific respirator protection factors is based on the above generic ratio of the concentration of airborne contaminant outside the respirator (in the ambient air) to the concentration inside the respirator (within the wearer's breathing zone). These specific protection factors will be discussed in more detail in the next section of this chapter.

a. Workplace protection factor (WPF) is the protection factor measured on an individual only while they are wearing a respirator continuously while at work in the wearer's usual workplace.
b. Effective protection factor (EPF) is the protection factor measured in a real workplace as above but includes time when the wearer takes the liberty to not wear the respirator continuously. Any period of time when the respirator is not worn will naturally reduce the EPF below what the WPF would have been had the respirator been worn continuously.
c. Assigned protection factor (APF) is determined by an administrative process after a review and statistical analysis of multiple studies of the protection afforded to people wearing respirators within a defined class or category. The same APF value is assigned to all respirators within each defined category or class. The contaminant concentration in the breathing zone of nearly all properly fitted and trained users of a properly functioning respirator should be reduced by at least the numeric value of the APF for that class of respirators.
d. Fit factor (abbreviated FF herein or "ff" by OSHA) is the protection factor determined on an individual wearing a particular respirator in a controlled (laboratory-like) setting, usually as part of selecting an acceptable tight-fitting facepiece for that individual. A quantitative fit test (QNFT) will result in a quantitative fit factor (QNFF) for that person and respirator. Someone who passes a qualitative fit test with a given respirator can be assumed to have an FF of at least 100.

2. "Penetration" has the same qualitative meaning for respirators as it had in the previous chapter, referring in this case to the portion of the contaminant that reaches the breathing zone by passing around the respirator rather than permeating through it or its air purifier. Equation 22.2 defines penetration in respirators quantitatively as the ratio of the contaminant concentration inside the respirator to the breathing zone concentration outside the respirator. Thus, penetration can be viewed as the inverse of the protection factor although it really implies a specific mechanism.

$$\text{“Penetration”} = \frac{C_{\text{inside}}}{C_{\text{outside}}} = \frac{1}{\text{the protection factor}} \tag{22.2}$$

[a] If oxygen is low, providing oxygen is vital and therefore not of secondary importance. OSHA requires virtually maximum protection in an oxygen deficient atmosphere, even though any atmosphere supplying a respirator provided with normal breathing air (oxygen content $= 20.8\%$) only needs to be capable of generating a protection factor of 16 to be able to provide a wearer with at least 19.5% oxygen in their breathing zone even with no oxygen outside the respirator.

Penetration is primarily the result of leakage. The latter term implies a process while the former term describes a result. The term penetration is not used often in respirators despite it being the most common pathway by which toxins reach the breathing zone of the wearer (in comparison to the chemical passing through the respirator's air purifier). Equation 22.2 will be mentioned in a subsequent discussion on Fit Factor (FF), circa Equation 22.3.[b]

3. A respiratory inlet covering is the broad term describing the part of a respirator that covers and/or protects the wearer's breathing zone. This basic term will be used many times before the six or seven different physical designs of such respiratory inlet coverings are discussed in subsections 8 and 9.

However, one cannot categorize a respirator by just its respiratory inlet covering alone. Respirators are categorized from multiple aspects that will be discussed below. One aspect is whether the source of the (breathing) air inside the respirator is clean air with oxygen (referred to as an atmosphere supplying respirator in Section I.4) or ambient air from which the toxic contaminants have been removed (referred to as an air purifying respirator in Section I.5). Another aspect is whether the air pressure inside the respiratory inlet covering, while the wearer is inhaling, is either below the ambient pressure outside of the respirator (referred to as a negative pressure respirator in Section I.6) or above it (referred to as a positive pressure respirator in Section I.7). A third aspect is whether the respiratory inlet covering is loose-fitting (such as a helmet or hood that at least covers the wearer's face in Section I.8) or tight-fitting (as in a mask or mouth insert to be described in Section I.9).

4. Atmosphere supplying respirators, as their name implies, provide an external supply of breathable air into the respirator. An atmosphere supplying respirator is the only respirator that provides oxygen, and thus it is the only respirator that can be used in an oxygen deficient atmosphere. There are two broad subtypes of atmosphere supplying respirators.
   a. A self-contained breathing apparatus (SCBA) is a respirator in which the supplied atmosphere can be carried or worn by the user. (Both the name and the technology are very similar to SCUBA gear used for underwater diving.) SCBA systems can be either open circuit (vastly the more common type of SCBA) or closed circuit.
      1. The atmosphere for an open circuit SCBA comes from a compressed air tank (2000 to 4500 psi) carried by the wearer, usually back-pack style (a small tank can be mounted on the wearer's belt). Figure 22.1 shows a wearer that has donned an SCBA from a walkaway mounting bracket; a second SCBA is ready to the right. The wearers of ensembles in Figure 21.7 and Figure 21.8 were also wearing SCBAs. The exhaled air in all open circuit SCBA's is exhausted back to the ambient atmosphere. That exhaust tends to inflate a Level A ensemble slightly. An open circuit SCBA will include a pressure regulator and a low pressure alarm to warn the user when the air supply tank is about to be exhausted.
      2. A closed circuit SCBA captures the wearer's exhaled air from which the carbon dioxide is scrubbed (removed) and to which either stored or chemically generated oxygen is added before it is rebreathed. Hence, a closed circuit SCBA is sometimes called a rebreather. Notice the two hoses in Figure 22.2 for recirculating the breathing air. Closed circuit SCBAs are used mostly for rescue or other special needs where the desired duration of the air supply would exceed

[b] The efficiency of a respirator is related to but different from any of the above terms and is no longer recommended as a term to describe respirator performance.

FIGURE 22.1 An example of an SCBA in use and another one ready for fast response. Courtesy of MSA.

FIGURE 22.2 A closed circuit SCBA, courtesy of Draeger Safety, Inc.

the maximum feasible size of an open circuit storage tank and/or where the urgency makes such a more costly system cost effective.

b. A supplied air respirator (SAR) or airline respirator provides breathing air under pressure via a hose to the user from a source that is not carried by the user.[c] Examples can be seen in Figure 22.4b and Figure 22.5c. For routine or continuous uses, the supplied air would come from a compressor located in clean (uncontaminated) air, although air can be supplied for a short time by one or more compressed air tanks too large to be worn as part of an SCBA (e.g., tanks mounted on a hand truck). The breathing air for a SAR supplied under a pressure of up to 125 psi must be reduced by a pressure regulator before reaching the respiratory inlet covering (similar to the regulator on an SCBA). As will be described shortly, the pressure may be regulated either in a continuous mode for helmets, hoods, and other loose-fitting respirators or in a pressure demand mode for tight-fitting respirators. If a SAR is used in an IDLH setting,[d] the wearer must also have an escape SCBA as a backup system.

5. An air purifying respirator (APR) removes the contaminant from the ambient air before it reaches the breathing zone. A purifier can be a filter for aerosols, a chemical absorber for a gas or vapor, or both a filter and absorber. There are also two broad subtypes of air purifying respirators.
   a. In a conventional air purifying respirator, the wearer's lungs do the work of drawing the air through the purifier and into a tight-fitting respiratory inlet covering. There is no widely accepted single term for the conventional tight-fitting air purifying respirators to be shown in Figure 22.7, Figure 22.8, and Figure 22.9. One term could be an unpowered air purifying respirator (cf., the powered air purifying respirator described below). The term "conventional air purifying respirator" could also work (and it would exclude mouthpiece respirators like those shown in Figure 22.10). Because the intra-mask pressure during inhalation into any conventional respirator is always less than the atmospheric pressure outside the respirator, any leakage between the mask and the wearer's face is inward. Thus, a conventional air purifying respirator is a negative pressure respirator (although the latter term also applies to an atmosphere supplying respirator used with a demand air flow regulator).
   b. In a powered air-purifying respirator (PAPR), a portable air pump or fan (either integral to the respirator or worn elsewhere on the user like on a belt) draws the air through the purifier and supplies it to the facepiece under a slight positive pressure. All existing PAPRs operate in a continuous flow mode (see Section I.10.a below). PAPRs are available with either, a loose-fitting facepiece (Figure 22.3a), a tight-fitting full face mask (Figure 22.3b and c), a hood (Figure 22.5b), or a tight-fitting half-mask facepiece (shown in conventional form in Figure 22.8). The slight positive pressure inside of a PAPR's respiratory inlet covering results in very little inward leakage and more protection (a higher protection factor), but the higher airflow rate through a PAPR's air cleaner results in a shorter service life for that air purifier than if the same air purifier only had to clean the air the wearer was going to inhale (as on a conventional air purifying respirator).

[c] A hose mask respirator is an old term for a tight fitting Type B respirator with no compressor or air blower. Air was drawn through ≤ 50 ft of hose with check valves just by the wearer's breathing; therefore, it needed no regulator. Again, such respirators are no longer marketed in the U.S.A. although some may still be legally in service.

[d] IDLH stands for a concentration that is immediately dangerous to life or health. NIOSH has assigned IDLH concentrations to many chemicals in its Pocket Guide to Chemical Hazards (accessible on-line at www.cdc.gov/niosh/npg/npg.html).[4]

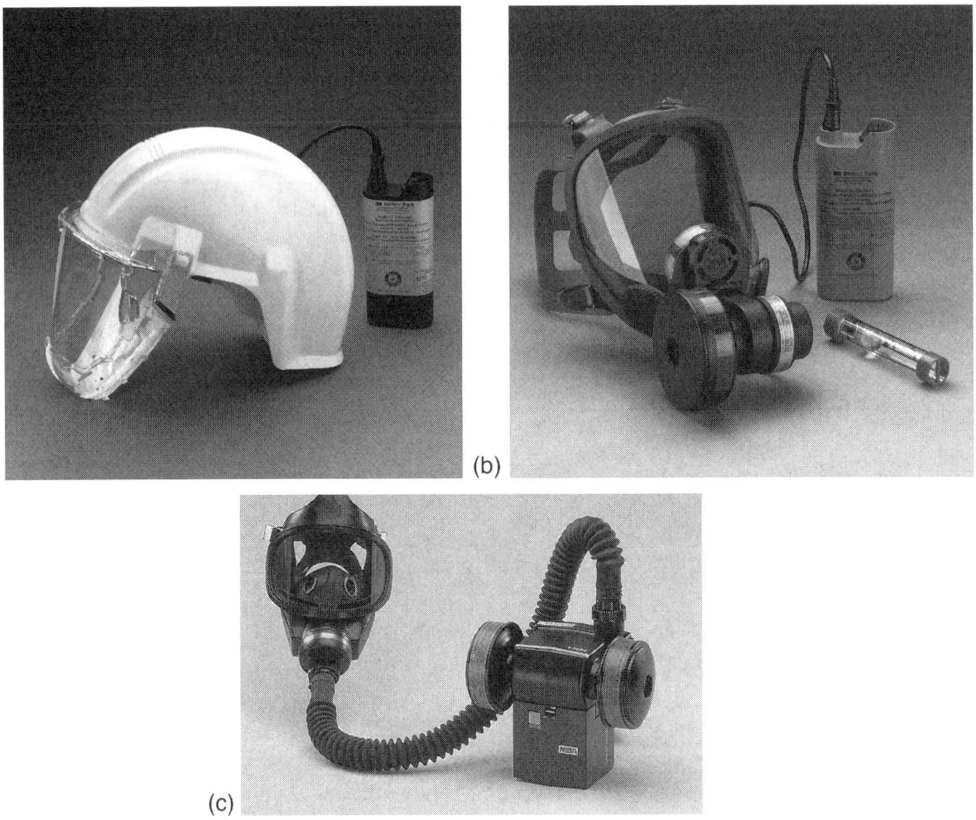

(a) (b) (c)

FIGURE 22.3a, b, c Examples of powered air purifying respirators configured as a loose-fitting facepiece with a filter built into the cap on the left, the air purifier mounted on a full face mask in the middle, and the air purifier mounted on a belt pack on the right. From left to right, courtesy of 3M™, 3M™, and MSA.

Respirators can achieve their protection factor in either of two distinctly different ways: either structurally (by creating a tight-fitting barrier between the ambient contaminant and the wearer's breathing zone that fits tightly to the wearer's face or mouth) or aerodynamically (by creating and maintaining an outward flow of air from the wearer's breathing zone back toward the contaminated ambient air). Some respirators use both methods to protect the wearer.

6. All negative pressure respirators rely completely on some form of structural barrier that physically separates the ambient air from the wearer's breathing zone. All negative pressure respirators are tight-fitting (held closely against the wearer's face or mouth). A conventional respirator is a tight-fitting mask that fits onto the wearer's face around their mouth and nose to create a pocket of safely breathable air inside of the mask (e.g., Figure 22.7, Figure 22.8, Figure 22.9, and Figure 22.11). Each time the wearer inhales through a conventional respirator, the pressure within the mask falls below the atmospheric pressure outside of the mask. Because of this negative pressure difference, any leakage between the mask and the wearer's face is inward. The protection factor created by any negative pressure respirator is directly proportional to the goodness of fit of that particular respirator to an individual wearer's face.
7. All positive pressure respirators rely in whole or at least in part on the outward flow of air to keep contaminants away from the wearer's breathing zone. A positive pressure respirator may be either tight-fitting (held against the face or mouth) or loose-fitting

(not held against the face or mouth). Positive pressure respirators require some form of pressure regulator to be discussed in Section I.10.

All loose-fitting respirators rely on the outward flow of air at a fairly high volume to keep contaminants out of the breathing zone. Loose-fitting respirators tend to be more comfortable than tight fitting respirators, but the large air flow rate that they require would soon deplete the amount of air that can be stored in a portable air tank, which therefore limits loose-fitting respirators to either a respirator connected to a remote compressor via an air line (hose) or to a respirator that blows ambient air through an air purifier before it reaches the breathing zone.

A good tight-fitting, positive pressure facepiece minimizes that outward air flow enough to allow an air tank on a backpack to supply a user with clean breathable air for 30 to 45 min. By relying on both a good seal and outward air flow, a tight fitting positive pressure respirator gives the maximum protection of any respirator.

8. A loose-fitting respirator always relies for chemical protection on the outward flow of excess air supplied into the respiratory inlet covering under positive pressure. A loose-fitting respirator may be either a SAR or PAPR, it is unlikely to be an SCBA (because it would use up air too fast), and it can never be negative pressure respirator (because it fits loosely).
   a. A helmet is a rigid covering (that also protects the head) with a transparent shield in front of the face. Virtually all helmet respirators include a shroud, bib, or blouse that extends down from the helmet, below the shoulders, as low as the waist, and sometimes includes sleeves. Breathable air is supplied into the helmet and is eventually exhausted out from beneath the shroud, bib, or blouse. The extended path of the outward flowing air provides much more protection to the wearer's breathing zone than would a helmet without a shroud, bib, or blouse (the latter comprises a loose-fitting facepiece that will be defined in subsection c). The helmet respirator in Figure 22.4a is shown sitting on its bib which is normally worn as seen in

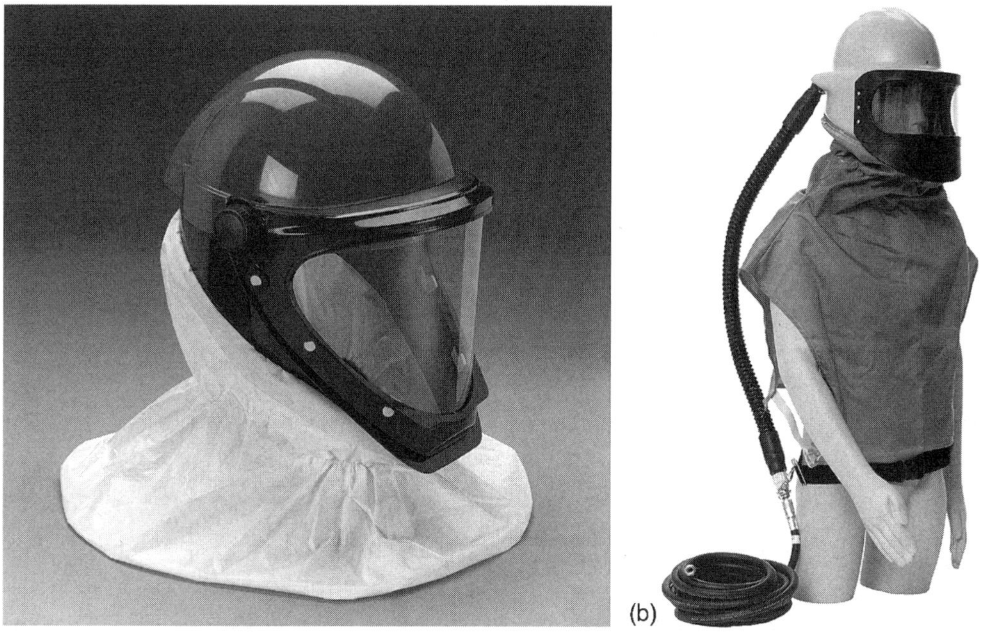

(a) (b)

FIGURE 22.4a, b A helmet respirator, courtesy of 3M™, and a helmet respirator with supplied air hose, courtesy of E.D. Bullard.

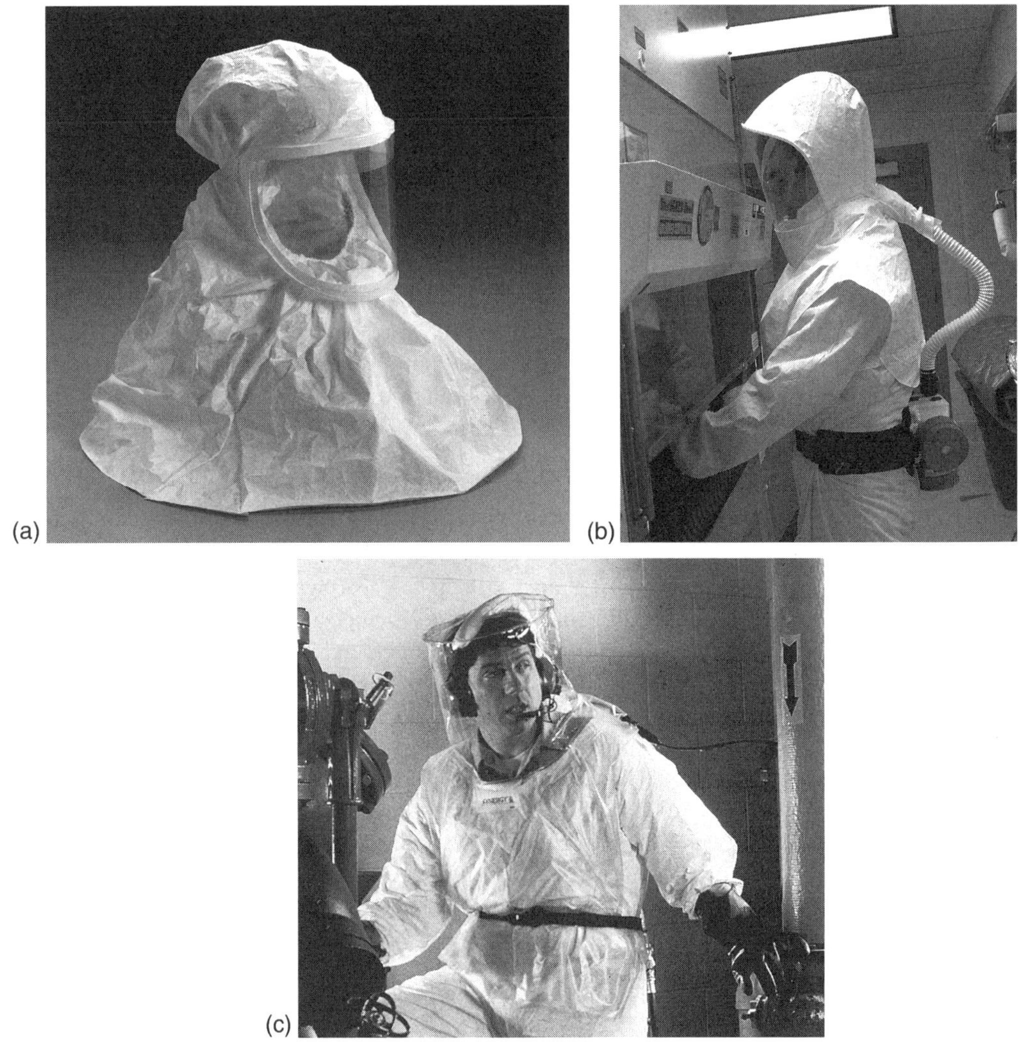

FIGURE 22.5a, b, c Examples of hood respiratory inlet covers that can be configured as atmosphere supplied or powered air purifying. From left to right, courtesy of 3M™, E.D. Bullard, and MSA.

Figure 22.4b. Helmets tend to be made commercially as atmosphere supplying respirators (cf., air purifying respirators).

b. A hood is a respiratory inlet covering that fully encloses the head and neck, similar to, but lighter and less rigid, than a helmet. A hood is a loose-fitting respirator that also has a shroud or blouse extending well below the shoulders under which the excess air escapes (similar to a helmet but different from a loose-fitting facepiece that does not have the extension). The protection factor offered by hoods and helmets with the extension are virtually the same, and both offer much better protection than the loose-fitting facepiece. The hood respirator shown in Figure 22.5a is normally worn with an extended blouse such as seen in Figure 22.5b and c. Hoods tend to be used as a powered air purifying respirator although most can also be configured to operate as an airline system.
c. A loose-fitting facepiece (shown in Figure 22.3a and Figure 22.6) blows breathing air down behind a transparent shield that covers the entire face, as depicted in Figure 22.6.

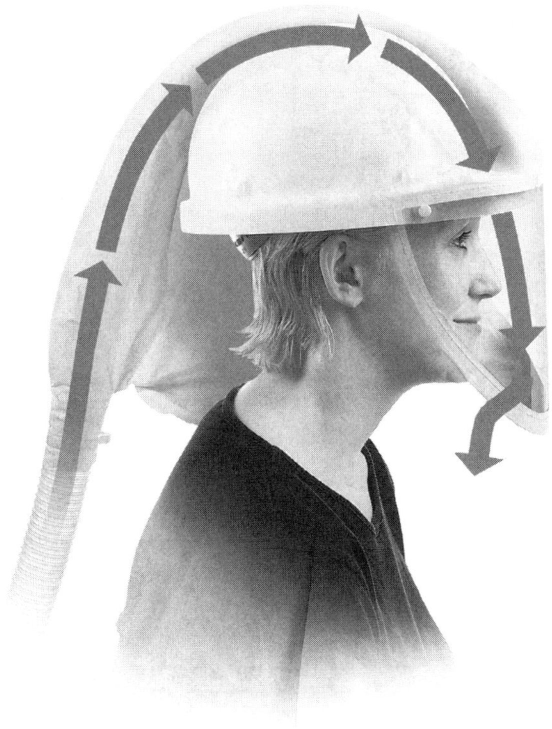

FIGURE 22.6 A depiction of the air flow with a loose-fitting facepiece. Courtesy of 3M™.

A loose-fitting facepiece may slightly constrain the escape of excess air by soft barriers along the sides of the face shield (barely visible in Figure 22.3a) that direct most of the air out the open bottom, but it does not have an extended bib or blouse like the helmets and hoods in Figure 22.4 and Figure 22.5 have. Thus, the protection factor expected from a loose-fitting facepiece is much less than a hood or helmet with the bib or blouse. Its simpler design makes it easier to don and doff than a helmet with a bib or blouse. Be sure to notice the important distinction between the broad group of loose-fitting respirators and the specific respirator called a loose-fitting facepiece.

9. A tight-fitting respirator always creates and relies upon a tight seal between the face (or mouth) and the respirator to protect the wearer. At least four different categories of respiratory inlet covers can be categorized as tight-fitting respirators. Those shown below all happen to be negative-pressure respirators; however, the first two respiratory inlet coverings can also be operated in a positive pressure mode (the full-face version was shown in Figure 22.1, Figure 22.3b, and Figure 22.3c), which will greatly increase their protection factors.
   a. A full-face mask (see Figure 22.7) covers from under the wearer's chin to above their eyes (approaching but not reaching the wearer's hair line) with an integral face shield. This design allows a better face-to-mask seal (less leakage) than the masks listed below and protects the wearer's eyes from irritants. The conventional (negative pressure) air purifying full-face mask has an assigned protection factor of 50. The same type of mask (without fittings for cartridges) operated in a positive pressure mode will have a protection factor of at least 1000.

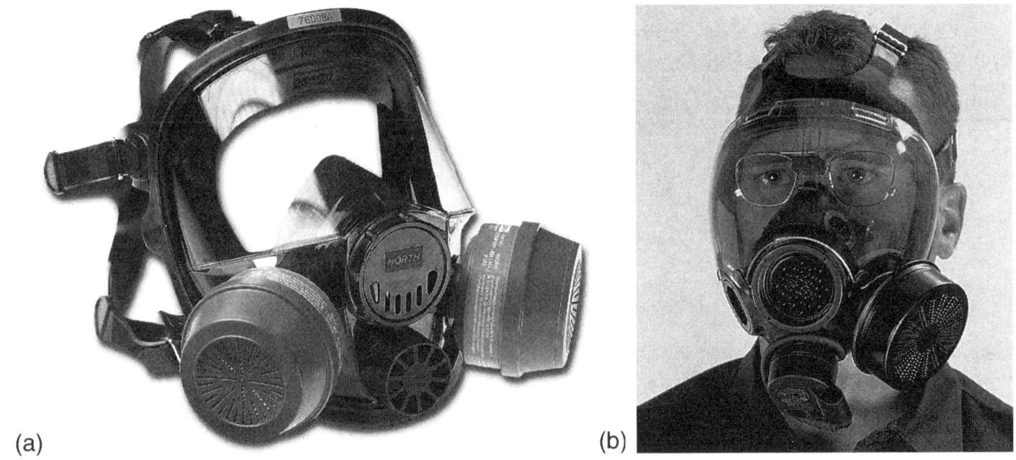

FIGURE 22.7 Two examples of a full-facepiece as a part of an air purifying respirator, one with two purifying cartridges on the left (courtesy of North Safety) and one cartridge and voice diaphragm on the right (courtesy of MSA).

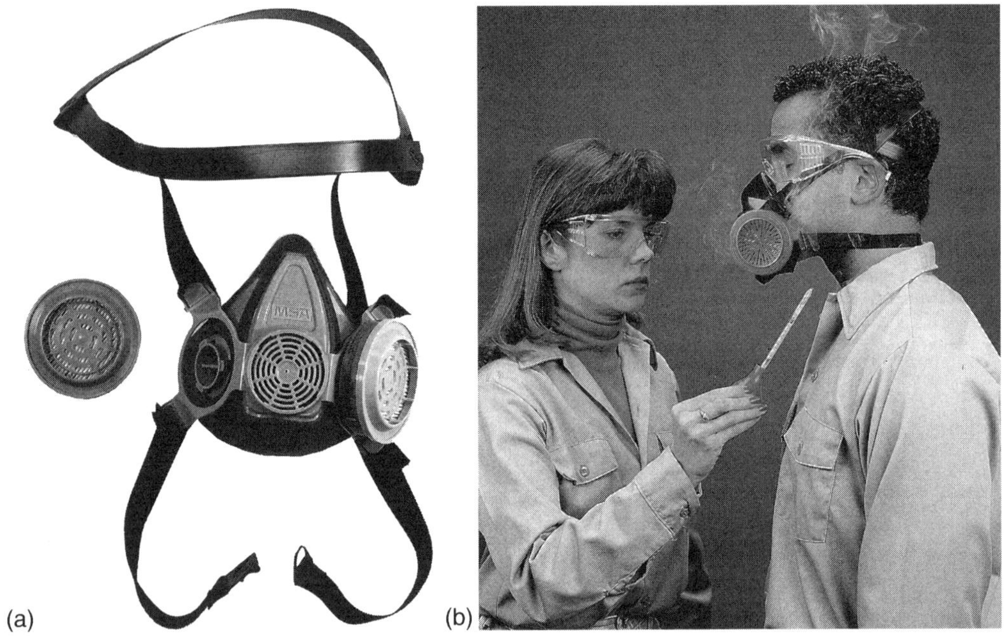

FIGURE 22.8 Two examples of a half-mask respirator. One cartridge is shown to the side an MSA half-mask. The other mask is being fit tested (courtesy of North Safety).

b. A half-mask (see Figure 22.8) covers from under the wearer's chin (like a full-face mask) to the bridge of the wearer's nose. This design has more difficulty than a full-face mask sealing the concave curve between the cheek and nose, often also the corners of the mouth, and sometimes the chin. The conventional air purifying half-mask respirators shown have an assigned protection factor of 10. However, the

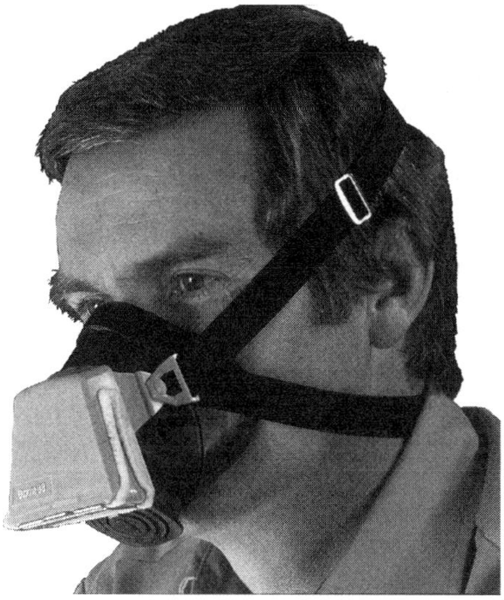

FIGURE 22.9 A quarter-mask air purifying respirator resting above the wearer's chin. Courtesy of MSA.

same type of mask (again without fittings for cartridges) operated in a positive pressure mode would have a protection factor of at least 50.

c. A quarter-mask covers from the bridge of the nose (like a half-mask) but stays above the chin (but below the mouth), as can be seen in Figure 22.9. This less common design is lighter, more comfortable,[5] but may provide a slightly less good fit than a half-mask face piece (OSHA has assigned quarter-masks a protection factor of 10, but NIOSH only thinks they deserve a factor of 5). Quarter-mask respirators are commercially available only as negative pressure respirators with dust filters.

d. The mouthpiece respirator comprises a tube (usually oval) that is inserted into the user's mouth and held between the teeth, a nose clamp to prevent nasal inhalation, and either an air purifier or an external pressurized air supply. The designs shown in Figure 22.10 are intended for emergency or escape use only. Because of this use, a

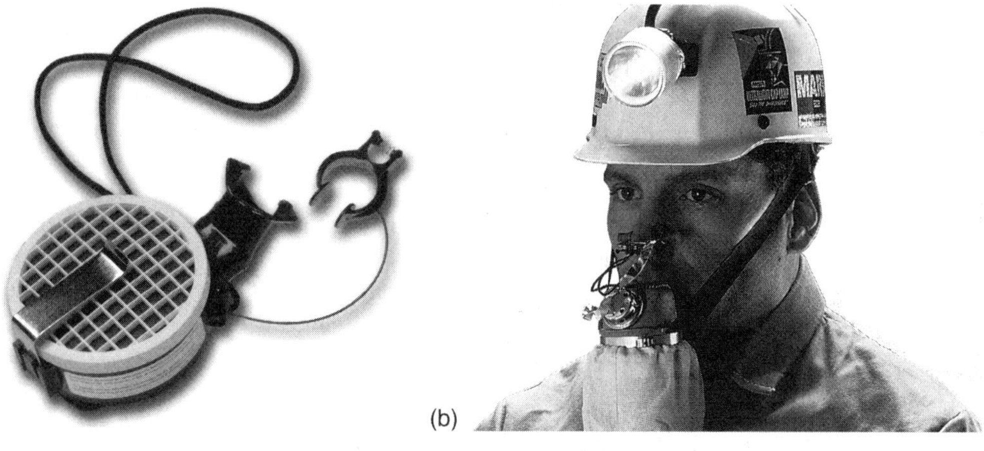

(a) (b)

FIGURE 22.10a, b Two examples of mouthpiece air-purifying respirators for escape use only. The one on the left from North Safety Products is for acid gases; the one on the right from MSA is for carbon monoxide.

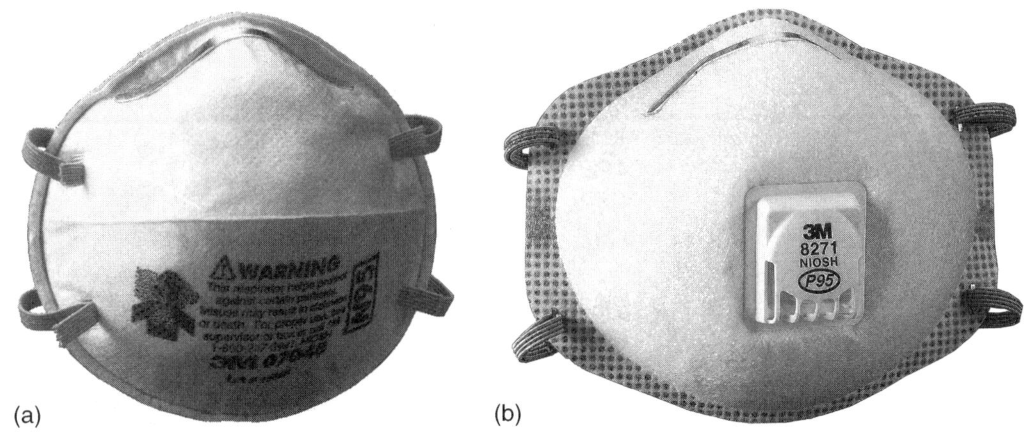

FIGURE 22.11a, b Two examples of a filtering facepiece or plain dust masks with and without exhaust valves. Courtesy of 3M™.

mouthpiece respirator is sometimes called an "escape only" respirator (although escape respirators are also designed with a slip-over hood that looks much like that in Figure 22.5c). Because of its limited use and almost universal ability for a trained user to get a good fit around the mouthpiece for the short period of escape, the mouthpiece respirator is not required to be fit tested.

e. A filtering facepiece (or what is often called a dust mask) is a single piece, negative pressure, particulate aerosol respirator in which virtually the entire half-mask facepiece is made of a filtering material. Such respirators may or may not have an exhaust valve (cf., Figure 22.11a and Figure 22.11b), they have no replaceable parts, and they are made to be disposable. Their simplicity has gained them some special treatment within the OSHA respiratory protection standard.

10. All positive pressure respirators use some form of a pressure regulator, either to maintain a constant flow rate of air into the respiratory inlet covering or to modulate the flow rate of compressed air in response to the wearer's breathing. Thus, positive pressure respirators can be differentiated by the type of flow or pressure regulator that they have.

a. Continuous flow regulators are the simplest. They maintain a constant, high-volume (low pressure) flow rate at all times. To be certifiable, they must maintain 115 Lpm for a tight-fitting facepiece and 170 Lpm for a loose-fitting helmet or hood. Such regulators may be used to provide that flow either from a compressed air line (at 10 to 40 psig) or in conjunction with a small (often belt-mounted) fan or blower. A loose-fitting respiratory inlet covering with a continuous flow regulator relies on the outward flow of a large volume of air for its protection. The outward-leakage of air from a tight-fitting respiratory inlet covering with a continuous flow regulator is an inexpensive supplement to a face mask's protection, but it would be a distinct limitation to an SCBA respirator with only a fixed volume of compressed air in a tank worn by the user. Thus, continuous flow regulators are normally only found on airline and powered air purifying respirators with a loose-fitting respiratory inlet covering. Most wearers feel that a loose-fitting positive pressure respirator with a continuous flow regulator is more comfortable than a tight-fitting mask.[5]

Both a pressure-demand regulator and the older (virtually obsolete) demand regulator vary their flow rates in response to the cyclic breathing demands of the wearer.

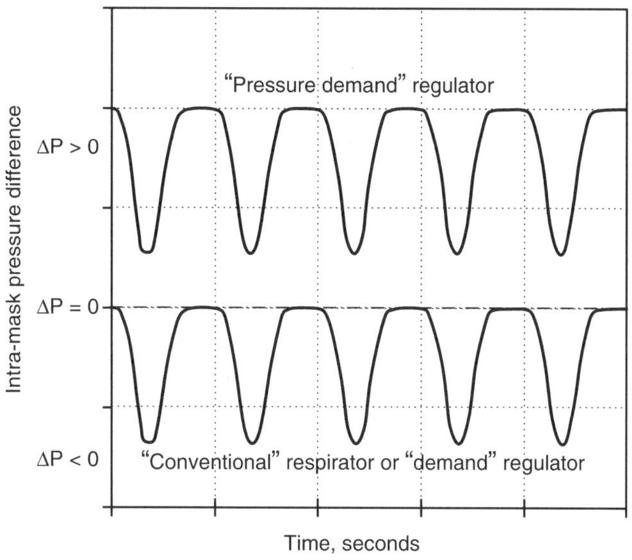

FIGURE 22.12 The pattern of intra-mask pressure over time with two kinds of pressure regulators.

These regulators reduce the air pressure supplied either by a compressor at up to 125 psig or from an air pressure tank at several thousand psig to just above the ambient atmospheric pressure (or just to P_{ambient} in the case the demand regulator).

b. A pressure-demand regulator tries to keep the pressure inside the respirator slightly positive (approximately 1.5″wg) with respect to the ambient pressure outside the respirator at all times (see the upper line in Figure 22.12). The flow rate between breaths (just enough to balance any outward leakage through the mask-to-face seal) is automatically increased when the intra-respirator pressure drops below this positive value as the wearer starts to inhale (the demand portion of the breathing cycle). A pressure-demand respirator also has special exhaust valves that do not open until that additional positive pressure is exceeded during exhalation. As long as the intra-respirator pressure is always positive, any leakage is outward. The maximum protection is obtained by combining both the physical seal of a tight-fitting mask and the aerodynamic flow from a positive pressure regulator. A respirator with a pressure-demand regulator still needs a reasonably good seal of the mask to the face to limit the rate of outward air loss. Minimizing this leakage is especially important if the supplied air is coming from a tank on the user's back (as in an SCBA). It is also possible upon maximal exertion (e.g., while fighting a fire) for a wearer to breathe faster than the regulator can provide air; such over-breathing creates a momentary negative pressure inside the mask that reduces its protection.

c. A demand regulator used to be common but is now virtually extinct (or at least it should be in terms of the protection afforded to the wearer). Think of it as a footnote.[e]

[e] A demand regulator only allows air to flow when the intra-mask pressure drops below the atmospheric pressure as the wearer starts to inhale (see lower line in Figure 22.12). This regulator conserves the air supply by not forcing breathing air to leak outward, but the negative pressure created inside the mask during inhalation allows contaminants to leak inward. The resulting protection factor for a mask with a demand regulator relies only on a good seal (just like a conventional negative pressure air-purifying mask) and is much lower than the same mask with a positive pressure demand regulator.

TABLE 22.1
Feasible Combinations of Respiratory Inlet Coverings within Functional Categories of Respirators with or without Various Kinds of Pressure Regulators

		Negative Pressure Respirators	Positive Pressure Respirators	
			Continuous Flow Regulator	Pressure-Demand Regulator
Air purifying respirators	Conventional respirators	Escape mouthpiece; filtering facepiece; half or full facepiece		
	Powered air-purifying respirators (PAPRs)		Half or full facepiece; loose-fitting facepiece; helmet or hood w. cape	
Atmosphere supplying respirators	Airline respirators (SAR)	Obsolete demand regulator with either half or full facepiece	Half or facepiece; helmet or hood w. cape	Half or full facepiece
	SCBA			Half or full facepiece

Table 22.1 reviews all of the previous categories of respirator systems, types of respirators, air regulators, and respiratory inlet coverings. Any respirator will either purify ambient air or have fresh air supplied to it. Similarly, the air pressure inside of any respirator will either be negative or positive with respect to the ambient air outside the respirator. Several combinations of respirator category and pressure regulator are either incompatible or no longer desirable as indicated by the blank spaces in Table 22.1. Practically speaking, only conventional tight-fitting respirators have a negative pressure inside the mask (demand regulators are shown but are obsolete). The blowers on PAPRs do not create enough positive pressure to operate a pressure demand regulator. This leaves only approximately five categories of respirators from which to select.

11. All air purifying respirators will have either an aerosol filter and/or a chemical absorbing air purifier.
    a. A mechanical filter can remove aerosols. Respirator filters vary in how efficiently they remove particles and the effect of oil in that aerosol on the removal efficiency of the filter. Thus, the selection of which filter to use in a particular setting is made on the basis of the removal efficiency desired and whether or not the aerosol contains oil. Section VI.6.b of Chapter 22 will discuss filter selection in detail.
    b. A chemical absorber placed in a cartridge or canister through which the air passes can remove certain gases and vapors. The selection of which absorber to use is made on the basis of either the specific chemical (e.g., ammonia or carbon monoxide) or the class of the chemical contaminant (e.g., organic vapor or acid gas). The selection of the size of the container is made primarily on the basis of the service life needed, secondarily upon the concentration of the contaminant to be removed, and tertiarily upon comfort (the smaller cartridge being much lighter than the larger canister). Section VI.6.a of this chapter will discuss cartridge service life and change-out schedules.
    c. A combination of a mechanical filter and a chemical absorber can be used in series, (e.g., for spray paint or pesticides), and a universal canister can remove organic vapors, dusts, acid gases and carbon monoxide.

Both respiratory protection program managers and respirator users must be made aware that all air purifiers are subject to some serious limitations. These limitations underlie the philosophy that atmosphere supplying respirators are always the default choice for respiratory protection. Air purifying respirators are only allowed when conditions are known to not violate the following limitations.

- ✍ No air purifier can make up for a deficiency of oxygen. Air purifiers only remove contaminants; they do not add anything to the air. An atmosphere supplying respirator must be used if oxygen is deficient.
- ✍ No chemical air purifier can remove all chemical hazards. Activated charcoal is the most common chemical adsorbent, but it does not remove many gases or some vapors. Aerosol filters remove only particles and no gases or vapors. Because some airborne contaminants cannot be removed by any purifier, an atmosphere supplying respirator must be used when the contaminant is unknown and/or its removal cannot be assured.
- ✍ No air purifier can last forever. All air purifiers have a maximum capacity of contaminants that it can hold. This capacity determines an air purifier's service life. An air purifier's service life determines its change-out schedule (the time when it should be changed). Thus, an atmosphere supplying respirator must be used if the service life cannot be predicted or if it is incompatible with the respirator's expected duration of use.

12. The following two other components of respirators are especially important for tight-fitting masks.
    a. Most respirators require a harness to hold the respiratory inlet covering in place. A 2-strap harness (seen in Figure 22.8 and Figure 22.10) holds any tight-fitting mask against the face at four points better than a 1-strap harness that only holds the mask at two points (or even the 2-straps in Figure 22.9). One of the two straps passes around the neck, and the other passes over the back of the head. Many head straps now bifurcate (split as shown in Figure 22.8) to be more comfortable and to hold a mask in a more stable position. The straps of full-face masks are sometimes either attached at five points (with an additional strap running straight-back across the top of the head, visible in Figure 22.3c and Figure 22.7a), six points (as in Figure 22.7b), or are attached to a netting made into a cap that holds the mask in position more securely than straps. Helmets and hoods rest on the head or shoulders, respectively, and might only have either a chin or a chest strap.
    b. Most tight-fitting masks have one or more valves that are usually simple flaps as depicted in Figure 22.13.[f] Exhalation valves prevent contaminated air from entering into the mask through the exhaust outlet during periods when the interior of the respirator is at a negative pressure. Normally air only enters either through the air purifier or from an external source via a pressure regulator. An exhalation valve that either sticks open (as in the right side of Figure 22.13), is bent and unable to seal, or is missing altogether will allow contaminated air to enter into the mask via the exhalation port. A tight fitting air purifying respirator may also have one or two similar inhalation valves to prevent exhaled air from exiting through the air purifier. The most damage likely to be caused by a failed inlet valve is to increase a respirator's dead air volume (by rebreathing some exhaled air pushed into the air purifier) and perhaps adding humidity to the air

[f] The lowest cost filtering facepieces (single-piece dust masks such as in Figure 22.11a) do not have an exhalation valve, in which case the exhaled air must pass directly back out through the filtering facepiece, causing some weakening of the mask's shape. Of course, loose-fitting facepieces, hoods, and helmets with continuous flow regulators do not have any valves.

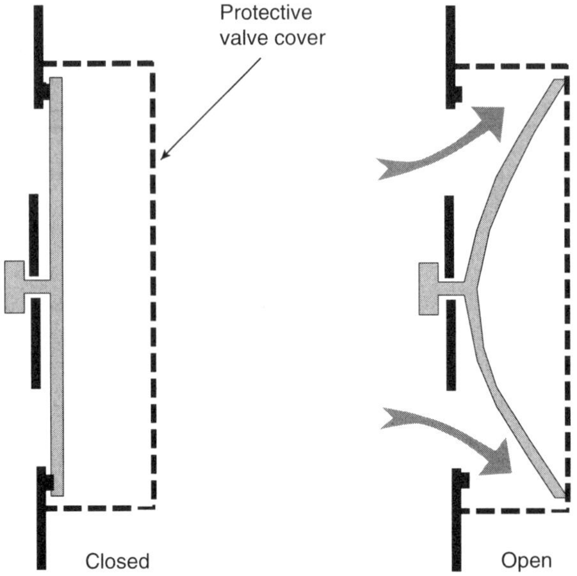

FIGURE 22.13 An exhaust valve in the open and closed positions.

purifier. Thus, the failure of an inhalation valve, while undesirable, is not as consequential as a failed exhalation value. A valve failure can be detected by a seal check (to be discussed in Section IV.3).

II. DIFFERENT KINDS OF PROTECTION FACTORS

The following four specific protection factors are all based on the concept embodied in the previously stated definition of a generic protection factor, repeated below.

$$\text{protection factor} = \frac{\text{contaminant concentration outside the respirator}}{\text{contaminant concentration inside the respirator}}$$

(repeat of Equation 22.1)

The sequence in which the following four specific protection factors are discussed parallels their historic development beginning with measuring respirator effectiveness in a controlled (lab-type) setting. Many years of studies in various field (workplace) settings were necessary to create the assigned protection factors which are a key focal point of those today who select respirators and/or manage a respiratory protection program.

1. The quantitative fit factor (QNFF) is simply the protection of a given tight-fitting, negative-pressure respirator worn by a given individual as measured using a quantitative fit test (QNFT) in a controlled test setting, usually an office setting. OSHA officially has only one term called simply a fit factor or ff in their formulas. According to the standard testing protocol described in Appendix A of OSHA 1910.134 (and discussed below in Section IV), the overall fit factor is calculated, as shown by Equation 22.3a, from seven individual fit factors, each measured while conducting one of seven specified exercises for 1 minute.

$$\text{Overall Fit Factor} = \frac{\text{Number of exercises} = 7}{1/\text{ff}_1 + 1/\text{ff}_2 + 1/\text{ff}_3 + 1/\text{ff}_4 + 1/\text{ff}_5 + 1/\text{ff}_6 + 1/\text{ff}_7} \quad (22.3a)$$

The reciprocal of each ff_i in the denominator is equivalent to a penetration value as defined by Equation 22.2. Thus, the overall fit factor in Equation 22.3a is equivalent to the QNFF in Equation 22.3c.

$$\text{Overall Fit Factor} = \frac{7}{\text{penetration}_1 + \text{penetration}_2 + \cdots + \text{penetration}_7} \tag{22.3b}$$

$$\text{QNFF} = \frac{7}{[C_{\text{in}}/C_{\text{out}}]_1 + [C_{\text{in}}/C_{\text{out}}]_2 + \cdots + [C_{\text{in}}/C_{\text{out}}]_7} \tag{22.3c}$$

In the ideal world, every wearer would be assured by a quantitative fit test that the selected mask provides a good fit. Unfortunately, the cost of the test apparatus is too expensive to justify requiring a quantitative fit test (QNFT) for all respirator users. The alternative to a QNFT is a qualitative fit test (QLFT) with the assurance that a respirator that passes a QLFT can be assumed to have a fit factor of at least 100. Protocols for both tests are described in Appendix A of OSHA 1910.134 and outlined in Section IV.

2. The workplace protection factor (WPF) has been defined as the protection factor achieved by a properly selected, fit tested, and functioning respirator, measured only while the respirator is properly worn and used during normal work activities. Functionally, the WPF is the ratio described by Equation 22.4 of time weighted average concentration from parallel samples are collected both inside and outside of the respirator being worn at work.

$$\text{WPF} = \frac{[C_1 \times \text{time}_1]_{\text{outside}} + [C_2 \times \text{time}_2]_{\text{outside}} + \cdots + [C_n \times \text{time}_n]_{\text{outside}}}{[C_1 \times \text{time}_1]_{\text{inside}} + [C_2 \times \text{time}_2]_{\text{inside}} + \cdots + [C_n \times \text{time}_n]_{\text{inside}}} \tag{22.4}$$

Experience has shown that the protection afforded to a wearer during normal work activities is less than the FF achieved during a fit test in a controlled setting. NIOSH, ANSI, and OSHA all use a factor of 10 to approximate the worst degradation of an individual's FF to their WPF in the same mask. Logically, one could hypothesize that an individual who achieves a larger QNFF should have a larger WPF. While that hypothesis could well be true, as will be explained below, there is no administrative advantage to obtaining a QNFF beyond 10× the APF for the class of respirator being tested. Thus, measuring WPF is largely relegated to a research procedure used to set APF values, as would be essential for new classes or brands of respirators for which an APF is not already established.

3. The effective protection factor (EPF) is similar to WPF but measured when the respirator is worn for only some fraction of the total exposure period in the workplace. As you might expect, if a user wears the respirator for less than 100% of the exposed time, the EPF will be less than the WPF. There is virtually no incentive for anyone to go out and measure EPFs; it is simply the default result of trying to measure WPF without being able to enforce the wearing of respirators continuously. However, understanding the relationship between EPF and WPF can help the practitioner to anticipate how rapidly a user's actual level of protection will drop if they do not wear their respirator all of the time and to judge how much effort should be expended to achieve 100% user compliance.

The following discussion will show how the effective protection factor can be predicted from the workplace protection factor for any level of noncompliance. While the cursory reader can jump directly to Equation 22.7, the student is encouraged to follow the algebraic derivation of that equation. Equation 22.7 will be used to find how long one can not wear a respirator before it can become a serious problem. Its derivation uses basic industrial hygiene (IH) principles that should be understood by the competent hygienist. The logic used in this derivation can be used to solve other time-weighted average problems that you may face some time in the future.

Suppose that a respirator is worn for a time ($time_{on}$) and not worn for a (hopefully shorter) time ($time_{off}$). Let "$fraction_{not\ worn}$" be the fraction of the time that the respirator is not worn as defined by Equation 22.5.

$$fraction_{not\ worn} = \frac{time_{off}}{time_{on} + time_{off}} \tag{22.5}$$

Just like WPF above, EPF needs to be expressed in terms of time weighted averages. The simplest assumption is that the ambient contaminant concentration ($C_{outside}$) is constant whether the respirator is worn or not worn. This assumption of one constant outside concentration is the simplest, but not the only possible, assumption (see Popendorf for more complicated assumptions).[6]

$$EPF = \frac{C_{outside}(time_{on} + time_{off})}{C_{inside}\ time_{on} + C_{outside} time_{off}} \tag{22.6a}$$

Let the measured ambient $C_{outside}$ be further simplified to just plain C, the high concentration of concern for which the respirator is being worn in the first place. C_{inside} during $time_{on}$ while the respirator is worn, can be simplified to C/WPF if the respirator functions to reduce C by its full potential WPF. Making these substitutions into Equation 22.6a allows both C's in Equation 22.6a to cancel each other out.

$$EPF = \frac{C(time_{on} + time_{off})}{C\,time_{on}/WPF + C\,time_{off}} = \frac{time_{on} + time_{off}}{time_{on}/WPF + time_{off}} \tag{22.6b}$$

Finally, by rearranging Equation 22.5 to form $time_{off} = time_{on} \times fraction_{not\ worn}/(1\text{-}fraction_{not\ worn})$, first $time_{off}$ then time itself can be removed from the above equation, allowing EPF to be expressed in terms of only WPF and the fraction of time that the respirator is not worn.

$$EPF = \frac{time_{on} + time_{on} fraction_{not\ worn}/(1 - fraction_{not\ worn})}{time_{on}/WPF + time_{on} fraction_{not\ worn}/(1 - fraction_{not\ worn})} \tag{22.6c}$$

$$= \frac{(1 - fraction_{not\ worn}) + fraction_{not\ worn}}{[(1 - fraction_{not\ worn})/WPF] + fraction_{not\ worn}} \tag{22.6d}$$

$$= \frac{WPF}{(1 - fraction_{not\ worn}) + (fraction_{not\ worn} WPF)} \tag{22.6e}$$

$$EPF = WPF\left(\frac{1}{1 + [fraction_{not\ worn}(WPF - 1)]}\right) \tag{22.7}$$

The term within the large parentheses of Equation 22.7 is the factor by which EPF is reduced below its ideal WPF. An analysis by inspection shows that the term in the large parentheses will decrease when either the not-worn fraction of time or the ideal WPF increase (they are both in the denominator). The resulting EPF values for potential WPF values of 10, 100, 1000, and 10,000 are plotted as Figure 22.14a and b. The effect of short fractions of noncompliance (near the left side) are hard to see in Figure 22.14a, but this

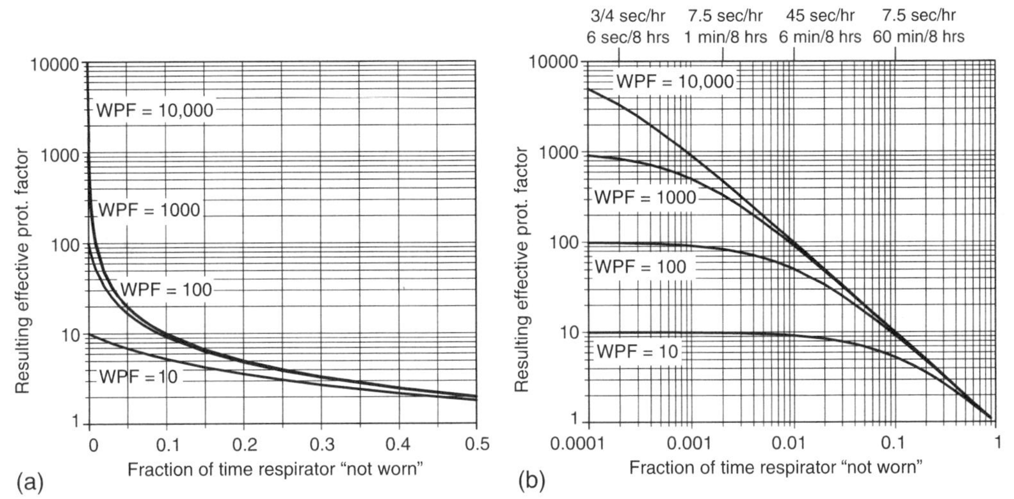

FIGURE 22.14a, b Plots of EPF as a function of the fraction of time the respirator is not worn. The figure on the left shows fraction on a linear scale, while that on the right shows fraction on a logarithmic scale.

figure clearly shows how the EPF values converge toward "$1/\text{fraction}_{\text{not worn}}$" at noncompliance fractions above approximately 0.1. Thus, if an exposed person does not wear his or her respirator very much, their protection is determined predominantly by the level of noncompliance no matter what kind of respirator they are wearing. For instance, if a respirator is not worn for 25% of the time, EPF values will only range from approximately 3 to 4 as the WPF ranges from 10 to 10,000, respectively.

How quickly not wearing a respirator for short intervals will affect the resulting EPF can be seen more clearly in Figure 22.14b. Table 22.2 shows the same trend for three nominal reductions in WPF. Each reduction in WPF increases the TWA exposures as defined by the inverse ratio of WPF/EPF (in other words, the wearer's exposure level will increase as much as the EPF is decreased below what the WPF would have been if the respirator were worn continuously). For instance, the wearer of a respirator capable of a WPF of 10,000 (typical of a pressure demand SCBA) would have to be noncompliant for only 2.9 sec out of an 8-h day to reduce the EPF by 50% and get a twofold increase in

TABLE 22.2
The Fraction and Actual Time Out of 8 h That a User Would Have to Not Wear Their Respirator to Decrease Their EPF Below the Potential WPF and to Increase Their TWA Exposure by WPF/EPF

		Potential WPF (protection if it were worn 100% of the time)				
		10,000	**1000**	**100**	**50**	**10**
EPF = 50% of WPF	$\text{Fraction}_{\text{not worn}}$	0.0001	0.001	0.010	0.020	0.111
(WPF/EPF = 2)	Time_{off}	2.9 sec	29 sec	5 min	10 min	53 min
EPF = 25% of WPF	$\text{Fraction}_{\text{not worn}}$	0.0003	0.003	0.030	0.061	0.33
(WPF/EPF = 4)	Time_{off}	9 sec	1.4 min	15 min	29 min	160 min
EPF = 12.5% of WPF	$\text{Fraction}_{\text{not worn}}$	0.0007	0.007	0.071	0.143	0.78
(WPF/EPF = 8)	Time_{off}	20 sec	3.4 min	34 min	69 min	373 min

exposure. If that notworn time continued for 20 sec, the result would be a reduction in the EPF down to 12.5% of the WPF and an eightfold increase in exposure. On the other hand, a wearer of a respirator capable of a WPF of only 10 (typical of a negative pressure, half-mask respirator) would have to not wear their respirator for 53 min to experience a similar 50% reduction in protection, and continue to not wear it for more than 75% of the time to get an eightfold drop in protection.

Two lessons can be learned from this analysis. One is that the user of a respirator with a very high WPF (>500) can only be noncompliant for a matter of a few seconds (or a $\text{fraction}_{\text{not worn}}$ as small as 0.001) before significantly reducing their protection, while the user of a respirator with a WPF of 10 to 50 can be noncompliant for several to many minutes (or a $\text{fraction}_{\text{not worn}}$ as much as 0.1) before significantly reducing their protection. The other lesson is that at high levels of noncompliance ($\text{fractions}_{\text{not worn}}$ in excess of 0.1), all EPF values eventually converge toward a value of $1/\text{fraction}_{\text{not worn}}$ no matter what kind of respirator the person was supposed to be wearing.

4. The assigned protection factor (APF) is the most commonly used protection factor in the routine management of a respiratory protection program (and a factor commonly misused by others not versed in the field). The APF is not measured; it is used to select a respirator.

   Each defined class of respirators has been given an assigned protection factor based on a statistical analysis of previously measured workplace protection factors (primarily WPFs or some SWPFs). Its original definition was the minimum level of respiratory protection that a properly functioning respirator is expected to provide to at least 95% of properly fitted and trained users.[2] Opinions vary on the quality, grouping, and analysis of the available data. In the past, two groups published two sets of APF values for somewhat different classes of respirators.[7–10] Table 22A.1 in the appendix to this chapter lists both sets of APFs. The most striking difference between the two sets of APF values is that the ANSI committee separated the performance of hoods and helmets with a cap or blouse from other loose-fitting facepieces while NIOSH continues to pool these two together. In June 2003, OSHA proposed a set of APF values for their 1910.134 Table I summarized herein as Table 22.3.[11] Their APF values differ slightly from both of the above groups, but they have the same classes of respirators as ANSI.

   The assigned protection factors listed in Table 22.3 include an important caveat. OSHA makes the point that "the assigned protection factors listed in (their) Table I are effective only when the employer has a continuing, effective respiratory protection program as specified by 29 CFR 1910.134, including training, fit testing, maintenance and use requirements." The agency added this caveat to emphasize the requirement that employers must select a respirator in the context of a comprehensive respiratory protection program. Users should not expect to achieve workplace protection factors as large as these APFs when any of the program elements (such as fit testing, maintenance, selection, use, or training) required by OSHA's Respiratory Protection Standard are absent from an employer's respirator program. Furthermore, these assigned protection factors also do not apply to respirators used solely for escape.

   An APF would be used after measuring the ambient concentration of a contaminant using normal IH air sampling practices, finding employee exposure to be excessive, and deciding that other means of source or pathway control are not feasible.[g] The assigned protection factor of an acceptable respirator must be equal to or greater than the ratio of

[g] Another important scenario in practice is when a respirator is used to investigate or to control an unknown chemical or a known chemical at an unknown concentration. In both situations it is prudent to select the most protective respirator available, the one with the highest APF.

TABLE 22.3
Assigned Protection Factors (APFs) Proposed by OSHA for Its 1910.134 Table I[11]

	Negative Pressure	Positive Pressure		
	Conventional or Unpowered APR and Demand Mode SAR or SCBA[a]	Continuous Flow	Pressure Demand	
		PAPR[b] or SAR	SAR	SCBA[c]
Filtering facepieces and quarter masks	10			
Loose-fitting facepiece		25		
Half-mask	10	50	50	
Hood/helmet with cap or blouse		1000		10,000
Full facepiece	50	1000	1000	10,000

[a] A demand mode regulator on an atmosphere supplying respirator results in a negative pressure within the respirator during inhalation and would have the same APF as a conventional air purifying respirator.
[b] Only high efficiency filters (N-, R-, or P-100) are certified for use in PAPRs against aerosols.
[c] This level of protection may not be achieved at high work rates (where a wearer can over-breathe the respirator).

the ambient hazard concentration to the exposure limit for that hazard, as defined by Equation 22.8. This ratio will be used in one of the last two steps when selecting a respirator, to be described in detail in Section VI.

$$\mathrm{APF}_{\text{of an acceptable respirator}} \geq \frac{\text{ambient contaminant concentration}}{\text{the contaminant exposure limit}} \quad (22.8)$$

The above ratio could apply in principle to either the PEL, TLV®, or any other chosen exposure limit; however, OSHA would apply it strictly to their PEL. The industrial hygienist or respirator program administrator is responsible to see that the APF for the respirator selected for each wearer meets the above selection criterion. Failure to meet this criterion would increase the likelihood of the user's breathing-zone concentration to exceed the exposure limit.

Another potential use of APFs is to determine the maximum use concentration (MUC) in which any given respirator should (safely) be used. In principle, a maximum use concentration could be found (or set by an agency) by multiplying the chemical's PEL or other exposure limit times the APF recommended for that class of respirator; this is the ideal approach described by Equation 22.9.

$$\text{ideal Maximum Use Concentration [MUC]} = \text{PEL} \times \mathrm{APF}_{\text{respirator}} \quad (22.9)$$

Unfortunately, such an ideal approach disregards important limitations that apply to half-face masks and most air purifiers and escape criteria that apply to many chemicals. For example, the MUC for an air purifying respirator should not exceed either a chemical's IDLH or the capacity of the air purifier to remove a particular chemical. NIOSH has a list of IDLH values.[12] A chemical can be an eye irritant at a level below 10 × the chemical's exposure limit; thus, wearers of a half-face mask would be suffering below the MUC. An MUC calculated via Equation 22.9 might exceed an air purifier's capacity to remove it (a level typically determined by the respirator's manufacturer). Thus, a more realistic MUC should not exceed either the value determined by

Equation 22.9, the chemical's IDLH, the chemical's eye irritation threshold if wearing a half-mask, or a value recommended by the manufacturer, whichever is less. This more comprehensive definition should help to avoid the misuse of a respirator beyond its intrinsic capability to protect the wearer.

III. VARIABLES AFFECTING PROTECTION FACTORS

This section begins with a short discussion on the important influence of the pressure difference between the inside and outside of a respirator as the most important factor affecting a respirator's protection factor. The variability in the fit of negative pressure respirators is the second most important factor, followed by differences among the types of respiratory inlet coverings. Wearer activities and environmental conditions can affect the seal of tight-fitting respirators and therefore decrease their protection factors. Of course, a wearer's noncompliance will reduce the protection of any respirator.

1. The presence (or absence) of a positive intra-mask pressure difference has more influence on protection than variations in fit within any type of negative pressure respirator and at least as much influence as differences between types of tight-fitting respirators (viz., between half-masks vs. full-face masks). In any negative pressure respirator (viz., a conventional air purifying respirator or any atmosphere supplying respirator with a demand type air-flow regulator), air does not flow into the mask until the wearer starts to inhale. Just like a local exhaust collection hood, the slight negative intra-mask pressure created by inhaling will allow some inward leakage of contaminants. However, if the intra-mask pressure can be maintained even slightly above the ambient pressure outside the mask (as in the top portion of Figure 22.12), then all leakage flows outward.[h] A comparison between the APFs within Table 22.3 for half-mask and full-face respirators shows that their protection factors can be increased by a factor of 5× and 20×, respectively, when the intra-mask pressure is converted from a negative into a positive difference, whether with a pressure demand or a continuous flow regulator. Loose fitting hoods and helmets rely on the continuous outward air flow to keep contaminants out of the breathing zone.
2. The best negative pressure respirator is no better than its poorest fit. The variability in fit of any mask to any person could be construed as the greatest determinant of protection, which is why fit-testing is so important before a tight-fitting respirator is assigned to an individual. However, after a person has passed a fit-test for a given respirator, the variability in the fit of any negative pressure respirator among wearers is the second most important determinant of protection. This conclusion stems from an analysis of the log-normally distributed data among past protection factor studies that found the typical interpersonal variability in protection is approximately one order of magnitude less than the difference in protection due to intra-mask pressure, and yet it is still approximately two to three times more important than the next variable (the type of mask being used).[5]

   Factors contributing to variations in fit include characteristics such as face size and shape, facial hair, and skin condition that are intrinsic to the individual, and personal practices of the wearer such as strap tension that are amenable to user training.

   a. Face length (nose to mouth) and face width (mouth) strongly affect fit but are not sufficient predictors to preclude fit testing. Most commercial respirators now come in two or three sizes; however, 90–95% of users will generally fit the medium size. Within the same size, significant variations exist among manufacturers in their mask dimensions, proportions, and internal contours. So, if someone cannot be fitted with

[h] Keep in mind the possibility of someone at maximal exertion to "over-breathe" a respirator, meaning they are breathing faster than the regulator can supply air to it, causing its normally positive intra-mask pressure to become negative.

one brand, they may find an acceptable or better fit with another model of the same brand (manufacturer) or with another brand of the same or different size. In fact §1910.134 addresses respirator size and selection in two locations:

- §134(d)(1)(iv) states, "The employer shall select respirators from a sufficient number of respirator models and sizes so that the respirator is acceptable to, and correctly fits, the user."
- Part A to Appendix A of §134 states, "The test subject shall be allowed to pick the most acceptable respirator from a sufficient number of respirator models and sizes so that the respirator is acceptable to, and correctly fits, the user."

b. Because most data indicate that facial hair (especially beards) can interfere with the mask-to-face seal and decrease QNFF and WPF,[13,14] OSHA §134(g)(1) bans "facial hair that comes between the sealing surface (of a tight-fitting) facepiece and the face …" §134 Appendix A. part A.9 even bans fit testing "if there is any hair growth between the skin and the facepiece sealing surface." §134(g)(1) more broadly precludes an employer from permitting "respirators with tight-fitting facepieces to be worn by employees who have … any condition that interferes with the face-to-facepiece seal or valve function." Missing dentures, facial scars, a strong dimple, a cleft chin, or even pimples or bug bits will make obtaining an acceptable fit more difficult if not impossible. However, such characteristics should not preclude the person from wearing a loose-fitting powered or atmosphere supplying respirator. §134(f)(3) reminds the employer that an additional fit test is required whenever changes affecting fit can be visually detected, such as might be caused by cosmetic surgery or an obvious change in body weight.

c. The ear-pieces of prescription glasses, goggles, sunglasses or any similar object that would protrude through the mask-to-face seal are also prohibited under §134(g)(1).[i] Glasses with either no or modified ear pieces can be mounted on clips or sockets inside some full-face masks, such as shown in Figure 22.15 and is visible in Figure 22.7b. The external interference between the nose-piece of prescription or safety glasses and the top of a half-mask respirator (see Figure 22.8b or Figure 22.9) can cause the glasses to slide down, but this problem is only an annoyance, not a hazard.

d. The strap tension of tight-fitting respirators is an important variable both between people and even within the same person from day-to-day or throughout the day. Many users have the tendency to pull the straps only as tight as needed to feel that the mask will not move. While that may be an acceptable way to wear such respirators for many hours per day, users who know they are at high-risk tend to pull the straps tighter (but short of being uncomfortable) to increase wearer protection. The difference in strap tightness between a 15-min fit test and a long workday probably contributes to the difference between QNFF and WPF. One of the challenges to good respirator training and program supervision is to impart the importance of enough strap tightness to achieve good protection without increasing discomfort and user noncompliance.

3. The type of mask (half or full-face) is the least important factor in comparison to either the relative pressure inside the respirator or inter- and intra-personal variability. Note in Table 22.3 the comparatively small (5×) difference between the APF of a full facepiece vs the APF of a half or quarter mask for the same intra-mask pressure or flow regulator. Despite masks differing in the material from which they are made (e.g., filtering

[i] OSHA used to prohibit someone from wearing contact lenses with a respirator, but that rule was not in the 1998 revision of 29 CFR 1910.134.

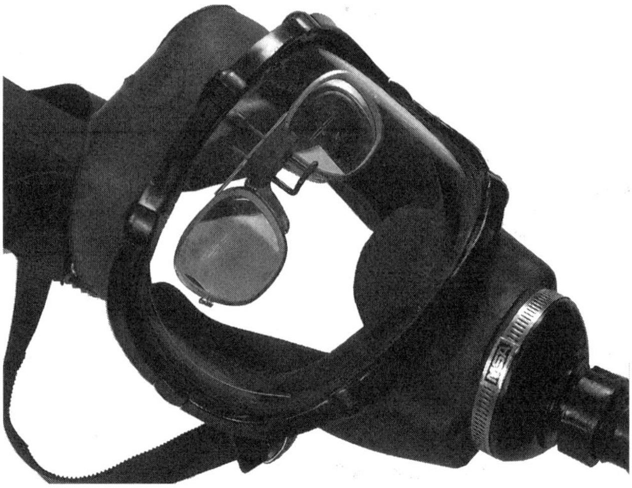

FIGURE 22.15 Eyeglass mounting clips in an MSA full face mask.

facepieces are less rigid than rubber), sufficient field data supports giving both kind of half-masks the same protection factor.

4. Work activities and thermal conditions will affect fit and therefore the protection afforded to any wearer of a negative pressure respirator. The wearer's activities are largely dictated by the work and are only weakly amenable to user training. Temperature and humidity also vary among workplaces (perhaps not so widely as activities) but are usually not easily controlled where respirators are used (if the environment were easy to control, respirators would likely not be needed).
   a. Any activity that can momentarily break the seal of a tight-fitting facepiece to the wearer's face will affect respiratory protection adversely. Certain activities such as cleaning ones mask and changing air purifying cartridges in the use-area are specifically prohibited by §134(g). Several other normally common activities such as eating, drinking, and smoking are not specifically addressed by the standard, but clearly should not be allowed in areas requiring respirator use. Talking is another common activity that not only enhances work efficiency but can also be a safety requirement of many tasks and is a social benefit if not a necessity. Based on the EPF that results from short periods of noncompliance predicted by Equation 22.7 and shown in Figure 22.14b, talking while wearing a respirator (especially for safety) need not necessarily be prohibited but should be minimized. Where frequent communication is required, respirators are commercially available with either a passive speaking diaphragm, an electronically amplified speaker, or wireless telecommunications.
   b. Temperature and physiologic factors can also degrade fit and protection from tight-fitting respirators but are sometimes an unavoidable part of the particular working environment.
      - Heat stress causes increased sweat, slippage of the mask on the face, a weakening of the mask-to-face seal, and the consequent loss of protection for the wearer. Sweat and exhaled moisture will also cause many filtering facepiece respirators to wilt, again weakening the seal and losing protection. This latter problem is aggravated if the facepiece is without an exhaust port.

- Cold conditions may cause a full-face facepiece or eyeglasses to fog up when worn with half-face mask respirators. Moving from a cold into a warmer and a sufficiently more humid condition will cause condensation (fogging) on glasses or the outside of a full facepiece visor. Such fogging can be alleviated by applying anti-fogging compounds onto one's glasses or facepiece. Fogging on the inside of a full-facepiece respirator can be alleviated by using nose-cups (internal baffles that direct exhaled air toward the exhaust ports and therefore away from the facepiece; just viewable in Figure 22.7). Nose-cups also reduce the dead-air volume.
- The high humidity in exhaled air may cause exhaust valves to freeze shut in very cold conditions. Stuck exhaust valves can usually be cleared with some extra pressure on the next exhalation (the wearer can feel the valves "pop" open). An exhaust port freezing open is rare but more dangerous because it will allow the direct entrance of contaminated air. Users and their supervisors should be warned about the potential for valves to freeze. Heating the workspace may be an additional management option to deal with such cold temperatures.

  Extreme temperature conditions beyond those accounted for by routine policies should trigger additional care and precautions to be taken by both the on-site supervisor and the user. Using loose-fitting powered air-purifying respirators can only alleviate a few of these problems.

5. As previously shown, user noncompliance with respirator use requirements can outweigh everything else. The previous factors such as facial features, strap tightness, work activities, and heat stress only affect protection when the respirator is worn, but if the respirator is not used when needed, it will not provide any protection at all. The effect of intermittent use on EPF predicted by Equation 22.7 was shown in Figure 22.14 and Table 22.2. User compliance is greatly aided by employee education, training, and self-motivation. User compliance without supervisory support is almost certainly doomed to failure. The importance of user compliance and the unwillingness of some organizations to make the long-term commitment to all of the elements of PPE management program are major reasons that respirators have been placed at the bottom of the IH control paradigm. However, with adequate program management (eventually meaning funding), respirator technology can provide effective control.

IV. RESPIRATOR FIT AND SEAL TESTING

Testing is needed to assure that a tight-fitting respirator adequately fits onto each wearer's face. OSHA §1910.134(f) requires either a qualitative fit test (QLFT) or a quantitative fit test (QNFT) in each of the following four circumstances:

1. prior to the initial use of any tight-fitting respirator and at least annually thereafter,
2. whenever a different respirator facepiece is used,
3. whenever changes in the employee's physical condition that could affect respirator fit are reported or noticed, and
4. if after passing a QLFT or QNFT an employee notifies the employer that the fit of the respirator is unacceptable. (This last circumstance implies an imminent change in respirator going back to circumstance number 2 above.)

Protocols for both QLFT and QNFT are described in part B of Appendix A to OSHA 1910.134, portions of which are quoted below. Both kinds of protocols include measuring or assessing the respirator's protection during (or after in one case) the following series of activities conducted for 1 min each: normal breathing, deep breathing, moving one's head from side to side, moving one's head up and down, bending over, and talking.[j] The QNFT also includes grimacing for 15 sec.

1. A QLFT is simply a pass or fail test. A given respirator on an individual wearer passes a QLFT if the wearer cannot detect the presence of a test substance inside the mask. If the wearer detects the test substance inside the mask, the test fails and another respirator should be tried. Each QLFT starts by screening the individual's odor or taste detection threshold for the test substance without a respirator, followed by a challenge at a higher level while wearing the respirator and exercising, as described above. The wearer in Figure 22.8b is undergoing a qualitative fit test.
   a. Acceptable test substances and their corresponding air purifier against which they are tested include:
      1. isoamyl acetate: detecting the odor from isoamyl acetate (also called banana oil or "IAA") is prevented from reaching inside the respirator by an organic vapor cartridge.
      2. saccharin: detecting the sweet taste from an aerosolized saccharin solution is prevented by a particulate filter. (Saccharin is a sugar substitute.)
      3. Bitrex™: detecting the bitter flavor of denatonium benzoate (used as a taste aversion agent for children) is prevented by a particulate filter.
      4. irritant fume: detecting irritation due to hydrochloric acid mist from $SnCl_4$ ventilation smoke tubes (available from MSA; Draeger makes a substantially similar $TiCl_2$ tube) is prevented specifically by a HEPA or P100 filter.
   b. §134(f)(6) states, "QLFT may only be used to fit test negative pressure air-purifying respirators that must achieve a fit factor of 100 or less." After applying the OSHA policy of dividing an individual's fit factor in a controlled test by a safety factor of 10, this paragraph implies that a respirator passing a QLFT can only assure that it will achieve a WPF of 10. Thus, it should only be used to protect against a contaminant that exceeds its exposure limit by no more than a factor of 10. Notice that the APF for a negative pressure or demand mode half-mask respirator in Table 22.3 is 10. Thus, by far the most common use of QLFT is to test the acceptability of a conventional half mask air purifying respirator.
   c. Although it is not explicitly stated in 1910.134, it is nonetheless true that a full facepiece that passes a QLFT can also be used in a negative pressure mode as if it had an APF of 10. A common instance of using a QLFT for a full facepiece respirator is if it were chosen because of its ability to prevent eye irritation rather than because of its higher APF. Thus, a full facepiece respirator passing a QLFT should also only be used where the amount of protection needed (as defined by Equation 22.8) is no greater than 10.
   d. A third use of a QLFT is to qualify any tight-fitting positive pressure respirator. ANSI Z88.2-1992 states (and OSHA agrees although it was again left unstated in 1910.134) that an acceptable individual fit factor for a respirator to be used in the positive pressure mode needs to be only 100. §134(f)(6) clearly states that a fit factor of 100 is achievable via the QLFT. Thus, any tight-fitting facepiece passing a QLFT qualifies it to be used in one of the positive pressure modes (either continuous or pressure-demand) with the higher APF stated in Table 22.3.

[j] Appx. A.14 to §1910.134 suggests reading the "Rainbow Passage" contained therein because speaking it supposedly creates a wide range of oral and facial movements.

   e. Finally, Appendix A to OSHA 1910.134 suggests that a QLFT may also be used informally as a screen for respirator selection prior to a QNFT (although a pressure seal test is also an adequate screen) and potentially as a field check (although most people think a QLFT is too cumbersome for routine field use in comparison to a seal test as described in Appendix Bto 1910.134 and in Section IV.3 below).
2. A QNFT generates a quantitative fit factor (QNFF). §134(f)(7) states "If the (overall) fit factor, as determined through an OSHA-accepted QNFT protocol, is equal to or greater than 100 for tight-fitting half facepieces, or equal to or greater than 500 for tight-fitting full facepieces, the QNFT has been passed with that respirator." OSHA acceptable QNFT protocols in Appendix A of 1910.134 further specify that the QNFT fails if any peak penetration in Equation 22.3 exceeds 5% for a half mask (equivalent via Equation 22.2 to a minimum fit factor of less than 20) or 1% for a full facepiece (equivalent to a minimum fit factor of less than 100). The above quantitative fit factors are minimally acceptable values; most people with a good-fitting mask score much higher.
   a. Again left unsaid but implied by §134(f)(7), the only respirator required to pass a QNFT is a full facepiece that is to be used in a negative pressure mode with an APF of more than 10. The only APF for a negative pressure respirator in OSHA's proposed 1910.134 Table I (Table 22.3 herein) is the full-face mask used as either an air purifying or demand mode atmosphere supplying respirator with an APF of 50. However, recall as just stated above that a full-face mask can be used for eye protection up to an APF of 10 based only on a QLFT. Similarly, a half-mask passing a QNFT gives it no operational benefit, although there may be some psychological benefits or liability assurance to this higher level of testing.
   b. Part C of Appendix A of 1910.134 specifies the following three quantitative test methods that are currently acceptable to OSHA:
      1. A generated aerosol quantitative fit testing protocol describes how a particle detector can be used to simultaneously measure the concentration of a generated test aerosol in the air both outside and inside a mask worn by someone inside a test booth. The aerosel may be either sodium chloride (NaCl), corn oil, polyethylene glycol (PEG 400), or di-2-ethyl hexy sebacate (DEHS). A key criterion for any particular testing apparatus to use this protocol is that it must be capable of quantifying a personal fit factor of at least 2000.
      2. The ambient aerosol condensation nuclei counter (CNC) quantitative fit testing protocol is specific to the TSI Inc. Portacount® that can simultaneously measure the concentration of very small particles (condensation nuclei) in ambient air outside and inside a respirator worn by someone in any room of average dustiness.
      3. The controlled negative pressure (CNP) quantitative fit testing protocol is reasonably specific to equipment manufactured by Dynatech Nevada, Inc. that can measure the air flow rate needed to maintain a constant intra-mask pressure of 15 mm (0.58 in.) of water. Dynatech Nevada's system translates that flow rate into a fit factor. This test can be used on a respirator worn in any environment, but it requires the wearer to hold their breath for 10 sec at a time, and it uses a slight modification of the test exercises used in all of the other protocols.
3. A seal check is a simple yet important field procedure used to ensure an adequate seal, a properly functioning valve, and general mask integrity. (A seal check is sometimes called either a pressure seal test or a pressure check but should never be called or confused with a fit test.) §1910.134(g)(1) states "For all tight-fitting respirators, the

employer shall ensure that employees perform a user seal check each time they put on the respirator ..."[k]

Two procedures are described in Appendix B-1 of 1910.134 for conducting a seal check. Both procedures are largely subjective but can detect a gross failure, changes in general fit, and/or the need for a user to make an adjustment to the respirator. Manufacturer's recommendations are allowed as a third alternative where they have been shown to be equally effective.

a. In a negative pressure check, the wearer covers or closes off the inlet, tries to inhale for 10 sec, and feels for leakage. This option parallels how leakage normally occurs during use.
b. In a positive pressure check, the wearer covers or closes off the exhaust port, tries to exhale, and feels for where and how easy it is (or difficult if it is really tight) for the mask to leak. To implement this option on some mask designs, the exhalation valve cover would have to be removed.

V. RESPIRATORY PROTECTION PROGRAM REQUIREMENTS

While OSHA requirements do not always equate to good IH practice and their requirements are subject to sporadic updates, the OSHA Respiratory Protection Standard (29 CFR 1910.134), as adopted in 1998, is organized in a way that largely parallels a good respiratory protection program, progressing from defining management responsibility, through equipment selection and supervision, to program recordkeeping and periodic evaluation requirements. §134(a) briefly restates the IH control paradigm from Chapter 1 in terms specific to respirators. §134(b) lists many of the definitions described above in more detail. §134(c) to §134(m) describe the following six generic PPE program elements plus three specialize elements required to manage a respirator program acceptable to OSHA.

1. As with most other personal protective equipment programs, employers using respiratory control are required to have a written respirator program. OSHA 1910.134(c) describes the required contents of that written program. §134(c)(1) "In any workplace where respirators are necessary to protect the health of the employee or whenever respirators are required by the employer, the employer shall establish and implement a written respiratory protection program with worksite-specific procedures." A 28 page example of a generic written program for a hypothetical XYZ Company is available from OSHA in their 328 page "Small Entity Compliance Guide" via OSHA's homepage.[l]
   a. It is essential that the person responsible to manage the respirator program be specifically named within this plan. §134(c)(3) "The employer shall designate a program administrator who is qualified by appropriate training or experience that is commensurate with the complexity of the program to administer or oversee the respiratory protection program and conduct the required evaluations of program effectiveness." While the program administrator is not required to be an industrial hygienist, an industrial hygienist is not only qualified but probably the most qualified person within the enterprise to be the program administrator.

[k] Does a QLFT or QNFT constitute "a use"? Should a person pressure test a respirator before being fit tested? Part A of Appendix A to 1910.134 states that a respirator must pass a user seal check before being subjected to a fit test. In this case, the seal check can also function as a screening tool for respirator selection.

[l] See either p. 260-288 of the whole www.osha-slc.gov/OshDoc/Additional.html#SECG-RPS document and/or p. 125-148 of a downloadable www.osha-slc.gov/Publications/secgrev-current.pdf file.

b. §134(c)(1) further spells out the requirements of that written program: "The employer shall include in the program the following provisions of this section, as applicable:"
   1. procedure for selecting respirators; §134(d)
   2. medical evaluations of employees required to use respirators; §134(e)
   3. a fit testing procedure to be used for tight-fitting respirators; §134(f)
   4. supervisory procedures to ensure the proper use of respirators; §134(g)
   5. procedures and scheduling for maintenance and care of respirators; §134(h)
   6. procedures that ensure quality breathing air is used for ASRs; §134(i)
   7. training employees in the hazards, use, and care of their respirators; §134(k) and
   8. procedures to regularly evaluate the effectiveness of the program. §134(l)

c. Section §134(c)(4) spells out the responsibility of employers to cover all the costs of a respirator program where respirators are required, "The employer shall provide respirators, training, and medical evaluations at no cost to the employee."

d. Section §134(c)(2) discusses the special case of voluntary usage where respirators are not required but an employee wants to wear one anyway. "An employer may provide respirators at the request of employees or permit employees to use their own respirators, if the employer determines that such respirator use will not in itself create a hazard. If the employer determines that any voluntary respirator use is permissible, the employer shall provide the respirator users with the information contained in Appendix D (of 1910.134)."

This section goes on to say that "... the employer must establish and implement those elements of a written respiratory program necessary to ensure that any employee using a respirator voluntarily is medically able to use that respirator, and that the respirator is cleaned, stored, and maintained so that its use does not present a health hazard to the user." However, it also states an important exemption: "Employers are not required to include in a written respiratory protection program those employees whose only use of respirators involves *the voluntary use of filtering facepieces (dust masks).*" (emphasis added)

2. A selection protocol is an important element of any PPE program. As a first step in that protocol, §134(d)(1) states that "The employer shall select and provide an appropriate respirator based on the respiratory hazard(s) to which the worker is exposed and workplace and user factors that affect respirator performance and reliability." A hazard assessment usually involves measuring exposures, but OSHA §134(d)(1)(iii) also allows the use of predictive physical models (such as presented in Chapters 5 and 20). "The employer shall identify and evaluate the respiratory hazard(s) in the workplace; this evaluation shall include a reasonable estimate of employee exposures to respiratory hazard(s) and an identification of the contaminant's chemical state and physical form. Where the employer cannot identify or reasonably estimate the employee exposure, the employer shall consider the atmosphere to be IDLH." Thus, a hazard assessment for respirator selection will require at least the following three steps:

a. Measure or predict the percent oxygen and anticipate the level to which it could change during the worst-case scenario that one can foresee. One should physically measure the oxygen level before measuring the chemical concentration to avoid the interaction of an oxygen deficiency degrading the accuracy of the chemical measurement method.

b. Measure or predict the concentration of the chemical compound(s) of interest and anticipate its maximum potential concentration during the worst foreseeable case scenario (e.g., from complete evaporation of an accidental spill or a container rupture or breakage; recall the models and use of a chemical's vapor pressure in Chapter 5).

c. Find (look up) the following toxicologic information on the chemical contaminant(s). Some of this data may be on an MSDS, but all of it probably will not be.
   1. The IDLH: a hard copy of the NIOSH *Pocket Guide to Chemical Hazards* is difficult to obtain between printings but is available at http://www.cdc.gov/niosh/npg/npg.html.[4] IDLH values are also a listed output in some commercial respirator manufacturer's websites. If an IDLH has not been established for the chemical contaminant, find its LFL, LC_{50} or LC_{low}, TCLo, etc. from its MSDS, NIOSH RTECS (the *Registry of Toxic Effects of Chemical Substances* is described but is not available at http://www.cdc.gov/niosh/rtecs.html), or/and other compendium.
   2. The exposure limit: the TLV or the AIHA WEEL® are guidelines; the PEL is mandatory. Any limit may be used, but exposures inside the respirator cannot exceed the PEL.
   3. Whether the compound is an eye irritant and at what level: of particular interest is whether the chemical is an eye irritant at the measured or expected concentration level.[15]
   4. The odor or other physiological warning threshold (not essential): sources of odor thresholds include published summaries[16,17] and respirator and chemical manufacturer data.[m]

3. The next step is to use that exposure assessment in a logical way to select a respirator. The decision logic described within the written plan need not necessarily cover all of the following topics, but the selection protocol that is described must comply with the broad outline of an acceptable (objective) decision logic contained within the rest of §134(d).[n] The most important concept or philosophical perspective that is common to all respirator selection protocols is that an atmosphere supplying respirator is always the default (safe) option. Thus, an air purifying respirator is only allowed if an atmosphere supplying respirator is not required for one or more specified reasons.

   The following respirator selection protocol is organized into three health-based criteria followed by four nonhealth-based criteria. The health-based selection criteria are

[m] The ability to detect a chemical's odor below its allowable exposure limit was, for many years, an acceptable means of detecting breakthrough of an air purifier. OSHA's new policy effectively deleted even good warning properties and objective tests on wearers as an indicator of breakthrough. However, independent of its use to detect breakthrough, an argument can be made against using a conventional negative pressure air purifying respirator for a hazardous gas or vapor without adequate warning properties because the wearer would then be unable to detect an over-exposure due to an improper fit, a stuck respirator exhaust valve, or other malfunction. Although this logic is not part of anyone's published criteria, a chemical's warning properties could be deemed inadequate for selecting an air purifying respirator in any of the following scenarios:

1. If the odor detection limit (or the detection limit for a significant portion of wearers) is greater than the PEL or other exposure limit, e.g., CH_3Cl and CO. In this scenario, the wearer would be over-exposed before a malfunction could be detected.
2. If the odor detection limit is so far below the exposure limit that it is likely to be detected even behind a properly functioning respirator. In this scenario, odor detection would be continuous, and the user would be required to detect a malfunction by judging when the intensity of the penetrating odor increases, a daunting task and dubious expectation.
3. The chemical causes dulling or numbing of the olfactory senses (e.g., H_2S above 50 ppm). In this scenario, the affected wearer would be unable to detect the odor if the chemical concentration were to increase to IDLH levels.
4. The odor detection limit is highly variable among people (e.g., the threshold for CH_3OH reportedly varies among people from 0.03 ppm to >9000 ppm), although an objective validation of the detection threshold of each individual wearer might be able to overcome this last objection.

[n] NIOSH released a new respirator decision logic as NIOSH Publ. No. 2005-100 that complies with the new OSHA 1910.134 standard. ANSI Z88.2-1992 is being revised and is expected out in 2005. Details of a decision logic are described in Section VI of this chapter. These decision logics are all based only on health-protection criteria.

presented in summary form here as subsections a to c, but will be presented in more detail as nine decision logic questions with yes or no answers within the next main Section VI of this Chapter.

a. The nature of the use setting or the seriousness of the chemical hazard that is present are sufficient to dictate that an atmosphere supplying respirator is required and (simultaneously) that an air purifying respirator is not allowed. An atmosphere supplying respirator is automatically required if any one of the following five use-conditions are met:
   1. where the respirator will be used to fight interior structural fires (see §134(g)(4)),
   2. where an atmosphere supplying respirator is specified by another standard (the other current OSHA standards that specify a particular respirator are listed in the appendix to this chapter),
   3. where the employer cannot identify or reasonably estimate the employee exposure,
   4. where either the contaminant concentration is above a level considered to be immediately dangerous to life or health (IDLH) or the oxygen level is below a level that OSHA considers to be IDLH,
   5. where oxygen is depleted below 19.5% (simply oxygen deficient).

   Some form of quantitative assessment of the workplace air quality prior to making a decision to use respirators for employee protection is the only way to avoid the third of the above conditions, and is necessary to assess the last two conditions, which is why a hazard assessment (as described above in Section V.2) is mandatory before an air purifying respirator can safely be used.
b. If an atmosphere supplying respirator is not required by the above criteria, the lack of an adequate air purifier can still preclude the use of an air purifying respirator. Any given gas or vapor air purifier will only remove certain contaminants for a limited time before the absorber will become saturated. After the absorber is saturated (shortly after breakthrough), virtually all of the gaseous contaminants will pass through allowing the contaminant concentration inside the respirator to exceed its exposure limit. Even if an atmosphere supplying respirator is not automatically required by the setting or concentration (step number 3.a above), one may still be required if an effective air purifier is not available from a manufacturer or its end of service life can neither be indicated nor predicted (an air purifier's end of service life is normally predicted by testing or a model).
c. The selected respirator must have an assigned protection factor (APF) at least as large as the ratio by which the contaminant concentration exceeds its exposure limit, as defined by Equation 22.8. The APF values in Table 22.3 are grouped by the relative pressure inside their respiratory inlet covering (negative, continuous flow, or pressure demand) and by the style of their respiratory inlet covering. The acceptability of the APF of a half-mask (or even a quarter mask) can be overridden by the need for eye protection that can only be provided by either a full facepiece mask, a loose-fitting facepiece, or a hood/helmet.

The three preceding health-based respirator selection criteria are explicit OSHA requirements within 1910.134. The following additional respirator selection criteria are very practical but neither health-related nor necessarily required by OSHA. Where one of these nonhealth criteria is implied within 1910.134, that paragraph is cited below. Cost, while clearly permeating all management decisions, is put at the end of the following list as much for convenience as for its relative priority.

d. Comfort of the selected respirator: comfort is an important subjective determination that can have a direct and strong bearing on the acceptability of any respirator by the users (and therefore on their compliance and the resulting effective protection factor predicted by Equation 22.7) and how tightly they wear the mask (and therefore their workplace protection factor). §134(d)(1) states that "The employer shall select respirators from a sufficient number of respirator models and sizes so that the respirator is acceptable to, and correctly fits, the user."

e. Usage pattern and setting: the frequency and duration of use, the climatic environment and work rate, and mobility and communication requirements affect both the costs of the program and the physiologic demands placed on the user.
   1. Frequency of use affects the economic choice of single-use (disposable) vs. multiple-use respirators, the storage facilities and maintenance level needed (§134(h)), and the physiologic demand and therefore the prerequisite health of the user (§134(e)).
   2. Duration of use affects the tank size and pressure if an SCBA is used, the cartridge or canister size if an air purifying respirator is used, the physiologic stress on the user, and again the prerequisite health of the user.
   3. Temperature and humidity affects not only the fit in the field of tight-fitting respirators, but can also affect the performance of air purifying cartridges and canisters (especially charcoal absorbers) and adds to the physiologic heat stress (in conjunction with the work rate).
   4. Work requiring the user to move over a wide distance is usually not compatible with an airline respirator, especially where the hose can become entangled with equipment. Selection of a self-contained breathing apparatus limits the duration of use for each charged bottle. If the wearer must frequently communicate orally, a respirator with a communication device may be needed; and if the communication distance is large, that device will probably need to be electronically enhanced. Each of these factors adds to the cost of respirators.
f. User capabilities: the capabilities of the user imply that both the wearer's health must be assessed and that subtle criteria (related to motivation) should not be overlooked.
   1. The health (physiologic ability) of the user is normally based on a medical assessment as discussed in §134(e) and the next subsection herein. The health of a user can affect the acceptability of the added physiologic demands placed on them by working while wearing a respirator; however, NIOSH has stated "if a worker is physically able to do an assigned job while not wearing a respirator, the worker will in most situations not be at increased risk when performing the same job while wearing a respirator".[18] An employer should balance work practice or workplace modifications with worker selection for respirator use to be consistent with reasonable accommodation within the American with Disabilities Act; however, §134(e)(6)(ii) also states in essence, that if the PLHCP finds that the use of a negative pressure respirator may place the employee's health at increased risk, the employer shall provide a PAPR if that is acceptable to the PLHCP.
   2. The motivation of the users affects their eventual compliance with use requirements, the allowable complexity of the equipment, and the necessary reliability of the equipment. The user's skill level and education often parallel their motivation but are neither the exclusive, sufficient, nor necessary criteria by which the chosen respirator should match the user.
g. While cost cannot be an overriding factor in the selection of PPE, the cost of control is always important to management. In general, the larger, the more complex, and the more reliable the respirator, the more costly it will be, although cost is not equally or consistently proportional to each of these qualities. With regard to reliability, §134(d)(1) states "The employer shall select a NIOSH-certified respirator." Uncertified respirators are often (but not always) cheaper. A respirator program's annual costs are based on more than the initial purchase price. The costs for respirator supplies can easily be \$100 to >\$500 per year per respirator user. Total costs

including a medical evaluation, training, and supervision can be two to four times that amount. Cost effectiveness calculations will generally show that control via respirators will eventually (if they are used long enough) cost the employer more than other options in the IH control hierarchy. Finding the best balance of cost and acceptability within the overriding constraint of protecting the user is truly an art within the science of industrial hygiene.

4. The medical evaluation of users is one of those "special" program management elements not common to all forms of PPE that is a requirement of respirators.
   a. §134(e)(1) "The employer shall provide a medical evaluation to determine the employee's ability to use a respirator, before the employee is fit tested or required to use the respirator in the workplace." This evaluation may be as simple as a questionnaire administered in conjunction with a physician or other licensed health care professional. §134(e)(2) "The employer shall identify a physician or other licensed health care professional (PLHCP) to perform medical evaluations using a medical questionnaire or an initial medical examination that obtains the same information as the medical questionnaire," such as provided in Appendix C to 1910.134.
   b. The above medical evaluation is a screening procedure. "The employer shall ensure that a follow-up medical examination is provided" that includes "any medical tests, consultations, or diagnostic procedures that the PLHCP deems necessary to make a final determination" under the following four conditions:
      1. per §134(e)(3): to an employee whose medical history or symptom responses to the questionnaire in Appendix C to §1910.134 or whose initial medical evaluation demonstrates the need for a follow-up medical examination;
      2. per §134(e)(7)(i): when an employee reports medical signs or symptoms that are related to their ability to use a respirator;
      3. per §134(e)(7)(ii and iii): when a PLHCP, a supervisor, the respirator program administrator, or a fit test technician informs the employer that an employee needs to be reevaluated;
      4. per §134(e)(7)(iv): when conditions change in the workplace that may result in a substantial increase in the physiological burden placed on an employee (e.g., metabolic work rate).

      There is no specific requirement for the frequency of periodic reevaluations. One guidance document suggests 1 to 5 years depending upon wearer age and the strenuousness of the work.[18]
   c. §134(e)(5): "The following information must be provided to the PLHCP before the PLHCP makes a recommendation concerning an employee's ability to use a respirator:
      1. The type and weight of the respirator to be used by the employee;
      2. The duration and frequency of respirator use (including use for rescue and escape);
      3. The expected physical work effort;
      4. Additional protective clothing and equipment to be worn; and
      5. Temperature and humidity extremes that may be encountered.
      6. ... a copy of the written respiratory protection program and a copy of this section."

      This information would usually be provided by the program administrator who is often the industrial hygienist.
   d. While §134(e)(6) states that the employer determines the employee's ability to use a respirator based on a written recommendation from the PLHCP, in practice there is a strong incentive to simply follow the PLHCP's recommendation. Several published references are available to guide an IH or PLHCP in evaluating the medical qualifications of someone to wear a respirator, although none very recently.[18–23]

Supervision of PPE has multiple dimensions including the supervision of the users, the use conditions, the respirators themselves, and the supervision of the supervisors. The next two subsections parallel the way the standard describes supervision.

5. Supervision needs to be provided over the users and the conditions in which respirators are being used.
   a. §134(g)(1) relates specifically to tight-fitting respirators:
      1. "The employer shall not permit respirators with tight-fitting facepieces to be worn by employees who have: facial hair ... or any (other) condition that interferes with the face-to-facepiece seal or valve function.
      2. The employer shall ensure that corrective glasses or goggles do not interfere with the seal.
      3. The employer shall ensure that employees perform a user seal check each time they put on the respirator.
   b. §134(g)(2)(i) "Appropriate surveillance shall be maintained of work area conditions and degree of employee exposure or stress" to detect changes that might affect the adequacy of the protection selected.
   c. §134(g)(2)(ii) relates to good user behaviors such as leaving the respirator use area to wash their face and respirators periodically, to replace air purifiers, to replace a cartridge if they detect the contaminant, or to respond appropriately if they detect a leak or other malfunction.
6. Supervision is also needed to assure the maintenance and care of respirators.
   a. §134(h)(1) states, "The employer shall provide each respirator user with a respirator that is clean, sanitary, and in good working order." Acceptable cleaning and disinfecting procedures are either those specified in Appendix B-2 of 1910.134 or those recommended by the respirator manufacturer (e.g. warm water, detergent, and a 1:1000 bleach solution). OSHA even specifies the following intervals at which respirators shall be cleaned and disinfected:
      1. Respirators issued for the exclusive use of an employee shall be cleaned and disinfected as often as necessary to be maintained in a sanitary condition;
      2. Respirators issued to more than one employee, shall be cleaned and disinfected before being worn by different individuals;
      3. Respirators maintained for emergency use or used in fit testing and training shall be cleaned and disinfected after each use (see also 22.E.6.c.3 below).
   b. §134(h)(2) states, "The employer shall ensure that respirators are stored ... to protect them from damage, contamination, dust, sunlight, extreme temperatures, excessive moisture, and damaging chemicals, and ... to prevent deformation of the facepiece and exhalation valve." Boxes or sealable plastic bags can suffice. The storage location should balance convenience with protection, supervision, and maintenance facilities. In addition, emergency respirators shall be accessible to the work area and clearly marked. Improper storage of respirators is a common OSHA citation.
   c. §134(h)(3) specifies the frequency of respirator inspections.
      1. All respirators used in routine situations shall be inspected before each use and during cleaning.
      2. All respirators maintained for use in emergency situations shall be inspected at least monthly and ... shall be checked for proper function before and after each use; and
      3. Emergency escape-only respirators shall be inspected before being carried into the workplace for use.
      4. SCBAs shall be inspected monthly, and gas cylinders recharged when the pressure falls to 90% of the manufacturer's recommended level.

   Subparts of this section describe further details that could easily be developed into a checklist.

d. §134(h)(4) contains some rather specific maintenance requirements. "The employer shall ensure that respirators that fail an inspection or are otherwise found to be defective are removed from service, and are discarded or repaired or adjusted ... only by persons appropriately trained to perform such operations and ... according to the manufacturer's recommendations and specifications ..." These general qualifications for most respirators are more specific for SCBAs: "Reducing and admission valves, regulators, and alarms shall be adjusted or repaired only by the manufacturer or a technician trained by the manufacturer."

The next two subsections covering atmosphere supplying and air purifying respirators, respectively, are also special requirements (along with Section V.4 above) beyond the six standard program elements.

7. The air quality required in atmosphere supplying respirators is specified in §134(i) along with other special requirements on compressors and air storage tanks.
   a. "Compressed breathing air shall meet at least the requirements for Grade D breathing air described in ANSI/Compressed Gas Association Commodity Specification for Air, G7.1-1989" that include:
      1. oxygen content (v/v) of between 19.5 and 23.5%
      2. hydrocarbon (condensed) content of 5 mg/m^3
      3. carbon monoxide (CO) content of 10 ppm
      4. carbon dioxide content of 1000 ppm
      5. the lack of a noticeable odor.

      In addition, breathing air stored in cylinders shall a moisture dew point of not more than -50°F (-45.6°C) at one atmosphere or 10°F lower than the coldest temperature expected in the area.
   b. Oil-lubricated compressors may use a high temperature alarm as a continuous surrogate for periodic monitoring of CO; however, the air supply must still be monitored periodically to prevent carbon monoxide from exceeding 10 ppm (the same as Grade D breathing air above).
   c. §134(i)(8) adds an additional precaution: "The employer shall ensure that breathing air couplings are incompatible with outlets for nonrespirable worksite air or other gas systems." It may be surprising how often employees have been asphyxiated by nitrogen (used for instance, to inert storage tanks) inadvertently cross-connected to a respirator inlet.[24]
8. §134(j): "The employer shall ensure that all (air purifying) filters, cartridges, and canisters used in the workplace are labeled and color coded with the NIOSH approval label and that the label is not removed and remains legible." ANSI/AIHA Z88.7-2001 was written (in parallel to the new 42 CFR 84) to redefine the color coding of all air purifying canisters, cartridges, and filters.[25] A selected range of these color codes is shown in Table 22.4.
9. Respirator users must be trained in the use, care, and limitations of their particular respirators. The industrial hygienist is likely to provide this training, although operations with a large number of respirator users might have either a full-time respirator technician and/or educator.
   a. OSHA §134(k) states that "The employer shall provide the training prior to requiring the employee to use a respirator in the workplace," annually thereafter, when changes in the workplace or respirator render previous training obsolete, or when it appears that the employee has not retained the requisite understanding or skill.[o]

[o] That user training appears as far back as §134(k) in the OSHA standard puts this one element quite out-of-synchrony with both its own requirement to train before use and with user training within a real respirator program.

TABLE 22.4
Extract of Colors Assigned to Canisters, Cartridges, and Filters within ANSI Z88.7-2001

Contaminant to Be Protected against	Color	Comments
Organic vapors	Black	
Ammonia and methyl amine gases	Green	
Acid gases	White	For example, SO_2, H_2S, HCl, Cl_2, ClO_2
Acid gas and organic vapors	Yellow	
P100 or HEPA filter	Purple	NIOSH still mandates magenta which is the same as the purple in ANSI Z88.7-2001
P95, P99, R95, R99, R100 filters	Orange	
N95, N99, N100 filters	Teal	Oil-free particulates[a]
Carbon monoxide[b]	Blue	
Universal or type N[b]	Red	For acid gas, ammonia, CO, and organic vapors
Any other gases or vapors	Olive	

[a] Teal is defined as either dark grayish blue or greenish blue, depending upon the dictionary.
[b] CO is oxidized into CO_2 by a MnO_2 + CuO catalyst called Hopcalite. Because Hopcalite can be poisoned by water vapor, a desiccant must precede the Hopcalite within a CO respirator. A viewable change in the color of that desiccant indicates the imminent contamination of Hopcalite by water.

b. Such training must cover the relevant respiratory hazards but goes well beyond the usual employee hazard communication required by OSHA 1910.1200. §134(k)(1) states that: "The employer shall ensure that each employee can demonstrate knowledge of at least the following:
   1. Why the respirator is necessary and how improper fit, usage, or maintenance can compromise the protective effect of the respirator;
   2. What the limitations and capabilities of the respirator are;
   3. How to use the respirator effectively in emergency situations, including situations in which the respirator malfunctions;
   4. How to inspect, put on and remove, use, and check the seals of the respirator;
   5. What the procedures are for maintenance and storage of the respirator;
   6. How to recognize medical signs and symptoms that may limit or prevent the effective use of respirators; and
   7. The general requirements of this (respirator standard)."

10. Personal protective equipment programs need to be evaluated periodically to ensure that they are still needed, that receiver controls are still the best approach, and that the particular program is adequate. §134(l)(1) states that: "The employer shall conduct evaluations of the workplace as necessary to ensure that the provisions of the current written program are being effectively implemented and that it continues to be effective." Evaluations shall include input from employee respirator users regarding respirator fit, selection, proper use, and maintenance. Most programs schedule annual evaluations.
11. Receiver controls should have good recordkeeping to support the periodic evaluations, to comply with OSHA, to support any claims against the employer in court, and to facilitate employee involvement in the respirator program. Retention of written records is required of medical evaluations, fit testing, and the respirator program itself. §134(m)(1) states that: "Records of medical evaluations required by this section must be retained and made available in accordance with 29 CFR 1910.1020" (which is Access to Employee Exposure and Medical Records). Beyond that however, §134(m)(2) states that only records of the most recent fit testing and the current respirator program are required to be

retained. On the other hand, 1910.1020 specifies that employee medical and environmental exposure data shall be retained for 30 years.

VI. A RESPIRATOR SELECTION PROTOCOL

A decision logic is a sequence of questions whose yes or no answers invariably lead the user to the correct action. A variety of ostensibly related respirator decision logics have been published.[7–9,26] The following decision logic is based on several but tailored to follow the new OSHA proposed requirements.[1,11] Details of the nine numbered questions listed below will be covered in the corresponding nine numbered subsections that follow, but first a short re-introduction to that logic.

The philosophy that should underlie any respirator selection protocol is that an air purifying respirator is allowed only if an atmosphere supplying respirator is not required. Also, an atmosphere supplying respirator is required if the answer to any of the first five questions is yes. The first three questions are qualitative. The answer to most of the questions beyond 3 depends upon having conducted a quantitative exposure assessment such as described above in Section V.2.

1. Is the respirator-use intended for fighting interior, structural fires (fire fighting)?
2. Is exposure to the appropriate respirator specified by another complete OSHA standard?
3. Is the exposure unknown (in terms of either the chemical or its concentration)?
4. Is the setting IDLH in terms of either its contaminant concentration or its oxygen level?
5. Is the oxygen level where the respirator would be used deficient (but not IDLH)?

If an atmosphere supplying respirator has not been required by one of the first five questions, an air purifying respirator will still be acceptable only if the answer to both of the next two questions is yes. The answers to these two questions usually depend upon a manufacturer's recommendation.

6. Does an air purifier exist that can remove the chemical? and
7. Does an objective method exist that will either predict or detect an air purifier's end of service life from which a purifier replacement schedule can be determined?

Although the basic decision of using either an atmosphere supplying or an air purifying respirator has usually been made by this point, the selection of an acceptable kind of respiratory inlet covering and pressure regulator (if any) will depend upon two more questions.

8. Is the contaminant an eye irritant at the expected exposure concentration?
9. How much protection (what magnitude of APF) is required?

   The answer to question 9 can still obligate one to choose an atmosphere supplying respirator, despite any of the above criteria.

The nonhealth criteria described in the last half of Section V.3 above (including the comfort of the selected respirator, the usage pattern and setting, the capabilities of the user, and cost) are not included within this selection protocol. It is reiterated here that while these nonhealth criteria involve important management decisions, the program administrator should never lose sight of the overriding priority to protect the health of the users.

The following decision logic parallels the decision flow chart in Figure 22.16 that begins when one is about to choose the kind of respirator for a given task or person. Figure 22.16c is a "legend" that depicts standard symbols to be used in any decision flow chart. The facepiece pressure and respiratory inlet covering are chosen by a series of decisions within question 9 leading to the respirator with the minimally acceptable APF in each case. The dashed lines shown near the bottom of both Figure 22.16a and b are nonstandard symbols used to depict acceptable upgrades (horizontal in each case) beyond the minimally acceptable respirator. Do not jump vertically to decisions that do not follow a line.

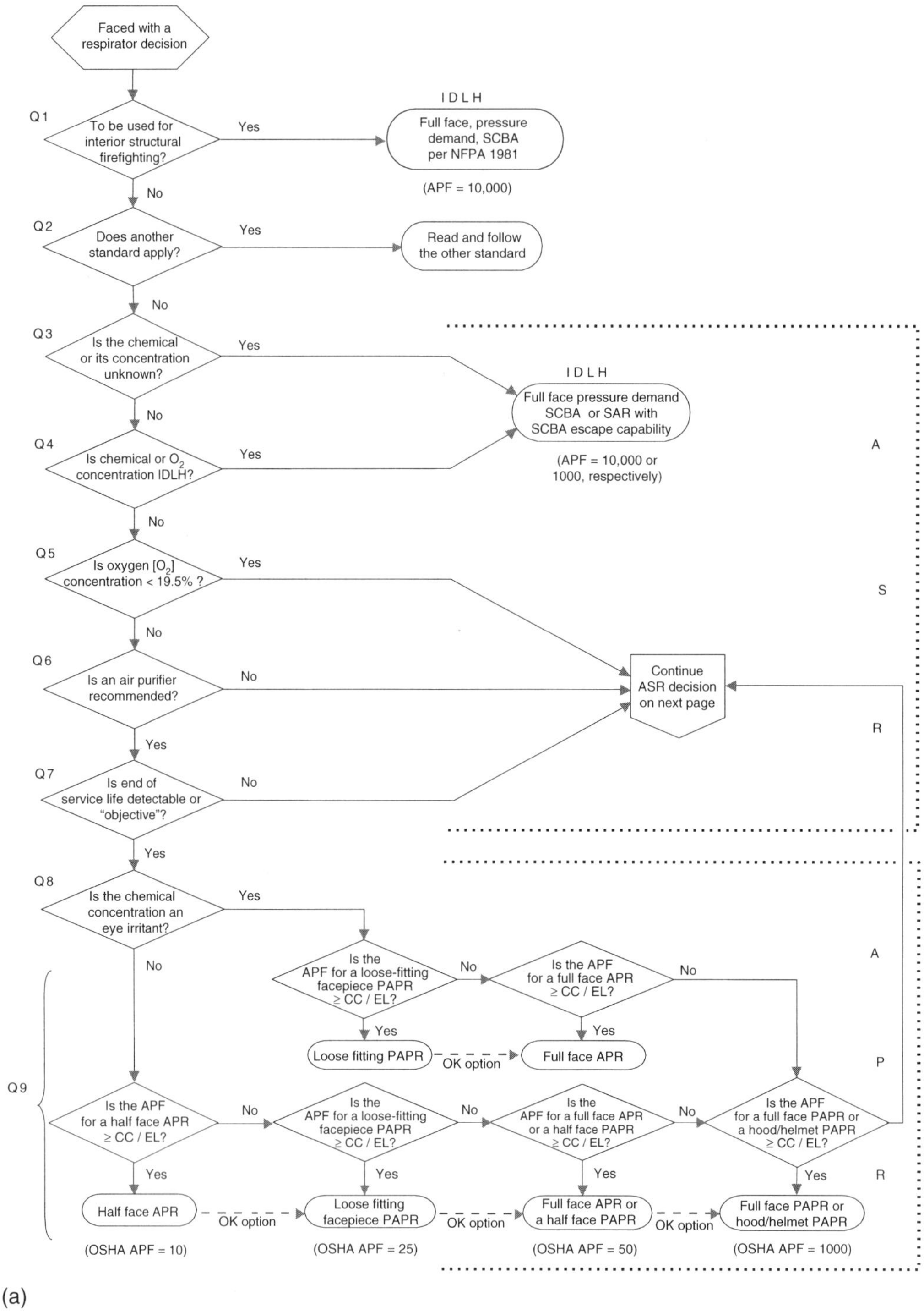

FIGURE 22.16a A decision logic flow chart leading to an air purifying respirator (see continuation in (b) for atmosphere supplying respirators). CC/EL = chemical concentration/exposure limit.

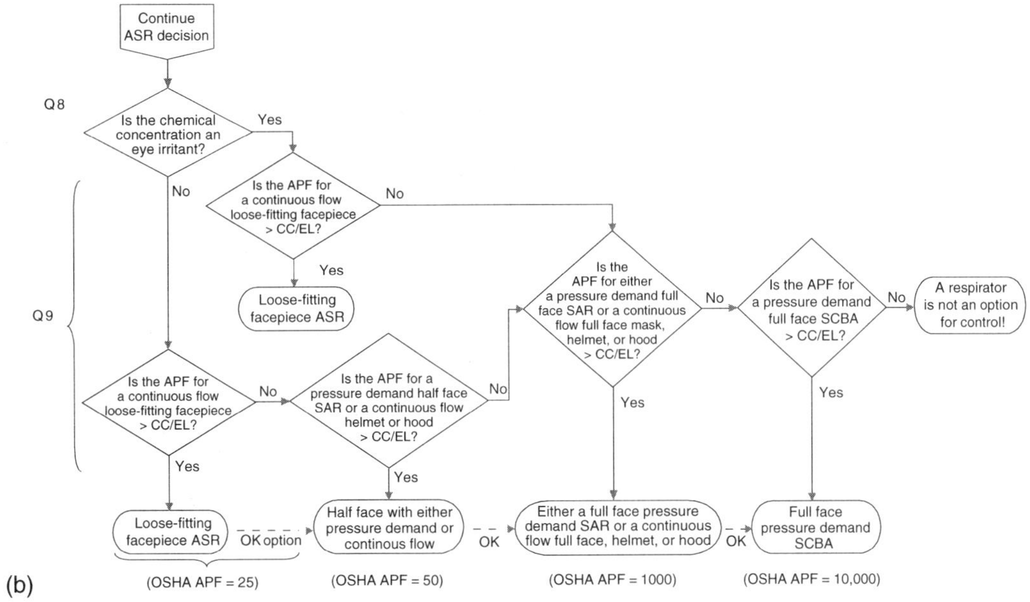

FIGURE 22.16b A continuation of the decision logic flow chart for questions 8 and 9 for atmosphere supplying respirators initiated in (a).

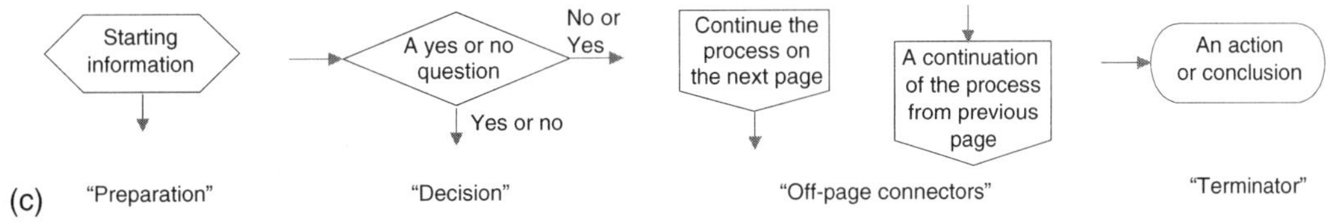

FIGURE 22.16c A legend of four standard decision logic symbols.

1. Is the respirator to be used for interior structural firefighting?

Both §134(a) and §134(g)(4) consider interior structural firefighting an IDLH environment that requires a self-contained breathing apparatus with a full facepeice and pressure demand regular as specified by NFPA 1981.

Firefighting involves both unknown chemicals and unpredictable concentrations and requires mobility. The decision logic flow chart goes directly to a pressure demand SCBA with requirements set by another organization (external to OSHA).[27]

2. Is the appropriate respirator specified within another standard?

§134(d)(3)(i) states (in part) "The employer shall … ensure compliance with all other OSHA statutory and regulatory requirements …" Other OSHA standards that specify which respirator to use include those for at least the two tasks and chemicals listed below.

The type of respirator is specified if the operation is either abrasive blasting (1910.94) or welding in a confined space (1910.252). Respirators are also specified for various levels of exposure to twenty-eight chemicals in standards 1910.1001 through 1052 (as listed in the second part of the appendix to this chapter).

3. Is either the airborne agent or its concentration unknown?

§134(d)(1) states "Where the employer cannot identify or reasonably estimate the employee exposure, the employer shall consider the atmosphere to be IDLH." The subsequent decisions are the same as those for a known IDLH condition.

The chemical is generally unknown in the initial phases of a community spill response or a chemical waste dump investigation. The concentration may be unknown during unplanned maintenance or if a chemical's vapor pressure is unknown. The decision chart leads one to the same respirator and procedures for an unknown condition that is considered (by default) to be IDLH as it would if the environment were known to be IDLH. Some of the special IDLH procedures are described in relation to question 5 below.

4. Is the environment immediately dangerous to life or health (IDLH)? More specifically, is either the chemical concentration greater than its IDLH or is the oxygen concentration less that its IDLH?

Working in any IDLH atmosphere automatically precludes an air purifying respirator, but OSHA §134(d)(2) further limits the choice of an ASR to use in an IDLH condition to either an SCBA or an airline respirator with an escape SCBA, either of which must be equipped with a full facepiece and pressure demand regulator.

OSHA has identified five conditions that can constitute an IDLH environment; three come from NIOSH, the last two are their own. IDLH values have been set by NIOSH for most of the chemicals with a PEL. See their Pocket Guide to Chemical Hazards or their web page http://www.cdc.gov/niosh/idlh/idlh-1.html.[4] The NIOSH definition of IDLH is the lowest chemical concentration that either:

a. is capable of causing severe eye or respiratory irritation sufficient to prevent escape (within a 30 min time frame),
b. has been reported to cause a lethal or irreversible health effect, or
c. is 10% or more of the compound's lower flammable limit (LFL).

OSHA added two more conditions that it considers IDLH for the purposes of respirator selection.

d. OSHA §134(d)(1)) states that settings shall be considered IDLH "Where the employer cannot identify or reasonably estimate the employee exposure, ..." This is the same criterion that in Selection Question 3 required the use of an atmosphere supplying respirator, but §134(d)(2) further limits the type of ASR by declaring an unknown hazard to be an IDLH setting.
e. While OSHA §1910.134(b) considers any ambient oxygen concentration that is less than 19.5% O_2 to be oxygen deficient, §1910.134(d)(2) Table II (reproduced below as Table 22.5) sets oxygen concentrations below which OSHA considers the deficiency to be IDLH. Recall from Chapter 2 that pure air is 20.95% oxygen and the discussion about oxygen defiencies in Chapter 4.

If the atmosphere is considered IDLH, then NIOSH recommends, and OSHA §134(d)(2) now requires, that one of two types of atmosphere supplying respirator systems must be used:

- Either "a combination full facepiece pressure demand supplied-air respirator (SAR) with auxiliary self-contained air supply." The auxiliary SCBA is a back-up respirator

TABLE 22.5
OSHA 1910.134 Table II Defines the Following Ranges of Oxygen Deficient Atmospheres at Various Altitudes That OSHA Does Not Consider IDLH

Altitude (feet)	Percent Oxygen (% O_2) for which the Employer May Rely on Any Atmosphere Supplying Respirator
Less than 3001	16.0–19.5
3001–4000	16.4–19.5
4001–5000	17.1–19.5
5001–6000	17.8–19.5
6001–7000	18.5–19.5
7001–8000[a]	19.3–19.5

[a] Above 8000 ft, the exception does not apply. Oxygen-enriched breathing air must be supplied above 14,000 ft.

for escape in case of an emergency in which case the user can disconnect from the external supplied airline source and breath for a short time on the small compressed air bottle such as that on the belt of the wearer in Figure 22.17.

- Or a "full facepiece pressure demand SCBA certified by NIOSH for a minimum service life of 30 min." The SCBA (such as that in Figure 22.1) does not require a separate back-up but is always limited in its duration of use by the capacity of its compressed air storage tanks.

OSHA §134(g)(3) further specifies that certain operational procedures (work practices) must be implemented whenever a respirator is used in an IDLH atmosphere, including the following:

- One or more standby employee (trained and equipped to provide effective emergency rescue) shall be located outside the IDLH atmosphere who can maintain visual, voice, or signal line communication with the employee(s) in the IDLH atmosphere.
- "The employer or designee is notified before the employee(s) located outside the IDLH atmosphere enters the IDLH atmosphere to provide emergency rescue."

5. Is the atmosphere oxygen deficient?

OSHA regulations preclude the use of an air purifying respirator if the oxygen is less than 19.5%, which leaves the ASR as the default respirator.

OSHA §1910.134(b) defines less than 19.5% oxygen as an oxygen deficient atmosphere requiring an atmosphere supplying respirator (again see Chapter 4). As explained in the previous question, the type of atmosphere supplying respirator (ASR) will depend upon how far below 19.5% the oxygen is. While this question can appear redundant with Question 4, logic demands that one cover the scenario where oxygen is deficient but not IDLH. If oxygen can be maintained between 19.5% and the lower limit within Table 22.5 for the altitude at which the respirator will be used, then any NIOSH approved ASR may be used but an air purifying respirator is insufficient and prohibited. This oxygen deficiency criterion provides protection against both inert gases that do not have a PEL (or TLV) but can displace oxygen and where oxygen has been consumed (as from microbial metabolism or chemical oxidation), possibly without any chemical contaminant.

FIGURE 22.17 An example of an airline respirator with auxiliary SCBA. Courtesy of MSA.

A "yes" answer to either of the first two questions virtually ended the decision-making process because the choice is dictated by another regulation (the short lines in Figure 22.16a). A "yes" to either of the next three questions (numbers 3–5) led either to the highest protection for an IDLH environment or to the selection of an atmosphere supplying respirator to be chosen in Figure 22.16b. Even if the decision logic has been followed to this point without an atmosphere supplying respirator having been mandated, an air purifying respirator is still allowed only if two more criteria are met. The next two questions relate to the availability of both an air purifier for the particular toxic chemical and an objective method to set the interval at which the air purifier should be changed. The answers to both of these questions for most users depend heavily upon the respirator manufacturer or their vendors.

6. Is there an adequate air purifier for the contaminant? In a more practical sense, does a manufacturer recommend an air purifier for the chemical contaminant?

If no air purifier is recommended for the contaminant, then one must again use an atmosphere supplying respirator. If an air purifier is recommended, go directly to the next question.

a. This criterion is primarily directed at gases and organic vapors. Although most gaseous contaminants can be absorbed by some air purifier, a few are poorly adsorbed by all of the cartridge materials currently available (e.g., $COCl_2$, PH_3, AsH_3, TDI, CH_2Cl_2, HCN, NO_x). Some of the causes for poor adsorption will be discussed in relation to estimating the chemical cartridge breakthrough time as part of the next respirator selection question. Answering this question generally means perusing the respirator manufacturer's literature or web site. If no cartridge is recommended, discuss the reasons with them. It may be that they have not tested your chemical, in which case they might test your chemical if they can sell you cartridges. Or they may know your chemical is incompatible with the adsorbents in any of their cartridges. If no absorbent is compatible, then an air purifying respirator is not an option, and the decision chart leads to an ASR decision (Figure 22.16b).

b. Finding an adequate filter for an aerosol is rarely a problem. From a strictly regulatory perspective, all filters certified by NIOSH under 42 CFR 84 are appropriate for use with any conventional APR to protect against any occupational aerosols except for hazards specifically regulated by another OSHA standard.[p] The problem for aerosols can be deciding which filter to use. Manufacturers can have their air-purifying, aerosol filters for conventional respirators certified by NIOSH under 42 CFR 84 within one of nine classes as summarized in Table 22.6.[q] Each class is defined by its minimum filtration efficiency against 0.3 μm particles and its ability to tolerate the presence of oil in the aerosol. Practicality and economics suggest that not all nine classes will be popular.

   The choice of using either the N, R, and P-series of filters should be based simply on the presence of oil within the aerosol. N filters are Not resistant to oil; oil will reduce the service life of N filters quickly (within one 8-hour shift). R filters are Resistant to oil, but oil will slowly degrade their filtration efficiency; thus, an R filter should not be used for more than one 8-h shift when oil is in the air. P filters are oil Proof, and oil will hardly degrade their filtration efficiency at all.

   Choosing the appropriate filter efficiency (95, 99, or 100) is less clear cut than choosing the N, R, or P series. Some guidance can be obtained by comparing the penetration through one of the filters to a perfectly fitting face mask (one with no leakage). Penetration through an N95 filter would give such a respirator a protection factor of at least 20 (equivalent to 100% of the concentration outside $\div$ (100% $-$ 95%) inside), which is better than the APF for a negative pressure half-mask but not as good as that for a full facepiece. Using the same rationale, a perfectly fitting respirator with an N99 filter would have a protection factor of at least 100, which is better than the APF for a negative pressure full facepiece mask. Thus, the protective effect of a 95% filter would exceed a functional APF of 10 and thus would be adequate for a half-mask respirator or for a full facepiece only passing a QLFT. Similarly, at least a 99% filter might be needed for the functional protection of a full facepiece respirator to achieve its APF of 50.

   More efficiency is not necessarily better. There are some disadvantages to arbitrarily selecting a higher efficiency filter than is necessary, besides the fact that "100" filters probably cost more than the lower efficiency filters.

[p] Recall that "conventional" means an unpowered, tight-fitting, air purifying respirator; PAPRs are only certified to use high efficiency filters. Similarly, abrasive blasting 1910.94(a), asbestos 1910.1001, or cadmium 1910.1027 that all require the use of either a HEPA or a 99.97% efficient N100, R100, and P100 filter.

[q] §134(d)(3)(iv) was added to allow the continued use of HEPA filters certified under NIOSH's old procedures but not allow the use of old "dust or dust/mist/fume filters" for protection against small aerosol particles ("particles with mass median aerodynamic diameters (MMAD) [below] 2 micrometers") as discussed in Chapter 3.

TABLE 22.6
The Aerosol Filter Classifications and the Challenge Aerosol Specified in 42 CFR 84

Minimum Filter Efficiency against a 0.3 μm $d_{aerodynamic}$ Aerosol (%)	Category of Filter (and the Challenge Aerosol)			Old Terms for These Filters
	N (NaCl) Not Oil Resistant	R (DOP[a]) Resistant to Oil (for One Shift)	P (DOP[a]) Oil Proof	
99.97	N100	R100	P100	HEPA
99	N99	R99	P99	Dust/mist/fume
95	N95	R95	P95	Dust

[a] Dioctyl phthalate is physically similar to mineral oil.

- The collection efficiency of any of these filters against polydisperse aerosols will be better than the efficiency when they were tested in compliance with 42 CFR 84 because 0.3 μm particles are the most penetrating (as mentioned in Section I.4 of Chapter 3).
- The pressure drop as air passes through a filter is likely to start higher and/or increase faster with dust loading for a high efficiency (99.97%) filter than for a low efficiency (95%) filter. A higher pressure drop means that any leakage through the face seal will be greater, more respiratory effort is required, and user comfort and compliance may decrease.

7. Is there a reliable means either to detect air purifier saturation and/or breakthrough or to predict an air purifier's service life or change schedule to prevent overexposing the wearer?

Recall from Section IV.5 of Chapter 21 that service life means the longest period of time that a respirator, filter, sorbent, or other temporally limited component provides adequate protection to the wearer. A change schedule needs to be set to replace any air purifier before it ceases to provide adequate protection. As in the previous question, the answer to this question depends on whether the airborne hazard is gaseous or particulate in nature, but the complexity of the answers is pretty much in the reverse order from the answers to the previous question. Subsections a and b below could almost be viewed as subquestions.

a. Aerosol filters do not get saturated and have breakthrough in the same way chemical cartridges do, but particulate loading will increase the filter's resistance to flow and eventually require that the filter be replaced. As a filter loads up with particles, both its particle collection efficiency and its resistance to air flow increase (except if the aerosol is oil-based and the filter is not oil proof or P-series). An increase in air flow resistance will make breathing more difficult and will eventually either increase the leakage through a tight-fitting mask's face seal or cause a PAPR's fan flow rate to decrease below that needed for adequate protection. Thus, a noticeable increase in breathing resistance, a decrease in air flow rate, or physical damage to the filter should each signal the need to change out the filter before the respirator's protection is compromised. A noticeable increase in breathing resistance is the primary determinant of the service life for N and P filters. The use of an R-series filter exposed to an oil-based aerosol is limited to one shift.

The following supplemental information merely parallels the new filter certification requirements.

1) The use of an N-series filter is not specifically time limited. To be certifiable, the resistance of N filters cannot increase noticeably when exposed to up to 200 mg of solid or water-based particulates. Thus, the change schedule should be at least as long as the minimum time given by Equation 22.10. However, N-series filters should be assumed to have lost their collection efficiency upon exposure to an oil-based aerosol.
2) The use of an R-series filter is similar to an N-series filter for an aerosol without oil, but an R-series filter will retain its collection efficiency of an oil-based aerosol for no more than one 8-h shift or a filter loading of 200 mg (although the latter is unlikely to occur in one shift).
3) The use of a P-series filter is not specifically time limited whether exposed to an oil-based aerosol or not. Continuing to use a P-series filter beyond a loading of 200 mg puts it beyond the range of certification testing conditions. If not changed per Equation 22.10, a P-series filter should be replaced whenever a noticeable increase in breathing resistance is detected.

Equation 22.10 can be used to predict the minimum service life of filter respirators based on the time necessary to accumulate 200 mg of dust, the maximum loading to which filters are challenged during testing. This 200 mg is a conservative limit in that the air flow resistance of any certified respirator should not become noticeably more difficult until some time past this point.[r]

$$\text{Minimum filter service life [minutes]} \geq \frac{200\ \text{mg} \times 10^3[\text{L/m}^3]}{C[\text{mg/m}^3] \times \text{RMV[Lpm]}} \quad (22.10)$$

Common values for the respiratory minute volumes can be 15, 25, and 35 Lpm corresponding to the upper edge of work rates typically categorized as low, moderate, and high, respectively. For example, Equation 22.10 predicts the minimum service life of a half-mask filter for use in an atmosphere of 10 mg/m^3 at a moderate work rate would be at least 13 h. This result is conservative in comparison to rule-of-thumb guidance to replace a filter at least weekly in heavy dust, bi-weekly for medium dust, and every 3 weeks in light dust.

$$\text{Minimum time} \geq \frac{200\ \text{mg} \times 10^3[\text{L/m}^3]}{10[\text{mg/m}^3] \times 25[\text{Lpm}]} = 800\ \text{min} \quad \text{(using Equation 22.10)}$$

b. A chemical absorber (for a gas or vapor) should be replaced before its purifying capacity has become saturated by gas and vapor molecules filling all of its adsorption or reaction sites. §134(d)(3) states that "For protection against gases and vapors, the employer shall provide (either) an atmosphere-supplying respirator, or an air-purifying respirator, provided that: (either) the respirator is equipped with an end-of-service-life indicator (ESLI) certified by NIOSH for the contaminant; or If there is no ESLI appropriate for conditions in the employer's workplace, the employer implements *a change schedule for canisters and cartridges that is based on objective*

[r] To be certified, a new filter cannot exceed 35 mm of water pressure drop at a continuous flow rate of 85 Lpm (cf. limits of 40 to 50 mm of water for a new chemical cartridge and 45 to 70 mm for their final resistance). Raven et al. (1981) tested a "demand" respirator with an inspiratory resistance of up to 85 mm of water and found that the physiological responses were the same for mildly pulmonary impaired subjects as normal controls at work rates up to 63% of maximal capacity.[20]

information or data that will ensure that canisters and cartridges are changed before the end of their service life" (emphasis added).

The above statement simply means that an air purifying respirator can be used for vapors and gases only if there is a reliable way either to detect or to predict when the chemical purifier has been exhausted and can no longer remove the contaminant from the inhaled air. There are very few ways to detect breakthrough chemically. Thus, most programs establish a change schedule supported by information from the manufacturer, laboratory data generated elsewhere, or an acceptably reliable predictive model. OSHA does not consider a warning property (such as a detectable odor below the allowable exposure limit) to be either reliable or objective.

OSHA appears to have a hierarchy to its preferred forms of objective information or data regarding a respirator's service life. While this hierarchy may eventually affect policy, at the moment, any of the following methods is acceptable to OSHA.

1) An ESLI can theoretically be installed behind a cartridge. An ESLI is a positive (usually visual) indication of breakthrough. However, almost no chemical-specific ESLIs exist at this time (except for Hg and the indication when the water vapor absorbent that protects the "Hopcalite" that normally removes CO in a Type-N universal canister has been exhausted; a picture of the latter is shown in Figure 22.18).
2) The service life (time to breakthrough) can be measured in laboratory tests at the maximal concentration that could be encountered. Such custom data can be generated by a respirator manufacturer if there is a good prospect of selling more respirators. Customized testing can also be contracted to a commercial lab or conducted in an employer's own lab.
3) Most laboratory tests are conducted at concentrations that rarely are exactly what a particular customer or user needs. Therefore, the service life is most often predicted by interpolation or extrapolation from a small number of breakthrough times measured experimentally on a given cartridge at other concentrations. The descriptive extrapolation model by Yoon and Nelson[28] has

FIGURE 22.18 One form of ESLI.

shown good reliability when based on at least two and preferably three or more chemical concentrations.

$$\log\,[\text{time}_{\text{breakthrough}}] = A - B \log C_i \qquad (22.11)$$

where

$\text{time}_{\text{breakthrough}}$ = the breakthrough time at a given concentration (C_i).

C_i = the contaminant concentration being tested.

A and B = statistical regression constants derived from paired $\text{time}_{\text{breakthrough}}$ and C_i data for at least two experimental concentration conditions. At least three conditions are needed to yield an R^2-value as an indicator of the goodness of fit of the model for that chemical.

4) The breakthrough time can also be predicted by a model using only physical chemistry data, e.g., the mathematical predictive model described by Wood[29] that uses the adsorption capacity and rate of adsorption of organic vapors from organic liquids. However, OSHA recommends that at least some experimental confirmation of such a model be conducted.

Excerpts from the OSHA Technical Links' web site describing both the "Yoon" and the "Wood" models are included as The third part of the appendix to this chapter. Commercial vendors have developed various on-line (web based) service life prediction tools to aid buyers in properly using their air purification products. The following are three examples. Note that the low, medium, and high breathing rates in these tools are all higher than the corresponding work rates used in heat stress conditions and aerosol physics research.[30,31]

- MSA: go to http://www.msanet.com then select North America, Resources (at the top of the page), Tools and Cartridge Life Calculator. Their input parameters include the chemical, its concentration (ppm or mg/m^3), the ambient temperature, relative humidity, and atmospheric pressure (or elevation above sea level), the wearer's breathing rate (Lpm), the percent of the PEL upon which you want to base your change-out schedule, and the cartridge type (the latter must be chosen by the user).[s]
- 3M: go to either http://csrv.3m.com/csrv/ or http://www.3m.com then select United States, then Products and Services (near the top of the page), Occupational Health & Environmental Services, and look for Cartridge Change Schedule. The 3M Service Life Software is also down-loadable. Both versions use the Wood method to determine the service life for organic vapor cartridges.
- North Safety Products: go to http://www.northsafety.com, then select United States, "Selection Guide for Respiratory and Hand Protection," and "Cartridge Service Life Estimation." This site allows the viewer to choose the OEL value, the ambient temperature and relative humidity, the wearer's breathing rate (among categories), the percent of the PEL upon which you want to base your change-out schedule, and a safety factor. The program provides values for all of their cartridge types.

c. The following rules of thumb were suggested in Chapter 36 of the AIHA *White book* to help anticipate organic vapor cartridge air purifier service life and change-out schedules when a computer is not available.[32]

[s] MSA has the site user choose either a full-face or half-mask and cartridge type, although the mask style has no direct effect on service life since the same cartridges fit both kinds of their masks.

✓ 1. The service life should be at least 8 h at a normal work rate as long as the chemical's boiling point is >70°C (roughly equivalent to $P_{vapor} < 100$ mmHg) and its concentration is less than 200 ppm.

✓ 2. Service life is inversely proportional to work rate (due to the increase in ones breathing rate).

✓ 3. Reducing the exposed concentration by a factor of 10 × will increase an adsorbent's service life by a factor of 5 × (based on Equation 22.11).

✓ 4. As the humidity increases above 85%, the service life will likely be reduced by 50%.

✓ 5. Used organic vapor cartridges should be changed daily even if their prior use was for less than their predicted service life because adsorbed vapors can migrate throughout the charcoal bed during an overnight period of nonuse.

Do not forget the following general admonitions against using a chemical cartridge air purifier to control an airborne chemical exposure even if it is allowed. These admonitions form the basis of why an atmosphere supplying respirator is the default respirator (the safest choice). The first three of these admonitions were made in Section I.5 of this chapter, but after seeing the above nuances of air purifying respirators, the value (and liability) of adding the fourth admonition should be obvious.

NO air purifier can protect against:
1. insufficient *oxygen*
2. all contaminants,
3. any contaminant forever, or
4. failure to see and follow the manufacturer's recommendations.

The complexity and management requirements of a respiratory protection program make the previous options to control a chemical exposure hazard via source and pathway controls look better and better, and there are still two respirator selection criteria to go. The final two questions, taken together, determine the type of respiratory inlet covering and in-mask pressure required. They apply whether an air purifying respirator or an atmosphere supplying respirator is used (Figure 22.16a or Figure 22.16b, respectively).

8. Is the agent an eye irritant at the concentration(s) to be encountered?

A full-face mask (or hood or helmet) is recommended if the contaminant is an eye irritant at the exposure concentration (even if a half-mask APF is adequate).

Eye irritation is an overriding justification for specifying a full facepiece mask, even though a half or quarter mask might provide an adequate APF. Although MSDS sheets etc. might say a given chemical is an eye irritant, they almost never give quantitative guidance regarding the threshold concentration at which eye irritation may begin. One way to estimate an eye irritation level is when a commercial respirator's selection guidance recommends using a full-face mask for a contaminant concentration that is less than the half-mask's APF × the exposure limit (if it were not an eye irritant, a half-mask respirator should be adequate for up to 10 × the exposure limit, per Equation 22.9). Grant's *Toxicology of the Eye* (1993) can also help.[15]

9.

Which respirator will reduce the exposure to an acceptable concentration?
Select a respirator whose APF is $\geq \frac{\text{Contaminant Concentration}}{\text{Exposure Limit}} = \frac{CC}{EL}$ (similar to Equation 22.8)

Recall that one primary goal of selecting a respirator in the first place was to reduce the contaminant concentration in the breathing zone of the exposed person to or below some acceptable concentration. This goal implies that the following constraint must be satisfied in order for the wearer's breathing zone concentration to be below the acceptable exposure limit (EL, typically either the PEL or TLV):

$$\text{Breathing concentration} \leq \text{EL} \leq \frac{\text{ambient Contaminant Concentration}}{\text{Assigned Protection Factor [APF]}} \quad (22.12)$$

To meet this constraint operationally, it is necessary to select a respiratory inlet covering (and perhaps a pressure regulator) that together create an APF greater than the ratio of the ambient contaminant concentration over its exposure limit, "CC/EL" as shown in Equation 22.8. A condensed set of proposed OSHA APF values was compiled into Table 22.3 earlier in this chapter. The decision flow chart asks the above question repeatedly in a series of decision boxes read from left to right with increasing levels of respiratory protection aligned vertically, until the first "yes" answer defines the minimally acceptable respirator whose APF exceeds the necessary reduction in exposure. The series of questions are arranged in order of increasing APF under the assumption that the cost of both the respirator and the program will increase for more protective respirators. The dashed lines (between some but not all decision boxes) indicate where the program administrator can choose a more protective respirator. These dashed lines can move the decision horizontally but never vertically between Questions 8 and 9.

Example 22.1

Suppose a maintenance employee occasionally has to disassemble and repair part of a transfer pump carrying cyclohexene. Recent measurements found airborne exposures of about 900 ppm. The TLV and PEL are both 300 ppm. What type of respirator would be appropriate?

The following solution parallels the decision logic in Figure 22.16:

Q1 Is the respirator to be used for structural firefighting? Simple answer, no.

Q2 Is the appropriate respirator specified within another OSHA standard? The PEL for cyclohexene is one of the old TLVs within 1910.1000 Table Z-1. So the answer is no.

Q3 Is either the airborne agent or its concentration unpredictable or unknown? While cyclohexene's vapor pressure in Appendix A indicates that a saturated vapor would exceed the 900 ppm by over 100 ×, the measurements are considered a good predictor of concentration as long as the ventilation and work practice remain unchanged. So the answer to this question is also no.

Q4 Is the environment immediately dangerous to life or health? More specifically, is either the chemical concentration greater than its IDLH or is the oxygen concentration less than its IDLH? Or to pose the question in short-hand, is C > IDLH? To answer this question, one would have to look up the IDLH. The *NIOSH Pocket Guide*[4] lists the IDLH for cyclohexene as 2000 ppm. None of the measured exposure concentrations (C) were above the IDLH, so a fifth no answer. Oxygen was not IDLH, as answered in the next question.

Q5 Is the atmosphere oxygen deficient? The work is not in a confined space and the vapor concentration is not nearly in the 7% range found in Section IV.4.a of Chapter 4 to be necessary to displace oxygen below 19.5%. So, the answers to this and the IDLH question are no.

A quick review of Figure 22.16a would show that a yes answer to any of the above questions would have taken the decision directly to some form of supplied air respirator.

Q6 Is there an adequate air purifier for the contaminant? Again in a more practical sense, does a manufacturer recommend an air purifier for the chemical contaminant? One must go to a manufacturer, their web page, or a vendor to answer this question. While any of the vendors listed under question number 7 would do (and all will recommend an activated charcoal cartridge), the author prefers to use the MSA site because it offers the viewer more choices over the use and endpoint conditions being modeled.

Q7 Is there a reliable means either to detect air purifier saturation and/or breakthrough or to predict an air purifier's service life or change schedule to prevent overexposing the wearer? For all organic vapors, a change schedule will be determined from the manufacturer's information. The following results were obtained from MSA's web site.

Chemical name:	Cyclohexene
Chemical PEL (ppm):	300 OSHA PEL
Temperature:	22 °C
Relative humidity:	50%
Pressure:	760 Torr
Breathing rate:	30 LPM
Use concentration:	900 ppm
Breakthrough concentration:	100% OSHA PEL
The estimated breakthrough time at which cartridges need to be replaced is: 258 min.	

The breakthrough concentration chosen by the viewer is important when translating breakthrough time into an actual change schedule. The breakthrough concentration can be allowed to exceed the 8-TLV or PEL for a short period. In this case, 258 min is approximately half a workday. A change schedule is easier to enforce if it is linked to the workpattern. Thus, this cartridge should be changed every 4 h at a mid-shift break.

Q8 Is the agent an eye irritant at the concentration(s) encountered? For purposes of eye irritation, both the average and the maximum credible concentration can be considered. As an alternative to Grant's *Toxicology of the Eye*,[15] one can use the manufacturer's data or recommendations to determine the eye irritation threshold. MSA's Response® respirator selector is an example (found elsewhere at the MSA website). An eye irritation threshold can be implied by the lowest concentration at which a half-mask respiratory inlet covering is not recommended. In the case of cyclohexene, that concentration is 2000 ppm (which also happens to be its IDLH). Thus, cyclohexene is not expected to be an eye irritant at the concentrations of about 900 ppm.

Q9 Which respirator will reduce the exposure to an acceptable concentration? This question brings the decision to the last lines on Figure 22.16a which are really a series of questions concerning the acceptability of each respirator's APR in comparison to the agent's compliance ratio (the ratio of the agent's concentration to the exposure limit from Equation 5.18). The question is answered by Equation 22.8 (or some variation of it):

$$\text{APF}_{\text{of an acceptable respirator}} \geq \frac{\text{contaminant concentration}}{\text{the contaminant exposure limit}} = \frac{900\text{ ppm}}{300\text{ ppm}} = 3$$

Thus, an APF of only three is required. A glance at Table 22.3 will confirm that the APF for a half-mask air purifying respirator is 10. Thus, it is acceptable. One can follow the dashed lines to higher levels of protection, but the simplest respirator on this line will suffice as long as the organic vapor cartridges are changed at least every 4 h.

Taking all of the limitations of respirators into consideration (from the words of warning in subsection 7 above for simple air purifying respirators to the complexity, training, and maintenance required for atmosphere supplying respirators), there should be no doubt about the wisdom of the industrial hygiene control paradigm. We all owe a debt of gratitude to the pioneers, innovators, and leaders of the past who wrote down or otherwise tried to put the knowledge from their old heads into young bodies and those who will follow. We can all be learners and teachers in our time.

APPENDIX TO CHAPTER 22

TABLE 22A.1
A Comparison of Assigned Protection Factors from NIOSH,[7,8] ANSI,[9] and OSHA[11]

			Assigned Protection Factors		
			NIOSH	ANSI	OSHA
Conventional air purifying respirator					
Quarter mask			5	10	10
Half-mask			10	10	10
Full facepiece w. low efficiency filter			10	100	n.a.
Full facepiece/HEPA or gas/vapor			50	100	50
Powered air purifying respirators (PAPR)					
Loose-fitting facepiece (face shield)			25	25	25
Hood or helmet with shroud (ANSI varies w. filter)			25	100–1000	1000
Half mask			50	50	50
Full facepiece (ANSI varies w. filter)			50	100–1000	1000
Compressed air respirator (airline)		Regulator (see alternative sequence below)			
Half-mask (tight fitting)		Demand mode (rare)	10	10	10
Loose-fitting facepiece		Continuous flow	25	25	25
Hood or helmet with shroud		Continuous flow	25	1000	1000
Half-mask (semi-tight)		Continuous flow	50	50	50
Full-facepiece (tight fitting)		Demand mode	50	100	50
Full-facepiece (semi-tight)		Continuous flow	50	1000	1000
Half-mask (tight fitting)		Pressure demand	1000	50	50
Full-facepiece (tight fitting)		Pressure demand	2000	1000	1000
Self-contained breathing apparatus (SCBA)					
Full-facepiece w. escape		Demand mode	50	100	50
Full-facepiece w. escape		Pressure demand	10,000	10,000	10,000
Alternative sequence of atmosphere supplying respirators listed in order of increasing OSHA APF:					
Half-mask (tight fitting)	SAR	Demand mode (rare)	10	10	10
Loose-fitting facepiece	SAR	Continuous flow	25	25	25
Half-mask (semi-tight)	SAR	Continuous flow	50	50	50
Half-mask (tight fitting)	SAR	Pressure demand	1000	50	50
Full-facepiece (tight fitting)	–	Demand mode (rare)	50	100	50
Hood or helmet with shroud	SAR	Continuous flow	25	1000	1000
Full-facepiece (semi-tight)	SAR	Continuous flow	50	1000	1000
Full-facepiece (tight-fitting)	SAR	Pressure demand	2000	1000	1000
Full-facepiece (tight-fitting)	SCBA	Pressure demand	10,000	10,000	10,000

TABLE 22A.2
The Following Is a List of Workplace Contaminants and Processes that Have OSHA Specific Standards with Respirator Requirements that Take Precedence over the Normal Respirator Selection Decision Logic

Substance	General Industry	Shipyards	Construction	Agriculture
Acrylonitrile	1910.1045 (h)	1915.1045	1926.1145	
Arsenic (inorganic)	1910.1018 (h)	1915.1018	1926.1118	
Asbestos	1910.1001 (g)	1915.1001 (h)	1926.1101 (h)	
Benzene	1910.1028 (g)	1915.1028	1926.1128	
1,3-Butadiene	1910.1051 (h)			
Cadmium	1910.1027 (g)	1915.1027	1926.1127 (g)	1928.1027
Coke oven emissions	1910.1029 (g)		1926.1129	
Cotton dust	1910.1043 (f)			
1,2-Dibromo-3-chloropropane	1910.1044 (h)	1915.1044	1926.1144	
Ethylene oxide	1910.1047 (g)	1915.1047	1926.1147	
Formaldehyde	1910.1048 (g)	1915.1048	1926.1148	
Lead	1910.1025 (f)	1015.1025	1926.62 (f)	
Methylene chloride	1910.1052 (g)	1915.1052	1926.1152	
Methylenedianiline	1910.1050 (h)	1915.1050	1926.60 (i)	
Vinyl Chloride	1910.1017 (g)	1915.1017	1926.1117	
13 Carcinogens*	1910.1003 (c)(4)(iv)	1915.1003	1926.1103	

Process	Standard	Summary
Welding in confined spaces	1910.252	Airline or SCBA respirators shall be used where "adequate ventilation" cannot be provided to welding and cutting operations carried on in confined spaces.
Abrasive blasting	1910.94	"Abrasive blasting respirators" shall be worn by all abrasive blasting operators: (a) When working inside of blast-cleaning rooms, or (b) When using silica sand in manual blasting operations where the nozzle and blast are not physically separated from the operator in an exhaust ventilated enclosure, or (c) Where concentrations of toxic dust may exceed the PELs.

Notes: OSHA part 1910 standards apply to General Industry; Part 1915 standards apply to Shipyards; Part 1926 standards apply to Construction; and Part 1928 standards apply to Agriculture.

* 4-Nitrobiphenyl, alpha-naphthylamine, methyl chloromethyl ether, 3,3′-dichlorobenzidine (and its salts), bis-chloromethyl ether, beta-naphthylamine, benzidine, 4-aminodiphenyl, ethyleneimine, beta-propiolactone, 2-acetylaminofluorene, 4-dimethylaminoazo-benezene, and N-nitrosodimethylamine).

Modeling Respirator Change-out Schedules (Adapted from the OSHA-slc.gov/SLTC/respiratoryprotection)

Experimental Testing[33]

Protocols for laboratory testing of respirator cartridges have been well established. These protocols call for strict control of test conditions. The data obtained is reliable and accurate (especially when controlling for multiple simultaneous contaminants). Such testing may be conducted by the manufacturer. Ideally respirator cartridges should be tested under conditions which accurately reproduce the work environment or are carried out in the workplace which generic testing may not. Problems with this approach include:

- Test equipment is bulky and sophisticated.
- A large number of individual tests are required in order to generate the required breakthrough curves.
- Personnel conducting the test must be knowledgeable of the workplace and of the test equipment and protocols.
- Costs associated with the tests may be excessive.

A limited number of studies have been completed which examine more convenient approaches to estimating service life for respirator cartridges. One approach developed at the University of Michigan (Cohen, 1991) uses a glass tube filled with sorbent from the respirator cartridge (Respirator Charcoal Tube or RCT) through which the test vapor is drawn until breakthrough is detected using either a continuous monitor or the analysis of a short-term chemical sample.[33] The data obtained from the RCT is extrapolated to the cartridge (perhaps using the Wood model described below).

Yoon–Nelson Descriptive Extrapolation Model[28,34]

The determination of breakthrough from field data does not require the determination of the full breakthrough curve for the respirator cartridge. The Yoon–Nelson model is a descriptive extrapolation model which uses experimental data to calculate parameters which are then entered into the model. The basic equation for the model is:

$$t = \tau + \frac{1}{k'}\ln\left[\frac{P}{1-P}\right] \qquad \text{(Yoon–Nelson Equation 1)}$$

where

$t =$ breakthrough time (min)
$\tau =$ 50% contaminant breakthrough time (min)
$k' =$ rate constant (min^{-1})
$P =$ fraction of the assault contaminant concentration that breaks through $= \dfrac{C_{\text{behind the cartridge}}}{C_{\text{in front of cartridge}}}$

The value of τ can be determined from experimental data; τ and k' can also be determined from the regression of at least three points on a breakthrough curve (plotting C behind the cartridge vs. time). The value of k' has been shown to be related to τ by the following formula:

$$k' = \frac{k}{\tau} \qquad \text{(Yoon–Nelson Equation 2)}$$

where

k = proportionality constant that is independent of concentration and varies only slightly with humidity.

The value of τ is related to the contaminant concentration by the equation:

$$\log \tau = K'' - a \log C_I \qquad \text{(Yoon–Nelson Equation 14)}$$

where

K'', a = constants which can be derived from experimental data. They vary with humidity, but for humidities $\leq 50\%$ they are essentially constant.

C_I = contaminant assault concentration or concentration in front of the (ppm).

It is possible to determine the constants k, K'', and a from a minimum of three experimental determinations of τ at different C_I. However, the inclusion of additional data points increases the accuracy of the model.

The Wood Mathematical Predictive Model[29,35,36]

This predictive model should not be relied upon without experimental confirmation of the calculation. The basic equation for service life is:

$$t_b = \left[\frac{W_e W}{C_o Q}\right] - \left[\frac{W_e \rho_b}{k_v C_o}\right] \ln\left[\frac{C_o C_x}{C_x}\right] \qquad \text{(Wood Equation 1)}$$

where

t_b = breakthrough time (min)
W_e = equilibrium adsorption capacity (g/g carbon)
W = weight of carbon adsorbent
ρ_b = bulk density of the packed bed (g/cm^3)
C_o = inlet concentration (g/cm^3)
C_x = exit concentration (g/cm^3)
Q = volumetric flow rate (cm^3/min)
k_v = absorption rate coefficient (min^{-1}).

The parameter W_e can be estimated using the following equation:

$$W_e = W_o d_L \exp\left[-b' W_o P_e^{-1.8} R^2 T^2 \left(\ln\left(\frac{\rho}{\rho_{sat}}\right)\right)^2\right] \qquad \text{(Wood Equation 4)}$$

where

W_o = carbon micropore volume (cm^3/g)
d_L = liquid density of adsorbate (g/cm^3)
T = absolute temperature (°K = °C + 273)
P_x = partial pressure corresponding to concentration C_x
P_{sat} = saturation vapor pressure at temperature T
P_e = molar polarization
R = ideal gas constant (1.987)
b' = an empirical coefficient with value 3.56×10^{-5}

The parameter P_e can be estimated using the following equation:

$$P_e = \left(\frac{n_D^2 - 1}{n_D^2 + 2}\right)\frac{\text{MW}}{d_L} \qquad \text{(Wood Equation 6)}$$

where
MW = molecular weight
n_D = refractive index

The parameter k_v has been estimated by Wood from experimental data to be equivalent to the following equation:

$$\frac{1}{k_V} = \left(\left(\frac{1}{V_L}\right) + 0.027\right)\left(\text{I} + \frac{S}{P_e}\right) \qquad \text{(Wood Equation 7)}$$

where
I = calculated to be 0.000825
S = 0.036 for 1% breakthrough
V_L = linear airflow velocity (cm/sec)

Wood uses an example of hexane with the following information for a person with a work rate of 53.3 l/min to calculate 94 min for a 1% breakthrough time through a pair of charcoal cartridges:

T = 22°C (295°K)
W_o = 0.454 (determined from experimental data)
d_L = 0.6603 (available from scientific handbooks)
P_e = 29.877 (calculated from available data)
P_{sat} = 121 torr (available from scientific handbooks)
P_x = 0.38 torr (500 ppm challenge concentration) (calculated from available data)
V_L = 11.22 cm/sec (calculated from available data)
W = 70.6 g (calculated from available data)
C_o = 0.00178 g/cm^3 (calculated from available data)
k_v = 4242 min^{-1}

REFERENCES

1. Respiratory Protection, OSHA Regulations, Title 29 of the Code of Federal Regulations, Part 1910 Standard 134.
2. Myers, W. R., Lenhart, S. W., Campbell, D., and Provost, G., Letter to the editor: respirator performance terminology, *Am. Ind. Hyg. Assoc. J.*, 44(3), B25–B26, 1983.
3. Myers, W. R., Respiratory protective equipment, In *Patty's Industrial Hygiene*, 5th ed., Vol. 2, V: Engineering Control and Personal Protection, Harris, R. L., Ed., Wiley, New York, pp. 1489–1550, chap. 32, 2000.
4. NIOSH *Pocket Guide to Chemical Hazards*, DHHS (NIOSH), 97-140 1997, Also on-line at http://www.cdc.gov/niosh/npg/npg.html.
5. Popendorf, W., Merchant, J. A., Leonard, S., Burmeister, L. F., and Olenchock, S. A., Respiratory protection and acceptability among agricultural workers exposed to organic dust, *J. Appl. Occup. Environ. Hyg.*, 10(7), 595–605, 1995.
6. Popendorf, W., Error analysis in assessing respiratory protection factors, *J. Appl. Occup. Environ. Hyg.*, 10(7), 606–615, 1995.

7. NIOSH *Guide to Industrial Respiratory Protection*, DHHS (NIOSH) Publication No. 87–116, 1987, http://www.cdc.gov/niosh/87-116.html.
8. NIOSH *Respirator Decision Logic* 2004, DHHS (NIOSH) Publication No. 2005-100 2004, See http://www.cdc.gov/niosh/docs/2005-100/.
9. ANSI: American National Standard for Respiratory Protection, ANSI Z88.2-1992, A.N.S.I., NY, 1992.
10. Nelson, T. J., The assigned protection factor according to ANSI, *Am. Ind. Hyg. Assoc. J.*, 57(8), 735–740, 1996.
11. Department of Labor, Occupational Safety and Health Administration, 29 CFR Parts 1910, 1915, and 1926 [Docket No. H049C] Assigned Protection Factors; Proposed Rule, Federal Register, 68(109), Friday, June 6, pp. 34036–34119, 2003.
12. NIOSH: *Documentation for Immediately Dangerous to Life or Health Concentrations (IDLH)*, NTIS Publication No. PB-94-195047, Also available at http://www.cdc.gov/niosh/idlh/intridl4.html.
13. Holt, G. L., Employee facial hair vs. employer respirator policies, *J. Appl. Ind. Hyg.*, 2(5), 200–203, 1987.
14. Stobbe, T. J., daRoza, R. A., and Watkins, M. A., Facial hair and respirator fit: a review of the literature, *Am. Ind. Hyg. Assoc. J.*, 49(4), 199–204, 1988.
15. Grant, W. M. and Schuman, J. S., In *Toxicology of the Eye*, Thomas, C. C., Ed., 4th ed, Springfield, IL, 1993.
16. Reist, P. C. and Rex, F., Odor detection and respirator cartridge replacement, *Am. Ind. Hyg. Assoc. J.*, 38: 563–566, 1977.
17. Ruth, J. H., Odor thresholds and irritation levels of several chemical substances: a review, *Am. Ind. Hyg. Assoc. J.*, 47: A142–A151, 1986.
18. NIOSH: Medical Aspects of Wearing Respirators in Appendix H to *Criteria for a Recommended Standard: Occupational Exposure to Ethylene Glycol Monomethyl Ether, Ethylene Glycol Monoethyl Ether, and Their Acetates*), DHHS (NIOSH) Publication, No. 91-119, 1991.
19. Raven, P. B., Dodson, A. T., and Davis, T. O., The physiological consequences of wearing industrial respirators: a review, *Am. Ind. Hyg. Assoc. J.*, 40(6), 517–534, 1979.
20. Raven, P. B., Jackson, A. W., Page, K., Moss, R. F., Bradley, O., and Skaggs, B., The physiological responses of mild pulmonary impaired subjects while using a "demand" respirator during rest and work, *Am. Ind. Hyg. Assoc. J.*, 42(4), 247–257, 1981.
21. Harber, P., Medical evaluation for respirator use, *J. Occup. Med*, 26: 496–502, 1984.
22. ANSI: American National Standard for Respiratory Protection/Respirator Use: Physical Qualifications for Personnel, ANSI Z88.6-1984, A.N.S.I., NY, 1984.
23. Hodous, T. K., Screening prospective workers for the ability to use respirators, *J. Occup. Med.*, 28(10), 1074–1080, 1986.
24. OSHA: Deaths Involving the Inadvertent Connection of Air-line Respirators to Inert Gas Supplies, Safety and Health Information Bulletin 04-27-2004, U.S. Department of Labor, 2004, See http://www.osha.gov/dts/shib/shib042704.html.
25. ANSI/AIHAZ, *88.7-2001 American National Standard Color Coding of Air-Purifying Respirator Canisters, Cartridges, and Filters*, American Industrial Hygiene Association, Fairfax, VA, 2001.
26. AIHA, *Respiratory Protection: A Manual and Guideline*, 3rd ed., American Industrial Hygiene Association, Fairfax, VA, 2001.
27. NFPA, *1981 Standard on Open-Circuit Self-Contained Breathing Apparatus for Fire and Emergency Services*, National Fire Protection Association, Quincy, MA, 2002.
28. Yoon, Y. H. and Nelson, J. H., Breakthrough time and adsorption capacity of respirator cartridges, *Am. Ind. Hyg. Assoc. J.*, 53: 303–316, 1992.
29. Wood, G. O., Estimating service lives of organic vapor cartridges, *Am. Ind. Hyg. Assoc. J.*, 55: 11–15, 1994.
30. Human respiratory tract model for radiological protection, ICRP Publication 66, *Health Phys.*, 24(1–3), 1–201, 1994.
31. Ramsey, J. D. and Bishop, P. A., Hot and cold environments, In *The Industrial Environment, its Evaluation and Control*, DiNardi, S., Ed., 2nd ed., American Industrial Hygiene Association, Fairfax, VA, chap. 24, 1997.

32. Colton, C. E. and Nelson, T. J., Respiratory protection. In *The Industrial Environment, its Evaluation and Control*, DiNardi, S., Ed., 2nd ed., American Industrial Hygiene Association, Fairfax, VA, chap. 36, 1997.
33. Cohen, H. J., Levine, S. P., and Garrison, R. P., Development of a field method for calculating the service lives of organic vapor cartridges — part IV, Results of field validation trials, *Am. Ind. Hyg. Assoc. J.*, 52(7), 263–270, 1991.
34. Yoon, Y. H., Nelson, J. H., and Lara, J., Respirator cartridge service-life: exposure to mixtures, *Am. Ind. Hyg. Assoc. J.*, 57(9), 809–819, 1996.
35. Wood, G. O. and Lodewyckx, P., Extended equation for rate coefficients for adsorption of organic vapors and gases on activated carbons in air-purifying respirator cartridges, *Am. Ind. Hyg. Assoc. J.*, 64(5), 646–650, 2003.
36. Wood, G. O., Estimating service lives of organic vapor cartridges, Part II, a single vapor at all humidities, *J. Occup. Environ. Hyg.*, 1(7), 472–492, 2004.

Appendix A

The vapor hazard ratio (VHR = $P_{vapor} \times 10^6/\text{TLV} \times 760$) for selected chemicals in alphabetical order. The 2005 TLV® and the compound's vapor pressure [P_{vapor}] at 25°C are used unless otherwise indicated.[a]

Chemical Name	CAS #	TLV (ppm)	Ceiling	Skin	P_{vapor} (mmHg)	VHR	VHI
Acetic acid	64-19-7	10			15.7	2066	3.3
Acetic anhydride	108-24-7	5			0.498	131	2.1
Acetone	67-64-1	500			231.0	608	2.8
Acetonitrile or methyl cyanide	75-05-8	20		x	88.8	5842	3.8
Acetylene tetrabromide	79-27-6	1			0.02	26	1.4
Acrolein	107-02-8	0.1	C	x	274.0	3,605,263	6.6
Acrylamide (0.03 mg/m^3)[b]	79-06-1	0.010		x	0.0070	893	3.0
Acrylic acid	79-10-7	2		x	4.0	2632	3.4
Acrylonitrile; see vinyl cyanide	107-13-1	2		x	109.0	71,711	4.9
Adiponitrile	111-69-3	2		x	0.00068	0.45	−0.3
Aldrin (0.25 mg/m^3)[b]	309-00-2	0.0168		x	0.00012	9	1.0
Allyl alcohol	107-18-6	0.5		x	26.1	68,684	4.8
Allyl chloride	107-05-1	1			368.0	484,211	5.7
Allyl glycidyl ether (at 20°C)[c]	106-92-3	1			2.0	2632	3.4
2-Aminopyridine	504-29-0	0.5			0.8	2105	3.3
n-Amyl acetate	628-63-7	50			3.5	92	2.0
sec-Amyl acetate	626-38-0	50			8.03	211	2.3
iso-Amyl acetate (see Pentyl acetate)	123-92-2	50			5.6	147	2.2
Aniline and homologs	62-53-3	2		x	0.49	322	2.5
o-Anisidine (0.5 mg/m^3)[b]	90-04-0	0.1		x	0.08	1053	3.0
p-Anisidine (0.5 mg/m^3)[b]	104-94-9	0.1		x	0.019	250	2.4
ANTU (α-Naphthylthiourea 0.03 mg/m^3)[b]	86-88-4	0.0363			6.6×10^{-6}	2.4×10^{-1}	−0.6
Atrazine (5 mg/m^{3}at 20°C)[b,c]	1912-24-9	0.566			3.0×10^{-7}	1.0×10^{-3}	−3.2
Azinphos methyl (Guthion 0.2 mg/m^3)[b]	86-50-0	0.0154		x	7.5×10^{-9}	6.4×10^{-4}	−3.2
Benomyl (10 mg/m^3)[b]	17804-35-2	0.842			3.7×10^{-9}	5.8×10^{-6}	−5.2
Benzene	71-43-2	0.5		x	95.2	250,526	5.4
Benzotrichloride	98-07-7	0.1	C	x	0.41	5395	3.7
Benzoyl chloride	98-88-4	0.5	C		0.7	1842	3.3
Benzoyl peroxide (5 mg/m^3)[b]	94-36-0	0.505			0.00042	1	0.0
Benzyl acetate	140-11-4	10			0.177	23	1.4
Benzyl chloride	100-44-7	1			1.3	1711	3.2
Biphenyl or Diphenyl	92-52-4	0.2			0.00893	59	1.8
Bromine (at 20°C)[c]	7726-95-6	0.1			172.0	2,263,158	6.4
Bromine pentafluoride (at 20°C)[c]	7789-30-2	0.1			328.0	4,315,789	6.6
Bromoform	75-25-2	0.5		x	5.4	14,211	4.2

Chemical Name	CAS #	TLV (ppm)	Ceiling	Skin	P_{vapor} (mmHg)	VHR	VHI
1-Bromopropane	106-94-5	10			111.0	14,605	4.2
n-Butanol (Butyl alcohol)	71-36-3	20			6.7	441	2.6
sec-Butanol (Butyl alcohol)	78-92-2	100			18.3	241	2.4
tert-Butanol (Butyl alcohol)	75-65-0	100			41.7	549	2.7
2-Butoxyethanol (EGBE)	111-76-2	20			0.88	58	1.8
2-Butoxyethyl acetate (EGBEA)	112-07-2	20			0.38	25	1.4
n-Butyl acetate	123-86-4	150			11.5	101	2.0
sec-Butyl acetate	105-46-4	200			17.0	112	2.0
tert-Butyl acetate	540-88-5	200			47.0	309	2.5
n-Butyl acrylate	141-32-2	2			5.45	717	2.9
n-Butyl glycidyl ether (BGE)	2426-08-6	3		x	4.44	1947	3.3
n-Butyl lactate (at 20°C)[c]	138-22-7	5			0.4	105	2.0
n-Butyl mercaptan	109-79-5	0.5			45.5	119,737	5.1
p-*tert*-Butyl toluene	98-51-1	1			0.67	883	2.9
n-Butylamine	109-73-9	5	C	x	91.5	24,079	4.4
n-Butylated hydroxytoluene (2 mg/m^3)[b]	128-37-0	0.22			0.005	30	1.5
o-*sec*-Butylphenol	89-72-5	5		x	0.05	13	1.1
Camphor, synthetic	76-22-2	2			0.65	428	2.6
Carbaryl (Sevin) (5 mg/m^3)[b]	63-25-2	0.608			1.4×10^{-6}	2.9×10^{-3}	−2.5
Carbofuran (0.1 mg/m^3 at 20°C)[b,c]	1563-66-2	0.011			5.0×10^{-6}	1	−0.2
Carbon disulfide	75-15-0	10		x	358.0	47,105	4.7
Carbon tetrachloride	56-23-5	5			115.0	30,263	4.5
Cellosolve acetate (2-Ethoxyethyl acetate)	111-15-9	5		x	2.34	616	2.8
Chlordane (0.5 mg/m^3)[b]	57-74-9	0.0298		x	9.8×10^{-6}	4.3×10^{-1}	−0.4
Chlorinated camphene (0.5 mg/m^3)[b]	8001-35-2	0.0295		x	6.7×10^{-6}	3.0×10^{-1}	−0.5
1-Chloro-1-nitropropane	600-25-9	2			6.0	3,947	3.6
Chloroacetaldehyde	107-20-0	1	C		64.3	84,605	4.9
Chloroacetone	78-95-5	1	C	x	0.12	158	2.2
2-Chloroacetophenone or Phenacyl chloride	532-27-4	0.05			0.0054	142	2.2
Chloroacetyl chloride (at 20°C)[c]	79-04-9	0.05		x	19.0	500,000	5.7
Chlorobenzene	108-90-7	10			12.0	1579	3.2
Chlorobromomethane (Halon 1011)	74-97-5	200			143.0	941	3.0
Chlorodiphenyl or PCB (0.5 mg/m^3)[b]	11097-69-1	0.0372		x	7.7×10^{-5}	3	0.4
bis(2-Chloroethyl) ether (see Dichloroethyl ether)	111-44-4	5		x	1.5	395	2.6
Chloroform (trichloromethane)	67-66-3	10			197.0	25,921	4.4
bis-(Chloromethyl) ether	542-88-1	0.001			30.0	39,473,684	7.6
Chloropicrin or Nitrotrichloromethane	76-06-2	0.1			23.8	313,158	5.5
β-Chloroprene (or 2-Chloro-1,3-butadiene)	126-99-8	10			216.0	28,421	4.5
2-Chloropropionic acid	598-78-7	0.1		x	1.06	13,947	4.1
o-Chlorostyrene	2039-87-4	50			0.96	25	1.4
o-Chlorotoluene	95-49-8	50			3.4	89	2.0
Chlorpyrifos (0.1 mg/m^3)[b]	2921-88-2	0.007		x	1.7×10^{-5}	3	0.5
Chromyl chloride (at 20°C)[c]	14977-61-8	0.025			20.0	1,052,632	6.0

Chemical Name	CAS #	TLV (ppm)	Ceiling	Skin	P_{vapor} (mmHg)	VHR	VHI
Clopidol (10 mg/m^3)[b]	2971-90-6	1.273			0.00066	6.8×10^{-1}	−0.2
Cobalt carbonyl (0.1 mg/m^3 at 20°C)[b,c]	10210-68-1	0.0072			0.7	128,812	5.1
Cresol (mixed isomers at 20°C)[c,d]	1319-77-3	5		x	0.18	47	1.7
Crotonaldehyde	4170-30-3	0.3	C	x	30.0	131,579	5.1
Cumene	98-82-8	50			4.5	118	2.1
Cyclohexane	110-82-7	100			96.9	1275	3.1
Cyclohexanol	108-93-0	50		x	0.8	21	1.3
Cyclohexanone	108-94-1	20		x	4.33	285	2.5
Cyclohexene	110-83-8	300			89.0	390	2.6
Cyclohexylamine	108-91-8	10			10.1	1329	3.1
Cyclonite or RDX (0.5 mg/m^3)[b]	121-82-4	0.055			4.1×10^{-9}	9.8×10^{-5}	−4.0
Cyclopentadiene	542-92-7	75			435.0	7632	3.9
Cyclopentane	287-92-3	600			318.0	697	2.8
2,4-D (Dichlorophen-oxyacetic acid) (10 mg/m^3)[b]	94-75-7	1.106			6.0×10^{-7}	7.1×10^{-4}	−3.1
Di-(2-ethylhexyl) phthalate (DEHP 5 mg/m^3)[b]	117-81-7	0.3130			9.7×10^{-6}	4.1×10^{-2}	−1.4
Diacetone alcohol	123-42-2	50			1.71	45	1.7
Diazinon (0.01 mg/m^3)[b]	333-41-5	0.0008		x	0.0112	1834	3.3
Diazomethane	334-88-3	0.2			760.0	5,000,000	6.7
Dibutyl phosphate	107-66-4	1			9.6×10^{-5}	1.3×10^{-1}	−0.9
Dibutyl phthalate (5 mg/m^3)[b]	84-74-2	0.44			7.3×10^{-5}	2.2×10^{-1}	−0.7
1,1-Dichloro-1-nitroethane	594-72-9	2			16.9	11,118	4.0
1,4-Dichloro-2-butene	764-41-0	0.005		x	3.0	789,474	5.9
Dichloroacetic acid	79-43-6	0.5		x	0.18	474	2.7
o-Dichlorobenzene	95-50-1	25			1.36	72	1.9
p-Dichlorobenzene	106-46-7	10			1.0	132	2.1
1,1-Dichloroethane (or Ethylidene chloride)	75-34-3	100			227.0	2987	3.5
1,2-Dichloroethane (or Ethylene dichloride)	107-06-2	10			78.9	10,382	4.0
Dichloroethyl ether	111-44-4	5		x	1.55	408	2.6
1,2-Dichloroethylene (mixed isomers)[d]	156-59-2	200			250.0	1645	3.2
Dichloromethane or Methylene chloride	75-09-2	50			435.0	11,447	4.1
1,3-Dichloropropene	542-75-6	1		x	34.0	44,737	4.7
2,2-Dichloropropionic acid (5 mg/m^3 inhal.)[b]	75-99-0	0.85			0.19	294	2.5
Dichlorvos (DDVP) (0.1 mg/m^3)[b]	62-73-7	0.011		x	0.053	6303	3.8
Dichrotophos (0.05 mg/m^3 at 20°C)[b]	141-66-2	0.0052		x	8.6×10^{-5}	4.4	0.6
Dicyclopentadiene	77-73-6	5			2.3	605	2.8
Dieldrin (0.25 mg/m^3)[b]	60-57-1	0.0160		x	5.9×10^{-6}	4.8×10^{-1}	−0.3
Diethanolamine (2 mg/m^3)[b]	111-42-2	0.465		x	0.00028	1	−0.1
Diethyl ketone	96-22-0	200			37.7	248	2.4
Diethyl phthalate	84-66-2	0.55			0.0016	4	0.6
Diethylamine	109-89-7	5		x	237.0	62,368	4.8
2-Diethylaminoethanol	100-37-8	2		x	1.4	921	3.0
Diethylene triamine	111-40-0	1		x	0.23	303	2.5

Chemical Name	CAS #	TLV (ppm)	Ceiling	Skin	P_{vapor} (mmHg)	VHR	VHI
Diisobutyl ketone or 2,6-Dimethyl-4-heptanone	108-83-8	25			1.65	87	1.9
Diisopropylamine	108-18-9	5		x	79.4	20,895	4.3
Dimethyl acetamide	127-19-5	10		x	2.0	263	2.4
Dimethyl aniline (*N,N*-Dimethylaniline)	121-69-7	5		x	0.7	184	2.3
Dimethyl formamide	68-12-2	10		x	3.87	509	2.7
1,1-Dimethyl hydrazine	57-14-7	0.01		x	157.0	20,657,895	7.3
Dimethyl phthalate (5 mg/m^3)[b]	131-11-3	0.63			0.00165	3	0.5
Dimethyl sulphate	77-78-1	0.1		x	0.68	8947	4.0
meta-Dinitro benzene	99-65-0	0.15		x	0.0009	8	0.9
ortho-Dinitro benzene	528-29-0	0.15		x	4.6×10^{-5}	4.0×10^{-1}	−0.4
para-Dinitro benzene	100-25-4	0.15		x	2.6×10^{-5}	2.3×10^{-1}	−0.6
Dinitro-*o*-cresol (0.2 mg/m^3)[b]	534-52-1	0.0247		x	0.00032	17	1.2
Dinitrotoluene (0.2 mg/m^3)[b]	25321-14-6	0.027		x	0.0004	17	1.2
1,4-Dioxane (or Diethylene dioxide)	123-91-1	20		x	38.1	2507	3.4
Diquat (Resp.) (0.1 mg/m^3)[b]	2764-72-9	0.0071		x	1.0×10^{-8}	1.9×10^{-3}	−2.7
Endosulfan (0.1 mg/m^3)[b]	115-29-7	0.0060		x	1.0×10^{-5}	2	0.3
Endrin (0.1 mg/m^3)[b]	72-20-8	0.0064		x	3.0×10^{-6}	6.2×10^{-1}	−0.2
Epichlorohydrin (1-Chloro-2,3-epoxypropane)	106-89-8	0.5		x	16.4	43,158	4.6
EPN (0.1 mg/m^3)[b]	2104-64-5	0.0076		x	9.5×10^{-7}	1.7×10^{-1}	−0.8
Ethanethiol; see Ethyl mercaptan	75-08-1	0.5			529.0	1,392,105	6.1
Ethanol or Ethyl alcohol	64-17-5	1000			59.3	78	1.9
Ethanolamine or 2-Aminoethanol	141-43-5	3			0.404	177	2.2
Ethion (0.05 mg/m^3)[b]	563-12-2	0.0032		x	1.1×10^{-6}	0.06	−1.2
2-Ethoxyethanol or Cellosolve or EGEE	110-80-5	5		x	5.31	1397	3.1
2-Ethoxyethyl acetate or Cellosolve acetate	111-15-9	5		x	2.34	616	2.8
Ethyl acetate	141-78-6	400			93.7	308	2.5
Ethyl acrylate	140-88-5	5			38.6	10,158	4.0
Ethyl alcohol (see Ethanol)	64-17-5	1000			59.3	78	1.9
Ethyl amyl ketone or 5-Methyl-3-heptanone	541-85-5	25			5.03	265	2.4
Ethyl benzene	100-41-4	100			9.6	126	2.1
Ethyl bromide	74-96-4	5		x	467.0	122,895	5.1
Ethyl butyl ketone (3-Heptanone)	106-35-4	50			2.6	68	1.8
Ethyl ether or diethyl ether	60-29-7	400			537.0	1766	3.2
Ethyl formate	109-94-4	100			245.0	3224	3.5
Ethyl mercaptan or Ethanethiol	75-08-1	0.5			529.0	1,392,105	6.1
Ethyl silicate	78-10-4	10			1.88	247	2.4
Ethylamine	75-04-7	5		x	1.0	263	2.4
Ethylene chlorohydrin	107-07-3	1	C	x	7.18	9447	4.0
Ethylene diamine or 1,2-Diaminoethane	107-15-3	10		x	12.0	1579	3.2
Ethylene dibromide (EDB) PEL	106-93-4	20		x	11.0	724	2.9
Ethylene dichloride or (1,2-Dichloroethane)	107-06-2	10			78.9	10,382	4.0
Ethylene glycol (100 mg/m^3)[b]	107-21-1	39	C		0.092	3	0.5

Chemical Name	CAS #	TLV (ppm)	Ceiling	Skin	P_{vapor} (mmHg)	VHR	VHI
Ethylene glycol dinitrate	628-96-6	0.05		x	0.072	1895	3.3
Ethyleneimine	151-56-4	0.5		x	213.0	560,526	5.7
n-Ethylmorpholine	100-74-3	5		x	5.03	1324	3.1
Fenamiphos (0.1 mg/m^3)[b]	22224-92-6	0.008		x	1.0×10^{-6}	0.16	−0.8
Fenthion (0.2 mg/m^3 at 20°C)[b,c]	55-38-9	0.018		x	3.0×10^{-5}	2.3	0.4
Fonofos (Difonate) (0.1 mg/m^3)[b]	944-22-9	0.0099		x	0.0002	27	1.4
Formalin (37% using Henry's law)	50-00-0	0.3	C		3.16	13,841	4.1
Formamide	75-12-7	10		x	0.06	8	0.9
Formic acid	64-18-6	5			42.6	11,211	4.0
Furfural	98-01-1	2		x	2.21	1454	3.2
Furfuryl alcohol	98-00-0	10		x	0.609	80	1.9
Glutaraldehyde (at 30°C)[c]	111-30-8	0.05	C		0.6	15,789	4.2
Glycerin (10 mg/m^3 mist)[b]	56-81-5	2.65			0.00017	8.3×10^{-2}	−1.1
Glycidol or 2,3-Epoxy-1-propanol	556-52-5	2			5.59	3678	3.6
Halothane	151-67-7	50			302.0	7947	3.9
Heptachlor (0.05 mg/m^3)[b]	76-44-8	0.0033		x	0.0004	161	2.2
Heptane (*n*-Heptane)	142-82-5	400			46.0	151	2.2
Hexachlorobenzene (0.002 mg/m^3)[b]	118-74-1	0.0002		x	1.8×10^{-5}	138	2.1
Hexachlorobutadiene	87-68-3	0.02			0.22	14,474	4.2
Hexachlorocyclopentadiene	77-47-4	0.01			0.06	789	2.9
Hexachloroethane	67-72-1	1		x	0.21	276	2.4
Hexamethylene diisocyanate (HDI)	822-06-0	0.005			0.03	7895	3.9
n-Hexane	110-54-3	50		x	151.0	3974	3.6
Hexane, other than *n*- (at 20°C)[c]		500			208.0	547	2.7
1-Hexene	592-41-6	50			186.0	8158	3.9
sec-Hexyl acetate	108-84-9	50			4.0	105	2.0
Hexylene glycol	107-41-5	25	C		0.013	1.0	0.0
Hydrazine (at 20°C)[c]	302-01-2	0.01		x	10.0	1,315,789	6.1
Hydrogen cyanide	74-90-8	4.7	C	x	91.5	25,616	4.4
Hydrogen peroxide (at 30°C)[c]	7722-84-1	1			5.0	6579	3.8
Hydroquinone (2 mg/m^3)[b]	123-31-9	0.444			0.00067	2	0.3
Isoamyl acetate (see Pentyl acetate)	123-92-2	50			5.6	147	2.2
Isoamyl alcohol (primary and secondary)	123-51-3	100			2.37	31	1.5
Isobutanol (isobutyl alcohol)	78-83-1	50			10.5	276	2.4
Isobutyl acetate	110-19-0	150			17.8	156	2.2
Isooctyl alcohol	26952-21-6	50		x	0.5	13	1.1
Isophorone	78-59-1	5	C		0.4	105	2.0
Isopropanol (Isopropyl alcohol or IPA)	67-63-0	200			45.4	299	2.5
Isopropyl acetate	108-21-4	100			59.2	779	2.9
Isopropyl amine	75-31-0	5			580.0	152,632	5.2
Isopropyl ether	108-20-3	250			0.149	7.8×10^{-1}	−0.1
Lindane (0.5 mg/m^3)[b]	58-89-9	0.042		x	0.00041	13	1.1
Malathion (1 mg/m^3)[b]	121-75-5	0.074		x	7.9×10^{-6}	1.4×10^{-1}	−0.9
Maleic anhydride	108-31-6	0.1			0.25	3289	3.5
Mercury (0.025 mg/m^3)[b]	7439-97-6	0.003		x	0.00185	799	2.9
Mesityl oxide	141-79-7	15			8.21	720	2.9
Methacrylic acid	79-41-4	20			1.0	66	1.8
Methanol or methyl alcohol	67-56-1	200		x	127.0	836	2.9
1-Methoxy-2-propanol (PGME)	107-98-2	100			12.5	164	2.2

Chemical Name	CAS #	TLV (ppm)	Ceiling	Skin	P_{vapor} (mmHg)	VHR	VHI
Methoxychlor (10 mg/m^3)[b]	72-43-5	0.7074			2.6×10^{-6}	4.8×10^{-3}	−2.3
2-Methoxyethanol or Methyl cellosolve	109-86-4	5		x	9.5	2500	3.4
2-Methoxyethyl acetate (Methyl cellosolve acetate)	110-49-6	5			4.01	1055	3.0
2-Methoxymethylethoxy propanol or Dipropylene glycol methyl ether (at 20°C)[c]	34590-94-8	100		x	0.5	7	0.8
4-Methoxyphenol (5 mg/m^3)[b]	150-76-5	0.99			0.0083	11	1.0
Methyl acetate	79-20-9	200			216.0	1421	3.2
Methyl acrylate	96-33-3	2		x	86.0	56,579	4.8
Methyl acrylonitrile	126-98-7	1		x	71.2	93,684	5.0
Methyl alcohol (see Methanol)	67-56-1	200		x	127.0	836	2.9
Methyl choloroform or 1,1,1-Trichloroethane	71-55-6	350			124.0	466	2.7
Methyl cyclohexane	108-87-2	400			46.0	151	2.2
Methyl cyclohexanol (at 30°C)[c]	25639-42-3	50			2.0	53	1.7
o-Methyl cyclohexanone	583-60-8	50		x	2.25	59	1.8
Methyl ethyl ketone (2-Butanone or MEK)	78-93-3	200			95.3	627	2.8
Methyl formate	107-31-3	100			586.0	7711	3.9
Methyl hydrazine (Monomethyl hydrazine)	60-34-4	0.01			50.0	6,578,947	6.8
Methyl iodine	74-88-4	2		x	405.0	266,447	5.4
Methyl isoamyl ketone	110-12-3	50			5.2	137	2.1
Methyl isobutyl carbinol	108-11-2	25		x	5.3	279	2.4
Methyl isobutyl ketone (Hexone or MIBK)	108-10-1	50			19.9	524	2.7
Methyl isocyanate (at 20°C)[c]	624-83-9	0.02		x	348.0	22,894,737	7.4
Methyl methacrylate	80-62-6	50			38.4	1011	3.0
Methyl *n*-amyl ketone	110-43-0	50			3.86	102	2.0
Methyl *n*-butyl ketone or 2-Hexanone	591-78-6	5			0.116	31	1.5
Methyl parathion (0.2 mg/m^3)[b]	298-00-0	0.0186		x	1.7×10^{-5}	1.2	0.1
Methyl propyl ketone or 2-Pentanone	107-87-9	200			35.4	233	2.4
α-Methyl styrene	98-83-9	50			0.29	8	0.9
Methyl *tert*-butyl either (MTBE)	1634-04-4	50			249.0	6553	3.8
Methyl vinyl ketone	78-94-4	0.2	C	x	152.0	1,000,000	6.0
Methyl-2-cyanoacrylate	137-05-3	0.2			0.796	5237	3.7
Methylal or Dimethoxy-methane	109-87-5	1000			398.0	524	2.7
o-Methylcyclohexanone (at 20°C)[c]	583-60-8	50		x	2.0	53	1.7
Methylene bisphenyl isocyanate (MDI)	101-68-8	0.005			1.2×10^{-5}	3	0.5
Methylene chloride (see Dichloromethane)	75-09-2	50			435.0	11,447	4.1
Mevinphos (Phosdrin) (0.01 mg/m^3)[b]	7786-34-7	0.0011		x	0.00013	157	2.2
Monomethyl aniline	100-61-8	0.5		x	0.453	1192	3.1
Morpholine	110-91-8	20		x	10.1	664	2.8
Naled (0.1 mg/m^3 at 20°C)[b,c]	300-76-5	0.0064		x	0.002	410	2.6
Naphthalene	91-20-3	10		x	0.085	11	1.0

Chemical Name	CAS #	TLV (ppm)	Ceiling	Skin	P_{vapor} (mmHg)	VHR	VHI
Nicotine (0.5 mg/m^3)[b]	54-11-5	0.075		x	0.038	664	2.8
Nitrapyrin (10 mg/m^3)[b,e]	1929-82-4	1.0588			0.0028	3	0.5
p-Nitroaniline (3 mg/m^3)[b]	100-1-6	0.53		x	3.2×10^{-6}	7.9×10^{-3}	−2.1
Nitrobenzene	98-95-3	1		x	0.245	322	2.5
p-Nitrochlorobenzene	100-00-5	0.1		x	0.0219	288	2.5
Nitroethane	79-24-3	100			20.8	274	2.4
Nitroglycerin	55-63-0	0.05		x	0.0002	5	0.7
Nitromethane	75-52-5	20			35.8	2355	3.4
1-Nitropropane	108-03-2	25			10.1	532	2.7
2-Nitropropane	79-46-9	10			17.2	2263	3.4
m-Nitrotoluene	99-08-1	2		x	0.205	135	2.1
o-Nitrotoluene	88-72-2	2		x	0.188	124	2.1
p-Nitrotoluene	99-99-0	2		x	0.164	108	2.0
Nonane	111-84-2	200			4.45	29	1.5
Octachloronaphthalene (0.1 mg/m^3)[b]	2234-13-1	0.0061		x	1.5×10^{-8}	3.3×10^{-3}	−2.5
Octane	111-65-9	300			14.1	62	1.8
Osmium tetroxide (as Os at 20°C)[c]	20816-12-0	0.0002			7.0	46,052,632	7.7
Oxalic acid (1 mg/m^3)[b]	144-62-7	0.33			0.00023	1	0.0
Paraquat (0.1 mg/m^3)[b]	4685-14-7	0.0095			1.0×10^{-7}	0.01	−1.9
Parathion (0.05 mg/m^3)[b]	56-38-2	0.0042		x	9.7×10^{-6}	3	0.5
PCB (see Chlorodiphenyl 0.5 mg/m^3)[b]	11097-69-1	0.037		x	7.7×10^{-5}	3	0.4
Pentaborane (at 20°C)[c]	19624-22-7	0.005			171.0	45,000,000	7.7
Pentachlorophenol (0.5 mg/m^3)[b]	87-86-5	0.046		x	0.00011	3	0.5
Pentaerythritol (10 mg/m^3)[b]	115-77-5	1.80			15.1	11,064	4.0
iso-Pentane	78-78-4	600			689.0	1511	3.2
n-Pentane	109-66-0	600			514.0	1127	3.1
Pentyl acetate (all isomers including Amyl acetate)	123-92-2	50			5.6	147	2.2
Perchloroethylene (see Tetrachoroethylene)	127-18-4	25			18.6	979	3.0
Perchloromethyl mercaptan	594-42-3	0.1			3.0	39,474	4.6
Phenol (Carbolic acid)	108-95-2	5		x	0.35	92	2.0
Phenyl ether (as vapor)	101-84-8	1			0.0225	30	1.5
Phenyl glycidyl ether (PGE)	122-60-1	0.1		x	0.01	132	2.1
Phenyl mercaptan (Thiophenol)	108-98-5	0.1		x	1.9	25,000	4.4
Phenylenediamine (mixed 0.1 mg/m^3)[b,d]		0.023			0.013	744	2.9
Phenylhydrazine	100-63-0	0.1		x	0.026	342	2.5
Phosphoric acid (1 mg/m^3 at 20°C)[b]	7664-38-2	0.2495			0.03	158	2.2
Phosphorus, yellow (0.1 mg/m^3 at 20°C)[b,c]	7723-14-0	0.0197			0.026	1734	3.2
Phthalic anhydride	85-44-9	1			0.00052	6.8×10^{-1}	−0.2
Picloram (10 mg/m^3)[b]	1918-02-1	1.01			7.2×10^{-11}	9.4×10^{-8}	−7.0
Picric acid (0.1 mg/m^3)[b]	88-89-1	0.0107			2.3×10^{-7}	0.03	−1.6
Propargyl alcohol	107-19-7	1		x	15.6	20,526	4.3
β-Propiolactone	57-57-8	0.5			3.4	8947	4.0
Propionaldehyde	123-38-6	20			317.0	20,855	4.3
Propionic acid	79-09-4	10			3.53	464	2.7
n-Propyl acetate	109-60-4	200			33.7	222	2.3
n-Propyl alcohol (propanol)	71-23-8	200		x	21.0	138	2.1

Chemical Name	CAS #	TLV (ppm)	Ceiling	Skin	P_{vapor} (mmHg)	VHR	VHI
n-Propyl nitrate	627-13-4	25			23.5	1237	3.1
Propylene dichloride or 1,2-Dichloropropane	78-87-5	75			53.3	935	3.0
Propylene glycol monomethyl ether (see 1-Methoxy-2-propanol)	107-98-2	100			12.5	164	2.2
Propylene imine	75-55-8	2		x	112.0	73,684	4.9
Propylene oxide or 1,2-Epoxypropane	75-56-9	2			538.0	353,947	5.5
Pyridine	110-86-1	5			20.8	5474	3.7
Quinone	106-51-4	0.1			0.9	11,842	4.1
Resorcinol	108-46-3	10			0.0005	0.07	−1.2
Ronnel (10 mg/m^3)[b]	299-84-3	0.76			7.5×10^{-5}	1.3×10^{-1}	−0.9
Rotenone (5 mg/m^3)[b]	83-79-4	0.31			2.6×10^{-9}	1.1×10^{-5}	−5.0
Rubber solvent (naphtha at 20°C)[c]	8030-30-6	400			4.0	13	1.1
Sodium fluoroacetate (0.05 mg/m^3)[b]	62-74-8	0.012		x	6.5×10^{-7}	7.0×10^{-2}	−1.2
Stoddard solvent	8052-41-3	100			4.0	53	1.7
Strychnine (0.15 mg/m^3)[b]	57-24-9	0.011			35.4	4,247,035	6.6
Styrene (vinyl benzene monomer)	100-42-5	20			6.4	421	2.6
Sulfotep (TEDP 0.1 mg/m^3)[b]	3689-24-5	0.015		x	0.00017	15	1.2
2,4,5-T (2,4,5-trichlorophenoxyacetic acid 10 mg/m^3)[b]	93-76-5	0.96			3.8×10^{-5}	5.2×10^{-2}	−1.3
Temephos (1 mg/m^3)[b]	3383-96-8	0.52			7.9×10^{-8}	2.0×10^{-4}	−3.7
1,1,2,2-Tetrachloro-1,2-difluoroethane	76-12-0	500			220.0	579	2.8
1,1,1,2-Tetrachloro-2,2-difluoroethane	76-11-9	500			54.9	144	2.2
1,1,2,2-Tetrachloroethane	79-34-5	1		x	4.6	6079	3.8
Tetrachoroethylene (or Perchloroethylene)	127-18-4	25			18.6	979	3.0
Tetraethyl lead (as 0.1 mg/m^3 Pb)[b]	78-00-2	0.0076		x	0.39	67,886	4.8
Tetraethyl pyrophosphate (TEPP 0.05 mg/m^3)[b]	107-49-3	0.0042		x	0.47	146,802	5.2
Tetrahydrofuran	109-99-9	200			162.0	1066	3.0
Tetranitromethane	509-14-8	0.005			8.42	2,215,789	6.3
Tetryl (1.5 mg/m^3)[b]	479-45-8	0.128			5.7×10^{-8}	5.8×10^{-4}	−3.2
Thiram (1 mg/m^3)[b]	137-26-8	0.102			1.7×10^{-5}	2.2×10^{-1}	−0.7
Toluene	108-88-3	50		x	28.4	747	2.9
Toluene-2,4-diisocyanate (TDI)	584-84-9	0.005			0.008	2105	3.3
Toluidine, mixed isomers		2		x	0.28	184	2.3
Toxaphene (see Chlorinated camphene)	8001-35-2	0.0295		x	0.4	17,824	4.3
Tributyl phosphate	126-73-8	0.2			0.00012	7.9×10^{-1}	−0.1
1,1,2-Trichloro-1,2,2-trifluoroethane	76-13-1	1000			363.0	478	2.7
Trichloroacetic acid	76-03-9	1			5.0×10^{-9}	6.6×10^{-6}	−5.2
1,2,4-Trichlorobenzene	120-82-1	5	C		0.29	76	1.9
1,1,2-Trichloroethane	79-00-5	10		x	23.0	3026	3.5
1,1,1-Trichloroethane (see Methyl chloroform)	71-55-6	350			124.0	466	2.7
Trichloroethylene	79-01-6	50			69.0	1816	3.3
1,2,3-Trichloropropane	96-18-4	10		x	3.69	486	2.7
Triethanolamine (5 mg/m^3)[b]	102-71-6	0.8193			3.6×10^{-6}	0.01	−2.2
Triethylamine	121-44-8	1		x	57.1	75,132	4.9

Chemical Name	CAS #	TLV (ppm)	Ceiling	Skin	P_{vapor} (mmHg)	VHR	VHI
Trimethyl benzene (mixed)	25551-13-7	25			2.1	111	2.0
2,4,6-Trinitrotoluene (TNT 0.1 mg/m^3)[b]	118-96-7	0.011		x	8.0×10^{-6}	9.8×10^{-1}	−0.0
Triphenyl amine	603-34-9	0.5			0.0004	1.05	0.0
Triphenyl phosphate (3 mg/m^3)[b]	115-86-6	0.225			6.3×10^{-6}	3.7×10^{-2}	−1.4
Turpentine (at 20°C)[c]	8006-64-2	20			4.0	263	2.4
Vinyl acetate (at 20°C)[c]	108-05-4	10			90.2	11,868	4.1
Vinyl toluene (P_{vapor} at 20°C)[c,f]	25013-15-4	50			1.0	48	1.7
n-Vinyl-2-pyrrolidone	88-12-0	0.05			0.114	3000	3.5
Vinylidene chloride	75-35-4	5			600.0	157,895	5.2
VM&P Naphtha	8032-32-4	300			10.0	44	1.6
Warfarin (0.1 mg/m^3)[b]	81-81-2	0.0079			1.2×10^{-7}	1.9×10^{-2}	−1.7
m-Xylene	108-38-3	100			8.3	109	2.0
o-Xylene	95-47-6	100			6.6	87	1.9
p-Xylene	106-42-3	100			8.8	116	2.1
Xylenes[d]	1330-20-7	100			8.0	105	2.0

[a] VHI = $\log_{10}$ (VHR) via Equation 8.11.

[b] The TLV in ppm is derived from the mass concentration in parentheses via ppm = $C \times 24.45/\text{MW}$.

[c] The P_{vapor} at the temperature indicated (which is other than at 25°C).

[d] The vapor pressure of "mixed" isomers will vary with the ratio of the isomers in a mixture.

[e] OSHA lists Nitrapyrin as 2-Chloro-6 (trichloromethyl) pyridine.

[f] The P_{vapor} of vinyl toluene is assumed to be midway between that of *p*- and *o*-methylstyrene.

Appendix B

The vapor hazard ratio ($VHR = P_{vapor} \times 10^6/(TLV \times 760)$) for selected chemicals rank ordered by their vapor hazard ratio. The 2005 TLV® and the compound's vapor pressure [P_{vapor}] at 25°C are used unless otherwise indicated.[a]

Chemical Name	CAS #	TLV (ppm)	Ceiling	Skin	P_{vapor} (mmHg)	VHR	VHI
Osmium tetroxide (as Os at 20°C)[c]	20816-12-0	0.0002			7.0	46,052,632	7.7
Pentaborane (at 20°C)[c]	19624-22-7	0.005			171.0	45,000,000	7.7
Bis-(chloromethyl) ether	542-88-1	0.001			30.0	39,473,684	7.6
Methyl isocyanate (at 20°C)[c]	624-83-9	0.02		x	348.0	22,894,737	7.4
1,1-Dimethyl hydrazine	57-14-7	0.01		x	157.0	20,657,895	7.3
Methyl hydrazine (Monomethyl hydrazine)	60-34-4	0.01			50.0	6,578,947	6.8
Diazomethane	334-88-3	0.2			760.0	5,000,000	6.7
Bromine pentafluoride (at 20°C)[c]	7789-30-2	0.1			328.0	4,315,789	6.6
Strychnine (0.15 mg/m^3)[b]	57-24-9	0.011			35.4	4,247,035	6.6
Acrolein	107-02-8	0.1	C	x	274.0	3,605,263	6.6
Bromine (at 20°C)[c]	7726-95-6	0.1			172.0	2,263,158	6.4
Tetranitromethane	509-14-8	0.005			8.42	2,215,789	6.3
Ethanethiol; see Ethyl mercaptan	75-08-1	0.5			529.0	1,392,105	6.1
Ethyl mercaptan or Ethanethiol	75-08-1	0.5			529.0	1,392,105	6.1
Hydrazine (at 20°C)[c]	302-01-2	0.01		x	10.0	1,315,789	6.1
Chromyl chloride (at 20°C)[c]	14977-61-8	0.025			20.0	1,052,632	6.0
Methyl vinyl ketone	78-94-4	0.2	C	x	152.0	1,000,000	6.0
1,4-Dichloro-2-butene	764-41-0	0.005		x	3.0	789,474	5.9
Ethyleneimine	151-56-4	0.5		x	213.0	560,526	5.7
Chloroacetyl chloride (at 20°C)[c]	79-04-9	0.05		x	19.0	500,000	5.7
Allyl chloride	107-05-1	1			368.0	484,211	5.7
Propylene oxide or 1,2-Epoxypropane	75-56-9	2			538.0	353,947	5.5
Chloropicrin or Nitrotrichloromethane	76-06-2	0.1			23.8	313,158	5.5
Methyl iodine	74-88-4	2		x	405.0	266,447	5.4
Benzene	71-43-2	0.5		x	95.2	250,526	5.4
Vinylidene chloride	75-35-4	5			600.0	157,895	5.2
Isopropyl amine	75-31-0	5			580.0	152,632	5.2
Tetraethyl pyrophosphate (TEPP 0.05 mg/m^3)[b]	107-49-3	0.0042		x	0.47	146,802	5.2
Crotonaldehyde	4170-30-3	0.3	C	x	30.0	131,579	5.1
Cobalt carbonyl (0.1 mg/m^3 at 20°C)[b,c]	10210-68-1	0.0072			0.7	128,812	5.1
Ethyl bromide	74-96-4	5		x	467.0	122,895	5.1
n-Butyl mercaptan	109-79-5	0.5			45.5	119,737	5.1
Methyl acrylonitrile	126-98-7	1		x	71.2	93,684	5.0
Chloroacetaldehyde	107-20-0	1	C		64.3	84,605	4.9
Triethylamine	121-44-8	1		x	57.1	75,132	4.9
Propylene imine	75-55-8	2		x	112.0	73,684	4.9
Acrylonitrile; see Vinyl cyanide	107-13-1	2		x	109.0	71,711	4.9

Chemical Name	CAS #	TLV (ppm)	Ceiling	Skin	P_{vapor} (mmHg)	VHR	VHI
Allyl alcohol	107-18-6	0.5		x	26.1	68,684	4.8
Tetraethyl lead (as 0.1 mg/m^3 Pb)[b]	78-00-2	0.0076		x	0.39	67,886	4.8
Diethylamine	109-89-7	5		x	237.0	62,368	4.8
Methyl acrylate	96-33-3	2		x	86.0	56,579	4.8
Carbon disulfide	75-15-0	10		x	358.0	47,105	4.7
1,3-Dichloropropene	542-75-6	1		x	34.0	44,737	4.7
Epichlorohydrin (1-Chloro-2,3-epoxypropane)	106-89-8	0.5		x	16.4	43,158	4.6
Perchloromethyl mercaptan	594-42-3	0.1			3.0	39,474	4.6
Carbon tetrachloride	56-23-5	5			115.0	30,263	4.5
β-Chloroprene (or 2-Chloro-1,3-butadiene)	126-99-8	10			216.0	28,421	4.5
Chloroform (Trichloromethane)	67-66-3	10			197.0	25,921	4.4
Hydrogen cyanide	74-90-8	4.7	C	x	91.5	25,616	4.4
Phenyl mercaptan (Thiophenol)	108-98-5	0.1		x	1.9	25,000	4.4
n-Butylamine	109-73-9	5	C	x	91.5	24,079	4.4
Diisopropylamine	108-18-9	5		x	79.4	20,895	4.3
Propionaldehyde	123-38-6	20			317.0	20,855	4.3
Propargyl alcohol	107-19-7	1		x	15.6	20,526	4.3
Toxaphene (see Chlorinated camphene)	8001-35-2	0.0295		x	0.4	17,824	4.3
Glutaraldehyde (at 30°C)[c]	111-30-8	0.05	C		0.6	15,789	4.2
1-Bromopropane	106-94-5	10			111.0	14,605	4.2
Hexachlorobutadiene	87-68-3	0.02			0.22	14,474	4.2
Bromoform	75-25-2	0.5		x	5.4	14,211	4.2
2-Chloropropionic acid	598-78-7	0.1		x	1.06	13,947	4.1
Formalin (37% using Henry's law)	50-00-0	0.3	C		3.16	13,841	4.1
Vinyl acetate (at 20°C)[c]	108-05-4	10			90.2	11,868	4.1
Quinone	106-51-4	0.1			0.9	11,842	4.1
Dichloromethane or Methylene chloride	75-09-2	50			435.0	11,447	4.1
Methylene chloride (see Dichloromethane)	75-09-2	50			435.0	11,447	4.1
Formic acid	64-18-6	5			42.6	11,211	4.0
1,1-Dichloro-1-nitroethane	594-72-9	2			16.9	11,118	4.0
Pentaerythritol (10 mg/m^3)[b]	115-77-5	1.80			15.1	11,064	4.0
1,2-Dichloroethane (or Ethylene dichloride)	107-06-2	10			78.9	10,382	4.0
Ethylene dichloride or (1,2-Dichloroethane)	107-06-2	10			78.9	10,382	4.0
Ethyl acrylate	140-88-5	5			38.6	10,158	4.0
Ethylene chlorohydrin	107-07-3	1	C	x	7.18	9447	4.0
Dimethyl sulfate	77-78-1	0.1		x	0.68	8947	4.0
β-Propiolactone	57-57-8	0.5			3.4	8947	4.0
1-Hexene	592-41-6	50			186.0	8158	3.9
Halothane	151-67-7	50			302.0	7947	3.9
Hexamethylene diisocyanate (HDI)	822-06-0	0.005			0.03	7895	3.9
Methyl formate	107-31-3	100			586.0	7711	3.9
Cyclopentadiene	542-92-7	75			435.0	7632	3.9
Hydrogen peroxide (at 30°C)[c]	7722-84-1	1			5.0	6579	3.8
Methyl *tert*-butyl either (MTBE)	1634-04-4	50			249.0	6553	3.8

Chemical Name	CAS #	TLV (ppm)	Ceiling	Skin	P_{vapor} (mmHg)	VHR	VHI
Dichlorvos (DDVP) (0.1 mg/m^3)[b]	62-73-7	0.011		x	0.053	6303	3.8
1,1,2,2-Tetrachloroethane	79-34-5	1		x	4.6	6079	3.8
Acetonitrile or Methyl cyanide	75-05-8	20		x	88.8	5842	3.8
Pyridine	110-86-1	5			20.8	5474	3.7
Benzotrichloride	98-07-7	0.1	C	x	0.41	5395	3.7
Methyl-2-cyanoacrylate	137-05-3	0.2			0.796	5237	3.7
n-Hexane	110-54-3	50		x	151.0	3974	3.6
1-Chloro-1-nitropropane	600-25-9	2			6.0	3947	3.6
Glycidol or 2,3-Epoxy-1-propanol	556-52-5	2			5.59	3678	3.6
Maleic anhydride	108-31-6	0.1			0.25	3289	3.5
Ethyl formate	109-94-4	100			245.0	3224	3.5
1,1,2-Trichloroethane	79-00-5	10		x	23.0	3026	3.5
n-Vinyl-2-pyrrolidone	88-12-0	0.05			0.114	3000	3.5
1,1-Dichloroethane (or Ethylidene chloride)	75-34-3	100			227.0	2987	3.5
Acrylic acid	79-10-7	2		x	4.0	2632	3.4
Allyl glycidyl ether (at 20°C)[c]	106-92-3	1			2.0	2632	3.4
1,4-Dioxane (or Diethylene dioxide)	123-91-1	20		x	38.1	2507	3.4
2-Methoxyethanol or Methyl cellosolve	109-86-4	5		x	9.5	2500	3.4
Nitromethane	75-52-5	20			35.8	2355	3.4
2-Nitropropane	79-46-9	10			17.2	2263	3.4
2-Aminopyridine	504-29-0	0.5			0.8	2105	3.3
Toluene-2,4-diisocyanate (TDI)	584-84-9	0.005			0.008	2105	3.3
Acetic acid	64-19-7	10			15.7	2066	3.3
n-Butyl glycidyl ether (BGE)	2426-08-6	3		x	4.44	1947	3.3
Ethylene glycol Dinitrate	628-96-6	0.05		x	0.072	1895	3.3
Benzoyl chloride	98-88-4	0.5	C		0.7	1842	3.3
Diazinon (0.01 mg/m^3)[b]	333-41-5	0.0008		x	0.0112	1834	3.3
Trichloroethylene	79-01-6	50			69.0	1816	3.3
Ethyl ether or Diethyl ether	60-29-7	400			537.0	1766	3.2
Phosphorus, yellow (0.1 mg/m^3 at 20°C)[b,c]	7723-14-0	0.0197			0.026	1734	3.2
Benzyl chloride	100-44-7	1			1.3	1711	3.2
1,2-Dichloroethylene (mixed isomers)[d]	156-59-2	200			250.0	1645	3.2
Chlorobenzene	108-90-7	10			12.0	1579	3.2
Ethylene diamine or 1,2-Diaminoethane	107-15-3	10		x	12.0	1579	3.2
iso-Pentane	78-78-4	600			689.0	1511	3.2
Furfural	98-01-1	2		x	2.21	1454	3.2
Methyl acetate	79-20-9	200			216.0	1421	3.2
2-Ethoxyethanol or Cellosolve or EGEE	110-80-5	5		x	5.31	1397	3.1
Cyclohexylamine	108-91-8	10			10.1	1329	3.1
n-Ethylmorpholine	100-74-3	5		x	5.03	1324	3.1
Cyclohexane	110-82-7	100			96.9	1275	3.1
n-Propyl nitrate	627-13-4	25			23.5	1237	3.1
Monomethyl aniline	100-61-8	0.5		x	0.453	1192	3.1
n-Pentane	109-66-0	600			514.0	1127	3.1
Tetrahydrofuran	109-99-9	200			162.0	1066	3.0
2-Methoxyethyl acetate (Methyl cellosolve acetate)	110-49-6	5			4.01	1055	3.0
o-Anisidine (0.5 mg/m^3)[b]	90-04-0	0.1		x	0.08	1053	3.0

Chemical Name	CAS #	TLV (ppm)	Ceiling	Skin	P_{vapor} (mmHg)	VHR	VHI
Methyl methacrylate	80-62-6	50			38.4	1011	3.0
Perchloroethylene (see Tetrachoroethylene)	127-18-4	25			18.6	979	3.0
Tetrachloroethylene (or Perchloroethylene)	127-18-4	25			18.6	979	3.0
Chlorobromomethane (Halon 1011)	74-97-5	200			143.0	941	3.0
Propylene dichloride or 1,2-Dichloropropane	78-87-5	75			53.3	935	3.0
2-Diethylaminoethanol	100-37-8	2		x	1.4	921	3.0
Acrylamide (0.03 mg/m^3)[b]	79-06-1	0.010		x	0.0070	893	3.0
p-tert-Butyl toluene	98-51-1	1			0.67	883	2.9
Methanol or Methyl alcohol	67-56-1	200		x	127.0	836	2.9
Methyl alcohol (see Methanol)	67-56-1	200		x	127.0	836	2.9
Mercury (0.025 mg/m^3)[b]	7439-97-6	0.003		x	0.00185	799	2.9
Hexachlorocyclopentadiene	77-47-4	0.01			0.06	789	2.9
Isopropyl acetate	108-21-4	100			59.2	779	2.9
Toluene	108-88-3	50		x	28.4	747	2.9
Phenylenediamine (mixed 0.1 mg/m^3)[b,d]		0.023			0.013	744	2.9
Ethylene dibromide (EDB) PEL	106-93-4	20		x	11.0	724	2.9
Mesityl oxide	141-79-7	15			8.21	720	2.9
n-Butyl acrylate	141-32-2	2			5.45	717	2.9
Cyclopentane	287-92-3	600			318.0	697	2.8
Morpholine	110-91-8	20		x	10.1	664	2.8
Nicotine (0.5 mg/m^3)[b]	54-11-5	0.075		x	0.038	664	2.8
Methyl ethyl ketone (2-Butanone or MEK)	78-93-3	200			95.3	627	2.8
Cellosolve acetate (2-Ethoxyethyl acetate)	111-15-9	5		x	2.34	616	2.8
2-Ethoxyethyl acetate or Cellosolve acetate	111-15-9	5		x	2.34	616	2.8
Acetone	67-64-1	500			231.0	608	2.8
Dicyclopentadiene	77-73-6	5			2.3	605	2.8
1,1,2,2-Tetrachloro-1,2-difluoroethane	76-12-0	500			220.0	579	2.8
tert-Butanol (Butyl alcohol)	75-65-0	100			41.7	549	2.7
Hexane, other than *n*- (at 20°C)[c]		500			208.0	547	2.7
1-Nitropropane	108-03-2	25			10.1	532	2.7
Methyl isobutyl ketone (Hexone or MIBK)	108-10-1	50			19.9	524	2.7
Methylal or Dimethoxy-methane	109-87-5	1000			398.0	524	2.7
Dimethyl formamide	68-12-2	10		x	3.87	509	2.7
1,2,3-Trichloropropane	96-18-4	10		x	3.69	486	2.7
1,1,2-Trichloro-1,2,2-trifluoroethane	76-13-1	1000			363.0	478	2.7
Dichloroacetic acid	79-43-6	0.5		x	0.18	474	2.7
Methyl chloroform or 1,1,1-Trichloroethane	71-55-6	350			124.0	466	2.7
1,1,1-Trichloroethane (see Methyl chloroform)	71-55-6	350			124.0	466	2.7
Propionic acid	79-09-4	10			3.53	464	2.7
n-Butanol (Butyl alcohol)	71-36-3	20			6.7	441	2.6

Chemical Name	CAS #	TLV (ppm)	Ceiling	Skin	P_{vapor} (mmHg)	VHR	VHI
Camphor, synthetic	76-22-2	2			0.65	428	2.6
Styrene (Vinyl benzene monomer)	100-42-5	20			6.4	421	2.6
Naled (0.1 mg/m^3 at 20°C)[b,c]	300-76-5	0.0064		x	0.002	410	2.6
Dichloroethyl ether	111-44-4	5		x	1.55	408	2.6
Bis(2-chloroethyl) ether (see Dichloroethyl ether)	111-44-4	5		x	1.5	395	2.6
Cyclohexene	110-83-8	300			89.0	390	2.6
Phenylhydrazine	100-63-0	0.1		x	0.026	342	2.5
Aniline and homologs	62-53-3	2		x	0.49	322	2.5
Nitrobenzene	98-95-3	1		x	0.245	322	2.5
tert-Butyl acetate	540-88-5	200			47.0	309	2.5
Ethyl acetate	141-78-6	400			93.7	308	2.5
Diethylene triamine	111-40-0	1		x	0.23	303	2.5
Isopropanol (Isopropyl alcohol or IPA)	67-63-0	200			45.4	299	2.5
2,2-Dichloropropionic acid (5 mg/m^3 inhal.)[b]	75-99-0	0.85			0.19	294	2.5
p-Nitrochlorobenzene	100-00-5	0.1		x	0.0219	288	2.5
Cyclohexanone	108-94-1	20		x	4.33	285	2.5
Methyl isobutyl carbinol	108-11-2	25		x	5.3	279	2.4
Hexachloroethane	67-72-1	1		x	0.21	276	2.4
Isobutanol (isobutyl alcohol)	78-83-1	50			10.5	276	2.4
Nitroethane	79-24-3	100			20.8	274	2.4
Ethyl amyl ketone or 5-Methyl-3-heptanone	541-85-5	25			5.03	265	2.4
Dimethyl acetamide	127-19-5	10		x	2.0	263	2.4
Ethylamine	75-04-7	5		x	1.0	263	2.4
Turpentine (at 20°C)[c]	8006-64-2	20			4.0	263	2.4
p-Anisidine (0.5 mg/m^3)[b]	104-94-9	0.1		x	0.019	250	2.4
Diethyl ketone	96-22-0	200			37.7	248	2.4
Ethyl silicate	78-10-4	10			1.88	247	2.4
sec-Butanol (Butyl alcohol)	78-92-2	100			18.3	241	2.4
Methyl propyl ketone or 2-Pentanone	107-87-9	200			35.4	233	2.4
n-Propyl acetate	109-60-4	200			33.7	222	2.3
sec-Amyl acetate	626-38-0	50			8.03	211	2.3
Dimethyl aniline (*N,N*-Dimethylaniline)	121-69-7	5		x	0.7	184	2.3
Toluidine, mixed isomers		2		x	0.28	184	2.3
Ethanolamine or 2-Aminoethanol	141-43-5	3			0.404	177	2.2
1-Methoxy-2-propanol (PGME)	107-98-2	100			12.5	164	2.2
Propylene glycol monomethyl ether (see 1-Methoxy-2-propanol)	107-98-2	100			12.5	164	2.2
Heptachlor (0.05 mg/m^3)[b]	76-44-8	0.0033		x	0.0004	161	2.2
Chloroacetone	78-95-5	1	C	x	0.12	158	2.2
Phosphoric acid (1 mg/m^3 at 20°C)[b]	7664-38-2	0.2495			0.03	158	2.2
Mevinphos (Phosdrin) (0.01 mg/m^3)[b]	7786-34-7	0.0011		x	0.00013	157	2.2
Isobutyl acetate	110-19-0	150			17.8	156	2.2
Heptane (*n*-Heptane)	142-82-5	400			46.0	151	2.2
Methyl cyclohexane	108-87-2	400			46.0	151	2.2
iso-Amyl acetate (see Pentyl acetate)	123-92-2	50			5.6	147	2.2
Isoamyl acetate (see Pentyl acetate)	123-92-2	50			5.6	147	2.2

Chemical Name	CAS #	TLV (ppm)	Ceiling	Skin	P_{vapor} (mmHg)	VHR	VHI
Pentyl acetate (all isomers incl. Amyl acetate)	123-92-2	50			5.6	147	2.2
1,1,1,2-Tetrachloro-2,2-difluoroethane	76-11-9	500			54.9	144	2.2
2-Chloroacetophenone or Phenacyl chloride	532-27-4	0.05			0.0054	142	2.2
Hexachlorobenzene (0.002 mg/m^3)[b]	118-74-1	0.0002		x	1.8×10^{-5}	138	2.1
n-Propyl alcohol (propanol)	71-23-8	200		x	21.0	138	2.1
Methyl isoamyl ketone	110-12-3	50			5.2	137	2.1
m-Nitrotoluene	99-08-1	2		x	0.205	135	2.1
p-Dichlorobenzene	106-46-7	10			1.0	132	2.1
Phenyl glycidyl ether (PGE)	122-60-1	0.1		x	0.01	132	2.1
Acetic anhydride	108-24-7	5			0.498	131	2.1
Ethyl benzene	100-41-4	100			9.6	126	2.1
o-Nitrotoluene	88-72-2	2		x	0.188	124	2.1
Cumene	98-82-8	50			4.5	118	2.1
p-Xylene	106-42-3	100			8.8	116	2.1
sec-Butyl acetate	105-46-4	200			17.0	112	2.0
Trimethyl benzene (mixed)	25551-13-7	25			2.1	111	2.0
m-Xylene	108-38-3	100			8.3	109	2.0
p-Nitrotoluene	99-99-0	2		x	0.164	108	2.0
n-Butyl lactate (at 20°C)[c]	138-22-7	5			0.4	105	2.0
sec-Hexyl acetate	108-84-9	50			4.0	105	2.0
Isophorone	78-59-1	5	C		0.4	105	2.0
Xylenes[f]	1330-20-7	100			8.0	105	2.0
Methyl *n*-amyl ketone	110-43-0	50			3.86	102	2.0
n-Butyl acetate	123-86-4	150			11.5	101	2.0
n-Amyl acetate	628-63-7	50			3.5	92	2.0
Phenol (carbolic acid)	108-95-2	5		x	0.35	92	2.0
o-Chlorotoluene	95-49-8	50			3.4	89	2.0
Diisobutyl ketone or 2,6-Dimethyl-4-heptanone	108-83-8	25			1.65	87	1.9
o-Xylene	95-47-6	100			6.6	87	1.9
Furfuryl alcohol	98-00-0	10		x	0.609	80	1.9
Ethanol or Ethyl alcohol	64-17-5	1000			59.3	78	1.9
Ethyl alcohol (see Ethanol)	64-17-5	1000			59.3	78	1.9
1,2,4-Trichlorobenzene	120-82-1	5	C		0.29	76	1.9
o-Dichlorobenzene	95-50-1	25			1.36	72	1.9
Ethyl butyl ketone (3-Heptanone)	106-35-4	50			2.6	68	1.8
Methacrylic acid	79-41-4	20			1.0	66	1.8
Octane	111-65-9	300			14.1	62	1.8
Biphenyl or Diphenyl	92-52-4	0.2			0.00893	59	1.8
o-Methyl cyclohexanone	583-60-8	50		x	2.25	59	1.8
2-Butoxyethanol (EGBE)	111-76-2	20			0.88	58	1.8
Methyl cyclohexanol (at 30°C)[c]	25639-42-3	50			2.0	53	1.7
o-Methylcyclohexanone (at 20°C)[c]	583-60-8	50		x	2.0	53	1.7
Stoddard Solvent	8052-41-3	100			4.0	53	1.7
Vinyl toluene (P_{vapor} at 20°C)[c]	25013-15-4	50			1.0	48	1.7
Cresol (mixed isomers at 20°C)[c,d]	1319-77-3	5		x	0.18	47	1.7
Diacetone alcohol	123-42-2	50			1.71	45	1.7
VM&P Naphtha	8032-32-4	300			10.0	44	1.6
Isoamyl alcohol (primary and sec.)	123-51-3	100			2.37	31	1.5

Chemical Name	CAS #	TLV (ppm)	Ceiling	Skin	P_{vapor} (mmHg)	VHR	VHI
Methyl *n*-butyl ketone or 2-Hexanone	591-78-6	5			0.116	31	1.5
n-Butylated hydroxytoluene (2 mg/m^3)[b]	128-37-0	0.22			0.005	30	1.5
Phenyl ether (as vapor)	101-84-8	1			0.0225	30	1.5
Nonane	111-84-2	200			4.45	29	1.5
Fonofos (Difonate) (0.1 mg/m^3)[b]	944-22-9	0.0099		x	0.0002	27	1.4
Acetylene tetrabromide	79-27-6	1			0.02	26	1.4
2-Butoxyethyl acetate (EGBEA)	112-07-2	20			0.38	25	1.4
o-Chlorostyrene	2039-87-4	50			0.96	25	1.4
Benzyl acetate	140-11-4	10			0.177	23	1.4
Cyclohexanol	108-93-0	50		x	0.8	21	1.3
Dinitro-*o*-cresol (0.2 mg/m^3)[b]	534-52-1	0.0247		x	0.00032	17	1.2
Dinitrotoluene (0.2 mg/m^3)[b]	25321-14-6	0.027		x	0.0004	17	1.2
Sulfotep (TEDP 0.1 mg/m^3)[b]	3689-24-5	0.015		x	0.00017	15	1.2
o-sec-Butylphenol	89-72-5	5		x	0.05	13	1.1
Isooctyl alcohol	26952-21-6	50		x	0.5	13	1.1
Lindane (0.5 mg/m^3)[b]	58-89-9	0.042		x	0.00041	13	1.1
Rubber solvent (naphtha at 20°C)[c]	8030-30-6	400			4.0	13	1.1
4-Methoxyphenol (5 mg/m^3)[b]	150-76-5	0.99			0.0083	11	1.0
Naphthalene	91-20-3	10		x	0.085	11	1.0
Aldrin (0.25 mg/m^3)[b]	309-00-2	0.0168		x	0.00012	9	1.0
meta-Dinitro benzene	99-65-0	0.15		x	0.0009	8	0.9
Formamide	75-12-7	10		x	0.06	8	0.9
α-Methyl styrene	98-83-9	50			0.29	8	0.9
2-Methoxymethylethoxy propanol or dipropylene glycol methyl ether (at 20°C)[c]	34590-94-8	100		x	0.5	7	0.8
Nitroglycerin	55-63-0	0.05		x	0.0002	5	0.7
Dichrotophos (0.05 mg/m^3 at 20°C)[b]	141-66-2	0.0052		x	8.6×10^{-5}	4.4	0.6
Diethyl phthalate	84-66-2	0.55			0.0016	4	0.6
Dimethyl phthalate (5 mg/m^3)[b]	131-11-3	0.63			0.00165	3	0.5
Ethylene glycol (100 mg/m^3)[b]	107-21-1	39	C		0.092	3	0.5
Methylene bisphenyl isocyanate (MDI)	101-68-8	0.005			1.2×10^{-5}	3	0.5
Parathion (0.05 mg/m^3)[b]	56-38-2	0.0042		x	9.7×10^{-6}	3	0.5
Chlorpyrifos (0.1 mg/m^3)[b]	2921-88-2	0.007		x	1.7×10^{-5}	3	0.5
Nitrapyrin (10 mg/m^3)[b,e]	1929-82-4	1.0588			0.0028	3	0.5
Pentachlorophenol (0.5 mg/m^3)[b]	87-86-5	0.046		x	0.00011	3	0.5
PCB (see Chlorodiphenyl 0.5 mg/m^3)[b]	11097-69-1	0.037		x	7.7×10^{-5}	3	0.4
Fenthion (0.2 mg/m^3 at 20°C)[b,c]	55-38-9	0.018		x	3.0×10^{-5}	2.3	0.4
Endosulfan (0.1 mg/m^3)[b]	115-29-7	0.0060		x	1.0×10^{-5}	2	0.3
Hydroquinone (2 mg/m^3)[b]	123-31-9	0.444			0.00067	2	0.3
Triphenyl amine	603-34-9	0.5			0.0004	1.05	0.0
Methyl parathion (0.2 mg/m^3)[b]	298-00-0	0.0186		x	1.7×10^{-5}	1.2	0.1
Benzoyl peroxide (5 mg/m^3)[b]	94-36-0	0.505			0.000423	1	0.0
Hexylene glycol	107-41-5	25	C		0.013	1.0	0.0
2,4,6-Trinitrotoluene (TNT 0.1 mg/m^3)[b]	118-96-7	0.011		x	8.0×10^{-6}	9.8×10^{-1}	0.0
Oxalic acid (1 mg/m^3)[b]	144-62-7	0.33			0.00023	1	0.0
Diethanolamine (2 mg/m^3)[b]	111-42-2	0.465		x	0.00028	7.9×10^{-1}	−0.1
Tributyl phosphate	126-73-8	0.2			0.00012	7.9×10^{-1}	−0.1
Isopropyl ether	108-20-3	250			0.149	7.8×10^{-1}	−0.1
Clopidol (10 mg/m^3)[b]	2971-90-6	1.273			6.6×10^{-4}	6.8×10^{-1}	−0.2
Phthalic anhydride	85-44-9	1			0.00052	6.8×10^{-1}	−0.2
Endrin (0.1 mg/m^3)[b]	72-20-8	0.0064		x	3.0×10^{-6}	6.2×10^{-1}	−0.2

Chemical Name	CAS #	TLV (ppm)	Ceiling	Skin	P_{vapor} (mmHg)	VHR	VHI
Carbofuran (0.1 mg/m^3 at 20°C)[b,c]	1563-66-2	0.011			5.0×10^{-6}	6.0×10^{-1}	−0.2
Dieldrin (0.25 mg/m^3)[b]	60-57-1	0.0160		x	5.9×10^{-6}	4.8×10^{-1}	−0.3
Ethion (0.05 mg/m^3)[b]	563-12-2	0.0032		x	1.1×10^{-6}	4.5×10^{-1}	−0.3
Adiponitrile	111-69-3	2		x	0.00068	4.5×10^{-1}	−0.3
Chlordane (0.5 mg/m^3)[b]	57-74-9	0.0298		x	9.8×10^{-6}	4.3×10^{-1}	−0.4
ortho-Dinitro benzene	528-29-0	0.15		x	4.6×10^{-5}	4.0×10^{-1}	−0.4
Chlorinated camphene (0.5 mg/m^3)[b]	8001-35-2	0.0295		x	6.7×10^{-6}	3.0×10^{-1}	−0.5
ANTU (α-Naphthylthiourea 0.03 mg/m^3)[b]	86-88-4	0.0363			6.6×10^{-6}	2.4×10^{-1}	−0.6
para-Dinitro benzene	100-25-4	0.15		x	2.6×10^{-5}	2.3×10^{-1}	−0.6
Dibutyl phthalate (5 mg/m^3)[b]	84-74-2	0.44			7.3×10^{-5}	2.2×10^{-1}	−0.7
Thiram (1 mg/m^3)[b]	137-26-8	0.102			1.7×10^{-5}	2.2×10^{-1}	−0.7
EPN (0.1 mg/m^3)[b]	2104-64-5	0.0076		x	9.5×10^{-7}	1.7×10^{-1}	−0.8
Fenamiphos (0.1 mg/m^3)[b]	22224-92-6	0.008		x	1.0×10^{-6}	1.6×10^{-1}	−0.8
Malathion (1 mg/m^3)[b]	121-75-5	0.074		x	7.9×10^{-6}	1.4×10^{-1}	−0.9
Ronnel (10 mg/m^3)[b]	299-84-3	0.76			7.5×10^{-5}	1.3×10^{-1}	−0.9
Dibutyl phosphate	107-66-4	1			9.6×10^{-5}	1.3×10^{-1}	−0.9
Glycerin (10 mg/m^3 mist)[b]	56-81-5	2.65			0.00017	8.3×10^{-2}	−1.1
Sodium fluoroacetate (0.05 mg/m^3)[b]	62-74-8	0.012		x	6.5×10^{-7}	7.0×10^{-2}	−1.2
Resorcinol	108-46-3	10			5.0×10^{-4}	6.6×10^{-2}	−1.2
2,4,5-T (2,4,5-Trichlorophenoxyacetic acid 10 mg/m^3)[b]	93-76-5	0.96			3.8×10^{-5}	5.2×10^{-2}	−1.3
Di-(2-ethylhexyl) phthalate (DEHP 5 mg/m^3)[b]	117-81-7	0.3130			9.7×10^{-6}	4.1×10^{-2}	−1.4
Triphenyl phosphate (3 mg/m^3)[b]	115-86-6	0.225			6.3×10^{-6}	3.7×10^{-2}	−1.4
Picric acid (0.1 mg/m^3)[b]	88-89-1	0.0107			2.3×10^{-7}	2.8×10^{-2}	−1.6
Warfarin (0.1 mg/m^3)[b]	81-81-2	0.0079			1.2×10^{-7}	1.9×10^{-2}	−1.7
Paraquat (0.1 mg/m^3)[b]	4685-14-7	0.0095			1.0×10^{-7}	1.4×10^{-2}	−1.9
p-Nitroaniline (3 mg/m^3)[b]	100-1-6	0.53		x	3.2×10^{-6}	7.9×10^{-3}	−2.1
Triethanolamine (5 mg/m^3)[b]	102-71-6	0.8193			3.6×10^{-6}	5.8×10^{-3}	−2.2
Methoxychlor (10 mg/m^3)[b]	72-43-5	0.7074			2.6×10^{-6}	4.8×10^{-3}	−2.3
Octachloronaphthalene (0.1 mg/m^3)[b]	2234-13-1	0.0061		x	1.5×10^{-8}	3.3×10^{-3}	−2.5
Carbaryl (Sevin) (5 mg/m^3)[b]	63-25-2	0.608			1.4×10^{-6}	2.9×10^{-3}	−2.5
Diquat (Resp.) (0.1 mg/m^3)[b]	2764-72-9	0.0071		x	1.0×10^{-8}	1.9×10^{-3}	−2.7
2,4-D (Dichlorophen-oxyacetic acid) (10 mg/m^3)[b]	94-75-7	1.106			6.0×10^{-7}	7.1×10^{-4}	−3.1
Azinphos methyl (Guthion 0.2 mg/m^3)[b]	86-50-0	0.0154		x	7.5×10^{-9}	6.4×10^{-4}	−3.2
Tetryl (1.5 mg/m^3)[b]	479-45-8	0.128			5.7×10^{-8}	5.8×10^{-4}	−3.2
Atrazine (5 mg/m^3 at 20°C)[b,c]	1912-24-9	0.566			3.0×10^{-7}	7.0×10^{-4}	−3.2
Temephos (1 mg/m^3)[b]	3383-96-8	0.52			7.9×10^{-8}	2.0×10^{-4}	3.7
Cyclonite or RDX (0.5 mg/m^3)[b]	121-82-4	0.055			4.1×10^{-9}	9.8×10^{-5}	−4.0
Rotenone (5 mg/m^3)[b]	83-79-4	0.31			2.6×10^{-9}	1.1×10^{-5}	−5.0
Benomyl (10 mg/m^3)[b]	17804-35-2	0.842			3.7×10^{-9}	5.8×10^{-6}	−5.2
Trichloroacetic acid	76-03-9	1			5.0×10^{-9}	6.6×10^{-6}	−5.2
Picloram (10 mg/m^3)[b]	1918-02-1	1.01			7.2×10^{-11}	9.4×10^{-8}	−7.0

[a] VHI = $\log_{10}$ (VHR) via Equation 8.11.

[b] The TLV in ppm is derived from the mass concentration in parentheses via ppm $= C \times 24.45/MW$.

[c] The P_{vapor} at the temperature indicated (which is other than at 25°C).

[d] The vapor pressure of "mixed" isomers will vary with the ratio of the isomers in a mixture.

[e] OSHA lists Nitrapyrin as 2-Chloro-6 (trichloromethyl) pyridine.

[f] The P_{vapor} of vinyl toluene is assumed to be midway between that of *p*- and *o*-methylstyrene.

Appendix C

Heating and cooling degree days at a base temperature of 65°F at about 350 U.S. locations selected from about 5500 reporting stations to achieve a wide geographic coverage with a bias towards large cities, university towns, and airports (the latter have an FAA designated call sign) as reported by the National Climatic Data Center in *CLIMATOGRAPHY OF THE UNITED STATES NO.* 81, Suppl 2, Annual Degree Days to Selected Bases 1971–2000 (available at www5.ncdc.noaa.gov/climatenormals/clim81_supp/CLIM81_Sup_02.pdf).

Station Name	(Call Sign)	State Abb.	AHDD	ACDD
BIRMINGHAM INTL	BHM	AL	2823	1881
HUNTSVILLE INTL	HSV	AL	3262	1671
MOBILE RGNL AP	MOB	AL	1681	2539
MONTGOMERY DANN	MGM	AL	2194	2252
TUSCALOOSA ACFD	TCL	AL	2371	2239
ANCHORAGE INTL	ANC	AK	10470	3
BARROW AP	BRW	AK	19674	0
FAIRBANKS INTL	FAI	AK	13980	74
JUNEAU INTL AP	JNU	AK	8574	0
KODIAK AP	ADQ	AK	8862	0
NOME AP	OME	AK	13674	2
FLAGSTAFF PULLI	FLG	AZ	6999	126
PHOENIX SKY HRB	PHX	AZ	1125	4189
PRESCOTT		AZ	4849	742
TEMPE A S U		AZ	1390	3655
TUCSON INTL AP	TUS	AZ	1578	3017
YUMA MCAS		AZ	852	4472
FAYETTEVILLE EX	FYV	AR	4166	1439
JONESBORO 4 N		AR	3737	1858
LITTLE ROCK ADA	LIT	AR	3084	2086
MONTICELLO 3 SW		AR	2937	1960
TEXARKANA WEBB		AR	2421	2280
ANAHEIM	FUL	CA	1286	1294
BAKERSFIELD KER	BFL	CA	2120	2286
BERKELEY		CA	2857	142
BISHOP AP	BIH	CA	4314	1003
BURBANK VALLEY	BUR	CA	1575	1455
CRESCENT CITY 3	CEC	CA	4687	6
FRESNO YOSEMITE	FAT	CA	2447	1963
LANCASTER ATC	WJF	CA	3241	1733
LIVERMORE		CA	2755	858
NEEDLES AP	EED	CA	1227	4545
OXNARD (CAMARIL)	CMA	CA	1935	404
PALM SPRINGS	PSP	CA	951	4224

Station Name	(Call Sign)	State Abb.	AHDD	ACDD
PALO ALTO		CA	2584	452
PASO ROBLES MUN	PRB	CA	2789	1038
REDDING MUNICIP	RDD	CA	2961	1741
SACRAMENTO AP	SAC	CA	2666	1248
SAN DIEGO LINDB	SAN	CA	1063	866
SAN FRANCISCO I	SFO	CA	2862	142
SONOMA		CA	2647	717
TAHOE VALLEY AP	TVL	CA	8300	38
UKIAH	UKI	CA	3083	843
UCLA		CA	1364	893
ASPEN 1 SW		CO	8835	35
BOULDER		CO	5687	552
COLORADO SPRING	COS	CO	6480	404
CRAIG 4 SW		CO	8351	209
DENVER INTL AP	DEN	CO	6128	695
DURANGO		CO	6779	238
FORT COLLINS		CO	6238	497
GRAND JUNCTION	GJT	CO	5700	1091
LAMAR		CO	5556	1088
PUEBLO AP	PUB	CO	5598	922
BRIDGEPORT SIKO	BDR	CT	5466	789
DANBURY		CT	6159	597
GROTON		CT	5799	511
HARTFORD BRADLE	BDL	CT	6104	759
DOVER		DE	4212	1262
MILFORD 2 SE		DE	4622	1058
NEWARK UNIVERSI		DE	4746	1047
WILMINGTON NEW	ILG	DE	4888	1125
DAYTONA BEACH I	DAB	FL	815	2961
FORT MYERS (PAG)	FMY	FL	302	3957
GAINESVILLE RGN	GNV	FL	1143	2659
JACKSONVILLE IN	JAX	FL	1354	2627
MIAMI INTL AP	MIA	FL	149	4361
ORLANDO INTL AP	MCO	FL	580	3428
PENSACOLA RGNL	PNS	FL	1498	2650
TALLAHASSEE MUN	TLH	FL	1604	2551
TAMPA INTL AP	TPA	FL	591	3482
ALBANY 3 SE	ABY	GA	2108	2264
ATLANTA HARTSFI	ATL	GA	2827	1810
AUGUSTA BUSH FI	AGS	GA	2525	1986
COLUMBUS METRO	CSG	GA	2154	2296
MACON MIDDLE GA	MCN	GA	2364	2115
SAVANNAH MUNICI	SAV	GA	1799	2454
HILO INTL AP	ITO	HI	0	3228
HONOLULU INTL A	HNL	HI	0	4561
LAHAINA 361		HI	0	4105
LANAI AP 656		HI	16	2261
BOISE AIR TERMI	BOI	ID	5727	807
COEUR D'ALENE		ID	6540	426

Station Name	(Call Sign)	State Abb.	AHDD	ACDD
MOSCOW U OF I		ID	6706	254
POCATELLO RGNL	PIH	ID	7109	387
REXBURG RICKS C		ID	8325	236
TWIN FALLS 6 E	TWF	ID	6715	364
CARBONDALE SEWA		IL	4930	1179
CHICAGO UNIVERS		IL	5787	994
MOLINE QUAD CIT	MLI	IL	6415	969
NORMAL		IL	6190	998
ROCKFORD AP	RFD	IL	6933	768
SPRINGFIELD CAP	SPI	IL	5596	1165
URBANA		IL	5916	979
BLOOMINGTON IND		IN	5348	1017
EVANSVILLE INTL	EVV	IN	4617	1422
FORT WAYNE BAER	FWA	IN	6205	830
INDIANAPOLIS IN	IND	IN	5521	1042
MUNCIE BALL STA		IN	6022	895
SOUTH BEND RGNL	SBN	IN	6294	812
TERRE HAUTE IN		IN	5433	1107
WEST LAFAYETTE	LAF	IN	5732	1024
AMES 8 WSW		IA	6791	830
DES MOINES AP	DSM	IA	6436	1052
DUBUQUE AP	DBQ	IA	7270	656
IOWA CITY		IA	6052	1134
MASON CITY AP	MCW	IA	7765	655
SIOUX CITY AP	SUX	IA	6900	914
WATERLOO MUNICI	ALO	IA	7348	758
COLBY 1 SW		KS	6302	891
DODGE CITY RGNL	DDC	KS	5037	1481
LAWRENCE		KS	4685	1582
SALINA MUNICIPA	SLN	KS	4952	1600
TOPEKA BILLARD	TOP	KS	5225	1357
WICHITA MID-CON	ICT	KS	4765	1658
BOWLING GREEN F	BWG	KY	4243	1413
LEXINGTON BLUE	LEX	KY	4713	1154
LOUISVILLE STAN	SDF	KY	4352	1443
MURRAY		KY	3861	1628
OWENSBORO 3 W		KY	4159	1565
WILLIAMSBURG		KY	4637	940
ALEXANDRIA INTL	AEX	LA	1908	2602
BATON ROUGE RYA	BTR	LA	1689	2628
LAKE CHARLES AP	LCH	LA	1546	2705
MONROE RGNL AP	MLU	LA	2190	2517
NEW ORLEANS INT	MSY	LA	1417	2773
SHREVEPORT AP	SHV	LA	2251	2405
AUGUSTA AP	AUG	ME	7358	388
BANGOR INTL AP	BGR	ME	7676	313
BELFAST		ME	7432	248
PORTLAND INTL A	PWM	ME	7318	347

Station Name	(Call Sign)	State Abb.	AHDD	ACDD
ANNAPOLIS POLIC		MD	4695	1162
BALTIMORE CITY		MD	3807	1774
COLLEGE PARK		MD	4377	1325
FREDERICK POLIC		MD	4430	1272
HAGERSTOWN		MD	5249	902
MECHANICSVILLE		MD	4299	1093
AMHERST		MA	6856	452
BOSTON LOGAN IN	BOS	MA	5630	777
FRAMINGHAM		MA	6060	651
LOWELL		MA	6575	532
NEW BEDFORD		MA	5734	740
WORCESTER RGNL	ORH	MA	6831	371
ANN ARBOR UNIV		MI	6503	691
DETROIT CITY AP		MI	5898	920
FLINT BISHOP IN	FNT	MI	7005	555
GRAND RAPIDS IN	GRR	MI	6896	613
KALAMAZOO STATE	AZO	MI	6235	773
LANSING CAPITAL	LAN	MI	7098	558
SAGINAW TRI STA	MBS	MI	7099	548
BEMIDJI		MN	9540	349
DULUTH INTL AP	DLH	MN	9724	189
INTL FALLS AP	INL	MN	10269	233
MANKATO		MN	8029	650
MINNEAPOLIS INT	MSP	MN	7876	699
ROCHESTER MUNIC	RST	MN	8308	473
ST CLOUD MUNICI	STC	MN	8815	443
BILOXI		MS	1645	2517
HATTIESBURG 5 S		MS	2024	2327
JACKSON THOMPSO	JAN	MS	2401	2264
MERIDIAN KEY AP	MEI	MS	2352	2173
NATCHEZ		MS	1910	2293
TUPELO RGNL AP	TUP	MS	3086	1884
COLUMBIA RGNL A	COU	MO	5177	1246
KANSAS CITY DOW	MKC	MO	4734	1676
KIRKSVILLE	IRK	MO	6134	988
ROLLA UNI OF MI		MO	4876	1359
ST LOUIS INTL A	STL	MO	4758	1561
SPRINGFIELD REG	SGF	MO	4602	1366
WARRENSBURG 4 N		MO	5063	1361
BILLINGS INTL A	BIL	MT	7006	583
BOZEMAN MONTANA		MT	7729	298
BUTTE MOONEY AP	BTM	MT	9399	127
GREAT FALLS INT	GTF	MT	7828	288
HELENA AP	HLN	MT	7975	277
KALISPELL GLACI	FCA	MT	8193	142
MISSOULA INTL A	MSO	MT	7622	256
GRAND ISLAND CT	GRI	NE	6385	1027
LINCOLN AP	LNK	NE	6242	1154

Station Name	(Call Sign)	State Abb.	AHDD	ACDD
MCCOOK	MCK	NE	5967	1037
NORTH PLATTE RG	LBF	NE	6766	750
OMAHA EPPLEY AP	OMA	NE	6311	1095
SCOTTSBLUFF AP	BFF	NE	6742	690
VALENTINE MILLE	VTN	NE	7255	779
CARSON CITY		NV	5661	419
ELKO		NV	7181	412
ELY	ELY	NV	7561	196
FALLON EXPERIME		NV	5586	624
LAS VEGAS AP	LAS	NV	2239	3214
RENO CANNON INT	RNO	NV	5600	493
WINNEMUCCA MUNI	WMC	NV	6271	526
BERLIN		NH	8515	169
KEENE		NH	7505	328
LEBANON MUNICIP	LEB	NH	7694	307
NASHUA 2 NNW		NH	6834	445
PLYMOUTH		NH	8253	201
ATLANTIC CITY A	ACY	NJ	5113	935
HIGHTSTOWN 2 W		NJ	5357	751
JERSEY CITY		NJ	5367	882
MILLVILLE MUNIC	MIV	NJ	4835	1009
MORRIS PLAINS 1		NJ	5908	544
NEWARK INTL AP	EWR	NJ	4843	1220
TOMS RIVER		NJ	5173	858
ALAMOGORDO		NM	3108	1715
ALBUQUERQUE INT	ABQ	NM	4281	1290
CARLSBAD AP	CNM	NM	2935	2031
CLOVIS		NM	3955	1305
GALLUP MUNICIPA	GUP	NM	6588	357
LORDSBURG 4 SE		NM	3280	1756
LOS ALAMOS		NM	6613	240
RATON FILTER PL		NM	6006	351
ROSWELL AP	ROW	NM	3332	1814
SANTA FE 2		NM	6073	414
ALBANY INTL AP	ALB	NY	6860	544
BINGHAMTON BROO	BGM	NY	7237	396
BUFFALO NIAGARA	BUF	NY	6692	548
ELMIRA		NY	6806	446
ITHACA CORNELL		NY	7182	312
NEW YORK CITY C	NYC	NY	4754	1151
PLATTSBURGH AFB		NY	7817	387
POUGHKEEPSIE	POU	NY	6438	550
ROCHESTER MONRO	ROC	NY	6728	576
SYRACUSE HANCOC	SYR	NY	6803	551
WATERTOWN AP	ART	NY	7681	301
ASHEVILLE RGNL	AVL	NC	4326	818
CHARLOTTE DGLAS	CLT	NC	3162	1681
GREENSBORO RGNL	GSO	NC	3848	1332
GREENVILLE		NC	3112	1636

Station Name	(Call Sign)	State Abb.	AHDD	ACDD
FAYETTEVILLE PW		NC	3097	1721
RALEIGH DURHAM	RDU	NC	3465	1521
WILMINGTON NEW	ILM	NC	2429	2017
BISMARCK MUNICI	BIS	ND	8802	471
DICKINSON AP		ND	8558	512
FARGO HECTOR AP	FAR	ND	9092	533
GRAND FORKS INT	GFK	ND	9489	420
MINOT AP		ND	8990	492
AKRON CANTON AP	CAK	OH	6154	678
ATHENS 2 N		OH	5690	705
BOWLING GREEN W		OH	6492	690
CHILLICOTHE MOU		OH	5570	863
CINCINNATI LUNK	LUK	OH	4883	1131
CLEVELAND HOPKN	CLE	OH	6121	702
COLUMBUS INTL A	CMH	OH	5492	951
DAYTON INTL AP	DAY	OH	5690	935
FINDLAY AP	FDY	OH	6242	704
TOLEDO EXPRESS	TOL	OH	6460	715
YOUNGSTOWN MUNI	YNG	OH	6451	552
DURANT		OK	3244	1963
ELK CITY		OK	4143	1618
ENID		OK	4269	1852
OKLAHOMA CITY A	OKC	OK	3663	1907
TULSA INTL AP	TUL	OK	3642	2049
BAKER		OR	7186	267
BEND		OR	7042	147
BURNS MUNICIPAL	BNO	OR	7785	218
CORVALLIS STATE		OR	4715	247
EUGENE MAHLON S	EUG	OR	4786	242
MEDFORD AP	MFR	OR	4539	711
NORTH BEND AP		OR	4464	12
PORTLAND INTL A	PDX	OR	4400	390
ALLENTOWN LEHIG	ABE	PA	5830	787
BRADFORD RGNL A	BFD	PA	7666	217
ERIE AP	ERI	PA	6243	620
HARRISBURG CAPI	MDT	PA	5201	955
PHILADELPHIA IN	PHL	PA	4759	1235
PITTSBURGH INTL	PIT	PA	5829	726
STATE COLLEGE		PA	6345	538
WILKES BRE SCTN	AVP	PA	6234	611
KINGSTON		RI	5993	441
PROVIDENCE GREE	PVD	RI	5754	714
CHARLESTON INTL	CHS	SC	2005	2306
CLEMSON UNIVERS		SC	3194	1584
COLUMBIA METRO	CAE	SC	2594	2074
FLORENCE RGNL A	FLO	SC	2524	2029
GRNVL SPART AP	GSP	SC	3272	1526
ABERDEEN RGNL A	ABR	SD	8348	626

Station Name	(Call Sign)	State Abb.	AHDD	ACDD
BROOKINGS 2 NE		SD	8490	523
PIERRE RGNL AP	PIR	SD	7282	919
RAPID CITY RGNL	RAP	SD	7211	598
SIOUX FALLS AP	FSD	SD	7812	747
BRISTOL TRI CIT	TRI	TN	4445	956
CHATTANOOGA AP	CHA	TN	3427	1608
KNOXVILLE AP	TYS	TN	3690	1450
MEMPHIS INTL AP	MEM	TN	3041	2187
NASHVILLE INTL	BNA	TN	3677	1652
ABILENE MUNICIP	ABI	TX	2659	2386
AMARILLO INTL A	AMA	TX	4318	1344
AUSTIN-BERGSTRO	AUS	TX	1509	2982
BROWNSVILLE AP	BRO	TX	644	3874
COLLEGE STATION	CLL	TX	1616	2938
CORPUS CHRISTI	CRP	TX	950	3497
DALLAS-FT WORTH	DFW	TX	2370	2568
DEL RIO INTL AP	DRT	TX	1417	3226
EL PASO INTL AP	ELP	TX	2543	2254
HOUSTON BUSH IN	IAH	TX	1525	2893
LUBBOCK RGNL AP	LBB	TX	3508	1769
MIDLAND INTL AP	MAF	TX	2716	2139
SAN ANTONIO INT	SAT	TX	1573	3038
WACO RGNL AP	ACT	TX	2164	2840
BLANDING		UT	5482	921
DELTA		UT	6198	809
LOGAN UTAH STAT	LGU	UT	6723	684
MOAB		UT	4455	1743
PROVO BYU		UT	5264	1028
ST GEORGE		UT	3103	2471
SALT LAKE CITY	SLC	UT	5631	1066
VERNAL AP	VEL	UT	7347	499
BURLINGTON INTL	BTV	VT	7665	489
MONTPELIER AP	MPV	VT	8245	225
RUTLAND		VT	7304	317
SAINT JOHNSBURY	1V4	VT	7806	296
CHARLOTTESVILLE		VA	4103	1212
DANVILLE		VA	3970	1418
FREDERICKSBURG		VA	4455	1182
NORFOLK INTL AP	ORF	VA	3368	1612
RICHMOND BYRD I	RIC	VA	3919	1435
ROANOKE WOODRUM	ROA	VA	4284	1134
WASHINGTON REAG	DCA	VA	4055	1531
ABERDEEN		WA	5161	34
BELLINGHAM INTL	BLI	WA	5400	67
OLYMPIA AP	OLM	WA	5531	97
PULLMAN 2 NW		WA	6688	276
RICHLAND		WA	5133	739
SEATTLE TACOMA	SEA	WA	4797	173
SPOKANE AP	GEG	WA	6820	394

Station Name	(Call Sign)	State Abb.	AHDD	ACDD
WENATCHEE AP	EAT	WA	5950	775
BLUEFIELD AP	BLF	WV	4891	663
CHARLESTON YEAG	CRW	WV	4644	978
ELKINS AP	EKN	WV	6036	416
MORGANTOWN HART	MGW	WV	5312	759
PARKERSBURG AP	PKB	WV	4966	966
WHEELING		WV	5313	926
EAU CLAIRE RGNL	EAU	WI	8196	554
GREEN BAY STRBL	GRB	WI	7963	463
LA CROSSE MUNIC	LSE	WI	7340	775
MADISON DANE CO	MSN	WI	7493	582
MILWAUKEE MITCH	MKE	WI	7087	616
PLATTEVILLE		WI	7478	603
WAUSAU AP	AUW	WI	8237	464
CASPER NATRONA	CPR	WY	7571	428
CHEYENNE MUNICI	CYS	WY	7388	273
JACKSON		WY	9529	50
LARAMIE RGNL AP	LAR	WY	9038	71
ROCK SPRINGS AP		WY	8670	230
SHERIDAN AP	SHR	WY	7721	398

Index

B

D

E

M

S

T

U

V